Economy

Critical Essays in Human Geography

Edited by

Ron Martin

Department of Geography and St Catharine's College,
University of Cambridge, UK

 Routledge
Taylor & Francis Group

LONDON AND NEW YORK

First published 2008 by Ashgate Publishing

Reissued 2018 by Routledge
2 Park Square, Milton Park, Abingdon, Oxon, OX14 4RN
711 Third Avenue, New York, NY 10017, USA

Routledge is an imprint of the Taylor & Francis Group, an informa business

First issued in paperback 2018

A Library of Congress record exists under LC control number: 2007941052

Notice:
Product or corporate names may be trademarks or registered trademarks, and are
used only for identification and explanation without intent to infringe.

Publisher's Note
The publisher has gone to great lengths to ensure the quality of this reprint but
points out that some imperfections in the original copies may be apparent.

Disclaimer
The publisher has made every effort to trace copyright holders and welcomes
correspondence from those they have been unable to contact.

ISBN 13: 978-0-815-38874-6 (hbk)
ISBN 13: 978-1-138-61954-8 (pbk)
ISBN 13: 978-1-351-15920-3 (ebk)

Contents

Acknowledgements

The editor and publishers wish to thank the following for permission to use copyright material.

Blackwell Publishing for the essays: Ash Amin (1999), 'An Institutionalist Perspective on Regional Economic Development', *International Journal of Urban and Regional Research*, **23**, pp. 365–78. Copyright © 1999 Joint editors and Blackwell Publishers Ltd; Ash Amin and Nigel Thrift (1992), 'Neo-Marshallian Nodes in Global Networks', *International Journal of Urban and Regional Research*, **16**, pp. 571–87; Neil M. Coe, Martin Hess, Henry Wai-chung Yeung, Peter Dicken and Jeffrey Henderson (2004), '"Globalizing" Regional Development: A Global Production Networks Perspective', *Transactions of the Institute of British Geographers*, **29**, pp. 468–84. Copyright © 2004 Royal Geographical Society (with The Institute of British Geographers); Gordon L. Clark and Neil Wrigley (1995), 'Sunk Costs: A Framework for Economic Geography', *Transactions of the Institute of British Geographers*, **20**, pp. 204–23; Jamie Peck and Adam Tickell (2002), 'Neoliberalizing Space', *Antipode*, **34**, pp. 380–404. Copyright © 2002 Editorial Board of Antipode; Kevin R. Cox (2004), 'Globalization and the Politics of Local and Regional Development: The Question of Convergence', *Transactions of the Institute of British Geographers*, **29**, pp. 179–94. Copyright © 2004 Royal Geographical Society (with The Institute of British Geographers).

Economic Geography for the essays: Ron Martin and Peter Sunley (1996), 'Paul Krugman's Geographical Economics and Its Implications for Regional Development Theory: A Critical Assessment', *Economic Geography*, **72**, pp. 259–92; Ann Markusen (1996), 'Sticky Places in Slippery Space: A Typology of Industrial Districts', *Economic Geography*, **72**, pp. 293–313; Henry Wai-chung Yeung and George C.S. Lin (2003), 'Theorizing Economic Geographies of Asia', *Economic Geography*, **79**, pp. 107–28. Copyright © 2003 Clark University; Jamie Peck (1992), 'Labor and Agglomeration: Control and Flexibility in Local Labor Markets', *Economic Geography*, **68**, pp. 325–47; Meric S. Gertler (1995), '"Being There": Proximity, Organization, and Culture in the Development and Adoption of Advanced Manufacturing Technologies', *Economic Geography*, **71**, pp. 1–26.

Hodder Education for the essay: Ron Martin and Peter Sunley (1997), 'The Post-Keynesian State and the Space Economy', in Roger Lee and Jane Wills (eds), *Geographies of Economies*, London: Edward Arnold, pp. 278–89; 289a–c.

Macmillan Publishers Ltd for the essay: Edward E. Leamer and Michael Storper (2001), 'The Economic Geography of the Internet Age', *Journal of International Business Studies*, **32**, pp. 641–65.

Oxford University Press for the essays: Allen J. Scott (2004), 'A Perspective of Economic Geography', *Journal of Economic Geography*, **4**, pp. 479–99. Copyright © 2004 Oxford

Series Preface

This series collects together some of the most significant articles from the major fields of human geography published over the past forty years. During this time something of a renaissance has occurred in the thinking that explores facets of human society using the concepts of space and place (and associated ones such as territory, geopolitics, mobility, diffusion, and locality). This reflects both the rediscovery of cultural difference and local knowledge as claims to universal, objective knowledge have come under critical scrutiny, and the exhaustion of historicist narratives owing to the failure of various projects to end global economic and political inequality. Thinking in terms of fixed territories of statehood has also become problematic owing to increased human mobility in a technologically-driven world in which religious and other, often non-national identities, appear to be in the political ascendancy. The period from 1970 until today has been one of particularly dramatic worldwide geopolitical change. Consider, for example, the impact of the political collapse of the Soviet Union, the economic rise of China, and the spread of globalization around the world. Many of the central tenets of Western social thought on nationalism, the primacy of economic change, and the centrality of states that were established during the Cold War are now subject to revision.

Of course, concepts of space and place as they are applied to human society have old philosophical roots, although these were undoubtedly obscured, particularly in the United States, for many years. Since the seventeenth century 'space' has meant the plane on which events and objects are located, while 'place', the older of the two terms, dates back to the ancient Greeks and refers to the 'lived' or 'occupied' space. What is new in recent years is the varied ways in which these concepts and related ones are now being used. This involves blending older understandings, drawing from the three classic geographic traditions that emphasize the environmental (physical-human), spatial (distributional), and regional (clustering) approaches to geographic sameness and difference, with new sensibilities and concerns about social and political divisions (from social class, ethnicity, disability, and gender to the new political divisions of a post-Cold War world). This sense of older ideas adapting to new circumstances accounts for the seemingly paradoxical title of the series: Contemporary Foundations of Space and Place. If the foundations of recent thinking are rooted in the past, such thinking also mirrors recent imperatives. What is also new is a more self-consciously 'critical' tenor to use of concepts compared to a past when much research was driven less by theoretical or methodological considerations and much more by just an interest in a given phenomenon in itself (landscape features, settlement types, borderlands, etc.) as if the conditions for its study were self-evident.

The ten volumes are divided, using fairly conventional distinctions, between those that are more general or cover large parts of the field (theory and methods, regions) and those that have relatively more constrained empirical subject matter (rural, economic, etc.). Unsurprisingly, there is considerable potential overlap across the volumes. Within the field as a whole, there is no simple division of labor between articles that are just 'theoretical' or 'methodological' and those that employ concepts in more mundanely 'empirical' ways. Nevertheless, there are definitely some articles that have had greater theoretical or methodological influence across the field as a whole. Within their respective areas, however, editors do attempt to cover the main conceptual differences and controversies of the past forty years, paying particular attention to the scope of the influence in question. Each volume is comprised of articles in English. Of course, English dominates global academic production today, but this does not mean that only work written by native English-speakers is significant. In choosing volume editors and volume content, an active effort has been made to ensure the representation of standpoints and perspectives from all over the world. The series as a whole should be particularly helpful to those looking for substantial overviews of the course of human geography between the 1970s and the early 2000s but who lack access to large library collections of academic journals. The individual volumes are crafted by specialists who have expansive rather than narrow conceptions of the purpose of the overall enterprise. The volumes are thus also designed to appeal to those looking for artful reviews of recent developments in different substantive areas of human geography.

JOHN AGNEW
General Editor
UCLA

Introduction

Space, Place and the Economy

How should we think about 'the economy'? What do we mean by the concept? How is 'the economy' best examined and understood? The economist's traditional approach to answering such questions has been to argue that there are just two levels at which economic processes operate, theoretical exposition is needed, and policy intervention is appropriate: the 'macro-economy' of the nation-state, and the 'micro-economy' of individual firms, industries and households. These two levels are taken to constitute 'the economy'. The macro-economy is viewed either as a set of aggregate-level forms and relationships (macro-economics), or as the sum of a myriad of micro-level agents (firms and households) and their actions and behaviours (micro-economics). There is no stopping point between these two levels, little or no sense that economic activity occurs in and across geographic space, that the 'everyday business of economic life' (to use Alfred Marshall's phrase) does not occur on the head of the proverbial pin, but in particular places and locations, in local communities, towns, cities and regions. It is within these various geographical spaces that people live and work, that businesses operate, wealth is produced, and services and products consumed. These local and regional economies are themselves linked into complex networks of flows and relationships, both national and global in reach (Hudson, Chapter 5). Yet even when economists talk of the 'international economy', or the 'global economy', these are typically conceived of in terms of interactions between spaceless national economies. Recently, Hodgson (2001) has criticized economists for having forgotten the importance and role of history. For the most part, economists have also neglected the importance and role of geography, space and place.[1]

The aim of economic geography is precisely to demonstrate why space, place and location are central to a full understanding of 'the economy' and 'the economic process'. Geography would be unimportant if economic growth and prosperity occurred equally everywhere. Patently, from the evidence of the world around us, that is not the case: economic development does not occur uniformly across geographic space, but rather differs in degree and form as between different locations and places, producing spatial inequalities in economic growth, prosperity and well-being as a result. Indeed, geographically uneven development appears intrinsic to capitalism, for it occurs at a whole variety of spatial scales within nations, between regions, cities and localities. For economic geographers, then, the economy does not reside on the head of the proverbial pin, but is spatially structured in ways that have different outcomes and consequences in different places.

---

1   Of course, there have been some notable exceptions to this general rule. For example, Gunnar Myrdal, Jane Jacobs and certain other 'heterodox' economists argued for a regional or city-based perspective in economics. However, these excursions into geography had little impact on the discipline's overwhelming pre-occupation with the spaceless national economy.

Geography is in fact integral to capitalism in a dual sense. At one level, capitalism uses space in its constant search for new market opportunities and more profitable locations, what David Harvey once described as new 'spatial fixes', and others have called 'spatial divisions of labour' (Massey, 1984; Storper and Walker, 1984). In so doing, capitalism continuously creates and destroys economic landscapes. Harvey's depiction of this process is still one of the most graphic, when he describes how

> Capitalism perpetually strives ... to create a social and physical landscape in its own image and requisite to its own needs at a particular point in time, only just as certainly to undermine, disrupt and even destroy that landscape at a later point in time. The inner contradictions of capitalism are expressed through the restless formation and re-formation of geographical landscapes. (1985, p.150)

But the causal arrow between economic process and geographic space points in both directions. Places are not simply produced and transformed by capitalism: capitalism is also produced and transformed within specific places. For the most part, economic processes are neither exogenous macro-level forces, 'out there' as it were, impinging differently on different places, nor ubiquitous micro-level mechanisms operating similarly everywhere and anywhere. Rather, many economic processes are spatially constituted, that is to say, they depend upon and are shaped by the specific conditions and relationships that exist in certain places, but not in others. In other words, the micro-level behaviours and decisions of individual economic agents are, in part at least, *place dependent*. And this localization itself often has unintended *emergent* effects,[2] such as the development of locally-embedded externalities, that cannot easily be replicated or reproduced in other places. The role of space in the economy is thus not merely a passive or simply mediative one, a pre-given cartography across which economic forces and developments are played out. Rather, space is integral to and constitutive of economic development. Just as the economy is a historical process, so it is also a geographical process (Lee and Wills, 1997).

Conceptualizing Economic Landscapes

Understanding the economy as a spatial process presents major conceptual and theoretical challenges. Over the past fifty years economic geographers have brought various approaches to bear on this task (see Barnes, 1996, 2001; Scott, 2000, and Chapter 1), but three key issues have recurred in this search for an explanatory framework. First, there is the matter of what sort of economic (or other) theory economic geographers should use as the basis for their explanations. Second, there is the question of how far and in what ways it is possible to achieve the seemingly contradictory ambition of constructing general explanations given the spatial complexity and particularity of the economic landscape. David Harvey has recently expressed this latter challenge thus:

---

2 Unintended in the sense that such effects are neither planned nor imposed, but rather arise or 'emerge' out of the multifarious and complex interactions between individual 'agents' (be these workers, households, firms or other collectivities). For an account of the idea of socio-economic emergence see Sawyer, 2005.

Any theory of uneven geographical development must be simple enough to aid comprehension and complex enough to embrace the nuances and particularities that call for interpretation. (2006, p. 75)

And third, there is the issue of how we conceptualize space itself; how we think economic space is structured. Over the past quarter of a century the intellectual orientation of the discipline has changed on all three fronts (see Table 1).[3]

Every theory, or explanatory schema, abstracts from phenomena to focus on certain facts, aspects and attributes to the neglect of others. A theory simplifies in order to bring out the structural and causal aspects of a complex, underlying reality. Different theoretical schemas involve different types of abstraction and different ambitions with respect to explanatory generalization (one of the most useful discussions of abstraction in economic geography is still that of Sayer, 1982). Consider two rather extreme positions on this matter. On the one hand we may believe that beneath the apparent spatial specificity and chaotic complexity of the economic landscape, general mechanisms and 'laws' are in fact at work, and that our task is therefore to render such complexity explicable by identifying that which is 'universal' or general in the particular. The quest for such universal or general causes typically requires high-level abstractions that ignore spatial particularities in favour of focusing on and emphasizing spatial similarities and commonalities. On the other hand, we might argue that spatial economic differentiation and specificity permit of no universal explanatory generalizations, that the spatially particularizing aspect of economic processes and outcomes is the existence of infinitely diverse mechanisms and phenomena that concern us more for what distinguishes them than for what they share. In this approach, explanation is highly context dependent, constructed through place-specific narrative-interpretative accounts, and not via high-level abstractions. The options, it would seem, are that we must either downplay the locally unique quality of economic life, or relinquish the effort to generalize across space. Of course, this is to state the explanatory challenge far too starkly and simply. Nevertheless, since the late 1970s one of the fundamental shifts in economic geography has been a move away from the search for or use of universal and general concepts and theory to a much greater focus on contextualized conceptualizations and explanations.

The positivistic location-analytic economic geography practised during the 1960s and 1970s (Harvey, 1969; Smith, 1971; Dicken and Lloyd, 1972; Amadeo and Golledge, 1975) largely revolved round constructing explanatory accounts of the location of economic activity by drawing theoretical inspiration from neoclassical economics and Germanic 'location theory'. Coupled with an absolute view of economic space – as a pre-given, fixed, and often idealized Euclidean geometry of locations or regional 'containers' – the aim was to explain the geographical distribution of economic activity across such landscapes in terms of general principles of optimal locational decision-making by hypothetical (or 'representative') firms and households, that is via abstractions intended to have wide spatial-temporal scope. Inductive variants, based on empirically-estimated models, sought to do much the same thing by searching for general statistical regularities in location patterns.

---

3 For a related discussion of the treatment of space (and time) in economic theories, see Corportaux and Crevoisier, 2007.

Table 1 Shifts in Dominant Theoretical Perspective in Economic Geography Since the mid-1970s

	Location-Theoretic Late 1960s to Mid-1970s	Marxist Late 1970s to Late 1980s	Post-Marxist Late 1980s Onwards
Form of economics drawn upon	Explicitly or implicitly based on Germanic location theory, and on neoclassical economic assumptions of optimizing behaviour on part of firms and households, leading to equilibrium spatial economic outcomes	Heavily dependent on Marxian political economy, with its emphasis on dynamics of capital accumulation, the labour process and social relations of production; focus on uneven regional development and spatial divisions of labour	Two main strands: one drawing on institutional, social and cultural economics, the other on neo-Schumpeterian and evolutionary economic theory; focus on local socio-cultural embeddedness of economy, and on spatialities of knowledge creation and transfer
Type of theorizing	Focus on high level generalization via models that abstract from spatial particularity, and which emphasize universal laws of spatial distribution and interaction	Based on historical materialist approach to abstraction, involving appeal to systemic 'laws of motion' of the economy, and the constitution of the latter in terms of antagonistic class conflict over the ownership of, and the distribution of the surplus from, production	Mostly partial and contextualized using theoretical concepts and ideas drawn from a range of frameworks, including institutional economics, economic sociology, cultural theory and evolutionary economics
Conceptions of space	Conceived of in absolute (Euclidean) terms, as fixed geometry of locations and regions, pre-given, unchanging and unproblematic	Absolute and relative in nature, not pre-given but constantly produced and reproduced by capitalist relations of production	Absolute, relative and relational, where both spatial proximity and relational proximity (propinquity without proximity) are important; regional spaces consequentially complex to define and not pre-given or fixed; increasing focus on spatial networks and flows

The Marxist economic geography that superseded industrial location theory in the 1980s was also based on a quest for explanatory generalization, though from a quite different perspective (Marxist political economy) and based on a different mode of abstraction (historical materialism). In this case, the economic landscape was explained not in terms of universal forms of idealized economic behaviour assumed to apply everywhere and anywhere, but by appeal to the fundamental 'deep-structure' processes of capital accumulation, surplus value extraction, and class relations that Marxist theory defines as the quintessential features of capitalism as a dynamic mode of production. The key argument was that the economic landscape is not as spatially particularistic nor as fixed as it outwardly seems, that geographically uneven development itself is not only the inevitable product of the immutable (and often contradictory) imperatives of capitalism's inner expansionist logic (Harvey, 1982; Smith, 1984), but is constantly being reconfigured in the process (witness Harvey's statement, quoted above). In this conception, geographic-economic space has a relative character: it is not simply a pre-given empty container or cadastral entity in which capitalism occurs, but takes on particular forms by virtue of the use that capitalism makes of it (Massey, 1984; Harvey, 2006).

Since the early 1990s, however, the mode of theorization and abstraction employed by economic geographers has moved distinctly away from the pursuit of high-level generalizations, whether of a positivistic location-theoretic or historical-materialist Marxist kind, to an increasing emphasis on the locally contextual, contingent and embedded (see Barnes, 1995; Sunley, 1996). In part this has been associated with the exploration of alternative, heterodox schools of economic theory – including institutional, neo-Schumpeterian and evolutionary economics – and even social and cultural theory, as sources of conceptual inspiration. These institutional, cultural and evolutionary 'turns' in the discipline have stimulated ongoing debates about what sort of economic theory is appropriate for economic geography, about the nature of the 'economic' in economic geography, and about appropriate modes of explanation (see for example, Sayer, 1994, 1997; Barnes, 1995; Thrift and Olds, 1996; Amin, Chapter 3; Boschma and Lambooy, 1999; Markusen, 1999; Amin and Thrift, Chapter 7; Lee, 2000; Martin, 2000a; Martin and Sunley, 2001; Rodríguez-Pose, 2001; Bathelt and Glückler, Chapter 4; Scott, Chapter 1; Boschma and Frenken, Chapter 6). What has emerged from these discussions and developments is a more pragmatic, relativistic, and open-ended approach to explanation and a downplaying of general 'laws' and concepts. Ontologically speaking, the economic landscape is now seen as the 'product of multiple separate determinations' (Massey, 1992), operating at many different spatial scales, including many processes that are highly localized and particularistic, but others that embrace larger geographical scales, including of course the global. Explanations and theories are regarded as spatially conditioned and hence spatially contingent. There has been a move, one might say, away from seeking generalizing 'grand theory' and meta-narratives to something more akin to 'contextualized theory' and 'particularistic narratives', to a mode of abstraction, theory construction and explanation that works as much 'bottom upwards', from intensive study of specific local cases, instances, practices and contexts to uncover causal explanations, as 'top downwards' by imposing high-level generalizations, 'laws' and statistical tendencies on the local (Yeung, 2003; Peck, Sheppard, Tickell, 2006). This is not to suggest that theoretical ideas and concepts have been subordinated; to the contrary, they remain key. But it is to recognize that our concepts and explanations may have specific (and thus limited) spatial-temporal scope and reach. At the same time, our conceptions of space have changed to recognize this, in an age when networks and flows have assumed greater significance and complexity, and economic space has a relational character (Bathelt and Glückler, Chapter 4): that in the economic landscape the relations between firms, households, institutions and the like need not be spatially proximate but geographically stretched, discontinuous and dispersed. The emphasis, we might say, is on the *spatiality of connectedness* and not just the spatially contiguous.[4]

Yet, interestingly, while economic geographers have been busy exploring alternative forms of economics, economic sociology and cultural economics, and downplaying explanatory generalization, another and quite different 'new economic geography' has developed within economics. The central goal of this so-called 'new economic geography', referred to as NEG by its exponents, is precisely to construct formal models that explain the economic landscape in terms of a limited set of highly generalized principles of location and agglomeration that supposedly apply at all spatial scales, from the intra-urban to the global (Krugman, 1991, 1995, 1996, 1998; 2000; Fujita, Krugman and Venables, 1999; Brakman, Garretsen and Marrevijk, 2000; Fujita and Thisse, 2002; Henderson, 2005). Because they are mathematical in form, these models inevitably involve a high level of idealization of the economic landscape and abstract away the detailed messiness of everyday spatial particularity. Even though a key focus of NEG – namely the widely observed tendency for economic activities to exhibit spatial agglomeration and clustering – is shared by 'proper' economic geography (or PEG, as we might refer to it), the former is based on an explanatory method and a mode of abstraction that differ markedly from those that currently dominate the latter. This combination of a shared interest but markedly different approaches, together with a fear on the part of some economic geographers of NEG's encroachment into their 'academic space', laced with not a little resentment that economists should be so provocative as to call their neophyte field the 'new economic geography', has sparked off lively commentaries and assessments by economic geographers (see Dymski, 1996; Martin and Sunley, Chapter 2; Boddy, 1999; Martin, 1999, 2001;). The scope for interaction and dialogue between PEG and the NEG has itself become a topic of debate (Overman, 2004; Sheppard and Plummer, 2001; Sjöberg and Sjöholm, 2002; Duranton and Rodríguez-Pose, 2005; Duranton and Storper, 2006;), and the rise of NEG has even prompted some economic geographers to argue that perhaps PEG has in fact moved too far away from quantitative analysis and explanatory generalization (Sheppard and Plummer, 2001; Scott, Chapter 1). Whilst a gulf of difference separates the conceptual and methodological dispositions of economic geographers and the new geographical economists,

---

4   It is the connective relational structures of the economy that render it a complex, open dynamic system. The geographies of the economy are thus defined not simply in terms of fixed spatial geometries of locations and places (cities, regions, etc), but by the multifarious and ever shifting geometries of economic connections that constitute such places and their interrelationships. In the new complexity economics, for example, the ontology of methodological individualism, social atomism and equilibrium of traditional economics is replaced by an ontology of 'connectivity':

> Generally connections are specific direct relationships between elements and are ubiquitous in the economic system. They exist in the structure of interdependencies and interactions between agents. They exist in the modalities of technology and the forms of organization and competence. They exist as contracts. They exist in the structure of decision rules and the way that information is processed, In all such events, the dynamics of economic systems can be seen to occur most in the space of connections. (Potts, 2000, p. 3)

nevertheless both groups agree that space and place have become more, not less, important in understanding today's process of increasing global economic integration.[5]

The Localization of Global Economic Space

One of the key processes driving contemporary globalization is technological change, and especially rapidly evolving advances in information and communications technologies, or ICT (Castells, 1996). Some suggest that such is the radical nature of this ICT revolution that, for perhaps the first time in history, geography is no longer the primary constraint on the boundaries of social and economic organization. Authors such as O'Brien (1992), Cairncross (1997) and Reich (2001) and others have all but declared the 'death of distance'. Most recently, Friedman (2005) has suggested that the development of a global communications and technology platform has 'flattened the world' by allowing unprecedented exchanges of knowledge across the globe, opening up almost limitless possibilities for contact, collaboration, alliances, off-shoring and outsourcing.

Not surprisingly, many economic geographers have strongly resisted this 'end of geography' prognosis (for example, see Martin, 1994; Yeung, 1998, Morgan, 2004), and have argued that the removal of certain physical geographical constraints on economic and social organization does not mean that cities and regions are less important as economic entities. To the contrary, they take on new forms and gain new significance. Thus Scott (2006) has dismissed the end of geography view as a profound misunderstanding, and argued just the opposite, that the world economy is being reconstituted as 'an integrated system of differentiated locations based on little more than functional divisions of labour and the ways they mould the competitive advantages of different places', and that 'far from being dissolved away in this process, the economic contrasts between distinct places have consistently been strengthened by it' (p. 2). Thus the challenge for economic geography in the contemporary era of rapid globalization is not to explain the 'flattening of the world', but to explain 'why regions keep emerging as centers for new rounds of growth even as our capacities for transcending the frictions of space continue to improve' (Storper, 1999, p. 45), that is, why a (re)localization of economic activity seems to be underway.

Over the past two decades a variety of explanations of this economic (re)localization have appeared, although almost all are in one way or another based on some notion of localized increasing returns or externalities. In the 1980s and early 1990s the argument was that the new market opportunities, production technologies and competitive imperatives of the emerging post-Fordist global economy were spawning new industrial spaces of 'flexible specialization' in areas previously untouched by the post-war large Fordist mass manufacturing complexes (Scott, 1988). One version of this thesis – based on the neo-Marshallian notions of localization economies – has highlighted the resurgent craft and neo-artisanal districts of the 'Third Italy' and argued that their competitive advantage has been due to a combination of flexibly specialized production, external economies and a dense supportive institutional fabric and co-operative culture (Piore and Sabel, 1984; Beccattini, 1990; dei Ottati, 1994). A second

---

5   As several authors have stressed, globalization is not a new phenomenon, but a process that has been unfolding, in nature, speed and geographical complexion since at least the nineteenth century, if not before (see for example, Scholte, 2000, chapter 3.)

approach, the so-called 'Californian School', argued that post-Fordist flexible production involves the agglomeration of vertically disintegrated and networked forms of production, and that this is driving not only the resurgence of craft districts but also high-technology regions and financial clusters (Scott, 1988; 1998; Storper and Walker, 1989). Most importantly, it is argued that agglomeration not only provides important external economies, such as skilled labour markets, but it also reduces the input-output transaction costs of these production systems and thus allows them to realize the advantages of flexibility and specialization. Thus vertical disintegration, out-sourcing and contracting are symbiotically fused with agglomeration.

But, arguably, the major influence has been the work of Michael Porter on business clusters. Arguing from a business strategy perspective, Porter's interest in clusters grew out of his research into international competitiveness where he concluded that the geographical clustering of leading industries within nation-state borders was a near-universal phenomenon (1990; 1998). He argues that clustering provides key advantages in terms of boosting firms' productivity and innovation, and raising rates of spin-offs and start-ups (Porter, 2000). Although Porter's cluster theory is eclectic, drawing on various conceptual ideas, it too has distinct neo-Marshallian aspects in that it stresses the highly beneficial localized externalities that clusters confer on the firms concerned.

The cluster concept has stimulated several vigorous debates (Benneworth and Henry, 2004).[6] Some critics have questioned the universality of the concept as alleged by Porter and others, and have warned of the dangers in applying it to every kind of local production system (Martin and Sunley, 2003). Relatedly, others have argued that we should be careful to examine the important differences between different types of industrial district (Markusen, Chapter 8). Still others have stressed the role of clusters as knowledge networks (Maskell and Malmberg, 1999; Pinch et al., 2003). In this context, a continuing debate centres on the whether it is the 'local buzz' of tacit knowledge or specialized access to external, more codified global knowledge networks and 'pipelines' that gives clusters their competitive edge (Amin and Thrift, Chapter 7; Cooke, 2001; Bathelt et al., 2004). While most researchers now agree that we need to examine the articulation and co-constitution of local and global channels and practices, there continues to be discussion on how these should be conceived and explained. Thus, two unresolved issues in the cluster literature revolve around the question of how far clusters genuinely do provide locational advantages to their component firms, and how they function as growth nodes in their respective regional settings (see Asheim, Cooke and Martin, 2006).

Weaving their way through these conceptual explorations are three issues in particular: first, the recognition that the dynamics of geographically uneven development are produced by mutually constitutive interactions between places and global flows and networks (a co-evolutionary dynamic between the 'space of flows' and the 'spaces of places'); second, the need to hold onto the fact that forms of societal and territorial embeddedness at a range of scales continue to be enormously important; and third, that in the new digital and information age, the creative, innovative and knowledge base of local areas is fundamental to their global competitiveness and successful development.

---

6   Despite these debates, and the limitations of the notion, Porter's cluster concept has become one of the most influential spatial policy tools across both advanced and developing countries (see Asheim, Cooke and Martin, 2006).

In the explication of the new space of global flows several key frameworks or metaphors have wrestled for dominance. One such is the idea of global commodity or value chains, which seeks to trace the stages and locations where economic value is being added, controlled and appropriated (Gereffi, 1996; Bair, 2005; Gereffi et al., 2005). A second approach focuses more on global production circuits and networks in which linkages are intricate and multilayered, horizontal and diagonal as well as vertical (Dicken and Malmberg, Chapter 12; Henderson et al., 2002; Coe et al., Chapter 9). Research has sought to clarify exactly what these networks mean and how their configuration differs between sectors and industries, as well as how they interact with different institutional systems and regional production complexes, including under different national contexts (see, for example, Yeung, 2003).[7]

While it is clear that globalization has increased the availability of some types of economic knowledge, and made certain information ubiquitous, several studies have documented how various knowledge spillovers, particularly those relevant to technological innovations such as patent citations, continue to show a marked localization (Audretsch and Feldman, 1996). In order to explain this paradox, many have argued that tacit knowledge, which is difficult to articulate and codify, is much more difficult to standardize and communicate outside of close social relations. It is developed within close 'communities of practice' which are characterized by a shared understanding of conventions of work and ways of problem solving (Wenger, 1998). It has been argued that the exchange of tacit knowledge depends on high degrees of trust and face to face personal contacts, so that it is more readily available to firms in local clusters and unavailable or available at much greater costs to firms outside of these agglomerations (Maskell and Malmberg, 1999; Malmberg and Maskell, 2002).

This is not to assume that tacit knowledge is inevitably local or, conversely, that codified knowledge is necessarily non-local, even global (Allen, 2000). Not only are the precise definitions of tacit and codified knowledge contentious (Gertler, 2003), the relationships between them are far from straightforward. Both tacit and codified knowledge are shared and travel within epistemic communities (Amin and Cohendet, 1999; Håkanson, 2005). Many scientific communities tend, of course, to be international and some argue that modern communications technologies are also making economic epistemic communities increasingly insensitive to distance. But the geographies of such communities are shaped by their institutional contexts so that their effect depends on how they operate at different scales and how their flows of knowledge are integrated and mixed in particular regions and innovation systems.

The concept of the regional or territorial innovation system has assumed considerable prominence in economic geography in recent years. Knowledge and innovation have been conceived as associational, interactive and distributed, so that innovation is a networked activity in which firms' research activities are closely linked to collaborating firms, and to scientific and research institutions. It has been argued that these innovation infrastructures are not simply nationally variable but that they often have important regional dimensions so that distinctive constellations and networks of firms, clusters and institutions can be identified at regional

---

7   Work on Asian and other non-Western corporations and multinationals has revealed significant cultural and organization differences from their North American and European counterparts, characteristics that are reflective of corresponding national differences in the institutional and cultural foundations of their respective capitalist systems (see Hall and Soskice, 2001).

and sub-regional scales. Economic geographers have attempted to systematize the different dynamics and forms of governance shown by different types of regional and local innovation system, and even to differentiate such systems on the basis of the types of knowledge on which they are based. Thus it is argued that science-based industries rely on forms of 'analytic knowledge', while engineering industries depend more on 'synthetic' knowledge, whereas creative industries rely on forms of 'symbolic knowledge', all requiring quite different local socio-economic-institutional conditions (Asheim and Gertler, 2005). Whether and how far these categorizations of knowledge help us to better understand knowledge flows and the emergence of regional and local innovation systems remains to be seen, but what is clear is that understanding the geographies of innovation is crucial to any appreciation of the new global knowledge economy. This in turn has highlighted the importance of the firm and skilled labour in shaping the innovative capabilities of regions and localities.

Firms, Workers and Places

Modern economic geography has long been aware of the firm, and of differences in labour force characteristics across space, but both the importance accorded to the geographies of firms and workers, and how these spatialities have been theorized and explained has changed significantly over the past two decades or so. For much of its initial post-war development, during its location-theoretic phase from the mid-1950s to the mid-1970s, economic geography tended to regard the firm as something of a 'black box', an entity of considerable importance for understanding the geographies of industry and employment, but for the most part viewed simply as a singular entity or 'agent' that makes location decisions in response to the relative attractiveness of alternative possible sites. There was relatively little investigation of the nature of the firm, how it is organized, or the specific form of its production process, or how and why firms innovate. Somewhat akin to the economist's notion of a production function, the model used to specify how specific inputs – labour, capital and technology – are combined in a firm (or industry) to produce an output, so in economic geography firms were deemed to have 'location functions', by which they trade-off the relative attractiveness of different possible locations to choose the 'optimal' place where production will yield maximum profits. In this approach, a key aspect of the relative attractiveness of different locations to firms was the local availability and skills and cost of labour. Just as there was scant discussion of the nature of the firm, of what actually goes on inside firms, how investment and disinvestment decisions are made, or how innovation occurs, so little attention was directed to the actual use of labour in the firm, to the qualities of the workforce, to the employment relation or to social conditions in firms.

Over the past twenty years or so, however, economic geographers have devoted increasing effort to recasting the way they think about firms and workers (for example, Clark and Wrigley, Chapter 11; Schoenberger, 1997; Maskell, 2001; Gertler, 2004). The growth and transformation of the large firm, and of multinational, transnational and truly global corporations, on the one hand, and the resurgence of independent small enterprises on the other, have helped stimulate this new interest. At the same time, the emergence of new theories of the firm within economics, and management, business and organizational studies has presented economic geographers with a richer conceptual canvas from which to draw ideas. The list of theories, many of which overlap, includes transaction cost theories, managerial theories, evolutionary theories, resource based theories, competence theories and entrepreneurial theories. Which of these, then, are of potential use in economic geography? According to Maskell (2001) the selection criteria rest largely on how far and in what ways a given theory of the firm gives theoretical significance to the spatial context in which the firm is placed. He suggests that of the various theories on offer, competence-based accounts of the firm would seem particularly relevant to economic-geographic work, since they offer most scope for linking firms to their specific local environments and settings, and allow geographers to identify and explain the ways in which a given location provides the firm with resources (externally derived competences, including tacit knowledge) that are not equally available elsewhere. There have also been arguments that even corporate cultures and cultural practices within firms are often place based (Schoenberger, 1997; Gertler, 2004).

This line of reasoning – of linking the nature, performance and relations of firms to the characteristics of their local and regional environments – is taken further in the notion of socio-spatial 'embeddedness', where the latter refers to the various locally-specific social, cultural, inter-firm and institutional networks of practices, alliances and knowledge that act as resources for, or constraints, on local firms.[8] In particular, the concept of local embeddedness emphasizes the importance of local networks and local knowledge for understanding the firm. It suggests that just as local and regional economies cannot be properly understood without understanding the firm, so the firm cannot be properly understood without understanding its relation to place, to the local and regional setting in which it is located and operates (Conti et al., 1995; Dicken and Malmberg, Chapter 12). It might be thought that whilst local embeddedeness should be important for small firms, it would be of much less significance for large, multinational and transnational firms, which are able to transcend local conditions by accessing and themselves providing resources across a global space. On the other hand, even large multidivisional corporations may take advantage of and often respond to specific local knowledge and social networks, and many such firms themselves claim to play to such local knowledge in tailoring their product or service to the specificities of each local market (witness HSBC's boast to be the 'world's local bank'). And as the trend towards radical decentralization and hierarchical fragmentation within many large corporations continues, the local embeddedness of individual divisions and units assume heightened significance even for such firms.

Nevertheless, it is important not to over-stress the role of local networks and embeddedness, even for small firms. In an increasingly global economy, firms often find it necessary and advantageous to establish links and interdependencies with other firms located at some spatial remove from themselves. Global networks can matter as much for the small firm as for the large multinational enterprise. These may be supply links, collaborative relationships, strategic alliances, product or technology licensing, or flows of codified information and knowledge. In short, both the formal and informal, traded and untraded networks that firms deploy and on which they depend need not be spatially proximate, or may be local for some aspects of a firm's functioning, but more distant, even global, for others. Propinquity does not necessarily depend on proximity: firms often are embedded in more distantiated networks. In thinking about the firm in network terms (Yeung, 2005), as a collaborative or 'relational'

---

8   The concept of 'embeddedness' is taken from the work of the economic sociologist, Granovetter (1985, 1993), and has been very influential in economic geography.

entity, therefore, the 'boundaries' of the firm are likely to extend over multiple spatial scales. Such blurred operational and knowledge boundaries clearly differ from the conventional legal boundaries of the firm.

The challenge facing economic geographers researching the firm is thus not only to understand firms in their appropriate local, global, institutional and sectoral contexts, but also to find the concepts and categories to connect micro-economic theorizations of the firm with broader conceptions of the economic, social and institutional construction of the geographical spaces and networks that shape a firm's access to resources, knowledge and finance. Making these connections is important because firms configure, and constantly reconfigure, the landscape of economic development, wealth creation and, importantly, of employment and work.

With regard to the world of work, here too economic geography has undergone substantial shifts in theoretical and empirical orientation. Unsurprisingly, the dramatic transformations that have recast the nature, composition and distribution of work over the past quarter of a century have compelled economic geographers to rethink the notion of labour. De-industrialization, tertiarization, globalization and the information technology revolution, have in combination reconfigured the occupational, skill, gender and reward structures of employment, and as a consequence the geographies of work, wages and welfare at all spatial scales, from the intra-urban to the global (Martin and Morrison, 2003). Compared to two or three decades ago, the landscape of labour-market inequality is now a much more rugged terrain.

These new forms of spatial differentiation of the labour market present economic geographers with a series of important empirical and theoretical questions, not only in the terms of making sense of the new patterns that are emerging, but also in terms of understanding how local labour markets function in today's rapidly changing and globalizing economic environment, and what form local labour market policies and interventions should take. A major breakthrough in the way economic geographers think about labour and labour markets came in the early 1980s, and was associated – in part at least – with the Marxist turn that affected the subject more generally. The focus of this Marxist approach was on the labour process, which is the way in which – technically and socially – labour is utilised and regulated within the firm, as part of the organization and techniques of production. The argument was that as the labour process changes – for example because of major developments in technology, or in the organizational structure of the firm itself – so the geographies of work are likely to be recast accordingly, in ways that depend on the specific labour requirements of firms on the one hand, and the existing geographies of employment, social relations and class structures on the other. In this way different labour processes produce different 'spatial divisions of labour' (Storper and Walker, 1984, Massey, 1984). Thus, while some industries may concentrate in particular regions, producing sectoral 'spatial divisions of labour', increasingly firms, and especially multinational firms, have located technically separate different stages of their activities (research and design, production, assembly, distribution, marketing) across different locations (regions and countries). Clearly various combinations are possible, and with the rise of the global corporation, spatial divisions of labour now often take on complex international dimensions.

Over the past decade and a half, the treatment of labour in economic geography has widened still further. One stream of work has focused on the conceptualization of local labour markets (Peck, 1989, and Chapter 13, 1996; Martin 2000). It is within specific spatial settings

– 'local labour markets' – that workers seek employment and employers hire and fire their workers, that particular forms of jobs, employment practices, work cultures, labour relations and worker and employer networks become established, and where particular forms of labour regulation become socialized and institutionalized. To be sure, local labour markets are often difficult to define with empirical precision, are intrinsically open and of course are interlinked in complex ways with other local labour markets, often in systems of indirect relations with other markets at considerable geographical distance, even at the other side of the globe. In an increasingly global economy, local labour has become ever more vulnerable to the shifting contours of international competition and the attendant investment and disinvestment strategies of transnational and global firms. These issues raise important questions concerning how and why local labour markets differ: are local labour markets based on common processes with local variations, or are the processes involved inherently different between localities? The jury is still out on this debate.

A second stream of work has sought to stress that local labour is not merely a passive entity at the mercy of the imperatives and exigencies of capital, but is itself an active agent. As Herod (1997, p. 3) argues 'the economic geography of capitalism does not simply evolve around workers who themselves are disconnected from the process. They are active participants in its very creation.' To be sure, this does not mean that workers are free to construct economic landscapes just as they please, for the scope for labour's agency is restricted; but such a perspective does highlight the key fact that labour can and does influence those landscapes through workers' social and spatial practices, resistance and reproduction (Herod, 1997, 2001, 2003). Nor is such local agency confined to organized workers in advanced economies, but is also becoming apparent among the low-wage workers voicing dissent over their pay and employment conditions in the overseas mass production factories of Western companies.

Indeed, the changes in the labour market and world of work over the past quarter of a century connect with a third stream of research in labour geography, concerned with the nature, practice and performance of post-industrial work. Three shifts in particular have combined to produce major disparities amongst different workers in the constraints on and opportunities for agency: marked de-unionization and loss of worker power in the industrial workplace, the drive by firms (and governments) to increase the 'flexibility' of work and workers (as part of the perceived need to remain 'competitive' in the new global economy), and the increasing importance of high-level skills, advanced education and creative talent in determining access to good quality jobs and decent incomes (Florida, 2002) . These developments have curtailed the scope for agency for many workers, whilst enhancing it for others. Similarly, as the service economy has expanded, so have the numbers of part-time, low-wage, and temporary jobs, especially in contract and other similar services. Flexibility in these jobs often means awkward hours, little or no job security, and low and variable wages (Carnoy, 2000). And while there has been an increase in the number of women in high-skilled and professional occupations, gender divisions within and between workplaces produce and reproduce inequalities in the type and experience of work, in pay and in career advancement, as McDowell (1997) showed most clearly for the City of London. For others, working in skilled jobs in high-technology sectors, in computing, in software, and in many of the cultural industries, what Reich (2000) calls 'symbolic' workers, flexibility means greater scope for individual creativity, high pay, and self-determined flexible working patterns, though self exploitation and adverse impacts on work–life balance are often attendant side effects (Benner, 2002). In this world of greater labour market 'flexibility', polarization, increasingly individualized employment contracts, and of widespread insecurity, unions seem to find themselves at a distinct disadvantage. Historically, trade unionism has been one of the most geographically prominent aspects of the labour market: most unions have their origins in the industrial and social specificities of particular areas. The sharp decline of trade unionism over the past two decades or so has thus challenged these geographical 'heartlands', and assessing the extent and ways in which this process has unfolded has spawned a number of conceptual and methodological enquiries by economic geographers (for example, Martin, Sunley and Wills, 1996; Herod 2000, 2001, 2003). At the same time, the transition to the so-called 'New Economy', with its 'new rules of work' (Freeman, 1994; Herzenberg, Alic and Wial, 1999; Reich, 2000; Ellwood et al., 2000; Beck, 2002; Benner, 2002; Head, 2003), poses new challenges for unions. The response of unions to such developments will require strategies that are not simply economic or political in nature, but also geographical. In the new global economy, the future of labour unions will depend in part on how far they manage not only to find new ways of appealing to and representing local labour, but also linking with and relating to similar workers on a wider global landscape (Waterman and Wills, 2001).

Culture, Technology and the Geographies of Knowledge

The rise of the so-called 'New Economy' has itself stimulated exciting developments in economic geography in recent years. For one thing there has been debate over the precise meaning and significance of the 'New Economy', and different interpretations of its key characteristics and spatial implications (see, for example, Gadrey, 2001; Thrift, 2005; Martin, 2006b; Daniels et al., 2006). Though associated by many with the dazzling – but as it turned out, short-lived - boom of internet 'dotcom' companies at the end of the 1990s, the concept has since taken on a number of more substantive, if debated, connotations (see Table 2). For many it symbolizes a profound shift towards an information-orientated, knowledge-driven economy, where constant innovation is the secret to competitive success in today's global marketplace (Castells, 1996; Cohen, 2003, Breschi and Malerba, 2005). This perspective highlights the spatialities of the creation and genesis of new knowledge within 'territorial innovation systems', and the networks by which such new knowledge flows from place to place (Warf, Chapter 15). Economic geographers have sought to determine what makes for innovative-rich and 'learning' regions, and how different forms of knowledge - such as codified, tacit, analytic, symbolic and synthetic – have different spatial logics and dynamics. In such discussions, emphasis is frequently put on local institutional, social and cultural determinants of knowledge creation and circulation (Cooke and Morgan, 1998).

Others focus more on what they see as the increasing 'dematerialization' of economic accumulation, on the 'weightless' nature of an expanding array of information-based services and products, many of which are characterized by a unique 'non-exclusivity' (for example the same internet service can be used by more than customer simultaneously). The internet economy is seen as not only reconstitutive of the geographies of economic activity, work, wealth and consumption, but by some of these commentators as rendering distance and even geography increasingly irrelevant: the argument is that because of advanced information-telecommunications technologies, economic proximity – between and amongst producers

Table 2 Conceptions of the so-called 'New Economy'

Concept	Key Characteristics
A new information-based, knowledge-driven networked economy	Revolutionizing effects of new information-communications technologies and industries of the social and spatial organization of economic life. 'Space of flows rather than space of places'; territorial innovation systems.
A new form of 'dematerialized capitalism'	New model of economic organization, accumulation, work, consumption and regulation based around (increasingly global) digital information, products and processes. The 'death of distance' and the rise of the 'global bazaar' (Coyle, 1997; Leadbeater, 1997; Reich, 2001; Friedman, 2005)
A new 'cultural capitalism'	Driven by prominence of cultural production and consumption, a 'New Economy' of symbols, signs and sensations. Cities as the new sites of cultural production (Scott, 2000, 2007; Power and Scott, 2004; Thrift, 2005)
A new era of 'creative capitalism'	The rise of new creative and entrepreneurial classes as the key drivers of innovation and growth. Cities and regions as creative spaces (Florida, 2003; Cooke and Schwartz, 2007).

and customers – no longer requires spatial propinquity (O'Brien, 1992; Coyle, 1997; Leadbeater, 1997; Reich, 2001; Friedman, 2005). Such extreme versions of this prognosis of the 'end of geography', 'death of distance' and 'flattening of the world', do not square, however, with the observation that many of these knowledge-producing and knowledge-intensive activities are highly spatially clustered (examples include biotechnology, computer software, internet service providers, film production, fashion design and financial services). Whilst economic geographers themselves increasingly recognize the importance of 'relational' as against 'geographical' proximity, nevertheless, spatial clustering suggests that while the networks of economic relationships may well now be globally extensive and diffuse, and the transmission of certain (codified) information all but instantaneous, spatial proximity is still an import factor in the locational logics of many 'New Economy' activities (Leamer and Storper, Chapter 20; Oinas and Malecki, 2002). The positive externalities that spatial clustering and agglomeration confer are not just 'hard' economic benefits – such as specialist suppliers, skilled labour, demand-supply linkages, and so on – but also include 'soft' and 'untraded' interdependencies amongst firms, such as locally embedded networks of social relations and cultural practices, norms and conventions that act as conduits for information and knowledge exchange and hence innovation and learning, as well as the bases of trust and cooperation (Gertler, Chapter 17; Maskell, 2001).

A third view of the 'New Economy' sees it as driven by the very prominence of an expanding array of 'cultural' activities. Culture itself has become commodified, and an increasing part of the economy is now concerned with the production and consumption of cultural products and services: the so-called 'cultural industries'. Products and services increasingly embody cultural signifiers and markers ('we are what we consume'), so that the modern economy is as much about symbols and signs as about 'hard' commodities.[9] The spaces of consumption and retailing both reflect and promote this cultural economy. Cities, in particular, it is argued, have taken on a new role and significance as the loci of cultural production and consumption (Scott, Chapter 16, 2001b, 2007; Power and Scott, 2004; Thrift, 2005). From film, music, art and theatre, to fashion, design, media, sport, restaurants, museums and tourism, the economies and physical landscapes of cities are being regenerated and revived around the production and consumption of culture, spectacle and leisure. According to some observers these activities, along with those knowledge based industries that support and facilitate them (such as architecture, ICT providers, marketing, exhibition agencies and the like) signal a new 'creative capitalism' driven by a new 'creative class' of innovative, entrepreneurial individuals (Scott, Chapter 16; Florida, 2003; Cooke and Schwartz, 2007) who flourish in the socio-economic diversity and 'buzz' that cities and certain city-regions can offer, and who themselves contribute to the 'cultural capital' of such places. But whilst the so-called 'creative industries' are indeed growing rapidly in terms of employment and output, they have also been over-hyped, and seized upon too unthinkingly by policymakers as some sort of panacea for urban and regional development. Just as the very notion of the 'New Economy' conjures up a misleading image of an unproductive and outmoded 'old economy', so the focus on a new set of culturally-based 'creative industries' seems to suggest that more traditional production and service industries are 'non-creative', non- innovative and neither require nor generate much (new) knowledge. Such representations would be most misleading, however. New knowledge and innovation – in short 'creativity' – have always been fundamental to the economic success of firms, industries and places, as Alfred Marshall emphasized in his discussion of the industrial districts of the nineteenth century.

In any case, these debates surrounding the 'New Economy' and its spatialities are only one facet of the growing interest in the 'cultural economy' by economic geographers. For over the past decade and a half, a more basic 'cultural turn' has been underway within the discipline. In some respects, the rise of 'cultural economic geography' is one of the most contentious developments in the discipline's recent history. Received wisdom has long regarded 'culture' and 'economy' as self-determining entities, each with its own discrete set of institutions, rationalities and conditions of existence, indeed each defined by what the other is not: 'economy' as irreducibly instrumental, materialistic, vulgar, tangible and devoid of morality; 'culture' as non-instrumental, intrinsic, aesthetic, normative and intangible (see Jackson, 2002). However, since the early 1990s economic geographers increasingly have embraced a range of more fluid and hybrid conceptions of 'the economic' that emphasize its mutual constitution by, and hence fundamental inseparability from, 'the cultural' (see for example Lee, 1989; Thrift and Olds, 1996; Barnett, 1998; Crang, 1997; Lee and Wills, 1997; Massey, 1997; Leyshon, Lee, and Williams, 1997; Sayer, 1994, 1997). Under the broad banner of 'cultural economy' geographers have, on the one hand, examined how 'traditional' economic concerns of capital, production, exchange, valuation and consumption, operate within, and impact on, the spatially variable sets of socio-cultural conventions, norms, attitudes, values and beliefs of the societies within which economic decisions and practices take place and,

---

9   Some would argue that the rise of the 'cultural economy' is even more fundamental, and has to do with the very meaning, nature and operation of 'markets' in post-industrial capitalism (du Gay and Pryke, 2002).

on the other hand, how these economic categories are themselves discursively as well as materially constructed, practised and performed at different spatial scales.

The 'cultural turn' in economic geography has involved not only a series of shifts in the substantive empirical research concerns of economic geographers, but also a related set of methodological, theoretical and conceptual shifts, themselves inseparable from an ontological shift at the deepest level in which geographers have called into question the very nature of 'the economic' itself. At one level, the 'cultural turn' is a direct response to the new economic realities that have accompanied the shift since the late 1970s to a post-industrial, knowledge-based, global capitalist economy, in which the socio-cultural foundations of economic success (and failure) have become increasingly apparent at multiple spatial scales, ranging from the national scale (Whitley, 1999) to the regional (Sayer and Walker, 1992) to the level of the firm (Schoenberger, 1997). In addition to the stimulus provided by the rise of the so-called 'cultural industries' mentioned above, geographers have turned to 'conventions theory' to reveal the role of socio-cultural conventions, codes, and norms in organizing and co-ordinating particular learning behaviours (for example, Lundvall, 1992; Storper, 1992; Storper, 1997). At the same time, the nature of contemporary service work has forced scholars to question previous 'culture'/'economy' binaries: because the production and consumption of the service occur simultaneously, the cultural identity, personality, appearance, embodied personality and social characteristics of the worker become part of the product consumed (see McDowell, 1997). Given the changed nature of the economy, many geographers have argued for a re-orientation and rethinking of economic geography that moves consideration of culture to centre stage (Thrift and Olds, 1996).

On a second level, the development of cultural economic geography also represents an epistemological critique of structurally determinist accounts of economic change, particularly ideas from Marxian political economy, regulation theory and neoclassical economics, the very approaches that, as we noted above, have variously dominated the discipline since the early 1970s, and in which singular notions of 'economy' invariably take precedence over the social and the cultural. Some have also criticized the modes of abstraction inherent in these previous approaches, arguing that they fail to account fully for complex sociocultural geographies of economic change. This is one of the reasons why economic geographers have shifted their focus to so-called 'background' factors, to the 'soft' socio-cultural, relational and contextual aspects of economic behaviour previously sidelined. The result has been a much more fluid and hybrid interpretation of the 'economic', in economic geography, fuelled at a broader level by postmodern and poststructuralist critiques that have developed within human geography more generally over the last decade or so.

Relatedly, the rise of cultural economic geography also represents a response to a call to revitalize and broaden out economic geography in order to retain its viability as an exciting, socially-relevant academic discipline. This call has stimulated an engagement with a variety of heterodox schools of economic thought (Amin and Thrift, 2000), including new economic sociology, feminist economics, evolutionary and institutionalist economics, economic anthropology, as well as other social sciences such as cognitive psychology, sociology, and management and organization studies. In so doing, cultural economic geographers have revisited, and sought inspiration from, previous work by a select group of 'celebrity scholars', including Karl Polanyi's work on embeddedness, Michael Polanyi's work on tacit knowledge, Thorstein

Veblen's work on institutions, Joseph Schumpeter's work on innovation and creative destruction, and the work of Michel Callon and Bruno Latour on actor networks to name but a few.

The cultural turn has had many positive effects, not least in escaping from previous reductionist treatments of culture as a mere reflection of material circumstances (Ray and Sayer, 1999). Another key effect has been the opening up of the discipline to a much broader range of conceptual, methodological and analytical frameworks. Ethnographic research methods and reflexive strategies of writing and authorship have assumed prominence over quantitative models and meta-theoretical schemas, and the contingent, open-ended and situated nature of the economic landscape is increasingly emphasized, together the unavoidable positionality of researchers and the situatedness of their knowledge claims (for example, Schoenberger, 1991; Clark, 1998; Barnes, 2001).

All this is not to suggest that the cultural turn in economic geography has been unproblematic or gone uncontested. In some versions, the cultural has been elevated to the point of dismissing (and denigrating) the economic altogether, and with it any idea of meta-theory, a tendency that Harvey (2006) complains has been especially evident amongst British geographers. It has also been associated with what some see as a move too far away from rigorously conceptualized, analytically based and detailed empirical work, in favour of fuzzy concepts, impressionistic story telling and thin empirics (Markusen, 1999; Martin and Sunley, 2001; Rodríguez-Pose, 2001; Scott, Chapter 1). As others have sought to stress, however, the aim of the new cultural economic geography is most certainly not to reject or downgrade the economic logics of production, allocation and distribution, but instead to reveal the diversity of ways in which 'cultural' and 'economic' practices are interwoven and mutually constitutive at a range of spatial scales and in different institutional contexts. In this way, cultural economy and political economy should most appropriately be seen as complementary rather than competing perspectives (Lee, 2002; Hudson, 2004).

Regulating Economic Spaces

If the nature and trajectory of economic development amongst the advanced countries (to say nothing of the global economy more generally) have been changing dramatically over the past quarter of a century, so too have the regulatory frameworks and structures within which that development has taken place. Geography is inextricably bound up with economic regulation. Economic regulation is a multi-scalar process. National states are obviously key agents and institutions of economic regulation, and shape the process and geographies of economic development through a wide range of fiscal, monetary, welfare and other economic policies, as well as legal and other arrangements (Painter, 2000). Above national states exist the various supra-national, international and global bodies and institutions (such as the WTO, IMF, NAFTA, European Commission) that also exert a powerful regulatory influence. Below the national state are numerous regional, city and local authorities and agencies that also have varying regulatory powers and responsibilities. The relationships and interactions between these different regulatory levels are complex and not infrequently a source of tension. At the same time, these different sources of regulation exert a key influence over and impact differently on the economic development of individual regions, cities and localities, and play a crucial role in shaping the economic landscape. In this dual sense, therefore, the study of economic regulation is an integral part of economic geography, and in recent years has become increasingly recognized as such.

One line of enquiry has focused on theorizing and empirically interrogating the economic-geographic implications of major historical shifts in the form and mode of national economic regulation, and in particular on the economic-geographic consequences of the breakdown of the post-war Fordist- or Keynesian-welfare state form of socio-economic regulation in the late 1970s and the subsequent rise and global diffusion of a neoliberal form. How to conceptualize these historical shifts in 'regulatory regime', and their implications for the economic landscape was a primary factor in the adoption and development of the so-called regulationist approach in economic geography in the 1980s and 1990s. This approach, itself a critical reaction to, and outgrowth from, the Marxist perspective of the mid-1970s to mid-1980s, views the capitalist economy as comprising two inextricably enmeshed spheres – a 'regime of accumulation' (the forms and organization of production, investment, consumption) and a 'mode of regulation' (the panoply of both formal and informal mechanisms and institutions, including the state and its policies, the monetary system, welfare system, education system, business conventions, and social norms and rules) that guides, steers and co-ordinates the social relations of economic accumulation (see Jessop, 1997, for a survey of regulation theory).

Whilst not all OECD states adopted the Keynesian-welfare model to the same extent or in the same precise form – and such differences were themselves both a reflection and a determinant of different national variants of capitalism (Hollingsworth and Boyer, 1997; Hall and Soskice, 2001) – it is possible to identify a distinct set of spatial-economic logics and outcomes of this historical form of state-led economic regulation and coordination (Martin and Sunley, Chapter 21). Not least among these were the spatially stabilizing effects of aggregate demand management, significant levels of state ownership of industry, the spatially redistributive effects of a generous and automatic benefit system, and various explicit policies to redistribute economic activity between regions and cities. In general, a key outcome of the Keynesian-welfare mode of regulation was a tendency for both income inequalities and regional economic disparities to narrow over the course of the 1950s, 1960s and 1970s.[10]

According to the regulation-theoretic approach the rise of the neo-liberal mode of regulation from the early 1980s onwards has been inextricably bound up with the collapse of Fordism and the transition to a 'post-Fordist' regime of economic accumulation (Jessop, 1994a; Tickell and Peck, 1992; Peck and Tickell, 1994). Again, whilst this model has also taken on different national variants (Jessop, 1994b, 1995; Peck, 2001), the spatial-economic implications have become all too apparent. By dismantling many of the main pillars of the Keynesian-welfarist model – especially by abandoning the pursuit of full employment via demand management, deregulating financial systems, privatizing large sections of the public sector, restricting the power of labour unions, replacing welfare by workfare, and driving through policies aimed at 'flexibilizing' the economy – the new regulatory model has intentionally exposed regions and cities to the full discipline of market forces. In so doing, it has also been complicit in the widening of income inequalities and regional disparities that has occurred over the past two

---

10 Interestingly, whereas the long post-war boom was widely associated with a steady narrowing of regional economic disparities, the boom period since the early 1990s has not seen further narrowing of regional inequalities. Indeed, in some cases regional disparities have widened, especially at the sub-regional scale.

Economy: Critical Essays in Human Geography

decades. The spatial-economic logic of the neoliberal model is founded on a belief that to survive in the new global market place, regions and cities must become more competitive and that this requires the supply-side of such local economies to be flexible, adaptive and innovative (Peck and Tickell. 2002). This political-ideological imperative has been implemented in part by decentralizing and devolving various aspects of economic regulation and intervention from the central state down to regional and city levels, on the grounds that this is where supply-side policy action is best implemented (Cox, Chapter 23). Set adrift in the harsh realities of the global economy, regions and cities have thus become busy with devising economic strategies aimed at raising their competitiveness, jockeying for position in the international league tables of local economic performance and well-being (Malecki, 2004; Martin, 2006; Martin, Kitson and Tyler, 2006; Bristow, 2006).

For some observers this rise of the region and the city as a new locus of economic regulation and governance is in fact linked to, and is a manifestation of, the decline of the nation state as an economic actor, a shift in the cartography of economic and regulatory power towards a new 'borderless world' in which major global city-regions are the major engines of economic growth and policy formulation (Ohmae, 1995; Scott, 1998, 2001). It might even be argued that the more that states pursue neoliberalism the more they will end up ceding economic and regulatory power to such global city-regions. However, it is important not to exaggerate such claims, for not only do nation states still retain distinctive key regulatory roles and functions, even in the face of accelerating globalization (see Cox, Chapter 23), economic geographers have only just begun to assess the dividends and disadvantages associated with regional decentralization and devolution (Rodríguez-Pose and Gill, Chapter 24, 2005). What is clear, however, is that the spaces of economic regulation are being reconfigured in historic ways, with major implications for how we think about the geographies of economies.

Future Challenges

Over the past quarter of a century, economic geography has undergone what is arguably the most intense phase of intellectual development in its century-old history. The theoretical bases of the discipline have broadened and deepened, and are now richer in scope and relevance than ever before. Likewise, challenged by unprecedented changes and transformations in the economic landscape, its subject matter has expanded apace. The essays collected below attempt to capture a flavour of these developments. They provide but a partial glimpse into the huge literature that has appeared in recent years: inevitably, several important areas of theoretical and empirical enquiry, and numerous equally formative contributions, have not been included. Nevertheless, the collection conveys a sense of the excitement and intellectual fervour that has come to energize the subject. It is probably not an exaggeration to claim that over the past two decades economic geography has undergone something of a renaissance – both inside geography itself, and across the social sciences more generally.

Whilst there is no space here to offer extensive promissory comments on the likely directions in which the subject will develop in the coming years, a few suggestions can be offered. Almost certainly, there will be continued engagement with social knowledges other than the economic, as geographers seek to capture the role of the cultural, social, institutional and political in shaping the economic landscape. For another, there is a need to deepen our understanding of how economic space is constituted and organized at different geographical scales, and how these different scales interact. What is required here is a major effort to bring together ideas on networks, relational proximity and the relationship between the local and the global to inform our conceptions of regions and cities. There is also considerable scope for developing the evolutionary approach that some economic geographers have begun to explore, using ideas from evolutionary economics, and perhaps complexity theory (see, for example, Boschma and Frenken, Chapter 6; Martin and Sunley, 2006). Perhaps most obviously, and certainly most urgently, there is a pressing need to bring environmental issues and considerations to the fore in economic geography. With the evidence for global warming mounting, we need to identify and analyse the environmental impacts of particular forms and patterns of spatial economic growth and development (for an example, see Hudson, 2000). For example, how do we conceptualize and measure the carbon 'footprint' of regions and cities? Yet another, and not unrelated, issue that is beginning to attract attention amongst policy makers is the measurement of regional and local 'well-being', in recognition of the limits of conventional measures of regional and spatial economic disparities (in terms of per capita incomes and GDP), to take account of the environmental, social and cultural factors that also determine people's quality of life. For, ultimately, this is surely what economic geography is about: understanding why socio-economic well-being varies from place to place, and thereby helping to inform policies to redress such variations.

References

Allen, J. (2000), 'Power/Economic Knowledge: Symbolic and Spatial Formations', In J.R. Bryson, P.W. Daniels, N. Henry and J.Pollard (eds), *Knowledge, Space, Economy*, London: Routledge, pp. 15–33.

Amadeo, D. and Golledge, R.G. (1975), *Introduction to Scientific Reasoning in Geography*, New York: Wiley.

Amin, A. and Cohendet, P. (1999), 'Learning and Adaptation in Decentralized Business Networks', *Society and Space*, **17**, pp. 87–104.

Amin, A. and Thrift, N. (2000), 'What Kind of Economic Theory for What Kind of Economic Geography?', *Antipode*, **32**, pp. 4–9.

Asheim, B., Cooke, P. and Martin, R. (eds) (2006), *Clusters and Regional Development: Critical Reflections and Explorations*, London: Routledge.

Asheim B. and Gertler. M. (2005), 'The Geography of Innovation: Regional Innovation Systems', in J. Fagerberg, D.C. Mowery and R.R. Nelson (eds), *The Oxford Handbook of Innovation*, Oxford: Oxford University Press, pp. 291–317.

Audretsch, D.B. and Feldman, M.P. (1996), 'R&D Spillovers and the Geography of Innovation and Production', *American Economic Review*, **86**, pp. 630–40.

Bachi-Sen, S. and Lawton Smith, H. (2006), *Economic Geography: Past, Present and Future*, London: Routledge.

Bair, J. (2005), 'Global Capitalism and Commodity Chains: Looking Forward, Going Forward', *Competition and Change*, **9**(2), pp. 153–80.

Barnes, T.J. (1995), 'Political Economy, I: "The Culture Stupid", *Progress in Human Geography*, **19**, pp. 423–31.

Barnes, T.J. (1996), *Logics of Dislocation: Models, Metaphors and Meanings of Economic Space*, New York: Guilford Press.

Barnes, T.J. (2001), 'Re-theorizing Economic Geography: From the Quantitative Revolution to the "Cultural Turn"', *Annals of the Association of American Geographers*, **91**, pp. 546–65.

Barnett, C. (1998), 'The Cultural Turn: Fashion or Progress in Human Geography?', *Progress in Human Geography*, **30**, pp. 379–94.

Bathelt, H., Malmberg, A. and Maskell, P. (2004), 'Clusters and Knowledge: Local Buzz, Global Pipelines and the Process of Knowledge Creation', *Progress in Human Geography*, **28**(1), pp. 31–56.

Beccattini, G. (1990), 'The Marshallian Industrial District as a Socioeconomic Notion, in F. Pyke, G. Beccattini and W. Sengenberger (eds), *Industrial Districts and Inter-firm Co-operation in Italy*, Geneva: ILO, pp. 37–51.

Benner, C. (2002), *Work in the New Economy: Flexible Labour Markets in Silicon Valley*, Oxford: Blackwell.

Benneworth, P. and Henry, N. (2004), 'Where is the Value added in the Cluster Approach: Hermeneutic Theorising, Economic Geography and Clusters as a Multiperspectival Approach', *Urban Studies*, **41**, pp. 1011–1023.

Boddy, M. (1999), 'Geographical Economics and Urban Competitiveness: A Critique', *Urban Studies*, **36**, pp. 811–42,

Boschma, R.A. and Lambooy, J.G (1999), 'Evolutionary Economics and Economic Geography', *Journal of Evolutionary Economics*, **9**(4), pp. 411–29.

Brakman, S., Garretsen, H. and Marrevijk, C. van (2001), *Introduction to Geographical Economics*, Cambridge: Cambridge University Press.

Breschi, S. and Malerba, F. (eds) (2005), *Clusters, Networks and Innovation*, Oxford: Oxford University Press.

Bristow, C. (2006), 'Everyone's a Winner: Problematising the Discourse of Regional Competitiveness', *Journal of Economic Geography*, **5**, pp. 285–304.

Cairncross, F. (1997), *The Death of Distance: How the Communications Revolution will Change Our Lives*, London: Orion Books.

Carnoy, M. (2000), *Sustaining the New Economy: Work, Family and Community in the Information Age*, Cambridge, MA: Harvard University Press.

Castells, M. (1996), *The Rise of the Networked Society*, Oxford: Blackwell.

Clark, G.L. (1998), 'Close Dialogue and Stylized Facts: Methodology in Economic Geography', *Annals of the Association of American Geographers*, **88**, pp. 73–87.

Clark, G.L., Feldman, M. and Gertler. M. (eds) (2000), *Oxford Handbook of Economic Geography*, Oxford: Oxford University Press.

Cohen, D. (2003), *Our Modern Times: The New Nature of Capitalism in the Information Age*, Cambridge, MA: MIT Press.

Conti, S., Malecki, E.J. and Oinas, P. (1995), *The Industrial Enterprise and its Environment: Spatial Perspectives*, Aldershot, England: Avebury.

Cooke, P. (2001), 'Regional Innovation Systems, Clusters, and the Knowledge Economy', *Industrial and Corporate Change*, **10**, pp. 945–74.

Cooke, P. and Morgan, K. (1998), *The Associational Economy: Firms, Regions and Innovation*, Oxford: Oxford University Press.

Cooke, P. and Schwartz, D. (eds) (2007), *Creative Regions: Technology, Culture and Knowledge Entrepreneurship*, London: Routledge.

Corportaux, J. and Crevoisier, O. (2007), 'Economic Theories and Spatial Transformations: Clarifying the Space-Time Premises and Outcomes of Economic Theories', *Journal of Economic Geography*, **7**, pp. 285–309.

Coyle, D. (1997), *The Weightless World*, Oxford: Capstone Publishing.

Crang, P. (1997), 'Cultural Turns and the (Re)constitution of Economic Geography', in R. Lee and J. Wills (eds), *Geographies of Economies*, London: Edward Arnold, pp. 3–15.

Daniels, P., Leyshon, A., Bradshaw, M. and Beaverstock, J. (eds) (2006), *Geographies of the New Economy*, London: Routledge.

Dei Ottati, G. (1994), 'Cooperation and Competition in the Industrial Districts as an Organizational Model', *European Planning Studies*, **2**, pp. 463–83.

Dicken, P., Kelly, P., Olds, C. and Yeung, H. W-C. (2001), 'Chains and Networks, Territories and Scales: Towards a Relational Framework for Analysing the Global Economy', *Global Networks*, **1**(2), pp. 80–112.

Dicken, P. and Lloyd, P. (1972), *Location and Space: A Theoretical Approach to Economic Geography*, London: Harper and Row.

du Gay, P. and Pryke, M. (eds) (2002), *Cultural Economy: Cultural Analysis and Commercial Life*, London: Sage.

Duranton, G. and Rodríguez-Pose, A. (2005), 'When Economists and Geographers Collide, or the Tale of the Lions and the Butterfly', *Environment and Planning, A*, **37**, pp. 1695–705.

Duranton, G. and Storper, M. (2006), 'Agglomeration and Growth: A Dialogue Between Economists and Geographers', *Journal of Economic Geography*, **6**, pp. 1–7.

Dymski, G. (1996), 'On Paul Krugman's Model of Economic Geography', *Geoforum*, **27**, pp. 439–52.

Ellwood, D.T., Blanck, R.M., Blasi, J., Kruse, D., Niskansen, W.A. and Lynn-Dyson, K. (2000), *A Working Nation: Workers, Work and Government in the New Economy*, New York: Russell Sage Foundation.

Florida, R. (2002), *The Rise of the Creative Class*, New York: Basic Books.

Freeman, R. (ed.) (1994), *Working Under Different Rules*, New York: NBER-Russell Sage Foundation

Friedman, R. (2005), *The World is Flat: The Globalised World in the Twenty-First Century*, London: Penguin.

Fujita, M., Krugman, P. and Venables, A. (1999), *The Spatial Economy: Cities, Regions and International Trade*, Cambridge, MA: MIT Press.

Fujita, M., and Thisse, J. (2002), *Economics of Agglomeration: Cities, Industrial; Location and Regional Growth*, Cambridge: Cambridge University Press.

Gadrey, J. (2001), *New Economy, New Myth*, London: Routledge.

Gereffi, G. (1996), 'Global Commodity Chains: New Forms of Co-ordination and Control among Nations and Firms in International Industries', *Competition and Change*, **1**(4), pp. 427–39.

Gereffi, G., Humphrey, J. and Sturgeon, T. (2005), 'The Governance of Global Value Chains', *Review of International Political Economy*, **12**(1), pp. 78–104.

Gertler, M. (2003), 'Tacit Knowledge and the Economic Geography of Context, or The Undefinable Tacitness of Being There', *Journal of Economic Geography*, **3**(1), pp. 75–99.

Gertler, M. (2004), *Manufacturing Culture: The Institutional Geography of Industrial Practice*, Oxford: Oxford University Press.

Granovetter, M. (1985), 'Economic Action and Social Structure: The Problem of Embeddedness', *American Journal of Sociology*, **93**, pp. 481–510.

Granovetter, M. (1993), 'Economic Institutions as Social Constructions: A Framework of Analysis', *Acta Sociologica*, **35**, pp. 3–12.

Håkanson, L. (2005), 'Epistemic Communities and Cluster Dynamics: On the Role of Knowledge in Industrial Districts', *Industry and Innovation*, **12**(4), pp. 433–63.

Hall, P. and Soskice D. (eds) (2001), *Varieties of Capitalism: The Institutional Foundations of Comparative Advantage*, Oxford: Oxford University Press.

Harvey, D. (1969), *Explanation in Geography*, London: Edward Arnold.

Harvey, D. (1982), *Limits to Capital*, Oxford: Blackwell.

Harvey, D. (1985), 'The Geopolitics of Capitalism', in D. Gregory, and J. Urry (eds), *Social Relations and Spatial Structures*, London: Macmillan, pp. 128–63.

Harvey, D. (2006), *Spaces of Global Capitalism: Towards a Theory of Uneven Geographical Development*, London: Verso.

Head, S. (2003), *The New Ruthless Economy: Work and Power in the Digital Age*, Oxford: Oxford University Press.

Henderson, J. Dicken, P. Hess, M. Coe, N. and Yeung, H. W-C. (2002), 'Global Production Networks and the Analysis of Economic Development', *Review of International Political Economy*, **9**(3), pp. 436–64.

Henderson, J.V. (ed.) (2005), *New Economic Geography*, Cheltenham: Edward Elgar.

Herod, A. (1997), 'From a Geography of Labour to a Labour Geography: Labour's Fix and the Geography of Capitalism', *Antipode*, **29**, pp. 1–31.

Herod, A. (2000), 'Labour Unions and Economic Geography', in E. Sheppard and T. Barnes (eds), *A Companion to Economic Geography*, Oxford: Blackwell, pp. 341–58.

Herod, A. (2001), *Labour Geographies: Workers and the Landscapes of Capitalism*, New York: Guilford Press.

Herod, A. (2003), 'Workers, Space and Labour Geography, *International Labour and Working Class History*, **64**, pp. 112–38.

Herzenberg, S.A., Alic, J.A. and Wial, H. (1999), *New Rules for a New Economy: Employment and Opportunity in Post-Industrial America*, Ithaca, NY: Cornell University Press.

Hodgson, G. (2001), *How Economics Forgot History: The Problem of Historical Specificity in Social Science*, London: Routledge.

Hollingsworth, J.R. and Boyer, R. (1997), *Contemporary Capitalism: The Embeddedness of Institutions*, Cambridge: Cambridge University Press.

Hudson, R. (2000), *Production, Places and Environment*, Harlow: Prentice Hall.

Hudson, R. (2004), 'Conceptualising Economies and their Geographies: Spaces, Flows and Circuits', *Progress in Human Geography*, **28**, pp. 447–471.

Jackson, P. (2002), 'Commercial Cultures: Transcending the Cultural and the Economic', *Progress in Human Geography*, **26**, pp. 3–18.

Jessop, B. (1994a), 'Post-Fordism and the State', in A. Amin (ed.), *Post-Fordism: A Reader*. Oxford: Blackwell, pp. 251–79.

Jessop, B. (1994b), 'The Transition to Post-Fordsim and the Schumpeterian Workfare State', in R. Burrows and B. Loader (eds), *Towards a Post-Fordist Welfare State*, London: Routledge, pp. 13–37.

Jessop, B. (1995), 'The Regulation Approach, Governance and Post-Fordism: Alternative Perspectives on Economic and Political Change', *Economy and Society*, **24**, pp. 307–33.

Jessop, B. (1997), 'Survey Article: The Regulation Approach', *The Journal of Political Philosophy*, **5**(3), pp. 287–326.

Krugman. P. (1991), *Geography and Trade*, London: MIT Press.

Krugman, P. (1995), *Development, Geography and Economic Theory*, London: MIT Press.

Krugman, P. (1996), *The Self-Organising Economy*, Oxford: Blackwell.

Krugman, P. (1998), 'What's New about the New Economic Geography?', *Oxford Review of Economic Policy*, **14**(2), pp. 7–17.

Krugman, P. (2000), 'Where In the World is the New Economic Geography?', in G.L. Clark, M. Feldman and M. Gertler (eds), *Oxford Handbook of Economic Geography*, Oxford: Oxford University Press, pp. 49–60.

Leadbeater, C. (1999), *Living on Thin Air: The New Economy*, London: Viking Books.

Lee, R. (1989), 'Social Relations and the Geography of Material life', in D. Gregory and R. Walford, R. (eds), *Horizons in Human Geography*, London: Macmillan, pp. 152–69.

Lee, R. (2002), 'Nice Maps, Shame about the Theory: Thinking Geographically About the Economic', *Progress in Human Geography*, **26**, pp. 333–55.

Lee, R. and Wills, J. (1997), *Geographies of Economies*, London: Edward Arnold.

Leyshon, A., Lee, R. and Williams, R. (2003), *Alternative Economic Spaces: Rethinking the Economic in Economic Geography*, London: Sage.

Lundvall, B. (ed.) (1992), *National Systems of Innovation: Towards a Theory of Innovation and Interactive Learning*, London: Pinter.

McDowell (1997), *Capital Culture: Gender at work in the City*, Oxford: Blackwell.

Malecki, E. (2004), 'Jockeying for Position: What It Means and Why It Matters to Regional Development Policy When Places Compete', *Regional Studies*, **38**(9), pp. 1101–120.

Malmberg, A. and Maskell, P. (2002), 'The Elusive Concept of Localisation Economies: Towards a Knowledge-Based Theory of Industrial Clustering', *Environment and Planning, A*, **34**, pp. 429–49.

Markusen, A. (1999), 'Fuzzy Concepts, Scanty Evidence and Policy Distance: the Case for Rigour and Policy Relevance in Critical Regional Studies', *Regional Studies*, **33**, pp 869–86.

Martin R.L. (1994), 'Stateless Monies, Global Financial Integration and National Economic Autonomy: the End of Geography?', in S. Corbridge, R. Martin and N. Thrift (eds), *Money, Power and Space*, Oxford: Blackwell, pp. 253–78.

Martin, R.L. (1999), 'The New "Geographical Turn" in Economics: Some Critical Reflections', *Cambridge Journal of Economics*, **23**, pp. 65–91.

Martin, R.L. (2000a), 'Institutionalist Approaches to Economic Geography', in T. Barnes and E. Sheppard (eds), *A Companion to Economic Geography*, Oxford: Blackwell, pp. 77–94.

Martin, R.L. (2000b), 'Local Labour Markets: Their Nature, Performance and Regulation', in G.L. Clark, M. Feldmann and M. Gertler (eds), *The Oxford Handbook of Economic Geography*, Oxford: Oxford University Press, pp 455–76.

Martin, R.L. (2006a), 'Economic Geography and the New Discourse of Regional Competitiveness', in S. Bagchi-Sen and H. Lawton-Smith (eds), *Economic Geography: Past, Present and Future*, London: Routledge, pp. 159–72.

Martin, R.L. (2006b), 'Making Sense of the New Economy? Realities, Myths and Geographies', in P. Daniels, A. Leyshon, M. Bradshaw and J. Beaverstock (eds), *Geographies of the New Economy*, London: Routledge, pp. 15–48.

Martin, R.L. and Morrison (eds) (2003), *Geographies of Labour Market Inequality*, London: Routledge.

Martin, R.L. and Sunley, P.J. (2001), 'Rethinking the 'Economic' in Economic Geography: Broadening Our Vision or Losing our Focus?', *Antipode*, **33**(2), pp.148–61.

Martin, R.L. and Sunley, P. (2003), 'Deconstructing Clusters: Chaotic Concept or Policy Panacea?', *Journal of Economic Geography*, **3**, pp. 5–35.

Martin, R.L. and Sunley, P.J. (2006), 'Path Dependence and Regional Economic Evolution', *Journal of Economic Geography*(1999), **6**, pp. 395–438.

Martin, R.L., Sunley. P.J. and Wills, J. (1996), *Union Decline and the Regions*, London: Jessica Kinglsey.

Martin R.L., Kitson, M. and Tyler, P. (eds) (2006), *Regional Competitiveness*, London: Routledge.

Maskell, P. (2001), 'The Firm in Economic Geography', *Economic Geography*, **77**, pp. 329–44.

Maskell P. and Malmberg, A. (1999), 'Localised Learning and Industrial Competitiveness', *Cambridge Journal of Economics*, **23**, pp. 167–86.

Massey, D. (1984), *Spatial Divisions of Labour*, London: Macmillan.

Massey, D. (1992), 'Politics and Space/Time', *New Left Review*, **196**, pp. 65–84.

Massey, D. (1997), 'Economic/Non-economic', in R. Lee and J. Wills (eds), *Geographies of Economies*, London: Arnold, pp. 27–36.

Morgan, K. (2004), 'The Exaggerated Death of Geography: Learning, Proximity and Territorial Innovation Systems', *Journal of Economic Geography*, **4**, pp. 3–22.

O'Brien, R.O. (1992), *Global Financial Integration; The End of Geography*, London: Royal Institute of International Affairs.

Ohmae, K. (1995), *The End of the Nation State: The Rise of Regional Economies*, New York: Free Press.

Overman, H. (2004), 'Can we Learn Anything from Economic Geography Proper?', *Journal of Economic Geography*, **4**, pp. 501–16.

Painter, J. (2000), 'State and Governance', in E. Sheppard and T.J. Barnes (eds), *A Companion to Economic Geography*, Oxford: Blackwell, pp. 359–76.

Peck, J. (1989), 'Reconceptualizing the Local Labour Market: Space, Segmentation and the State', *Progress in Human Geography*, **13**, pp. 42–61.

Peck, J. (1996), *Work Place: The Social Regulation of Labour Markets*, New York: Sage.

Peck, J. (2001), 'Neoliberalizing States', *Progress in Human Geography*, **25**, pp. 445–55.

Peck, J. and Tickell, A. (1994), 'Searching for a New Institutional Fix: the After-Fordist Crisis and the Global-local Disorder', in A, Amin (ed.), *Post-Fordism: A Reader*, Oxford: Blackwell, pp. 280–15.

Peck, J. and Tickell, A. (2002), 'Neoliberalising Space', *Antipode*, **34**, pp. 380–404.

Peck, J., Sheppard, E. and Tickell, A. (eds) (2006), *Politics and Practice in Economic Geography*, London: Sage.

Peet, R. (1997), 'The Cultural Production of Economic Forms', in R. Lee. and J. Wills (eds), *Geographies of Economies*, London: Arnold, pp. 37–46.

Pinch, S., Henry, N., Jenkins, M. and Tallman, S. (2003), 'From "Industrial Districts" to "Knowledge Clusters": A Model of Knowledge Dissemination and Competitive Advantage in Industrial Agglomerations, *Journal of Economic Geography*, **3**, pp. 373–88.

Piore, M. and Sabel, C. (1984), *The Second Industrial Divide: Possibilities for Prosperity*, New York: Basic Books.

Porter, M. (1990), *The Competitive Advantage of Nations*, London: Macmillan.

Porter, M. (1998), *On Competition*, Boston, MA: Harvard Business Review.

Porter, M. (2000), 'Location, Competition and Economic Development: Local Clusters in the Global Economy', *Economic Development Quarterly*, **14**, pp. 15–31.

Potts, J. (2000), *The New Evolutionary Microeconomics*, Cheltenham: Edward Elgar.

Power, D. and Scott, A.J. (eds) (2004), *Cultural Industries and the Production of Culture*, London: Routledge.

Ray, L. and Sayer, A. (eds) (1999), *Culture and Economy After the Cultural Turn*, London: Sage.

Reich, R.B. (2001), *The Future of Success; Work and Life in The New Economy*, London: Heinemann.

Rodríguez-Pose, A. (2001), 'Killing Economic Geography with a "Cultural Turn" Overdose', *Antipode*, **33**, pp 176–82.

Rodríguez-Pose, A. and Gill, N. (2005), 'On the "Economic Dividend" of Devolution, *Regional Studies*, **39**, pp 405–20.

Sawyer, R.K. (2005), *Social Emergence: Societies as Complex Systems*, Cambridge: Cambridge University Press.

Sayer, A. (1982), 'Explanation in Economic Geography: Abstraction versus Generalization', *Progress in Human Geography*, **6**, pp. 68–88.

Sayer, A. (1994), 'Cultural Studies and "the Economy Stupid"', *Environment and Planning, D: Society and Space*, **12**, pp. 635–7.

Sayer, A. (1997), 'The Dialectics of Culture and Economy', in R. Lee and J. Wills (eds), *Geographies of Economies*, London: Arnold, pp. 16–26.

Sayer, A. and Walker, R. (1992), *New Social Economy: Reworking the Division of Labour*, Oxford: Blackwell.

Schoenberger, E. (1991), 'The Corporate Interview as Research Method in Economic Geography', *Professional Geographer*, **43**, pp. 183–9

Schoenberger, E. (1997), *The Cultural Crisis of the Firm*, Oxford: Blackwell.

Scholte, J.A. (2000), *Globalisation: A Critical Introduction*, London: Macmillan.

Scott, A. (1988), *New Industrial Spaces*, London: Pion.

Scott, A.J. (1998), *Regions and the World Economy: The Coming Shape of Global Production, Competition and Political Order*, Oxford: Oxford University Press.

Scott, A.J. (2000), 'Economic Geography: The Great Half-Century', in G.L. Clark, M. Feldman and M. Gertler (eds), *Oxford Handbook of Economic Geography*, Oxford: Oxford University Press, pp. 1–44. Reprinted from: *Cambridge Journal of Economics*, 1999, **24**, pp. 483–504.

Scott, A.J. (ed.) (2001a), *Global City Regions: Trends, Theory, Policy*, Oxford: Oxford University Press.

Scott A. J. (2001b), 'Capitalism, Cities and the Production of Symbolic Forms', *Transactions of the Institute of British Geographers*, **26**, pp. 11–23

Scott. A.J. (2006), *Geography and Economy*, Oxford: Oxford University Press.

Sheppard, E. and Barnes, T. (eds) (2000), *A Companion to Economic Geography*, Oxford: Blackwell.

Sheppard, E. and Plummer, P. (2001), 'Must Emancipatory Economic Geography be Qualitative', *Antipode*, **33**, pp. 194–9.

Sjöberg, O. and Sjöholm, F. (2002), 'Common Ground? Prospects for Integrating the Economic Geography of Geographers and Economists', *Environment and Planning, A*, **34**, pp. 467–86

Smith, D. (1971), *Industrial Location: An Economic Geographical Analysis*, New York: Wiley.

Smith, N. (1984), *Uneven Development: Nature, Capital and the Production of Space*, Oxford: Oxford University Press.

Storper, M. (1992), 'Regional Worlds of Production', *Regional Studies*, **27**, pp.433–55.

Storper, M. (1997), *The Regional World: Territorial Development in a Global Economy*, New York: Guilford Press.

Storper, M. (1999), 'The Resurgence of Regional Economics: Ten Years Later', in T. Barnes and M. Gertler (eds), *The New Industrial Geography: Regions, Regulations and Institutions*, London: Routledge, pp. 23–51.

Storper, M. and Walker, R. (1984), 'The Spatial Division of Labour: Labour and the Location of Industries', in L. Sawyers and W. Tabb (eds), *Sunbelt Snowbelt: Urban Development and Regional Restructuring*, Oxford: Oxford University Press, p. 19–47.

Storper, M. and Walker, R. (1989), *The Capitalist Imperative: Territory, Technology and Industrial Growth*, Oxford: Blackwell.

Sunley, P. (1996), 'Context in Economic Geography: The Relevance of Pragmatism', *Progress in Human Geography*, **20**, pp. 338–55.

Thrift, N. (2005), *Knowing Capitalism*, London: Sage.

Thrift, N. and Olds, K. (1996), 'Refiguring The Economic in Economic Geography', *Progress in Human Geography*, **20**, pp. 311–37.

Tickell, A. and Peck, J. (1992), 'Accumulation, Regulation and the Geographies of Post-Fordism: Missing Links in Regulationist Research', *Progress in Human Geography*, **16**, pp. 190–218.

Waterman, P. and Wills, J. (2001), *Place, Space and the New Labour Internationalism*, Oxford: Blackwell.

Wenger, E. (1998), *Communities of Practice: Learning, Meaning and Identity*, Cambridge: Cambridge University Press.

Whitley, R. (1999), *Divergent Capitalisms*, Oxford: Oxford University Press.

Yeung, H. W-C. (1998), 'Capital, State and Space: Contesting the Borderless World', *Transactions of the Institute of British Geographers*, **23**, pp. 291–309.

Yeung, H. W-C. (2003), 'Practising New Economic Geographies: A Methodological Examination', *Annals of the Association of American Geographers*, **93**, pp. 445–66.

Yeung, H. W-C. (2005), 'The Firm as Social Networks: an Organizational Perspective', *Growth and Change*, **36**(3), pp.307–28.

Part I
Conceptual Developments
in Economic Geography

[1]

A perspective of economic geography

*Allen J. Scott**

Abstract

The paper opens with a statement on the social embeddedness of knowledge. The disciplinary situation and practices of economic geographers are reviewed in the light of this statement. The rise of a new geographical economics is noted, and its main thrust is summarized in terms of a description of the core model as formulated by Krugman. The geographers' reception of the new geographical economics is described, and some key aspects of this reception are assessed. I then subject the core model itself to critical evaluation. Its claims about pecuniary externalities in the context of Chamberlinian competition provide a number of useful insights. However, I argue that the model is deficient overall in the manner in which it tackles the central problem of agglomeration. The discussion then moves on to consideration of the recent interest shown by many economic geographers in issues of culture. After a brief exposition of what this means for economic geography, I offer the verdict that this shift of emphasis has much to recommend it, but that in some of its more extreme versions it is strongly susceptible to the temptations of philosophical idealism and political voluntarism. In the final part of the paper, I attempt to pinpoint some of the major tasks ahead for economic geography in the phase of post-'late capitalism'. I suggest, in particular, that a new cognitive map of capitalist society as a whole is urgently needed, and I offer some brief remarks about how its basic specifications might be identified.

Keywords: conceptual bases of economic geography; cultural turn; geographical economics; methodology; philosophy of geography
JEL classifications: R10, Z19

1. In search of perspective

In this paper, I attempt to evaluate a number of prominent claims put forward in recent years by both geographers and economists about the methods and scope of economic geography. Much of the paper revolves around two main lines of critical appraisal. First, I seek to the strong and weak points of geographical economics as it has been formulated by Paul Krugman and his co-workers (though I also acknowledge that geographical economics is now moving well beyond this initial point of departure). Second, I provide a critique of the version of economic geography that is currently being worked out by a number of geographers under the rubric of the cultural turn, and

---

* Department of Geography, University of California–Los Angeles, Los Angeles, CA 90024, USA.
email < ajscott@ucla.edu>

480 • *Scott*

here I place special emphasis on what I take to be its peculiar obsession with evacuating the economic content from economic geography. On the basis of these arguments, I then make a brief effort to identify a viable agenda for economic geography based on an assessment of the central problems and predicaments of contemporary capitalism. This assessment leads me to the conclusion that the best bet for economic geographers today is to work out a new political economy of spatial development based on a full recognition of two main sets of circumstances: first, that the hard core of the capitalist economy remains focused on the dynamics of accumulation; second, that this hard core is irrevocably intertwined with complex socio-cultural forces, but also that it cannot be reduced to these same forces. In order to ground the line of argument that now ensues, we need at the outset to establish a few elementary principles about the production and evaluation of basic knowledge claims.

A large recent body of work in the theory of knowledge and social epistemology has made us increasingly accustomed to the notion that research, reflection, and writing are not so much pathways into the transcendental, as they are concrete social phenomena, forever rooted in the immanence of daily life. By the same token, knowledge is in practice a shifting patchwork of unstable, contested, and historically-contingent ideas shot through from beginning to end with human interests and apologetic meaning (Barnes, 1974; Rorty, 1979; Latour, 1991; Shapin, 1998). Mannheim (1952), an early exponent of the sociology of knowledge, expressed something of the same sentiments in the proposition that the problems of science in the end are mediated outcomes of the problems of social existence.

Postmodernists, of course, have picked up on ideas like these to proclaim the radical relativism of knowledge and the dangers of 'totalization' (cf. Dear, 2000), though the first of these claims carries the point much too far in my opinion, and the second turns out on closer examination to be largely a case of mistaken identity. I accept that knowledge is socially constructed and not foundational, but not that it is purely self-referential, for although knowledge is never a precise mirror of reality, it does not follow—given any kind of belief that some sort of external reality actually exists—that one mirror is as good as another (Sayer, 2000). The aversion to so-called totalization among many geographers today seems to translate for the most part, in a more neutral vocabulary, into the entirely sensible principle that theories of social reality should not claim for themselves wider explanatory powers than they in fact possess. However, the principle strikes me as pernicious to the degree that it is then used to insinuate that small and unassuming concepts are meaningful and legitimate whereas large and ambitious concepts are *necessarily* irrational. This in turn has an unfortunately chilling effect on high-risk conceptual and theoretical speculation.

These brief remarks set the stage for the various strategies of assessment of economic geography that are adopted in what follows. We want to be able to account for the shifting substantive emphases and internal divisions of the field in a way that is systematically attentive to external contextual conditions, but which does not invoke these conditions as mechanical determinants. We must, in particular, be alert to the social and institutional frameworks that encourage or block the development of ideas in certain directions, as well as to the professional interests that drive choices about research commitments. Moreover, since science is (either consciously or unselfconsciously) a vehicle for the promotion of social agendas, we need to examine the wider ideological and political implications of any knowledge claims. A basic question in this regard is: whose interests do they ultimately serve, and in what ways? The simple posing of this question implies already that the form of appraisal that follows entails a degree of partisan engagement (Haraway, 1991; Yeung, 2003), though in a way, I hope (given my preceding critical comments on relativism), that maintains a controlled relationship to an underlying notion of coherence and plausibility. Last but by no means least, then, we must certainly pay close attention to the logical integrity, the scope of reference, the correspondence between ideas and data, and so forth, of the various versions of economic geography that are on offer.

2. Economic geographers at work

2.1. Geography and the disciplinary division of labor

Geographers long ago gave up trying to legislate in a priori terms the shape and form of their discipline. In any case, from what has gone before, we cannot understand geography, or any other science for that matter, in relation to some ideal normative vision of disciplinary order. Geography as a whole owes its current standing as a distinctive university discipline as much to the inertia of academic and professional institutions as it does to any epistemological imperative. The geographer's stock-in-trade, nowadays, is usually claimed to revolve in various ways around questions of space and spatial relations. This claim provides a reassuring professional anchor of sorts, but is in practice open to appropriation by virtually any social science, given that space is intrinsically constitutive of all social life. In fact, geographers and other social scientists regularly encounter one another at points that lie deep inside each other's proclaimed fields of inquiry, and this circumstance reveals another of modern geography's peculiarities, namely its extreme intellectual hybridity. It is perhaps because of this hybridity that geography is so susceptible to rapidly shifting intellectual currents and polemical debate, but also—and this is surely one of its strengths—an unusual responsiveness to the burning practical issues of the day.

2.2. The wayward course of economic geography in the last half-century

Economic geography reproduces these features of geography as a whole in microcosm. On the one side, it is greatly influenced by issues of social and political theory. On the other side, given its substantive emphases, it has particularly strong areas of overlap with economics and business studies. At any given moment in time, it nevertheless functions as a more or less distinctive intellectual and professional community that brings unique synthetic perspectives to the tension-filled terrain(s) of investigation that it seeks to conquer. At the same time, economic geography has been greatly susceptible to periodic shifts of course over the last several decades, often in surprising ways, and equally often with the same dramatis personae, as it were, appearing and re-appearing in different costumes in different acts of the play.

The period of the 1950s and 1960s was especially important as a formative moment in the emergence of economic geography as a self-assertive subdiscipline within geography as a whole. This was a period of great intellectual and professional struggle in geography between traditionalists and reformers, with the latter seeking to push geography out of its perceived idiographic torpor and—on the basis of quantitative methodologies and formal modeling—into a more forthright engagement with theoretical ideas (Gould, 1979). Economic geographers were in the vanguard of this movement, and they were able to push their agendas vigorously, partly because of their strategic affiliation with a then-powerful regional science, partly because the questions they were posing about the spatial organization of the economy were of central concern to much policy making in the capitalism of the era, with its central mass-production industries and its activist forms of social regulation as manifest in Keynesian economic policy and the apparatus of the welfare state (Benko, 1998; Scott, 2000).

This early moment of efflorescence was succeeded by a sharp turn toward political economy as the crises of the early 1970s mounted in intensity, and as the general critique of capitalism became increasingly vociferous in academic circles. This was a period in which geographers developed a deep concern about the spatial manifestations of economic crisis generally, as reflected in a spate of papers and books on topics of regional decline, job loss, regional inequalities, poverty, and so on (Carney et al., 1980; Bluestone and Harrison 1982; Massey and Meegan, 1982). It was also a period in which much of economic geographers' portrayal of basic social realities was cast either openly and frankly in Marxian terms or in variously *marxisant* versions. The first stirrings of a vigorous feminist encounter with economic geography also began to take shape at this time.

As the initial intimations of the so-called new economy made their appearance in the early 1980s, and as the crisis years of the 1970s receded, economic geography started to go through another of its periodic sea changes. A doubly-faceted dynamic of economic and geographic transformation was now beginning to push geographers toward a reformulated sense of spatial dynamics. On the one hand, new spatial foci of economic growth were springing up in hitherto peripheral or quasi-peripheral regions in the more economically-advanced countries, with neo-artisanal communities in the Third Italy and high-technology industrial districts in the US Sunbelt doing heavy duty as early exemplars of this trend (Scott, 1986; Becattini, 1987). In this connection, geographers' interests converged intently on the theoretical and empirical analysis of spatial agglomeration. On the other hand, a great intensification of the international division of labor was rapidly occurring, especially under the aegis of the multinational corporation (Fröbel et al., 1980). In this connection, the main issues increasingly crystallized around globalization and its expression in international commodity chains, cross-border corporate linkages, capital flows, foreign branch plant formation, and so on (e.g., Dicken, 1992; Johnston et al., 1995; Taylor et al., 2002). The themes of agglomeration and international economic integration more or less continue to dominate the field today, though many detailed changes of emphasis have occurred as research has progressed. Indeed, of late years, these two themes have tended increasingly to converge together around the notion of the local and the global as two interrelated scales of analysis within a process of economic and political rescaling generally (Swyngedouw, 1997).

These thematic developments represent only a thumb-nail sketch of the recent intellectual history of economic geography. We must recognize that there have been many additional twists and turns within this history, both of empirical emphasis and of theoretical debate. As it stands, however, this account now serves as a general point of entry into a detailed examination of some of the major conceptual tensions that run through the field today, including a number of claims, which if they can be sustained, presage some quite unexpected new directions of development.

2.3. Turbulence and challenge

Economic geography, then, has been marked over its post-War history by a great susceptibility to turbulence. A notable recent sign of this tendency is the various 'turns' that the field is said to have taken or to be about to take. A cursory count reveals an empirical turn (Smith, 1987), an interpretative turn (Imrie et al., 1996), a normative turn (Sayer and Storper, 1997), a cultural turn (Crang, 1997), a policy turn (Martin, 2001), and a relational turn (Boggs and Rantisi, 2003), among others.

In some instances, the proclamation of these turns has been no more than an attempt to test the waters. In others, it has registered some real underlying tendency in geographic research. Today, the field is subject to particularly strong contestation from two main sources. One of these lies largely outside geography proper and is being energetically pushed by economists under the rubric of a new geographical economics. It represents a major professional challenge to economic geographers by reason of its threatened appropriation and theoretical transformation of significant parts of the field. The other is represented by the cultural turn that comes in significant degree from within geography itself, but also reflects the wider politicization of cultural issues and the rise of concerns about identity in contemporary society. The cultural turn represents a very different kind of challenge to economic geography on account of its efforts to promote within the field a more highly developed consciousness of the role of culture in the eventuation of economic practices. Much of the rest of this paper is concerned with investigating the nature of this current conjuncture in the light of the arguments already marshalled.

3. Geographical economics: accomplishments and deficits

3.1. The core model

Krugman's *Geography and Trade*, published in (1991) rang a tocsin in the ears of geographers, with its twofold proclamation that the project of economic geography was now at last beginning, and that economic geographers (of the variety found in geography departments) had hitherto been more or less sleeping at the wheel.

The new geographical economics did not, as we might expect, reach back to regional science, but appeared quite unexpectedly from another quarter: the new growth and trade theories that had been taking shape in economics over the previous decade or so (Thisse, 1997; Meardon, 2000). The core model is built up around the idea of monopolistic competition as originally propounded by Chamberlin (1933) and subsequently formalized by Dixit and Stiglitz (1977). The model also has some points of resemblance to an older tradition of heterodox economics focussed on increasing returns and cumulative causation, as represented by Hirschman (1958), Myrdal (1959), and Kaldor (1970). Strictly speaking, Krugman's model, and the surge of research activities that it has sparked off, are not neoclassical, for it firmly eschews any notion of constant returns to scale and perfect competition. That said, the model retains a strong kinship with mainstream economics by reason of its commitment to methodological individualism, full information, utility-maximizing individuals and profit-maximizing firms, and an exclusive focus on socially disembedded relationships of exchange (Dymski, 1996).

The model itself is an ingenious if convoluted piece of algebra. Imagine a set of regions[1] with production represented by immobile farmers and mobile manufacturing workers

---

1   Because of the model's complexity, it is typically defined in terms of just two regions for expository purposes.

484 • *Scott*

and firms. Manufacturing firms engage in product differentiation (monopolistic competition) with increasing returns to scale, or better yet, unexhausted economies of scale. Thus, each firm produces a unique or quasi-unique variety in its given product class. Consumers in all regions (both farmers and manufacturing workers) purchase some portion of every firm's output. Wages are determined endogenously. Market prices always reflect the transport costs incurred in product shipment. Consumers in regions with many producers will therefore pay less than those less favorably situated. Any individual's 'utility' is a function of both nominal wages and price levels. Mobile manufacturing workers will migrate from (peripheral) regions with lower utility to (core) regions with higher utility. Nominal wages in any region whose manufacturing labor force is increasing in this way will tend to fall (though corresponding utilities will increase because of the decreasing cost of final goods). More and more manufacturing firms will therefore be attracted to the region, which will in turn induce further in-migration of labor. The net result will be a path-dependent process of spatial development leading to a stable core-periphery pattern. Eventually an equilibrium of production, wages, prices, and demand will be attained, and the final result will exhibit market-driven pecuniary externalities (i.e., overall real price reductions) derived from intra-firm increasing returns under conditions of Chamberlinian competition. In a later formulation of the core model, Krugman and Venables (1995) showed that core-periphery contrasts will tend to be relatively subdued (or even to disappear entirely) in situations where transport costs are uniformly very high or very low, whereas core-periphery contrasts will be maximized where transport costs are contained within some intermediate range of values.

Depending on the distribution of immobile workers, transport costs, elasticities of demand and substitution, and other basic parameters, the model is capable of generating widely varying locational outcomes. Numerous modifications and extensions of the basic model have been proposed since its first formulation. For example, Krugman and Venables (1995, 1996) and Venables (1996) introduce inter-industrial linkages into the model. Abdel-Rahman and Fujita (1990) have suggested that the production functions of downstream industries are sensitive to the variety of available upstream inputs. In this case, agglomeration and pecuniary externalities are brought about by the productivity effects of input variety. In another variation of the model, Baldwin (1999) has shown how demand-linked circular causality can induce agglomeration, even in the absence of labor mobility.[2]

3.2. The geographers' reception of geographical economics

Rather predictably, the geographers' first reaction to the new geographical economics was one of virtually unqualified rejection. In a brief review of Krugman's book, Johnston (1992, p.1066) dismisses it with the comment 'not recommended'. Martin (1999, p.67) writes that

---

2   Other work in contemporary geographical economics (not all of it strictly in line with the core model) include regional income inequalities (Barro and Sala-i-Martin, 1995; Quah, 1996), the dynamics of city systems (Ellison and Glaeser, 1997; Duranton and Puga, 2000), regional productivity and growth (Henderson, 2003), and so on. My remarks in this paper are focused on the narrower (Krugmanian) view of geographical economics, both because of the extremely large claims that have been made on its behalf and because of its current centrality in the entire project of geographical economics (but see my later assessment of how this situation may change).

geographical economics, 'is not that new and it most certainly is not geography'. In a similar vein, Lee (2002, p.353) rebuffs the entire enterprise with the comment that 'there are ... precious few grounds for some mutually beneficial conversation here'.

Something of these reactions can no doubt be ascribed to geography's endemic professional anxieties reflecting its relatively low standing on the academic totem pole, certainly by comparison with economics. Krugman's tasteless self-promotion as the creative genius par excellence of economic geography, and the champion of four-square thinking generally, did nothing to assuage those anxieties. The geographers' main complaints about this work have tended to revolve around their concern that it is unduly cut off from the wider social and political frameworks within which economic issues are actually played out. Many economic geographers' perception of their own work, as well, points to practices of research that are grounded, open, polycentric, focussed on rich empirical description, and deeply conscious of the contingency and complexity of things (Thrift and Olds, 1996; Boddy, 1999). I shall take issue with this particular line of self-justification later, but let us for the moment simply note some of its basic modulations. Thus, Clark (1998, p.75) suggests that 'a fine-grained substantive appreciation of diversity, combined with empirical methods of analysis like case studies are the proper methods of economic geography'. Martin (1999, p.77) castigates geographical economics for its neglect of 'real communities in real historical, social and cultural settings'. In a similar vein, Barnes (2003), picking up on the work of Geertz (1983), proclaims that all knowledge is local and that locational analysis in geography is (or should be) born out of specific contextual settings.

Some of these comments point to significant deficiencies of geographical economics, and they need to be taken seriously (see also David, 1999). In my opinion, however, they provide at best only peripheral glosses on the main issues at stake and they fail signally to grapple with the target's central weakness, which, as I hope to demonstrate, reside in its limited *analytical* grasp of agglomeration economies and locational processes. I would also argue that the geographers' critique has tended to veer too enthusiastically in favor of the virtues of the empirical and the particular and too forcefully against theoretical systematization and formal analysis, thereby implicitly abdicating from far too much that is of value on their own side (though I suspect that most of the geographers mentioned earlier would not consider this to be a fair judgment of what they are saying). In any case, a scientifically meaningful and politically progressive economic geography can scarcely allow itself to be reduced merely to close dialogue with endless empirical relata (Sayer, 2000; Plummer and Sheppard, 2001). A more penetrating engagement with the internal theoretical structure of geographical economics, it seems to me, is more than overdue.

3.3. An evaluation of the Krugman model

One of the obvious failures of earlier neoclassical theories in economic geography and regional science is that their commitment to perfect competition and constant returns induced them to overemphasize the divisibility of economic activities, leading in turn to a radical underemphasis of agglomeration as a force in shaping the economic landscape. Fujita and Thisse (2002) describe the space-economy as seen through neoclassical spectacles as a tending to a system of 'backyard capitalism'.

The originality and value of the Krugman model as an approach to spatial analysis is its formulation of the problem in terms of monopolistic competition and increasing returns within the firm. The notion that agglomeration has its roots at least partially in

486 • *Scott*

monopolistic competition is particularly interesting, and corresponds well with the character of much industry today. Modern sectors such as high-technology manufacturing, business and financial services, cultural products, and so on, are especially prone to form distinctive clusters, and it is exactly in such sectors that we find the high levels of product variety, intra-sectoral trade, and the drive to market extension that characterize monopolistic competition. At the same time, the core model breathes new life into the notion of pecuniary externalities as originally formulated by Scitovsky (1954). The emphasis on agglomeration as an outcome of the complex pecuniary effects of Chamberlinian competition and internal economies of scale is unquestionably the model's principal claim to theoretical significance, and it is all the more interesting because it sets these within a framework of multi-region interdependencies.

Once all of this has been said, many reservations remain. At the outset let me state that I do not share the inclination of some geographers to discard the new geographical economics simply on the basis of its commitment to a priori forms of deductive theorizing. In practice, of course, such theorizing sometimes turns in upon itself in highly dysfunctional ways, and economists are notorious for their cultivation of an ingrown professional culture focussed on displays of bravura but vacuous analytics (cf. McClosky, 2002). The Krugman model and its derivative expressions certainly suffer from this syndrome, especially in view of the implausibility and arbitrariness of many of its assumptions, where enormous compromises with reality are made so as to ensure that numerical solutions can be generated,[3] and it is tempting to reject the model out of hand on the grounds of its unrealistic assumptions alone. However, I think it better to issue the main challenge from the basis of a related but slightly different perspective. In other words, *what is this a model of?* It may well be a description of life on some planet somewhere in the universe, but what exactly is its relevance to an understanding of economic realities on planet earth at any time in the past or the foreseeable future?

This question is underlined by the fact that the core model puts the emphasis on market-driven pecuniary relationships in an equilibrium Chamberlinian framework. In so far as it goes, this point of departure has the merit of making inter-regional competitive forces an explicit element of the analysis (as befits its intellectual origins in international trade theory). Its principal deficiency is that it fails adequately to grasp at the notion of the region as a *nexus of production relationships and associated social infrastructures* from which streams of external economies of scale and scope continually flow, even in

---

3   Some sectors are taken to be monopolistically competitive; others are deemed subject to perfect competition where this is analytically convenient. Elasticities of demand and substitution are always held constant. Firms have no opportunities for strategic interaction. The multidimensional character of business transactions is reduced to the fiction of iceberg transport costs. Labor is mobile when the algebra demands it; labor is immobile otherwise. Wages are adjustable in some cases, sticky in others. Generalizations of the model to more than two regions result in a world that is shaped like a doughnut, not because this makes any sense in substantive terms but because the mathematics are otherwise intractable. The model contains fixed costs but no sunk costs, and there are thus no inertial barriers to adjustment in the model (where adjustment proceeds relatively rapidly whereas adjustment of the economic landscape in reality tends to be extremely slow). And what exactly is the model's appropriate spatial scale of resolution? Ottaviano and Thisse (2001) appear to feel, with some justification I believe, that the model of pecuniary externalities works best at a level of resolution where regions approximate the size of the US Manufacturing Belt. The model certainly does not seem to have much relevance to cases where regions are defined at small scales of spatial resolution. For a more extended discussion of the problem of scale in the new geographical economics, see Olsen (2002).

single-region economies, and even in cases where competition in final product markets is non-monopolistic. Equally, the model diverts attention away from the fact that for any kind of regional development to occur, productive assets need to be physically mobilized and integrated with one another on the ground in specific regions (Hirschman, 1958). In fact, the model, as such, has virtually nothing to say about the *endogenous* intra-regional organization and dynamics of production, and almost as little about the region as a motor (as opposed to a receptacle) of economic activity (cf. Scott, 2002b). In more specific terms, and despite the fact that Krugman and his co-workers make frequent reference to Marshall, the model actually gives short shrift to any meaningfully Marshallian approach to regional development and agglomeration.[4]

Four specific lacunae of the core model merit further attention in this connection:

First, the model identifies productive activity only in terms of monopolistically-competitive firms with fixed and variable costs. In its initial formulation it makes no reference whatever to the dynamics of the social division of labor and the networks of transactional relations that flow from this process. In later formulations (e.g., Krugman and Venables, 1996; Venables, 1996) an intermediate goods industry is assumed by fiat to exist in the model. However, the model is silent on the endogenous relations that exist in reality between the vertical structure of production and spatially dependent transactions costs. These relations tend to be of special interest and importance in clustered economic systems where intra- and inter-firm transactional structures are usually extremely complex (see, for example, Scott, 1983). Accordingly, the model pays inadequate attention to the wider logic of locational convergence/divergence, and, in particular, it is deficient in its grasp of the individual regional economy as a source of competitive advantage (cf. Porter, 2001).

Second, these failings are compounded by the model's neglect of local labor market processes, such as information flows, job search patterns, labor-force training, and so on (Peck, 1996). True enough, Krugman pays lip service to the existence of processes like these, but makes no effort to incorporate them into the workings of the core model.

Third, region-based learning and innovation processes are conspicuous by their absence from the core model. A consequence of this absence is that the core model pays little or no attention to patterns of temporal change in the qualitative attributes and competitive advantages of regional production systems. The rich parallel literature by economists such as Jaffe et al. (1993), Audretsch and Feldman (1996), or Acs (2002) on regional innovation systems compensates in some degree for this omission, but the model itself remains more or less impervious to conceptions of technology-led growth (Acs and Varga, 2002).

Fourth, given its resolute commitment to microeconomic forms of analysis, the model actively suppresses the possibility that collective region-based strategies of economic adjustment might play a role in the construction of localized competitive advantages (Neary, 2001). In practice, such strategies are often highly developed in regions with active production systems, both in the private sphere (e.g., inter-firm collaboration), and in the public sphere (e.g., local economic development and training programs under the aegis of regional agencies). Numerous researchers have shown time and again that strategies like

---

4   Sheppard (2000) makes much the same point. For examples of the Marshallian approach as developed by geographers see Amin and Thrift (1992), Cooke and Morgan (1998), Gertler (2003b), Rigby and Essletzbichler (2002), Scott (1998), Storper (1997).

these are critical to the creation of regional competitive advantages and an important tool in the search for improved rates of local economic growth (Bianchi, 1992; Saxenian, 1994; Storper and Scott, 1995; Cooke, 1999).

Some of the lacunae pointed out here can no doubt be dealt with in part by appropriate reformulations of the model (such as the introduction of commuting costs to reflect the spatial organization of local labor markets, or explicit reference to coalition formation processes), but at the cost of enormous increases of algebraic complexity. The Krugman model is for the most part a black box that occludes what by many accounts must be seen as some of the most important aspects of regional economic growth and development. As such, it casts only a very limited light on the full the play of externalities, competitive advantage, and locational agglomeration in economic geography. Needless to say, the model is silent on wider social and political issues of relevance to the analysis of agglomeration, such as, for example, region-specific forms of worker socialization and habituation, the emergence of local governance structures, or the historical shifts that occur periodically in technical-organizational structures of accumulation, and that greatly impact regional trajectories of development.

By its elevation of atomistic exchange relations to an exclusive ontology of the economic and the geographic (albeit in a Chamberlinian context), the core model provides only a very partial account of the genesis and logic of the economic landscape. In the end, the model can be seen more as an effort to codify atomized market processes in simple spatial frameworks than it is an attempt to understand spatial relations in any thorough-going sense of the term. The strong point of the model is its description of pecuniary externalities in multi-region systems; the weak point is its account of locational adjustment in which units of capital and labor move like billiard balls (except for the ones that have been nailed to the table) from one equilibrium to another, and then simply fall magically into a fully functional economic system as they accumulate in receiving regions. Nevertheless—and to be fair to the wider body of urban and regional economics generally—there is an obvious and encouraging trend in much of the current literature to move beyond the limitations of the core model as expounded by Krugman and to deal in a more flexible and open-ended manner with many of the issues where it is most vulnerable to criticism (see, for example, the essays collected together in Cheshire and Mills, 1999; and Henderson and Thisse, 2004).

4. Capitalism, culture, and geography

4.1. Economic geographers discover culture

Just as geographical economics was making its appearance on the academic scene in the early 1990s, a number of economic geographers were transferring their attentions to an altogether different set of approaches rooted in issues of culture. The emergence of this interest coincided with a growing conviction that not only were certain earlier generations of geographers and other social scientists incorrect to regard culture simply as an outcome of underlying economic realities, but that these realities themselves are in fundamental ways subject to the play of cultural forces.

Any casual scrutiny of contemporary capitalism reveals at once that it is inflected with different social and cultural resonances in different localities, and that these resonances are directly implicated in the organization of economic life and modalities of economic calculation. American, Japanese, and Chinese capitalism, for example, are at once generically similar and yet are marked by socio-cultural idiosyncrasies with significant effects on the ways in which they function. In addition, production systems in contemporary capitalism, while still obviously highly focused on the mechanical manufacture of things, are shifting more and more into the processing of information and symbols, from business advice to cultural services. This trend is leading to dramatic changes in the form and function of commodified goods and services, and much research is now moving forward on the reception, interpretation, and social effects of these outputs (Jackson, 1999; Thrift, 2000; Bridge and Smith, 2003). This research points to important theoretical issues concerning the hermeneutics of the commodity, and the functions of capitalism generally as a fountainhead of symbolic representation in modern life (e.g., Harvey, 1989; Lash and Urry, 1994). Striking changes are also occurring in the social and psychological make-up of the workplace. Over large areas of the new economy of capitalism, dress, mannerisms, forms of speech, self-presentation, and so on have become essential elements of workers' performance. Equally, gender, race, ethnicity, and so on, together with specific forms of empathy associated with them are being actively exploited and managed in various ways in the workplace (cf. McDowell, 1997). More generally, markets as a whole could not work in the absence of a socio-cultural system regulating the conventions and behaviors that sustain them.

These brief remarks, schematic as they may be, already underline the obvious and pressing need for economic geographers to pay close attention to the ways in which culture and economy intersect with one another in mutually constitutive ways. The urgency of this need is reinforced by the observation that the economic and the cultural come together with special intensity in place (Shields, 1999), and that many of the key agglomerations constituting the focal points of the new economy around the world are critically dependent on the complex play of culture. Thus, to an ever-increasing degree, the productive performance of agglomerations like the City of London (Thrift, 1994), Hollywood (Scott, 2002a), or Silicon Valley (Saxenian, 1994) can only be understood in relation to their joint economic and cultural dynamics. Each of these places is shot through with distinctive traditions, sensibilities, and cultural practices that leave deep imprints on phenomena such as management styles, norms of worker habituation, creative and innovative energies, the design of final outputs, and so on, and these phenomena in turn are strongly implicated in processes of local economic growth and development.

4.2. For culture; against the cultural turn

In view of the discussion above, it seems fairly safe to say that only a few die-hards and philistines are likely to make strenuous objections to attempts to bring culture more forcefully into the study of economic geography. In spite of the neologisms and cliché-ridden prose that Martin and Sunley (2001) rightly complain about, there is obviously a significant nexus of ideas in a more culturally-inflected economic geography that responds in a very genuine way to major problems posed by contemporary capitalist society. Once this point has been made, however, a number of the reforms of economic geography that have been most strenuously advocated under the rubric of the cultural turn are rather less obviously acceptable, and have recently been subject to heated debate by economic geographers (see, for example, Sayer, 1997; Martin and Sunley, 2001; Plummer and Sheppard, 2001; Rodríguez-Pose, 2001; Storper, 2001).

This debate has tended to find its sharpest expression in relation to the curious reluctance by some proponents of the cultural turn to make any concession to the play of economic processes in economic geography except insofar as they are an expression of underlying cultural dynamics. In a number of their more fervent statements, indeed, some of these proponents occasionally verge on an inversion of the classical Marxian conceit to the effect that culture flows uni-causally from the economy, by offering equally exaggerated claims about the influence of culture on the economy. In a statement that displays much enthusiasm about the study of culture and much acrimony in regard to the discipline of economics, Amin and Thrift (2000) essentially recommend withdrawal from economic analysis, as such, and a wholesale re-description of economic realities in terms of cultural points of reference. Thus, in writing about the problem of eventuation they one-sidedly argue that 'acting into the words confirms the discourse and makes a new real' (p.6), so that in their formulation, the economy becomes nothing more than a series of 'performances' derived from a script. Elsewhere, Thrift (2001) further proclaims that the new economy of the 1990s was fundamentally a rhetorical phenomenon. The argument here starts off promisingly enough with an examination of the role of the press, business consultants, financial advisors, and the like, in helping to foment the fast-paced, high-risk economic environment of the period, but then it veers into the blunt assertion that the new economy as a whole can be understood simply as a discursive construct. In formulations like these, basic economic realities—the state of technology, the rhythms of capital accumulation and investment, the rate of profit, the flow of circulating capital, etc.—become just so much inert plasma to be written upon this way or that as cultural shifts occur and as revisions of the script are introduced. Certainly, words are a critical moment in the circuit of mediations through which economic reality operates, and there can be no doubt that many unique effects are set in motion at this particular level of analysis. Conversely, and it is puzzling that so trivial and obvious a point should need to be made, there are also deeply-rooted *economic* logics and dynamics at work in the contemporary space-economy, and at least some of these (such as the dynamics of industrial organization, or the increasing returns effects that lie at the root of industrial districts), require investigation on their own terms above and beyond invocations of the causal powers of discourse and culture.

In a series of recent writings, Barnes (e.g., 1996; 2001; 2003), has pursued a related line of investigation opened up by the cultural turn. Barnes' work is much influenced by Derrida and Rorty, and is centrally focussed on the metaphorical and narratological character of geographical writing. There is actually much of interest in the approach Barnes takes. He has many useful things to say about the ideologies and working habits of economic geographers, as well as about the rhetorical devices that they deploy in their written reports. This helps among other things to keep us focussed on the critical idea that our intellectual encounters with the real are always deeply theory-dependent (Sunley, 1996). But as the plot thickens—or thins, according to your taste—we steadily lose sight of economic geography as a discipline with concrete substantive concerns (such as regional development or income inequalities), for these simply dissolve away into the primacy of the text and its metaphorical perplexities. I am perfectly prepared to admit that there may be strong elements of metaphor in, for example, a geography of hunger, but I certainly have no sympathy for the idea that hunger is just a metaphor, if only on the ad hominem grounds that it has painful physical manifestations and morbid long-term effects. Here, the legitimate claim that we can only know the world through socially-constructed codes of reference seems to have given way to the sophism that all we can know about the world is the codes themselves.

An even more extreme case of the solipsism that haunts much of the cultural turn can be found in the book by Gibson-Graham (1996) about strategic possibilities for progressive social change in contemporary capitalism. The central arguments of the book hinge upon the proposition that the criteria for validating a theory are purely internal to the theory to be validated. As Gibson-Graham writes (p.60): 'We cannot argue that our theory has more explanatory power or greater proximity to the truth than other theories because there is no common standard which could serve as the instrument of such a metatheoretical validation process'. If this proposition were indeed true it would presumably undermine much of the point in Gibson-Graham proceeding any further in her argument, though she does in fact continue on for another 200-odd pages. In the course of this discussion, the relativism of her main thesis is steadily transformed from a merely academic exercise into a political agenda of sorts. Thus, she announces (p.260), 'the way to begin to break free of capitalism is to turn its prevalent representations on their heads'. Presto. Not even a hint about a possible transitional program, or a few suggestions about, say, practical reform of the banking system. The claim is presented in all its baldness, without any apparent consciousness that attempts to break free of any given social system are likely to run into the stubborn realities of its indurated social and property relations as they actually exist. More generally, Gibson-Graham's argument leads inexorably beyond the perfectly acceptable notion that all intellectual work is theory-dependent and into those murky tracts of idealist philosophy where reality is merely a reflection of theory, and where theory produces social change independently of concrete practice and disciplined attention to the refractory resistances of things as they really are.

So, quite apart from its dysfunctional depreciation of the role of economic forces and structural logics in economic geography, the cultural turn also opens a door to a disconcerting strain of philosophical idealism and political voluntarism in modern geography. The net effect is what we might call economistic grand theory in reverse: a remarkable failure to recognize sensible boundaries as to just what precisely a cultural theory of the economy can achieve, and a concomitant over-promotion of the notion that social and economic transformation involves nothing more than the unmediated power of theoretical ideas. Again, nothing in this argument is intended to deny the important continuities and intersections between culture and economy or the significance of the economy as a site of cultural practices; neither is it in any sense an attempt to eject the study of cultural economy from geography. The problem is not 'culture' but the cultural turn as it has emerged out of cultural studies with its militant project of reinterpreting all social relations as cultural relations, and its naïve, if understandable, attempt to humanize the iron cage of capitalist accumulation by unwarranted culturalization of its central economic dynamics (Rojek and Turner, 2000; Eagleton, 2003).

5. Toward a re-synthesis

As I have tried to show in this paper, Krugman-style geographical economics offers at best an extremely narrow vision of the dynamics of the economic landscape, and is in any case, less preoccupied with geography as such than it is with geography as just another domain within which markets unfold. Geographers and economists certainly occupy much common ground at the present time, but encounters between them on this shared terrain are endemically susceptible to deeply-seated disputes about theoretical priorities. My guess is that the influence of the Krugman model will in any case soon wither away as geographical economics comes up against the model's inner and outer limits, just as neoclassical regional science began to show signs of enervation after the mid-1970s in part as a consequence of its commitment to the strait-jacket of convexity and constant returns to scale (cf. Thisse, 1997; Neary, 2001). As it happens, any such retreat may just possibly help to open up opportunities for more fruitful future encounters between geographers and economists over issues of space (see also Sjöberg and Sjöholm, 2002). For the present, the undisputed major contribution of geographical economics to our understanding of spatial problems has been its resuscitation of the notion of pecuniary externalities in a world of Chamberlinian competition.

The cultural turn, for its part, has sought to take economic geography in an altogether different direction. In some degree, of course, the clashing claims of economic geographers and cultural geographers over the last decade or so can be interpreted as expressions of an internal power struggle for status and influence in the profession of geography as a whole. This struggle owes much to the unquestioned intellectual re-invigoration and consequent self-assertion of cultural geography that occurred over the 1990s as cultural studies expanded in the academy at large. Despite the clashes, there remains, as I have indicated, much useful work to be accomplished by cooperation between economic and cultural geographers in any effort to comprehend the spatiality and locational dynamics of modern capitalism (Gregson et al., 2001; Bathelt and Glückler, 2003; Gertler 2003a; Yeung, 2003). At the same time, there will undoubtedly continue to be strong points of divergence between the two subdisiciples; lines of investigation opened up by economic geographers where cultural geographers hesitate to tread, and vice versa. A degree of mutual tolerance (though certainly not automatic and uncritical mutual endorsement) is no doubt called for in this situation.

Notwithstanding all the theoretical turbulence of the last few decades, there is probably still wide agreement among economic geographers, as such, that one of the main tasks we face is in the end some sort of transformative understanding of the historical geography of capitalist society (Harvey, 1982; Harvey and Scott, 1989). I suspect, as well, that most economic geographers would agree with the proposition that we need some sort of new synthesis in order to pursue this task more effectively (cf. Castree, 1999), i.e., a revised cognitive map that can help us make sense of all the complex contemporary tendencies that have turned what critical theorists used hopefully to call 'late capitalism' into the triumphant and rejuvenated juggernaut that it is today. I make this claim about the need for a new synthesis in full cognizance of the reductionist dangers that it opens up (cf. Amin and Robbins, 1990; Sayer, 2000). Equally, I want to avoid the self-defeating conclusion that because of these dangers we must always downsize our theoretical ambitions. One of the truly disconcerting aspects of much geographical work today is that it preaches a doctrine that privileges the small, the piecemeal, and the local, even as capital plays out its own grandiose saga of expansion and recuperation at an increasingly globalized scale.

A prospective economic geography capable of dealing with the contemporary world must hew closely, it seems to me, to the following programmatic goals if it is to achieve a powerful purchase on both scientific insight and progressive political strategy.

1. To begin at the beginning: economic geography needs to work out a theoretical re-description of capitalism as a structure of production and consumption and as an engine of accumulation, taking into account the dramatic changes that have occurred in recent decades in such phenomena as technology, forms of industrial and corporate organization, financial systems, labor markets, and so on. This theoretical re-description must be sensitive to the generic or quasi-generic forms of capitalist development that occur in different times in different places, which, in turn, entails attention to the kinds of issues that regulation theorists have

      identified under the general rubric of regimes of accumulation (Aglietta, 1976; Lipietz, 1986).

2.   In addition to these economic concerns, we must recognize that contemporary capitalism is intertwined with enormously heterogeneous forms of social and cultural life, and that no one element of this conjoint field is necessarily reducible to the other. Directions of causality and influence across this field are a matter of empirical investigation, not of theoretical pre-judgment. Note that in this formulation, class becomes only one possible dimension of social existence out of a multiplicity of other actual and possible dimensions.

3.   This nexus of economic, social, and cultural relationships constitutes a creative field or environment within which complex processes of entrepreneurship, learning, and innovation occur. Geographers have a special interest in deciphering the spatial logic of this field and in demonstrating how it helps to shape locational dynamics.

4.   In combination with these modalities of economic and social reality, we need to reserve a specific analytical and descriptive space for collective action and institutional order at many different levels of spatial and organizational scale (the firm, the local labor market, the region, the nation, etc.), together with a due sense of the political tensions and rivalries that run throughout this sphere of human development. By the same token, a vibrant economic geography will always not only be openly policy-relevant (Markusen, 1999), but also politically engaged. A key question in this context is how to build local institutional frameworks that promote both economic success and social justice.

5.   We must recognize that social and economic relations are often extremely durable, and that they have a propensity to become independent in varying degree of the individuals caught up within them. This means that any normative account of social transformation and political strategy, must deal seriously with the idea that there are likely to be stubborn resistances to change rooted in these same relations. The solutions to this problem proposed by sociologists like Bourdieu (1972) and Giddens (1979) strike me as providing reasonable bases for pushing forward in this respect, for they explicitly recognize the inertia of social structures while simultaneously insisting on the integrity of individual human volition. Unfortunately, these solutions (most especially the structure-agency formulation of Giddens) have been much diluted in recent years by reinterpretations that lean increasingly heavily on the agency side of the equation, partly as a reflection of the cultural turn, partly out of a misplaced fear of falling into the pit of determinism.[5] Invocations of unmediated agency (or, for that matter, neoclassical utility) as an

---

5   It is useful here to recall the early argument of Martin (1951) to the effect that any self-respecting determinism insists on a direct mechanistic link from matter or the external world to mind so that what passes for free will is (so the determinist would say) nothing more than a cause-effect relationship. The existence of structural constraints on human action, or even the emergence of common social predispositions and habits, do not, by this standard of judgment, amount to any form of determinism. Nor can determinism in this rigorous sense necessarily be equated with the existence of macro-social outcomes that occur independently of any explicit decision that the world should be structured thus and so, or with situations where these outcomes assume 'laws of motion' without our explicit permission, as it were (e.g., the pervasive separation of home and work in the modern metropolis and the daily waves of commuting that are a result of this circumstance). Mutatis mutandis, when geographers invoke unmediated 'agency' or 'volition' as an explanatory variable, they are implicitly confessing to a failure of analysis, even though agency and volition are always a component of any human action.

explanatory variable in social science are often little more than confessions of ignorance, in the sense that when we are unable to account for certain kinds of relationships or events, we are often tempted to fall back on the reassuring notion that things are thus and so for no other reason than because that's the way we want them to be, irrespective of any underlying structural conditions.

6. A corollary of the structured organization and sunk costs of social life is that economic relationships (especially when they are locationally interrelated, as in the case of a regional production system) are likely to be path-dependent. This observation suggests at once that an evolutionary perspective is well suited to capture important elements of the dynamics of the economic landscape (cf. Nelson and Winter, 1982; Boschma and Lambooy, 1999). It follows that any attempt to describe the economic landscape in terms of instantaneous adjustment and readjustment to a neoclassical optimum optimorum is intrinsically irrelevant.

7. All of these moments of economic and social reality occur in a world in which geography has not yet been—and cannot yet be—abolished (Leamer and Storper, 2001). The dynamics of accumulation shape geographic space, and equally importantly, geographic space shapes the dynamics of accumulation. This means, too, that capitalism is differentiated at varying levels of spatial resolution, from the local to the global, and that sharp differences occur in forms of life from place to place. Indeed, as globalization now begins to run its course, geographic space becomes more important, not less important, because it presents ever-widening possibilities for finely-grained locational specialization and differentiation. Critical analysis of these possibilities must be one of modern economic geography's principal concerns.

8. Finally, I want to enter a plea for methodological variety and openness. One corollary of this plea is that economic geographers need to recover the lost skills of quantitative analysis, not out of some atavistic impulse to reinstate the economic geography of the 1960s, but because of the proven value of these skills in the investigation of economic data. The steady erosion of geographers' capabilities in this regard over the last couple of decades is surely a net loss to the discipline.

These remarks still leave open a wide range of alternative research strategies and theoretical orientations in economic geography, including approaches marked variously by heavy doses of algebraic formalization or cultural commentary. A particular point of focus, however, is provided by the continuing commitment by significant numbers of economic geographers to *critical* analysis and to the search for progressive social change. The pursuit of some sort of social democratic agenda and the fight against global neoliberalism, it seems to me, must stand high in any set of priorities in this regard at the present time, and all the more so as our world remains an arena in which tremendous variations in living standards, economic opportunity, and possibilities for cultural self-realization persist tenaciously from place to place and country to country. More than anything else, the great testing ground for economic geography, now and in the foreseeable future, must surely be identified in one way or another in relation to the central question of development, not only in its expression as a problem in historical geography, but as a normative project of global significance.

Economic geographers have much work to do in dealing with the multiple challenges of this evolving situation. But they also need periodically to take critical soundings of their tools, their practices, and their theoretical commitments if they are to remain equal to the daunting tasks ahead.

Acknowledgements

I am grateful to Jeffrey Boggs, Steven Brakman, Harry Garretsen, David Rigby, Michael Storper, and Jacques Thisse for their comments on an earlier draft of this paper. None of these individuals bears any responsibility for the opinions expressed here.

References

Abdel Rahman, H., Fujita, M. (1990) Product variety, Marshallian externalities, and city sizes. *Journal of Regional Science*, 30: 165–183.

Acs, Z. J. (2002) *Innovation and the Growth of Cities*. Cheltenham: Edward Elgar.

Acs, Z. J., Varga, A. (2002) Geography, endogenous growth and innovation. *International Regional Science Review*, 25: 132–148.

Aglietta, M. (1976) *Régulation et Crises du Capitalisme*. Paris: Calmann-Lévy.

Amin, A., Robbins, K. (1990) The re-emergence of regional economies? The mythical geography of flexible accumulation. *Environment and Planning D: Society and Space*, 8: 7–34.

Amin, A., Thrift, N. (2000) What kind of economic theory for what kind of economic geography? *Antipode*, 32: 4–9.

Amin, A., Thrift, N. J. (1992) Neo-Marshallian nodes in global networks. *International Journal of Urban and Regional Research*, 16: 571–581.

Audretsch, D. R., Feldman, M. P. (1996) R&D spillovers and the geography of innovation and production. *American Economic Review*, 86: 630–640.

Baldwin, R. E. (1999) Agglomeration and endogenous capital. *European Economic Review*, 43: 253–280.

Barnes, B. (1974) *Scientific Knowledge and Sociological Theory*. London: Routledge and Kegan Paul.

Barnes, T. J. (1996) *Logics of Dislocation: Models, Metaphors, and Meanings of Economic Space*. New York: Guilford.

Barnes, T. J. (2001) Retheorizing geography: from the quantitative revolution to the cultural turn. *Annals of the Association of American Geographers*, 91: 546–565.

Barnes, T. J. (2003) The place of locational analysis: a selective and interpretive history. *Progress in Human Geography*, 27: 69–95.

Barro, R. J., Sala-i-Martin, X. (1995) *Economic Growth*. New York: McGraw-Hill.

Bathelt, H., Glückler, J. (2003) Toward a relational economic geography. *Journal of Economic Geography*, 3: 117–144.

Becattini, G. (1987) *Mercato e Forze Local : il Distretto Industriale*. Bologna: Il Mulino.

Benko, G. (1998) *La Science Régionale*. Paris: Presses Universitaires de France.

Bianchi, P. (1992) Levels of policy and the nature of post-fordist competition. In M. Storper and A. J. Scott (eds) *Pathways to Industrialization and Regional Development*. London: Routledge, 303–315.

Bluestone, B., Harrison, B. (1982) *The Deindustrialization of America*. New York: Basic Books.

Boddy, M. (1999) Geographical economics and urban competitiveness: a critique. *Urban Studies*, 36: 811–842.

Boggs, J. S., Rantisi, N. M. (2003) The relational turn in economic geography. *Journal of Economic Geography*, 3: 109–116.

Boschma, R. A., Lambooy, J. G. (1999) Evolutionary economics and economic geography. *Journal of Evolutionary Economics*, 9: 411–429.

Bourdieu, P. (1972) *Esquisse d'une Théorie de la Pratique*. Geneva: Librairie Droz.

Bridge, G., Smith, A. (2003) Intimate encounters: culture—economy—commodity. *Environment and Planning D: Society and Space*, 21: 251–268.

496 • *Scott*

Carney, J., Hudson, R., Lewis, J. (1980) *Regions in Crisis: New Perspective in European Regional Theory*. New York: St Martin's Press.

Castree, N. (1999) Envisioning capitalism: geography and the renewal of Marxian poliltical economy. *Transactions of the Institute of British Geographers.*, 24: 137–158.

Chamberlin, E. (1933) *The Theory of Monopolistic Competition*. Cambridge, MA: Harvard University Press.

Cheshire, P., Mills, E. S. (1999) *Handbook of Regional and Urban Economics, Volume 3, Applied Urban Economics*. Amsterdam: North Holland.

Clark, G. L. (1998) Stylized facts and close dialogue: methodology in economic geography. *Annals of the Association of American Geographers*, 88: 73–87.

Cooke, P. (1999) The co-operative advantage of regions. In T. J. Barnes and M. S. Gertler (eds) *The New Industrial Geography: Regions, Regulation and Institutions*. London: Routledge, 54–73.

Cooke, P., Morgan, K. (1998) *The Associational Economy: Firms, Regions, and Innovation*. Oxford: Oxford University Press.

Crang, P. (1997) Introduction: cultural turns and the (re)constitution of economic geography. In J. Wills and R. Lee (eds) *Geographies of Economies*. London: Arnold, 3–15.

David, P. A. (1999) Krugman's economic geography of development: NEGs, POGs and naked models in space. *International Regional Science Review*, 22: 162–172.

Dear, M. J. (2000) *The Postmodern Urban Condition*. Oxford: Blackwell.

Dicken, P. (1992) *Global Shift: The Internationalization of Economic Activity*. New York: Guilford.

Dixit, A. K., Stiglitz, J. E. (1977) Monopolistic competition and optimum product diversity. *American Economic Review*, 67: 297–308.

Duranton, G., Puga, D. (2000) Diversity and specialization in cities: why, where and when does it matter? *Urban Studies*, 37: 533–555.

Dymski, G. (1996) On Krugman's model of economic geography. *Geoforum*, 27: 439–452.

Eagleton, T. (2003) *After Theory*. New York: Basic Books.

Ellison, G., Glaeser, E. L. (1997) Geographic concentration in US manufacturing industries: a dartboard approach. *Journal of Political Economy*, 105: 889–927.

Fröbel, F., Heinrichs, J., Kreye, O. (1980) *The New International Division of Labor*. Cambridge: Cambridge University Press.

Fujita, M., Thisse, J.-F. (2002) *Economics of Agglomeration: Cities, Industrial Location, and Regional Growth*. Cambridge: Cambridge University Press.

Geertz, C. (1983) *Local Knowledge: Further Essays in Interpretive Anthropology*. New York: Basic Books.

Gertler, M. S. (2003a) A cultural economic geography of production. In K. Anderson, M. Domosh, S. Pile, and N. Thrift (eds) *Handbook of Cultural Geography*. London: Sage, 131–146.

Gertler, M. S. (2003b) Tacit knowledge and the economic geography of context, or, the undefinable tacitness of being (there) *Journal of Economic Geography*, 3: 75–99.

Gibson-Graham, J. K. (1996) *The End of Capitalism (As We Knew it): A Feminist Critique of Political Economy*. Oxford: Blackwell.

Giddens, A. (1979) *Central Problems in Social Theory: Action, Structure and Contradiction in Social Analysis*. London: Macmillan.

Gould, P. (1979) Geography 1957–1977: the Augean period. *Annals of the Association of American Geographers*, 53: 290–297.

Gregson, N., Simonsen, K., Vaiou, D. (2001) Whose economy for whose culture? Moving beyond oppositional talk in European debate about economy and culture. *Antipode*, 33: 616–646.

Haraway, D. J. (1991) *Simians, Cyborgs and Women: the Reinvention of Nature*. New York: Routledge.

Harvey, D. (1982) *The Limits to Capital*. Oxford: Blackwell.

Harvey, D. (1989) *The Condition of Postmodernity: An Enquiry into the Origins of Cultural Change*. Oxford: Blackwell.

Harvey, D., Scott, A. J. (1989) The practice of human geography: theory and empirical specificity in the transition from fordism to flexible accumulation. In B. Macmillan (ed.) *Remodelling Geography*. Oxford: Blackwell, 217–229.

A perspective of economic geography • **497**

Henderson, V. (2003) The urbanization process and economic growth: the so-what question. *Journal of Economic Growth*, 8: 47–71.

Henderson, V., Thisse, J. F. (2004) *Handbook of Regional and Urban Economics, Volume 4, Cities and Geography*. Amsterdam: North-Holland.

Hirschman, A. O. (1958) *The Strategy of Economic Development*. New Haven, CT: Yale University Press.

Imrie, R., Pinch, S., Boyle, M. (1996) Identities, citizenship and power in the cities. *Urban Studies*, 33: 1255–1262.

Jaffe, A. B., Trajtenberg, M., Henderson, R. (1993) Geographic localization of knowledge spillovers as evidenced by patent citations. *Quarterly Journal of Economics*, 108: 577–598.

Johnston, R. J. (1992) Review of 'Geography and Trade'. *Environment and Planning A*, 24: 1066.

Johnston, R. J., Taylor, P. J., Watts, M. J. (1995) *Geographies of Global Change*. Oxford: Blackwell.

Kaldor, N. (1970) The case for regional policies. *Scottish Journal of Political Economy*, 17: 337–348.

Krugman, P. (1991) *Geography and Trade*. Leuven: Leuven University Press.

Krugman, P., Venables, A. J. (1995) Globalization and the inequality of nations. *Quarterly Journal of Economics*, 110: 857–880.

Krugman, P., Venables, A. J. (1996) Integration, specialization, and adjustment. *European Economic Review*, 40: 959–967.

Lash, S., Urry, J. (1994) *Economies of Signs and Space*. London: Thousand Oaks: Sage

Latour, B. (1991) *Nous n'avons jamais été modernes: essai d'anthropologie symétrique*. Paris: La Découverte.

Leamer, E. E., Storper, M. (2001) The economic geography of the Internet age. *Journal of International Business Studies*, 32: 641–665.

Lee, R. (2002) Nice maps, shame about the theory: thinking geographically about the economic. *Progress in Human Geography*, 26: 333–355.

Lipietz, A. (1986) New tendencies in the international division of labor: regimes of accumulation and modes of social regulation. In A. J. Scott and M. Storper (eds) *Production, Work, Territory: The Anatomy of Industrial Capitalism*. Boston: Allen and Unwin, 16–40.

Mannheim, K. (1952) *Essays in the Sociology of Knowledge*. Henley-on-Thames: Routledge and Kegan Paul.

Markusen, A. (1999) Fuzzy concepts, scanty evidence, policy distance: the case for rigour and policy relevance in critical regional studies. *Regional Studies*, 33: 869–884.

Martin, A. F. (1951) The necessity for determinism. *Transactions of the Institute of British Geographers*, 17: 1–11.

Martin, R. (1999) The new geographical turn in economics: some reflections. *Cambridge Journal of Economics*, 23: 65–91.

Martin, R. (2001) Geography and public policy: the case of the missing agenda. *Progress in Human Geography*, 25: 189–210.

Martin, R., Sunley, P. (2001) Rethinking the 'economic' in economic geography: broadening our vision or losing our focus? *Antipode*, 33: 148–161.

Massey, D., Meegan, R. (1982) *Anatomy of Job Loss: The How, Why, Where, and When of Employment Decline*. London: Methuen.

McClosky, D. (2002) *The Secret Sins of Economists*. Chicago: Prickly Paradigm Press.

McDowell, L. (1997) *Capital Culture: Gender at Work in the City*. Oxford: Blackwell.

Meardon, S. J. (2000) Eclecticism, inconsistency, and innovation in the history of geographical economics. In R. E. Backhouse and J. Biddle (eds) *Toward a History of Applied Economics*. Durham, NC: Duke University Press, 325–359.

Myrdal, G. (1959) *Economic Theory and Under-Developed Regions*. London,: Gerald Duckworth.

Neary, J. P. (2001) Of hype and hyperbolas: introducing the new economic geography. *Journal of Economic Literature*, 39: 536–561.

Nelson, R. R., Winter, S. G. (1982) *An Evolutionary Theory of Economic Change*. Cambridge, MA.: Belknap Press.

Olsen, J. (2002) On the units of geographical economics. *Geoforum*, 33: 153–164.

Ottaviano, G. I. P., Thisse, J. F. (2001) On economic geography in economic theory: increasing returns and pecuniary externalities. *Journal of Economic Geography*, 1: 153–179.

Peck, J. (1996) *Work-Place: the Social Regulation of Labor Markets.* New York: Guilford Press.

Plummer, P., Sheppard, E. (2001) Must emancipatory economic geography be qualitative? *Antipode*, 33: 194–199.

Porter, M. E. (2001) Regions and the new economics of competition. In A. J. Scott (ed.) *Global City-Regions: Trends, Theory, Policy.* Oxford: Oxford University Press, 139–157.

Quah, D. T. (1996) Regional convergence clusters across Europe. *European Economic Review*, 40: 951–958.

Rigby, D. L., Essletzbichler, J. (2002) Agglomeration economies and productivity differences in US cities. *Journal of Economic Geography*, 2: 407–432.

Rodríguez-Pose, A. (2001) Killing economic geography with a 'cultural turn' overdose. *Antipode*, 33: 176–182.

Rojek, C., Turner, B. (2000) Decorative sociology: towards a critique of the cultural turn. *The Sociological Review*, 48: 629–648.

Rorty, R. (1979) *Philosophy and the Mirror of Nature.* Princeton, NJ: Princeton University Press.

Saxenian, A. (1994) *Regional Advantage: Culture and Competition in Silicon Valley and Route 128.* Cambridge, MA: Harvard University Press.

Sayer, A. (1997) The dialectic of culture and economy. In R. Lee and J. Wills (eds) *Geographies of Economies.* London: Arnold, 16–26.

Sayer, A. (2000) *Realism and Social Science.* London: Sage.

Sayer, A., Storper, M. (1997) Ethics unbound: for a normative turn in social theory. *Environment and Planning D: Society and Space*, 15: 1–17.

Scitovsky, T. (1954) Two concepts of external economies. *Journal of Political Economy*, 62: 143–151.

Scott, A. J. (1983) Industrial organization and the logic of intra-metropolitan location: I. Theoretical considerations. *Economic Geography*, 59: 233–250.

Scott, A. J. (1986) High technology industry and territorial development: the rise of the Orange County complex, 1955–1984. *Urban Geography*, 7: 3–45.

Scott, A. J. (1988) *New Industrial Spaces: Flexible Production Organization and Regional Development In North America and Western Europe.* London: Pion.

Scott, A. J. (2000) Economic geography: the great half-century. *Cambridge Journal of Economics*, 24: 483–504.

Scott, A. J. (2002a) A new map of Hollywood: the production and distribution of American motion pictures. *Regional Studies*, 36: 957–975.

Scott, A. J. (2002b) Regional push: the geography of development and growth in low- and middle-income countries. *Third World Quarterly*, 23: 137–161.

Shapin, S. (1998) Placing the view from nowhere: historical and sociological problems in the location of science. *Transactions of the Institute of British Geographers*, 23: 5–12.

Sheppard, E. (2000) Geography or economics? Conceptions of space, time, interdependence, and agency. In G. L. Clark, M. P. Feldman, and M. S. Gertler (eds) *The Oxford Handbook of Economic Geography.* Oxford: Oxford University Press, 99–119.

Shields, R. (1999) Culture and the economy of cities. *European Urban and Regional Studies*, 6: 303–311.

Sjöberg, O., Sjöholm, F. (2002) Common ground? Prospects for integrating the economic geography of geographers and economists. *Environment and Planning A*, 34: 467–486.

Smith, N. (1987) Dangers of the empirical turn: some comments on the CURS initiative. *Antipode*, 19: 59–68.

Storper, M. (1997) *The Regional World: Territorial Development in a Global Economy.* New York: Guilford Press.

Storper, M. (2001) The poverty of radical theory today: from the false promises of Marxism to the mirage of the cultural turn. *International Journal of Urban and Regional Research*, 25: 155–179.

Storper, M., Scott, A. J. (1995) The wealth of regions: market forces and policy imperatives in local and global context. *Futures*, 27: 505–526.

Sunley, P. (1996) Context in economic geography: the relevance of pragmatism. *Progress in Human Geography*, 20: 338–355.

Swyngedouw, E. (1997) Neither global nor local: 'glocalization' and politics of scale. In K. R. Cox (ed.) *Spaces of Globalization: Reasserting the Power of the Local.* New York: Guilford, 137–166.

Taylor, P. J., Catalano, G., Walker, D. R. F. (2002) Measurement of the world city network. *Urban Studies*, 39: 2367–2376.

Thisse, J. F. (1997) L'oubli de l'espace dans la pensée économique. *Région et Développement*, 13–39.

Thrift, N. (1994) On the social and cultural determinants of international financial centres. In S. Corbridge, N. Thrift, and R. Martin (eds) *Money, Power and Space*. Oxford: Blackwell, 327–355.

Thrift, N. (2000) Pandora's box? cultural geographies of economies. In G. L. Clark, M. P. Feldman, and M. S. Gertler (eds) *The Oxford Handbook of Economic Geography*. Oxford: Oxford University Press, 689–704.

Thrift, N. (2001) 'It's the romance, not the finance, that makes the business worth pursuing': disclosing a new market culture. *Economy and Society*, 30: 412–432.

Thrift, N., Olds, K. (1996) Refiguring the economic in economic geography. *Progress in Human Geography*, 20: 311–337.

Venables, A. J. (1996) Equilibrium locations of vertically linked industries. *International Economic Review*, 37: 341–359.

Yeung, H. W. (2003) Practicing new economic geographies: a methodological examination. *Annals of the Association of American Geographers*, 93: 445–466.

[2]

Paul Krugman's Geographical Economics and Its Implications for Regional Development Theory: A Critical Assessment*

Ron Martin

Department of Geography, University of Cambridge, Cambrige CB2 3EN, U.K.

Peter Sunley

Department of Geography, University of Edinburgh, Edinburgh EH8 9XP, U.K.

Abstract: Economists, it seems, are discovering geography. Over the past decade, a "new trade theory" and "new economics of competitive advantage" have emerged which, among other things, assign a key importance to the role that the internal geography of a nation may play in determining the trading performance of that nation's industries. Paul Krugman's work, in particular, has been very influential in promoting this view. According to Krugman, in a world of imperfect competition, international trade is driven as much by increasing returns and external economies as by comparative advantage. Furthermore, these external economies are more likely to be realized at the local and regional scale than at the national or international level. To understand trade, therefore, Krugman argues that it is necessary to understand the processes leading to the local and regional concentration of production. To this end he draws on a range of geographical ideas, from Marshallian agglomeration economies, through traditional location theory, to notions of cumulative causation and regional specialization. Our purpose in this paper is to provide a critical assessment of Krugman's "geographical economics" and its implications for contemporary economic geography. His work raises some significant issues for regional development theory in general and the new industrial geography in particular. But at the same time his theory also has significant limitations. We argue that while an exchange of ideas between his theory and recent work in industrial geography would be mutually beneficial, both approaches are limited by their treatment of technological externalities and the legacy of orthodox neoclassical economics.

Key words: Krugman, trade, external economies, regional industrial concentration, regional industrial policy.

The relationship between economic geography and economics has long been an asymmetric one. In constructing their theories and explanations of regional development, economic geographers have drawn freely on the concepts and perspectives of different schools of economics; but, for their part, economists have tended to accord little if any attention to the role of geography in the economic process. The case of trade theory admirably illustrates this point. Regional devel-

* This is a revised version of a paper presented at the Special Session on Economic Geography held at the Annual Conference of the Institute of British Geographers, Newcastle, January 1995. The authors gratefully acknowledge the many useful and constructive comments made on that occasion, and we have tried to respond to as many of these as possible. We also wish to thank the three anonymous referees for their suggestions, which have also helped to sharpen our argument.

opment theory has always been concerned with the question of interregional trade, because a region's ability to export goods and services is one of the foundations of local economic growth and employment (Erickson 1989). The typical approach to the study of interregional trade has been to borrow and adapt the ideas and models of comparative advantage (factor endowment) trade theory from economics. Trade economists, however, have invariably regarded the national economy as spaceless, and even international trade typically has been seen as an exchange system devoid of any geography, a world where goods and services move between dimensionless points at zero or uniform transport costs. This lack of a sensitivity to geography by trade theorists partly explains why there is no overall theoretical framework guiding geographical research on international trade (Grant 1994). The absence of such a framework is particularly evident at a time when the "globalization" of economic relations and the continental regionalization of trade are challenging the territorial and regulatory significance of national economic spaces and giving greater prominence to the nature and performance of individual regional and local economies within nations (Dunford and Kafkalas 1992; Anderson and Blackhurst 1993; Gibb and Michalak 1994).

Recently, however, there have been developments within economics which may mark the beginning of a closer relationship with economic geography in general and regional development theory more particularly. Over the past decade, a "new" trade theory and a new economics of competitive advantage have emerged which, among other important features, assign a key significance to the role that the internal geography of a nation may play in determining the trading performance of that nation's industries.[1] Econo-

mists, it seems, are discovering geography. In particular, Paul Krugman, the leading and extraordinarily prolific exponent of the "new" trade theory,[2] has sought to show how trade is both influenced by and in turn influences the process of geographical industrial specialization within nations (for example, Krugman 1991a). In his view, the importance of regional industrial specialization and concentration is such that economic geography should be accepted as a major subdiscipline within economics, "on a par with or even in some senses encompassing the field of international trade" (Krugman 1991a, 33). Likewise, from a different, but ultimately related perspective, Michael Porter, the eminent business economist, has argued that the degree of geographical clustering of industries within a national economy plays an important role in determining which of its sectors command a competitive advantage within the international economy (Porter 1990). In a similar vein to Krugman, Porter also argues that there are strong grounds for making economic geography a "core discipline in economics" (Porter 1990, 790).

Paul Krugman's work, especially, is worthy of closer interest by geographers. Krugman has written on a wide range of issues that impinge on the regional development question: trade, externalities, the localization of industry, strategic industrial policy, globalization, the role of history and "path dependence," and the implications of economic and monetary integration for regional growth. One of the key thrusts of his work is that in order to understand trade we need to understand the process of regional development within nations. A number of his writings have thus sought to explain why industrial development is likely to be geographically

---

[1] The set of ideas referred to as the "new" trade theory" was originally expounded in a series of papers by Dixit and Norman (1980),

Lancaster (1980), Krugman (1979, 1980, 1981), Ethier (1982), and Helpman (1984).

[2] Such has been Krugman's influence within the economics profession that Paul Samuelson (1994, vii) refers to him as "the rising star of this century and the next."

uneven. To this end, he draws on a range of economic and geographical ideas, from Alfred Marshall's account of localization economies, through traditional location theories, to notions of cumulative causation. For Krugman, economic geography—by which he means uneven regional development—is a central part of the process by which national economic prosperity and trade are created and maintained.

Our aim in this paper is to provide a critical assessment of Krugman's "geographical economics" and its implications for contemporary economic geography. An exchange of ideas between his theory and recent work in economic geography would be mutually beneficial. Such an exchange is not easy to engineer, however, as there are several significant obstacles, on both sides. First, Krugman's ideas are far from static. Indeed, his views seem to change continuously over time—sometimes in a self-criticizing way—so it is important to base any evaluation on a range of his works. Second, throughout Krugman's writings there is a strong distinction between what is theoretically possible and what is empirically and practically important, so that his conclusions have to be read carefully and closely. Most important, however, Krugman's geographical economics and contemporary economic geography are very different academic genres, with different methodological styles and conventions of analysis and writing (Krugman 1993a). Krugman's method is to start with a real world problem and then build a model to capture the "essence" of that problem (Krugman 1989, 1992). The model, which is usually mathematically specified, is made as simple as possible to remove unnecessary clutter, although in most cases he also gives a highly readable narrative account of the model. The mathematical aspect of his methodology may well explain the strong location-theory flavor to much of his geographical

economics.[3] However, this methodological and theoretical disposition is unlikely to appeal to many economic geographers, who have abandoned formal models and rigorous exegesis for a more discursive approach, in which broad master concepts (like "flexible specialization" and "post-Fordism") are mingled with anecdotal spatial stereotypes ("industrial spaces" and "industrial districts").

These differences probably largely explain why Krugman's writings have thus far had a limited impact on economic geography, and why they have been summarily dismissed by certain geographers. Johnston (1992) dismisses Krugman's *Geography and Trade* (1991a) as "highly simplistic" in its treatment of "geography" and "patronizing" in its comments on the work of economic geographers.[4] In a somewhat similar vein,

---

[3] This location-theory orientation has if anything become even more pronounced in his two most recent books on "spatial economics," *Development, Geography and Economic Theory* (Krugman 1995) and *The Self-Organising Economy* (Krugman 1996). In the former, lamenting economic geographers' "retreat" from quantitative models into Marxist and regulationist concerns with "post-Fordism," Krugman resurrects what he calls the five "exiled traditions" of economic geography: location theory, social physics, cumulative causation, land use modeling, and Marshallian local external economies. In the latter, von Thunen's model and central place theory occupy a key role in his theorization of the "self-organizing" space economy.

[4] It is not difficult to see how economic geographers might take offense at Krugman's view of their work. In *Geography and Trade* (1991a, 3–4) he writes: "The decision by international economists to ignore the fact that they are doing geography wouldn't matter so much if someone else were busy . . . looking at localization and trade within countries. Unfortunately, nobody is. That is, of course an unfair statement. There are excellent economic geographers out there. . . . However, . . . economic geographers proper are almost never found in economics departments, or even talking to economists. . . . They may do excellent work,

in his review of the same book, Hoare (1992, 679) criticizes the particular economic geography used by Krugman as "dated, historically and intellectually" and his analysis as based on the "flimsiest of empirical support." However, Krugman's remarks are leveled primarily at his own colleagues' failure to admit that "space matters" (Krugman 1991a, 8), and he should at least be congratulated for wanting "to bring geography back into economic analysis," even if the particular form of geography he uses—essentially a form of regional science—is open to criticism. Furthermore, *Geography and Trade* gives only a partial glimpse into Krugman's analyses, and any considered judgment as to the significance of Krugman's work for economic geography must also be based on his numerous other writings in the field.

We too have criticisms to make of Krugman's treatment of economic geography, although we also believe that his work raises some interesting issues for contemporary regional theory. We begin by outlining what we take to be the essential arguments and components of his "geographical economics," focusing on his interpretation of the relationships between location and trade, the role of increasing returns and externalities in the localization of industry, and the signifi-

---

but it does not inform or influence the economics profession." It could equally be argued, of course, that it is the economists who have failed to talk to economic geographers and that, as a result, like Krugman they are largely ignorant of the major developments that have taken place in economic and industrial geography over the past decade or so. Equally irritating is Krugman's comment, in *Development, Geography and Economic Theory* (1995, 88), that "in the end, we [i.e., economists] will integrate spatial issues into economics through clever models (preferably but not necessarily mine) that make sense of the insights of the geographers in a way that meets the standards of the economists." Whether economists have any such monopoly over analytical or theoretical standards may most certainly be questioned.

cance of history, "lock-in," and path dependence for regional development. The subsequent section examines these ideas in closer, more critical detail, and compares Krugman's theories with those that have emerged from the "new industrial geography" in the past few years. We then examine his arguments about the impact of economic integration on regional development, especially his prognoses of the regional implications of integration within the European Union and his views on regional stabilization and industrial policy. We conclude the paper by drawing together the main strengths and weaknesses of Krugman's approach to economic geography.

Trade, Externalities, and Industrial Localization: The Bases of Krugman's "Geographical Economics"

The New Trade Theory and Location

Krugman's geographical economics and theorization of uneven regional development are firmly rooted in his contributions to the "new trade theory." Conventional trade economics is based on Ricardian comparative advantage theory (especially in its Heckscher-Ohlin-Samuelson versions), which argues that under conditions of perfect competition, and given the relative immobility of one or more factors of production, nations will specialize in those industries in which they have comparative factor advantages (favorable resources of raw materials, cheaper labor, and so forth). The relative factor endowments of different nations is thus the main reason for international trade and specialization. The principle of comparative advantage, then, predicts that countries with dissimilar resource endowments will exchange dissimilar goods. The theory does not and cannot, however, predict what sort of goods will be exchanged by countries that have similar resource endowments. But much of world trade, and most of Organization for Economic Cooperation and Development (OECD) trade, is between countries with similar factor endowments, and they exchange predominantly similar products. Such *intraindustry* trade has been expanding rapidly in recent decades, even though countries have been converging in skill levels and per capita endowments of capital (OECD 1994). The "new trade theory" is an attempt to account for this form of trade. The new trade theory acknowledges that differences among countries are one reason for trade, but it goes beyond the traditional view in four main ways (Krugman 1990).[5]

First, it argues that much trade between countries, especially intraindustry trade between similar countries, represents specialization to take advantage of *increasing returns to scale* rather than to capitalize on inherent differences in national factor endowments. Contrary to the assumptions of perfect competition and constant returns to scale that underpin the basic Ricardian theory of comparative advantage and trade, according to the new theory *imperfect* competition and *increasing* returns are pervasive features of contemporary industrial economies.[6] If specialization and trade are driven by increasing returns and economies of scale rather than by comparative advantage, the gains from trade arise because production costs fall as the scale of output increases. Second, with this view of the world, specialization is to some extent a historical

accident. The specific location of a particular microindustry is to a large degree indeterminate, and history-dependent. But once a pattern of specialization is established, for whatever reason, that pattern gets "locked in" by the cumulative gains from trade. There is thus a strong tendency toward "path dependence" in the patterns of specialization and trade between countries: *history matters.* Third, the patterns of demand for and rewards to factors of production under conditions of imperfect competition and intraindustry trade will depend on the *technological conditions* of production at the micro level, and nothing can be said a priori about the evolution of factor demands. Fourth, whereas under the Ricardian model free trade is assumed to be the appropriate policy stance, the new trade theory argues that the existence of imperfect competition and increasing returns opens up the possibility of using trade policies *strategically* to create comparative advantage by promoting those export sectors where economies of scale—and particularly external economies—are important sources of rent. In other words, strategic trade policy may enable a nation to shift the pattern of international economic specialization in its own favor (Krugman 1980).

In Krugman's view, these developments in the "new trade theory" both necessitate and facilitate a rapprochement between trade theory and location theory. In recent work he has compared the contrasting assumptions underlying these two, hitherto largely separate, sets of economic literature (Krugman 1993a). His geographical economics is a hybrid of the two. It combines the models of imperfect competition and scale economies used in new trade theory with location theory's emphasis on the significance of transport costs. The interaction of external economies of scale with transport costs is the key to his explanation of regional industrial concentration and the formation of regional "centers" and "peripheries" (Krugman 1991a; Krugman and Venables 1990). His model suggests that high

[5] There are in fact several different versions of the new trade theory, but the various strands all subscribe to the basic elements elaborated by Krugman in his *Rethinking International Trade* (1990).

[6] Of course, the idea that increasing returns and economies of scale could be alternatives to comparative advantage as explanations of international specialization and trade goes back to Ohlin (1933), if not to Adam Smith. But while their importance has been recognized in principle, they invariably have been assigned a subsidiary or supplementary role in formal trade theory. The novelty of the "new trade theory" is that increasing returns and economies of scale are moved into the mainstream.

transport costs will act to prohibit the geographical concentration of production. With some reduction in transport costs, however, firms will want to concentrate in one site to realize economies of scale both in production and in transport. In Krugman's words, "Because of the costs of transacting across distance, the preferred locations for each individual producer are those where demand is large or supply of inputs is particularly convenient—which in general are the locations chosen by other producers" (1991a, 98). If transport costs continue to fall, the model suggests that the need to locate near to markets will disappear and production may disperse. However, given that some transport costs will remain, the circular relation, or positive feedback, between production and demand means that regions which have a head start in manufacturing, typically as a result of accidental good fortune, will attract industry and growth away from regions with less favorable initial conditions. Krugman (1991a, 1991d) argues that this model explains the rise of the manufacturing belt in the Northeastern United States during the nineteenth century. It has also been applied to the discussion of the likely fate of peripheral regions in the European Union (Krugman and Venables 1990). Recently, Krugman (1993d) has developed a new variant of the approach which argues, not that successful regions systematically attract industry away from peripheral areas, but that trade and external economies produce more-specialized regions, which are then more vulnerable to random "shocks."

On the basis of this location model, Krugman (1993b) argues that large-scale regions are more significant economic units than nation-states. He writes that a satellite image of the world at night shows regional agglomerations rather than national concentrations. Furthermore, in his view, "The best evidence for the practical importance of external economies is so obvious that it tends to be overlooked. It is the strong tendency of both economic activity in general and of particular

industries or clusters of industries to concentrate in space" (Krugman 1993b, 173). This tendency, he argues, provides a decisive refutation of the competitive model of economic equilibrium, for when one turns to the location of production in space the "irrelevance of equilibrium economics" is compelling and there are multiple possible equilibria. An economy's form is determined by contingency, path dependence, and the initial conditions set by history and accident. Forward and backward linkages mean that once an initial regional advantage is established it may become cumulative. There is therefore no automatic tendency toward an optimum solution, as apparently "irrational" economic distributions may be "locked in" through increasing returns. So that while Krugman associates economic geography with path dependence, or what he calls "the economics of qwerty,"[7] he does not neglect the possibility of reversal and change. Rather, he argues that when change in regional fortunes occurs it will be sudden and unpredictable. He repeatedly uses the example of Massachusetts as a regional economy that has gone into a tailspin. Krugman (1991c) suggests that, under certain conditions, self-fulfilling expectations may outweigh accumulated advantages, and pessimistic expectations about a region's prospects may become self-justifying.

Increasing Returns and Imperfect Competition

According to Baldwin (1994), Krugman's analysis represents a genuinely new location theory. In actuality, not only does

---

[7] This term derives from the first line of letters (QWERTYUIOP) on the keyboard of a typewriter or word processor. That this order is the same today as it was on the first mechanical typewriters of the nineteenth century, even though more efficient sequences are possible, represents a form of "lock-in" and persistence that has analogous parallels in the economy.

it echo Ohlin (1933), Hirschman (1958), and Myrdal (1957), it strongly resembles Weber's (1929) model of the overlaying of transport costs on agglomeration economies. Whereas Weber identified spatial overlaps, the Krugman-Venables model adds the general level of transport costs as a variable that can fluctuate over time. Given these predecessors, we should consider whether there is anything really new in Krugman's geographical economics. In several places he himself states that he is simply retelling an old story in a more rigorous way. It would be tempting to conclude, as some critics have done, that there is nothing new in this. However, this conclusion overlooks the way in which Krugman's reading of agglomeration has been shaped both by the developments in trade theory and by recent models of industrial organization. One of the main reasons for trade theory's traditional neglect of the advantages which arise from increasing returns and economies of scale was the difficulty of modeling market structure. In one sense, recent developments in modeling market structure with nonconstant returns have facilitated the new trade theory (Helpman 1984; Krugman 1983a; Buchanan and Yoon 1994; Smith 1994). Hence the best place to start, in order to understand Krugman's interpretation of increasing returns, is with these models. Two approaches are particularly relevant to Krugman's account of geographical concentration, namely the Marshallian and Chamberlinian models.

The Marshallian approach to understanding increasing returns is already familiar in economic geography. It is based in a long tradition that sees economies of scale as primarily external, as arising from the specialization of the social division of labor (Young 1928; Stigler 1951). Typically, economies of scale have been taken to be purely external, so that the assumptions of perfect competition may be retained (Chipman 1970). While Krugman is aware of this long tradition, he suggests that recent advances in the modeling of such

external economies (see, for example, Romer 1990) have given them a new tractability. He argues, in one paper (Krugman 1981), that external economies at a national level are the key to the uneven development of countries. Yet, increasingly, Krugman has been reluctant to treat nations as economic units and has emphasized the significance of external economies at a local and regional scale. Indeed, in *Geography and Trade* his account of the localization of industries and agglomeration at a relatively small scale is based on Marshall's three types of external economy: labor market pooling, the availability of specialist suppliers, and the presence of technological knowledge spillovers. However, he places greater emphasis on the first two of these and deals only briefly with local technological externalities. This might seem strange given that elsewhere he has argued that, empirically, the most plausible source of positive externalities from trade is the inability of innovative firms to monopolize the knowledge they create (Krugman 1987a, 137). But the arguments are not incompatible, for Krugman argues that these externalities are difficult to measure and track and that many of them are national or international in scope (see also Ethier 1982). So while local technological externalities are important in some high-tech districts, he considers their general locational significance to be limited.

Further reasons for Krugman's lack of emphasis on technological spillovers become apparent when we turn to the second model of market structure that has been influential in new trade theory, namely the Chamberlinian model (Chamberlin 1949). This model of market equilibrium envisages competition among similar firms producing differentiated products which are close but not perfect substitutes. Each firm faces a downward sloping demand curve and has some monopoly power. The entry of new firms producing slightly different products eliminates monopoly profits and means that there are many little monopolists. Many explanations of intraindustry trade by new trade theorists have been developed from this model, with the assumption of economies of scale that are internal to firms.[8] According to Helpman and Krugman (1985), these internal economies are easy to justify. They argue that firms could both achieve economies of scale and meet a demand for differentiated products from other producers and consumers by locating at one site and engaging in intraindustry trade (Krugman 1989). Krugman sees this approach as especially relevant to intermediate products and components, where the scope for differentiation is high and the market often too small for an exhaustion of scale economies. Moreover, he argues that

> where intermediate goods produced with economies of scale are not tradeable, the result will be to induce the formation of "industrial complexes"—groups of industry tied together by the need to concentrate all users of intermediate goods in the same country. In this case the pattern of specialisation and trade in the Chamberlinian world will come to resemble the pattern in the Marshallian world described previously. (Krugman 1987c, 319; compare Losch 1967, 109)

Both of these models of competition are implicit in Krugman's discussion of regional and local externalities.

The Role and Implications of Externalities

Krugman uses Marshall's theorization of external economies to explain geographical clustering at a relatively small scale, that of urban specialization and city agglomerations. However, the Chamberlinian approach has been equally important to his account of externalities. In Krugman's view, the presence of increasing returns implies that the orthodox

divide between "technological" and "pecuniary" externalities is misleading and unhelpful.[9] In the competitive equilibrium model, technological externalities are defined as those consequences of activity which directly influence the production function in ways other than through the market. They have real welfare and efficiency consequences (Meade 1952; Mishan 1971). In a situation of perfect competition and constant returns, however, pecuniary externalities which arise through buying and selling in the market are scarce (Scitovsky 1954). Krugman argues that this type of distinction is misleading. In his words, "It is valid only when there are constant returns and perfect competition; in a world of increasing returns and imperfect competition, the range of significant external economies is much larger. In particular, there are true external economies associated with a variety of market-size effects" (Krugman 1993b, 166). In summary, there are increasing returns in production, so that the size of the market matters and pecuniary economies also have real welfare significance. Elsewhere Krugman writes,

> Over the past decade . . . it has become a familiar point that in the presence of imperfect competition and increasing returns, pecuniary externalities matter; for example, if one firm's actions affect the demand for the product of another firm

---

[8] Hanink (1988, 1994) describes these approaches as the theory of differentiated markets, and uses an extended Linder model to explain the consequences of geographical product differentiation.

[9] The term "pecuniary externalities" was used by Scitovsky (1954) to refer to externalities arising from market imperfections of both demand and supply. Market-size effects are an important form of pecuniary external economy; the larger the market, the more individual firms can increase their output without having to cut prices. Increasing market size permits increasing returns. Such market-size effects may operate at various geographical scales, from the international to the local. Technological externalities refer to the situation where there are spillovers from the production function of one firm into those of other firms, for example when a firm makes an innovation that other firms can imitate.

whose price exceeds marginal cost, and this is as much a "real" externality as if one firm's research and development spills over into the general knowledge pool. At the same time, by focusing on pecuniary externalities, we are able to make the analysis much more concrete than if we allowed external economies to arise in some invisible form. (Krugman 1991b, 485)

This focus on pecuniary externalities shapes Krugman's interpretation of Marshall (see also David and Rosenbloom 1990). For instance, he claims that both labor market pooling and the availability of nontraded intermediate goods are examples of market-size effects.[10] Moreover, he emphasizes "pecuniary" externalities which are derived from both external and internal economies of scale. As he notes, "Even if economies of scale are internal to firms, internal economies in the production of intermediate inputs can behave like external economies for the firms which buy them" (Krugman 1981, 151). Furthermore, in a recent paper on metropolitan location (Krugman 1993c), he demonstrates that the "centripetal forces" which hold a city together can be derived from the interaction of economies of scale at the plant level with transport costs. Thus, he argues, local external economies do not have to be assumed; instead, they are again derived from market-size or market-potential effects. The key point to note about these arguments is that internal economies of scale, by increasing the incentive for firms to concentrate on one site, intensify the tendency toward the geographical concentration of production. Thus Krugman associates the rise of the North American manufacturing belt with the rise of the Chandlerian corporation.

Krugman's analysis emphasizes that externalities operating within and between industries in these regional agglomerations make a difference to the competitive advantage of the constituent firms. In this sense, then, Krugman's work conveys a sense of regional competitiveness.[11] At the same time, his recent writings on the international economy have criticized certain popular definitions and uses of competitiveness, and it is important to set his regional work within the context of his more general understanding of the consequences of trade (Krugman 1994a, 1994b, 1994c, 1994d). The primary issue here is that Krugman sees all forms of international economic integration, including trade and capital and labor mobility, as essentially beneficial. For example, the specialization produced by trade raises the efficiency of the world economy as a whole and produces mutual benefits to the trading nations (Krugman 1994d). This view is partly founded on his belief that comparative advantage still remains important and useful. It is also based on his belief that the "new" trade theory's recognition of externalities and imperfect competition highlights the *potential gains* from economic integration. Increased trade may allow greater economies of scale through rationalization and, in other situations, it may have a beneficial effect on oligopolistic markets by increasing competition.

The complexity of much of Krugman's work also reflects the fact that the existence of significant externalities and nonconstant returns also opens up possible ways in which increased trade and integration may have adverse effects. Krugman (1989) highlights two main sources of adverse effects. The first is the possibility of the uneven distribution of benefits associated with the existence of excess returns in imperfectly competitive industries. A country that gains a disproportionate share of high-returns industries can gain at others' expense, raising the possibility that trade policies designed to

---

[10] Krugman also praises Fujita (1989) for his emphasis on market-size effects as explanations of urban agglomeration.

[11] In a similar fashion, Porter (1990) argues that the geographical concentration of leading industries often reinforces and intensifies their competitive advantage.

foster these industries will lead to trade conflict. Hence, "While the possibility of actual losses from trade is probably purely academic, there is a real issue of conflict over the division of the gains" (Krugman 1989, 361). (For a global example, see Krugman and Venables 1994.) There is clearly also a regional dimension to this problem of uneven distribution. As a result of the importance of external economies and the accumulated, path-dependent advantages of certain regions, it is possible that these leading regions will capture a disproportionate share of the benefits of increased integration. A major obstacle to integration, in this view, is that its benefits are not equally shared across regions *within countries*.

This forms the basis of the second set of adverse consequences identified by Krugman, namely adjustment costs. He argues that although it is costly for capital and labor to shift into new industries, these costs represent a type of investment. They may be deserving of compensation, but they are not reasons to prevent or delay change. On the other hand, where these adjustments involve significant social costs, most importantly unemployment, he concedes that this may provide a case against moving too fast. The possibility of adjustment costs becoming real social costs should not be dismissed lightly. One ameliorating factor that Krugman notes is that the growth of trade between the industrialized nations since the Second World War, especially within Europe, has largely been intraindustry trade. "Thus the specialisation that took place as trade in manufactured goods grew tended to involve concentration on different niches within sectors rather than wholesale concentration of different countries on different industries" (Krugman 1989, 364). There is clearly a tension in Krugman's work between his positive evaluation of trade and integration in general and his demonstration that significant adverse consequences are possible. As we have seen, the question of uneven distribution of the benefits and costs is probably the most important of these. Indeed, while

Krugman's "geographical economics" shows the positive effects of agglomeration on productivity, it is also particularly well suited to explaining these possible adverse consequences.

Krugman's Geographical Economics and Economic Geography: A Critical Comparison

Clearly, Krugman shares an interest in regional agglomeration and the geographical consequences of trade with many economic geographers. At the same time, his treatment of these issues has been significantly different from the approaches pursued in economic geography in recent years. In this part of the paper we shall examine the most important of these differences and consider the lessons that Krugman and economic geographers can learn from each other. As we have noted already, a fundamental difference between Krugman's geographical economics and the various schools of contemporary economic geography is one of method. Krugman's reliance on formal models means that his work is rigorous and supported by mathematical proofs. In his view, the dependence of many of these models on unrealistic assumptions is not a grave problem nor a serious limitation. Instead, he appears to regard these models as rough metaphors or representations of the core of real world problems (Krugman 1995). When the models' results are found to be inadequate, their assumptions can be modified. In contrast, most contemporary economic geography has abandoned the use of formal modeling and is dominated instead by various types of political economy, which aim, above all, to be "realistic." From this perspective, Krugman's models have an inadequate sense of geographical and historical context. Knox and Agnew (1994), for example, argue that Krugman's core-periphery model in *Geography and Trade* differs from other location models in that it does not suggest a long-term process of conver-

KRUGMAN'S GEOGRAPHICAL ECONOMICS   269

gence. Instead, "the long term never arrives" (Knox and Agnew 1994, 83). There are multiple equilibria as concentrations persist for long periods of time but may then be unraveled by new patterns of concentration. However, Knox and Agnew insist that

> concentration somewhere . . . is the perpetual rule. So though apparently attentive to historical change, this model is static in its assumptions about the operation of economic-locational principles. The same principles of increasing returns, imperfect competition, and agglomeration are at work *in the same way* all the time. From this point of view, geographical outcomes can change but the process driving them does not. (1994, 83; original emphasis)

While this phrasing may be too strong,[12] Krugman (1991a) clearly states that the patterns of concentration that he describes are typical only of some industries under certain conditions; nevertheless, it does identify an important weakness in Krugman's work. He claims that the same broad locational forces which explain the growth of nineteenth-century concentrations also underlie the continued tendency to agglomeration. Indeed, this is one reason why he is reluctant to emphasize technological spillovers as a key determinant of contemporary clusters. At the same time, however, Krugman makes several passing references to the way in which the nature of agglomeration has changed over time. Thus he suggests, in one paper, that the railway and steamship were responsible for the emergence of core-periphery distinctions and that the age of this type of divergence may have passed. But such a "throwaway" suggestion requires a great deal more explanation. The historical ground-

ing of Krugman's approach remains unclear and clouded by ambiguity. What is clear is that his emphasis on continuity in the forces responsible for capital's agglomeration contrasts with economic geographers' focus on historical patterns of restructuring. However, the relative merits of this more historical approach depend on precisely how change is theorized and explained. It is impossible here to talk about economic geography as a whole; we have therefore selected two relevant areas of work, namely the recent literature in industrial geography on regional agglomeration and recent writing on theorizing the geography of trade.[13]

The Resurgence of Regional Economies

During the past decade, the most influential approach to industrial organization within economic geography has been the notion of a fundamental transition from Fordist mass production to more flexible production methods, such as flexible specialization. Scott and Storper (1992a, 1992b), Scott (1988), Storper and Walker (1989), and others have argued that internal economies of scale and scope have been undermined by increased market uncertainty and technological change. They argue that the response has been horizontal and vertical disintegration, or an externalization of production, which enables a greater ability to meet differentiated demand and a greater adaptability to market forces. Where a multiplicity of linkages are created which have geographically sensitive transaction costs, externalization is positively related to agglomeration. In this view, "Agglomeration is a strategy whereby producers ease the tasks of transactional interaction because proximity translates into lower

---

[12] After all, most of the main schools of economic geography and political economy may be criticized on similar grounds. All, for example, assume that the basic laws of economic development (as they perceive them) remain essentially unchanged as capitalism evolves over historical time.

[13] These are two of the leading fields in contemporary (post-Marxist) economic geography. It would require another paper to consider Krugman's work in relation to the complete corpus of geographical work on uneven regional development.

costs and wider opportunities for match-ing needs and capabilities" (Scott and Storper 1992b, 13). In summary, the shift to flexible specialization has been respon-sible for the rise of new industrial districts and for the new, or renewed, significance of regional agglomeration (Sabel 1989). While there are many contrasts between this "new industrial geography" and Krugman's geographical economics (see Table 1), we shall focus on three issues: the treatment of industrial and market structure, of externalities, and of nonmar-ket transactions and relations.

As we have seen, Krugman tends to rely on several abstract models of monop-olistic and oligopolistic market structure. The assumptions of these models are in some ways unrealistic, but the underlying rationale is that they are useful because of the pervasive presence of imperfect com-petition. In contrast, the flexible special-ization approach has envisaged a new type of competition involving downsizing and

disintegration and therefore a movement back toward perfect competition. How-ever, the idea that corporate disintegra-tion is a necessary response to uncertainty can be criticized (see, for example, Lover-ing 1990; Phelps 1992). Moreover, as Phelps (1992) has argued, Scott's analysis of the causes of agglomeration pertains primarily to situations approximating to that of perfect competition. In Phelps's view, "The assumption of near-perfect competition is otherwise implicit in an analysis which applies to single plant firms and neglects considerations of differential economic power embodied in linkage structures" (1992, 41). This is especially problematic when the analysis is applied to international trade. As Markusen (1993, 287) writes, "Most important internation-ally traded industries are now multinucle-ated, with large national firms thrust into more spirited competition with similarly sized and politically well-endowed firms from other nations." Hence, "the number of players is relatively small, their sizes and clout are varied, and none of them is

Table 1

A Comparison of Krugman's "Geographical Economics" with the "New Industrial Geography"

	Krugman	New Industrial Geography
Externalities	Marshallian, especially labor pooling, specialist suppliers "Pecuniary" market-size effects	Marshallian trio Labor market Specialist suppliers Technological and knowledge spillovers
Agglomeration	Local clusters Interregional center-periphery pattern	Industrial districts Craft-based High-tech Financial centers
Competition	Imperfect: monopolistic and oligopolistic; economies of scale	Competitive flexible specialization; economies of scope
Transfer costs	Transport, including trade barriers	Transactions costs
Technological spillovers	Not typical, but important in some industries; local and international	Local and fundamental to innovatory success in high-tech clusters
Labor market pooling	Strategy of insurance against risk (both employers and employees)	Form of local social embeddedness
Social and cultural characteristics of clusters	Difficult to formalize and assumed a priori; best left to sociologists	Key preconditions for successful localization

unaware of the behaviour of its neighbors. These are characteristics of oligopolized markets, not perfectly competitive ones."

Several arguments have been used to support the association of near-perfect competition with agglomeration. One is the finding that in some industries and places larger producers are located away from local clusters of industry (Hoare 1975; Scott 1986), and another is the observation that the decline of some industrial districts has been associated with the concentration of production into larger firms (Steed 1971). These are contingent findings, however. For example, Scott (1992a) notes that large producers are integral to the Southern Californian computer districts. Even where large firms are not found in local industrial clusters, they may be central to the regional and metropolitan concentrations modeled by Krugman. It cannot be assumed that internal economies of scale and scope act against agglomeration. Indeed, the intraindustry trade literature implies that, with increasing product diversity, internal economies and agglomeration become more closely linked. There is clearly a need to research the relations between market structure and locational dynamics in more detail.

This difference on the issue of competition has important consequences for understanding externalities. In order to compare Krugman's representation of externalities with that used in the "new industrial geography," it is helpful to set both approaches within a general framework. De Melo and Robinson (1990) argue that three main approaches to externalities are apparent in recent economic literature. The first is the Marshallian externalities approach, which we discussed above. They suggest that some parts of endogenous growth theory fall within this approach. For example, in an article on the externalities arising from human capital formation, Lucas (1988) talked about increasing returns at an economywide level. The second type of externality that De Melo and Robinson identify are those that result in uneven

rates of growth and occur with imperfect competition. Again, there are examples from endogenous growth theory: Romer (1990) sees investment in R&D in a situation of monopolistic competition as generating externalities in disembodied knowledge. The third type of externality arises from demand spillovers between sectors and industries. Murphy, Schleifer, and Vishny (1989), for example, argue that there are low-level equilibrium traps, where industrialization remains unprofitable. Industrial production only becomes profitable for individual firms in the context of more general demand linkages.

This framework provides a means of comparing the two approaches to agglomeration (Table 2). Krugman's focus on market size effects is clearly closest to, and draws most heavily on, pecuniary externalities.[14] As we have seen, his explanation of local clustering also invokes certain types of Marshallian external economy. Importantly, he has tended to downplay the significance of externalities based on spillovers in technological knowledge. Krugman (1987c) describes these as "elusive," preferring to concentrate on externalities that can be modeled. The difference between his approach and that of the "new industrial geography" is apparent. In accordance with the reliance on situations of near-perfect competition, Marshallian external economies have been at the forefront of the industrial districts literature. As Phelps has argued, and as Krugman's account demonstrates, the external economies that can be used in this approach are only a subset of the range available (Phelps 1992). To a certain extent this limitation has been weakened by those revisionist studies of industrial

---

[14] Krugman (1993b, 1995) describes this type of external economy as similar to those envisaged in "Big Push" interpretations of industrialization. He argues that a large-scale program of industrialization can take advantage of external economies and complementarities and so reduce the risk of investment (see Rosenstein-Rodan 1943).

Table 2

Comparison of the Treatment of Externalities in Krugman's
"Geographical Economics" and the "New Industrial Geography"

Type of Externality	Application to Agglomeration	
	Krugman	New Industrial Geography
Marshallian external economies	Local clusters of industry associated with market-size effects (labor pooling and specialist suppliers) and with internal economies	Districts associated with vertical distintegration and transaction costs; technological spillovers important
Knowledge and technological spillovers under imperfect competition	Important in some industries, but not typical and difficult to model—"too fashionable"	(Not typical; where present large producers tend to adopt decentralized and flexible organizational forms)
Pecuniary externalities (demand and supply spillovers)	Regional specialization and concentration on a grand scale (center-periphery) through interaction of market size, demand, and transport costs	(Typically regarded as Marshallian; much more emphasis on nonmarket conditions)

districts which argue that large producers can imitate decentralization. However, the contradiction between a commitment to perfect competition and the dependence of Schumpeterian models of "creative destruction" and local technological spillovers on imperfect competition cannot be resolved easily.[15]

The differences between Krugman's geographical economics and the recent work in economic geography on regional development are not confined to industrial structure and externalities, but also extend to the question of nonmarket transactions. Thus another important contrast between Krugman's approach and those of economic geographers is the manner in which the increasing power of larger producers has been related to contemporary localizations of industry. In economic geography there has been some dissatisfaction with the way in which the flexible specialization literature has ignored the increasing internationalization of firm structures and globalization more generally (Amin and Robins 1990; Gertler 1992). Consequently, there has been an interest in the way in which large firms interact with industrial districts. In contrast to Krugman's market-size effects, however, the main emphasis has been on the intermingling of firm and local networks (Amin and Thrift 1992; Grabher 1993). Networks have usually been defined as types of organizational relation that are neither market transactions nor hierarchies, and the term has been used to refer to cooperative and mutually beneficial relationships among producers (Cooke and Morgan 1993). Using this definition, the boundaries of firms become blurred, and firms and districts become intermingled. On the one hand,

[15] Schumpeterian models of "creative destruction," technological spillovers, and endogenous growth depend on imperfect competition. Typically, the incentive for firms to develop new products and processes stems from the temporary monopoly profits which they can earn (Grossman and Helpman 1991; Aghion and Howitt 1993). This sits uneasily with the new industrial geography's emphasis on near-perfect competition.

Krugman's contrary emphasis on pecuniary relations is a reminder to geographers not to lose sight of market effects. But on the other hand, Krugman's neglect of externalities that are intangible and leave no paper trail appears too restrictive. As Jaffe, Trajtenberg, and Henderson (1993) have pointed out, knowledge flows do sometimes leave a paper trail, in the form of citation of patents.

The interest in network forms of organization in economic geography reflects a more general concern to examine the ways in which economic activities are "embedded" in, and made possible by, social and cultural conditions. This has been applied with particular force by Storper (1992a) to high-technology districts. As Harrison (1992) notes, this interest in embeddedness has been the distinctive contribution of the recent geographical literature on industrial districts.[16] This stands in complete contrast to Krugman's rejection of invisible externalities. However, as Storper (1992b) argues, in the context of increasing market contestability, it is difficult to explain the continuing competitive advantage of certain districts if their conventions, rules of behavior, and implicit accords are not taken into account. Conversely, the decline of other regions appears to be partly a result of the "lock-in" of outmoded conventions and rules of behavior (Grabher 1993). Krugman's rejection of nonmarket linkages seems to be made primarily on the grounds that if externalities cannot be modeled then they have to be assumed a priori, so that the analyst can say anything she or he likes about types of spillover. But this rules out other methods of research and more sociological approaches. Moreover, Krugman's reluc-

tance to envisage nonmarket linkages seems to conflict with his commitment to new-Keynesian economics, where expectations and conventions are central. He himself shows (Krugman 1991c) that, under certain conditions, expectations may affect the course of regional development. But, if they are to be understood, expectations cannot be treated as exogenous "animal spirits." Rather, they are an integral part of social conventions and meanings, and their formation should be an important area for regional research. Our conclusion, then, is that there is a need for a greater exchange of ideas between Krugman's work and the geographical literature. But this applies not only to research on regional agglomeration but also to that on trade more generally.

The New Political Economy of Trade

Recent years have seen a growing interest by economic geographers in the spatial patterns of international trade. While this work has lacked a comprehensive theoretical framework, it has nevertheless been characterized by shared themes revolving around the inability of conventional geographies, based on Ricardian comparative advantage, to explain fully the complex character of contemporary international patterns. In accord with the "new trade theory," this geographical revival has stressed the importance of shifts in the world economy and the rise of intraindustry and intracorporate trade. One of the defining features of this revival has been a call to study the ways in which the geography of trade is shaped by states and by trade regimes. This emphasis on state policy, and the interpretation of trade on which it is based, contrasts with Krugman's approach in ways that raise fundamental questions about the effects of trade and its policy implications.

Some years ago, Johnston (1989) called for trade to be explained as part of a holistic theory of uneven development that combines the logics of capitalism and the policies of states. To some degree, his

---

[16] Amin and Thrift (1994) describe this embedding as best summed up by the phrase "institutional thickness." This is defined by a strong institutional presence in a local area, high levels of interaction among these institutions, strong social structures, and a collective awareness of common enterprise.

plea for an enlarged research agenda has
begun to be recognized. Grant summarizes recent developments as follows:

> The unifying theme in newer approaches is
> their study of the interactions between
> governments and firms and their connec-
> tions to trade and industrial policy within
> the context of a politically and economically
> competitive world economy, one in which
> governments attempt to "create" the most
> advantageous environment for national
> business. Accordingly, approaches recon-
> textualize comparative advantage to include
> an understanding of developments in the
> trade-industrial policy arena. (1994, 301)

In line with this theme, Grant focuses
on the role of governments, especially the
formation of regional blocs, and the role of
firms as the bases of a more comprehensive theory. Moreover, he argues that
high-technology trade occupies a key
place in any new theory, as success in
high-technology bestows national benefits
on productivity and high-wage job creation (see also Drache and Gertler 1991).
Likewise, in their recent study of trade in
textiles and clothing, Glasmeier, Thompson, and Kays (1993) contend that it is
necessary to understand how the actions
of the state influence the structure of
global competition. Indeed, they conclude
that state actions have superseded market
forces as the regulator of the industry's
geographical evolution.

The conceptual movement away from
orthodox comparative advantage explanations has been most fully spelled out in
Trading Industries, Trading Regions, edited by Noponen, Graham, and Markusen
(1993). Here again it is argued that
success in trade is fundamentally shaped
by government intervention. In a chapter
in that volume, Howes and Markusen
claim that governments have played a key
role in creating and maintaining industrial
leadership, and that "in a world with
governments successfully conducting such
industrial and trade policies, open economies without such efforts will find themselves the targets of import penetration
and potential export market shrinkage"
Howes and Markusen 1993, 4). In this

view, factor endowments can be used to
explain trade in minerals, agricultural
goods, and some labor-intensive consumer goods, but the majority of trade
between developed market economies
can only be explained by a "dynamic
revisionist" theory. This has four major
tenets that contradict orthodox trade
theory. First, the mix of sectors matters,
as some industries have greater growth
and productivity differentials. Second,
growth is not constrained by factors but
by demand for the product. Third, in
some industries rapid growth leads to
continuing success due to increasing
returns. Fourth, because of the existence
of increasing returns, comparative advantage may conceivably be created by
strategic intervention on the part of
nation-states and regional authorities. On
this basis they argue that the orthodox
view that free trade means growth for all
regions is mistaken; instead, "there is
some danger that the unfettered pursuit
of free trade will actually depress wages
and employment and lower world living
standards" (Howes and Markusen 1993,
35). Furthermore, Markusen (1993) argues that in the United States free trade
and laissez-faire strategies have produced
persistent unemployment and a waste of
infrastructure.

While this "dynamic revisionist" theory
shares an emphasis on "new trade theory"
with Krugman, it more closely resembles
the strategic trade views of authors such
as Tyson (1992) and Reich (1991), whom
Krugman (1994a, 1994c) has recently
criticized.[17] As we have noted, Krugman
remains convinced that the mutual benefits of greater international trade outweigh
the costs. Moreover, in his opinion,
comparative advantage is not just a
sector-specific theory, it remains a general
principle that explains the beneficial

---

[17] For a debate on Krugman's critique see
the discussion on "The Fight over Competitiveness" in *Foreign Affairs* (1994a), Friedman
(1994), and *The Economist* ("The Economics of
Meaning" 1994).

consequences of trade. The concept makes clear that absolute productivity advantage in some areas is not necessary for a country to gain from economic integration. Trade, therefore, is not a zero-sum game, so that concerns about national competitiveness are misplaced and unfounded. Krugman (1987b) concedes that the intellectual case for free trade has been weakened and that it is not an absolute ideal, but he believes that it is still the best general policy or rule of thumb. But Krugman's position faces several key questions. The first is the extent to which this continuing use of comparative advantage is compatible with his own emphasis on the pervasive presence of increasing returns. Kaldor (1985), for example, argues that the presence of increasing and diminishing returns conflicts with the basic tenets of Ricardian comparative advantage. Simply put, he contends that diminishing returns may mean that the resources released by trade will not necessarily be employed in other sectors, so that there is a real possibility of absolute loss (a "negative sum" game). Conversely, increasing returns in some industries may inhibit the transfer of resources elsewhere. Krugman's economic geography pays insufficient attention to these problems. This is reflected by his insistence that it is pointless to try to identify high-return sectors, so that the mix of sectors does not really matter.[18] Given his insistence on the importance of productivity, it is surprising that he devotes little attention to the extent to which high-technology sectors do generate the productivity spill-

overs which some authors have suggested (for example, Hanink 1994).

The second question is whether Krugman underestimates the significance of adjustment costs and the obstacles to regional adjustment. On the one hand, Krugman is committed to a nonequilibrium view of economic geography in which there is no process of convergence to a spatial equilibrium where all factors are equally rewarded. He rejects the neoclassical faith in the efficiency of markets on the grounds that the collective result of individual choices may be to "lock-in" a bad result. On the other hand, in a methodological sense, Krugman (1993a) insists that all economic models should contain a well-specified equilibrium. By this he means that they should specify how individuals behave and show how market outcomes emerge from the interaction of these individual behaviors (Krugman 1993a, 115–16). He holds these two opposing convictions together, it seems, through a commitment to a "new Keynesian" brand of economics. According to this, economic trends and patterns are the products of innumerable individual decisions, but these decisions are not perfectly rational and informed. Instead they are frequently both near-rational and individually reasonable and sensible. However, in imperfectly competitive markets the aggregate result will be unstable and irrational. In his words, "What look like highly irrational outcomes in the marketplace are caused by the interaction between imperfectly competitive markets and slightly less than perfectly rational individuals" (Krugman 1994c, 213). But if emphasis is placed on the second of these factors, then the position is readily reassimilated into a neoclassical view of the economy. It lends itself to the view that markets would adapt efficiently and rapidly if only people would behave rationally. This is exemplified, perhaps, by Krugman's (1993e) argument that Eurosclerosis, or the problem of a persistently high level of unemployment in Europe, has been caused by the effects of welfare states on labor markets. The whole

---

[18] Krugman's (1994c) argument is that it is wrong to assume that high-technology sectors such as computers and aerospace are the sectors with highest value added per worker. In fact, he notes that in the United States the real high-value industries are extremely capital-intensive sectors, such as cigarettes and oil refining. This says nothing, however, about the possibility of positive spillovers from the high-technology sectors.

question of adjustment to the effects of trade is one that Krugman has recently considered explicitly in terms of the impact of economic integration on regional development, particularly in the European Union, and it is to this aspect of his work that we now turn.

Krugman's Model of Economic Integration and Regional Development: The Lessons of the United States for Europe?

The regional consequences of European economic integration is an issue that has attracted surprisingly little attention from economic geographers. At the heart of this issue is the question of what the impact of progressive economic and monetary integration in the European Union (EU) will be on regional patterns of economic growth, employment, and income across member states. Economists have offered two opposing answers to this question. On the one hand, there are those who believe that the free movement of goods, services, and capital associated with European economic and monetary integration (EMU) should lead to regional convergence, not only in factor returns and economic performance but also in economic structure. To the extent that wages and other costs are lower in the less productive and slower-growing regions, the removal of barriers to trade and factor movements, it is argued, should enable industries and services in these regions to better exercise this comparative advantage and to attract increased flows of capital investment.[19] This optimistic scenario is, on balance, the view taken by the

European Commission (Commission of the European Communities 1991, 1994). In contrast, others argue that economic integration will intensify rather than reduce regional imbalances in growth and income across the European Union. Instead of leading to equalizing centrifugal movements of firms and investment toward depressed and peripheral regions within the European Union, economic integration is likely to stimulate a spatial reconfiguration of economic activity in favor of growth regions precisely because these are the areas that already enjoy greater comparative advantage in terms of access to markets, inputs, expertise, and business infrastructure.[20]

Krugman falls into the second of these two camps, although he appears to subscribe to two somewhat different models of regional divergence. In an earlier paper (Krugman and Venables 1990), he follows a core-periphery argument not unlike that in *Geography and Trade*. Although the removal of barriers to trade and movement of capital and labor within the European Union will increase the inflow of capital into, and the relative competitiveness of, the low-wage peripheral regions, given transport costs this centrifugal process is on balance likely to be outweighed by further concentration of industry and employment in the high-wage core regions, because these areas have the largest markets, well-developed external economies and infrastructures, and a comparative advantage in terms of

[19] Additionally, economic integration represents a major supply shock to such regions, since it exposes them to the full force of competition elsewhere in the system. Such shocks, the argument continues, should (allowing for adjustment lags) eliminate inefficient firms, work practices, and products in depressed regions and improve their supply-side competitiveness and flexibility.

[20] Because the gains foreseen from completion of the internal market are thought to be generated mainly endogenously, the various processes of resource allocation are bound to cumulate resources in the leading core regions. It is the historically established competitive advantage of the growth regions which enables them to capture a disproportionate share of the benefits of economic integration. As for the depressed and lagging regions, economic integration is seen as bringing prolonged problems of adjustment and the need for greater levels of spending on regional policies.

relative accessibility. His second approach is more emphatic, but different in its specific arguments. In his paper on the "Lessons of Massachusetts for EMU," he supports the movement toward European economic integration as "a generally good thing," but argues that it will lead to greater regional instability and divergence of regional growth rates (Krugman 1993d, 241). In developing this thesis he begins by drawing on his earlier ideas on trade and the localization of industries that we have discussed above:

> For regional issues . . . in the EC, . . . the key aspect of regional specialisation is the dependence of regional economies on export clusters held together by Marshallian external economies. . . . Are such regional clusters more likely to form in a more integrated economy? The answer is definitely yes. (Krugman 1993d, 244)

These ideas are then used in a somewhat different way from his earlier work to produce a theoretical account that not only carries over some of the problems we have already highlighted, but also introduces additional elements of contention.

The gist of this second model may be summarized as follows. First, given the existence of increasing returns, the expansion of interregional trade that EMU will bring about will lead to greater regional industrial concentration and specialization along essentially arbitrary lines. Once under way, there will be a tendency for this regional specialization process to become "locked in" by the operation of location-specific external economies. Second, Krugman argues that this increased regional specialization will render the European regions much more subject to random, idiosyncratic demand and technology shocks, so that region-specific recessions and crises will be more likely to occur.[21] Third, when combined with

the increased factor mobility that integration will promote, such region-specific shocks will lead to divergent long-term regional growth paths. Thus, fourth, given that under EMU member states will no longer be able to use the exchange rate mechanism as a policy instrument (see also Krugman 1989), the only way regional adjustment problems can be ameliorated is by transferring a significant part of national budgets to the European Union to allow fiscal federalism to function as an automatic stabilizer.

Thus, in contrast to his previous work— for example, in *Geography and Trade* (1991a) and Krugman and Venables (1990)—Krugman argues that the process of uneven regional development that EMU may be expected to produce will not be one of cumulative divergence into a core-periphery pattern. He believes that the forces generating this form of uneven regional development have probably reached their limit in advanced industrial nations; indeed, he suggests that in both the United States and Europe industrial activity is becoming much more evenly distributed geographically (Krugman 1993d). Rather, in his view the process will be one of increasing regional export specialization, with the result that the pattern of regional growth and decline will be more unpredictable, dependent on the particular incidence of random demand shocks. Hence, unlike the argument in *Geography and Trade*, past regional success need not be self-reinforcing, and even prosperous regions may experience sudden reversals of fortune.

Another distinctive feature of Krugman's exposition is the method he uses to support his theory empirically. The United States is taken to be the sort of integrated economic and monetary unit which the European Union is seeking to emulate, so that regional experiences in the former are considered to be a good guide as to what to expect in the latter.

---

[21] In an earlier paper, Krugman (1989) stressed that increasing interdependence in Europe acts as a buffer against regional and national shocks, but this buffering effect only

acts against locally generated recessions such as those caused by investment slumps.

Using simple measures of the dispersion of economic structure, Krugman (1991a, 1993d) finds that the broad regions of the United States are more industrially specialized than are European countries. Furthermore, a comparison of Belgium with the state of Ohio is used to suggest that the regional employment growth rates in the United States are more unstable than in the European Union. He then examines the disparities in long-term growth rates between certain states in the United States and among the main EU countries and finds that these disparities are larger in the United States than in Europe. In addition, he uses the recent economic slump in the New England region of the United States as an illustration of how, in a monetary union, regional industrial specialization can give rise to pronounced local instability in the face of region-specific demand shocks, and how such shocks can lead to permanently lower levels of employment (Krugman 1993d). These various empirical results are taken as lending support to his thesis that increased market integration in the European Union will lead to more regional specialization and unequal growth. In our view, however, Krugman's empirical examples and findings are not of themselves sufficient to prove his case, and several features of his analysis are problematic.

Economic Integration and Regional Specialization

The first problem concerns the evidence on regional specialization. What is the "regional" scale being referred to? The "regions" used by Krugman in his comparisons of regional specialization and regional growth rate disparities in the United States and the European Union are extremely aggregate ones: the four "Great Regions" and individual states in the former and whole countries in the latter. Krugman argues that these spatial units are of roughly similar size, and thus broadly comparable. That may be so, but they do not necessarily represent the

geographical scale at which local external economies and the processes leading to industrial clustering actually operate. The basic point is that the analysis of localization economies requires an identification of the relevant regions as economic areas and the relevant level of industrial disaggregation at which to measure the extent of geographical concentration and specialization. The geographical literature on "new" flexible industrial districts indicates that such clusters are in fact quite localized, and far smaller than the broad spatial units used by Krugman. Certainly in the European Union, local differences in economic structure and economic growth rates *within* member countries (for example, at the so-called NUTS1 and NUTS2 level regions) are much larger than the disparities *between* countries (Collier 1994; Dunford 1993; Dunford and Kafkalas 1992). Likewise, as von Hagen and Hammond (1994) argue, the metropolitan rather than the state or broad regional level is the most meaningful one for analyzing geographical differences in industrial concentration and localization economies in the United States, a view to which, as we have already noted, Krugman has elsewhere subscribed. These findings imply that Krugman's method of comparing the European Union with the United States will generate different results according to the geographical scale used to define economic regions in the two areas. Indeed, it may even be that at some geographical scales regional specialization is not in fact greater in the more-integrated economy of the United States than in the European Union.

In any case, is increasing regional industrial specialization an inevitable outcome of economic integration? While the existence of external economies and localization economies in the European Union could well lead to the increased regional specialization that Krugman predicts (Baldwin and Lyons 1990; Cabellero and Lyons 1990, 1991; Martin and Rogers 1994a, 1994b), some observers have argued that product market integration in the European Union will increase the scope of *intra*industry trade there still further, and that this is likely to render regional industrial structures increasingly *similar* over time (Commission of the European Communities 1991; Eichengreen 1993; Emerson, Anjean, and Catinat 1988). Indeed, possible evidence for this effect is provided for the United States by Krugman: as he shows, U.S. statistics indicate that regional specialization there has actually been declining since the Second World War (Krugman 1991a, Chap. 3). He suggests that this may be a statistical illusion, in that specialization may have become more difficult to measure but may not necessarily be less in fact. However, there is also evidence from Europe that economic integration and increased trade lead to regional industrial *diversification* rather than specialization (Peschel 1982). Indeed, both the definition of regional economic "specialization" and the question of how specialization actually influences regional instability are not straightforward issues.[22] As a number of writers have shown, the empirical patterns of regional shocks in both the United States and the European Union appear to be more complex than those posited by Krugman (see, for

---

[22] There is a sizable literature on this topic, although it is not referred to by Krugman (for example, see Barth, Kraft, and Wiest 1975; Conroy 1975; Brewer 1984; Jackson 1984; Kurre and Weller 1989). Much of this is based on what is called a "portfolio" approach to the analysis of regional industrial specialization. This type of analysis, first applied to regional economics by Conroy (1975), borrows the concepts of expected return and risk from theories of the optimal diversification of financial portfolios developed by Markowitz (1959). The regional industrial structure may be conceptualized as a "portfolio" which provides "returns" to the region in the form of employment, income, and tax revenues. These returns are associated with risk—arising from demand and technology shocks—as represented by the variance and covariance in the returns. It is this measure of risk, "the portfolio variance," which measures the degree of instability of the region.

example, Bayoumi and Eichengreen 1993; Palmini and Cray 1992; von Hagen and Hammond 1994). There seems to be no simple relationship between economic integration, regional specialization, and regional shocks. Both the pattern and severity of shocks will depend not only on the degree and geography of regional industrial specialization, but also on how such shocks are transmitted between regions (for example, through interregional input-output linkages and the impact of government policies) and on how flexible regional labor markets are in adjusting to disturbances. In short, much more theoretical and empirical analysis of regional industrial specialization within both the United States and the European Union is required before the former can be taken as a guide of what to expect in the latter.

Economic Integration and Divergent Regional Growth

This last point links with the third element of Krugman's thesis, that demand shocks in an integrated Europe will have permanent regional growth effects, in the same way that temporary policies may have long-term implications (Krugman 1987a). Suppose a region experiences a decline in the demand for its clusters of export industries. This would put downward pressure on relative wages and other factor costs in the region. If relative wages and other costs fall, this would help to restore the region's competitiveness vis-à-vis other, higher-cost regions, so that new industries would be attracted there and demand and growth should be restored. As Krugman puts it,

> Regions that have been unlucky in their heritage of industries from the past will have lower costs than lucky regions, and will therefore be more likely to break into industries in the future. We would expect this process to put limits on the extent of regional divergence in growth. (Krugman 1993d, 248)

Unfortunately, however, according to

Krugman labor mobility prevents the wage flexibility mechanism from bringing regional growth rates into balance in this self-correcting way. To the contrary:

> An unfortunate region will not have lower factor prices for very long: capital and labour will move to other regions until payments are equalized. This means, however, that there is no particular reason to expect a region whose traditional industries are faring badly to attract new industries. It can simply shed people instead. The implication is that relative output and employment of regions should look more like a random walk than like a process that returns to some norm. (Krugman 1993d, 248)

In developing this argument, Krugman draws on Blanchard and Katz's (1992) study of patterns of growth among U.S. states. According to these authors, while employment growth rates differ consistently across U.S. states, unemployment rates and wages vary much less, suggesting that when states are hit by demand shocks workers react by relocating (see also Barro and Sala-i-Martin 1992). There is no discernible tendency for states to recover lost jobs: relative regional unemployment returns to normal through the out-migration of workers. This would seem to be in contrast to the adjustment process in the European Union, where historically factor mobility has tended to be far lower than in the United States and regional unemployment disparities appear to be characterized by greater hysteresis (Eichengreen 1993). Krugman draws the obvious implication that if Europe moves toward U.S. levels of regional specialization *and* factor mobility, disparities in economic growth rates among countries and regions may be expected to increase.

Labor mobility is thus central to Krugman's model of divergent regional growth. In this respect his analysis is similar to local "endogenous growth" models, in which labor mobility intensifies local disparities in the accumulation of human capital and hence long-term development (Grossman and Helpman 1991; Bertola 1993). In this respect we find it somewhat

curious that Krugman is at pains to distinguish his model of uneven regional development in the European Union not only from "core-periphery" models of cumulative concentration but also from "local endogenous growth" models (Krugman 1993d). His own model implies a similar cumulative divergent growth mechanism, at least in the sense that interregional shifts in labor prevent the reequilibration of regional growth rates. The question mark over his analysis is exactly how far labor mobility will increase in an integrated Europe. Although interregional migration across national borders will in principle be unrestricted, there are further reasons to doubt whether labor mobility will ever reach the levels found in the United States. The marked cultural and language differences across Europe will continue to be a significant barrier to migration for many groups of workers. But if this form of adjustment to regional shocks remains slow, where does this leave Krugman's view of regional long-term growth differences in an integrated European Union?

The implication of his model is that if labor mobility is low, then local (downward) relative wage flexibility will serve to restrain the degree of divergence between regional growth rates. Unfortunately, wages in the European Union do not seem to be particularly flexible: European labor markets appear to be more rigid or "sclerotic" than their American counterparts, a point highlighted by Krugman (1993e). In the European regions, adverse sectoral demand shocks trigger greater unemployment, without the equilibrating mechanisms of labor migration or downward relative wage movements (the rigidity of the latter possibly reflects the considerably higher rates of institutionalized wage setting among workers and the availability of more generous unemployment benefits in the EU countries compared to the United States). The Commission of the European Communities (1990) argues that EMU, by increasing the credibility of fiscal authorities' commitment not to bail out depressed regions, should encourage workers in such areas to moderate their wage claims, thus imparting greater local wage flexibility. In practice, little is known about how far regional relative wages would have to fall in order to stimulate capital inflows and the restoration of employment. Equally, we still know little about interregional productivity and technology spillovers, which may offset the need for wage reductions (Jaffe, Trajtenberg, and Henderson 1993; Audretsch and Feldman 1994). In short, it is by no means obvious whether increasing integration in the European Union will lead to convergence or divergence of regional growth. The evidence so far would seem to suggest that "club convergence" may be the most likely outcome, with convergence within the northern, core regions, on the one hand, and within the southern and peripheral regions, on the other, but little or no convergence between these subsets (Button and Pentecost 1993; Chatterji 1993; Neven and Gouyette 1994).

Thus, though suggestive, Krugman's arguments about the impact of economic integration on regional trade, specialization, instability, and long-term growth disparities in the European Union are problematic and limited. Comparison between the United States and the European Union in terms of "regions" and their structures, shocks and reactions to them is not, perhaps, as valid as Krugman and others (such as Eichengreen) assume. We do not have a counterfactual history for the United States—that is, a picture of what regional development would be like if the United States was not an economic and monetary union. Nor do we know what would have happened to the regions of the European countries in the absence of the formation of the European Community and its recent movement toward EMU. Finally, what of Krugman's views on the regional policy implications of European integration? To assess this aspect of his analysis we need to look at the policy debate within the new trade theory more generally.

Trade and the Regional Policy Issue

Strategic Trade Policy

Like much of his other work, Krugman's views on the role of trade and industrial policy have shifted over time. In his early writings he reacted against the idea of targeted industrial policies, on the grounds that they were based on crude misconceptions and that even if more-sophisticated theorizations could be found such policies were still unlikely to be effective in practice (Krugman 1983, 1983b, 1984). Not long after, however, he had constructed a sophisticated theoretical argument for "strategic trade policy" (Krugman 1986). One of the most contentious aspects of the new trade theory has been the debate it has generated over the question of strategic industrial policy. Whereas conventional trade theory denies there is any case for "activist" trade or industrial policies, the new trade theory directly challenges the conventional view. According to Krugman (1986) and other new trade theorists (for example, Brander and Spencer 1983, 1985), an "activist" trade policy can benefit a country relative to free trade in two ways. The first is through "rent creation." If a government can promote a new or expanded monopoly position for domestic factors of production in industries that trade internationally, then a targeted industrial policy can in principle raise a country's income at foreign expense. Second, targeting can raise income if there are certain industries in which the resources committed by individual firms indirectly raise the earnings of other firms' resources—that is, where external economies can be generated. In both instances, the argument is that it may well be possible to identify some "strategic sectors" that at the margin are more valuable than others, and that the promotion of these sectors through protection, export subsidies, support of R&D, and so on could raise national income.

More recently, however, Krugman has reacted against strategic trade policy. In

Peddling Prosperity (1994c), he questions the theoretical validity of strategic industrial policy and goes on to berate leading American politicians (especially President Clinton) and their economic advisors (especially Thurow and Reich) for misappropriating strategic trade theory and applying it in a "simple-minded way." Both Reich (1991) and Thurow (1994) are criticized for peddling the view that if the United States is to compete in the global economy, the government should abandon its notions of free trade and minimalist industrial intervention and instead pursue a more activist stance aimed at promoting the shift of American industry into "high value" (Reich) and "sunrise" (Thurow) sectors. Krugman believes that such views are based on fallacious theory, impractical politics, and an erroneous obsession with the idea of "competitiveness": "While competitive problems could arise in principle, as a practical, empirical matter the major nations of the world are not to any significant degree in economic competition with each other" (Krugman 1994a, 35). In his view, competitiveness relies on the metaphor of a country being a big corporation, when, in fact, countries (and regions) are nothing like corporations. Hence it is very difficult to establish a meaningful definition of national or regional competitiveness. Furthermore, he argues that it is wrong to see international trade as competition—as a sort of zero-sum game—when it is a process of exchange involving mutual benefit. By the early 1990s, then, Krugman had come to denounce strategic trade policy as "bad economics." Yet, while Krugman has vigorously attacked the whole ensemble of policies that have come to be labeled as "strategic trade policy," it appears that he now sees a role for a limited and focused industrial policy.

Geographical Clustering and Strategic Industrial Policy

In a recent paper Krugman (1993b, 160) states that he has "now changed his mind and . . . gone, at least slightly, soft on

industrial policy." His initial skepticism of the theoretical credentials and practical applicability of the external economies rationale for targeted industrial policies was on the grounds that only technological, not pecuniary external economies are of concern, that technological externalities in any case are of limited significance, and that they are international rather than national in scope (Krugman 1983b, 1984). But, as we have seen, he now believes that external economies associated with market-size effects are substantial and demonstrable (and often indistinguishable from technological external economies), and this means that targeted industrial policies have a potential role after all (Krugman 1987a, 1993b). Moreover,

> many of the important market-size effects apply not at the level of the international or even the national economy, but at a regional or local level. The argument that the gains from support of industries that generate external economies will be dissipated abroad is therefore mostly wrong. (Krugman 1993b, 167)

In this revised view of industrial policy, not only are regional and local industrial clusters considered to provide empirical proof of the importance of external economies, such clusters help to *define* what industries should be supported. Geographical clustering provides the justification for industrial intervention, and the aim of that intervention should be to foster local externalities. In effect, what Krugman seems to be suggesting, though he does not use the term explicitly, is that the only justifiable form of industrial (trade) policy is in fact regional industrial development policy. In line with strategic trade theory, the underlying premise is that national industrial comparative advantage can be created through supportive and targeted industrial policies which aim to create and facilitate key sectoral specializations. The twist in Krugman's argument, however, is that the most effective scale at which to create that advantage is at the level of *regional clusters*. Essentially the same argument is implicit in Porter's (1990) major work on national competitive advantage. Indeed, he now sees local and economic development policies as having an instrumental role in fostering national industrial competitiveness (Porter 1994).

It is only a small step from this to argue that the promotion of specialized export clusters is also the most viable approach to reviving and regenerating old industrial regions. This is, in fact, what certain writers on flexible specialization and industrial districts have implied. These authors use the success of certain well-known specialized (usually export-orientated) industrial districts as a model for "indigenous" local economic regeneration more generally (see, for example, Hirst and Zeitlin 1989; Pyke, Becattini, and Sengenberger 1989; Sabel 1989; Stohr 1989; Cooke 1990; Scott 1992b). The path to the reindustrialization of economically and structurally depressed regions is seen to be via the promotion and support of neo-Marshallian small-firm, flexibly specialized production complexes involving dense local networks of cooperation, competition, and horizontal interdependencies. In some ways this support for a local industrial strategy based on multiple externalities is similar to Krugman's rationale for industrial policy. However, the advocacy of regional export specialization either as a local economic development strategy or as a form of trade policy is contentious.

The key question over the promotion of regional industrial specialization is whether the potential advantages are outweighed by the likelihood of greater regional instability and shocks, and the risk of structural depression. As Krugman (1993d) notes for the case of Massachusetts, regional industrial specialization is a double-edged sword: it can be the basis of a high rate of export-led local economic growth in one period, but the source of prolonged local economic depression if that demand subsequently collapses or is captured by other competing regions (often in other countries). This is precisely what happened to many of the specialized

industrial districts celebrated by Alfred Marshall early this century (for example, see Sunley 1992). An equally persuasive case can be made that industrial *diversification* rather than specialization is the most appropriate regional development policy route, that diversifying the regional industrial "portfolio" reduces the susceptibility of the regional economy to adverse demand shocks and localized structural crisis (this is the general conclusion of the portfolio studies referred to earlier; see also Geroski 1989).

Krugman, on the other hand, appears to believe that the most important policy response to the possibility of regional instability in more-specialized regions is fiscal stabilization. In the case of European economic integration, for example, Krugman suggests that national budgets will have to be substantially centralized so that automatic federal European fiscal transfers can perform the required stabilization role when asymmetric regional shocks occur.[23] He notes the way in which the U.S. federal budget tends automatically to redistribute resources toward regions affected by negative economic shocks (via compensating adjustments in the tax-take and in welfare payments across regions):

> While the US does not cope with the problems perfectly (as the current travails not only of New England but of the North East, in general, and increasingly of California, demonstrate), a highly federalized fiscal system helps a good deal. The lack of such a system in Europe therefore is a real problem. (Krugman 1993d, 258)

[23] In the European case, although fiscal federalism is indeed a natural corollary to EMU, national governments in a future European monetary union would not lose all of their instruments of economic policy. National budgetary policies would continue to have some, even if constrained, automatic stabilizing role (see Boonstra (1991) on the limits that EMU will impose on national budgetary autonomy). In this sense member states in a European EMU would be somewhat different from the individual states in the United States.

Krugman is at pains to distinguish this form of regional policy response from that needed to ameliorate "core-periphery" patterns of uneven regional development or the regional decline that stems from specialization in outmoded industries and products. The policy response to these sorts of regional issues, he says, "is much less related to EMU than the stabilization problem" (Krugman 1993d, 259).

However, while automatic fiscal transfers may well help to alleviate and stabilize the income and growth problems associated with economically depressed regions, they are not an adequate response to uneven regional development. By themselves, they are not sufficient to recast the structure and dynamics of regional development so as to improve the long-term economic performance and wealth of the regions concerned. This is why, of course, historically many European countries have developed elaborate systems of *region-specific* developmental aid and assistance, and why the European Union has been strengthening and reforming its own centrally administered regional structural funds in the context of the movement toward increasing economic integration (and enlargement) of the Union (see, for example, Martin 1993; Collier 1994). Krugman's distinction between regional instabilities due to idiosyncratic demand shocks and regional problems of a more "core-periphery" and "structural" nature is surprising and questionable. For if, as he argues, short-term regional shocks have long-term effects on regional growth, then interregional fiscal stabilization is an inadequate policy response, and other, more strategic forms of regional policy are required.

In our view, the proposal of fiscal federalism does not temper the worry that regional clusters of specialized industry will be unstable and fragile. The basic tension in Krugman's argument remains—namely, how to reconcile his suggestion that the aim of industrial policy should be to promote industrially specialized regional clusters with his thesis that increased regional industrial concentra-

tion and specialization leads to regional economic instability and divergent long-term growth paths. The response of the new industrial geography to this dilemma, of course, is to insist that flexibly specialized industrial districts are more adaptable to economic and technological change by virtue of the dynamism and networking of the small enterprises of which they are (invariably assumed to be) composed. However, this claim remains far from proven. In addition, the number of flexibly specialized districts remains small, and their origins and dynamics are matters of debate (see Markusen 1993; Markusen and Park 1993). This is not to dismiss the new "indigenous" approach to regional policy based on arguments of (flexibly) specialized industrial development; but it is to signal that this approach is no more of a general panacea for uneven regional development than was the old model of redistributive regional policy. Nor do we wish to imply that increasing returns and external economies are unimportant in the regional policy debate. To the contrary, not only is there evidence from Europe that increasing returns industries are more concentrated in regions with better infrastructures, especially technological and educational infrastructures (Martin and Rogers 1994a, 1994b), endogenous growth theory also suggests that external economies and technological spillovers are likely to play a key role in the local growth process in an integrated Europe. But in our judgment, there is an urgent need for much more thought on how local and regional policies can foster and support these externalities without simultaneously narrowing the industrial base and increasing the vulnerability of regions to demand shocks.

Conclusions

A few years ago, Neil Smith (1989) argued for a rebuilding of regional theory within geography based on a synthesis of ideas from location theory and uneven development theory. More recently, Krugman (1993a) has argued for a similar synthesis of location theory and trade theory, for using economic geography as a key component in the construction of a new "geographical economics" of trade. In this paper we have sought to provide a critical assessment of Krugman's ideas on economic geography and his attempts to use these ideas to forge a "geographical economics." Because of the enormous volume and breadth of his writings we have had to skate across many of his ideas, and as a result we have no doubt failed to accord some of them the full attention they deserve. In addition, Krugman's tendency to constantly revise and even reject his earlier ideas renders the task of assessment akin to tracking a moving target. Nevertheless, we believe we have succeeded in isolating the core components of his arguments sufficiently to be able to identify some of their main strengths and weaknesses, particularly as they relate to the question of regional growth and development. In many ways, Krugman's approach to economic geography is a regional science one, a reworking of traditional location theory concepts and models. The new economic and industrial geography, of course, has moved well away from that tradition. For this reason, it might well be questioned whether Krugman's work contains anything that is new or useful for economic geographers. It would be wrong to be so readily dismissive, however; Krugman's work is not as simplistic as Johnston and others have suggested (nor for that matter is geographical work always as sophisticated as its practitioners appear to believe). For it is perhaps less the specific results of Krugman's analyses that are important for economic geography than the general stimulus they provide for further inquiry. In this respect we concur with the view that Krugman's work "is rich in ideas, seductive in taking us though simple logical arguments to surprising conclusions and so self-confident in the discussion of its assumptions and its premises that reading it is at the same time great fun and a continuous challenge" (Casella 1993, 261–62). The challenge, as we see it,

is to pursue a closer exchange between Krugman's "geographical economics" and the new industrial and economic geography. Neither can claim to have a monopoly of insight, but an exchange of ideas between the two would, we believe, be beneficial.

One strength of Krugman's work, without doubt, is that his linking of external economies and regional industrial agglomeration with trade provides an important corrective to the flexible specialization thesis of the new industrial geography, in which regional industrial development is viewed overwhelmingly as an indigenous process and the role of trade is typically either subordinated or neglected altogether. Furthermore, Krugman's emphasis on imperfect competition and pecuniary externalities likewise exposes the limitations of the conceptions of externalities now prevalent within the geographical literature. The thrust of flexible specialization ideas in economic geography is that agglomeration is associated with the shift from vertical integration to the horizontal integration of related activities among small, competitive firms which cluster together to minimize transaction costs. Williamsonian transaction costs economics—itself a neoclassical-oriented form of institutional economics—has been used to give a new theoretical underpinning to Marshall's notion of industrial localization. Krugman's focus on pecuniary externalities, especially market-size effects, and the role that large, oligopolistic producers can play in industrial agglomeration suggests that industrial geographers need to reassess their theoretical accounts accordingly. Yet, at the same time, one of the most important limitations of Krugman's geographical economics is his stubborn concentration only on those externalities that can be mathematically modeled, and thus his reluctance to discuss the geographical impacts of technological and knowledge spillovers. Although the recent geographical literature has begun to assign key importance to technical change and technological externalities in shaping and transforming the space economy, and hence to some extent holds some important lessons for Krugman's analysis in this respect, it too has yet to explore fully the cumulative and spillover effects associated with technological change.

A second significant aspect of Krugman's geographical is the recognition that regional economic development is a historical, path-dependent process. His repeated exhortation that "history matters," both in terms of the arbitrary initial conditions and accidental events that set in motion particular patterns of industrial development over time and space and in terms of the subsequent "locking in" of those patterns via self-reinforcing effects, is not of course particularly novel to economic geographers. Geographers have long recognized that a given pattern of uneven regional development, once established, tends to exhibit a high degree of persistence or "inertia" over time, and that this inertia can operate either to foster regional growth or to retard it. The more recent interest by geographers in the local socio-institutional "embeddedness" of economic activity also bears upon the issues of path dependence and lock-in. In Krugman's view the role played by geography in determining "lock-in" is strictly an increasing returns phenomenon, in the form of the Marshallian externalities associated with local industrial agglomeration (or, under certain circumstances, in the form of self-fulfilling expectations). What he fails to consider is the influence exerted by local institutional, social, and cultural structures in facilitating or constraining local economic development. This neglect would seem to stem in large part from Krugman's complaint that noneconomic or "social" factors are not easily modeled and that they should therefore be left to sociologists. But as recent studies in the new industrial and economic geography have begun to show, the "thickness" and nature of such socio-institutional "externalities" are fundamental to the initial emergence, trajectory, and adaptability of industrial districts and regional economies. Thus

Krugman is right to stress the role of geography in the historical, path-dependent nature of the economic process, but he fails to explicate the nature of that role.

A third aspect of Krugman's geographical economics that we want to highlight, and which also has both strengths and weaknesses, is his analysis of the way that region-specific shocks can have long-term growth consequences. How regions respond and adjust to demand and supply shocks, both in the short term and in the long run, in an increasingly deregulated, market-propelled, and uncertain world is an important research issue, but one that has been neglected by the new industrial geography. Krugman's analysis for the EU regions, using the U.S. regions for comparison, provides a useful basis for developing this research agenda. However, as we have seen, his analysis is far from unproblematic. Apart from being too American-centered (as exemplified by his emphasis on the central role of labor mobility, which is considerably higher in the United States than in Europe), his models do not adequately explain why a successful regional economy (like Massachusetts, for example) can suddenly go into reverse, or why the geography of uneven regional development can and does undergo significant reconfigurations ("spatial switching"), or why some regions seem better able than others to withstand or adjust to negative external shocks. Krugman singles out industrial specialization as the main factor shaping the relative stability of different regions and the labor market as the key determinant of the regional adjustment process. But a full account must surely also consider other reasons for regional crisis and restructuring and mechanisms other than labor market flexibility in order to explain the degree of and differences in regional adjustment.

There is, then, considerable scope for a potentially fruitful cross-fertilization of ideas between Krugman's geographical economics and the new industrial-economic geography, and for the elaboration of each. Both draw heavily on a Marshallian view of industrial localization. But whereas the new industrial geography has sought to reinterpret the Marshallian account in terms of transaction cost economics, Krugman instead has tried to link Marshallian industrial localization with the economics of imperfect competition, increasing returns, path dependence, and cumulative causation. These concerns were at the center of Nicholas Kaldor's (1978, 1981, 1985) earlier seminal work on trade, endogenous growth, and regional development, a debt that Krugman acknowledges:

> This clear dependence on history is the most convincing evidence available that we live in an economy closer to Kaldor's vision of a dynamic world driven by cumulative processes than to the standard constant returns model. (1991a, 9–10)

Krugman even goes so far as to admit that in a sense his own work is only "a repetition" of Kaldor's ideas. There are significant differences between the two, however. In his quest for economic rigor, Krugman's mathematical formalization of the processes of industrial agglomeration and uneven regional development has taken him away from the richness of Kaldor's original approach toward the limited abstract landscapes of regional science. Indeed, in *Development, Geography and Economic Theory* (Krugman 1995) and *The Self-Organising Economy* (1996), his role model seems to be that doyen of regional science Walter Isard, rather than Nicholas Kaldor, who, one suspects, would have been extremely skeptical of the unrealistic, deductive model-building that is the hallmark of the regional science tradition. And despite Krugman's apparent agreement with Kaldor's argument for the "irrelevance of equilibrium economics," the ghosts of constrained maximization and equilibrium solutions still haunt much of his analysis.[24]

There would be much to be gained, in our view, if both Krugman's geographical economics and the new industrial and economic geography revisited the method and the message of Kaldor's work. But that, as Krugman would say, is another story.

References

Aghion, P., and Howitt, P. 1993. A model of growth through creative destruction. In *Technology and the wealth of nations: The dynamics of constructed advantage*, ed. D. Foray and C. Freeman, 145–72. London: Pinter.

Amin, A., and Robins, K. 1990. Industrial districts and regional development: Limits and possibilities. In *Industrial districts and inter-firm co-operation in Italy*, ed. F. Pyke, G. Becattini, and W. Sengenberger, 185–219. Geneva: Institute for Labour Studies.

Amin, A., and Thrift, N. 1992. Neo-Marshallian nodes in global networks. *International Journal of Urban and Regional Research* 16:571–87.

———, eds. 1994. *Globalization, institutions and regional development in Europe*. Oxford: Oxford University Press.

Anderson, K., and Blackhurst, R., eds. 1993. *Regional integration and the global trading system*. London: Harvester Wheatsheaf.

Audretsch, D., and Feldman, M. 1994. *Knowledge spillovers and the geography of innovation and production*. Discussion Paper 953. London: Centre for Economic Policy Research.

Baldwin, R. 1994. *Towards An integrated Europe*. London: Centre for Economic Policy Research.

Baldwin, R., and Lyons, R. 1990. *External economies and European integration: The potential for self-fulfilling expectations*. Dis-

---

[24] The same is true of the new trade theory and the new endogenous growth theory more generally. Many of the ideas found in these theories were in fact anticipated by Kaldor. But whereas he eschewed the deductive and mathematical for the inductive and realistic (as expressed in his emphasis on "stylized facts" and nonequilibrium), the new theorists have deliberately sought to "systematize" his ideas through mathematical formalism and appeal to all-embracing principles of optimizing economic behavior (see the assessments by Kitson and Michie 1995; Skott and Auerbach 1995).

cussion Paper 471. London: Centre for Economic Policy Research.

Barro, R. J., and Sala-i-Martin, X. 1991. Convergence across states and regions. *Brookings Papers on Economic Activity* 1:107–82.

Barth, J.; Kraft, J.; and Wiest, P. 1975. A portfolio-theoretic approach to industrial diversification and regional unemployment. *Journal of Regional Science* 15 (1):9–15.

Bayoumi, T., and Eichengreen, B. 1993. Shocking aspects of European monetary integration. In *Adjustment and growth in the EMU*, ed. F. Torres and F. Giavazzi, 193–229. Cambridge: Cambridge University Press.

Bertola, G. 1993. Models of economic integration and localized growth. In *Adjustment and growth in the EMU*, ed. F. Torres and F. Giavazzi, 159–79. Cambridge: Cambridge University Press.

Blanchard, O. J., and Katz, L. F. 1992. Regional evolutions. *Brookings Papers on Eonomic Activity* 1:1–59.

Boonstra, W. W. 1991. The EMU and national autonomy on budget issues. In *Finance and the international economy*, Vol. 4., ed. R. O'Brien and S. Hewin, 208–24. Oxford: Oxford University Press.

Brander, J., and Spencer, B. 1983. International R and D rivalry and industrial strategy. *Review of Economic Studies* 50: 707–22.

———. 1985. Export subsidies and international market share rivalry. *Journal of International Economics* 18:83–100.

Brewer, H. 1984. Regional economic stabilization: An efficient diversification approach. *Review of Regional Studies* 14 (1):8–21.

Buchanan, J., and Yoon, Y., eds. 1994. *The return to increasing returns*. Ann Arbor: The University of Michigan Press.

Button, K., and Pentecost, E. 1993. *Testing for convergence of the EC regional economies*. Economics Research Paper 93/5. Loughborough: Department of Economics, Loughborough University of Technology.

Cabellero, R., and Lyons, R. 1990. Internal versus external economies in European manufacturing. *European Economic Review* 34:805–30.

———. 1991. External effects and European integration. In *The effect of 1992 on trade and industry*, ed. L. A.Winters and A. Venables, 34–53. Cambridge: Cambridge University Press.

Casella, A. 1993. Discussion of Krugman's "Lessons of Massachusetts for EMU." In *Adjustment and growth in the EMU*, ed. F. Torres and F. Giavazzi, 261–66. Cambridge: Cambridge University Press.

Chamberlin, E. H. 1949. *The theory of monopolistic competition*. 6th ed. London: Oxford University Press.

Chatterji, M. 1993. Convergence clubs and endogenous growth. *Oxford Review of Economic Policy* 8:57–69.

Chipman, J. 1970. External economies of scale and competitive equilibrium. *Quarterly Journal of Economics* 84:347–85.

Collier, J. 1994. Regional disparities, the Single Market and EMU. In *Unemployment in Europe*, ed. J. Michie and J. Grieve Smith, 145–59. London: Academic Press.

Commission of the European Communities. 1990. *An empirical assessment of factors shaping regional competitiveness in problem regions*. Main report. Luxembourg: Office for Official EC Publications.

———. 1991. *The regions in the 1990s*. Luxembourg: Office for Official EC Publications.

———. 1994. *Competitiveness and cohesion trends in the regions*. Luxembourg: Office for Official EC Publications.

Conroy, M. 1975. The concept and measurement of regional industrial diversification. *Southern Economic Journal* 41:492–505.

Cooke, P. 1990. Manufacturing miracles: The changing nature of the local economy. In *Local economic policy*, ed. M. Campbell, 25–42. London: Cassell.

Cooke, P., and Morgan, K. 1993. The network paradigm: New departures in corporate and regional development. *Environment and Planning D: Society and Space* 11:543–64.

David, P., and Rosenbloom, J. 1990. Marshallian factor market externalities and the dynamics of industrial localization. *Journal of Urban Economics* 2:349–70.

De Melo, J., and Robinson, S. 1990. *Productivity and externalities: Models of export-led growth*. Discussion Paper 400. London: Centre for Economic Policy Research.

Dixit, A. K., and Norman, V. 1980. *Theory of international trade*. Cambridge: Cambridge University Press.

Drache, D., and Gertler, M., eds. 1991. *The new era of global competition: State policy and market power*. Toronto: McGill-Queens University Press.

Dunford, M. 1993. Regional disparities in the EC. Evidence from the REGIO data bank. *Regional Studies* 27:727–43.

Dunford, M., and Kafkalas, G., eds. 1992. *Cities and regions in the new Europe.* London: Belhaven.

Economics of meaning, The. 1994. *The Economist,* 30 April, 13–14.

Eichengreen, B. 1993. European monetary unification. *Journal of Economic Literature* 31:1321–57.

Emerson, M.; Aujean, M.; and Catinat, M. 1988. *The economics of 1992.* Oxford: Oxford University Press.

Erickson, R. 1989. Export performance and state industrial growth. *Economic Geography* 65:280–92.

Ethier, W. 1982. National and international returns to scale in the modern theory of international trade. *American Economic Review* 72:389–405.

Friedman, B. 1994. Must we compete? *New York Review of Books,* 20 October, 14–20.

Fujita, M. 1989. *Urban economic theory: Land use and city size.* Cambridge: Cambridge University Press.

Geroski, P. A. 1989. The choice between scale and diversity. In *1992: Myths and realities,* ed. E. Davis, 29–45. London: Centre for Business Strategy, London Business School.

Gertler, M. 1992. Flexibility revisited: Districts, nation-states and the forces of production. *Transactions of the Institute of British Geographers n.s.* 17:259–78.

Gibb, R., and Michalak, W., eds. 1994. *Continental trading blocs.* Chichester: Wiley.

Glasmeier, A.; Thompson, J.; and Kays, A. 1993. The geography of trade policy: Trade regimes and location decisions in the textile and apparel complex. *Transactions of the Institute of British Geographers n.s.* 18:19–35.

Grabher, G. 1993. *The embedded firm: On the socioeconomics of industrial networks.* London: Routledge.

Grant, R. 1994. The geography of international trade. *Progress in Human Geography* 18: 298–312.

Grossman, G., and Helpman, E. 1991. *Innovation and growth in the global economy.* London: MIT Press.

Hanink, D. 1988. An extended Linder model of international trade. *Economic Geography* 64:322–34.

———. 1994. *The international economy: A geographical perspective.* London: Wiley.

Harrison, B. 1992. Industrial districts: Old wine in new bottles? *Regional Studies* 26:469–84.

Helpman, E. 1981. International trade in the presence of product differentiation, economies of scale and monopolistic competition. *Journal of International Economics* 11:305–40.

———. 1984. Increasing returns, imperfect markets and trade theory. In *Handbook of international economics,* Vol. 1, ed. R. Jones and P. Kenen, 325–65. Oxford: North Holland.

Helpman, E., and Krugman, P. 1985. *Market structure and foreign trade: Increasing returns, imperfect competition and the international economy.* London: MIT Press.

Hirschman, A. 1958. *The strategy of economic development.* New Haven: Yale University Press.

Hirst, P., and Zeitlin, J., eds. 1989. *Reversing industrial decline?* Oxford: Berg.

Hoare, A. 1975. Linkage flows, locational evaluation, and industrial geography: A case study of Greater London. *Environment and Planning A* 7:41–58.

———. 1992. Review of P. Krugman's "Geography and Trade." *Regional Studies* 26:679.

Howes, C., and Markusen, A. 1993. Trade, industry and economic development. In *Trading industries, trading regions,* ed. H. Noponen, J. Graham, and A. Markusen, 1–44. London: Guilford Press.

Jackson, R. 1984. An evaluation of alternative measures of regional industrial diversification. *Regional Studies* 8:103–12.

Jaffe, A.; Trajtenberg, M.; and Henderson, R. 1993. Geographical localization of knowledge spillovers as evidenced by patent citations. *Quarterly Journal of Economics* 108:577–98.

Johnston, R. J. 1989. Extending the research agenda. *Economic Geography* 65:338–47.

———. 1992. Review of P. Krugman's "Geography and Trade." *Environment and Planning A* 24:1006.

Kaldor, N. 1978. The case for regional policies. In *Further essays on economic theory,* 139–54. London: Duckworth.

———. 1981. The role of increasing returns, technical progress and cumulative causation in the theory of international trade and economic growth. *Economie Appliquee* 34 (4). Reprinted in *The essential Kaldor* (1989), ed. F. Targetti and A. Thirlwall, 327–50. London: Duckworth.

———. 1985. Interregional trade and cumulative causation. In *Economics without equilibrium,* 57–79. New York: Sharpe.

Kitson, M., and Michie, J. 1995. Conflict, cooperation and change: The political economy of trade and trade policy. *Review of International Political Economy* 2 (4):632–57.

Knox, P., and Agnew, J. 1994. *The geography of the world economy.* 2d ed. London: Edward Arnold.

Krugman, P. 1979. Increasing returns, monopolistic competition and international trade. *Journal of International Economics* 9:469–79.

———. 1980. Scale economies, product differentiation and the pattern of trade. *Amercan Economic Review* 70:950–59.

———. 1981. Trade, accumulation and uneven development. *Journal of Development Economics* 8:149–61.

———. 1983a. The "new theory" of international trade and the multinational enterprise. In *The multinational corporation in the 1980s,* ed. C. Kindleberger and D. Audretsch, 57–73. London: MIT Press.

———. 1983b. Targeted industrial policies: Theory and evidence. In *Structural change and public policy.* Kansas City: Federal Reserve Bank of Kansas City.

———. 1984. Foreign industrial targetting and the US economy. *Brookings Papers on Economic Activity* 1:77–121.

———, ed. 1986. *Strategic trade policy and the new international economics.* Cambridge: MIT Press.

———. 1987a. The narrow moving band, the "Dutch disease" and the competitive consequences of Mrs. Thatcher. *Journal of Development Economics* 26:42–55.

———. 1987b. Is free trade passe? *Economic Perspectives* 1 (2):131–44.

———. 1987c. Increasing returns and the theory of international trade. In *Advances in economic theory,* ed. T. Bewley, 301–28. Cambridge: Cambridge University Press.

———. 1989. Economic integration in Europe: Some conceptual issues. In *The European internal market—trade and competition,* ed. A. Jacquemin and A. Sapir, 357–80. Oxford: Oxford University Press.

———. 1990. *Rethinking international trade.* Cambridge: MIT Press.

———. 1991a. *Geography and trade.* Leuven: Leuven University Press.

———. 1991b. Increasing returns and economic geography. *Journal of Political Economy* 99:483–99.

———. 1991c. History versus expectations. *Quarterly Journal of Economics* 106:651–67.

———. 1991d. History and industrial location: The case of the manufacturing belt. *American Economic Review (Papers and Proceedings)* 81:80–83.

———. 1992. *Currencies and crises.* London: MIT Press.

———. 1993a. On the relationship between trade theory and location theory. *Review of International Economics* 1:110–22.

———. 1993b. The current case for industrial policy. In *Protectionism and world welfare,* ed. D. Salvatore, 160–79. Cambridge: Cambridge University Press.

———. 1993c. First nature, second nature and metropolitan location. *Journal of Regional Science* 33 (2):129–44.

———. 1993d. The lessons of Massachusetts for EMU. In *Adjustment and growth in the European Monetary Union,* ed. F. Torres and F. Giavazzi, 241–69. Cambridge: Cambridge University Press.

———. 1993e. *Inequality and the political economy of Eurosclerosis.* November. London: Centre for Economic Policy Research.

———. 1994a. Competitiveness: A dangerous obsession. *Foreign Affairs* (March–April): 28–44.

———. 1994b. *The age of diminished expectations: US economic policy in the 1990s.* Rev. ed. London: MIT Press.

———. 1994c. *Peddling prosperity: Economic sense and nonsense in the age of diminshed expectations.* New York: Norton.

———. 1994d. Does Third World growth hurt First World prosperity? *Harvard Business Review* (July–August):113–21.

———. 1995. *Development, geography and economic theory.* Cambridge: MIT Press.

———. 1996. *The self-organising economy.* Cambridge, Mass.: Blackwell Publishers.

Krugman, P., and Venables, A. 1990. Integration and the competitiveness of peripheral industry. In *Unity with diversity in the European Community,* ed. C. Bliss and J. Braga de Macedo, 56–75. Cambridge: Cambridge University Press.

———. 1994. *Globalization and the inequality of nations.* Discussion paper 1015. London: Centre for Economic Policy Research.

Kurre, J., and Weller, B. 1989. Regional cyclical instability. *Regional Studies* 24:318–29.

Lancaster, K. 1980. Intra-industry trade under perfect monopolistic competition. *Journal of International Economics* 10:151–76.

Losch, A. 1967. *The economics of location.* New York: John Wiley Science Edition.

Lovering, J. 1990. Fordism's unknown successor. *International Journal of Urban and Regional Research* 14:159–74.

Lucas, R. 1988. On the mechanics of economic development. *Journal of Monetary Economics* 22:3–42.

Markowitz, H. 1959. *Portfolio selection and efficient diversification of investment.* New Haven: Yale University Press.

Markusen, A. 1993. Trade as a regional development issue: Policies for job and community preservation. In *Trading industries, trading regions,* ed. H. Noponen, J. Graham, and A. Markusen, 285–302. London: Guilford Press.

Markusen, A., and Park, S. O. 1993. New industrial districts: A critique and extension from the developing countries. Working paper, Project on Regional and Industrial Economics, Rutgers University.

Martin, P. J., and Rogers, C. A. 1994a. *Industrial location and public infrastructure.* Discussion paper 909. London: Centre for Economic Policy Research.

———. 1994b. *Trade effects of regional aid.* Discussion paper 910. London: Centre for Economic Policy Research.

Martin, R. L. 1993. Reviving the economic case for regional policy. In *Spatial policy in a divided nation,* ed. R. T. Harrison and M. Hart, 270–90. London: Jessica Kinglsey.

Meade, J. 1952. External economies and diseconomies in a competitive situation. *Economic Journal* 62:54–67.

Mishan, E. 1971. Externalities. *Journal of Economic Literature* 9:1–28.

Murphy, A.; Schleifer, A.; and Vishny, R. 1989. Industrialization and the big push. *Journal of Political Economy* 97:1003–26.

Myrdal, G. 1957. *Economic theory and underdeveloped regions.* London: Duckworth.

Neven, D. J., and Gouyette, C. 1994. *Regional convergence in the European Community.* Discussion paper 914. London: Centre for Economic Policy Research.

Noponen, H.; Graham, J.; and Markusen, A., eds. 1993. *Trading industries, trading regions.* London: Guilford Press.

Ohlin, B. 1933. *Interregional and international trade.* Cambridge: Harvard University Press.

Organization for Economic Cooperation and Development (OECD). 1994. *The OECD jobs study: Part 1. Labour market trends and underlying forces of change.* Paris: OECD.

Palmini, D., and Cray, R. F. 1992. Convergence or divergence? A study of regional business cycle patterns among and within US census regions. *Regional Science Perspectives* 22(2):30–52.

Peschel, K. 1982. International trade, integration and industrial location. *Regional Science and Urban Economics* 12:247–69.

Phelps, N. 1992. External economies, agglomeration and flexible accumulation. *Transactions of the Institute of British Geographers* n.s. 17(1):35–46.

Porter, M. E. 1990. *The competitive advantage of nations.* London: Macmillan.

———. 1994. The role of location in competition. *Journal of the Economics of Business* 1(1):35–39.

Pyke, F.; Becattini, G.; and Sengenberger, W., eds. 1989. *Industrial districts and inter-firm co-operation in Italy.* Geneva: Institute for Labour Studies.

Reich, R. B. 1991. *The work of nations: Preparing ourselves for 21st century capitalism.* London: Simon and Schuster.

Romer, P. 1990. Endogenous technological change. *Journal of Political Economy* 98:71–102.

Rosenstein-Rodan, P. 1943. Problems of industrialization of Eastern and South-Eastern Europe. *Economic Journal* 53:202–11.

Sabel, C. 1989. Flexible specialization and the re-emergence of regional economies. In *Reversing industrial decline?* ed. P. Hirst and J. Zeitlin, 17–70. Oxford: Berg.

Samuelson, P. 1994. Preface. In *The age of diminished expectations,* ed. P. Krugman, vii. London: MIT Press.

Scitovsky, T. 1954. Two concepts of external economies. *Economic Journal* 62:52–67.

Scott, A. J. 1986. Industrial organization and location, divisons of labor and the firm's spatial process. *Economic Geography* 62:215–31.

———. 1988. *New industrial spaces.* London: Pion.

———. 1992a. The role of large producers in industrial districts: A case study of high-tech systems houses in Southern California. *Regional Studies* 26:265–75.

———. 1992b. The collective order of flexible production agglomeration: Lessons for local economic development policy and strategic choice. *Economic Geography* 68:219–33.

Scott, A. J., and Storper, M. 1992a. Regional development reconsidered. In *Regional development and contemporary industrial response: Extending flexible specialization,* ed.

H. Ernste and V. Meier, 1–24. London: Belhaven.

————. 1992b. Industrialization and regional development. In *Pathways to industrialization and regional development*, ed. M. Storper and A. Scott, 3–17. London: Routledge.

Skott, P., and Auerbach, P. 1995. Cumulative causation and the "new" theories of economic growth. *Journal of Post Keynesian Economics* 17(3):381–402.

Smith, A. 1994. Imperfect competition and international trade. In *Surveys in international trade*, ed. D. Greenaway and A. Winters, 43–65. Oxford: Basil Blackwell.

Smith, N. 1989. Uneven development theory and location theory: Towards a synthesis. In *New models in geography*, Vol. 1, ed. R. Peet and N. Thrift, 142–63. London: Unwin Hyman.

Steed, G. 1971. Internal organization, firm integration and locational change: The Northern Ireland linen complex, 1954–64. *Economic Geography* 47:371–83.

Stigler, G. 1951. The division of labour is limited by the extent of the market. *Journal of Political Economy* 59:185–93.

Stohr, W. B. 1989. Regional policy at the crossroads. In *Regional policy at the crossroads: European perspectives*, ed. L. Albrechts, F. Moulaert, P. Roberts, and E. Swyngedouw, 191–97. London: Jessica Kingsley Press.

Storper, M. 1992a. Regional "worlds" of production: Learning and innovation in the technology districts of France, Italy and the USA. *Regional Studies* 27:433–55.

————. 1992b. The limits to globalization: Technology districts and international trade. *Economic Geography* 68:60–93.

Storper, M., and Walker, R. 1989. *The capitalist imperative: Territory, technology and industrial growth*. Oxford: Basil Blackwell.

Sunley, P. J. 1992. Marshallian industrial districts: The case of the Lancashire cotton industry in the inter-war years. *Transactions of the Institute of British Geographers n.s.* 17:306–32.

Thurow, L. 1994. *Head to head: The coming economic battle among Japan, Europe and America*. 2d ed. London: Brealey Publishing.

Tyson, L. 1992. *Who's bashing whom? Trade conflict in high technology industries*. Washington, D.C.: Institute for International Economics.

von Hagen, J., and Hammond, G. 1994. *Industrial localization: An empirical test for Marshallian localization economies*. Discussion paper 917. London: Centre for Economic Policy Research.

Weber, A. 1929. *Theory of the location of industries*. Chicago: University of Chicago Press.

Young, A. A. 1928. Increasing returns and economic progress. *Economic Journal* 38: 527–42.

[3]

An Institutionalist Perspective on Regional Economic Development*

ASH AMIN

Introduction

Until recently, regional policy has been firm-centred, standardized, incentive-based and state-driven. This is certainly true in the case of the Keynesian legacy that dominated regional policy in the majority of advanced economies after the 1960s. It relied on income redistribution and welfare policies to stimulate demand in the less favoured regions (LFRs) and the offer of state incentives (from state aid to infrastructural improvements) to individual firms to locate in such regions. Paradoxically, the same principles apply also to pro-market neoliberal experiments which have come to the fore over the last fifteen years. The neoliberal approach, placing its faith in the market mechanism, has sought to deregulate markets, notably the cost of labour and capital, and to underpin entrepreneurship in the LFRs through incentives and investment in training, transport and communication infrastructure, and technology. The common assumption in both approaches, despite their fundamental differences over the necessity for state intervention and over the equilibrating powers of the market, has been that top-down policies can be applied universally to all types of region. This agreement seems to draw on the belief that at the heart of economic success lies a set of common factors (e.g. the rational individual, the maximizing entrepreneur, the firm as the basic economic unit and so on).

The achievements of both strands of such an 'imperative' approach (Hausner, 1995) have been modest in terms of stimulating sustained improvements in the economic competitiveness and developmental potential of the LFRs. Keynesian regional policies, without doubt, helped to increase employment and income in the LFRs, but they failed to secure increases in productivity comparable to those in the more prosperous regions and, more importantly, they did not succeed in encouraging self-sustaining growth based on the mobilization of local resources and interdependencies (by privileging non-indigenous sectors and externally-owned firms). The 'market therapy' has threatened a far worse outcome, by reducing financial transfers which have proven to be a vital source of income and welfare in the LFRs, by exposing the weak economic base of the LFRs to the chill wind of ever enlarging free market zones or corporate competition and by failing singularly to reverse the flow of all factor inputs away from the LFRs. In short, the choice has been that between dependent development or no development.

* The original draft of this paper was commissioned by the Territorial Development Services Unit of the OECD. I am grateful to Mario Pezzini for allowing me to develop it for publication in this journal. I thank also David Freshwater, Meric Gertler, Patrick Le Galès, Amy Glasmeier and Charlotte Sammelin for their thoughtful comments on the original draft.

Partly in response to these failings, more innovative policy communities have begun to explore a third alternative, informed by the experience of prosperous regions characterized by strong local economic interdependencies (e.g. Italian industrial districts, certain technopoles, Baden Württemberg). It is an alternative centred on mobilizing the endogenous potential of the LFRs, through efforts to upgrade a broadly defined local supply-base. It seeks to unlock the 'wealth of regions' as the prime source of development and renewal. This is not an approach with a coherent economic theory behind it, nor is there a consensus on the necessary policy actions. However, its axioms contrast sharply with those of the policy orthodoxy, in tending to favour bottom-up, region-specific, longer-term and plural-actor based policy actions. Conceptually, against the individualism of the orthodoxy (e.g. the centrality of *homo economicus*), it recognizes the collective or social foundations of economic behaviour, for which reason it can be described loosely as an institutionalist perspective on regional development.

This article seeks to develop the institutionalist perspective by bringing together strands of policy action scattered across the literature, as well as suggesting new strands. It claims that the new perspective, taken beyond its current emphasis on local interfirm complementarities, economies of association and tacit knowledge, does open up novel but challenging opportunities for policy action at the local level. It also claims, however, that the 'new regionalism' will amount to very little in the absence of sustained macro-economic support for the regions, notably a secure financial and income transfer base and expansionary programmes to boost overall growth at national and international level. The first section of the paper summarizes the axioms of economic action and governance which emerge from a theorization of the economy rooted in institutional economics and socio-economics. The second section discusses applications of institutionalist thought within regional development studies that seek to explain the importance of territorial proximity for economic competitiveness. Part of the purpose of these two sections is to demonstrate that the new policy orientations, outlined in the third section, are not just ex-post generalizations based on the experience of a small number of regions, but also ex-ante suggestions based on a particular conceptualization/abstraction of the economy and its territoriality. As such, the suggested actions are not crude (mis)translations of unique local experiences, nor are they — on the grounds that abstracted axioms never convert into mirrored practices — a policy recipe.

The economy and economic governance in institutional economics

The rise of institutional and evolutionary economics is now well documented (Hodgson, 1988; 1998; Samuels, 1995; Metcalfe, 1998), as is thought in economic sociology which stresses the influence of wider social relations in economic life (Smelser and Swedberg, 1994; Ingham, 1996). The two bodies of thought stress that economic life is both an instituted process and a socially embedded activity and therefore context-specific and path-dependent in its evolution. Against orthodox assumptions that the economy is equilibrium-oriented and centred on the rational individual or machine-like rules, the stress falls on processes of institutionalization as a means of stabilizing and interpreting an economy that is essentially non-equilibrating, imperfect and irrational.

My aim here is not to discuss institutionalist economic thought in any depth or detail. The arguments are well charted elsewhere and, more importantly, my purpose is simply to provide an abbreviated account of influences on economic behaviour which might be relevant in a discussion of novel actions necessary to encourage regional economic success. Three sets of ideas seem to be especially important in this regard.

First, from economic sociology comes the well-known idea that markets are socially constructed (Bagnasco, 1988) and that economic behaviour is embedded in networks of interpersonal relations. Crucially, therefore, economic outcomes are influenced by network properties such as mutuality, trust and cooperation, or their opposite (Dore, 1983; Granovetter, 1985; Grabher, 1993; Fukuyama, 1995; Misztal, 1996). Granovetter, for example, has suggested that networks of weak ties might be more dynamic than those dominated by strong ties (e.g. enforced loyalty) or easy escape (e.g. contract-based ties). While weak ties offer economic agents the benefits of both cooperation and access to a varied selection environment for new learning, strong ties, as in many crime networks, pose the threat of both lock-in and restricted selection, and contractual self-reliance poses very high search costs. In addition, the rising influence of actor-network theory has furthered analysis of the powers of networks by stressing the inseparability of people and things within them, producing distinctive properties that weave together actors, organizational cultures, knowledge environments, machines, texts and scripts (Latour, 1986; Callon, 1991).

Second, and against the rational actor model in standard economics, comes the idea from evolutionary and cognitive psychology (Cosmides and Tooby, 1994; Plotkin, 1994) and the behavioural tradition in economics (Simon, 1959) that different actor-network rationalities produce different forms of economic behaviour and decision-making. For example, an instrumentalist or substantive rationality is likely to favour reactive responses to problems, based on a largely rule-following behaviour. Reactive responses may prove to be adequate in relatively unchanging and predictable environments but their underlying rationality is not equipped for a varying environment. In contrast, a procedural actor rationality is one which seeks to adapt to the environment, drawing upon perceptive powers and generally more complex cognitive arrangements for solving problems. Cognitive actor networks behave intentionally and knowledgeably, searching for solutions based on perception and conscious design. They respond heuristically and procedurally in satisfying goals. Finally, while the latter two rationalities tend to assume an invariant environment (therefore largely problem-solving), a recursive rationality is problem-seeking and assumes that the environment can be anticipated and to a degree manipulated through such procedures as strategic monitoring, experimental games, group learning and so on (Delorme, 1997). It tends to generate creative actor networks with the ability to shape the environment, owing to the capacity to think and act strategically and multidimensionally (Orillard, 1997). Thus, the creative, learning and adaptive capacities of economic agents are centrally dependent upon the rationalities of their constituent actor networks.

Third, from the recent rediscovery of 'old' institutional economics (Hodgson, 1988; 1998; Hodgson *et al.*, 1993) comes the idea that the economy is shaped by enduring collective forces, which make it an instituted process as claimed by Polanyi, not a mechanical system or set of individual preferences. These forces include formal institutions such as rules, laws and organizations, as well as informal or tacit institutions such as individual habits, group routines and social norms and values. All of these institutions provide stability in the real economic context of information asymmetry, market uncertainty and knowledge boundedness, by restricting the field of possibilities available, garnering consensus and common understandings and guiding individual action. They are also, however, templates for, or constraints upon, future development. It is their endurability and framing influence on action by individuals and actor networks that forces recognition of the path- and context-dependent nature of economic life, or, from a governance perspective, the wide field of institutions beyond markets, firms and states which need to be addressed by policies seeking to alter the economic trajectory.

From these strands of institutionalist thought derives an understanding of the economy as something more than a collection of atomized firms and markets driven by rational preferences and a standard set of rules. Instead the economy emerges as a composition of collective influences which shape individual action and as a diversified and path-dependent entity moulded by inherited cultural and socio-institutional influences. In turn, the influences on economic behaviour are quite different from those privileged by the economic orthodoxy (e.g. perfect rationality, hedonism, formal rules etc.). Explanatory weight is given to the effects of formal and informal institutions, considered to be socially constructed and subject to slow evolutionary change; to values and rationalities of action ensconced in networks and institutions; to the composition of networks of economic association, especially their role in disseminating information, knowledge, and learning for economic adaptability; and to intermediate institutions between market and state which are relatively purposeful and participatory forms of arrangement.

On the basis of these principles, we can begin to derive a number of general axioms of economic governance associated with an institutionalist approach. First, there is a preference for policy actions designed to strengthen networks of association, instead of actions which focus on individual actors alone. Second, part of the purpose of policy action might be to encourage voice, negotiation and the emergence of procedural and recursive rationalities of behaviour, in order to secure strategic vision, learning and adaptation (Amin and Hausner, 1997). Third, emphasis is given to policy actions which aim to mobilize a plurality of autonomous organizations, since effective economic governance extends beyond the reach of both the state and market institutions (Hirst, 1994). Fourth, the stress on intermediate forms of governance extends to a preference for building up a broad-based local 'institutional thickness' that might include enterprise support systems, political institutions and social citizenship (Amin and Thrift, 1995). Finally, a key institutionalist axiom is that solutions have to be context-specific and sensitive to local path-dependencies.

These governance axioms and their underlying concepts are now beginning to filter into regional development studies and to generate new policy implications, as shown in the next two sections.

The institutional turn in regional development studies

In recent years, the region has been rediscovered as an important source of competitive advantage in a globalizing political economy (Scott, 1995; Cooke, 1997). In part, this rediscovery is based on studies of the success of highly dynamic regional economies and industrial districts which draw extensively upon local assets for their competitiveness. However, the rediscovery is also based on the insights of institutional economic theory, particularly its explanation of why territorial proximity matters for economic organization. Two conceptual strands stand out.

One strand — perhaps the closest to the economics mainstream — derives from renewed interest in endogenous growth theory, which acknowledges the economic externalities and increasing returns to scale associated with spatial clustering and specialization (Porter, 1994; Krugman, 1995). Places specializing in given industries and their upstream and downstream linkages benefit from the scale economies of agglomeration and advantages associated with specialization. The contention of Krugman and Porter is that the spatial clustering of interrelated industries, skilled labour and technological innovations offers some of the key elements of growth and competitiveness. These include increasing returns, reduced transaction costs and economies associated with proximity and interfirm exchange, as well as specialized know-how, skills and technological advancement.

This 'new economic geography' has gained considerable influence and is undoubtedly appealing, as it provides solid economic reasons for local agglomeration in a globalizing economy (reduced transaction costs, economies of specialization, externalities etc.). It fails, however, to properly investigate the sources of these local advantages (Martin and Sunley, 1996), which, according to a second conceptual strand developed largely by economic geographers, lie in the character of local social, cultural and institutional arrangements. More specifically, insight is drawn from institutional and evolutionary economics concerning ties of proximity and association as a source of knowledge and learning (Amin and Thrift, 1995; Sunley, 1996; Storper, 1997).

A leading exponent is Michael Storper (1997), who has suggested that a distinctive feature of places in which globalization is consistent with the localization of economic activity is the strength of their 'relational assets' or 'untraded interdependencies'. These are assets which are not tradable, nor are they easily substitutable, since they draw on the social properties of networks in which economic agents are implicated. They include tacit knowledge based on face-to-face exchange, embedded routines, habits and norms, local conventions of communication and interaction, reciprocity and trust based on familiarity and so on.

These relational assets are claimed to have a direct impact on a region's competitive potential insofar as they constitute part of the learning environment for firms. They provide the daily access to the relevant resources (information, knowledge, technology, ideas, training and skills) that are activated through the networks of interdependency and common understandings that surround individual firms. Many of the insights of the literature on the so-called learning regions (Cooke and Morgan, 1998), such as Silicon Valley, Baden Württemberg and Italian industrial districts, derive from analysis of the learning properties of local, industry-specialist, business networks. These networks of reciprocity, shared know-how, spillover expertise and strong enterprise support systems, according to Malmberg (1996), are sources of learning, facilitated through such advantages as reduced opportunism and enhanced mutuality within the relationships of interdependence.

Other observers who note the difference between formal and informal knowledge for economic competitiveness (e.g. Becattini and Rullani, 1993; Asheim, 1997; Maskell *et al.*, 1998; Blanc and Sierra, 1999; Nooteboom, 1999) suggest that geographical proximity plays a unique role in supplying informally-constituted assets. For instance, Maskell *et al.* argue that tacit forms of information and knowledge are better consolidated through face-to-face contact, not only due to the transactional advantages of proximity, but also because of their dependence upon a high degree of mutual trust and understanding, often constructed around shared values and cultures. Similarly, scholars (Becattini and Rullani, 1993; Asheim, 1997; Nooteboom, 1999) have distinguished between codified knowledge as a feature of trans-local networks (e.g. R&D laboratories or training courses of large corporations) and formally constituted institutions (e.g. business journals and courses, education and training institutions, printed scientific knowledge) and non-codified knowledge (e.g. workplace skills and practical conventions) as aspects locked into the 'industrial atmosphere' of individual places. The consensus among these commentators seems to be that in a world in which codified knowledge is becoming increasingly ubiquitously available, uncodified knowledge, rooted in relations of proximity, attains a higher premium in deriving competitive advantage owing to its uniqueness.

The new stress on the comparative advantages of face-to-face learning environments has some obvious limitations (Amin and Cohendet, 1999). One is the tendency to ignore the relational proximity provided by global links of reciprocity, such as the networks of transnational corporations, that also constitute a rich source of informal learning. Another is its inadequate appreciation of the power of organizational environments in combining the fruits of tacit knowledge with science and technology (e.g. design and quality-conscious high performance companies), as well as its underestimation of the normal transaction cost savings associated with size. A third limitation is the under-appreciation of the variety of possible learning outcomes (Odgaard, 1998), which are in part influenced by differences in actor-network rationalities and cognitive frames (e.g. craft networks of incremental learning based on trust and reciprocity versus networks of problem-seeking learning based on reflexive knowledge). Finally, at times it is not clear whether the strengths of dynamic networks derive from their learning capabilities or their ability to anticipate change and adapt (Hudson, 1996).

Notwithstanding these limitations, the institutionalist geographers offer a much richer understanding of territorial proximity than that offered by endogenous growth theory, which continues to stress well-known but rather tired agglomeration factors. Proximity has come to be considered in ways which acknowledge the territorial parameters of the institutional and social sources of economic action (Barnes, 1995; Thrift and Olds, 1996). This includes the power of local rationalities and traditions of behaviour, the properties of face-to-face networks and the quality of local institutions, social habits and conventions. But, beyond their important differences, the consensus shared across both strands is that regional-level industrial configurations, supply-side characteristics and institutional arrangements continue to play a critical role in securing economic success in a globalizing economy characterized by the increased transnational flows of factor inputs and global-level industrial and financial organization.

Regional policy orientations

Both strands of the new regionalism imply practical action which transcends the limits of traditional local economic development initiatives. The focus falls on building the wealth of regions (not the individual firm), with upgrading of the economic, institutional and social base considered as the prerequisite for entrepreneurial success. Thus, local effort might focus on developing the supply base (from skills through to education, innovation and communications) and the institutional base (from development agencies to business organizations and autonomous political representation) in order to make particular sites into key staging points or centres of competitive advantage within global value chains. In addition, the 'relational' strand of the new regionalism implies that attention might be paid to the nature of local interfirm dependencies and rationalities of behaviour that work to local advantage (e.g. loose ties, interactive decision-making, recursive knowledge).

In my view there are four novel areas of action which emerge from the 'wealth of regions' perspective. I should stress that the recommendations are not offered as templates for action, but as issues that policy-makers need to consider in devising practical solutions to encourage regional endogenous growth.

Building clusters and local economies of association
The experience of some of the most dynamic economies in Europe shows that supply-side upgrading of a generic nature (e.g. advanced transport and communications systems or provision of specialized training and skills), though desirable, is not sufficient to secure regional economic competitiveness. Instead, in small nations such as Denmark and successful regional economies such as Emilia-Romagna, Baden Württemberg and Catalonia, policy action is increasingly centred on supporting clusters of interrelated industries which have long roots in the region's skill- or capabilities-base. This helps not only to secure meaningful international competitive advantage, but also to reap the benefits of local specialization along the supply chain. Firm-specific initiatives, such as small-firm development programmes or incentives to attract inward investors, tend to be integrated within such cluster programmes in order to build up a system of local interdependencies. Institutional support, in the form of technology transfer, training and education and access to producer services such as market intelligence, business innovation and finance, tends to be sectorally specific so that help can be targeted to firms in specific clusters.

In addition, considerable policy attention is paid to building economies of association within clusters. This might include efforts to improve cultures of innovation within firms by encouraging social dialogue and learning based on shared knowledge and information exchange. It might include initiatives to encourage interfirm exchange and reciprocity through buyer-supplier linkage programmes, incentives for pooling of resources, joint ventures, task specialization and so on. Finally, in order to maximize the efficiency of collective resources, it might include conscious effort to establish contact between sector-specific organizations (e.g. trade associations, sectorally-based service centres) and other support organizations (e.g. large and small-firm lobbies, function-specific producer services agencies, trade unions, chambers of commerce, local authorities, regional development agencies). Building economies of association along these lines would help regions to consolidate local ties and encourage continual upgrading and capacity-building across sectoral networks of horizontal and vertical interdependency.

Cluster programmes are no longer new to the regional policy community. Indeed, following the spectacular translation of Michael Porter's ideas into policy action through his world famous consultancy group, Monitor, most regions seem to have a cluster programme of some sort. And, ironically, in contradiction to the institutionalist stress on context-specificity and path-dependency, the most common tendency beyond the selection of locally-sensitive industrial clusters has been to copy from the experience of successful regions or from some 'expert' manual. Cluster programmes are becoming as standardized a mantra as were the incentive packages of preceding regional policy (Enright, 1998). Very few regions have attempted to develop unique industrial strategies based on deep assessment of local institutional and cultural specificities. To a degree, this failing stems from the inability of the policy community to recognize the centrality of 'softer' influences, such as the three considered in turn below.

Learning to learn and adapt
The geographical strand of the new regionalism stresses learning as a key factor in dynamic competitiveness. Indeed, it is claimed that economically successful regions are 'learning' or 'intelligent' regions (Cooke and Morgan, 1998). It is their capacity to adapt around particular sectors and to anticipate at an early stage new industrial and commercial opportunities that enables them to develop and retain competitive advantage around a range of existing and future possibilities. Their strength lies in 'learning to learn' (Hudson, 1996). By contrast, a very large number of less favoured regions suffer from the problem of industrial and institutional lock-in and that of reactive adaptation to their economic environment, thus preventing the formation of a learning culture.

Vexing from a policy perspective is that there is no received wisdom on the factors which contribute to regional learning and adaptability. However, some of the contributing factors can be discerned from an observation of the relevant regions. One obvious factor is quite simply the scale and density of 'intelligent' people and institutions, as reflected in the skill and professional profile of the labour market, the volume and quality of training and education across different levels, the depth of linkage between schools, universities and industry, the quality and diversity of the research, science and technology base, and the availability of intermediate centres of information and intelligence between economic agents and their wider environment (e.g. commercial media, trade fairs, business service agencies). These are vital sources of codified knowledge, grounded in the regional milieu. Many LFRs display a discernible lack of most of these attributes, with policy actions often geared towards the production of low-grade skills and training or towards disembodied ventures such as university expansion, science parks and training schemes which fail to build the necessary connections.

Less obviously, the quality of ties associated with economies of association is another important source of learning and adaptation, through its impact on the circulation of informal information, innovation and knowledge. Networks of association in the economy facilitate the spread of information and capabilities and the prospect of economic innovation through social interaction. Of course, there is always the danger that ties which are too strong and long-standing might actually prevent renewal and innovation by encouraging network closure and self-referential behaviour (Grabher and Stark, 1997). On the other hand, in contexts where economic agents have the option of participating in many competing networks on the basis of loose ties and reciprocal relations, often through independent intermediaries, the prospect for learning through interaction is enhanced. Not only is variety in the selection environment for options increased, but so is the potential for uptake and adaptation, through the ties and obligations that bind economic agents, leading, ultimately, to network cultures of permanent innovation. The policy challenge in this regard for LFRs is to find a way of substituting their traditional ties of hierarchy and dependency (e.g. big firms, state provision, family connections) with links of mutuality between economic agents and institutions.

Third, as mentioned earlier, research has begun to appreciate the connection between rationalities of action and adaptive potential. It would appear that rule-based, substantive rationality, which encourages reactive responses to the external environment, is ill-equipped for learning and adaptation. Procedural rationality, on the other hand, based on cognitive and behavioural interpretation by economic agents of the external environment, favours incremental adjustment and adaptation. In contrast, a reflexive rationality, involving strategic and goal-monitoring behaviour (Sabel, 1994), encourages experimental anticipation and actions seeking to shape the external environment. The cognitive frame of regional actors and institutions, in short, is the central source of learning. The culture of command and hierarchy that characterizes so many LFRs has stifled the formation of a reflexive culture among the majority of its economic institutions and, consequently, prevented the encouragement of rationalities geared towards learning and adaptation. Considerable policy attention needs to be paid to the nature of organizational and management cultures and actor rationalities which circulate within a region's dominant institutions. Only too often, policy action has sought to introduce new players and institutions in a region, without giving due regard to the dominant 'mind set' and its effects on learning and adaptability.

Importantly, but rarely addressed by the policy community, the capacity to change lies centrally in the ability of actor networks to develop an external gaze and sustain a culture of strategic management and coordination in order to foresee opportunities and secure rapid response. The key factor is the ability to evolve in order to adapt (Amin and Hausner, 1997). The encouragement of this ability requires effort to identify the potential sources of behavioural alternatives — for example the preservation of diverse competencies (e.g. redundant skills and industrial slack — see Grabher and Stark, 1997); the scope for subaltern groups to break the grip of hegemonic interests which gain from preserving the past; the openness of organizations to external and internal influences; the scope for strategic decision-making through agent-environment interaction; and the encouragement of diversity of knowledge, expertise and capability, so that new tricks are not missed.

Broadening the local institutional base
The last point illustrates the need for wider institutional changes to tackle impediments to economic renewal rooted in institutional dominance and closure. Partly in recognition of this problem, it has become increasingly common to assume that region-building has to be about mobilizing independent political power and capacity. In the European Union this assumption lies at the centre of the discourse on 'Europe of the regions' and has led to strong endorsement for local fiscal and financial autonomy, together with enlargement of the powers of local government and the establishment of vigorous regional assemblies or parliaments. The linkage made with economic development is that local political power and voice facilitates the formation of a decision-making and decision-implementing community able to develop and sustain an economic agenda of its own.

The institutionalist perspective, however, suggests that region-building cannot stop at simply securing regional political autonomy. Equally — perhaps more important — are matters of who makes decisions and how. Let us recall two institutionalist governance axioms, namely, the desirability of decision-making through independent representative associations and the superiority of participatory decision-making. The added challenge for the regions, therefore, is to find ways of developing a pluralist and interactive public sphere that draws in both the state and a considerably enlarged sphere of non-state institutions. It would be an error if regional institutional reform became a matter of simply substituting government by the central state with a regional corporatism that relies on a small elite drawn from the regional government offices, local authorities, development agencies, the business leadership and perhaps even mayors wielding extraordinary powers.

Governance, especially in the institutionally thin regions, has always been in the hands of elite coalitions, and the resulting institutional sclerosis has been a source of economic failure by acting as a block on innovation and the wider distribution of resources and opportunity. In an increasingly global economy, these elites and their charismatic leaders may undoubtedly help regions to jostle for influence with national and international organizations (e.g. the European Commission or transnational corporations), but they will achieve little in terms of mobilizing a regional development path based on unlocking hidden local potential. This is why it is vital that regional actors ask whether their decision-making processes constitute an obstacle to institutional renewal, away from a culture of hierarchy and rule-following, towards one that focuses on informational transparency, consultative and inclusive decision-making, and strategy-building on the basis of reflexive monitoring of goals.

In the sphere of local state action, such deliberation might well lead towards very new institutional practices. For example, regional authorities — in the search for innovative ideas or unrecognized potential — could extend decision-making beyond the professional politician and draw in — perhaps through specialist committees — experts and representatives from the various professional and civic groups that make up local society. In addition, the principle of learning through social inclusion, taken seriously, might stimulate special effort to draw in minority and excluded interests. In turn, special attention might be paid to how business is conducted, in order to allow full and proper debate, potential for creative decisions, empowerment of the dialogically disadvantaged, and open and transparent interaction with the public and other representative institutions.

But, ultimately, the process of institutional reform has to go beyond the decentralization and democratization of a region's official organizations. Many of the prosperous regions of Europe are also regions of participatory politics, active citizenship, civic pride, and intense institutionalization of collective interests — of society brought back into the art of governance. Within them, associational life is active, politics is contested, public authorities and leaders are scrutinized, public space is considered to be shared and commonly owned, and a strong culture of autonomy and self-governance seeps through local society. They are regions of developed 'social capital' (Putnam, 1993), serving to secure many economic benefits, including public-sector efficiency in the provision of services; civic autonomy and initiative in all areas of social and economic life; a culture of reciprocity and trust which facilitates the economics of association; containment of the high costs of social breakdown and conflict; and potential for economic innovation and creativity based on social confidence and capability.

The LFRs face a daunting task in reconstructing local social capital, damaged as it may be by decades of economic hardship, state-dependency, elite domination and so on. But this is not an impossible task. Some catalyst projects might focus on popular projects which restore a pride of place and belonging (e.g. festivals, the recovery of local public spaces, cheap and efficient public transport), community development programmes, schemes involving public participation, investment in the social infrastructure, civic educational programmes, and initiatives in marginalized communities designed to rebuild confidence and capability. These are projects which need public involvement and imagination, constituting a small but necessary step towards reconstructing damaged civic identities.

Mobilizing the social economy

The preceding discussion implies that a regional culture of social inclusion and social empowerment is likely to encourage economic creativity by allowing diverse social groups and individuals to realize their potential. This reinforces the view that policies to stimulate regional entrepreneurship should recognize, oblique though it may appear, the centrality of policies to combat social exclusion in this process. This is especially relevant in the context of regions marked by problems of persistent structural unemployment and rudimentary entrepreneurship, both of which act as a severe constraint on economic renewal. In such regions, the depth and scale of unemployment and the trend towards jobless growth in the economy at large, makes a return to full employment highly unlikely through improvements in regional economic competitiveness (via, say, industrial upgrading, clusters and economies of association).

More direct action to stimulate job generation is required, but as a catalyst for building a social economy capable of nurturing skills, expertise and capabilities. The action might involve active labour market programmes targeted towards reintegrating vulnerable social groups such as young persons, the under-qualified or ethnic minorities. It might include sustained effort to monitor and understand the informal economy, perhaps expecting improvements in business practice in exchange for policy supports which firms on the margins of illegality might find acceptable. For example, regions in which the sweatshop economy thrives might consider providing firms with access to bridge-loans and specialized services in order to help firms upgrade and, through this process, emerge into the formal economy.

An interesting contemporary policy innovation in the European Union is experimentation with the social economy as a source of local renewal. In countries such as Germany, France, Belgium, the Netherlands, Italy and Ireland, there is growing public policy support (e.g. subsidies and indirect aids such as training, facilitating legislation, specialized services) for community projects that are run by the third sector and involve excluded groups either as providers or users of socially useful services. This might involve support for a community group that employs school leavers to offer affordable housing to low-income groups or for a cooperative through which the long-term unemployed provide domestic care or transport access to the elderly. In other words, the battle against social exclusion is being combined with reforms to the welfare state, towards building an intermediate economic sphere that serves to meet real local welfare needs. In turn, this intermediate sphere, sustained by both monetary and innovative non-monetary metrics of exchange (e.g. service vouchers or services in kind), is seen as a source of employment and entrepreneurship in 'markets' which are of limited interest to state organizations and private-sector firms. In the longer run, it is seen as a vital source for unlocking social confidence and creativity among the excluded.

The policy implication is that regions need to incorporate a social economy programme into their efforts to improve regional economic competitiveness. It is important, however, for the reasons given in the preceding section, that support is provided with a light governmental touch, leaving a great deal to local actors. For example, regional, or city-based, 'social inclusion commissions' could be established, with an elected chair from a widely-drawn membership of relevant local organizations. The commissions would audit local service needs, propose rules for action, invite and consider applications for funding, work with the local authorities and other economic interest groups and so on. The local authorities and the central government would play a facilitating role, providing, for example, resources and legislation, but they would not provide a direct steer on local priorities and projects.

Conclusion: back to the macro-economy

The new regionalism offers a solution based on the mobilization of local resources. But it does so on the basis of a very broad definition of what constitutes the economy and economic action. It is an approach that builds outwards from a new industrial policy and effort to strengthen local economies of association to actions to improve institutional reflexivity, learning potential and social creativity. The institutionalist agenda threatens the inherited policy approach in three ways: first, by placing faith in long-term, evolutionary actions which tend to span across normal planning and electoral cycles; second, by suggesting new actor rationalities to replace reliance on standardized, off-the-shelf formulae applied mechanically by an unreflexive policy community; and third, by expecting policy actors to considerably broaden their definition of the factors of economic success.

To a degree the focus on endogenous regional solutions has been forced by an uncomfortably pervasive agreement across the policy community around the neoliberal rejection of macro-economic actions in favour of LFRs which might hinder market forces. Thus, for example, little support can be marshalled these days for monetary or credit policies which are monitored for their regional effects on demand or entrepreneurship, or for competition policies which might veto a merger or takeover if proven to be damaging to an LFR. Across the political spectrum, the consensus has grown that national and regional competitiveness is the only pathway to prosperity and that redistributive measures alone will not suffice.

This is a perilous supposition, not least because of the institutionalist axiom that action has to be contextually relevant and medium- to long-term. The policy orientations outlined above are not equally applicable to all types of region, and where they are, they require time to be built up. The orientations are especially appropriate for regions characterized by certain impediments to economic renewal: fragile small-firm entrepreneurship; domination by externally owned or controlled firms with poor levels of local economic integration; restricted diversification, innovation and learning capacity; and state dependency and institutional closure. These are problems which are rather typical of old industrial regions and their particular institutional legacies. Lagging rural regions face a different set of impediments, and their institutional base might also be less equipped for experimenting with learning-based industrial clusters and reflexive goal-monitoring institutional behaviour. Region-building, in short, may not be an option for all regions, owing to the restrictions of context and time.

This is why it is vital that an approach based on mobilizing the wealth of regions does not degenerate into localist sentiment. There is a risk that the institutionalist turn in regional thought and practice reinforces a parochial optimism centred on the belief that building local capabilities might be sufficient for establishing a privileged position within global networks. There are two flaws in this assumption. First, as Hudson (1996) argues, drawing on the example of once-prosperous regions which too were learning regions, such internal connectivity unattended can quite easily end up reinforcing, through institutional lock-in, path-dependencies which are inappropriate for new economic circumstances. Second, and as a consequence, the critical factor for economic success is not the presence of local relations of association and institutional advancement but the ability of places to anticipate and respond to changing external circumstances. Thus it is the management of the region's wider connectivity that is of prime importance, rather than its intrinsic supply-side qualities.

To a degree, the responsibility for the management of this wider connectivity lies in the hands of non-regional actors, notably government. No amount of imaginative region-building will be able to sustain a spiral of endogenous economic growth in the absence of a conducive macro-economic framework. Interregional competition in a Europe in recession and dominated by restrictive macro-economic policies will continue to work in favour of the core regions. Therefore, something has to be done to secure the less-favoured regions sufficient time and resources to implement boot-strapping reforms. So entrenched is the recent history in the EU and other regional confederations of member-state commitment to macro-economic prudence — from monetary stability to reduced public expenditure — that manipulation of the rules in favour of the LFRs is a dim prospect. For example, inflationary or deficit-inducing expenditure programmes steered towards the less favoured regions are likely to be blocked.

Yet it is imperative that the European economy, with its alarmingly high levels of unemployment, be given an expansionary kick start. Historically, governments have implemented Keynesian, demand-led recovery programmes by financing public building and infrastructure programmes, as well as relaxing investment and credit restrictions in order to stimulate expenditure and consequently industrial expansion. With careful regulation of potential inflationary outcomes, there is no reason why controlled expansion of the economy along these lines is not possible. Without it, there can be little scope for redistributing jobs and economic opportunity to the regions.

Secondly, regional financial security, decoupled from the ideological whims of centralizing governments, needs to be secured across member states in order to adequately fund policy priorities and meet the income and welfare needs of the local population. Controversially, this might involve as bold a step as automatic fiscal transfers to the regions aligned to local income. In this way, tax revenue pooled at, say, the EU level can be automatically, and continually, redirected to the regions. Such a regionally equitable fiscal system would ensure that the less favoured regions are compensated for their inability to generate as high a level of local tax revenue as their more prosperous counterparts.

These are controversial suggestions which need further debate. However, the point of raising them here is that in the absence of a conducive macro-economic framework, it seems irresponsible to ask the regions to embark upon a long-term and comprehensive overhaul in pursuit of an endogenous pathway to prosperity.

Ash Amin (Ash.Amin@durham.ac.uk), Department of Geography, University of Durham, Durham DH1 3LE, UK.

References

Amin, A. and N. Thrift (1995) Institutional issues for the European regions: from markets and plans to socioeconomics and powers of association. *Economy and Society* 24.1, 41–66.
—— and J. Hausner (eds.) (1997) *Beyond market and hierarchy: interactive governance and social complexity.* Edward Elgar, Aldershot.
—— and P. Cohendet (1999) Learning and adaptation in decentralised business networks. *Environment and Planning D: Society and Space* 17, 87–104.
Asheim, B. (1997) 'Learning regions' in a globalised world economy: towards a new competitive advantage of industrial districts? In S. Conti and M. Taylor (eds.), *Interdependent and uneven development: global-local perspectives*, Avebury, London.
Bagnasco, A. (1988) *La costruzione sociale del mercato.* Il Mulino, Bologna.
Barnes, T. (1995) Political economy I: 'the culture, stupid'. *Progress in Human Geography* 19.3, 423–31.
Becattini, G and E. Rullani (1993) Sistema locale e mercato globale. *Economia e Politica Industriale* 80, 25–40.
Blanc, H. and C. Sierra (1999) The internationalisation of R&D by multinationals: a trade-off between external and internal proximity. *Cambridge Journal of Economics* 23.2, 187–206.
Callon, M. (1991) Techno-economic networks and irreversibility. In J. Law (ed.), *A sociology of monsters*, Routledge, London.

Cooke, P. (1997) Regions in a global market: the experiences of Wales and Baden-Württemberg. *Review of International Political Economy* 4.2, 349–81.

—— and K. Morgan (1998) *The associational economy: firms, regions and innovation*. Oxford University Press, Oxford.

Cosmides, L. and J. Tooby (1994) Better than rational: evolutionary psychology and the invisible hand. *American Economic Review* 84.2, 327–32.

Delorme, R. (1997) The foundational bearing of complexity. In A. Amin and J. Hausner (eds.), *Beyond market and hierarchy: interactive governance and social complexity*, Edward Elgar, Aldershot.

Dore, R. (1983) Goodwill and the spirit of market capitalism. *British Journal of Sociology* 34.3.

Enright, M. (1998) *The globalisation of competition and the localisation of competitive advantage*. Paper presented at the workshop on Globalisation of Multinational Enterprise Activity and Economic Development, University of Strathclyde, 15–16 May.

Fukuyama, F. (1995) *Trust: the social virtues and the creation of prosperity*. Free Press, New York.

Grabher, G. (ed.) (1993) *The embedded firm*. Routledge, London.

Grabher, A. and D. Stark (1997) (eds.) *Restructuring networks in postsocialism: linkages and localities*. Oxford University Press, Oxford.

Granovetter, M. (1985) Economic action and social structure: the problem of embeddedness. *American Journal of Sociology* 91, 481–510.

Hausner, J. (1995) Imperative vs. interactive strategy of systematic change in central and eastern Europe. *Review of International Political Economy* 2.2, 249–66.

Hirst, P. (1994) *Associative democracy*. Polity Press, Cambridge.

Hogdson, G.M. (1988) *Economics and institutions*. Polity Press, Cambridge.

—— (1998) The approach of institutional economics. *Journal of Economic Literature* 36.1, 162–92.

——, W.J. Samuels and M.R. Tool (eds.) (1993) *The Elgar companion to institutional and evolutionary economics*. Edward Elgar, Aldershot.

Hudson, R. (1999) 'The learning economy, the learning firm and the learning region': a sympathetic critique of the limits to learning. *European Urban and Regional Studies* 6, 59–72.

Ingham, G. (1996) Some recent changes in the relationship between economics and sociology. *Cambridge Journal of Economics* 20, 243–75.

Krugman, P. (1995) *Development, geography and economic theory*. MIT Press, London.

Latour, B. (1986) The powers of association. In J. Law (ed.), *Power, action and belief*, Routledge and Kegan Paul, London.

Malmberg, A. (1996) Industrial geography: agglomeration and local milieu. *Progress in Human Geography* 20.3, 392–403.

Martin, R. and P. Sunley (1996) *Slow convergence? Post-neoclassical endogenous growth theory and regional development*. Working Paper 44, ESRC Centre for Business Research, University of Cambridge.

Maskell, P., H. Eskelinen, I. Hannibalsson, A. Malmberg and E. Vatne (1998) *Competitiveness, localised learning and regional development*. Routledge, London.

Metcalfe, J.S. (1998) Evolutionary economics and creative destruction. Routledge, London.

Misztal, B. (1996) *Trust in modern societies*. Polity, Cambridge.

Nooteboom, B. (1999) Innovation, learning and industrial organisation. *Cambridge Journal of Economics* 23.2, 127–50.

Odgaard, M. (1998) *The 'misplacement' of learning into economic geography*. Working Paper 132, Department of Geography, University of Roskilde.

Orillard, M. (1997) Cognitive networks and self-organisation in a complex socio-economic environment. In A. Amin and J. Hausner (eds.), *Beyond market and hierarchy: interactive governance and social complexity*, Edward Elgar, Aldershot.

Plotkin, H.C. (1994) *Darwin machines and the nature of knowledge*. Penguin, Harmondsworth.

Porter, M. (1994) The role of location in competition. *Journal of the Economics of Business* 1.1, 35–9.

Putnam, R.. (1993) *Making democracy work*. Princeton University Press, Princeton, NJ.

Sabel, C.F. (1994) Learning by monitoring: the institutions of economic development. In N. Smelser and R. Swedberg (eds.), *Handbook of economic sociology*, Princeton University Press, Princeton, NJ.

Samuels, W. (1995) The present state of institutional economics. *Cambridge Journal of Economics* 19, 569–90.

Scott, A.J. (1995) The geographic foundations of industrial performance. *Competition and Change* 1.1, 51–66.

Simon, H.A. (1959) Theories of decision-making in economic and behavioral sciences. *American Economic Review* 49.2, 253–83.

Smelser, N. and R. Swedberg (1994) *Handbook of economic sociology*. Princeton University Press, Princeton, NJ.

Storper, M. (1997) *The regional world: territorial development in a global economy*. Guilford Press, New York.

Sunley, P. (1996) Context in economic geography: the relevance of pragmatism. *Progress in Human Geography* 20.3, 338–55.

Thrift, N. and K. Olds (1996) Refiguring the economic in economic geography. *Progress in Human Geography* 20.3, 311–37.

[4]

Toward a relational economic geography

Harald Bathelt* and Johannes Glückler**

Abstract

In this paper, we argue that a paradigmatic shift is occurring in economic geo-graphy toward a relational economic geography. This rests on three propositions. First, from a structural perspective economic actors are situated in contexts of social and institutional relations. Second, in dynamic perspective economic pro-cesses are path-dependent, constrained by history. Third, economic processes are contingent in that the agents' strategies and actions are open-ended. Drawing on Storper's holy trinity, we define four ions as the basis for analysis in economic geography: organization, evolution, innovation, and interaction. Therein, we employ a particular spatial perspective of economic processes using a geographical lens.

Keywords: Economic geography, regional science, relational economic geography, ions of economic geography

JEL classifications: B41, B52, D83, L22, Z13

1. Prelude: multiple 'turns' in economic geography?

During the 1990s, a controversial debate has emerged in economic geography and other social sciences, such as economics and sociology, focusing on the question of what research program, key focus and methodology a novel economic geography should embody (Perrons, 2001). This was, partially, a reaction to the work of Krugman (1991), Fujita et al. (2001), and others who claimed to have developed a new economic geography. This self-proclaimed new economic geography offers an interesting economic perspective on the conventional problems of spatial distribution and equilibrium, based on an analysis of increasing returns, transportation costs, and other traded interdependencies (Martin and Sunley, 1996; Bathelt, 2001). Yet it fails to develop a comprehensive research program as a basis for economic geography because '... the new economic geography ignores almost as much of the reality they study as old trade theory did' (Krugman, 2000, p. 50).[1] In following Martin and Sunley's (1996) suggestion, this approach is better classified as geographical economics.

While this literature brings economic geography closer to the core ideas of neoclassical economics, Amin and Thrift (2000) have recently suggested another fundamentally different direction for economic geography, capitalizing on concepts and theories from other social sciences. Amin and Thrift (2000, p. 4) provocatively claim

---

* Faculty of Geography, Philipps-University of Marburg, Deutschhausstraße 10, D-35037 Marburg, Germany. *email:* < bathelt@mailer.uni-marburg.de >

** Institute of Economic and Social Geography, Johann Wolfgang Goethe-University of Frankfurt/Main, Postfach 111932, Dantestraße 9, D-60054 Frankfurt/Main, Germany. *email:* < glueckler@em.uni-frankfurt.de >

1   Berry (2002), one of the great thinkers in regional science, also criticizes the new economic geographers for their refusal to acknowledge previous academic contributions on similar topics. Further, their concept of space would be rather limited and partial.

118 • *Bathelt and Glückler*

that economic geography is no longer able to 'fire the imagination' of researchers. Therefore, they ask for a critical reflection and renewal of this field's basic goals, concepts, and methods. The reactions to their contribution have stimulated a debate, parts of which have been published in a special issue of *Antipode* in 2001. This debate has unfortunately been dominated by discipline-political arguments, opinions, and claims. In essence, it focuses on the question of whether economic geography should be closely associated with economics or lean towards the social, political, and cultural sciences. In particular, Thrift (2000) has identified a growing interest in the cultural dimension of economic relations, as well as in economic issues of cultural studies.

While Amin and Thrift (2000) propose a *cultural turn* away from neoclassical economics, their critics emphasize existing linkages with and the importance of economic theories as a foundation of economic geography (Martin and Sunley, 2001; Rodriguez-Pose, 2001). We agree with Martin and Sunley (2001) that this debate is partly based on false dualisms, such as economics vs. sociology and quantitative vs. qualitative methodology. In our view, this discussion is unclear because it mixes normative accounts of the discipline's policy implications with epistemological and methodological arguments. The debate is also somewhat misdirected for it tries to separate those economic and social aspects that are inseparable. The decisive question cannot be whether economic geography should be *economized* or *culturalized*. Rather, the economic and the social are fundamentally intertwined. They are dimensions of the same empirical reality which should be studied in a dialogue of perspectives rather than in mutual exclusion and reductionist prioritization (Stark, 2000). This is what we emphasize in our conceptualization of relational economic geography which we will develop in the following sections.

2. Setting the scene: two transitions in economic geography

In this paper, we wish to contribute to the discussion aimed toward a re-conceptualization of economic geography. We aim at a conceptual rather than a strategic-political renewal of the field. Based on the traditions in German geography, three broad paradigmatic approaches towards economic geography are distinguished which have developed sequentially over time. These are Länderkunde (i.e. science of regional description and synthesis), regional science (or spatial analysis), and relational economic geography. Although discussing this shift from the perspective of German geography, our intent is not to instigate a debate merely focusing on the state of contemporary economic geography in Germany.[2] Apart from this, there is a much broader discussion about the conceptualization of economic geography, as has been

---

2   Even though our thinking has evolved within the context of German economic geography, we do not limit our argument to this particular context. We refer to the German context primarily to identify recent parallels between German and Anglo-American economic geography. Our plea for a re-conceptualization of economic geography, thus, addresses a wider audience within Anglo-American geography. Already in the 1980s, Clark (1983) has presented an early version of a relational perspective based on his analysis of labor market dynamics. This study demonstrates from within the Anglo-American context that neoclassical models of discrete exchange are not capable of explaining rigidities in local labor markets. As a consequence of this, Clark (1983) suggests a relational view of contracts which does not exclusively focus on economic exchange but also includes social elements. He particularly refers to the interdependencies of unilateral power (e.g. that of a single party based on laws and regulations) and bilateral power (i.e. based on negotiations between two or more parties), with the latter being built upon the former in a contingent manner. This conceptualization also requires a consideration of the specific context within which labor markets operate (Clark, 1982).

indicated in the work of Amin and Thrift (2000), Scott (2000), Barnes (2001), and others. It is this debate to which we aim to contribute. Our intention is not to oversimplify the rich history of economic geography and state that the discipline has been characterized by three clear-cut research programs. Further, we do not intend to capture all of the different and sometimes contradictory approaches to economic geography, which have developed over time (for an overview, see, for instance, Schamp, 1983; Scott, 2000), into these broad categories to which we refer. We are, however, convinced that the paradigmatic approaches distinguished here have had and still have an important influence on the way in which economic geographers have thought of their discipline and how economic geography has been structured over time.

The conclusions which we draw from this discussion serve as a basis for the development of a novel analytical framework which we refer to as relational economic geography (Bathelt and Glückler, 2000, 2002a,b).[3] We believe that the current changes in economic geography do not unfold incrementally to the extent that existing concepts are simply being improved or updated. Since the late 1980s, a wealth of new perspectives, methods, and conceptions has developed that involves compelling ideas to be used as a basis of a re-conceptualization of economic geography. The implications of this are so fundamental that we treat them as paradigmatic and view them as constituting the *second transition*[4] of the discipline: (i) we call it a *transition* because paradigmatic shifts are neither incommensurable nor do they occur abruptly. They evolve as phases of intense dialogue between different perspectives yielding new arguments which, in turn, become the impetus for a transition towards a new perspective. Over time, this gains dominance based on consensus and 'good reason' (Toulmin, 1972); (ii) we refer to it as the *second* transition because it follows a first major transition in Germany from Länderkunde to regional science and characterizes another transition from regional science to a new paradigm which we refer to as relational economic geography.

The stages and shifts in German economic geography have been part of larger movements and shifts within the discipline. Post-war German geography has been strongly influenced by two important paradigms, that of Länderkunde (the science of regional description and synthesis) and Raumwissenschaft (spatial science).[5] Of course, as mentioned above, not all research practices can be captured by these two idealized programs. Länderkunde and spatial science have, however, substantially impacted other research practices in terms of their view of space and have influenced their

---

3   In suggesting this framework, we do *not* intend to proclaim a new, relational turn in economic geography which emphasizes some aspects while neglecting others. The proclaimed cultural, social, economic, political, and institutional turns from recent years have unnecessarily served to split the discipline into exclusive sub-fields (Ettlinger, 2001). In contrast, our conceptualization aims to integrate the economic and social aspects of human action in space which are inseparably intertwined (see, also, Lee, 2002).

4   Similar to our notion of the second transition, Barnes (2001) identifies a shift in post-War Anglo-American economic geography from first-wave theory (i.e. the quantitative and theoretical revolution) towards new-wave theory (associated with the *cultural turn* or new economic geography). While the first approach aims at objectively formalizing an independent reality, the latter one describes an interpretative mode of theorizing which is open, critical and reflexive. In our view of relational economic geography, however, we conceptualize this shift differently from Barnes (2001).

5   In other countries, such as the US, similar shifts have primarily taken place within economic and urban geography and do not necessarily correspond with a larger movement of the whole discipline. In Germany, this was somewhat different because of the integrated nature of geography as expressed in its Länderkunde tradition.

120 • *Bathelt and Glückler*

particular programs, methods and objectives. Länderkunde was developed as a science of regional description and synthesis by Hettner (1927) and represents the first academic school within German geography. Hartshorne's (1939) approach of areal differentiation resembles this view. According to Hettner (1927), the primary objective of the discipline was to classify the earth into naturally-defined Landschaften (landscapes) and understand their diverse structures through a description and synthesis of its layers of different geographies, such as bio, social, economic, and other geographies. The main purpose of economic geography was to analyse the spatial order of the economy and detect Wirtschaftsräume (economic spaces) to be included into the wider Länderkunde framework. Economic landscapes, formations, and spaces (e.g. Lütgens, 1921; Waibel, 1933; Krauss, 1933) were identified, described, and synthesized with other layers of geography. Research in economic geography was largely descriptive, listing those economic activities which take place within naturally defined landscapes.

Much later than in Anglo-American geography, this research program was criticized and challenged by new ideas from spatial science. This resulted in open conflict and culminated in the debates and controversies at the Meeting of German Geographers in Kiel in 1969 (Meckelein and Borcherdt, 1970, pp. 191–232). In this critique, Länderkunde was heavily criticized for its lack of a sound epistemological basis and for having a largely ideographic, descriptive, holistic, and naturalistic program. Drawing upon the work of Anglo-American spatial science (e.g. Haggett, 1965; Chorley and Haggett, 1965, 1967), which developed during the 1960s, Bartels (1968) developed a new conceptualization, i.e. Raumwissenschaft. This became the basis of a new paradigm in geography which lasted at least throughout the 1970s and the 1980s.[6] In the program of this science of space, mere description was replaced by analytic explanation and the naturalistic conception of space as a particular landscape by an abstract conception of space as a formal geometry (Bartels, 1970; Bahrenberg, 1972).

In *economic* geography, this first transition gave rise to Raumwirtschaftslehre (Schätzl, 1998), i.e. regional science (or spatial analysis).[7] Its research program was strongly impacted by Isard's (1956, 1960) influential work in American economic geography. In the late 1950s, Isard (1956, 1960) established regional science as science of spatial order and organization of the economy and drew attention towards theories and models in neoclassical economics (von Böventer, 1962). Economic theories were incorporated into economic geography through the integration of spatial variables, i.e. the cost of transit and transport over distance. It was the objective of regional science to develop general theories and models of the spatial order of the economy. Location patterns, trade relations and processes of agglomeration were typically explained using spatial parameters, such as distance, catchment areas and their economic equivalents (Voppel, 1999).

In the following section, we illustrate some of the analytical and methodological limits of the regional science approach through the use of two simplified examples. From this, we develop some ideas for re-conceptualization. Section 3 lays out the foundations of an alternative research design for economic geography which is based

---

6   Unlike in the US, Marxist approaches did not gain much importance in post-war geography in Germany. Although many German geographers were aware of the work of Harvey (1982), it was not quoted and reflected upon in influential economic geography textbooks (e.g. Schätzl, 1998).

7   We use the term regional science (or spatial analysis) to refer to those views of economic geography which are associated with the quantitative and theoretical revolution and which took place in the late 1950s and the 1960s in American geography.

on a relational perspective of economic action. Related to recent work in economic and social geography, we discuss the fundamental dimensions of a relational economic geography. This also incorporates ideas from critical realism. We suggest this framework as an alternative to regional science. Section 4 discusses a recent conceptualization by Storper (1997a,b) in which he formulates the foundations and goals of economic geography. Drawing from this reformulation, we introduce four core concepts or ions of analysis in economic geography in Section 5, i.e. organization, evolution, innovation, and interaction. They are conceptualized as relational categories for a novel economic geography (Bathelt and Glückler, 2002a,b). We apply this conceptualization to those areas of economic geography that we are familiar with through our own personal experience. However, we would like to invite others to try to use this conceptualization in other areas of economic geography. Section 6 summarizes our main arguments and draws some concluding remarks.

3. Limits of regional science and ideas for re-conceptualization

Since the late 1980s, a new set of ideas, conceptualizations, and models has been published which have come to form a counterweight against the regional science approach in economic geography. This work is characterized by contrasting perspectives and enlarged complexity in the analysis of economic and social processes compared to that in regional science. Much of the critique expressed in this work can be illustrated through the simplified examples discussed below (for further examples of methodological problems in regional science, see Sayer, 2000).[8]

3.1. Example 1: spatial characteristics and the acting region

A classical research focus of regional science is to explain why some regions grow faster than others. In order to provide an answer to this question, a number of consecutive stages are conducted in a spatial analysis. First, spatial characteristics are identified, such as indicators of the regional infrastructure, labor force, and other resources, as well as cost factors, such as wage levels. In the second stage, statistical analyses (e.g. correlation and regression analysis) are conducted to reveal which features are typical for both growing and shrinking regions. Often, causal mechanisms are derived from this analysis. It is argued, for instance, that low regional costs stimulate high regional growth. A major problem of this approach is that regions are treated as if they were the economic actors themselves having their own particular characteristics. One policy conclusion which often results from this view is that regions should lower their costs in order to stimulate growth.

However, such an argument neglects the fact that regions are not real actors. They are socially constructed entities, dependent on the particular economic, social, cultural, and political settings and realities under which people in firms and other organizations act and interact (e.g. Maskell, 2001). Sometimes, even one large dominant firm might

---

8 We are aware that these examples provide a simplified view of regional science; i.e. one, which does not deal with all aspects and complexities of this literature. Our intention is not, of course, to deny the historical importance of this work and its path-breaking findings. In empirical studies, the contrast between regional science and relational economic geography may not always be that evident at first sight. We think, however, that the paradigmatic reconstruction that we draw is helpful to identify those problems and dimensions which are in need of re-conceptualization (for a more nuanced history of approaches in economic geography; see Scott, 2000).

122 • *Bathelt and Glückler*

cause regional growth or decline simply through its linkages with other regional agents (Romo and Schwartz, 1995). This might be a consequence of a change in the firm's overall global market strategy and may not be related to the actual locational characteristics identified in the region (e.g. Schamp, 2000). Massey (1985, p. 11) has criticized the inappropriateness of this science of the spatial as follows: 'There was an obsession with the identification of spatial regularities and an urge to explain them by spatial factors. The explanation of geographical patterns, it was argued, lay within the spatial. There was no need to look further.... This is an untenable position.... There are no such things as purely spatial processes; there are only particular social processes operating over space'.

3.2. Example 2: location analysis and spatial incentives

Another focus of regional science is the analysis of spatial distributions and location decisions (Isard, 1956; Bartels, 1988; Schätzl, 1998). The location analysis of a sector involves a particular methodology. In the first stage, the locational requirements of the firms of this sector are listed. In the next stage, location factors and particular features of each region are identified. Finally, the locational requirements of the firms and the locational characteristics of the regions are systematically compared with one another to find the best match. The assumption behind this approach is that firms, in order to maximize their profits, choose exactly that location which can satisfy their requirements best, based on its inherent characteristics.

In this example spatial attributes are again used as explanatory variables which are supposed to explain location decisions and spatial distributions. Barnes (2001, p. 550) refers to these practices in quantitative regional science as fetishization because 'the social processes that actually produce such figures are hidden'. Also those aspects of social power which lay behind quantitative representations remain concealed. In contrast, recent work in economic geography has come to realize that firms do not just act according to spatial attributes but that they themselves create such spatial attributes in the first place (Scott, 1998). In their model of geographical industrialization, Storper and Walker (1989) deal with the question of how industries create their respective regions through the regular training of employees, recruitment of expertise from outside, support of newly established suppliers, outsourcing to other local suppliers and services, and learning processes with nearby customers. In this view, spatial structures are seen as being socially constructed. They result from complex interactions between regional actors and groups of actors, such as firms. Through the process of reflexive knowledge-creation, the localized structures, in turn, exercise an influence on economic decision-makers and their respective actions and policies (Storper, 1997a).

Recent work in economic geography, such as that of Amin (1994), Lee and Wills (1997), Barnes and Gertler (1999), Bryson et al. (1999), Sheppard and Barnes (2000), and Clark et al. (2000), has resulted in new approaches and ways of thinking which contrast with the traditional regional science view.[9] One could argue that the new views and approaches do not yet establish a well-defined, fully coherent theoretical framework (e.g. Barnes, 2001). What these studies share, however, is a critique of the

---

9   The approach suggested in this paper might, however, not be the only alternative to regional science (for different paths, see Webber and Rigby, 1996; Sheppard and Barnes, 2000). It is, however, well-suited to overcome some of the shortcomings of the regional science approach and integrate a large number of studies from the 1980s and 1990s which have moved away from regional science.

traditional approach and a strong interest in understanding localized economic and social processes. In addition to economic theory, social theories and conceptualizations are also applied in these approaches to explain localized economic phenomena as an expression of both the economic and the social. From these trends, we argue for a comprehensive transition in economic geography toward a relational conceptualization which we aim to characterize in the next sections.

4. Research design for a relational perspective

The second transition is characterized by a reformulation of the core concepts of economic geography. In the following sections, discontinuities between relational economic geography and regional science will be identified according to five dimensions of the research design. These dimensions include the conception of space, object of knowledge, conception of action, epistemological perspective, and research goal. From this, we develop a relational framework for analysis which systematically focuses on economic actors and their action and interaction. The basic propositions of this framework will be developed in the remainder of this section (Table 1).

4.1. Conception of space

A relational view of economic geography is based on a relationship between space and economy which is contrary to that of regional science.[10] Specifically, regional science views space as a container which confines and determines economic action. It treats space as a separate entity which can be described and theorized independently from economic action. In contrast, a relational approach assumes that economic action transforms the localized material and institutional conditions of future economic action. Similar to Storper and Walker (1989), this approach emphasizes that the economic actors themselves produce their own regional environments. The way in which spatial categories and regional artifacts have an impact on economic action can only be understood if the particular economic and social context of that action is analysed (Bahrenberg, 1987). Spatial structures and processes have, however, been socially and economically under-conceptualized in regional science. We contend that space can neither be used as an explanatory factor for economic action nor be treated as a separate research object in isolation from economic and social structures and relations. Consequently, as space is not an object of causal power to explain social or economic action it cannot be theorized (Sayer, 1985; Saunders, 1989; Hard, 1993).[11] Of course, economic processes also have material outcomes (e.g. infrastructure) which are localized in certain places and territories and exist over longer time periods. Such structures clearly have an impact on economic action and interaction in these localities.

Nonetheless, economic actors and their action and interaction should be at the core

---

10 This shift in causality between space and human action has also been emphasized in human geography (Werlen, 1993, 1995, 2000).

11 One could argue that the economy, like space, is also a social construct and yet we have economic theories. Why then should there not also be spatial theories? The key to this question lies within the very nature of relational action. There are economic agents which develop strategies and act according to economic and non-economic goals. Their action and interaction has intended and unintended spatial outcomes. There are, however, no spatial agents which act according to spatial goals. Intentions for actions derive from the actors, not from spatial representations.

124 • *Bathelt and Glückler*

Table 1. Changing research designs in the paradigms of German economic geography

Dimensions of the research design	Economic geography in Länderkunde	Regional science (or spatial analysis)	Relational economic geography
Conception of space	Space as object and causal factor	Space as object and causal factor	Space as perspective (geographical lens)
Object of knowledge	Specific economic-space formations of a landscape	Spatially manifested consequences of action (structure)	Contextual economic relations (social practice, process)
Conception of action	Environmental determinism/possibilism	Atomistic: methodological individualism	Relational: network theory/embeddedness-perspective
Epistemological perspective	Realism/naturalism	Neo-positivism/critical rationalism	Critical realism/evolutionary perspective
Research goal	Ideographic understanding of the nature of a landscape	Discovery of spatial laws of economic behavior	De-contextualization of principles of socio-economic exchange in spatial perspective

of a theoretical framework of economic geography and not space and spatial categories. Spatial scientists, such as Bunge (1973), treat spatiality as the object of knowledge in economic geography. They aim to detect those spatial laws which govern human action without looking at the actors themselves. Instead of treating space as a container, we suggest a conception of space as *perspective* (Glückler, 1999). In other words, we use space as a basis for asking particular questions about economic phenomena but space is not our primary object of knowledge. It is this conception that we refer to as the geographical lens. As part of this, economic exchange becomes the focus of analysis and not space. Similarly, we do not seek to identify spatial laws but, instead, look for explanations of localized economic processes and their consequences.[12]

It is particularly through the application of a distinct perspective to the study of an object of knowledge that discipline-specific research problems can be formulated. The spatial perspective or geographical lens leads economic geographers to pose research questions about an economic phenomenon, different from those typically asked by economists or sociologists. We also suggest that the perspective applied helps mobilize a particular terminology and, over time, a set of tacit knowledge which entails an understanding of what it is that is being analysed and how this subject matter can be described and evaluated adequately.

---

12 The term 'localized' does not imply that all economic action is locally or regionally bound but, instead, physically materializes in place. All action and interaction takes place somewhere; for instance, within particular places, regions, nations and trading blocks but also between them. The key is that economic action includes people in some places and excludes others. Because any economic activity takes place somewhere, it necessarily interacts with other economic and social processes which take place in the same places. This is because the same agents participate in various processes at the same time and because different processes, in part, involve the same group of agents. It is therefore not that easy to isolate a particular process and neglect others. Processes are necessarily interdependent, either because they take place within the same region or exactly because they do not. It is exactly this fundamental spatial overlap of a multitude of social and economic processes that our geographical lens aims to capture.

A brief example may help to illustrate the idea of the geographical lens. Depending on the disciplinary perspective used, observing changes in a firm's division of labor due to a structural crisis may cause researchers to formulate different sets of questions: a sociologist would, for instance, focus on the consequences of this crisis on the distribution of responsibilities and competencies within the work organization. In contrast, an economist may analyse the effects of this crisis on the firm's strategy, production program, or new market opportunities. An economic geographer would employ a particular spatial perspective and thus typically investigate different localized aspects and consequences of this crisis. Through her geographical lens, she would for example analyse the impacts of this on the local labor market, supplier relations, or the division of labor between different sites and locations. Of course, this example is a simplification. The point is that each perspective can shed some light on parts of a phenomenon but necessarily neglects other issues. Any research findings which result from the use of a particular perspective can themselves be used as a point of departure for new research applying a different perspective. Thus, a sociologist might use the results from a geographer's study on the localized consequences of the labor market to develop further research questions about social inclusion or exclusion. An economist, in turn, might investigate it in terms of welfare and efficiency effects and so forth. This exemplifies that space is not, in and of itself, the object of knowledge in economic geography.

4.2. Object of knowledge (or research object)

In contrast to regional science, a relational approach does not accept the space-economy, spatial systems, or spatial categories as being the core focus of a research program in economic geography. Instead, economic action and interaction are the central object of knowledge in the analysis. Thrift (2000, p. 698) has criticized a lack of explanation in traditional economic geography and demands that '…economic geographers cannot just be tied to the locational dimension as under-labourers for economists, noting down the "wheres" whilst economists do the "whys"'. Consequently, any analysis in relational economic geography is based on an understanding of the intentions and strategies of economic actors and ensembles of actors and the patterns of how they behave. Economic action is viewed as being embedded in structures of social (and economic) relations and is thus conceptualized as a context-specific process. Research in relational economic geography thus focuses on processes, such as institutional learning, creative interaction, economic innovation, and inter-organizational communication, and investigates these through a geographical lens, rather than uncovering spatial regularities and structures. Economic processes and relations broadly defined are at the heart of this approach which integrates (and requires) both economic and social theory.

4.3. Conception of action

Similar to neoclassical economics, regional science employs an atomistic view of economic agents (for a discussion of agency and space, see Sheppard, 2000). Often, economic agents are seen to act in isolation from other agents and their institutional environments according to the rationale of a homo economicus. In contrast, a relational framework emphasizes the importance of contextuality of human action.

126 • *Bathelt and Glückler*

Economic action is embedded in structures of ongoing social relations (Granovetter, 1985; Grabher, 1993).

From the view of the geography of the firm, this means that firms are not independent entities but are closely interconnected in communication and adjustment processes with their suppliers, customers and institutions and must be analysed accordingly.[13] An atomistic view of economic agents thus leads to a very limited understanding of their activities as context is neglected (Granovetter, 1992b). The socio-institutional context creates opportunities for economic action and interaction that would otherwise not exist. From these opportunities, agents develop new goals and strategies for action. A relational approach in economic geography thus systematically includes context into its research program.[14] As a consequence of this, particular attention is paid to economic action as a social process and the structure of relations between agents and the creation of formal and informal institutions.[15] In sum, economic action is a process, situated in time and place (Philo, 1989; Giddens, 1990; Martin, 1994, 1999; Sunley, 1996; Bathelt and Glückler, 2002a,b; Glückler, 2001).

4.4. Epistemological perspective

Since human action is viewed as contextual, it cannot be explained by universal laws. Action in open systems is not fully predictable and, thus, cannot be adequately conceptualized through deterministic theory (see, also, Peck, 1996). This is, however, what approaches in regional science aim at when they apply existing or identify new spatial laws and regularities of economic action. In contrast, critical realism provides a fundamentally different epistemological perspective of causality in that it systematically accounts for context-specificity in human action (Archer et al., 1998).[16] Critical realism

---

13 Maskell (2001) has criticized that traditional economic geography often treats firms like a 'black box'. To overcome this he suggests the use of a resource-based or competence-based view of the firm (see, also, Wernerfelt, 1984; Prahalad and Hamel, 1990). Nonaka et al. (2000) explicitly include the particular role of knowledge and develop this further into a knowledge-creation view of the firm. These views correspond nicely with a relational perspective of economic action as portrayed in this article.

14 This is particularly important when analysing the growing importance of knowledge within the economy (Lundvall and Johnson, 1994). Knowledge is context-specific and bound to people. In order to understand the process of knowledge-creation, it is necessary to understand how people act and interact with one another (Bathelt and Glückler, 2000). New knowledge is being created through interactive processes of interpreting, integrating and transforming existing knowledge within a specific context. Nonaka et al. (2000) use the Japanese notion of 'ba' to refer to the organizational context of knowledge-creation, related to the particular time and place where interaction occurs.

15 The role of institutions in shaping the context of economic agents has been recognized in much of the recent literature on innovations. Institutional contexts which are determined at the level of the nation-state, such as the educational and research infrastructure, work rules, and organizational standards with respect to social division of labor and the like, have a substantial influence on the nature of economic problems and shortages identified and the learning processes applied. As a result, particular national systems of innovation develop (Lundvall, 1992; Nelson, 1993). In these systems, institutions have a strong impact on interaction and innovation and, thus, shape the national production structure. The production structure, in turn, strengthens and reshapes existing institutions and develops them further, thus driving national specialization patterns (Lundvall and Maskell, 2000). Even further, such approaches need to consider aspects of cultural context, difference, 'othering', etc. in order to explain different production patterns more carefully (e.g. Saxenian, 1994; Schoenberger, 1997; Thrift, 2000).

16 Critical realism is not necessarily the only methodological alternative toward a relational economic geography (see, also, Harvey, 1996). It provides, however, a particularly well-suited conceptualization to integrate the key concepts 'context' and 'contingency' into a coherent framework which allows to overcome a number of structural problems associated with the regional science approach.

was developed by the British philosopher Bhaskar (1975) and propagated in the social sciences by Sayer (1992, 2000). It serves as a pragmatic epistemological alternative which attempts to avoid the problems of both deductive-nomological determinism in logical empiricism and relativism in postmodern theory (Lovering, 1989; Thrift, 1990; Sayer, 2000). In contrast to postmodern approaches, critical realism maintains the assumption that an objective reality exists which is independent from the individual. The relation between reality and human knowledge about it is, however, asymmetrical. The fact that empirical observations are necessarily mediated through concepts does not imply that they are a product of these concepts only. Instead, these observations are also dependent on the structural properties of the real objects (Sayer, 2000, p. 41). Critical realism also aims at developing causal explanations for general mechanisms. In contrast to positivist approaches, however, causality is no longer implied from their universal co-occurrence.

Conventional causal analysis as employed in regional science is based on Hume's (1758) principle of regularity. In this principle, an event is the cause of another subsequent event, if its occurrence is always associated with the occurrence of the latter event. Constant conjunction here is used as an associative principle of causality (Sayer, 2000). This explanation claims to be universal for it assumes that an event has particular consequences which occur at any time and any place in association with this event. In contrast, critical realism establishes a contextual explanation based on the principle of contingency. This approach distinguishes two types of relations between events (Sayer, 1985):

1. *Necessary relations.* Relations are necessary, if two events always occur in association with one another, independent from a specific context. Such non-contextual relations or universal laws are, however, extremely rare in social and economic processes (Fleetwood, 2002).
2. *Contingent relations.* Relations are contingent, if two events occur in conjunction with one another only under specific circumstances. Such relations are quite typical in the analysis of economic action using a geographical lens.

The principle of contingency states that one event does not *necessarily* cause another particular event. Therefore, identical preconditions for human action do not necessarily have the same consequences at any time and place. This provides an epistemological basis for a context-specific conceptualization of the intentions and consequences of human action. At the same time, it is recognized that future actions and development are fundamentally open-ended.[17] Contingency, however, does not result from human action alone, but also arises from differences in the localized material structure, as well as from variations between places and territories in the institutional architecture.[18] This creates deviations between

---

17 One could argue that the inclusion of stochastic processes in regional science also allows for open-ended, contingent development (e.g. Curry, 1967; Cliff et al., 1981). While this is certainly the case we would like to emphasize that this is a more formalistic view of contingency which does not provide a closer insight into the nature of economic action and interaction at work.

18 Particularly during the 1980s and early 1990s when 'new' regional configurations of industrial districts and industrial spaces were discovered and discussed, this encouraged researchers and policy makers, at least implicitly, to seek for general models of regional development. In a way, this was like searching for universal forms and general laws of spatial economic development and, thus, created similar problems to those of some earlier regional science work. We owe this argument to one of the reviewers.

128 • *Bathelt and Glückler*

regions, nations, etc. and results in different sets of opportunities and restrictions for economic action. Overall, this leads to particular structure–agency dynamics.[19]

The application of this concept does not mean, however, that research ends with a contextual explanation of singular events in particular locations and circumstances at a given time. Instead, another important step of realist analysis is to go beyond individual events and their specific contexts in order to identify common aspects of the causal mechanisms that affect economic action. This involves the identification of the causal mechanisms which are at the heart of localized economic action and interaction as opposed to the formulation of spatial laws. This methodology aims to uncover basic conditions of specific contexts and relate them to others. In this way, de-contextualization provides a methodology to identify trans-contextual, more-or-less necessary circumstances and structures from contextualized events.

4.5. Research goal

Relational economic geography explicitly draws attention to the importance of economic agents and how they act and interact in space, instead of focusing on the description of spatial categories, processes, and regularities. This has been accomplished through the application of the following changes to the research design: first, this has been achieved by inverting the causality between space and economy and adopting a conception of space as perspective (i.e. the geographical lens); second, this approach views economic action as a relational process which is situated in structures of relations; third, the principle of contingency has been introduced into the analysis of causal relations to establish an epistemological basis which accounts for contextual action and development. The objective of relational economic geography is to formulate research questions which are associated with the analysis of economic relations using a geographical lens (Bathelt and Glückler, 2002a,b).

The transition from regional science toward a relational approach, which we have aimed to illustrate in this section, has fundamental consequences for analysis in economic geography. It rests on three propositions:

1. *Contextuality.* From a structural perspective, economic agents are situated in contexts of social and institutional relations (Granovetter, 1985, 1992a,b). Since this conceptualization views action as being embedded in specific contexts, it cannot be explained through the application of universal spatial laws.
2. *Path-dependence.* From a dynamic perspective, contextuality leads to path-dependent development because yesterday's economic decisions, actions and interactions enable and constrain the context of today's actions. They also direct future intentions and actions to some extent (Nelson and Winter, 1982; Nelson, 1995).
3. *Contingency.* Economic processes are at the same time contingent in that the agents' strategies and actions may deviate from existing development paths. Economic action in open systems is not fully determined and cannot be predicted through

---

19 Contingent action reproduces or transforms specific contextual structures which, in turn, shape the preconditions for future action. In this respect, context is related to structure and contingency associated with agency or, as Jessop (1992) would say, strategy. The resulting interdependence between contextual structure and contingent action corresponds with those reflexive mechanisms at the heart of structuration theories, as developed by Giddens (1984) and Bourdieu (1977).

universal spatial laws. Despite its path-dependent development which provides a particular history, economic action is subject to unforeseeable changes and is therefore fundamentally open-ended (Sayer, 1992, 2000).

Relational economic geography enables a complex understanding of economic action and its localized consequences. It focuses on those people, firms, institutions and other organizations which are involved in economic decision-making, as well as on those people and environments which are subject to the consequences of economic action. This relational perspective does not intend to identify spatial regularities and avoids treating regions and other spatial configurations as actors. Instead, the strategies and objectives of economic agents and their relations with other agents and institutions are the core of the analysis.[20] A contextual, path-dependent and contingent perspective is quite different from other theoretical programs which view economic geography in line with universal laws, linear developments, and closed systems. Therefore, a transition from regional science to relational economic geography requires a reformulation of the concepts used to understand economic structures and processes. In the following sections we lay out the basic propositions and concepts of an alternative framework for economic geography.

5. Storper's conceptualization of the holy trinity

The most sophisticated attempt to reformulate the foundations and goals of economic geography is that developed by Storper (1997a,b). His conceptualization of the holy trinity in economic geography serves as our reference and point of departure. We integrate his ideas into a general model of relational action and interaction as a basis for a re-conceptualized economic geography (Bathelt and Glücklcr, 2002a,b).

Storper (1993, 1997c) argues that localized production systems continue to play a decisive role in the global economy despite revolutionary improvements in information, communication, and transportation technologies. He suggests that this continued importance of proximity is due to the advantages from reduced transaction costs and enhanced capabilities for organizational and technological learning in specialized agglomerations of interrelated economic activities. Apart from traded interdependencies, which are the key variables in the regional science literature, untraded interdependencies play a decisive role to enable communication, adjustment, and learning processes between economic agents (Storper, 1997a). They are embodied in the role of relations between particular people and the existence of conventions as expressed in accepted norms, rules, and practices. Relations and conventions are also localized because they are bound to those people, firms, and places involved and cannot easily be transferred to other places. They become region-specific assets and form the basis for further concentration and specialization of economic activities (Maskell and Malmberg, 1999a,b).

To understand the complex nature of economic production and its geography, Storper (1997a,b) identifies technology, organization, and territory as overlapping

---

20 The use of regression analysis provides a good example to illustrate the problems with context and contingency in regional science. The use of such tools is based on the assumption that the same conditions always result in identical outcomes (i.e. one particular value of the explanatory variable necessarily produces the same estimate of the dependent variable over and over). However, reality is different from this. Firms which operate under similar conditions in the same sector and region might pursue rather different strategies because they are embedded into a different market logic and context.

130 • *Bathelt and Glückler*

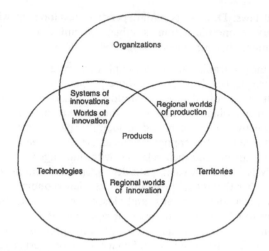

Figure 1. Storper's holy trinity (*Source:* Storper, 1997a, pp. 42 and 49).

constituent pillars of economic geography. They form a holy trinity through which economic and social processes and their interactions and power relations can be analysed.[21] This holy trinity serves as a conceptualization of economic geography radically different from regional science (Figure 1): (i) *Technology*. Technological change lies at the heart of the dynamics of the economy. It results in the rise of new and the decline of old products and processes; (ii) *Organization*. This pillar emphasizes the importance of the ways in which firms and networks of firms are organized and how these organizational structures are impacted by institutions; (iii) *Territory*. At the territorial level, it is possible to analyse the co-evolution of organizations and technologies. Through regional input–output linkages, knowledge transfers, and adjustments between firms, spillover effects and learning processes occur which enhance the collective competitiveness of those regional agents which are interrelated in the same value chain. Therein, untraded interdependencies play a decisive role in transforming technological and organizational worlds into regional worlds (Storper, 1993, 1997a).

An important aspect of Storper's (1995, 1997a) approach is that he emphasizes the role of context-specific institutions. He also views social interactions, as expressed in processes of organizing, learning, and knowledge-creation, as being central to analyses in economic geography. Storper (1997a,b) identifies mechanisms through which socio-institutional contexts stimulate processes of geographical clustering of industrial production and provides an explanatory framework which concentrates on economic agents as opposed to their spatial settings. We believe, however, that this conceptualization also bears the risk for misinterpretations with respect to its implicit spatial foundations:

1. *Overemphasis of the spatial dimension.* In Storper's (1997a,b) holy trinity, the territory forms a pillar of its own in addition to the dimensions of organization

---

21 Crevoisier (2001) uses a similar conceptualization when analysing creative milieus through the complex interrelationships and dynamics of co-existing technological, organizational and territorial paradigms.

and technology. This implies that the three have the same status and importance within this conceptualization of economic geography. We suggest that it is necessary to exercise care in this respect and not to treat economic and social processes in a similar way as spatial processes. While organization and technology can be conceptualized in a meaningful way, the metaphor of territory could provide a temptation for regional science-based interpretations which try to develop theories of space, in the way criticized by Massey (1985). We suspect this is not Storper's (1997a,b) intention. Instead of using the concept of territory as a constituent pillar of economic geography, we conceptualize space as perspective (Glückler, 1999). This perspective serves as a geographical lens which guides the analysis of economic and social processes in economic geography (Bathelt and Glückler, 2002a,b).

2. *Isolation of the geographical perspective from the economic and the social.* The conceptualization of territory as a separate pillar of economic geography is associated with further disadvantages. It serves to restrict the analysis of spatial processes and structures to a single analytical dimension, as well as its overlapping areas with those of other dimensions. This could lead some to view organization and technology as abstract dimensions which are not localized. To avoid such interpretations, we build our conceptualization of economic geography around the geographical lens because it enables us to contextualize all processes from the very beginning of our analysis. In this conceptualization, territory can hardly be treated as a separate entity.

6. Four ions of a relational economic geography

In this section, we aim to develop the criticisms, new ideas and approaches mentioned above into a novel, relational framework for analysis in economic geography based on Storper's (1997a,b) conceptualization of the holy trinity. This framework revolves around the basic concepts of organiza*tion*, evolu*tion*, innova*tion*, and interac*tion*. We refer to these as the four 'ions' of a relational economic geography. These concepts are founded on the relational perspective developed above. In our conceptualization, social institutions are of great importance in understanding and explaining context-specific behavior and action. We refer to numerous institutional concepts, such as those discussed by Storper (1997a), Schamp (2000), and others. They help to understand the mechanisms through which all ions are being constructed and reconstructed and serve as basic mechanisms to develop our relational framework.

The point of departure of our conceptualization is that those economic and social processes which drive the four ions are to be analysed and evaluated using a distinct geographical lens (Figure 2). This allows us to develop an interdisciplinary approach to economic geography which integrates both economic and social theories. In the following sections, we will indicate which research problems are associated with each of the ions. In applying a geographical lens, we will also demonstrate that these ions are closely interrelated with one another (Bathelt and Glückler, 2000, 2002a,b). The structure of the four ions, which is employed in Figure 3, serves as a heuristic framework to systematically apply the consequences of a relational perspective to the

132 • *Bathelt and Glückler*

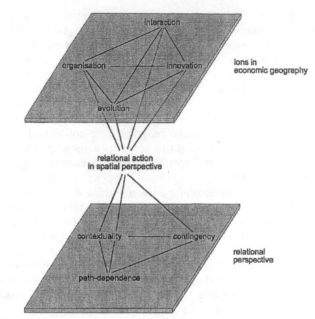

Figure 2. The four ions of economic geography in a relational perspective.

theoretical debates in much of the work of economic geographers today, particularly with respect to the geography of the firm.[22]

6.1. Organization

A basic problem of organizing industrial labor and production processes is to establish an efficient division and integration of labor (Sayer and Walker, 1992). This involves the coordination of the labor force, raw materials, intermediates, machinery, and equipment applied within and between workplaces and firms. In essence, it includes the establishment of a particular social and spatial division and integration of labor. This has to be done in such a way that sufficient control can be exercised over the production process to continuously produce the same goods at high quality standards according to customer needs (Bathelt, 2000). To solve the problem of industrial organization, decisions have to be made with respect to which process technologies will be used and the way in which the labor and production stages will be bundled together. This also includes decisions regarding which intermediate products will be produced in-house and which will be acquired from subcontractors and suppliers. If some vertical disintegration is intended, which is usually the case, the next questions to be answered are: which suppliers from which regions will be contacted; which competencies will they be given; and in which places, regions, and nations

---

22 The relational perspective proposed in this paper suggests that the research questions asked are themselves contextual, depending on the context of a particular discipline focus. Here, the geography of the firm (Dicken, 1990) serves as an example to demonstrate the consequences of a relational approach. This also means that a different pattern of ions could result if another discipline focus was applied. Different contexts might, for instance, be derived from feminist, labor market, or political economy literatures (for interesting accounts, see Harvey, 1982, 1996; Peck, 1996; McDowell, 2000).

will which parts of the production chain be located. These and other aspects of industrial organization can be analysed through the application of institutional theories, such as the transaction cost approach in economics (Coase, 1937; Williamson, 1975, 1985) and the embeddedness approach in social sciences (Granovetter, 1985, 1992a,b).

A firm's particular solution to this organization problem has a direct impact on the locational structure of the firm and the spatial organization of its production. In turn, the spatial distribution of potential suppliers and customers, as well as the strategies and decisions of major competitors, also have an impact on the resulting social and spatial division of labor. Overall, the organizational problem of industrial production is so complex that it is not possible to explain its outcome simply through the use of location factors and regional attributes. Spatial proximity and institutional affinity (or proximity) can in some technological and political contexts help to stabilize network relations between specialized firms because they reduce information costs, generate information spillovers and enable more efficient communication (Scott, 1988, 1998). Depending on the institutions which shape industrial relations and organization, the existing training and education system and other aspects of the capital–labor nexus, regional or national agglomerations can result.

These organizational structures are, however, not static ones. An evolutionary perspective is required to understand the dynamic nature of the organization of firms and value chains as a result of socio-institutional relations and their changes (Nelson and Winter, 1982; Swedberg and Granovetter, 1992). Whether a firm develops a vertically integrated production structure or whether it engages in vertical disintegration and, if so, in which regions subcontracting relationships will be established also depends on its experience and the particular sequence of organizational decisions made in the past. Learning from experience results in enhanced organizational reflexivity.

At the same time, organizational structures are embedded in social, cultural, and institutional structures and relations which cannot be separated from the economic sphere. The existence of accepted rules, habits, norms, and other institutional arrangements creates a reliable environment for interactive learning (Hodgson, 1988; North, 1991) and has a direct impact on the organization of innovation and production processes. The organizational structure of a firm and its development is also related to formal institutions and authorities which create societal standards and laws (e.g. Baum and Oliver, 1992). Therefore, the spatial organization of production is a result of complex negotiations and temporary compromises between firms and formal institutions and authorities and takes place within the context of particular power relations (e.g. Taylor, 1995; Allen, 1997; Berndt, 1999).

6.2. Evolution

The evolutionary dimension[23] is of great importance because it allows for the analysis of the impact of historical structures and processes on today's decisions. Evolutionary

---

23 The inclusion of this ion does not, of course, imply that we prioritize time over space. Even though both are treated in a similar way, the spatial perspective serves as a key concept for analysis in economic geography to relate all ions to one another. Despite its importance, the evolutionary dimension has often been neglected in neoclassical economics and regional science. We are aware, however, that early antecedents of evolutionary thinking exist, such as Myrdal's (1957) and Hirschman's (1958) work on cumulative causation in regional development and Kaldor's (1970) contribution on the role of increasing returns in regional development paths, albeit that this work does not reflect the relational approach presented here.

concepts of change assume that economic and social processes are experience-based, cumulative, and reflexive in nature. They are path-dependent in that they follow particular histories of decisions, actions, and their consequences. In this respect, there is a remarkable convergence of evolutionary perspectives in economics, sociology, and economic geography (Bathelt and Glückler, 2000).

Evolutionary economics supposes that techno-economic change defines a development path which follows particular routines and heuristics (Nelson and Winter, 1982; Dosi, 1982, 1988; Nelson, 1995). As part of this, the direction of technological change is pre-structured by existing technologies, albeit not in a deterministic way. Existing technologies are products of past decisions made about innovation and previous technologies. Through this, processes of selection, mutation, variation, and chance are initiated which aim at the creation of knowledge and new technologies to increase efficiency. Past choices generate potentials for present actions and at the same time limit the set of feasible solutions because old decisions cannot be easily reversed.

New approaches in economic sociology have extended this techno-economic view by applying aspects of socio-institutional embeddedness (Zukin and DiMaggio, 1990). Granovetter (1985) has pointed out that economic activities are deeply embedded in structures of social relations. This means that firms cannot be analysed as independent entities but must be viewed within their respective socio-economic contexts; that is, they are closely interconnected in communication and adjustment processes with their suppliers, customers, service providers, and state authorities (Grabher, 1993).

This is closely related to the role of institutions. Institutions do not only restrict the possibilities of economic action (North, 1991). More importantly, they also create a basis for mutual communication, collective learning, and joint problem-solving, without which a technical and social division of labor and economic interaction would not be possible (Giddens, 1984; Hodgson, 1988). Formal and particularly informal institutions, such as conventions, accepted rules, and habits, are of great significance because they invigorate and further stimulate localized production systems at different spatial levels. Embeddedness is not spontaneous but is experience-based and develops over time from a historical process. Contextual economic relations are the result of previous and ongoing experience in durable economic transactions. Through the same process, they also form the basis for future path-dependent and context-specific developments. In an evolutionary perspective, informal institutions can be materialized and transformed into formal institutions, such as laws. They are then typically integrated into an organizational context (e.g. a government agency) from which they are inseparable (Amin and Thrift, 1994).

Recent studies in economic geography which try to explain the rise of new industrial spaces and the process of geographical industrialization also tend to be evolutionary in their view (Scott, 1988; Storper and Walker, 1989). These studies integrate findings from evolutionary economics and the embeddedness approach into a specific spatial development perspective, albeit sometimes in an implicit way. In their model of geographical industrialization, Storper and Walker (1989) argue that novel industries have few specific locational requirements in their early growth stages. New firms are thus relatively free in their decision where to locate. This occurs because the particular inputs required do not exist anywhere and, instead, must be acquired from whatever materials are at hand. Thus, numerous regions exist which could house the industry. Later on, a handful of regions come to specialize in this industry, driving selective clustering. Due to their superior growth performance, they are able to actively shape their locational environment according to their needs. For instance, they stimulate the development of a specialized supplier sector and create a labor market which fits their particular skill requirements. The firms in these clusters thus develop a competitive advantage over those in other regions. This supports further agglomeration and specialization in the existing clusters. A more complex understanding of industrial and regional development paths and their socio-institutional contexts can be achieved if the effects of localized capabilities (Maskell and Malmberg, 1999a,b) and untraded interdependencies (Storper, 1995, 1997a) are included in the analysis (Bathelt, 2002).

6.3. Innovation

This dimension is closely associated with processes of knowledge-creation, the development of new technologies, and the effects of technological change, especially in a spatial perspective. Many traditional concepts in economics and geography fail to properly understand the processes of generating new products and processes and introducing innovations to established markets. Technological change is either viewed as a given, being external to the models used, or portrayed as a predictable outcome of a linear research process which follows a controlled sequence of research and development stages.

More recent evolutionary interpretations, such as those of Dosi (1988) and Storper (1997a), have realized that the process of generating new technologies has to be conceptualized with care. The creation of new technologies is viewed as an interactive social process, characterized by a particular social division of labor within a firm and between different firms of the value chain, as well as between firms and universities and governmental research facilities. This process is characterized by continuous feedback from various stages in research, testing, and production, reflexive patterns of economic behaviour, and interactive learning between the agents involved. Innovative activities are risky in that researchers do not know if they will succeed. Thus, uncertainty plagues the innovation process. Successful innovations are often associated with the creation of new or the modification of existing knowledge. The process of generating new technologies and knowledge is path-dependent in that it depends on the actors' experiences. Further, search activities are often limited to a particular class of techno-economic problem-solving mechanisms. Innovation processes of firms therefore follow particular technological trajectories in which search processes are stimulated and directed by existing routines, heuristics, and cognitive scripts (DiMaggio, 1997).

The particular spatial organization of innovation depends on a number of influences. These include the degree to which production processes are vertically disintegrated, the existence of near-by firms which could become partners in innovation and the type of knowledge which is needed in the innovation process, as well as the degree to which this knowledge is localized (or sticky). The spatial organization is also greatly impacted by experience from previous innovation processes. Many new technologies are incremental in character and have not been developed in integrated research processes in large globally organized firms. In addition, specialized industrial agglomerations support those innovation processes which rely on an extensive social division of labor. In this case, spatial proximity enables regular personal communication, joint problem-solving and adjustments between the people and firms involved. These interactions stimulate information spillovers and processes of knowledge-creation (Storper and Walker, 1989; Storper, 1997a; Maskell and Malmberg, 1999a,b; Bathelt and Glückler, 2000). This is particularly well developed if a large number of firms of the same value chain are involved in this social division of labor and if they share the same local socio-institutional context (Crevoisier and Maillat, 1991).

Nation-states in particular have a great impact on the structure of technology- and knowledge-creation because they define the primary institutional settings for the development of national innovation systems (Lundvall, 1992; Nelson, 1993). These national innovation systems are for instance characterized by different culture, organizational structures, varying degrees of vertical integration and centralization, different routines and habits in technological trajectories, and distinct ways in which they enable regional adjustments to localized capabilities. At the regional level, national innovation systems are being modified and adjusted to those local cultures (e.g. Saxenian, 1994; Schoenberger, 1997), institutions and production specificities at hand, and thus result in region-specific innovation and development paths.

6.4. Interaction

Interactions between actors and ensembles of actors in economic contexts are another crucial element of our relational framework. One important conclusion from the preceding discussion is that the particular organization of production and the processes of generating innovation are a result of ongoing interaction. This also operates as an enabling force for further interaction between people in various workplaces, firms, and formal institutions and authorities. An evolutionary approach helps to understand how the nature and extent of interaction changes over time according to ongoing experience between those organizations involved in innovation. The processes of interactive learning, creative variation and collective knowledge-production can thus be viewed as conceptual tools which link the organization and innovation ions and generate an evolutionary dynamic. Interactions of various kinds between economic agents at different spatial and organizational levels enable firms to modify and refine heuristics and routines along existing trajectories or to develop new technologies towards new development paths.

Interaction and learning are at the core of the reflexive economy (Lundvall and Johnson, 1994). This means that the outcomes of previous actions are recorded, checked, and evaluated in a systematic way in order to draw conclusions for further improvements in future actions. Indeed, empirical studies have shown that production and innovation are social processes, characterized by a particular social division of labor. This social character gains in importance as technologies become more complex and sophisticated and as specialization and segmentation increase. Through this, the process of learning by interacting has become a central issue of analyses in economics, sociology, and economic geography (Lundvall, 1988; Gertler, 1993, 1995). Learning by interacting refers to a process through which systematic communication and adjustment between producers and users results in mostly incremental improvements of product and process technologies and organizational routines.

The existence and acceptance of formal institutions and particularly informal institutions are important prerequisites which stimulate interactive learning between economic agents. Routines, conventions, and habits with respect to the technologies and resources used enable producers to communicate with their suppliers and customers and to collectively decide upon product and process adjustments in particular projects (Storper, 1997a). Interactive learning has also distinct consequences for the spatial organization of production and innovation because conventions and social relations are only preliminary, always restructured through feedback between agents. Their adjustments require the co-presence of agents, which is most efficiently conducted through co-location (Storper, 1997a). Some conventions cannot be easily transferred over large distances to other social contexts (Maskell and Malmberg, 1999a,b). As a consequence, industry clusters develop which are characterized by close inter-firm interaction, proximity, and learning. They enable efficient information and knowledge transfers (Storper, 1995; Asheim, 1999; Bathelt, 2002). Over time, this encourages the development of shared technological attitudes and expectations between the local actors and stimulates trust-based linkages (Crevoisier and Maillat, 1991). Therefore substantial interaction still takes place within national and regional socio-economic contexts despite the development of new forms of global organization of production (e.g. Dicken, 1994; Zeller, 2001).

7. Conclusion: implications of a relational perspective in economic geography

Since the late 1980s, new ideas, conceptions, and models, especially those originating from Anglo-American geography, have spread within the discipline and have initiated a reorientation of research practices in economic geography. This paper conceptualizes this as a shift toward a relational economic geography which is currently taking place in numerous countries despite their different academic histories. We have begun to illustrate the occurrence of this shift in both German geography and economic geography. In German geography, we have identified this as a second transition from spatial science towards a socially constructed, actor-centered, and process-oriented geography. This followed the first transition from Länderkunde towards spatial science which took place in the late 1960s and early 1970s. In economic geography, this change can be described as a transition from regional science (or spatial analysis) towards relational economic geography. In reviewing the literature, we argue that this transition corresponds with a wider shift in Anglo-American economic geography (e.g. Lee and Wills, 1997; Clark et al., 2000; Barnes, 2001).

The paradigmatic differences between these conceptualizations have been laid out systematically and five dimensions in the research design identified which explicate discontinuities between the paradigms discussed. These dimensions include the conception of space, object of knowledge, conception of action, epistemological perspective, and research goal. From this, a relational view of economic action has been developed which rests on three basic propositions. First, economic actors are structurally situated in contexts of social and institutional relations. Second, economic processes are path-dependent to the extent that future action is dependent on past decisions, structures, and processes. Third, economic processes are at the same time contingent and open-ended in that agents make choices which may deviate from existing development paths.

Drawing on Storper's (1997a,b) holy trinity, we define four ions as the basis for analysis in economic geography, i.e. organization, evolution, innovation, and interaction. These interrelated concepts serve as a heuristic framework to systematically apply the consequences of a relational perspective to the theoretical debates in much of the work of economic geographers throughout the 1990s. The point of departure for our conceptualization is that those economic and social processes which drive the four ions are analysed using a distinct geographical lens. This allows us to develop an approach to economic geography which integrates both economic and social theories. In applying the geographical lens, this allows us to formulate specific geographical research problems and implications, different from those in economics and sociology. Our conceptualization does not treat space and territory as entities which are research objects in themselves, being separate from both the economic and the social. Similarly, we do not intend to theorize space or identify spatial laws but look for explanations of human action in localized economic processes. Alternatively, we apply an understanding of space as perspective. This geographical lens draws our attention towards particular localized representations of economic processes and their outcomes.

Our conceptualization does not attempt to develop a comprehensive standard theory which is capable of explaining and predicting all possible outcomes of social and economic processes in space. Rather, the framework developed in this paper presents an interdisciplinary and multidimensional relational view which can be applied to a large number of research problems in economic geography, especially those that we have personally dealt with over the past decade. From a relational view, which rests on the assumptions of contextuality, path-dependence, and the contingency of economic action, a standard body of theory would be a virtual impossibility anyway. Instead, we aim to uncover a novel way to help formulate research questions in economic geography, different from those in traditional regional science (or spatial analysis), yielding subsequently different answers.

In focusing on the geography of the firm, an alternative relational view of economic geography would pose the following types of questions: how do firms interact with one another and what are the consequences for localized processes and structures? In which way are firms influenced by institutional and socio-cultural contexts in their home base and how does this change when they expand to other contexts? How are firms, networks of firms and production systems organized and how does this organization vary from place to place and which territorial consequences result from this? Through which processes do new institutions evolve and how are the localized?[24] Which interactive communication and adjustment processes with other firms and formal institutions allow firms to shape their environments and improve competitiveness? What are the impacts of global changes in technology, demand and competition on the organization of production and how do these effects vary between communities, regions, and nations?

---

24 One central issue on the agenda of relational economic geography is to provide a better understanding of how institutions shape contextual action, as well as the ways in which contingent action helps to transform existing and evolve new institutional forms. This is quite important because dynamic institutions continuously enable and constrain human action. Since they are sometimes territorial in character and vary between places, regions or nations, they are of great significance to understand different processes and structures in spatial perspective.

Towards relational economic geography • **139**

In our view, these questions are particularly relational for they systematically draw on the concepts of context, contingency and path-dependence of economic action and its localized consequences as a point of departure for analysis in economic geography. To answer these questions, we require a fundamental revision of traditional concepts in economic geography, such as that of the four ions which we have developed in the context of the geography of the firm. Our intention is to provide a starting point for analysis which can be further explored and applied in different contexts of economic geography.

Acknowledgements

We would like to thank Gerhard Bahrenberg, Christian Berndt, Gordon Clark, Ernst Giese, Eric Sheppard, Ute Wardenga, Benno Werlen, and Clare Wiseman for their stimulating comments and constructive suggestions on earlier versions of this paper, as well as the editors of this special issue. Parts of this paper were presented at the 2001 Meeting of German Economic Geographers in Rauischholzhausen near Marburg (Germany), at the 2002 Meeting of the Association of American Geographers in Los Angeles (USA), and at a Geography Colloquium at University of Zurich (Switzerland). We would like to thank the participants of these meetings for their helpful remarks. We are also grateful for the thought-provoking comments of four anonymous referees which have helped us to clarify some of our arguments. The usual disclaimers do of course apply. Parts of this research were funded through the Deutsche Forschungsgemeinschaft (German Science Foundation).

References

Allen, J. (1997) Economies of power and space. In R. Lee and J. Wills (eds) *Geographies of Economies*. London, New York, Sydney: Arnold, 59–70.

Amin, A. (ed.) (1994) *Post-Fordism*. Oxford, Cambridge, MA: Blackwell.

Amin, A., Thrift, N. (1994) Living in the global. In A. Amin and N. Thrift (eds) *Globalization, Institutions, and Regional Development in Europe*. Oxford, New York: Oxford University Press, 1–22.

Amin, A., Thrift, N. (2000) What kind of economic theory for what kind of economic geography. *Antipode*, 32: 4–9.

Archer, M., Bhaskar, R., Collier, A., Lawson, T., Norrie, A. (eds) (1998) *Critical Realism. Essential Readings*. London, New York: Routledge.

Asheim, B. (1999) Interactive learning and localised knowledge in globalising learning economies. *GeoJournal*, 49: 345–352.

Bahrenberg, G. (1972) Räumliche Betrachtungsweise und Forschungsziele der Geographie. *Geographische Zeitschrift*, 60: 8–24.

Bahrenberg, G. (1987) Über die Unmöglichkeit von Geographie als 'Raumwissenschaft'. Gemeinsamkeiten in der Konstituierung von Geographie bei A. Hettner und D. Bartels. In G. Bahrenberg, J. Deiters, M. Fischer, W. Gaebe, G. Hard, and G. Löffler (eds) *Geographie des Menschen. Dietrich Bartels zum Gedenken*. Bremer Beiträge zur Geographie und Raumplanung, 11. Bremen, 225–239.

Barnes, T. J. (2001) Retheorizing economic geography: from the quantitative revolution to the 'cultural turn'. *Annals of the Association of American Geographers*, 91: 546–565.

Barnes, T. J., Gertler, M. S. (eds) (1999) *The New Industrial Geography: Regions, Regulation and Institutions*. London, New York: Routledge.

Bartels, D. (1968) *Zur wissenschaftstheoretischen Grundlegung einer Geographie des Menschen*. Erdkundliches Wissen, 19. Wiesbaden: Steiner.

Bartels, D. (1970) Einleitung. In D. Bartels (ed.) *Wirtschafts- und Sozialgeographie*. Köln, Berlin: Kiepenheuer & Witsch, 13–45.

Bartels, D. (1988) *Wirtschafts- und Sozialgeographie*. Handwörterbuch der Wirtschaftswissenschaft, 9. Stuttgart, New York: Fischer; Tübingen: Mohr (Siebeck); Göttingen, Zürich: Vandenhoeck & Ruprecht, 44–54.

Bathelt, H. (2000) Räumliche Produktions- und Marktbeziehungen zwischen Globalisierung und Regionalisierung – Konzeptioneller Überblick und ausgewählte Beispiele. *Berichte zur deutschen Landeskunde*, 74: 97–124.

Bathelt, H. (2001) Warum Paul Krugmans Geographical Economics keine neue Wirtschaftsgeographie ist! Eine Replik zum Beitrag 'New Economic Geography' von Armin Osmanovic in *Die Erde* 131 (3): 241–257. *Die Erde*, 132: 107–118.

Bathelt, H. (2002) The re-emergence of a media industry cluster in Leipzig. *European Planning Studies*, 10: 583–611.

Bathelt, H., Glückler, J. (2000) Netzwerke, Lernen und evolutionäre Regionalentwicklung. *Zeitschrift für Wirtschaftsgeographie*, 44: 167–182.

Bathelt, H., Glückler, J. (2002a) *Wirtschaftsgeographie. Ökonomische Beziehungen in räumlicher Perspektive*. Stuttgart: UTB – Ulmer.

Bathelt, H., Glückler, J. (2002b) Wirtschaftsgeographie in relationaler Perspektive: Das Argument der zweiten Transition. *Geographische Zeitschrift*, 90: 20–39.

Baum, J. A., Oliver, C. (1992) Institutional embeddedness and the dynamics of organizational populations. *American Sociological Review*, 57: 540–559.

Berndt, C. (1999) Institutionen, Règulation und Geographie. *Erdkunde*, 53: 302–316.

Berry, B. J. L. (2002) Book review of 'The spatial economy: cities, regions, and international trade' by M. Fujita, P. Krugman, and A. Venables. Cambridge, MA: MIT Press, 2001 (2nd edition). *Annals of the Association of American Geographers*, 92: 359–360.

Bhaskar, R. (1975) *A Realist Theory of Science*. London, New York: Verso (reprint from 1997).

Bourdieu, P. (1977) *Outline of a Theory of Practice*. Cambridge: Cambridge University Press.

Bryson, J., Henry, N., Keeble, D., Martin, R. (eds) (1999) *The Economic Geography Reader. Producing and Consuming Global Capitalism*. Chichester, New York: Wiley.

Bunge, W. (1973) Ethics and logic in geography. In R. J. Chorley (ed.) *Directions in Geography*. London: Methuen, 317–331.

Chorley, R. J., Haggett, P. (eds) (1965) *Frontiers in Geographical Teaching*. London: Methuen.

Chorley, R. J., Haggett, P. (eds) (1967) *Models in Geography*. London: Methuen.

Clark, G. L. (1982) Rights, property, and community. *Economic Geography*, 58: 120–138.

Clark, G. L. (1983) Fluctuations and rigidities in local labor markets. Part 2: Reinterpreting relational contracts. *Environment and Planning A*, 15: 365–377.

Clark, G. L., Feldman, M. P., Gertler, M. S. (eds) (2000) *The Oxford Handbook of Economic Geography*. Oxford: Oxford University Press.

Cliff, A. D., Haggett, P., Ord, J. K., Versey, G. R. (1981) *Spatial Diffusion: A Historical Geography of Epidemics in an Island Community*. Cambridge: Cambridge University Press.

Coase, R. H. (1937) The nature of the firm. *Economica*, 4: 386–405.

Crevoisier, O. (2001): Der Ansatz des kreativen Milieus: Bestandsaufnahme und Forschungsperspektiven am Beispiel urbaner Milieus. *Zeitschrift für Wirtschaftsgeographie*, 45: 246–256.

Crevoisier, O., Maillat, D. (1991) Milieu, industrial organization and territorial production system: towards a new theory of spatial development. In R. Camagni (ed.) *Innovation Networks: Spatial Perspectives*. London, New York: Belhaven Press, 13–34.

Curry, L. (1967) Central places in the random space economy. *Journal of Regional Science*, 7/2 (suppl.): 217–238.

DiMaggio, P. J. (1997) Culture and cognition. *Annual Review of Sociology*, 23: 263–289.

Dicken, P. (1990) The geography of enterprise. Elements of a research agenda. In M. de Smidt and E. Wever (eds) *The Corporate Firm in a Changing World Economy. Case Studies in the Geography of Enterprise*. London: Routledge, 234–244.

Dicken, P. (1994) The Roepke lecture in economic geography: global-local tensions: firms and states in the global space-economy. *Economic Geography*, 70: 101–128.

Dosi, G. (1982) Technological paradigms and technological trajectories: a suggested reinterpretation of the determinants and directions of technical change. *Research Policy*, 2: 147–162.

Dosi, G. (1988) The nature of the innovative process. In G. Dosi, C. Freeman, R. R. Nelson, G. Silverberg, and L. L. G. Soete (eds) *Technical Change and Economic Theory*. London, New York: Pinter, 221–238.

Ettlinger, N. (2001) A relational perspective in economic geography: connecting competitiveness with diversity and difference. *Antipode*, 33: 216–227.

Fleetwood, S. (2002) Boylan and O'Gorman's causal holism: a critical realist evaluation. *Cambridge Journal of Economics*, 26: 27–45.

Fujita, M., Krugman, P., Venables, A. (2001) *The Spatial Economy: Cities, Regions, and International Trade*, 2nd edn. Cambridge, MA: MIT Press.

Gertler, M. S. (1993) Implementing advanced manufacturing technologies in mature industrial regions: towards a social model of technology production. *Regional Studies*, 27: 665–680.

Gertler, M. S. (1995) 'Being there': proximity, organization, and culture in the development and adoption of advanced manufacturing technologies. *Economic Geography*, 71: 1–26.

Giddens, A. (1984) *The Constitution of Society. Outline of the Theory of Structuration*. Cambridge: Polity Press.

Giddens, A. (1990) *The Consequences of Modernity*. Stanford, CA: Stanford University Press.

Glückler, J. (1999) *Neue Wege geographischen Denkens? Eine Kritik gegenwärtiger Raumkonzepte und ihrer Forschungsprogramme in der Geographie*. Frankfurt/Main: Verlag Neue Wissenschaft.

Glückler, J. (2001) Zur Bedeutung von Embeddedness in der Wirtschaftsgeographie. *Geographische Zeitschrift*, 89: 211–226.

Grabher, G. (1993) Rediscovering the social in the economics of interfirm relations. In G. Grabher (ed.) *The Embedded Firm. On the Socioeconomics of Industrial Networks*. London, New York: Routledge, 1–31.

Granovetter, M. (1985) Economic action and economic structure: the problem of embeddedness. *American Journal of Sociology*, 91: 481–510.

Granovetter, M. (1992a) Economic institutions as social constructions: a framework for analysis. *Acta Sociologica*, 35: 3–11.

Granovetter, M. (1992b) Problems of explanation in economic sociology. In N. Nohria and R. G. Eccles (eds) *Networks and Organisations: Structure, Form, and Action*. Cambridge, MA: Harvard Business School, 25–56.

Haggett, P. (1965) *Locational Analysis in Human Geography*. London: Arnold.

Hard, G. (1993) Über Räume reden. Zum Gebrauch des Wortes 'Raum' in sozialwissenschaftlichem Zusammenhang. In J. Mayer (ed.) *Die aufgeräumte Welt. Raumbilder und Raumkonzepte im Zeitalter globaler Marktwirtschaft*. Loccumer Protokolle, 74/92. Loccum: Evangelische Akademie Loccum, 53–78.

Hartshorne, R. (1939) *The Nature of Geography*. Lancaster, PA: Association of American Geographers.

Harvey, D. (1982) *The Limits of Capital*. Oxford: Basil Blackwell.

Harvey, D. (1996) *Justice, Nature and the Geography of Difference*. Oxford: Blackwell.

Hettner, A. (1927) *Die Geographie. Ihre Geschichte, ihr Wesen und ihre Methoden*. Breslau: Hirt.

Hirschman, A. O. (1958) *The Strategy of Economic Development*. New Haven: Yale University Press.

Hodgson, G. M. (1988) *Economics and Institutions: A Manifesto for a Modern Institutional Economics*. Cambridge: Polity Press.

Hume, D. (1758) *Eine Untersuchung über den menschlichen Verstand*. Stuttgart: Reclam (German edition from 1982).

Isard, W. (1956) *Location and Space-Economy: A General Theory Relating to Industrial Location, Market Areas, Land Use, Trade and Urban Structure*. New York, London: Wiley.

Isard, W. (1960) *Methods of Regional Analysis: An Introduction to Regional Science*. Cambridge, MA, London: MIT Press.

Jessop, B. (1992) Fordism and post-Fordism: a critical reformulation. In M. Storper and A. J. Scott (eds) *Pathways to Industrialization and Regional Development*. London, New York: Routledge, 46–69.

142 • *Bathelt and Glückler*

Kaldor, N. (1970) The case for regional policies. *Scottish Journal of Political Economy*, 17: 337–347.

Krauss, T. (1933) Der Wirtschaftsraum. Gedanken zu seiner geographischen Erforschung. In T. Krauss (ed.) *Individuelle Länderkunde und räumliche Ordnung*. Erdkundliches Wissen, 7. Wiesbaden: Steiner (1960 edition), 21–45.

Krugman, P. (1991) *Geography and Trade*. Leuven: Leuven University Press; Cambridge, MA, London: MIT Press.

Krugman, P. (2000) Where in the world is the 'new economic geography'? In G. L. Clark, M. P. Feldman, and M. S. Gertler (eds) *The Oxford Handbook of Economic Geography*. Oxford: Oxford University Press, 49–60.

Lee, R. (2002) 'Nice maps, shame about the theory'? Thinking geographically about the economic. *Progress in Human Geography*, 26: 333–355.

Lee, R., Wills, J. (eds) (1997) *Geographies of Economies*. London, New York, Sydney: Arnold.

Lovering, J. (1989) The restructuring debate. In R. Peet and N. Thrift (eds) *New Models in Geography 1*. London: Hyman, 198–223.

Lundvall, B.-Å. (1988) Innovation as an interactive process: from producer-user interaction to the national system of innovation. In G. Dosi, C. Freeman, R. R. Nelson, G. Silverberg, and L. L. G. Soete (eds) *Technical Change and Economic Theory*. London, New York: Pinter, 349–369.

Lundvall, B.-Å. (ed.) (1992) *National Systems of Innovation: Towards a Theory of Innovation and Interactive Learning*. London: Pinter.

Lundvall, B.-Å., Johnson, B. (1994) The learning economy. *Journal of Industry Studies*, 1: 23–42.

Lundvall, B.-Å., Maskell, P. (2000) Nation states and economic development: from national systems of production to national systems of knowledge creation and learning. In G. L. Clark, M. P. Feldman, and M. S. Gertler (eds) *The Oxford Handbook of Economic Geography*. Oxford: Oxford University Press, 353–372.

Lütgens, R. (1921) Grundzüge der Entwicklung des La Plata-Gebietes. *Weltwirtschaftliches Archiv*, 17: 359–374.

Martin, R. (1994) Economic theory and human geography. In D. Gregory, R. Martin, and G. Smith (eds) *Human Geography. Society, Space and Social Science*. Houndsmills: Macmillan, 21–53.

Martin, R. (1999) The 'new economic geography': challenge or irrelevance? *Transactions of the Institute of British Geographers*, 24: 387–391.

Martin, R., Sunley, P. (1996) Paul Krugman's geographical economics and its implications for regional development theory: a critical assessment. *Economic Geography*, 74: 259–292.

Martin, R., Sunley, P. (2001) Rethinking the 'economic' in economic geography: broadening our vision or losing our focus? *Antipode*, 33: 148–161.

Maskell, P. (2001) The firm in economic geography. *Economic Geography*, 77: 329–344.

Maskell, P., Malmberg, A. (1999a) The competitiveness of firms and regions: 'ubiquitification' and the importance of localized learning. *European Urban and Regional Studies*, 6: 9–25.

Maskell, P., Malmberg, A. (1999b) Localised learning and industrial competitiveness. *Cambridge Journal of Economics*, 23: 167–185.

Massey, D. (1985) New directions in space. In D. Gregory and J. Urry (eds) *Social Relations and Spatial Structures*. Basingstoke: Macmillan, 9–19.

McDowell, L. (2000) Feminists rethink the economic: the economics of gender/the gender of economics. In G. L. Clark, M. P. Feldman, and M. S. Gertler (eds) *The Oxford Handbook of Economic Geography*. Oxford: Oxford University Press, 497–517.

Meckelein, W., Borcherdt, C. (eds) (1970) *Tagungsberichte und wissenschaftliche Abhandlungen*. 37. Deutscher Geographentag Kiel 1969. Wiesbaden: Steiner.

Myrdal, G. (1957) *Economic Theory and Underdeveloped Regions*. London: Duckworth.

Nelson, R. R. (ed.) (1993) *National Innovation Systems: A Comparative Analysis*. Oxford: Oxford University Press.

Nelson, R. R. (1995) Evolutionary theorizing about economic change. *Journal of Economic Literature*, 23: 48–90.

Nelson, R. R., Winter, S. G. (1982) *An Evolutionary Theory of Economic Change*. Cambridge, MA: Harvard University Press.

Nonaka, I., Toyama, R., Nagata, A. (2000) A firm as a knowledge-creating entity: a new perspective on the theory of the firm. *Industrial and Corporate Change*, 9: 1–20.

North, D. C. (1991) Institutions. *The Journal of Economic Perspectives*, 5: 97–112.

Peck, J. (1996) *Workplace: The Social Regulation of Labour Markets*. New York: Guilford.

Perrons, D. (2001) Towards a more holistic framework for economic geography. *Antipode*, 33: 208–215.

Philo, C. (1989) Contextuality. In A. Bullock, O. Stallybrass, and S. Trombly (eds) *The Fontana Dictionary of Modern Thought*. London: Fontana Press, 173.

Prahalad, C., Hamel, G. (1990) The core competence of the corporation. *Harvard Business Review*, 68: 79–91.

Rodríguez-Pose, A. (2001) Killing economic geography with a 'cultural turn' overdose. *Antipode*, 33: 176–182.

Romo, F. P., Schwartz, M. (1995) The structural embeddedness of business decisions: the migration of manufacturing plants in New York State, 1960–1985. *American Sociological Review*, 60: 874–907.

Saunders, P. (1989) Space, urbanism and the created environment. In D. Held and J. B. Thompson (eds) *Social Theory of Modern Societies: Anthony Giddens and His Critics*. Cambridge: Cambridge University Press, 215–234.

Saxenian, A. L. (1994) *Regional Advantage: Culture and Competition in Silicon Valley and Route 128*. Cambridge, MA, London: Harvard University Press.

Sayer, A. (1985) The difference that space makes. In D. Gregory and J. Urry (eds) *Social Relations and Spatial Structures*. Basingstoke: Macmillan, 49–66.

Sayer, A. (1992) *Method in Social Science*. London: Routledge.

Sayer, A. (2000) *Realism and Social Science*. London: Sage.

Sayer, A., Walker, R. (1992) *The New Social Economy: Reworking the Division of Labor*. Cambridge, MA, Oxford: Blackwell.

Schamp, E. W. (1983) Grundansätze der zeitgenössischen Wirtschaftsgeographie. *Geographische Rundschau*, 35: 74–80.

Schamp, E. W. (2000) *Vernetzte Produktion: Industriegeographie aus institutioneller Perspektive*. Darmstadt: Wissenschaftliche Buchgesellschaft.

Schätzl, L. (1998) *Wirtschaftsgeographie 1: Theorie*, 7th edn. Paderborn, München, Wien: UTB – Schöningh.

Schoenberger, E. (1997) *The Cultural Crisis of the Firm*. Cambridge, MA, Oxford: Blackwell.

Scott, A. J. (1988) *New Industrial Spaces: Flexible Production Organization and Regional Development in North America and Western Europe*. London: Pion.

Scott, A. J. (1998) *Regions and the World Economy: The Coming Shape of Global Production, Competition, and Political Order*. Oxford, New York: Oxford University Press.

Scott, A. J. (2000) Economic geography: the great half-century. *Cambridge Journal of Economics*, 24: 483–504.

Sheppard, E. (2000) Geography or economics? Conceptions of space, time, interdependence, and agency. In G. L. Clark, M. P. Feldman, and M. S. Gertler (eds) *The Oxford Handbook of Economic Geography*. Oxford: Oxford University Press, 99–119.

Sheppard, E., Barnes, T. J. (eds) (2000) *A Companion to Economic Geography*. Oxford: Blackwell.

Stark, D. (2000) For a sociology of worth. Paper presented at the conference on economic sociology at the edge of the third millennium, Moscow.

Storper, M. (1993) Regional 'worlds' of production: learning and innovation in the technology districts of France, Italy and the USA. *Regional Studies*, 27: 433–455.

Storper, M. (1995) The resurgence of regional economics, ten years later. *European Urban and Regional Studies*, 2: 191–221.

Storper, M. (1997a) *The Regional World. Territorial Development in a Global Economy*. New York, London: Guilford.

Storper, M. (1997b) Regional economies as relational assets. In R. Lee and J. Wills (eds) *Geographies of Economies*. London, New York, Sydney: Arnold, 248–258.

Storper, M. (1997c) Territories, flows, and hierarchies in the global economy. In K. R. Cox (ed.) *Spaces of Globalization. Reasserting the Power of the Local*. New York, London: Guilford, 19–44.

Storper, M., Walker, R. (1989) *The Capitalist Imperative. Territory, Technology, and Industrial Growth*. New York, Oxford: Basil Blackwell.

Sunley, P. (1996) Context in economic geography: the relevance of pragmatism. *Progress in Human Geography*, 20: 338–355.

Swedberg, R., Granovetter, M. (1992) Introduction. In M. Granovetter and R. Swedberg (eds) *The Sociology of Economic Life*. Oxford: Westview Press, 1–26.

Taylor, M. (1995) The business enterprise, power and patterns of geographical industrialisation. In S. Conti, E. J. Malecki, and P. Oinas (eds) *The Industrial Enterprise and its Environment: Spatial Perspectives*. Aldershot: Avebury, 99–122.

Thrift, N. (1990) For a new regional geography 1. *Progress in Human Geography*, 14: 272–277.

Thrift, N. (2000) Pandora's box? Cultural geographies of economies. In G. L. Clark, M. P. Feldman, and M. S. Gertler (eds) *The Oxford Handbook of Economic Geography*. Oxford: Oxford University Press, 689–704.

Toulmin, S. (1972) *Human Understanding*, Volume I, *General Introduction and Part I: The Collective Use and Evolution of Concepts*. Princeton: Princeton University Press.

von Böventer, E. (1962) *Theorie des räumlichen Gleichgewichts*. Tübingen: Mohr (Siebeck).

Voppel, G. (1999) *Wirtschaftsgeographie: Räumliche Ordnung der Weltwirtschaft unter marktwirtschaftlichen Bedingungen*. Teubner Studienbücher der Geographie. Stuttgart, Leipzig: Teubner.

Waibel, L. (1933) Das System der Landwirtschaftsgeographie. In L. Waibel (ed.) *Wirtschaftsgeographische Abhandlungen – Nummer 1*. Leipzig: Hirt, 7–12.

Webber, M. J., Rigby, D. (1996) *The Golden Age Illusion: Rethinking Postwar Capitalism*. New York: Guilford.

Werlen, B. (1993) *Society, Action and Space. An Alternative Human Geography*. London: Routledge.

Werlen, B. (1995) *Sozialgeographie alltäglicher Regionalisierungen*. Band 1: *Zur Ontologie von Gesellschaft und Raum*. Erdkundliches Wissen, 116. Stuttgart: Steiner.

Werlen, B. (2000) *Sozialgeographie: Eine Einführung*. Bern, Stuttgart: UTB – Haupt.

Wernerfelt, B. (1984) A resource-based view of the firm. *Strategic Management Journal*, 5: 171–180.

Williamson, O. E. (1975) *Markets and Hierarchies: Analysis and Anti-Trust Implications*. New York: Free Press.

Williamson, O. E. (1985) *The Economic Institutions of Capitalism. Firms, Markets, Relational Contracting*. New York: Free Press.

Zeller, C. (2001) *Globalisierungsstrategien – Der Weg von Novartis*. Berlin, Heidelberg, New York: Springer.

Zukin, S., DiMaggio, P. (1990) Introduction. In S. Zukin and P. DiMaggio (eds) *Structures of Capital: The Social Organization of the Economy*. New York: Cambridge University Press, 1–36.

[5]

Conceptualizing economies and their geographies: spaces, flows and circuits

Ray Hudson

University of Durham, Department of Geography and International Centre for Regional Regeneration and Development Studies, Wolfson Research Institute, Durham DH1 3LF, UK

Abstract: The last decade or so has been one of ongoing, at times heated, debate in economic geography as to how best to conceptualize and theorize economies and their geographies. Reflecting on these debates, I identify six axioms that are central to conceptualizing economic geographies. I then go on to consider issues of culture and the economy and the relationships between them. The paper explores the links between political-economic and cultural-economic approaches, suggesting that they are most productively seen as complementary both/and approaches rather than as competitive either/or ones.

Key words: economic geographies, circuits, flows, spaces, capitalism, political economy, cultural economy.

I Introduction

The last decade or so has been one of ongoing, at times heated, debate in economic geography as to how best to conceptualize and theorize economies and their geographies, and, relatedly, how best to practise and carry out research on such economic geographies. This debate is reflected in a number of edited volumes that seek to define the current state-of-the-art and (re)define the conceptual boundaries of economic geography (for example, see Amin and Thrift, 2004b; Clark *et al.*, 2000; Lee and Wills, 1997; Sheppard and Barnes, 2000). During the 1970s and 1980s, in the wake of the critique of spatial science and views of the space economy that drew heavily on the orthodoxies of neoclassical economics, strands of heterodox political-economic approaches in general and Marxian political economy in particular rose to prominence. These were important in introducing concerns with issues of evolution, institutions and the state, alongside those of agency and structure, in seeking to develop more powerful and nuanced understandings of economies and their geographies. Much of the subsequent debate in the 1990s has been informed by poststructural critiques of such political-economic approaches, especially those that were seen (rightly or wrongly) to rely upon an overly deterministic and structural reading of the economy and its geographies (Hudson, 2001). These have been important in seeking to rethink relationships between categories such as consumption and production and to provoke more serious consideration of issues such as the relations between agency, practice and structure, the materiality of the economy and the relations between people, nature and things, and of the discursive construction and representation of 'the economy'.

Perhaps the focal point of these critiques and debates revolves around the issues of culture and the economy, and the relationships between them, both ontologically and epistemologically.[1] There is an as yet unresolved – and maybe irresolvable – debate as to the character of the relationships between culture and economy, with important differences within as well as between the advocates of culture/cultural economy and those of economy/political economy. Recognizing this, and so at the risk of some oversimplification, I want to suggest that the recent debates in economic geography can be represented in terms of a dialogue between the proponents of political economy and those of cultural economy (which I will elaborate below). However, rather than seeing these as competitive alternatives (which I think it is fair to say has been largely the case to date, and I will return briefly to the implications of this in the final section of the paper), I want to argue that they are most appropriately seen as complementary perspectives from which we can seek to understand more fully and in more subtle and nuanced ways economies and their geographies.

My primary focus in this paper, then, is to explore the possibilities for developing more subtle and nuanced conceptualizations of economies and their geographies rather than on the practice of empirical research in economic geography, although I will make a few remarks on the latter subject[2] and clearly the two are codependent. I therefore begin with two introductory questions and sets of issues. First, how do we best conceptualize the production of social life in general, in terms of relations between structures/practices/agents and between people and things? Secondly, and more specifically, how do we most appropriately conceptualize 'the economy' in capitalism, its temporalities and spatialities, its circuits and spaces and the links between them? By the 'economy' I refer to those processes and practices of production, distribution and consumption which are simultaneously discursive and material constructions through which people seek to create wealth, prosperity and well-being and so construct economies; to circuits of production, circulation, realization, appropriation and distribution of value. Value is *always* culturally constituted and defined. What counts as 'the economy' is, therefore, always cultural, constituted in places and distributed over space, linked by flows of values, monies, things and people that conjoin a diverse heterogeneity of people and things.

By 'capitalism' I refer to a particular mode of political-economic organization defined by socially produced structural relations and parameters, which are always – and necessarily – realized in culturally and time/space specific forms. The extent to which the contemporary phase of capitalism represents a break from past trajectories of capitalist development continues to be a matter for debate. Although there is now more emphasis on continuity than on radical ruptures between – say – Fordism and post-Fordism, or other dichotomous binaries, there are still claims and counterclaims about the extent to which the economy is characterized by greater 'flexibility' or has become more 'cultural'.

The prime focus of this paper is the second introductory question, the conceptualization of capitalist economies, but it is framed by the first. Capitalist economies are constituted via a complex mix of social relations, of understandings, representations and interpretations, and practices. Certainly the class relations of capital are decisive in defining such societies *as* capitalist but these are (re)produced in varying ways and in relation to noncapitalist class relations and nonclass social relationships of varying sorts (such as those of age, ethnicity, gender and territory). The social relationships of noncapitalist economies undoubtedly assume a great variety of forms, and occasional reference will be made to them. However, in order to allow some depth of analysis, the focus will be on the economies of capitalisms and the social relations of capital that define and dominate them.

II Guiding principles: six axioms and some of their implications

In seeking to answer the two introductory questions, I begin from six axioms. First, there is a need for concepts at a variety of levels of abstraction. This theoretical variety is necessary in order to describe and account for the diverse individual and collective practices, with varying temporalities and spatialities, involved in processes of production, distribution, exchange and consumption and in the spatiotemporal flows of materials, knowledge, people and value (variously defined) that constitute 'economies'.[3] All social life occurs in irreversible flows of time and has a necessary spatiality.

Secondly, however, the 'economy' must be conceptualized in such a way that these diverse practices are seen as necessarily interrelated and avoid fragmenting the economy into dislocated categories such as production and consumption, seeing these as at best unrelated and at worst hermetically sealed and self-contained. For a considerable period of time, much social scientific analysis of the economy – whatever its theoretical stripe – tended to separate the analysis of consumption from that of production[4] and explicitly or implicitly prioritized production over consumption. Consumption was simply seen as a necessary adjunct to production. Now this *is* the case in capitalist economies in one very precise sense. Both production and consumption – or, more accurately, exchange and sale – form moments in the totality of the production process and the point of sale is critical as this realizes the surplus value embodied in commodities and returns it to the monetary form. However, this is only a partial perspective on consumption. While services of necessity are (co)produced and consumed in the same time/space, the moment of sale of material commodities marks a shift in emphasis from their exchange to their use value characteristics, to what can be done with them postsale in a variety of spaces of private and public consumption in homes and civil society. For the life of commodities after they have been sold has important instrumental, material and symbolic connotations and dimensions (ranging from the creation of waste to the giving of gifts based on relations of family, friendship, love and reciprocity, and to the creation of identities).

Thirdly, knowledgeable and skilled subjects, motivated via various rationalities, undertake *all* forms of economic behaviour and practices.[5] Although people are certainly not the all-knowing one-dimensional rational automatons of neoclassical theory, what they do, how they do it and where they do it are the outcomes of purposeful behaviour, underpinned by knowledge and learning. People are not cultural dupes, not passive bearers of structures or habits, norms and routines. Conversely, flows of people in the course of their actions within the economy (and in other arenas, such as those of family and community) can become a mechanism and medium for flows of knowledge. Such flows can occur both in the form of embodied knowledge (often tacit) and that of the transmission of information in codified forms (written, spoken) via a variety of media (letter, telephone, fax, e-mail, for example).

Seen from this perspective, the economy is performed and (re)produced via meaningful and intentional human action but knowledge does not translate in any simple one-to-one relationship to behaviour. Knowledge is a necessary but not sufficient condition. Action is much more than simply a product of information and knowledge, shaped by diverse influences, from emotion to economic possibilities. Moreover, people and organizations have differential abilities to acquire and use knowledge in pursuit of their various projects (although this is not to equate such behaviour with generalized self-reflexivity and the continuous monitoring of individuals' life projects: see Giddens, 1991; Lash and Friedman, 1992). What people come to know and do depends in part upon their positionality in terms of class, ethnicity, gender and other dimensions of social differentiation and identity and the powers and resources available to them by virtue of their position within a given social structure, its organizations and institutions.

Furthermore, intention does not translate in any simple one-to-one relationship to outcome. Purposeful behaviour may have unavoidable and unintended as well as, or instead of, intended outcomes because people chronically act in circumstances in which they lack complete knowledge of the context, of other people and objects, and of the relationships between the people and objects on which they act. Miller (2002: 166) draws attention to 'the degree to which the political economy around us is the result of the unintended consequences of intentional actions'. There may be emergent properties because of the excess of practices, and the messy conjoining of people and things in heterogeneous networks and processes of ordering that produce emergence. Consequently, it is necessary to take seriously the unintended consequences of human action, at all levels from the individual to the formal organizations and institutions of the state (see Habermas, 1976; Offe, 1975). Complex change may be unrelated to agents actually seeking to produce change. They may simply recurrently perform the same actions but 'through iteration *over time* they may generate unexpected, unpredictable and chaotic outcomes. Often the opposite of what human agents may be seeking to realize' (Urry, 2000b: 4). Nevertheless, given these qualifications about uncertainty, ignorance and unintended outcomes, a concept of an economy that is not underpinned by intentional, purposeful behaviour, knowledge and learning is simply, literally, inconceivable. Economic practices are performed by knowledgeable, socially constituted subjects, although the outcomes of their actions may differ from those intended. However, the ways and forms in which knowledge and learning influence economic practice can and do vary over space and time.

The fourth axiom follows from the third: the economy is socially constructed, socially embedded, instituted in a Polyanian sense (with institutions ranging from the informality of habits to the formal institutions of government and the state: for example, see Hodgson, 1988).[6] These various institutions exhibit a degree of stability over the medium to long term, set within the *longue dureé* of structural parameters and necessary relationships that define a particular mode of political-economic organization (such as capitalism) As such, the economy can be thought of as a relatively stable *social* system of production, exchange and consumption.[7] However, institutional stability is always conditional and contingent, as there are processes that seek to disrupt and break out of established institutional forms as well as processes that seek to reproduce them. Hollingsworth (2000: 624) emphasizes that '. . . there is a great deal of path dependency to the way that institutions evolve'. Consequently institutional evolution is path dependent, as economic practices are performed in and create real, irreversible time. However, this is also a conditional dependence, for there are forces that seek to break path dependency as well as those that reproduce it. Therefore it would be more accurate to describe economic and institutional development trajectories as path contingent, with periodic cyclical crises along a given path and the potential for secular changes from one path to another.

The fifth axiom is that behaviour (individual and collective) is both institutionalized and enabled and constrained by structures, understood as stable yet temporary (albeit very long-term) settlements of social relationships in particular ways (Figure 1). Structural relations specify the boundary conditions and parameters that define a particular mode of political economic organization as *that* mode. For example, the class structural relation between capital and labour is a defining feature of the capitalist mode of production – if this was not present, then some other mode of production would exist. However, this relationship can be constituted in varying instituted forms and this is central to the possibilities of creating many capitalisms and their historical geographies. Whatever the specific form, however, economic agents behave in instituted ways that are shaped by, and at the same time help reproduce, such structural relations. There is a definite relationship between practices in the short term and in the long(er) term. This is not to say that such relationships may not be challenged – they often are. However, such challenges are typically folded into and absorbed in ways that alter, but do not radically break and transform, the defining structural characteristics and boundary conditions defined by capitalist social relations. Nonetheless, there is theoretical space for structural change, a point of immense political significance.

The sixth axiom follows from the previous two. 'The economy' is constructed via social relations and practices that are not natural and typically are competitive. Consequently, they must be politically and socially (re)produced via regulatory and governance institutions that ensure the more or less smooth reproduction of economic life. These range from very informal governance institutions such as habits and routines in a variety of spheres, including those of civil society, community, family and work, to the legal frameworks and formal regulatory mechanisms of the state. In short, there is a need to ensure the reproduction of the social relations of capitalism and not just those of *capital*, while acknowledging that the latter are both defining and dominant in capitalist economies and societies. However, while dominant, they are neither singular nor uncontested. Equally, there is a significant difference between the existence of rules and behavioural conformity with them. People may

452   Conceptualizing economies and their geographies

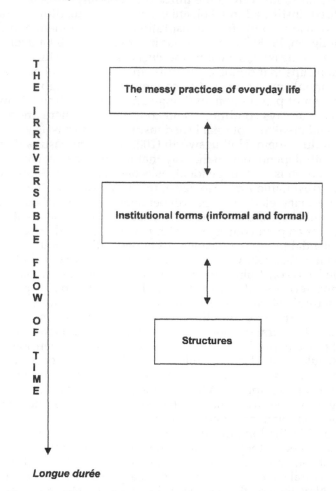

La vie quotidienne

The messy practices of everyday life

Institutional forms (informal and formal)

Structures

Longue durée

Figure 1   Temporalities of practices, institutions and structures

seek to break rather than obey rules, raising key questions as the circumstances in which they will do so (not least in terms of issues of predictability of behaviour). A distinction may be drawn between the formally regulated economy, the informal economy and the illegal economy (Figure 2). The formal economy consists of legal activities governed and regulated within the parameters of legislation. The informal economy consists of legal activities that are regulated by customary mechanisms and practices that fall outside the legal framework. Other activities are illegal but none-theless form part of the economy (the economy of criminality, of the Mafia, for example). However, the boundaries between formal, informal and illegal are fluid and vary over time/space.

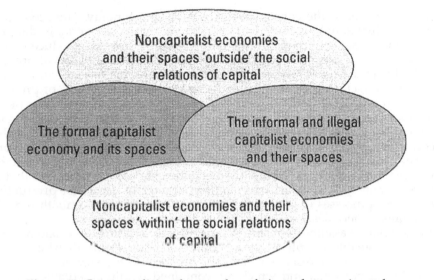

Figure 2 Conceptualizing the complex relations of economies and their spaces in capitalism

The variety of institutions leads to complex spatialities of governance and regulation. These combine the diverse spaces and spatial scales (national, supranational and subnational) of state organizations and institutions within civil society. Systems of governance and regulation are now more multiscalar (Brenner *et al.*, 2003) but national states retain a critical role within them (Sassen, 2003; Weiss, 1997; Whitley, 1999). While generally concerned with regulating the conditions that make markets possible, state activity can extend to supplementing or replacing market mechanisms in resource allocation, for example in the provision of welfare services or the production of key goods and materials. 'The economy' is chronically reproduced in situations of contested understandings, interests and practices because of the construction of governance and regulatory mechanisms that keep such potential disputes within 'acceptable' and 'workable' limits. However, such mechanisms themselves must be socially (re)produced, often via processes of conflict and struggle – and do not simply emerge automatically to meet the functional needs of capital. Thus the practices of government, governance and governmentality are of critical importance. Furthermore, within forms of capitalism that encompass formal political democracies, these mechanisms must be generally regarded as acceptable and legitimate, but in dictatorial capitalisms they may be more violently enforced as a result of state power. One way or another, however, modes of governance and regulation must be sufficiently held in place, at least for a time.

The requirement for a degree of admittedly contingent institutional – and even more so – structural stability reflects the need for a degree of predictability in the outcomes of economic practices and transactions. Such stability is a necessary condition for a required degree of predictability in performing the economy. This requirement is complicated precisely because of the dynamic character of the capitalist economy, the constant becoming of the economy. The economy is not something that simply is but always something that is *necessarily* in the processes of becoming (as, for example, companies constantly strive to produce new things, in new ways). The capitalist economy is performative, a practical order that is constantly in action (Thrift, 1999). Consequently, economic actors – workers, banks, manufacturing companies and so on – require a degree of predictability in order that the transactions and practices of the economy can be performed with some certainty as to outcome over varying time horizons. Companies need to be confident that customers will pay their bills on time, workers that they will receive their wages regularly, and governments that tax revenues will arrive at the due date. As such, there is an unavoidable tension between destabilizing processes that would undermine predictability, stabilizing processes that seek to assure it, and the necessarily dynamic character of capitalist production that complicates processes of governance and regulation and the smooth reproduction of capitalist economies.

In summary, I assume an instituted and structurally situated economy produced by knowledgeable people behaving purposefully in pursuit of different, and often competitive, interests, which can be pursued with a sufficient degree of pre-dictability of outcome, and which are contained within acceptable – or at least tolerable – limits via a range of governance and regulatory mechanisms. There is an unavoidable tension between processes of institutionalization that seek to create a degree of stability and predictability and the emergent outcomes of prac-tices that seek to disturb this, either deliberately or inadvertently. I therefore also acknowledge that there is no single totalizing metanarrative that can explain everything about economies and their geographies but that nonetheless metanarra-tives remain valuable – indeed are necessary – in seeking such explanations (cf. Massey, 1995: 303–304).

More specifically, I argue that there are, broadly speaking, two analytic strategies for understanding the economy and its geographies, with different but complemen-tary inflections. The first approach can be defined as '(political) economic', taking categories such as value, firms and markets as given, with these assumed to exist prior to their being observed and described from 'on high' and to using them in analysing the economy. However, different types of economics conceptualize and represent these in different ways and I draw on Marxian and other heterodox tra-ditions. The second can be thought of as 'cultural (economic)', with an epistemologi-cal focus on the discursive and practical construction and 'making-up' of these categories, while rejecting ontological claims that the economy has become more cul-tural. It emphasizes the ways in which the 'economy' is discursively as well as mate-rially constructed, practised and performed, exploring the ways in which economic life is built up, made up and assembled, from a range of disparate but always inten-sely cultural elements.

III   A caveat about consumption and production

I want to enter a qualification at this point and insist that any simple equation of production/economic and consumption/cultural, and of the primacy of the latter over the former, or *vice versa*, must be firmly rejected. While there is clearly a case for paying attention to consumption, there has been a tendency for the pendulum to swing too far, replacing one-sided accounts that were overly productionist in their emphasis with equally one-sided consumptionist accounts (Gregson, 1995), which, moreover, often conflate consumption with exchange and sale. This was especially so with 'first wave' consumption studies. For example, Bauman (1992: 49) asserts that 'in present day society consumer conduct (consumer freedom geared to the consumer market) moves steadily into the position of, simultaneously, the cognitive and moral focus of life, the integrative bond of society . . . in other words, it moves into the self-same position which in the past – during the "modern" phase of capitalist society – was occupied by work'. Baumann alludes here to the elision of (allegedly) postproductionist consumer society with postmodern society. Echoing this, Lash and Urry (1994: 296, emphasis in original) claim that 'the consumption of goods and services becomes *the* structural basis of western societies. And via the global media this [pleasure seeking] principle comes to be extended world-wide'.

There are two kinds of lessons to be drawn from this, which are reflected in 'second wave' consumption studies. First, politically, it clearly exemplifies the dangers of confusing fashions in academic thought based on the very class and socially specific experiences of an affluent minority with substantive changes in the living conditions and lifestyles of a much broader spectrum of the world's population. Even in the core territories of capitalism, at best only a fraction of the consumption activities of the vast majority of people could be said to be 'pleasure seeking'. McRobbie (1997) criticizes the political complacency of recent work on consumption that emphasizes pleasure and desire precisely because it marginalizes issues of poverty and social exclusions in its urge to reclaim the 'ordinary consumer' as a skilled and knowledgeable actor. For ordinary mundane consumption was (and indeed still is) not hedonistic, materialistic nor individualistic, but was above all the form by which 'capitalism was negated and through which labour brought its products back into the creation of humanity' (Miller, 2002: 182). For the vast majority of people living beyond the affluent core territories, hedonistic consumption and pleasure-seeking behaviour – let alone the attainment of pleasure – is a distant pursuit of the affluent minority, occasionally glimpsed on TV screens in a world characterized by perpetual hunger and malnutrition for the impoverished majority.

Secondly, theoretically, it illustrates the dangers of divorcing a concern with consumption from issues of production, both specifically in the context of capitalist economies but also more generally.[8] Understanding capitalist economies and their geographies requires a more nuanced and subtle stance in theorizing relations between the moments of consumption and production within the totality of the economic process.[9] Equally in terms of epistemology, there is, for example, a long history of rich ethnographic accounts of life in the workplace that seek to build an understanding of work and the social relations of the workplace in terms of the categories, understandings and practices of those engaged in the process (for instance, see Beynon, 1973[10]). Conversely, there are powerful political economies of consumption (for example, see Fine and Leopold, 1993; Miller, 1987). My argument is that such approaches are equally valid and should be seen as complementary. We need both to grasp the complexity of capitalist economies and their historical geographies, examining diverse practices of production, exchange and consumption from both political-economic and cultural-economic perspectives.

IV Cultural-economic approaches to understanding the economy

There has recently been a considerable resurgence of emphasis on 'cultural' approaches to understanding economies and their geographies, although these are far from uniform, broadly falling into ontological and epistemological concepts of a 'cultural economy' (Ray and Sayer, 1999). For example, Lash and Urry (1994) argue that there has recently been a significant 'culturalization' of economic life, which is expressed in three ways. First, there has been a growth in the numbers of innovative companies producing cultural hardware and software. Secondly, 'there is a growing aestheticization or "fashioning" of seemingly banal products whereby these are marketed to consumers in terms of particular clusters of meanings, often linked to lifestyles' (p. 7). Thirdly, there has been a growing 'turn to culture' in the worlds of business and organizations, precisely because maintaining or enhancing competitiveness requires companies to change the ways in which they conduct business and people to change the ways in which they behave within organizations.

However, the significance and validity of these epochal claims of 'increased culturalization' are far from assured. In certain limited respects, the economy may have become more cultural but to claim that the economy overall has become 'more cultural' is more problematic. The evidence in support of 'the exemplary oppositions between a more "use"-value-centred past and a more "sign"-value-centred present' is simply 'empirically insubstantial' (du Gay and Pryke, 2002: 7). Typically it is fragmentary, at times simply anecdotal. However, there is also an issue of adequate theorization and conceptualization of the links between 'economic' and 'cultural'. In practice, social actors cannot actually define a market or a competitor, '*except* through extensive forms of cultural knowledge' (Slater, 2002: 59, emphasis added). Producers cannot know what market they are in without extensive cultural calculation; and they cannot understand the cultural form of their product and its use outside a context of market competition. Understanding culture and (local) cultural difference is vital in order successfully to produce and sell globally (Franklin *et al.*, 2000: 146). In like fashion, the economic practices of advertizing, evocatively described as the 'magic system' (Williams, 1960), are intrinsically caught up with the cultural understanding of the role, functions and nature of advertisements (McFall, 2002: 161).

This draws attention to the way in which (to adopt a famous phrase from the cultural analysis of resources: Zimmerman, 1951) 'products and markets are not, they become'. This is perhaps most sharply emphasized by the iconic commodity of twentieth-century capitalism – the automobile – in which the cultural and economic are inextricably fused via the market segmentation and the symbolic meanings associated with automobiles and automobility (Sheller and Urry, 2000). Furthermore, in order to be(come) a particular kind of economic institution, a market must also be a certain kind of culturally defined domain, because it depends on the social categorization of things as (dis)similar (Slater, 2002: 68).[11] The dependence of markets upon such social categorization undermines propositions about the increasing 'culturalization of the economy' and the increasing, even complete, separation of the material and sign values of commodities.

Culturalism in its various forms reduces the product to its sign value and semiotic processes. As a result, the object becomes entirely dematerialized as a symbolic entity or sign, infinitely malleable and hence never stabilized as a sociohistorical object; its definition can be entirely accounted for in terms of the manipulations of codes by skilled cultural actors. As such, the materiality of the object and the material economy and social structures through which it is elaborated as a meaningful entity are ignored. Consequently, there is also a tendency to reduce market structures and relations to semiotic ones. It is rather difficult to imagine how markets could exist over time, as they patently do, if products actually underwent the kind of semiotic reduction that culturalists assume. As Slater (2002: 73) notes, 'markets are in fact routinely institutionalized, and are even stabilized, around enduring definitions of products, whereas the semiotic reduction would assume that – as sign value – goods will be redefined at will'.

However, while the definition of a commodity, or of a thing, cannot be resolved by drifting off into the realm of floating signifiers, neither can its definition be simply and solidly anchored in given material properties. In contrast, the meanings of things, and things themselves, are stabilized or destabilized, negotiated or contested, within complex asymmetrical power relations and resource inequalities. This emphasizes three things. First, the processes and interplay between the realms of the material, the symbolic and the social through which the meanings of commodities are created, fixed and reworked. Secondly, the instituted social field within which multiple actors seek to intervene to establish the meaning of things. Thirdly, the political-economic structural relations within which both actors and social fields are located – although these can get lost in the emphasis on meanings.

Moreover, Law (2002) argues that culture is located and performed in human and nonhuman material practices, which extend beyond human beings, subjects and their meanings, and implicate technical, architectural, geographical and corporeal arrangements.[12] As such, social production systems comprise a heterogeneity of people and things and links among and between them. That the social has an irreducible materiality is – or ought to be – old news: 'Perhaps Marx told us this. Certainly Michel Foucault and a series of feminist and non-feminist partial successors have done so' (Law, 2002: 24).[13] The reference to Marx is important, since one strand of the Marxian view of production centres on the labour process and transformation of elements of nature by people using artifacts and tools. In this regard, Law does no more – nor less – than restate a proposition from Marxian analysis that conceptualizes the economy as *always* a product of interactions between heterogeneous networks of people, nature (both animate subjects and inanimate objects) and things; of relationships between the social and the natural.

The conceptualization of the economy therefore remains contested terrain, a terrain that is now more complex and in some ways more slippery in its analysis of relationships than it used to be. This raises some important issues about the relationships between 'culture' and 'economy'. Miller (2002: 172–73) argues that it seems 'quite absurd' to suggest that we live within some new self-conscious, self-reflexive economy. There are undoubtedly powerful marketing discourses in the contemporary economy, but 'advertising and Hollywood were extraordinarily important' in the USA of half a century ago, and these made as much use as they could of the current psychological theories about how to create subjects (Williams, 1960). On the other hand, the economy was just as cultural 'at the time when most academics saw themselves as Marxists'. It is undeniably true that a small, affluent minority live more self-reflexive (self-centred) lifestyles. It is also certainly an exaggeration to claim that there was a time when 'most' academics were Marxists. Neither point, however, negates the force of Miller's argument about the limitations of claims about the culturalization of the economy.

In summary, positing a binary opposition between 'economy' and 'culture' is simply implausible and unhelpful. There is, however, considerable merit in an epistemological conception of cultural economy that envisages the 'cultural' as a 'bottom-up' method of analysis,[14] complementary to a more top-down political economy. In contrast, suggestions that somehow the 'economy' has (ontologically) become more 'cultural' are misconceived and deeply problematic. Miller (2002: 172–73) is particularly scathing in his comments about the 'culturalization of the economy' thesis. He suggests that there seems to be 'a sleight of hand' through which a shift in academic emphasis is supposed to reflect a shift in the world, an economy that is more cultural than in earlier times. In this, he echoes Hall (1991: 20), who cogently argues that 'we suffer increasingly from a process of historical amnesia in which we think just because we are thinking about an idea it has only just started'. It is important to avoid such amnesia and to avoid conflating changes in the economy and changes in academic fashion.[15] There is a need for eternal vigilance to guard against the constant danger of confusing new movements within thought (the (allegedly) new understanding that culture and economy cannot be theorized separately) from new empirical developments. The history of classical political economy (as evidenced, for example, in the writings of Smith and Marx) prior to the marginalist revolution and the rise of neoclassical economics and its claims to universal economic laws was one that recognized the cultural constitution of 'the economy' (see Amin and Thrift, 2004a). For 'culture is everywhere and little has changed in this respect... economically relevant activity has always been cultural' (Law, 2002: 21). 'Is it the case', Slater (2002: 78) asks rhetorically, 'that culture is more central to the economic process than it was before?... the answer, I think, is only in particular circumstances and instances but, in general, "no"'. Seeking to recover the ground conceded by the rise of neoclassicism in economics and acknowledge the long history of a cultural dimension within political economy is a very different matter to assuming that there has been a qualitative change involving the 'culturalization' of the economy – the hard realities (if not quite iron laws) of commodity production and the production of surplus-value remain.

V Political-economic approaches to understanding the economy

As with culture, the economic is a contested concept. There are several versions of 'the economic', based on differing theoretical presuppositions and forms and levels of abstraction (including neoclassical and mainstream orthodoxies, heterodoxies of various sorts including Marxian political economy and evolutionary and 'old' and 'radical' institutional approaches[16]). I reject the technicist conceptions of the economy and its geographies exemplified by neoclassical and mainstream orthodox economics, which persistently seek methodologically to fix economic categories as self-evident or natural (and which are central to the (allegedly) new 'geographical economics': for example, see Krugman, 2000; Glaeser, 2000).[17] Indeed, Slater (2002: 72) argues more generally that within economic analysis 'needs and goods appear as natural and self-evident'. In more critical theory, the use value/exchange value distinction within the commodity form 'has generally functioned as a proxy for the distinction for a "natural metabolism" between man and nature, and the warped social form taken by need and things within capitalist market relations'. While Slater's comments regarding neoclassical and mainstream orthodoxies are reasonable, his view of critical theory reveals a partial and warped understanding of Marxian political economy. For 'critical' heterodox positions embrace more than Marxism, while the notion of some 'unwarped' natural form is difficult to reconcile with *any* notion of 'the economy' as socially constituted and embedded.

Recognizing the heterogeneity of heterodox economics, I argue that a political-economic approach needs to combine the differing but complementary levels of abstraction of various heterodox positions – Marxian, institutional and evolutionary. This multiple approach is needed in order to begin to grasp the complexity of 'the economy' as constituted by labour processes, process of material transformation and processes of value creation and flow in specific time/space contexts.[18] Marxian analyses allow a specification of the structural features common to all capitalist economies that define them *as* capitalist. However, such structures do not exist independently of human practice; quite the contrary. They are both a condition for and an expression and a result of such practices and are always contingently reproduced. Practices may give rise to emergent effects that challenge the reproduction of these structures, although there are powerful social forces and institutions that seek to assure their continuation. In short, there is a permanent tension between processes that seek to destabilize these structural relations and those that seek to reproduce them, that is generally – but not inevitably – resolved in favour of the latter. This may involve folding disruptive processes into new institutional forms of capitalism while leaving the defining class structural relations unchanged.

Indeed, the distinction within Marxian political economy between modes of production and social formations recognizes that capitalism was – and is – constituted in variable ways. This insight has been considerably developed within other strands of heterodox political economy, in particular within evolutionary and institutional economics and sociology. Institutional approaches emphasize the ways in which these economies are constituted and embedded in specific cultural and time/space contexts. Evolutionary approaches foreground the path dependent character of development. At its most abstract level, the 'economy' in capitalism is certainly dominated, indeed defined, by the social relations of capital – and powerful analytic tools are needed to theorize these. At this level, Marxian political economy and its value-theoretic account of the social relations and structures of capital provides powerful conceptual tools to understand accumulation by, through and as commodity production and surplus-value production. However, this is a highly abstract approach and so it is necessary to develop less (or differently) abstract concepts to understand how capitalist production and the (re)production of capital are secured. This requires other theoretical constructs to capture the ways in which capitalism is instituted in specific time/space contexts, discursively and materially formed and concretized in and through specific informal and formal institutions. As such, it necessarily includes theorizing the state, regulation and governance within capitalism and also links between the formal and informal sectors of capitalist economies.[19] Put another way, it requires understanding how practices, institutions and structures interrelate in the reproduction of capital (understood as a social relationship).

This in turn, however, requires acknowledging that the commodity form within capitalism is a slippery one, temporally and spatially (Appadurai, 1986), and that the social structural relations of capital intersect with those of other social structures (such as ethnicity or gender) in varying ways. While there may be coevolution of structures, this is a variable and contingent process. Massey (1995: 303–304) recognizes that there are broad social structural relations – of class, gender and ethnicity, for example – which have determinate though nondeterministic effects. Recognition of such broad structures 'is not the same as the commitment to, or the adoption of, a metanarrative view of history. None of the structures . . . need to be assumed to have any inexorability in their unfolding . . . outcomes are always uncertain, history and geography have to be made'. These effects are determinate rather than deterministic precisely because of the multiplicity of structures, the conjunctural specificities of which combination of structures intersect and interact in a given time/space (which may also activate specific 'local' contingencies),[20] and the emergent properties of practices.

The process of commodification brings about, albeit unevenly, the extension and penetration of capitalist mechanisms and forms into aspects of the world and lifeworld from which they were previously absent. However, these processes result in uncertainty about the fate of commodities once they have been sold. The purchase of commodities depends (*inter alia*) upon the meanings that consumers attach to them. Consumption is one source of meaning and identity, both for those purchasing the commodity and those consuming it (for example, the recipient of a gift). There are claims that we are what we eat, what we wear and so on and, beyond those, that the body itself has become an accumulation strategy, with bodies worked on in terms of physical fitness, health clubs and plastic surgery to reshape various parts of human anatomies in socially sanctioned ways. Goods acquire meaning and value, becoming 'culturally drenched' and so taking on 'identity values', expressed in rituals around possession and the giving of gifts, for example (Featherstone, 1991). However, such identity values are subject to change and renegotiation. Not least this is because commodities are manufactured with their own preplanned trajectories, with built-in obsolescence within a product life cycle. As commodities reach the end of their socially useful lives to their original purchasers, they may be 'sold on', both formally and informally in a variety of spaces (such as street markets and car-boot sales). In this process, the meanings attached to commodities by their original purchasers are typically reworked (as, often, are the things themselves) so that there are recursive circuits of things and meanings rather than simply a linear path or a single circuit of meaning.

However, commodity production and consumption are also often complex processes of material transformations. The resultant 'environmental footprint' of these activities emphasizes the critical grounding of 'economies' in nature. Elements of 'first nature' become increasingly commodified while a 'second nature' is also increasingly produced from within the social relations of capital. There is a significant difference between the appropriation of an 'external' first nature into capitalist social relations and producing a second nature within those relations. With recent developments in biotechnologies even life itself has become capitalized and produced as part of second nature (Franklin *et al.*, 2000).

Finally, notwithstanding the increasing production of nature as second nature, within capitalist economies there remain 'economic activities' that are not under the direct sway of capitalist relations of production, both within and outside the spaces of capitalism (Figure 2). This raises questions as to how capitalist and noncapitalist economies relate to one another and about strategies of 'accumulation by dispossession' – that is, (forcibly) taking things/people not produced as commodities and commodifying them (Harvey, 2002).[21] Not least, *the* key requirement of any form of capitalist production – the availability of labour-power – requires that people produced in a noncommodity form become commodified as labour-power, selling their capacity to work on the labour market in exchange for a wage. This requires understanding of the processes whereby people are reproduced as sentient, thinking human beings, conscious agents with their own agendas, pathways and plans – that is, *not* as commodities – and the circumstances in which and the processes through which they become commodified as labour-power.[22]

The key point in terms of conceptualizing 'the economy', however, is that recognizing the existence of noncapitalist social relations within capitalist economies and noncapitalist economies alongside capitalist ones requires considering different concepts and theories of value and other economic categories to those appropriate to the mainstream, formal capitalist economy. It requires consideration of different processes of valuation, in which value is not defined as socially necessary labour time but in terms of some other metric, perhaps in a more multidimensional way that reflects a broader range of cultural and social concerns.[23] This raises the issue of how best to understand processes of production and consumption in these 'alternative' economies and their circuits, flows and spaces, both in themselves and in their (lack of) relationships to the mainstream.[24] This raises questions of political character of political economy and leads into a normative question of future alternatives, of 'sustainable economies' and their spaces.

VI   Cultural economy and political economy: complementary not alternative approaches

While some see cultural economy as an alternative to political-economy approaches, I prefer to see them as complementary perspectives: understanding geographies of economies necessarily needs to embrace both. This does no more than recover a position that was central to classical political economy but that was generally (there were exceptions) denied for many decades following the ascendancy of neoclassical orthodoxy (and that continues to be denied within the discipline of mainstream economics). Nonetheless, such recovery is vital to a more nuanced understanding of economies and their geographies. Thus the objects of analysis can be both taken 'as given' and can be problematized in terms of their discursive and material constitution. For example, consider the central concept of 'market'. A market 'is physically a place, a set of socio-technologies, and a set of practices. . . . Socially it is also a set of rules' (Law, 2002: 24). In contrast, du Gay and Pryke (2002: 2) suggest that 'the turn to culture' reversed the perception that markets exists prior to and hence independently of descriptions of them. A cultural approach indicates the ways in which objects are constituted through the discourses used to describe and to act upon them. As such, economic discourses format and frame markets and economic and organizational relations, '"making them up" rather than simply observing and describing them from a God's-eye vantage point'. This has critical analytical implications since it suggests that 'economic discourse is a form of representational and technological (that is, cultural) practice that constitutes the spaces within which economic action is formatted and framed'. Put slightly differently, the discursive space of the economic decisively shapes the practical spaces of the economy; and vice versa. Discursive and practical spaces are codetermining, coevolutionary.

As such, economic categories (for instance, firms or markets) need to be analysed in complementary ways that acknowledge the processes through which commodities are produced and the meanings of commodities created, fixed and reworked and the political-economic structural relations in which people are unavoidably located. What is required is a culturally sensitive political economy that begins from the assumption that the economy is – necessarily – always cultural and a politically sensitive cultural economy that is alert to the power geometries and dynamics of political economy. These provide complementary approaches, viewing the economy from different analytic windows rather than an 'either/or' ontological and epistemological choice. Indeed, these approaches in some respects interpellate one another rather than being discrete and self-contained. As such, the space currently occupied by cultural-economic divisions and reductions could be at least partially reconstructed by treating concepts such as competition, markets, products and firms as *both* lived realities *and* formal categories (cf. Slater, 2002: 76). Indeed, it could reasonably be argued that Marxian political economy has always contained strands of both approaches (Anderson, 1984).

VII Reconsidering the issues

Given the above, I now want further to explore in a preliminary way two sets of inter-related issues. First, the conceptualization of relations between agents, practices, representations and structures and their varying temporalities (Figure 1), using the notion of practice as what people do in the economy as a way of better grasping relationships between agency and structure by emphasizing doing rather than just thinking, the material and affective as well as the cognitive.[25] Law (2002: 21–23) defines practices as 'materially heterogeneous relations' that 'carry out and enact complex interferences between orders or discourses'. As such, economic practices in their various and multiple specificities interfere in different and specific performances with other, alternative strategies and styles. Moreover, this interference and multiplicity produce an 'irreducible excess', which is necessary to the survival of discourses and performances grounded in them. Secondly, there is the issue of the conceptualization of relations between spaces, flows and circuits, addressing the question of how to explain which parts of circuits are 'fixed' in which spaces for a given period of time. Three points can be made briefly in relation to this second question. First, spaces must be understood relationally, as socially constructed. Secondly, economic process must be conceptualized in terms of a complex circuitry with a multiplicity of linkages and feedback loops rather than just 'simple' circuits or, even worse, linear flows[26] (though for convenience the terminology of 'circuits' is used below: see Jackson, 2002, for a similar argument[27]). Thirdly, the economy must be conceptualized as a complex system, *a fortiori* given recognition that it involves material transformations and coevolution between natural and social systems.

There are two important implications of 'complexity' in this context. First, economic practices may have unintended as well as, or instead of, intended consequences, because people chronically act in circumstances of partial knowledge. Secondly, 'complexity' implies emergent properties that *may* lead to a change between developmental trajectories rather than simply path-dependent development along an existing trajectory. There is a danger that concepts of path dependency (especially if grounded in biological analogy) can lead to an underestimation of the role of agency and reduce actors to 'cultural dupes' (Jessop, 2001). People thus cease to be knowledgeable actors and come to be regarded as the passive bearers of habits, norms and routines (much as structuralist readings of Marx reduced them to passive bearers of structures). As a result, the concept of path contingency better expresses the possibilities of moving between as well as along developmental paths (Hardy, 2002). Actions and practices and systemic interactions may create emergent properties that alter, incrementally or radically, the direction of developmental trajectories. Consequently, evolutionary paths may be far from straightforward. As such, recognition of complexity and emergent properties can aid understanding of a shift from a simple evolutionary perspective of change along a given trajectory to evolution understood as a change from one trajectory to another.[28]

There has been a lively – at times, heated – debate as to the conceptualization of contemporary economy and society in terms of circuits, flows or spaces, and of the relations between them. Some argue that 'fixities' no longer matter, or at least matter less, in a world of flows and (hyper)mobilities (Castells, 1996; Urry, 2000a). There is undeniably evidence of greater mobility, albeit unevenly, across a wide range of activities and spatial scales.[29] Yet, for social life to be possible, for the economy to be performable, fluid sociospatial relations and flows require a degree of permanence, of fixity of form and identity – whether in terms of the boundaries of the firm, of national states or of local places.

However, there is also a dialectic of spaces and flows and circuits, centred on the *necessary* interrelations of mobilities and fixities, of spaces, circuits and flows. Circuits and flows require spaces in which their various 'stages/phases' can be performed and practised, while stretching social relations to create spaces of different sorts, fixing capital in specific time/space forms and ensembles (Hudson, 2001: Chapter 8). Spaces are both discursive and material. Material spaces are constituted as built environmental forms, a product of materialized human labour. Discursive spaces enable meanings to be both contested and established, permissible forms of action to be defined and sanctioned, and inadmissible behaviour to be disciplined. Recognizing that spaces are discursively as well as materially constructed implies that this process does not simply describe the economy. It is also in part constitutive of it, defining the economy as an object of analysis, constructing the spaces of meaning and the meaning of the spaces in which the economy is enacted and performed. These spaces of meaning then become guides to social and individual action. The same point can be made about concepts of circuits and flows, which are also constitutive rather than simply descriptive. As such, spaces, flows and circuits are socially constructed, temporarily stabilized in time/space by the social glue of norms and rules, and both enable and constrain different forms of behaviour.

Spaces, flows and circuits are thus both the medium and products of practices (over varying timescales), based on human understandings and knowledges. Moreover, flows, circuits and practices are also instituted, situated in specific time/space contexts. As such, they are socially constructed and shaped (but not mechanistically determined) by prevailing rules, norms, expectations, and habits and by dominant power relations. As Law (2002: 24) remarks of factories, markets, offices and other spaces of the economy, each is 'a set of socio-technologies and a set of practices. But socially it is also a set of rules'. Such spaces are thus simultaneously materially constructed, a fixation of value in built form, a product of and an arena for practices, defined and regulated by socially sanctioned rules which prescribe or proscribe particular forms of behaviour. In this sense there are structural limitations on action and understanding but, reciprocally, these limitations are a product of human action, beliefs and values: structures are both constraining and enabling. Structural constraints are most powerful when they are hegemonic, taking effect because they have become taken-for-granted, unquestioned determinants of everyday behaviour (Gramsci, 1971). Everyday routine then – even if unintentionally and unconsciously – reproduces these structural relations. Not least this is because of the existence of 'enabling myths' (Dugger, 2000), deeply embedded in the beliefs and meanings in which such routine is grounded, which have the effect of 'naturalizing' the social and reproducing the structural. However, as structures do not exist independently of human action and understanding but are always immanent, contingently reproduced, they are in principle changeable. This is a key theoretical point and – potentially – one of immense political importance.

Bourdieu catches this sense of hegemony via his concept of habitus. Habitus emphasizes the doxic (taken-for-granted, unthinking) elements of action, social classification and practical consciousness. He (1977: 72) argues that the structure of a particular constitutive environment produces 'habitus, systems of durable, transposable dispositions, as structured structures, that is, as principles of the generation of practices and representations which can be objectively regulated and "regular" without in any way being the product of obedience to rules'. They are 'objectively adapted to their goals without presupposing a consensus aiming at ends or an express mastery of the operation necessary to attain them, and being all this collectively orchestrated without being the product of the orchestrating actor of the conductor'. Bourdieu (1981: 309) later makes a critical point in insisting that habitus is 'an analytic construct, a system of "regulated improvization", or generative rules that represents the (cognitive, affective and evaluative) internalization by actors of past experience on the basis of shared typifications of social categories, experienced phenomenally as "people like us"' that varies by and is differentiated between social groups. Crucially, however, 'because of common histories, members of each "class fraction" share similar habitus, creating regularities of thought, aspirations, dispositions, patterns of action that are linked to the position that persons occupy in the social structure they continually reproduce'. While Bourdieu refers specifically to 'class fractions', commonality of experience and identity could as well be based on 'people like us' defined via other social attributes, such as ethnicity, gender or place of residence. Furthermore, historical processes of class formation will reflect the intersection of structures of class relations with those of other social structures (cf. Massey, 1995: 301–305).

VIII   Taking stock

We need to take what people do and their reasons for doing it, their actions and performances, seriously if we are to understand how structures are (un)intentionally (re)produced and constitute 'guides to action', informing social agents of appropriate ways of 'going on'. For example, capitalists and workers behave in particular ways because they understand the world in terms of a specific class structural representation of capital:labour relations. Nationalists and regionalists behave in particular ways because of their understanding of the world as principally organized around shared territorial interests and identities. Moreover, such behaviour may well be paradoxical precisely because social actors behave in circumstances beyond their control. For example, radical trades unionists go to work, even though they understand the capitalist labour process as exploitative, since on a quotidian basis they and their families need to eat, to have a place to live and so on.

The 'economy' is thus instituted, based on shared understandings, discursively established, regarding 'proper' behaviour and conduct by the owners and managers of capital and the vast variety of workers in factories, offices, shops, consumers and so on. However, these shared understandings and resultant practices/performances are structured by understandings of what capitalist production *necessarily* requires (a sufficient mass and rate of profit) and of how this can be produced. As such, they are shaped by and simultaneously help reproduce structural constraints and the materiality of the economy. Thus, capitalist business is based in a material culture of relations between people and things that ranges from the vast number of intermediaries required to produce trade, through the wide range of means of recording and summarizing business, to the different arrangements of buildings (spaces of work) that discipline workers' bodies. These devices and arrangements 'are not an aid to capitalism; they are a fundamental part of what capitalism is' (Thrift, 1999: 59). Of course, not least this is the case because a large part of these 'aids' is produced as commodities.[30]

The recognition of different arrangements of buildings as spaces of production that discipline workers' bodies touches on an important aspect of the spaces of economies and the ways in which these are both a medium for and product of human behaviour. More generally, economic spaces, circuits and flows both help produce and are (re)produced by performance. They both constrain and enable different forms of economic practice. In this way consumers and producers of these spaces both produce and consume their own (formally economic) citizenship. Those who cannot produce or consume in this way cease to be legitimate citizens. Spaces and practices are 'binding agents' in terms of how economies are performed and subject positions created and inflected (Thrift, 2000); the same point can be made about circuits and flows. Alternatively, and simultaneously, they are agents of social exclusion for those denied access to them.

However, relationships between agency, practice and structure are even more complicated because (as the Foucauldian comment about disciplining workers' bodies hints) there are typically contested and competing understandings of what is and what is possible in terms of action and change. For example, there is a struggle within workplaces between managers and workers, a contest to define and dominate the 'frontier of control' (Beynon, 1973). Equally, there are typically competitive struggles between capitalists for markets and profits and among groups of workers seeking to promote their interests in competition with other groups of workers (Herod, 2001; Hudson, 2001). All must also be disciplined to accept the 'rules of the game' of the commodity-producing market economy in conducting these struggles, though these rules vary through time/space. As a result, there is a complicated and multidimensional struggle for domination between competing views of the world and material interests. Consequently, the reproduction of structural constraints is a product of contested processes, unless, of course, one particular view becomes generally if not universally accepted as hegemonic.

IX Conclusions

My focus in this paper has been on conceptualizing capitalist economies and the spaces, circuits and flows through which they are constituted, especially in capitalism's late modern phase. There are claims that this represents a radical break with earlier phases, and that in particular it is marked by an enhanced 'culturalization' of the economy. This is an argument that I broadly reject, not least as it conflates changes in intellectual fashion and perspective with alleged changes in the economy. Practices of production, exchange and consumption, linked to flows of capital, commodities, information and people, are central to the constitution of the spaces of capitalist economies and are neither more nor less cultural than they previously have been (although they may be differently cultural) and neither more nor less material (Lee, 2002). As such, spaces of consumption, exchange and production are linked via a complex circuitry of flows and at the same time constitute the material and discursive spaces through which these flows can and must occur. Circuits, flows and spaces exist in relations of mutual determination, socially produced in historically-geographically variable ways. As Dicken and Yeung (1999: 125) put it, 'we need to recognize the organization and geographical diversity of internationalizing, regionalizing and globalizing processes and forms, . . . together with the multiple scales at which they are enacted'. A corollary of recognizing this diversity is a need for a variety of theoretical and methodological approaches in order to comprehend the economy: for example, political-economic and cultural-economic approaches.

Recognition of the coexistence of spaces of production, exchange and consumption also points to the ways in which the same individual may fulfil different roles and niches within capitalist economies at different time/spaces. People have different motives and rationalities, depending upon their positions in the economy: for example, as capitalists, managers of capital, workers or consumers. They may also participate simultaneously in the social relations of the economy in different ways in the same time/space – for example, as consumer, producer and indirect owner of capital via their participation in company pension schemes. Because of this multiplicity of positions in the social relations of the economy, people develop multiple understandings of capitalist economies and their implications, depending upon their own variable positionality, not just in terms of class relations within capitalism but also in terms of ethnicity, gender, age and so on. This has manifold implications for the creation of (multiple) identities within the circuits and spaces of economies. Once again, the cultural and economic fold into one another, with profound implications for political action.

Finally, I have claimed in this paper that cultural economy and political economy both can and should be regarded as complementary perspectives. It is the case, however, that there is comparatively little evidence that those practising economic geography operate in this way.[31] There seems to be a mild to strong tension on both sides, and quite a bit of mutual suspicion. Specifically in terms of research priorities, there is precious little agreement. Culturalists tend to see political-economic geographers' choices of research foci as unnecessarily centred on the formal economy/production/markets (and indeed certain industries) while political economists often see culturalists' research choices to be almost wilfully trivial (for example, car-boot sales, or local exchange and trading systems (LETS), which are certainly conceptually intriguing but hardly pivotal to contemporary capitalism). Seen in this sense, one group seeks to centre the economy, the other to decentre it. However, another way of approaching the issue is to recognize that what is at stake here is precisely what *is* defined as and taken to be 'the economy', and the ways on which different social relations constitutive of different versions of economies are thought of both in everyday practices of production, exchange and consumption and in social scientific practices that seek to make theoretical sense of them both from the 'bottom up' and from the 'top down'. Carrying out such research certainly requires theoretical sophistication and recognition of the need for different types of theory but, in addition, empirical research of a variety of sorts that is grounded in these different theorizations and their conceptions of valid evidence and knowledge. So, for example, more thoroughly understanding the geographies of the automobile industry would entail political-economic analysis of the production (both as a labour process and as a process of material transformation) and consumption of vehicles but also cultural analyses of work on the production line and of the automobile as one of the iconic commodities of consumption in modern capitalism (exploring its relations to advertising and the production of meanings), and including work on markets for secondhand, vintage and 'retro' cars and extending to the use of the car boot itself as a space of sale for a range of other 'recycled' commodities. Such work, grounded in different conceptions of theory and synthesizing the results of different sorts of research and evidence, will certainly not be easy, but is necessary if there is to be further progress in economic geography.

Acknowledgements

Nicky Gregson, Roger Lee, Jamie Peck and Henry Yeung commented fully and constructively on an earlier draft, while Peter Dicken and three anonymous referees for *Progress in Human Geography* provided equally full and valuable comments on a later one. I am grateful to all of them, and as will be clear at times have explicitly drawn on their comments and suggestions, although I have no doubt that I have failed adequately to deal with issues they raised. The usual disclaimers apply.

Notes

1.   As one of the referees of the paper rightly pointed out, however, there are strands of the economic geography literature that prefigure this more general concern with relationships between the cultural and the economic (for example, from Buchanan, 1935, to Harvey, 1973: 195–284, to Lee, 1989). I can only plead guilty to the charge of 'historical shortsightedness' that (s)he levels at me. In my (partial) defence, I would argue that I am not alone in this regard as economic geographers have generally sought inspiration in links with other social science disciplines rather than in the history of thought in their own subdiscipline (though see Barnes, 2001) but will be happy on a future occasion to seek to correct this shortcoming.

468   Conceptualizing economies and their geographies

2.   The paucity of my remarks on the latter partly reflect limitations of space and certainly should not be seen as suggesting that empirical research is somehow seen as less important. The value of theory lies in capacity to inform and guide empirical research rather than theory being a substitute for it.

3.   Gough (2003) takes me to task for not rigorously deducing such concepts from the value categories of capital (see Hudson, 2001). But to do so would be to seek a single totalizing metanarrative account that can explain anything and everything.

4.   Economic geography was handicapped for some time by the legacies of 1960s spatial sciences approaches to location theory that sought to develop specific partial equilibrium theorizations of the space economies of exchange, sale and production – consumption as such was simply ignored.

5.   The significance of knowledge and learning in the contemporary economy has recently been emphasized (Giddens, 1990). However, the key issue is the new ways that knowledge is important economically, not that knowledge has suddenly become economically important.

6.   Institutional approaches are discussed more fully in section V.

7.   Hollingsworth (2000: 614–15) suggests that a social system of production 'is the way that a society's institutions, its institutional arrangements and its institutional sectors are integrated into a social configuration. A society's modes of economic governance and co-ordination and its institutional sectors develop according to a particular logic . . . institutions and institutional arrangements within sectors are historically rooted'. While a useful elaboration, this emphasis on 'sectors' suggests only a partial grasp of the institutions and processes through which the social relations of capital are (re)produced.

8.   One of the strengths of the commodity chain approach is that is seeks to connect consumption and production practices: see Smith *et al.*, (2002).

9.   If this smacks of metanarrative, it is only one, not the only one, of relevance here.

10.   Though Beynon's work is cast more in the mould of political economy and economic sociology than cultural analysis.

11.   There are parallels here to the definition of an economic sector.

12.   The economy conceptualized as a heterogeneous networked association of people and things is a more general characteristic of Actant-Network Theorists such as Law and Latour (1987).

13.   Thus capitalist production is always more than just the production of commodities by means of commodities (as neo-Ricardians such as Sraffa, 1960, argue). Furthermore, classical political economy embraced relations between economy and environment as well as recognizing the cultural constitution of 'the economy', as noted above.

14.   Methodologically, this involves ethnographic and participant-observation approaches as well as interview-based approaches. However, in several respects such approaches are not new. Hermeneutic/ interpretative and ethnographic approaches have a long history, both generally in the social sciences and specifically in seeking to understand economic forms and practices (albeit not in the framework of cultural economy: for example, see Beynon, 1973).

15.   Law (2002) makes the same point about relationships between change in the world that social scientists analyse and changes in style and fashion of analysis that they deploy.

16.   The distinction between 'radical', 'old' and 'new' institutional analysis is important. 'New' institutionalism is close to the mainstream orthodoxies (Williamson, 1975). 'Old' institutionalism emphasizes the institutional and social embeddedness of 'the economy' (as expressed in the work of Common, Polyani or Veblen: see Hodgson, 1993). Radical institutionalism emphasizes issues such as asymmetrical power relations in shaping economic life (Dugger, 1989; 2000).

17.   Thrift (1999: 59) notes the irrelevance of most formal economic – that is mainstream orthodox – theory to business practices. Indeed, these seem more in tune – albeit only implicitly – with heterodox theories which emphasize the need to extract surplus-value from workers, to ensure the creation of monopoly rents via product and organizational innovation, and so on. Equally many trades' union practices have been shaped more by Marxian political economy than by mainstream orthodoxies. Much public-sector economic policy is guided by mainstream theoretical orthodoxies, which helps account for the chronic disjunction between policy intentions and outcomes, although there is an increasing influence of various heterodox strands of thought on local and regional economic development policy.

18. In turn, methodologically, it requires a variety of forms of evidence (quantitative and qualitative) relating to concepts such as firms, markets or labour productivity.

19. For comparable arguments, see McFall (2002).

20. Miller (2002: 166) likewise emphasizes the degree to which the political economy around us is the result of 'structural conjunctions', as different structures interact in specific time/space conjunctures.

21. Historically, within radical political economy this was reflected in literatures on the articulation of modes of production (for example, see Amin, 1977).

22. There are circumstances in which slavery and indentured labour become mechanisms to assure the supply of labour-power outside the normal sphere of market transactions.

23. There is a connection here between concepts of space (defined in terms of capital's one-dimensional interest in locations as a source of profit) and place (reflecting people's multidimensional attachments to a location via relations of family, friendship and community): see Hudson (2001: Chapter 8).

24. For example, conceptualizing the relationship between the mainstream formal economy and the social economy or 'Third Sector' is discussed in Amin *et al.* (2002).

25. Deploying practice in this way may well be anathema to nonrepresentational theorists.

26. Such as commodity chains (Gereffi and Korzeniewicz, 1994) or production seen as a linear series of materials transformations (Jackson, 1996).

27. 'My argument involves a move from linear commodity chains to more complex circuits and networks as a way of subverting dualistic thinking and unsettling the kind of linear logic that sees consumption at one end of a chain that begins with production' (Jackson, 2002: 5).

28. It is an open question as to whether emergent properties lead to changes within the parameters of capitalist social relations or to a shift onto alternative noncapitalist paths. This also raises key political questions about 'steering' changes along or between developmental trajectories (Hudson, 2001: Chapter 9).

29. Damette introduced the concept of hypermobility of capital in 1974 (see Damette, 1980). Thus the notion that the capitalist economy has suddenly 'speeded up' in the last decade or so requires some careful reconsideration.

30. There are strong echoes here of Sraffa's (1960) neo-Ricardian account of capitalist production as the production of commodities by means of commodities.

31. This point was made forcible, and fairly, by one the referees' of the paper. Much of this final paragraph is directly derived from his/her comments, a debt that I am happy to acknowledge.

References

Amin, A. and Thrift, N. 2004a: Cultural economy: the genealogy of an idea. In Amin, A. and Thrift, N., editors, *Cultural economy: a reader*, London: Sage, in press.

——, editors 2004b: *Cultural economy: a reader*. London: Sage, in press.

Amin, A., Cameron, A. and Hudson, R. 2002: *Placing the social economy*. London: Routledge.

Amin, S. 1977: *Unequal development*. Lewes: Harvester.

Anderson, P. 1984: *In the tracks of historical materialism*. London: Verso.

Appadurai, A., editor 1986: *The social life of things*. Cambridge: Cambridge University Press.

Barnes, T. 2001: 'The beginning was economic geography' – a science studies approach to disciplinary history. *Progress in Human Geography* 25, 521–44.

Baumann, Z. 1992: *Intimations of postmodernity*. London: Routledge.

Beynon, H. 1973: *Working for Ford*. Harmondsworth: Penguin.

Bourdieu, P. 1977: *Outline of a theory of practice*. Cambridge: Cambridge University Press.

—— 1981: Men and machines. In Knorr-Cetina, K. and Cicourcel, L., editors, *Advances in social theory and methodology*, Boston: Routledge and Kegan Paul, 304–18.

Brenner, N., Jessop, B., Jones, M. and MacLeod, G. 2003: *State/space: a reader*. Oxford: Blackwell.

Buchanan, R.O. 1935: The pastoral industries of New Zealand. *Institute of British Geographers Publication Number 2*. London: Philip and Son, 99 pp.

Castells, M. 1996: *The rise of the network society*. Oxford: Blackwell.

Clark, G.L., Feldman, M.P. and Getler, M.S., editors 2000: *The Oxford handbook of economic geography.* Oxford: Oxford University Press.

Damette, F. 1980: The regional framework of monopoly exploitation. In Carney, J., Hudson, R. and Lewis, J., editors, *Regions in crisis: new perspectives in European regional theory,* Beckenham: Croom Helm, 76–92.

Dicken, P. and Yeung, H. 1999: Investing in the future. In Olds, K., Dicken, P., Kelly, P., Kong, L. and Yeung, H.W.-C., editors, *Globalization and the Asia Pacific: contested territories,* London: Routledge, 107–280.

Du Gay, P. and Pryke, M. 2002: Cultural economy: an introduction. In du Gay, P. and Pryke, M., editors, *Cultural economy,* London: Sage, 1–20.

Dugger, W.M. 1989: Radical institutionalism: basic concepts. In Dugger, W.M., editor, *Radical institutionalism: contemporary voices,* Westport, CT: Greenwood Press, 1–20.

—— 2000: Deception and inequality; the enabling myth concept. In Pullin, R., editor, *Capitalism, socialism and radical political economy,* Cheltenham: Edward Elgar, 66–80.

Featherstone, M. 1991: *Consumer culture and postmodernism.* London: Sage.

Fine, B. and Leopold, E. 1993: *The world of consumption.* London: Routledge.

Franklin, S., Lury, C. and Stacey, J. 2000: *Global nature, global culture.* London: Sage.

Gereffi, G. and Korzeniewicz, M., editors 1994: *Commodity chains and global capitalism.* Westport, CT: Greenwood Press.

Giddens, A. 1990: *The consequences of modernity.* Cambridge: Polity.

—— 1991: *Modernity and self-identity: self and society in the late modern age.* Cambridge: Polity.

Glaeser, E.L. 2000: The new economics of urban and regional growth. In Clark, G.L., Feldman, M.P. and Gertler, M., editors, *The Oxford handbook of economic geography,* Oxford: Oxford University Press, 83–98.

Gough, J. 2003: Review of 'producing places'. *Economic Geography* 79, 96–99.

Gramsci, A. 1971: *Selections from the prison notebooks.* London: Lawrence and Wishart.

Gregson, N. 1995: And now it's all consumption? *Progress in Human Geography* 19, 135–41.

Habermas, J. 1976: *Legitimation crisis.* London: Heinemann.

Hall, S. 1991: The local and the global: globalization and ethnicity. In King, A.D., editor, *Culture, globalization and the world system,* London: Macmillan, 19–30.

Hardy, J. 2002: An institutionalist analysis of foreign Investment in Poland: Wroclaw's second great transformation. Unpublished PhD thesis, University of Durham.

Harvey, D. 1973: *Social justice and the city.* London: Arnold.

—— 2002: Reflecting on 'The Limits to Capital'. Paper presented to the Annual Conference of the Association of American Geographers, Los Angeles, 19–23 March.

Herod, A. 2001: *Labor geographies: workers and the landscapes of capitalism.* New York: Guilford.

Hodgson, G. 1993: *Economics and evolution: bringing life back into economics.* Cambridge: Polity.

—— 1988: *Economics and institutions: a manifesto for modern institutional economics.* London: Polity.

Hollingsworth, J. Rogers 2000: Doing institutional analysis. *Review of International Political Economy* 7, 595–640.

Hudson, R. 2001: *Producing places.* New York: Guilford.

Jackson, P. 2002: Commercial cultures: transcending the cultural and the economic. *Progress in Human Geography* 26, 3–18.

Jackson, T. 1996: *Material transformations.* London: Routledge.

Jessop, B. 2001: Institutional (re)turns and the strategic-relational approach. *Environment and Planning A* 33, 1213–35.

Krugman, P. 2000: Where in the world is the 'New Economic Geography'. In Clark, G.L., Feldman, M.P. and Gertler, M., editors, *The Oxford handbook of economic geography,* Oxford: Oxford University Press, 49–60.

Lash, S. and Friedman, J., editors 1992: *Modernity and identity.* Oxford: Blackwell.

Lash, S. and Urry, J. 1994: *Economies of signs and space.* London: Sage.

Latour, B. 1987: *Science in action: how to follow scientists and engineers through society.* Milton Keynes: Open University Press.

Law, J. 2002: Economics and interference. In du Gay, P. and Pryke, M., editors, *Cultural economy,* London: Sage, 21–38.

Lee, R. 1989: Social relations and the geography of material life. In Gregory, D. and Walford, R., editors, *Horizons in human geography,* London: Macmillan, 152–69.

—— 2002: 'Nice maps, shame about the theory'? Thinking geographically about the economic. *Progress in Human Geography* 26, 333–54

Lee, R. and Wills, J., editors 1997: *Geographies of economies.* London: Arnold.

Massey, D. 1995: *Spatial divisions of labour: social structures and the geography of production.* London: Macmillan.

McFall, L. 2002: *Advertising, persuasion and the culture/economy dualism.* In du Gay, P. and Pryke, M., editors, *Cultural economy,* London: Sage, 148–65

McRobbie, A. 1997: Bridging the gap: feminism, fashion and consumption. *Feminist Review* 55, 73–89.

Miller, D. 1987: *Material culture and mass consumption.* Blackwell: Oxford.

—— 2002: The unintended political economy. In du Gay, P. and Pryke, M., editors, *Cultural economy,* London: Sage, 166–84.

Offe, C. 1975: The theory of the capitalist state and the problem of policy formation. In Lindberg, L.N., Alford, R., Crouch, C. and Offe, C., editors, *Stress and contradiction in modern capitalism,* Lexington: DC Heath, 125–44.

Ray, L. and Sayer, A. 1999: Introduction. In Ray, L. and Sayer, A., editors, *Culture and economy after the cultural turn,* London: Sage.

Sassen, S. 2003: Globalization or denationalisation? *Review of International Political Economy* 10, 1–22.

Sheller, M. and Urry, J. 2000: The city and the car. *International Journal of Urban and Regional Research* 24, 737–57.

Sheppard, E. and Barnes, T., editors 2000: *A companion to economic geography.* Oxford: Blackwell.

Slater, D. 2002: Capturing markets from the economists. In du Gay, P. and Pryke, M., editors, *Cultural economy,* London: Sage, 59–77.

Smith, A., Rainnie, A., Dunford, M., Hardy, J., Hudson, R. and Sadler, D. 2002: Networks of value, commodities and regions: reworking divisions of labour in macro-regional economies. *Progress in Human Geography* 26, 41–64.

Sraffa, P. 1960: *The production of commodities by means of commodities.* Cambridge: Cambridge University Press.

Thrift, N. 1999: The globalisation of the system of business knowledge. In Olds, K., Dicken, P., Kelly, P., Kong, L. and Yeung, H.W.-C., editors, *Globalization and the Asia Pacific: contested territories,* London: Routledge, 57–71.

—— 2000: Performing cultures in the new economy. *Annals of the Association of American Geographers* 90, 674–92.

Urry, J. 2000a. *Sociology beyond societies: mobilities for the twenty first century.* London: Routledge.

—— 2000b: Time, complexity and the global. Available at http://www.comp.lancs.ac.uk/sociology/soc030ju.html (last accessed 14 August 2002).

Weiss, L. 1997: Globalisation and the myth of the powerless state. *New Left Review* 225, 3–27.

Whitley, R. 1999: *Divergent capitalisms: the social structuring and change of business systems.* Oxford: Oxford University Press.

Williams, R. 1960: Advertising: the magic system. *New Left Review* 4.

Williamson, O. 1975: *Markets and hierarchies.* New York: Free Press.

Zimmerman, E. 1951: *World resources and industries.* New York: Harper and Row.

[6]

Why is economic geography not an evolutionary science? Towards an evolutionary economic geography

Ron A. Boschma* and Koen Frenken**

Abstract

The paper explains the commonalities and differences between neoclassical, institutional and evolutionary approaches that have been influential in economic geography during the last couple of decades. By separating the three approaches in terms of theoretical content and research methodology, we can appreciate both the commonalities and differences between the three approaches. It is also apparent that innovative theorizing currently occurs at the interface between neoclassical and evolutionary theory (especially in modelling) and at the interface between institutional and evolutionary theory (especially in 'appreciative theorizing'). Taken together, we argue that Evolutionary Economic Geography is an emerging paradigm in economic geography, yet does so without isolating itself from developments in other theoretical approaches.

Keywords: evolutionary economic geography, new economic geography, institutional economic geography

JEL classifications: A12, B20, B25, B52, R0, R1

Date submitted: 14 Feburary 2005 **Date accepted:** 12 December 2005

1. Introduction

Since the 'Geographical Turn' in economics, a true *Methodenstreit* has been raging in the field of economic geography (Martin, 1999). From the 1980s onwards, economic geography moved away from traditional economic analysis and transformed into a more interdisciplinary approach using insights from social, cultural and political sciences. This turn has been characterized by the 'Cultural Turn' (Amin and Thrift, 2000; Barnes, 2001) or the 'Institutional Turn' (Martin, 2000) in economic geography.[1] A decade later, following a seminal contribution by Krugman (1991a), neoclassical economists have re-entered the field of economic geography (Fujita *et al.*, 1999;

---

* Section of Economic Geography, Urban and Regional research centre Utrecht (URU), Faculty of Geosciences, Utrecht University, PO Box 80115, NL-3508 TC, Utrecht, The Netherlands.
email: <r.boschma@geog.uu.nl>

** Section of Economic Geography, Urban and Regional research centre Utrecht (URU), Faculty of Geosciences, Utrecht University, PO Box 80115, NL-3508 TC, Utrecht, The Netherlands.
email: <k.frenken@geog.uu.nl>

1 A similar institutional approach exists in economics, yet by far has not gained the support within the community of economists as it did within the community of geographers.

274 • *Boschma and Frenken*

Brakman *et al.*, 2001; Fujita and Thisse, 2002; Puga, 2002), yet met harsh resistance from the side of economic geographers. Neoclassical economists are renewing their interest in geography while geographers are moving away from economics; the debate between economists and geographers has been little fruitful, and is probably best characterized by a 'dialogue between the deaf' (Martin, 2003).

Evolutionary economics can be considered a third approach in economic geography, yet has hardly drawn serious attention. Although it is noticeable that, to an increasing extent, lip service is paid to evolutionary thinking and concepts (e.g., Storper, 1997; Cooke and Morgan, 1998; Martin, 1999; Sjöberg and Sjöholm, 2002; Cooke, 2002; Scott, 2004), there are few systematic attempts to apply evolutionary economics into the realm of economic geography (Rigby and Essletzbichler, 1997; Boschma and Lambooy, 1999; Essletzbichler and Rigby, 2005). According to Martin (2003) evolutionary economics has not (yet) developed into 'a coherent body of theory and empirics' in economic geography. It is even fair to say that evolutionary economists themselves have been somewhat more active in linking evolutionary economics with geographical issues (Arthur, 1987, 1990; Swann and Prevezer, 1996; Antonelli, 2000; Caniëls, 2000; Breschi and Lissoni, 2001, 2003; Bresnahan *et al.*, 2001; Klepper, 2002a; McKelvey, 2004; Brenner, 2004; Werker and Athreye, 2004). Perhaps one of the reasons of the relatively minor impact of evolutionary economics in economic geography so far is that economic geographers tend to refer to evolutionary economics and institutional economics as being more or less indistinguishable.

As reflected in the title, we propose an evolutionary approach in economic geography paraphrasing Veblen's (1898) seminal article *Why is economics not an evolutionary science?* Our main objective is to outline the basic elements of Evolutionary Economic Geography. Before sketching the main contours of this new approach, we show that Evolutionary Economic Geography is reducible neither to the neoclassical approach nor to the institutional approach in economic geography. In order to do so, we first sketch two theoretical developments in economic geography that have been taken place in the last couple of decades; that is, the New Economic Geography around the 1990s and the 'cultural or institutional turn' in economic geography around the 1980s. We explain in Section 2 why the interface between these two strands of thought has shown to be a fertile ground for conflict rather than for exchange. In Section 3, we present three key issues that represent dividing lines within economic geography (and economics): the assumption debate, the use of mathematics, and statics versus dynamics. This framework will allow us to discuss the main similarities and differences between neoclassical, institutional and evolutionary approaches, because we argue that each key issue unites two approaches and differentiates them from the third. We also show the value added provided by the evolutionary approach and claim that Evolutionary Economic Geography indeed puts 'new wine in new bottles'. With this purpose in mind, we compare the Evolutionary Economic Geography approach with the Neoclassical Economic Geography and the Institutional Economic Geography in Sections 4 and 5, respectively. The exchanges along the interfaces are shown to be fruitful and should be further encouraged, although synthesis between the evolutionary approach and the neoclassical or institutional approach is not expected. Rather, an Evolutionary Economic Geography approach is unique in its core assumptions, units of analysis and type of explanations. To support this thesis, we briefly present, in a programmatic manner, the basic outlines of Evolutionary Economic Geography in the final section.

Before introducing the three approaches in economic geography, it should be reminded that our objective is not to discuss and compare each approach in all its details (for this see Nelson, 1995a; Hodgson, 1998; Marchionni, 2004). Consequently, we inevitably dispense some of the nuances. We refer mainly to 'textbook versions' of the three theories, without claiming that modern writings would all perfectly fit into one of the three categories. On the contrary, it should be reminded throughout the text that our stylized differentiation into three approaches primarily serves a heuristic use and ultimately aims to contribute to theorizing at the interfaces between the approaches.

2. *Methodenstreit* in economic geography

Economic geography has been subjected to a lot of turmoil during the last two decades or so (Martin and Sunley, 1996; Amin and Thrift, 2000; Barnes, 2001; Meardon, 2001; Overman, 2004; Scott, 2004). If any 'revolution' has hit economic geography recently, it must be the application of neoclassical economics in economic geography by Krugman (1991a) and others. Below, we refer to this new research programme as New Economic Geography, a term proposed by Krugman, although we share Martin's view that Krugman's models are better characterized as economics than as geography (Martin, 1999).[2] We will also make use of the term Neoclassical Economic Geography, by which we refer to both the pre-Krugman contributions in regional science and the more recent New Economic Geography, as both start from the neoclassical assumptions of utility maximization and the 'representative agent', and both derive model conclusions from equilibrium analysis, as in neoclassical economics.

Krugman's (1991a) approach can best be considered as a recent extension of neoclassical thinking to explain trade, specialization and agglomeration, relaxing the frequently used assumptions of perfect competition and constant returns to scale. It basically is a micro-economic theory that explains the existence and persistence of agglomerations in terms of rational decisions of economic agents. Assuming increasing returns to scale at the firm level and imperfect competition between firms, the contribution of Krugman has been to show that agglomeration can occur without having to assume regional differences or external economies. In particular, with transportation costs falling, a critical transition point is reached when both firms and workers find it more profitable to cluster in one region rather than to spread out over more regions. The transition point depends on the balance between internal scale economies for firms and economies of product variety for consumers related to clustering on the one hand and inter-regional transportation costs on the other hand. What is more is that the core model of Krugman has been shown to be extendable in many directions, including other factors such as congestion and unemployment (Fujita *et al.*, 1999; Brakman *et al.*, 2001; Puga, 2002; for a critical review see Neary, 2001).

Not long before Krugman and others set out their main ideas, the community of economic geographers itself had undergone an important reorientation. We refer to this change as the institutional turn in economic geography. One can view the institutional turn in economic geography as the successful development of the programme of

---

2   Krugman's approach fits within the regional science tradition in geography, which is based on general-equilibrium-analysis from neoclassical economics. Thus, one may better speak of the 'new regional science' or 'geographical economics' (Martin, 1999; Brakman *et al.*, 2001).

276 • *Boschma and Frenken*

institutionalism, which had little success within the boundaries of the economics profession.[3] Having said this, it is important to note that there is not (yet) a fully articulated 'institutional economic geography approach' (Martin, 2000). The same is true for institutional economics, which has never developed into a coherent, systematic paradigm (Hodgson, 1998). Both are better described as a collection of approaches that share common concepts and interests in explaining particular phenomena (Samuels, 1995). For most institutional scholars, the methodological and theoretical pluralism does not reflect incoherence. On the contrary, pluralism lies at the heart of methodology and is to be encouraged, at least if one accepts Institutional Economic Geography as an interdisciplinary and contextual science (Hodgson, 1988).

In its most stringent form, institutional approaches argue that differences in economic behaviour are primarily related to differences in institutions (Hodgson, 1988, 1998; Whitley, 1992, 2003; Saxenian, 1994; Gertler, 1997). Institutional differences can be present among firms (in terms of organizational routines and business cultures) and among territories (in terms of legal frameworks, informal rules, policies, values and norms). Comparative analysis between these units with different institutions can then be related to differences in economic outcomes, such as profit, growth, income distribution and conflicts. It should be noted that this definition of the institutional approach is only partial. One can distinguish between over- and under-socialized accounts, related to putting primacy to institutions and social class regulating individual behaviour or individuals whose rational actions result in institutions (Granovetter, 1985). In economics, for example, the 'old' institutional economics corresponds largely to the over-socialized account, while the 'new' institutional economics (Williamson, 1985) is in line with the under-socialized account (and, in this respect, is closer to neoclassical economics). Our characterization of institutional approaches in economic geography deals primarily with the over-socialized account, because a large part of economic geography research can fairly be characterized as being closer to that account, putting primacy at institutions rather than individual action (Gertler, 1997).[4]

The New Economic Geography and the Institutional Economic Geography have developed independently from each other. There has been some debate between the two (e.g., Amin and Thrift, 2000; Martin and Sunley, 2001), but we agree with Martin (2003) that it has led to little fruitful exchange of ideas so far. On the contrary, debates have been fierce and with little progress. This comes as no surprise, because the

---

3   An exception is transaction costs economics, which has become an important institutional theory in economics (Williamson, 1985). The success of transaction costs economics is most probably related to the fact that both transaction costs economics and neoclassical theory share a micro-economic atomistic view on economic agents. For that same reason, transaction costs economics has hardly found applications in economic geography, a notable exception being Scott (1993).

4   Still, it must be recognized that the division between the two accounts is no longer as sharp as before. In many cases, institutional analyses do no longer explain economic behaviour from institutions alone. In fact, we argue below that the interesting developments in economic geography take place exactly on the interfaces between different approaches; for example, on the institutional/evolutionary interface. Still, for heuristic reasons, we find it useful to characterize the institutional approach in economic geography as an over-socialized account. Central to this definition is the idea that institutions determine the larger part of economic behaviour, and, consequently, differences in economic behaviour and performance can be related more or less directly to differences in institutions. Accordingly, we define institutional approaches in economic geography as an archetype way of reasoning, rather than a coherent school of thought (which it is not).

two strands of thoughts differ in fundamental ways. We understand the clash between the two approaches as reflecting at least two important incommensurabilities.

First, institutional and neoclassical approaches differ in methodology and, they conceptualize space in very different ways. Institutional economic geographers dismiss *a priori* the use of formal modelling and econometric specifications derived from these. Instead, they apply an inductive, often, case-study research approach, signalling out the local specificity of 'real places'. One of the objectives of institutional analysis is to understand the effect of the local specificity of 'real places' on economic development, which is mainly attributed to place-specific institutions at different spatial scales. Thus, an institutional approach takes differences between localities as the starting point of the analysis and analyses how place-specific institutions affect local economic development. In contrast, the New Economic Geography approaches the matter deductively using formal models assuming utility maximization and representative agents, and using equilibrium analysis to come to theoretical conclusions or predictions. Proponents of the latter approach do not value or even reject altogether case-study research highlighting local specificity (e.g., Overman, 2004). The New Economic Geography does not even require differences between regions to exist, be it differences in factor prices or institutional set-ups. Rather, the models start from a 'neutral space' and aim to explain how agglomeration can *occur* from this. Their main goal is to show how uneven spatial patterns can emerge from an initially uniform world and, thus, they abstract from local specificity and different levels of spatial aggregation.

Second, the two approaches differ in their behavioural assumptions underlying explanations of economic phenomena. The New Economic Geography aims to explain geographical patterns in economic activity from utility-maximizing actions of individual agents. Institutional scholars start from the premise that economic behaviour is not described accurately as utility-maximizing but is better understood as being rule-guided. Agents are bounded rationally and rely heavily on the institutional framework they operate in, guiding their decisions and actions. Institutions are embedded in geographically localized practices, which imply that localities ('real places') are the relevant unit of analysis. By doing so, Institutional Economic Geography analyses how institutional specificity affects economic behaviour and thereby local patterns of economic development. In contrast, institutions play no role in neoclassical models, or do only in a loose and implicit sense (e.g., relating to particular parameters in the model) (Olsen, 2002). Local institutional and cultural factors are left out of the analysis, because these are not regarded as essential to an economic explanation and should therefore be 'best left to the sociologists', as Krugman once put it (Martin, 1999, p. 75).

Our argument holds that Evolutionary Economic Geography should be regarded as a third approach in economic geography that differs in turn from neoclassical and institutional approaches. Evolutionary Economic Geography applies core concepts and methodologies from evolutionary economics in the context of economic geography. It provides alternative explanations for the main *explananda* including agglomeration and regional growth differences. The starting point is to open the black box of organizations and to view organizations as competing on the basis of their routines that are built up over time (Nelson and Winter, 1982; Maskell, 2001). Evolutionary models of organizations' decision-making are based on the concept of bounded rationality and routine behaviour, rather than on utility maximization (Simon, 1955a). Routines can be understood as organizational skills, which cannot be reduced to the sum of individual skills (Nelson and Winter, 1982). Routines are manifested at the firm level due to

Economy: Critical Essays in Human Geography

division-of-labour and thereby due to division-of-skills between workers in a firm. Organizational routines, as for individual skills, consist of a large part of experience knowledge (learning-by-doing) and tacit knowledge, which are hard to codify. Both aspects of routines render them difficult to imitate by other firms (Teece *et al.*, 1997). Consequently, organizations are heterogeneous in their routines, and persistently so. Modelling organizations can thus no longer rely on assuming a 'representative agent'. It is this variety that fuels the selection process as an open-ended and out-of-equilibrium process of economic development (Hodgson, 1999). And, as organizations compete on the basis of their routines, and competition is driven by Schumpeterian innovation based on new products and technologies requiring new routines, rather than on production costs alone as assumed in neoclassical models.[5]

Basically, evolutionary economics explains the (changing) distribution of routines as the outcome of search behaviour and selection forces (Alchian, 1950). First, firms learn from their own mistakes through trial-and-error. When routines do not work well, failure induces active search for other routines (Nelson and Winter, 1982); for example, by investing in Research and Development. Evolutionary economics predicts most firms to innovate incrementally and to exploit their knowledge built up in the past. Empirical research shows that while innovations generally increase the life chances of firms (Cefis and Marsili, 2006), major organizational transformations tend to decrease the survival rates of firms (Anderson and Tushman, 1990; Carroll and Hannan, 2000). Organizations can also learn by networking while running the risk of competencies being copied by other firms (Cowan and Jonard, 2003), and by imitating, although imitation is failure-prone because the tacit components of routines are hard to copy (Teece *et al.*, 1997). Second, 'intelligence' also exists at the level of an industry as a whole, analogous to the population level in biology (Nelson and Winter, 1982). As long as firms show routinized behaviour, market competition acts as a selection device causing 'smart' fit routines to diffuse and 'stupid' unfit routines to disappear. In particular, differential profits leading to differential growth rates render fitter routines to become more dominant in an industry. This selection logic is in line with evidence that firm growth is temporally autocorrelated, meaning that some firms persistently grow over time (Bottazzi *et al.*, 2002; Cefis and Orsenigo, 2001; Cefis, 2003; Garnsey *et al.*, 2006).

Evolutionary Economic Geography aims to understand the spatial distribution of routines over time. It is especially interested in analysing the creation and diffusion of new routines in space, and the mechanisms through which the diffusion of 'fitter' routines occurs. Following this reasoning, the emergence of spatial agglomerations is to be analysed neither in terms of rational location decisions, as in neoclassical theory, nor in terms of the set-up of specific local institutions, as in institutional theory, but in terms

---

5 Our definition of evolutionary economics is closest to neo- or post-Schumpeterian economics as defined by Nelson and Winter (1982), Andersen (1994) and Nelson (1995a). We recognize that other evolutionary branches are distinguished in the literature. For example, there is a growing literature on evolutionary game theory, which is close to neoclassical economics in its reliance on equilibrium analysis (Friedman, 1998a, b). Other scholars include 'old institutionalism', which, confusingly, is often referred to as evolutionary economics in the United States (Hodgson, 1998; Martin, 2000). One could also mention complexity theory as a branch of evolutionary economics (or *vice versa*), with its explicit focus on modelling concepts such as path dependence and emergence (e.g., Foster and Holzl, 2004; Frenken, 2006). In particular, Colander (2000) argued that complexity theory is emerging as an alternative modelling paradigm in economics.

of the historically grown spatial concentration of knowledge residing in organizational routines. In this respect, there are several evolutionary mechanisms that may produce the spatial concentration of firms.

Agglomerations may be the result of a process in which chance events become magnified by positive feedbacks at the firm level (Arthur, 1990). As success breeds success through learning, some firms will be lucky and grow out into industry leaders while other firms are unlucky and have to exit. Successful firms also produce more spin-offs, and more successful spin-offs, which almost invariably remain in the region of the parent firm. The resulting industrial and spatial dynamics involve path dependence in firm and regional leadership, and once a spatial pattern has settled historically it becomes largely irreversible. In this case, evolutionary processes lead to spatial concentration in the absence of agglomeration economies (Klepper, 2002b). Spatial agglomeration may also be the result of increasing returns at the regional level. Knowledge not only is embodied in organizational routines in firms, but may also spill over from one firm to the other. As tacit knowledge is hard to be exchanged through contracts in global markets, knowledge spillovers occur more often among geographically proximate agents (Jaffe *et al*, 1993; Breschi and Lissoni, 2003; Verspagen and Schoenmakers, 2004). Agglomeration economies act both as an incentive and as a selection mechanism, explaining why economic activity become more and more concentrated in leading regions, driving out firms in other regions (Malmberg and Maskell, 2002; Boschma, 2004). It must be recognized, however, that the tacit nature of knowledge and routines implies that spillovers do not occur automatically ('in the air') but rely on transfer mechanisms, such as inter-firm collaborations, professional networks and labour mobility (Camagni, 1991; Capello, 1999; Breschi and Lissoni, 2003; Giuliani and Bell, 2005). Although they often are, these mechanisms are not tied to regional levels *per se*, and may even become increasingly detached from local contexts over time (Breschi and Lissoni, 2001).

In the following, we argue that Evolutionary Economic Geography is linking the neoclassical and institutional approaches in that it agrees with the neoclassical approach methodologically (using formal modelling), and it agrees with the institutional approach in terms of behavioural foundations (as captured by the concept of bounded rationality). Given these similarities between the evolutionary approach on the one hand and the neoclassical and institutional approaches on the other hand, one can expect the exchange of ideas along these two interfaces to be fruitful in economic geography. We will therefore explore in detail the interface between Evolutionary and Neoclassical Economic Geography (Section 4) and the interface between Evolutionary and Institutional Economic Geography (Section 5), respectively. In Section 3, though, we first start with a brief description of three key issues in economic geography that are helpful in understanding the nature of the interfaces between the three approaches in more depth.

3. Three key issues in economic geography

Since we plead for an Evolutionary Economic Geography approach that shares certain features and also differs in many ways from the Neoclassical and Institutional Economic Geography, we aim to clarify the similarities and differences with these two latter approaches. Though any attempt to describe and characterize the major theories in any discipline is inherently difficult and complex, we feel that it is useful as a way to differentiate a new approach from existing ones, as well as to show the linkages between the proposed approach and more familiar lines of thought. We will do so by introducing three key issues, which are positioned within the triangle depicted in Figure 1. Each of the issues unites two of the three approaches and differentiates them from the third. The three issues recurrently show up both in the history of economics and in the history of economic geography.

The first issue concerns the usefulness of formal modelling, which unites evolutionary and neoclassical scholars, and differentiates them from institutional scholars. As mentioned before, most institutionalists reject the use of formal modelling because it does not capture the contextual nature of economic and social life (Martin, 2000). According to institutional scholars, formal models take an anti-realist stance almost as a rule, because they exclude place-specific qualitative factors (such as culture and institutions) that are hard to put into 'Greek letter economics', but which are considered essential to the explanation of regional differences (Gertler, 1997).[6] In contrast, neoclassical and evolutionary scholars use formal modelling as a tool in theorizing albeit in slightly different ways.

The second issue centres on what might be called the assumption debate. Evolutionary and institutional approaches share a fundamental critique on the neoclassical assumption of utility-maximizing individuals. As Dosi (1984) once put it, 'we must abandon the neoclassical framework because we cannot assume an exogenous and given context and many God-like actors who behave in accordance with a uniform rationality' (p. 107). In contrast, evolutionary and institutional scholars claim that economic agents are bounded rationals and base their decisions on routines and institutions (Veblen, 1898; Simon, 1955a; Nelson and Winter, 1982). This is not to say that evolutionary and institutional approaches assume that agents do not strive to maximize utility, but that real-world agents are not able to do so due to bounded rationality. Instead, agents have to rely on routines (at the micro-level) and institutions (at the macro-level). Since routines and institutions are context-specific, with routines being specific to organizations, and institutions being specific to territories ('real places'), both approaches reject the atomistic view of neoclassical theory that ignores the contextuality of human action.

The third issue is about the conceptualization of time. Here evolutionary approaches take a critical stand towards static analysis in neoclassical and institutional approaches. Characteristic for evolutionary theory, be it as a theory of natural history in biology or as a theory of economic development in economics, is that it explains a current state of affairs from its history: 'the explanation to why something exists intimately rests on how it became what it is' (Dosi, 1997, p. 1531). Thus, the current state of affairs cannot be derived from current conditions only, since the current state of affairs has emerged from and has been constrained by previous states of affairs. Evolutionary theory deals

---

6   Though institutional scholars often take a realist stance on scientific explanation in social science, it is important to recognize that realist explanations do not exclude the use of mathematics *per se* even though many mathematical models take an instrumentalist stance. Interestingly, Marchionni (2004) claims that Krugman is best regarded as a realist who uses models as a research strategy to come closer to unravelling the complex mechanisms underlying the economy, rather than an instrumentalist who judges mathematical models primarily on the basis of its predictive value. Mäki (1992) and Mäki and Oinas (2004) also argue at length that the use of abstract modelling does not imply an anti-realist stance *per se*.

Towards an evolutionary economic geography • **281**

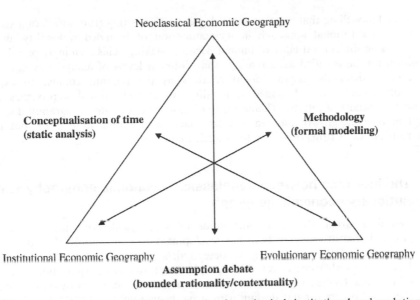

Neoclassical Economic Geography

Conceptualisation of time
(static analysis)

Methodology
(formal modelling)

Institutional Economic Geography

Evolutionary Economic Geography

Assumption debate
(bounded rationality/contextuality)

Figure 1. Three key issues within the triangle of neoclassical, institutional and evolutionary economic geography.

with *path dependent* processes, in which previous events affect the probability of future events to occur. In this view, small events can have large and long-lasting effects due to self-reinforcing processes (Arthur, 1989). In short, history matters (David, 1985).[7] In this respect, evolutionary approaches differ in a fundamental sense from those approaches in neoclassical and institutional thinking that share an interest in static analysis.[8]

Summarizing, the clash between Neoclassical and Institutional Economic Geography can be understood as a result of two fundamental differences, related to methodology (use of formal modelling) and key behavioural assumptions (bounded rationality and routines/institutions guiding decision-making). Evolutionary Economic Geography takes an intermediate position: it agrees with the neoclassical approach in the usefulness

---

7   See also the early critique by Atkinson and Stiglitz (1969) on neoclassical growth theory and the notion of production function. They argued that economic growth is essentially a historical process that cannot be understood without taking into account historical specificity.

8   We do not, however, claim that all institutional approaches make use of static analysis. On the contrary, the evolution of institutions is often an object of study. Hodgson (1998), for example, stresses that institutional economics does not only concern static comparative studies on different institutional regimes, but is also engaged in studies of institutional *change*, which is, very often, described as an evolutionary process (North, 1990). Some, including Samuels (1995), characterize institutionalism as an evolutionary approach, due to its emphasis on process and evolution: 'Veblenian evolutionism is Darwinian in having neither cause of causes nor predetermined end state; it is non-teleological and open-ended' (p. 580). Taking the evolution of institutions as object of study, institutional and evolutionary approaches have more in common than suggested in Figure 1. This proves again that new developments in research are often taking place at the interface of approaches. Still, when institutions are being explained and explanatory, it remains unclear what are the factors that drive institutional change, unless one adopts a teleological approach after all.

of formal modelling that requires some degree of abstracting from local contexts and with the institutional approach in its assumption of bounded rationality and its emphasis on the contextuality of human decision-making. This seemingly paradoxical position can be clarified as stemming from different levels of analysis: evolutionary economics views the organizational routines as the relevant context to explain decision-making under bounded rationality, while institutional approaches start from territorial institutions. Therefore, Evolutionary Economic Geography does not explain regional growth differences from macro-institutional differences, but from micro-histories of firms that operate in territorial contexts.

4. The interface between neoclassical economic geography and evolutionary economic geography

As described earlier, the main contribution of neoclassical economics to economic geography in recent years has been the development of a new family of models based on Krugman's (1991a) core model. As these models are better understood as economic models treating only some aspects of geography (in particular transportation costs), the New Economic Geography has been attacked on various occasions by economic geographers and others for not dealing with 'true' geography (e.g., Martin and Sunley, 1996; David, 1999; Amin and Thrift, 2000; Nijkamp, 2001). Nevertheless, the New Economic Geography can be considered an important contribution to our theoretical understanding of possible mechanisms creating uneven spatial development. We argue that, despite fundamental differences, the New Economic Geography shares some properties with Evolutionary Economic Geography, and can thus be considered to be located at the interface between Neoclassical Economic Geography and Evolutionary Economic Geography. At the same time, we make clear it would be wrong to assume that convergence between the two approaches will necessarily occur. As argued earlier, evolutionary and neoclassical approaches share a common methodology of modelling, including the usage of the concept of neutral space and the possibility of lock-in and irreversibility, yet the two approaches differ in key behavioural assumptions, units of analysis, treatment of time and their conceptualization of agglomeration economies.

The New Economic Geography can be considered as being part of a family of increasing-returns models in neoclassical economics, including growth theory, trade theory and economic geography. The new family of models has replaced the assumption of constant or decreasing returns to scale and perfect competition by the assumptions of increasing returns to scale and imperfect competition. These assumptions better capture the characteristics of most sectors in the modern economy, these being oligopolies with large firms realizing increasing returns to scale internally. As for evolutionary approaches, the New Economic Geography differs in important respects from the traditional neoclassical approaches that typically involve models of ahistorical and reversible processes with a unique optimal equilibrium. In contrast, both in evolutionary and New Economic Geography models, there is the possibility of multiple equilibria, path dependence in the process leading to one of the possible equilibria, irreversibility of outcomes leading the system to lock-in and sub-optimal outcomes.

Another feature both approaches share is that they are keen on explaining how uneven spatial patterns emerge from uniform or 'neutral space'. Even when assuming away regional differences, it is still possible to explain spatial concentration. In New

Economic Geography models, agglomeration occurs when both consumers and firms foresee that it is more advantageous to cluster in one location, thus minimizing transport costs and maximizing profits (increasing returns to scale) and utility (higher variety of consumption goods). The precise location, then, does not matter as long as agents cluster somewhere in space.[9] A similar question preoccupies evolutionary thinking. For example, assuming that new firms are spin-offs, and each firm has an equal probability to create a new firm by spin-off, the resulting locational dynamics can be modelled as a stochastic Polya urn process (Arthur, 1987), leading to skewed spatial distributions of firms. Similarly, Klepper (2002a) explains how Detroit became the capital of the U.S. car industry using a spin-off model assuming that routines are carried over from parent to spin-off, implying that survival rates of parents and spin-offs are correlated. From the 'industry life cycle' model, Klepper (1996, 2002b) derived that early entrants have a higher survival probability than late entrants, because they have more time available to improve their organizational routines than firms entering later in time. Only spin-off firms that enter later but stem from parent firms with fit routines are able to overcome the latecomer disadvantage, because these spin-offs inherit the fit routines of the parent firm. And as spin-offs locate in the same region as the parent firm, firms with fit routines will cluster in geographical space (Klepper 2002a).

The stochastic logic underlying evolutionary models has also been applied to the spatial evolution of networks where new nodes can occur anywhere in space, and connections between nodes are made dependent on both geographical space (negatively) and preferential attachment (positively). Preferential attachment means that a new node prefers to link with a node that is well connected as to profit from its connectivity (Barabasi and Albert, 1999; Albert and Barabasi, 2002). The resulting topology and spatial organization of a network can then be understood as a purely stochastic and myopic sequence (Andersson *et al.*, 2003, 2006) that may generate hubs-and-spokes networks observed in infrastructure networks (e.g., Guimerà and Amaral, 2004; Barrat *et al.*, 2005). Equally, the historically grown network patterns between cities in urban systems can be conceptualized as stemming from preferential attachment (Castells, 1996).

Thus, although the precise modelling techniques and underlying theoretical assumptions greatly differ between evolutionary and neoclassical approaches, both use formal models assuming 'neutral space' to explain the emergence of uneven distributions in an initially even world. Despite these common features, the New Economic Geography and the Evolutionary Economic Geography differ fundamentally on at least four grounds.

First, the New Economic Geography remains firmly within the neoclassical framework using the core assumptions of utility maximization of economic agents and homogeneity of agents ('the representative agent'). In this, it differs greatly from evolutionary theory that is based on a different set of assumptions including bounded rationality, routine behaviour and heterogeneity among agents. While neoclassical models assume a given market structure (monopolistic competition in the case of the New Economic

---

9 This has been called 'putty-clay geography' by Fujita and Thisse (1996): 'there is a priori considerable uncertainty and flexibility in where particular activities locate, but once spatial differences take shape they become quite rigid' (Martin, 1999, p. 70).

Geography), evolutionary models take into account entry, exit and innovation, and let market structure to evolve endogenously. Put differently, the New Economic Geography has rather weak foundations in modern industrial organization (Neary, 2001).

Second, the economic levels of aggregation in the two approaches differ. Neoclassical models address the spatial economy at the macro-level in terms of location decisions of agents (firms and consumers) at the micro-level assuming a given market structure. In this context, Martin (1999) is right in stating that the New Economic Geography is 'unable to tell where it (industrial localization and specialization) occurs, or why in particular places and not in others' (p. 78).[10] In contrast, evolutionary approaches aim to explain the spatial *evolution* of industries and networks at the meso-level of the economy. The spatial evolution of the economic system at the macro-level, then, is addressed in a framework of structural change, in which catching-up and falling-behind of territorial units is analysed in terms of the rise and fall of sectors and infrastructure networks in space (Hall and Preston, 1988), be it at the level of countries (Dosi and Soete, 1988), regions (Boschma, 1997) or cities (Hohenberg and Lees, 1995).[11]

Third, the treatment of dynamics in both theories is different. Although the New Economic Geography models are often interpreted as reflecting the formation of agglomerations in time, its conclusions are based on static equilibrium analysis, as in other neoclassical models. Model predictions are derived by computing the one-off locational choice of all individual agents, such that their joint actions are in equilibrium.[12] In these models, a change in equilibrium is 'caused' by a change in the exogenous parameters and not endogenously in time. For example, a fall in transportation costs or a removal in trade barriers may lead firms to cluster in one region rather than being uniformly distributed in space. It follows that true dynamics are only addressed in terms of comparative static analysis of different equilibrium states with different parameter settings.[13] This aspect of neoclassical models differs from

---

10   Furthermore, regarding the spatial unit of analysis in New Economic Geography models, Neary (2001, p. 551) rightly remarked that 'there is nothing intrinsic to the models that conclusively identifies these units.'

11   Note that analysing regional convergence and divergence in a multi-sector analysis also provides a straightforward theory of spatial leapfrogging (Martin and Sunley, 1998), in which regions specializing in new sectors take over regions locked in mature industries.

12   As noted by proponents of the New Economic Geography (Krugman, 1996, 1998; Brakman and Garretsen, 2003), model outcomes are derived from Nash-equilibria, as in game theory. In this respect, one can consider the New Economic Geography as dealing with location games involving many players. See especially Krugman (1998, p. 11) who stated that new economic geography models can be regarded as 'games in which actors choose locations rather than strategies—or rather in which locations *are* strategies—in which case one is engaged not in oldfashioned static expectations analysis but rather in state-of-the-art evolutionary game theory!' Krugman (1998, p. 11) continues by explaining that evolutionary game theory, as it is used in economic geography models, is just an alternative way to incorporate equilibrium analysis in models with maximizing agents: 'To middlebrow modellers like myself, it sometimes seems that the main contribution of evolutionary game theory has been to re-legitimize those little arrows we always wanted to draw on our diagrams.'

13   According to Martin (1999), history is not regarded as 'real history' in the New Economic Geography: 'there is no sense of the real and context-specific periods of time over which spatial agglomerations have evolved' (p. 76). It is relevant to distinguish between two different meanings of path dependence here. Path dependence may reflect a dynamic process in which small events, magnified by increasing returns, produce spatial outcomes. This meaning of path dependence has been adopted by

evolutionary models, in which economic dynamics only show temporary convergence towards equilibrium to be 'upset' by endogenously determined innovative firm behaviour (Nelson and Winter, 1982). The disequilibrium tendency caused by deviant firm behaviour is not regarded as 'noise' but as the fundamental driving force underlying economic development. Evolutionary economists view the search for supra-normal profits by innovation, called Schumpeterian competition, as the primary dynamic in the economy (moving away from equilibrium), while the erosion of profits due to price competition is only considered as a secondary dynamic (converging to equilibrium). In modelling terms, this implies that the growth and decline of firms, sectors and territories are modelled explicitly in time, assuming some underlying stochastic process to reflect innovation. In this vein, evolutionary economics increasingly makes use of interacting agent models from complexity theory (for a review, see Frenken, 2006). Within the context of economic geography, both simple stochastic models (Simon, 1955b; Arthur, 1987; Gabaix, 1999) and more elaborated models (Klepper, 2002a; Andersson *et al.*, 2003, 2006; Bottazzi *et al.*, 2004; Brenner, 2004; Guimerà and Amaral, 2004; Barrat *et al.*, 2005) have recently been developed.

A final difference between neoclassical and evolutionary approaches concerns the underlying theory of agglomeration economies. As described earlier, the New Economic Geography relies in their explanation of agglomerations on pecuniary rents (increasing returns to scale internal to the firm). Evolutionary approaches, instead, are more interested in agglomeration economies arising from knowledge externalities.[14] In an evolutionary perspective, knowledge spillovers contribute to the self-reinforcing nature of agglomeration economies in which firms locating in a region generate and attract new firms in the same region as knowledge spillovers rise with the number of firms (Arthur, 1990; cf. Myrdal, 1957). At the same time, knowledge spillovers may be responsible for sustained regional variety in technological trajectories as knowledge specific to each technology spills over primarily among proximate firms (Essletzbichler and Rigby, 2005).

A number of research questions follow from the concept of knowledge spillovers in an evolutionary perspective (Feldman, 1999; Schamp, 2002). First, as knowledge can spill over in more than one way (imitation, spin-offs, social networks, labour mobility, collaborative networking), one question is which of the mechanisms of knowledge spillovers are most important (Breschi and Lissoni, 2003). In the particular case of networks, one can ask the question to what extent networks of knowledge spillovers are different from other economic networks, and whether network centrality affects one's

---

New Economic Geography models and some evolutionary models, including the ones developed by Arthur (1989). Another notion of path dependence is employed by evolutionary (but also institutional) approaches, which interpret spatial outcomes as directed and channelled by structures (as embodied in routines and institutions) laid down in the past. Or, as Martin (1999) has put it, 'path dependence does not just "produce" geography as in the 'new economic geography' models; places produce path dependence' (p. 80). To be more precise, it is the dynamic interplay between agency and structure producing specific outcomes in particular places, and leading to real space that are put central in an evolutionary approach (Boschma, 2004).

14   Krugman (1991b) also criticized the notion of knowledge spillovers on empirical grounds when claiming that knowledge flows could hardly be measured: 'knowledge flows are invisible, they leave no paper trail by which they may be measured and tracked' (p. 53). Since, a number of scholars have developed methodologies to indicate knowledge spillovers, in particular, by making use of patent citations as pioneered by Jaffe *et al.* (1993).

ability to absorb such spillovers (Lissoni, 2001; Giuliani, 2005; Giuliani and Bell, 2005). Second, for each of these mechanisms one can analyse whether geographically close or more distant relationships are driving knowledge creation and spillovers (Rallet and Torre, 1999; Malmberg and Maskell, 2002; Bathelt *et al.*, 2004). Lastly, evolutionary theory is likely to contribute to a still unresolved issue about whether regional variety (Jacobs, 1969) or specialization is more favourable for knowledge spillovers (Glaeser *et al.*, 1992). Theoretically, evolutionary theory would predict variety to be more important for knowledge spillovers to occur, at least with regard to knowledge supporting radical innovation, involving a recombination of knowledge. It would also claim that some degree of related variety (defined as complementary capabilities among sectors) is needed to enable effective interactive learning and enhance regional growth (Frenken *et al.* 2005). In addition, evolutionary theory would expect that the effect of regional specialization depends on the stage of the product life cycle of the respective industry (Boschma and Wenting, 2005).

5. The interface between institutional economic geography and evolutionary economic geography

As stated in the introduction, it is quite common to share evolutionary approaches under the umbrella of institutional approaches (e.g., Martin, 2000, p. 83). This association has largely been based on the aforementioned common critiques on neoclassical economics, rather than on the fundamental principles that evolutionary and institutional approaches would share *per se*.[15] Both approaches reject utility maximization and equilibrium analysis, and both stress the important role of institutions in economic development. However, we claim that it is not only confusing but potentially misleading to equate institutional and evolutionary approaches in economic geography. Few people would agree that all studies gathered under the umbrella of institutional geography could equally be called evolutionary and *vice versa*. This is especially true for those studies that assess the impact of particular institutional arrangements on economic performance, but which tend to ignore the role of dynamics central to evolutionary approaches. Conversely, quite some influential evolutionary studies do not include the role of institutions in their analyses (e.g., Arthur, 1987; Klepper, 2002a; Bottazzi *et al.*, 2002). Having said this, it is clear that evolutionary and institutional approaches have more 'family resemblance' than evolutionary and neoclassical approaches, if only in that they both account for the historical and geographical context in the analysis of economic agency (Bathelt and Glückler, 2003; Martin, 2003).

One issue of disagreement, which has been explained earlier, holds that Institutional Economic Geography takes a critical stand towards formal modelling. Evolutionary Economic Geography uses formal modelling as a theoretical tool to derive testable hypotheses, while Institutional Economic Geography tends to dismiss the use of formal models *a priori*. In regional studies, for example, institutionalists call for anti-reductionist qualitative methodologies, in particular in-depth case-study research, to appreciate the complex and multi-faceted nature of regional development. The use of

---

15   Illustrative is that followers of the 'old' institutional economics in the US have somewhat confusingly
     called themselves evolutionary economists (Hodgson, 1998).

qualitative methodologies more or less follows from the nature of theorizing. However, in some cases their core concepts turn out to be hard to operationalize also in qualitative research designs. For example, the notion of 'institutional thickness' (Amin and Thrift, 1994; Keeble *et al.*, 1999) has been influential as a concept in economic geography, but has also been criticized for being a vague concept that can not be accurately measured, let alone that its impact on regional development can be determined and tested (Markusen, 1999). More generally, according to some criticasters, institutional and cultural approaches in economic geography show 'a lack of rigour, lack of hypothesis testing and ill-defined concepts' (Martin, 2003, p. 36).[16] The contributions of institutional approaches in economic geography have thus been, most importantly, theoretical, by suggesting new explanations and mechanisms underlying regional development, and in terms of policy implications, by opening up new discourses on the cultural meaning and heritage of places and the limited transferability of locally rooted economic production (e.g., Gertler, 1997).

Even if research methods often follow from theoretical premises, the use of qualitative research methods does not automatically follow from theoretical premises in Institutional Economic Geography in all instances. For instance, recent network approaches in Institutional Economic Geography could make use of statistical techniques from social network analysis (Wasserman and Faust, 1994) and modelling techniques from graph theory (Barabasi and Albert, 1999; Watts, 2004). However, in their programmatic contribution on relational economic geography, Boggs and Rantisi (2003, pp. 114–115) argue that 'doing relational economic geography' implies, as a rule, a case-study approach. Thus, some seem to have *a priori* objections to the use of quantitative tools, even if theoretical contributions allow for their fruitful application. The same observation has been made recently in Markusen's (2003) reply to institutional economic geographers, in which she pleas to go beyond the oppositional distinction between inductive and deductive research, and between qualitative and quantitative research. Her argument is in line with the methodological foundations of evolutionary economics that has combined what Nelson and Winter (1982) called 'appreciative theorizing' and 'formal modelling' from its very start.

A second more subtle issue in comparing evolutionary and institutional approaches is their treatment of context. While evolutionary approaches start from organizational routines at the firm level, institutional approaches start from institutions at some territorial level(s). Thus, both acknowledge the importance of context in economic decision-making and reject the framework of utility maximization central to the neo-classical paradigm, yet they differ in the precise context that is assumed to underlie economic behaviour. Organizational routines are specific to each firm providing a micro-context that results from the past experience and activities of the firm. Institutions, in contrast, are specific to communities and territories providing a macro context. This institutional context may exert considerable influence on the routines of firms. In this respect, it is meaningful to speak of varieties of capitalism, in the sense that the routines of firms will share many characteristics in one institutional system but will differ from one system to the other (Gertler, 1997; Hall and Soskice, 2001).

---

16   Though cultural studies have become well developed and established in sociology, anthropology and geography, some suggest that these studies suffer from 'conceptual imprecision, theoretical ambiguity and empirical open-endedness' (Martin and Sunley, 2001, p. 10).

Understanding the fitness of routines thus requires an analysis not only of markets but also of institutions as relevant constraining contexts. Having said that, from an evolutionary perspective, to take institutions as explanatory variables in economic analysis is not without conceptual difficulty. While institutions may indeed constrain economic behaviour, as routines should not conflict with territorial institutions, the presence of institutions still allows for heterogeneity in routines among firms. Accordingly, a territory as the unit of analysis is problematic, though not without meaning, as there is no strong reason to assume beforehand that routines are place-specific[17]. Some regions may be characterized by a strong degree of homogeneity in routines, while others may not. Conversely, many firms have multiple sites in different territorial contexts, yet these sites share corporate routines, even if some routines may be adapted to local contexts (Kogut and Zander, 1993; Cantwell and Iammarino, 2003). Thus, despite being a contextual approach, Evolutionary Economic Geography is mainly interested in determining whether, and if so in what way, geography matters, rather than theoretically preassuming that it matters in all cases.[18]

Let us illustrate the previous remarks when dealing with the innovation system approach, which is a good example of fruitful exchange between evolutionary and institutional concepts in geography (Freeman, 1987; Nelson, 1993; Edquist, 1997; Cooke *et al.*, 1998; Cooke, 2001; Asheim and Isaksen, 2002; Simmie, 2005). This approach has its historical roots in evolutionary economics, yet shares many characteristics of an Institutional Economic Geography approach. The initial concept of national systems of innovation, for example, aimed to uncover the institutional setting in a country affecting the interaction patterns between actors involved in the innovation process. As such, it takes the existence of institutions for granted and tries to link differential economic performances to different institutional settings. This approach has later been extended to the regional level (Cooke *et al.*, 1998; Cooke, 2001; Asheim and Isaksen, 2002). More recently, however, evolutionary scholars stress the specificity of sectoral innovation systems and the properties these innovation systems share across regions (Breschi and Malerba, 1997; Breschi, 2000). This sectoral approach suggests that the history of innovation systems, in specific places, should be understood from a *dynamic* perspective, by analysing how institutions have co-evolved with the emergence of a new sector.[19] In doing so, it acknowledges that the implementation

---

17   Boschma (2004) claims that territories can only be called relevant and meaningful units when the idea of routines and competences can be transferred from the organizational level to the regional level. In that respect, the region has become an entity on its own, providing intangible and non-tradable assets based on a unique knowledge and institutional base, which is not accessible for non-local firms. Only in those (quite exceptional) circumstances, one needs to understand the success and failure of firms through their local context (Lawson, 1999).

18   This also requires a multi-level analysis to test at which spatial levels behaviour and performance of firms are conditioned (Van Oort, 2004; Phelps, 2004). Within an evolutionary context, multi-level decomposition measures of selection using Price's equation (Frank, 1998; Andersen, 2004) and of variety using the entropy measure (Theil, 1972; Frenken *et al.*, 2005) are particularly useful.

19   While it may be true that institutions are primarily sector-specific, it may not be excluded that sector-specific institutional models may converge to some extent over time, due to evolutionary forces such as competition, selection and imitation. For instance, a key sector in a country may become so dominant that its institutions (e.g. research system, or property rights) become part of a national system (Hollingsworth, 2000). However, in practice, the transfer of institutional models between sectors is expected to be subject to many problems, due to, among other things, the systemic nature of institutions.

and diffusion of novelty often requires the restructuring of old institutions and the establishment of new institutions (Freeman and Perez, 1988; Galli and Teubal, 1997). A well-known example is the rise of the synthetic dye industry in the second half of the nineteenth century, which induced many institutional changes (such as new scientific and educational organizations and new patent laws), which Germany succeeded to implement, but the UK and the US did not (Murmann, 2003). Another example is a study of the evolution of the UK retail banking industry from the 1840s to the 1990s emphasizing the co-evolution of industrial organization, technology and institutions (Consoli, 2005). Consequently, in an evolutionary framework, the key issue is to analyse the extent to which institutions are flexible and responsive to changes in different places. Institutional differences between regions or nations, in this view, are part of the *explananda*, as institutions co-evolve with processes of technological innovation and industrial dynamics (Nelson, 1995b). When adopting such a co-evolutionary perspective, in which technology, markets and institutions mutually influence each other over time, it becomes apparent that institutional and evolutionary approaches converge.[20]

The question that is still to be answered is how Evolutionary Economic Geography can reconcile the notion of neutral space in formal models (similar to neoclassical approaches) with the concept of real places in real-world cases (as in institutional approaches). In an evolutionary perspective, neither can specific institutions in real places provide a sufficient explanation for differences in regional growth, nor can traditional determinants (e.g., factor prices) from neoclassical growth theory. While these factors certainly constrain the set of regions where growth may occur, they fail to explain why even regions with similar institutions and factor endowments can have different rates and patterns of growth. Consequently, factors related to institutions and factor endowments are to be supplemented by a dynamic analysis at the sector and network level, in which the path dependent and self-reinforcing nature of locational dynamics is at the core of a systematic explanation. As a result, Evolutionary Economic Geography claims that real places emerge from actions of economic agents, rather than fully determining their actions.[21]

When dealing with the emergence of new sectors and new networks in particular regions, Evolutionary Economic Geography has theoretical reasons to assume that

---

In that case, differences between sectoral systems of innovation are likely to co-exist and persist in one territory (Amable, 2000). What this example shows is that a dynamic perspective on institutions is highly relevant and exactly what an evolutionary approach is all about.

20 See also, as an example, a recent application of evolutionary economics in the field of transportation planning by Bertolini (2005).

21 Differences between territories can only be understood as the outcome of a long-term evolutionary process. Therefore, imitation of successful routines or institutions by other territories is inherently difficult and, more importantly, the effects are expected to be very different, depending on the set of routines and institutions in which it is introduced (Gertler, 2003). Consequently, comparative analysis, including benchmarking of regions, has its limitations, because a set of successful micro-routines and macro-institutions cannot simply be carried over to different historical contexts. Comparisons are useful to analyse which dimensions of an innovation system perform relatively poor and require adaptation, but they are less useful in providing solutions to fit the historical context of specific innovation systems. The core problem of policy by imitation concerns the high degree of tacitness and interdependencies that exist between the factors contributing to a successful model (Boschma, 2004). In sum, the trajectory of a territory sets limits on copying an external strategy that owed its success to its roots in an alien environment (Zysman, 1994; Rivkin, 2000).

firms operate in neutral space (rather than for reasons of modelling simplicity, cf. Krugman, 1991a). Place-specific features do not determine the location of new sectors, because the selection pressure of existing spatial structures is still rather weak when new industries emerge. That is, the environment is considered to be of minor importance at the initial stage of development of a sector, because a gap is likely to exist between the requirements of the new firms (in terms of knowledge, skills, etc) and its environment. Utmost, regional conditions may play a generic and rather unimportant role at the start of a new sector, such as providing generic knowledge and skills, functions that are often equally well provided in many other regions (Boschma and Lambooy, 1999). The crucial inputs, being sector-specific knowledge and skills, are to be developed by firms themselves as their organizational routines evolve over time. For this reason, one can expect firms in new sectors to emerge in many different locations. In this context, Storper and Walker (1989) have used the term open windows of locational opportunity to describe the locational dynamics of firms in new sectors, which comes close to the assumption of neutral space in evolutionary models.[22] Over time, windows close again, and, after a shake-out, the industry prospers in few regions only, while remaining marginal in most other regions. Similarly, the spatial evolution of networks can be understood as a process that starts off in neutral space, where many but probably not all locations are candidates to become new hubs. Yet, over time, only few locations will develop a central hub function with high connectivity, and consequently the windows of locational opportunity will close again (Castells, 1996).

Over time, the initial neutral space is transformed in real places as the new sectors and new infrastructure networks become spatially concentrated in some regions according to a path dependent process, and trigger the institutional base of these regions to transform and adapt. The renewal of institutions to become supportive of new economic activity is an outcome of a long process of co-evolution, rather than the initial determinant of new sectors locating in a region (recall the example of Germany's chemical industries at the end of the nineteenth century). Thus, regional development is more about path dependence than place dependence, although some places may be better in renewing their institutions than others. Institutions play only a generic role at the start of a new sector, and become more specific and better developed in those areas where a critical mass of firms locates. Thus, at one moment in time, the same institutional base

---

22   Such an evolutionary approach should, however, not take the notion of neutral space for granted, but, instead, should test it in empirical research. In doing so, neutral space is not confused with empty space, because it would be wrong to rule out the impact of regional conditions when a new industry emerges (Boschma, 1997; Boschma and Frenken, 2003). What we claim is that these regional structures only condition the range of possible behaviour of agents, but do not determine their actual behaviour and location. Consequently, the essence of an evolutionary approach applied to the spatial evolution of an industry is 2-fold: (i) to determine which territories are likely candidates (i.e. endowed with favourable conditions) and which territories can be excluded from the beginning. This provides an answer to what degree the windows of locational opportunity are open when a new industry emerges. (ii) to explore the mechanisms behind the path-dependent nature of the spatial evolution of a new industry. Here we answer the question that which of the candidate region(s) become the winner(s), and why. Such an approach has been adopted in a long-term study of the evolution of the British automobile industry (Boschma and Wenting, 2005). The study demonstrated that a local supply of related industries (such as bicycle and coach making) provided a basis for the emergence of the British automobile sector, but it was the success of early entrants and spin-off companies (especially the ones that had acquired experience in successful parent automobile firms) that contributed to the concentration of the automobile industry in the Coventry-Birmingham area.

Table 1. A comparison of the three approaches in economic geography

Key issues	Neoclassical	Institutional	Evolutionary
Methodology	Deductive	Inductive	Both
	Formal modelling	Appreciative theorizing	Both
Key assumptions	Optimising agent	Rule-following agent	Satisficing agent
	A-contextual	Contextual (macro)	Contextual (micro)
Conceptualization	Equilibrium analysis	Static analysis	Out-of-equilibrium analysis
of time	Micro-to-macro	Macro-to-micro	Recursive
Geography	Neutral space	Real place	Neutral space → real place
	Transport costs	Place dependence	Path dependence

Table 2. Summary of Evolutionary Economic Geography (EEG)

- EEG combines appreciative theorizing (inductive) and formal modelling (deductive)
- EEG takes firms, and their routines, as the basic but not the sole unit of analysis
- EEG assumes the behaviour and success of firms to be dependent primarily on the routines a firm (or its founder) has built up in the past (path dependence)
- EEG views the traditional determinants of firm (location) behaviour as being price signals (neoclassical) and place-specific institutions as conditioning the range of possible (location) behaviours and potential locations, but not determining actual (location) behaviour and locational outcomes
- EEG views institutions as primarily influencing innovation in a generic sense, and as co-evolving with technologies over time and differently so in different regions
- EEG describes the spatial evolution of sectors and networks as a dynamic co-evolutionary process transforming neutral space into real places
- EEG explains regional economic development from the dynamics of structural change at the level of sectors, networks and institutions at multiple territorial levels

of a region may be functioning well for mature industries and may be irrelevant, or even dysfunctional, for emerging sectors. Naturally, the paradox of regional policy holds that it can be effective in conserving economic activity, yet it has difficulties to trigger new economic activity necessary for long-term development (Pasinetti, 1993; Saviotti, 1996).

6. Towards an evolutionary economic geography

To sum up our discussion on neoclassical, institutional and evolutionary approaches in economic geography, we present in Table 1 the similarities and differences between them. The three categories of methodology, key assumptions and conceptualization of time correspond to the interfaces in the triangle presented in Figure 1. For reasons of clarity, we have included geography as an additional category to underline the notions of neutral space and real place. As a first attempt, we also listed in Table 2 the key propositions of the evolutionary approach in economic geography, as discussed throughout the paper.

Methodologically, we can conclude that Evolutionary Economic Geography disagrees with institutional approaches in their dismissal of formal modelling and their reluctance to test statistically theoretical propositions. However, different from neoclassical thinking, evolutionary scholars also acknowledge the value of case studies as tool in appreciative theorizing. Thus, Evolutionary Economic Geography strongly supports 'methodological variety and openness' in economic geography, as recently advocated by Plummer and Sheppard (2000), Markusen (2003) and Scott (2004). Following Nelson and Winter (1982), an evolutionary approach employs formal modelling (being more deductive) as well as 'appreciative' theorizing (being more inductive). Thus, Evolutionary Economic Geography makes use of formal theorizing grounded in more realistic assumptions (like bounded rationality), but it also conducts case-study approaches that analyse regional specificities from a dynamic perspective. In short, evolutionary scholars favour methodological pluralism.

Concerning key assumptions, Evolutionary Economic Geography is closer to the institutional approach in assuming economic action to be contextual rather than driven by maximization calculus. However, while institutional scholars tend to relate behaviour of agents to macro-institutions of territories, evolutionary scholars put primacy on micro-routines of organizations. In this view, price differentials (the neoclassical view) and place-specific institutions (the institutional view) only condition the range of possible behaviours and potential locations of firms, but the actual behaviour and location is largely determined by organizational routines acquired in the past. Having said this, firms are not only victims of their history in time and space: routines can be changed by innovation and relocation also. Accordingly, it is the dynamic interplay between structure and agency that produce the evolution of real places.

As far as the conceptualization of time is concerned, Evolutionary Economic Geography takes an explicit dynamic perspective, in which processes of birth and death of firms and sectors are put central, as well as the role of innovation and the co-evolution of firms/sectors with institutions. In contrast, the New Economic Geography is based on a static account of equilibrium analysis, while institutional approaches often focus, though not exclusively, on quite static analyses of institutions employing case studies and comparative studies. From this, it follows that the notions of neutral space (as assumed in neoclassical models for modelling simplicity) and real place (central to Institutional Economic Geography) can be reconciled in evolutionary thinking by viewing the spatial evolution of new sectors or new networks as a dynamic process transforming neutral space into real places.

To further underline and support our claim that Evolutionary Economic Geography potentially provides a comprehensive framework for theoretical and empirical research in economic geography, we propose a multi-layer scheme as depicted in Figure 2. The micro-unit of analysis in Evolutionary Economic Geography is the firm and its routines (Maskell, 2001). The location behaviour of firms is analysed from a historical perspective. One can make use of behavioural geography, in particular Pred (1967), to develop theoretically informed explanations of location decisions. Like evolutionary economists, adherents of behavioural geography start from bounded rationality, which implies that firms' location decisions are heavily constrained by the past. For example, most firms start from home, and spin-offs typically locate in the region of the parent firm. In both cases, previous decisions taken in a different historical context determine the location decision of a new firm. Furthermore, firms are expected to display a considerable degree of locational inertia. The probability of relocation decreases over time as a firm develops a stable set of relations with suppliers and customers and sunk costs accumulate *in situ* (Stam, 2003). In line with Nelson and Winter (1982) and

Towards an evolutionary economic geography • **293**

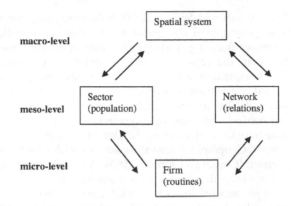

Figure 2. Evolutionary Economic Geography applied at different levels of aggregation.

Cohen and Levinthal (1990), Pred (1967) also emphasized that firms have different capabilities to absorb information about potential locations. Thus, firms are not only imperfectly informed about locations, but are also heterogeneous in their capability to use information in a meaningful way. The final spatial pattern, then, is the outcome of a selection process operating on heterogeneous firms and their location choices. When firms choose, intentionally or by accident, a location that falls within the so-called spatial margin of profitability, they have a better chance to survive and prosper (Smith, 1966).

Taking this one step further, one can assume that some firms develop sophisticated strategies to replicate their routines in different territorial contexts, while other firms continue to pursue strategies in an *ad hoc* manner. Kogut and Zander (1993), for example, argue that successful multinational corporations are those displaying a superior efficiency to transfer knowledge across borders. Also in service firms, systematic replication of routines in new branches constitutes an important part of firms' competitiveness (Winter and Szulanski, 2001). In general, the ability of a firm to replicate its routines in different geographical contexts is expected to contribute to a firm's performance.

Starting from the theory of the firm, Evolutionary Economic Geography applies to two meso-levels; that is, the spatial evolution of sectors and of networks. Firms' relations at the sector level are mainly of a competitive nature, which renders entry-and-exit models and survival analysis obvious techniques for analysis. The core models of the spatial evolution of industry are Simon's (1955b) model on stochastic growth and Arthur's (1987) models on spin-offs and agglomeration economies, while more elaborated methodologies have been developed by Klepper (2002a), Bottazzi *et al.* (2002), Maggioni (2002) and Brenner (2004). Taking a dynamic perspective, the spatial evolution of a new industry in these analyses is described in terms of locations of entry, spin-offs and exits driving the distribution of organizational routines in a population of firms over time (Boschma and Frenken, 2003).

Importantly, in an evolutionary context, spatial concentration (or its absence) is not only an outcome of a process of industrial evolution, but also affects an industry's further evolution. This recursive relationship has, at least, three dimensions

294 • *Boschma and Frenken*

(Hannan *et al.*, 1995; Stuart and Sorenson, 2003; Boschma and Wenting, 2005; van Wissen, 2004). First, geographical concentration of industrial activities can generate agglomeration economies fostering start-ups and innovation and, possibly, the birth of a related industry in the region. Second, geographical concentration of firms increases the level of competition and makes exits of firms raise the average fitness of routines. Third, spatial concentration of firms can also affect the opportunities of collective action as such initiatives are more likely to emerge among proximate agents that can more effectively control opportunistic behaviour.

Networks provide another unit of analysis. One important aspect of networks in Evolutionary Economic Geography is that these act as vehicles for knowledge creation and knowledge diffusion (Cowan and Jonard, 2003). A key research question is then to determine whether knowledge diffusion and innovation is more a matter of being in the right place or in the right network, or in both (cf. Castells, 1996). A recent study found that, using co-inventor data to indicate networks and patent citations to indicate knowledge flows, geographical localization of knowledge spillovers can be largely attributed to social networks and labour mobility (Breschi and Lissoni, 2003). Since most network relationships and job moves are local, with job mobility in turn strongly channelled through network structures, the understanding of knowledge diffusion requires a detailed understanding of the underlying social networks. Using social network analysis, success and failure of economic agents, and regions as an aggregate, can be related to the network centrality of agents within local and global networks of knowledge. As noticed before, this implies that empirical studies on the innovative performance of firms should not take for granted the impact of the region, but should also explore the impact of firm characteristics (competences, market power) and the network position of firms (Boschma and Weterings, 2005; Giuliani, 2005; Giuliani and Bell, 2005). Another important research question becomes to what extent regional and national institutions affect the propensity of agents to network locally and globally (Bathelt and Glückler, 2003).

Apart from analysing network structure, an issue shared with many institutional theories, Evolutionary Economic Geography also aims to explain the spatial evolution of networks. In evolutionary models of network formation, network evolution is understood as an entry process of new nodes connecting with certain probability to existing nodes depending on geographical distance and the latter's connectivity (Barabasi and Albert, 1999; Guimerà and Amaral, 2004; Barrat *et al.*, 2005; cf. Castells, 1996). Well-connected nodes become even better connected nodes rendering the final distribution of connections skewed: networks automatically evolve towards a hierarchy with some nodes becoming highly connected primary hubs, other secondary hubs, while most nodes evolve into poorly connected spokes. A powerful feature of models of network evolution holds that these equally apply to the spatial evolution of social networks between actors as to the spatial evolution of infrastructure networks among locations (e.g., transportation networks, ICT networks, trade networks).

Reasoning from the dynamics of sectors and networks, Evolutionary Economic Geography also applies to the macro-level of the spatial system as a whole. The economic development of cities and regions can be analysed as an aggregate of sectors and networks in a region, and its geographical position in a global system of trade and commerce. The sectoral logic underlying the evolution of spatial systems is better known as the process of structural change (Freeman and Perez, 1988; Pasinetti, 1993; Boschma, 1997). Cities and regions that are capable of generating new sectors with new product life cycles and expanding demand will experience growth, while cities and regions that are locked into earlier specializations with mature life cycles will experience decline. Importantly, there is no automatic economic or political mechanism that assures cities or regions to successfully renew themselves. Rather, one expects localities in most instances to experience decline after periods of growth due to vested interests, institutional rigidities and sunk costs associated with previous specializations.

Theorizing about the network logic underlying the evolution of spatial systems is more recent, in which geographers play a prominent role (Hohenberg and Lees 1995; Castells, 1996). According to these contributions, growth crucially depends on a city's or region's inclusion in global networks of trade and commerce. A central network position can be achieved by attracting corporate headquarters, developing specialized business services and functioning as transportation hubs. Again, one can expect central cities in one era (e.g., based on railways) to be less successful in the next era (e.g., based on airlines) due to institutional rigidities and sunk costs associated with previous infrastructures.

Following from the meso levels of sectors and networks, differential regional growth patterns and processes of convergence, divergence and leapfrogging can be modelled by simulation or econometrically. For example, research interest has been renewed in stochastic models of urban growth using time series on city size. These models investigate *sustained* growth and decline in urban growth thus going beyond the simple logic of Gibrat's Law stating that urban growth rates are stochastic and independent of city size (Pumain, 1997; Gabaix, 1999). Complementary to this research, historical analysis is required to understand the co-evolution of regional economic development and institutional structures underlying the individual regional histories of systematic growth or decline (Nelson, 1995b, 2002). In that respect, institutions can become an integral part of an Evolutionary Economic Geography framework when applied to the analysis of the dynamics of industries, networks and spatial systems.

Having said this, Evolutionary Economic Geography is still at an early stage of development. Some of its fundamental concepts, such as routines and path dependence, need more careful elaboration both theoretically and empirically (see, e.g., Martin, 2003; Becker, 2004). Furthermore, there are relatively few studies to date that can serve as 'Kuhnian exemplars' of this new approach. Notwithstanding these shortcomings, we believe that Evolutionary Economic Geography provides genuine new explanations for the main *explananda* in economic geography, such as location behaviour of the firm, the spatial evolution of sectors and networks, the co-evolution of firms, technologies and territorial institutions, and convergence/divergence in spatial systems. The comparison of evolutionary approach with neoclassical and institutional approaches shows that Evolutionary Economic Geography indeed offers value added to the field of economic geography. What is more, an evolutionary approach offers interfaces with neoclassical and institutional approaches that are potentially much more fertile than the uneasy interactions that we have witnessed between neoclassical and institutional scholars so far. We realize there is still a long way to go before Evolutionary Economic Geography becomes an established field. Having said this, we are convinced that evolutionary theory constitutes a truly new and promising paradigm in economic geography. Time will tell whether it will live up our expectations: it is evolution as usual.

296 • *Boschma and Frenken*

References

Albert, R., Barabasi, A. L. (2002) Statistical mechanics of complex networks. *Reviews of Modern Physics*, 74(1): 47–97.

Alchian, A. A. (1950) Uncertainty, evolution, and economic theory. *Journal of Political Economy*, 58: 211–221.

Amable, B. (2000) Institutional complementarity and diversity of social systems of innovation and production. *Review of International Political Economy*, 7(4): 645–687.

Amin, A., Thrift, N. (eds) (1994) *Globalization, Institutions and Regional Development in Europe*. Oxford: Oxford University Press.

Amin, A., Thrift, N. (2000) What kind of economic theory for what kind of economic geography? *Antipode*, 32(1): 4–9.

Andersen, E. S. (2004) Population thinking and evolutionary economic analysis: exploring Marshall's fable of the trees. *DRUID Working Paper* 2004-5. http://www.druid.dk.

Anderson, P., Tushman, M. L. (1990) Technological discontinuities and dominant designs: a cyclical model of technological change. *Administrative Science Quarterly*, 35: 604–633.

Andersson, C., Frenken, K., Hellervik, A. (2006) A complex network approach to urban growth. *Environment and Planning A*, in press.

Andersson, C., Hellervik, A., Lindgren, K., Hagson, A., Tornberg, J. (2003) Urban economy as a scale-free network. *Physical Review E*, 68(3): 036124 Part 2.

Antonelli, C. (2000) Collective knowledge communication and innovation: the evidence of technological districts. *Regional Studies*, 34(6): 535–547.

Arthur, W. B. (1987) Urban systems and historical path dependence. In J. H. Ausubel and R. Herman (eds) *Cities and Their Vital Systems*. Washington DC: National Academy Press), 85–97.

Arthur, W. B. (1989) Competing technologies, increasing returns, and lock-in by historical events. *The Economic Journal*, 99: 116–131.

Arthur W. B. (1990) Silicon Valley locational clusters: when do increasing returns imply monopoly? *Mathematical Social Sciences*, 19(3): 235–251.

Arthur, W. B. (1994), *Increasing Returns and Path Dependence in the Economy*. Ann Arbor: University of Michigan Press, 99–110.

Asheim, B. T., Isaksen, A. (2002) Regional innovation systems: the integration of local 'sticky' and global 'ubiquitous' knowledge. *Journal of Technology Transfer*, 27: 77–86.

Atkinson, A. B., Stiglitz, J. E. (1969) A new view on technological change. *The Economic Journal*, 79: 573–578.

Barabasi, A. L., Albert, R. (1999) Emergence of scaling in random networks. *Science*, 286(5439): 509–512.

Barnes, T. J. (2001) Retheorizing economic geography: from the quantitative revolution to the 'cultural turn'. *Annals of the Association of American Geographers*, 91(3): 546–565.

Barrat, A., Barthelemy, M., Vespignani, A. (2005) The effects of spatial constraints on the evolution of weighted complex networks. *Journal of Statistical Mechanics*, Art. No. P05003.

Bathelt, H., Glückler J. (2003) Toward a relational economic geography. *Journal of Economic Geography*, 3: 117–144.

Bathelt, H., Malmberg, A., Maskell, P. (2004) Clusters and knowledge: local buzz, global pipelines and the process of knowledge creation. *Progress in Human Geography*, 28(1): 31–56.

Becker, M. C. (2004) Organizational routines. A review of the literature. *Industrial and Corporate Change*, 13(4): 643–677.

Bertolini, L. (2005) Evolutionary urban transportation planning? An exploration. *Papers in Evolutionary Economic Geography* #05.12. http://econ.geog.uu.nl.

Boggs, J. S., Rantisi, N. M. (2003) The 'relational turn' in economic geography. *Journal of Economic Geography*, 3: 109–116.

Boschma, R. A. (1997) New industries and windows of locational opportunity. A long-term analysis of Belgium. *Erdkunde*, 51: 12–22.

Boschma, R. A. (2004) The competitiveness of regions from an evolutionary perspective. *Regional Studies*, 38(9): 1001–1014.

Boschma, R. A., Frenken, K. (2003) Evolutionary economics and industry location. *Review for Regional Research*, 23: 183–200.

Boschma, R. A., Lambooy, J. G. (1999) Evolutionary economics and economic geography. *Journal of Evolutionary Economics*, 9: 411–429.

Boschma, R. A., Wenting, R. (2005) The spatial evolution of the British automobile industry. *Papers in Evolutionary Economic Geography* #05.04. http://econ.geog.uu.nl.

Boschma, R. A. and Weterings, A. B. R. (2005) The effect of regional differences on the performance of software firms in the Netherlands. *Journal of Economic Geography*, 5: 567–588.

Bottazzi, G., Cefis, E., Dosi, G. (2002) Corporate growth and industrial structures: some evidence from the Italian manufacturing industry. *Industrial and Corporate Change*, 11(4): 705–723.

Bottazzi, G., Dosi, G., Fagiolo, G., Secchi, A. (2004) Sectoral and geographical specificities in the spatial structure of economic activities. *LEM Working Paper* 2004-21. http://www.lem.sssup.it/.

Brakman, S., Garretsen, H. (2003) Rethinking the 'new' geographical economics. *Regional Studies*, 37(6–7): 637–648.

Brakman, S., Garretsen, H., van Marrewijk, C. (2001) *An Introduction to Geographical Economics*. Cambridge, UK: Cambridge University Press.

Brenner, T. (2004) *Local Industrial Clusters. Existence, Emergence and Evolution*. London and New York: Routledge.

Breschi, S. (2000) The geography of innovation: a cross-sector analysis. *Regional Studies*, 34(3): 213–230.

Breschi, S., Lissoni, F. (2001) Knowledge spillovers and local innovation systems: a critical survey. *Industrial and Corporate Change*, 10(4): 975–1005.

Breschi, S., Lissoni, F. (2003) Mobility and social networks: Localised knowledge spillovers revisited. *CESPRI Working Paper* 142. http://www.cespri.unibocconi.it/.

Breschi, S., Malerba, F. (1997) Sectoral innovation systems: technological regimes, Schumpeterian dynamics, and spatial boundaries. In: C. Edquist (ed.) *Systems of Innovation. Technologies, Institutions and Organizations*, London/Washington: Pinter, 130–156.

Bresnahan, T., Gambardella, A., Saxenian, A. (2001) 'Old Economy' inputs for 'New Economy' outcomes: Cluster formation in the new Silicon Valleys. *Industrial and Corporate Change*, 10(4): 835–860.

Camagni, R. (ed.) (1991) *Innovation Networks. Spatial Perspectives*. London/New York: Bellhaven Press.

Caniëls, M. (2000) *Knowledge Spillovers and Economic Growth. Regional Growth Differentials across Europe*. Cheltenham: Edward Elgar.

Cantwell, J. A., Iammarino, S. (2003) *Multinational Corporations and European Regional Systems of Innovation*. London: Routledge.

Capello, R. (1999) Spatial transfer of knowledge in high technology milieux: learning versus collective learning processes. *Regional Studies*, 33(4): 353–365.

Carroll, G. R., Hannan, M. T. (2000) *The Demography of Corporations and Industries*. Princeton, NJ: Princeton University Press.

Castells, M. (1996) *The Rise of the Network Society*. Oxford: Blackwell.

Cefis, E. (2003) Is there persistence in innovative activities? *International Journal of Industrial Organization* 21(4): 489–515.

Cefis, E., Marsili, O. (2006) A matter of life and death: innovation and firm survival. *Industrial and Corporate Change*, in press.

Cefis, E., Orsenigo, L. (2001) The persistence of innovative activities: a cross-countries and cross-sectors comparative analysis. *Research Policy*, 30(7): 1139–1158.

Colander, D. (2000) The death of neoclassical economics. *Journal of the History of Economic Thought*, 22(2): 127–143.

Consoli, D. (2005) The dynamics of technological change in UK retail banking services. An evolutionary perspective, *Research Policy* 34: 461–480.

Cooke, P. (2001) Regional innovation systems, clusters, and the knowledge economy. *Industrial and Corporate Change*, 10: 945–974.

Cooke, P. (2002) *Knowledge Economies. Clusters, Learning and Cooperative Advantage*. London/New York: Routledge.

298 • *Boschma and Frenken*

Cooke, P., Morgan, K (1998) *The Associational Economy. Firms, Regions, and Innovation.* Oxford: Oxford University Press.

Cooke, P., Uranga M. G., Extebarria, G. (1998) Regional innovation systems: an evolutionary perspective. *Environment and Planning A*, 30: 1563–1584.

Cowan, R., Jonard, N. (2003) The dynamics of collective invention. *Journal of Economic Behavior and Organization*, 52(4): 513–532.

David, P. A. (1985) The economics of QWERTY. *American Economic Review (Papers and Proceedings)*, 75: 332–337.

David, P. A. (1999) Krugman's economic geography of development: NEGs, POGs, and naked models in space. *International Regional Science Review*, 22(2): 162–172.

Dosi, G. (1984) *Technical Change and Industrial Transformation.* Basingstoke and London: MacMillan.

Dosi, G. (1997) Opportunities, incentives and the collective patterns of technological change. *The Economic Journal*, 107(444): 1530–1547.

Dosi, G., Soete, L. (1988) Technical change and international trade. In: G. Dosi, C. Freeman, R. Nelson, G. Silverberg and L. en Soete (eds) *Technical Change and Economic Theory.* London: Pinter, 401–431.

Edquist, C. (ed.) (1997) *Systems of Innovation. Technologies, Institutions and Organizations.* London/Washington: Pinter.

Essletzbichler, J., Rigby, D. L. (2005) Competition, variety and the geography of technology evolution. *Tijdschrift voor Economische en Sociale Geografie*, 96(1): 48–62.

Feldman, M. (1999) The new economics of innovation, spillovers and agglomeration: a review of empirical studies. *Economics of Innovation and New Technology*, 8: 5–25.

Foster, J., Hölzl, W. (2004) (eds.) *Applied Evolutionary Economics and Complex Systems,* Cheltenham: Edward Elgar.

Frank, S. A. (1998) *Foundations of Social Evolution.* Princeton NJ: Princeton University Press.

Freeman, C. (1987) *Technology Policy and Economic Performance. Lessons from Japan.* London: Pinter.

Freeman, C., Perez, C. (1988) Structural crisis of adjustment, business cycles and investment behaviour. In: G. Dosi, C. Freeman, R. Nelson, G. Silverberg and L. Soete (eds) *Technical Change and Economic Theory.* London: Pinter, 38–66.

Frenken, K. (2006) Technological innovation and complexity theory. *Economics of Innovation and New Technology*, in press.

Frenken, K., Van Oort, F. G., Verburg, T., Boschma, R. A. (2005) Variety and regional economic growth in the Netherlands. *Papers in Evolutionary Economic Geography* #05.02. http://econ.geog.uu.nl.

Friedman, D. (1998a) On economic applications of evolutionary game theory. *Journal of Evolutionary Economics* 8(1): 15–43.

Friedman, D. (1998b) Evolutionary economics goes mainstream: a review of the theory of learning in games. *Journal of Evolutionary Economics* 8(4): 423–432.

Fujita, M., Krugman, P., Venables, A. J. (1999) *The Spatial Economy. Cities, Regions and International Trade.* Cambridge MA: MIT Press.

Fujita, M., Thisse, J. F. (1996) Economics of agglomeration. *Journal of the Japanese and International Economics*, 10: 339–378.

Fujita, M., Thisse, J. F. (2002) *Economic of Agglomeration. Cities, Industrial Location and Regional Growth.* Cambridge, MA: Cambridge University Press.

Gabaix, X. (1999) Zipf's law for cities: an explanation. *Quarterly Journal of Economics* 114: 739–767.

Galli, R., Teubal, M. (1997) Paradigmatic shifts in national innovation systems. In: C. Edquist (ed.) *Systems of Innovation. Technologies, Institutions and Organizations.* London/Washington: Pinter, 342–370.

Garnsey, E., Stam, E., Heffernan, P. (2006) New firm growth: exploring processes and paths. *Industry and Innovation*, 13(1), in press.

Gertler, M. S. (1997) The invention of regional culture. In: R. Lee and J. Wills (eds) *Geographies of Economies.* London: Arnold, 47–58.

Gertler, M. S. (2003) Tacit knowledge and the economic geography of context or the undefinable tacitness of being (there). *Journal of Economic Geography*, 3: 75–99.

Glaeser, E. L., Kallal, H. Scheinkman, J., Shleifer, A. (1992) Growth in cities. *Journal of Political Economy* 100: 1126–1152.

Granovetter, M. (1985) Economic action and social structure: the problem of embeddedness. *American Journal of Sociology*, 91: 481–510.

Giuliani, E. (2005) The structure of cluster knowledge networks: uneven and selective, not pervasive and collective. *DRUID Working Paper* 2005-11. http://www.druid.dk.

Giuliani, E., Bell, M. (2005), The micro-determinants of meso-level learning and innovation: evidence from a Chilean wine cluster. *Research Policy*, 34(1): 47–68.

Guimerà, R., Amaral, L. A. N. (2004) Modelling the world-wide airport network. *European Physical Journal B*, 38(2): 381–385.

Hall, P. G., Preston, P. (1988) *The Carrier Wave: New Information Technology and the Geography of Innovation 1846-2003*. London: Unwin Hyman.

Hall, P. A., Soskice, D. (2001) *Varieties of Capitalism. The Institutional Foundations of Comparative Advantage*. Oxford: Oxford University Press.

Hannan, M. T., Carroll, G. R., Dundon, E. A., Torres, J. C. (1995) Organizational evolution in a multinational context: entries of automobile manufacturers in Belgium, Britain, France, Germany, and Italy, *American Sociological Review* 60(4): 509–528.

Hodgson, G. M. (1988) *Economics and Institutions. A Manifesto for a Modern Institutional Economics*. Cambridge: Polity.

Hodgson, G. M. (1998). The approach of institutional economics. *Journal of Economic Literature* 36(1): 166–192.

Hodgson, G. M. (1999) *Evolution and Economics. On Evolutionary Economics and the Evolution of Economics*. Cheltenham: Edward Elgar.

Hohenberg, P. M., Lees, L. H. (1995) *The Making of Urban Europe 1000-1994*. Cambridge, MA: Harvard University Press.

Hollingsworth, J. R. (2000) Doing institutional analysis: implications for the study of innovations. *Review of International Political Economy*, 7(4): 595–644.

Jacobs, J. (1969). *The Economy of Cities*. New York: Vintage Books.

Jaffe, A. B., Trajtenberg, M., Henderson, R. (1993) Geographic localization of knowledge spillovers as evidenced by patent citations. *Quarterly Journal of Economics*, 108(3): 577–598.

Keeble, D., Lawson, C., Moore, B., Wilkinson, F. (1999) Collective learning processes, networking and 'institutional thickness' in the Cambridge region. *Regional Studies*, 33(4): 319–332.

Klepper, S. (1996) Entry, exit, growth and innovation over the product life cycle. *American Economic Review*, 86: 562–583.

Klepper, S. (2002a) The evolution of the U.S. automobile industry and Detroit as its capital. *Paper presented at 9th Congress of the International Schumpeter Society*, Gainesville, FL, March 27–30.

Klepper, S. (2002b) The capabilities of new firms and the evolution of the U.S. automobile industry. *Industrial and Corporate Change*, 11(4): 645–666.

Kogut, B., Zander, U. (1993) Knowledge of the firm and the evolutionary theory of the multinational corporation. *Journal of International Business Studies*, 24: 625–646.

Krugman, P. R. (1991a) Increasing returns and economic geography. *Journal of Political Economy*, 99(3): 483–499.

Krugman, P. R. (1991b) *Geography and Trade*. Cambridge, MA: MIT Press.

Krugman, P. R. (1996) What economists can learn from evolutionary theorists. *A talk given to the European Association for Evolutionary Political Economy*, November, downloadable at http://web.mit.edu/krugman/www/evolute.html.

Krugman, P. R. (1998) What's new about the new economic geography? *Oxford Review of Economic Policy*, 14: 7–17.

Lawson, C. (1999) Towards a competence theory of the region. *Cambridge Journal of Economics*, 23: 151–166.

Lissoni, F. (2001) Knowledge codification and the geography of innovation: the case of Brescia mechanical cluster. *Research Policy*, 30(9): 1479–1500.

300 • *Boschma and Frenken*

Maggioni, M. A. (2002) *Clustering Dynamics and the Location of High-Tech-Firms.* Springer: Heidelberg.

Mäki, U. (1992) Friedman and realism. *Research in the History of Economic Thought and Methodology,* 10: 171–195.

Mäki, U., Oinas, P. (2004) The narrow notion of realism in human geography. *Environment and Planning A,* 36: 1755–1776.

Malmberg, A., Maskell, P. (2002) The elusive concept of localization economies: Towards a knowledge-based theory of spatial clustering. *Environment and Planning A,* 34(3): 429–449.

Marchionni, C. (2004) Geographical economics versus economic geography: towards a clarification of the dispute. *Environment and Planning A,* 36: 1737–1753.

Markusen, A. (1999) Fuzzy concepts, scanty evidence, policy distance: the case for rigour and policy relevance in critical regional studies. *Regional Studies,* 33(9): 869–884.

Markusen, A. (2003) On conceptualization, evidence and impact: a response to Hudson, Lagendijk and Peck. *Regional Studies,* 37(6–7): 747–751.

Martin, R. (1999) The new 'geographical turn' in economics: some critical reflections. *Cambridge Journal of Economics,* 23(1): 65–91.

Martin, R. (2000) Institutional approaches in economic geography. In: E. Sheppard and T. J. Barnes (eds) *A Companion to Economic Geography.* Oxford and Malden, MA: Blackwell Publishing, 77–94.

Martin, R. (2003) Putting the economy back in its place: one economics and geography. *Paper presented at the Cambridge Journal of Economics Conference 'Economics for the Future: Celebrating 100 years of Cambridge Economics',* Cambridge, UK, September 17–19.

Martin, R., Sunley, P. (1996) Paul Krugman's geographical economics and its implications for regional development theory. A critical assessment. *Economic Geography* 72(3): 259–292.

Martin, R, Sunley P. (2001) Rethinking the 'economic' in economic geography: broadening our vision or losing our focus? *Antipode,* 33(2): 148–161.

Maskell, P. (2001) The firm in economic geography. *Economic Geography,* 77(4): 329–344.

McKelvey, M. (2004) Evolutionary economics perspectives on the regional–national–international dimensions of biotechnology innovations. *Environment and Planning C,* 22(2): 179–197.

Meardon, S. J. (2001) Modeling agglomeration and dispersion in city and country: Gunnar Myrdal, Francois Perroux, and the New Economic Geography. *The American Journal of Economics and Sociology,* 60(1): 25–57.

Murmann, J. P. (2003) *Knowledge and Competitive Advantage. The Co-evolution of Firms, Technology, and National Institutions.* Cambridge: Cambridge University Press.

Myrdal, G. (1957) *Economic Theory and Underdeveloped Regions.* London: Duckworth.

Neary, J. P. (2001) Of hype and hyperbolas: introducing the new economic geography. *Journal of Economic Literature,* 39(2): 536–561.

Nelson, R. R. (ed.) (1993) *National Innovation Systems. A Comparative Analysis.* Oxford and New York: Oxford University Press.

Nelson, R. R. (1995a) Recent evolutionary theorizing about economic change. *Journal of Economic Literature,* 33(1): 48–90.

Nelson, R. R. (1995b) Co-evolution of industry structure, technology and supporting institutions, and the making of comparative advantage. *International Journal of the Economics of Business,* 2(2): 171–184.

Nelson, R. R. (2002) Bringing institutions into evolutionary growth theory. *Journal of Evolutionary Economics,* 12: 17–28.

Nelson, R. R., Winter, S. G. (1982) *An Evolutionary Theory of Economic Change.* Cambridge, MA and London: The Belknap Press.

Nijkamp, P. (2001) The spatial economy: cities, regions, and international trade. By Fujita M, Krugman P, Venables AJ. *The Economic Journal* 111(469): F166–F168.

North, D. C. (1990) *Institutions, Institutional Change and Economic Performance.* Cambridge: Cambridge University Press.

Olsen, J. (2002) On the units of geographical economics. *Geoforum* 33: 153–164.

van Oort, F. G. (2004) *Urban Growth and Innovation. Spatially Bounded Externalities in the Netherlands.* Aldershot: Ashgate.

Overman, H. G. (2004) Can we learn anything from economic geography proper? *Journal of Economic Geography,* 4: 501–516.

Pasinetti, L. L. (1993) *Structural Economic Dynamics.* Cambridge: Cambridge University Press.

Phelps, N. A. (2004) Clusters, dispersion and the spaces in between. For an economic geography of the banal. *Urban Studies* 41(5–6): 971–989.

Plummer, P., Sheppard, E. (2000) Must emancipatory economic geography be qualitative? A response to Amin and Thrift. *Paper presented at the Global Conference on Economic Geography,* National University of Singapore, December 5–9.

Pred, A. R. (1967) *Behavior and Location. Foundations for a Geographic and Dynamic Location Theory.* Lund: Lund Studies in Geography, 27.

Puga, D. (2002) European regional policy in light of recent location theories. *Journal of Economic Geography* 2(4): 372–406.

Pumain, D. (1997), City size distributions and metropolisation. *Geojournal,* 43(4): 307–314.

Rallet, A., Torre, A. (1999) Is geographical proximity necessary in the innovation networks in the era of global economy? *Geojournal* 49(4): 373–380.

Rigby, D. L., Essletzbichler, J. (1997) Evolution, process variety, and regional trajectories of technological change in US manufacturing. *Economic Geography,* 73(3): 269–284.

Rivkin, J. W. (2000) Imitation of complex strategies. *Management Science,* 46(6): 824–844.

Samuels, W. J. (1995) The present state of institutional economics. *Cambridge Journal of Economics,* 19: 569–590.

Saviotti, P. P. (1996) *Technological Evolution, Variety and the Economy.* Cheltenham and Brookfield: Edward Elgar.

Saxenian, A. (1994) *Regional Advantage.* Cambridge MA: Harvard University Press.

Schamp, E. W. (2002) Evolution und Institution als Grundlagen einer Dynamischen Wirtschafts-geographie: Die Bedeutung von Externen Skalenertragen fur Geographische Konzentration. *Geographische Zeitschrift,* 90(1): 40–51.

Scott, A. J. (1993) *Technopolis. High-technology Industry and Regional Development in Southern California.* Berkeley: University of California Press.

Scott, A. J. (2004) A perspective of economic geography. *Journal of Economic Geography,* 4: 479–499.

Simmie, J. (2005) Innovation and space: a critical review of the literature. *Regional Studies,* 39(6): 789–804.

Simon, H. A. (1955a) A behavioral model of rational choice. *Quarterly Journal of Economics,* 6: 99–118.

Simon, H. A. (1955b) On a class of skew distribution functions. *Biometrika* 42(3–4): 425–440.

Sjöberg, O., Sjöholm, F. (2002) Common ground? Prospects for integrating the economic geography of geographers and economists. *Environment and Planning A,* 34(3): 467–486.

Smith, D. (1966) A theoretical framework for geographical studies of industrial location. *Economic Geography,* 42: 95–113.

Stam, E. (2003) *Why Butterflies Don't Leave. Locational Evolution of Evolving Enterprises.* Dissertation, Utrecht University.

Stuart, T., Sorenson, O. (2003) The geography of opportunity: spatial heterogeneity in founding rates and the performance of biotechnology firms. *Research Policy,* 32(2): 229–253.

Storper, M. (1997) *The Regional World. Territorial Development in a Global Economy.* London: Guilford Press.

Storper, M., Walker, R. (1989) *The Capitalist Imperative. Territory, Technology and Industrial Growth.* New York: Basil Blackwell.

Swann, P., Prevezer, M. (1996) A comparison of the dynamics of industrial clustering in computing and biotechnology. *Research Policy,* 25: 1139–1157.

Teece, D., Pisano, G., Shuen, A. (1997) Dynamic capabilities and strategic management. *Strategic Management Journal,* 18(7): 509–533.

Theil, H. (1972) *Statistical Decomposition Analysis.* Amsterdam: North-Holland.

Veblen, T. (1898) Why is economics not an evolutionary science? *Quarterly Journal of Economics*, 12: 373–397.

Verspagen, B., Schoenmakers, W. (2004) The spatial dimension of patenting by multinational firms in Europe. *Journal of Economic Geography*, 4(1): 23–42.

Wasserman, S., Faust, K. (1994) *Social Network Analysis: Methods and Applications*. Cambridge: Cambridge University Press.

Watts, D. (2004) The 'new' science of networks. *Annual Review of Sociology* 30: 243–270.

Werker, C., Athreye, S. (2004) Marshall's disciples: knowledge and innovation driving regional economic development and growth. *Journal of Evolutionary Economics*, 14: 505–523.

Whitley, R. (1992) *Business Systems in East Asia: Firms, Markets and Societies*. London: Sage.

Whitley, R. (2003) Developing innovative competences: the role of institutional frameworks. *Industrial and Corporate Change*, 11(3): 497–528.

Williamson, O. E. (1985) *The Economic Institutions of Capitalism*. New York: The Free Press.

Winter, S. G., Szulanski, G. (2001) Replication as strategy. *Organization Science* 12: 730–743.

van Wissen, L. (2004) A spatial interpretation of the density dependence model in industrial demography. *Small Business Economics*, 22 (3–4): 253–264.

Zysman, J. (1994) How institutions create historically rooted trajectories of growth. *Industrial and Corporate Change*, 3: 243–283.

Part II
The Localization of Global Economic Space

[7]

Neo-Marshallian Nodes in Global Networks*

ASH AMIN AND NIGEL THRIFT

Introduction

The literature on industrial districts seems to have reached something of an impasse. On one side the proponents of industrial districts sit around their camp fires, supposedly wild-eyed with enthusiasm, talking flexible specialization and postfordism. On the other side are a series of supposedly grim-faced critics, shouting destructive comments about globalization and corporate networks from out of the mist. This paper is an attempt to break out of this often acrimonious impasse. We want to take the emergence of new localized industrial complexes seriously, but we want to set them firmly within a context of expanding global corporate networks.

Accordingly, the paper is in four parts. In the first part of the paper, we summarize the key arguments of the localization thesis which predicts a return to industrial districts, and some of the major criticisms that have been made of the claim that there is a resurgence of the regional economy on a pervasive scale. In the second part of the paper, we attempt to reformulate the localization and globalization theses so as to provide a space for local agglomeration within growing global production filieres. In particular, we want to focus on Marshall's idea of industrial atmosphere, indicating a set of socio-cultural characteristics which are still crucial in global production filieres and which can lead to a degree of localization. The third part of the paper attempts to illustrate these and other contentions via a consideration of the history of two industrial districts. The first, Santa Croce in Tuscany, has become a Marshallian industrial district of the old kind over the last 20 years. The second, the City of London in London, finally stopped being a conventional Marshallian industrial district at the same time that Santa Croce was becoming one. But the City did not, therefore, cease to be a localized complex. Rather, we argue, it heralds a new form of localization. Finally, in the fourth part, we address some of the local economic development policy implications of the previous parts of the paper.

The localization thesis

The most powerful case for the possibility of a major return to the regional economy comes from a group of writers speculating on the rise of locally agglomerated production systems

* The section on the localization thesis draws on a paper written by Ash Amin with Anders Malmberg. We are grateful to Anders for allowing us to use some of the material. The section on Santa Croce is based on research funded by the Nuffield Foundation. Special thanks are due to Fabio and Anna Sforzi, without whose contacts, support and research input the project would not have taken place. The section on the City of London is based on work with Andrew Leyshon. Useful comments were received from Neill Marshall, Diane Perrons and two anonymous referees. Both authors also wish to acknowledge the support of the European Science Foundation's research programme on Regional and Urban Restructuring in Europe (RURE). It is their membership of the programme which gave rise to this paper.

out of the crisis of mass production. What is envisaged, to put it somewhat reductively, is a return to a division of labour between self-contained, product specialist regional economies as first conceived by Adam Smith at a national level. This is a thesis which draws upon the work of Michael Piore and Charles Sabel (Piore and Sabel, 1984; Sabel, 1989), Allen Scott and Michael Storper (Scott, 1988; Storper and Scott, 1989; Storper, 1989), Paul Hirst and Jonathan Zeitlin (1989; 1991) and others borrowing the concepts of 'flexible specialization' or 'flexible accumulation'[1] to describe the transition to a new era of vertically disintegrated and locationally fixed production.

The key argument is that the irreversible growth in recent decades of consumer sovereignty, market volatility and shortened product life-cycles requires production to be organized on an extremely flexible basis. Size, scale, hierarchy, vertical integration and task dedication on the part of machinery and employees are deemed to be too inflexible to turn out short runs of better quality and differentiated goods with the minimum of time and effort. Instead, the market is said to require decentralized coordination and control; the 'deverticalization' of the division of labour between independent but interlinked units; numerical and task flexibility among the workforce; greater reliance on innovation, ingenuity and skills; the deployment of multi-purpose and flexible tools and machinery; and the elimination of time and wastage in supply and delivery.

Such a change is said to be particularly evident in industries which face pronounced volatility and product innovation in their niche markets. Examples include electronics, designer clothing, craft products and other light industrial consumer products. In organizational terms, the new market circumstances are said to require a radical transformation of the production system towards flexible intra-firm and inter-firm arrangements which can simultaneously combine the economies of scale, scope and versatility.

This change, it is argued, implies a return to place — a dependence on locational proximity between different agents involved in any production filiere. Agglomeration is said to offer a series of advantages upon which a system of vertically disintegrated production can draw. Echoing the factors first identified by Alfred Marshall in his work on small-firm districts in Lancashire and Yorkshire during the nineteenth century, these advantages are said to include the build-up of a local pool of expertise and know-how and a culture of labour flexibility and cooperation resulting from dense social interaction and trust; lowered transport and transaction costs; and the growth of a local infrastructure of specialized services, distribution networks and supply structures. Via the consolidation of particular product specialisms in different regions a federation of self-contained regional economies is anticipated, each with its own cumulative causation effects drawing upon strong external economies of agglomeration.

Empirical verification of this thesis comes from the claim that over the last few decades the most dynamic and competitive examples of industrial restructuring have been 'Marshallian' in their spatial dynamics. The examples which are quoted have now become almost too familiar. They include high-tech, R & D and innovation-intensive areas such as Silicon Valley, Boston, the M4 corridor, Grenoble and other successful technopoles. They also include industrial districts in both semi-rural contexts such as those in the Third Italy regions and those in inner-city environments (e.g. motion pictures in Los Angeles, the furniture industry in inner London), in which networks of specialist small firms produce craft or quality consumer goods. Finally, also cited is the example of areas such as Baden-Württemburg in Germany, where leading-edge large companies such as Bosch are said to

1. These two concepts, it should be noted, are not interchangeable. Flexible specialization is a concept deployed to describe transformations in the production process stimulated by new technological, skill and market developments. In contrast, the term flexible accumulation, drawing upon the Regulation Approach developed in France as well as a recent essay of David Harvey (Harvey, 1989), refers to a broader macroeconomic design for the twenty-first century, transcending the 'fordist' regime of accumulation which was built upon the pillars of mass consumption, mass production and Keynesian regulation of the economy.

rely on local subcontracting and supply networks for their flexibility and innovative excellence.

The significance of this thesis should not be underestimated, equating, as it does, industrial renovation with territorial development. The cited examples are very real cases of success, and their experiences could inform policy measures in other areas. The novel conceptual aspect of the thesis is the (re)discovery of the locational importance of patterns of linkages and the formation of inter-firm relationships, notably in relation to the exchange of information and goods between buyer and seller and its influence on linkage costs. The new literature makes the interesting proposition that negotiations involved in producing and exchanging certain types of commodity are less conveniently carried out at a distance. Customer-specific supplies, for instance, are often based on extensive technical cooperation between the seller and the buyer. Therefore such cooperation requires reliable and rapid communication, usually best conveyed through personal contacts. Production of customized goods and services under conditions of 'dynamic' competition, the hallmark of the post-mass-production economy, will therefore tend to bring with it agglomeration and local networking.

To anticipate a pervasive, perhaps even total, return to local production complexes in the postfordist economy is nonsensical, for a number of reasons.[2] First, it is inaccurate to refer to the conditions and areas cited by the localization thesis as the only examples of success. Others must surely include the reconsolidation of the major metropolitan areas such as London, Milan, Frankfurt and Paris as centres of growth through their magnet-like pull on finance, management, innovation, business services and infrastructure. Reasons related to their status as core metropolitan areas and size as centres of consumption have far more to do with their economic success than the rediscovery of Marshallian tendencies. Should any citation of success not also include the resurgence of major provincial cities such as Birmingham, Turin and Grenoble, which have managed to carve out a niche as intermediate centres of agglomeration within global financial, corporate and service networks? Indeed, why not also include, as an example, the growing concentration of wealth in certain rural areas characterized by an odd combination of capital intensification in agriculture, the decentralization of offices and service industries and in-migration by commuters looking for a pleasant lifestyle? These additions to the geography of 'postfordist' success have little in common with the logic of flexible specialization.

Second, in proposing local agglomeration as the symbol of a future regime of capital accumulation, the localization thesis effectively rules out the possibility of transformation and change within the very areas cited as examples of postfordist growth. These areas, too, are likely to evolve, and perhaps fragment internally, in much the same way as did, for example, Alfred Marshall's cutlery district in Sheffield in the course of the twentieth century. Evidence of such change is already apparent in 'mature' production complexes such as Silicon Valley, now being drawn into a wider spatial division of labour as a result of intense inward investment by overseas multinational corporations, and the export of assembly and intermediate production functions respectively to areas of cheap labour and growing market demand. Some Italian industrial districts, too, are undergoing change, as local linkages begin to replace external ones, owing to either the threat of takeover of local banks by foreign financial institutions, or the increase in international rather than local intra-firm and inter-firm linkages (see the section on Santa Croce below, and Amin, 1989; Harrison, 1990; Bianchini, 1990).

Third, taking the ingredients for local success identified by the localization thesis seriously, it has to be concluded that the proliferation of localized production complexes

2. Some of the exponents of the localization thesis, for example Allen Scott and Michael Storper (Scott, 1991; Storper, 1991), have begun in their more recent work to talk of the future as a complex juxtaposition of local and global production networks. However, this development still does not draw them sufficiently far away from the localization thesis to warrant a reformulation of their position in the debate on the geography of flexible accumulation.

is likely to be restricted in so far as these ingredients are not readily transferable to other areas. Local containment of the division of labour requires a gradual build-up of know-how and skills, cooperative traditions, institutional support, specialist services and infrastructure. These not only take time to consolidate, but also escape the traditional instruments of spatial policy owing to their ephemeral and composite nature (Amin and Robins, 1990). In addition, there remains the problem that 'new' growth cannot, as on a *tabula rasa*, sweep aside local traditions which might resist such change (Glasmeier, 1991). The dismal failure of strategies to promote technopoles in different European less favoured regions (LFRs), as well as efforts to encourage greater local networking among and between small and large firms within the depressed industrial regions, bears witness to this difficulty.

A final problem with the notion of a pervasive spread of local production complexes is related to the observation made by a number of critics that there is no conclusive evidence of the demise of fordist principles of mass production and consumption and of the multitude of labour processes which co-existed under fordism (e.g. customization, batch production, mass assembly, continuous flow). The idea of a clean break between one macrosystem dominated by one way of doing things and another regime with its own distinctive organizational structure is too simple a caricature of historical change and a denial of the ebb and flow, the continuity and discontinuity and the diversity and contradiction that such change normally suggests (see Gertler, 1988; Sayer, 1989; Thrift, 1989). Sensitivity to diversity is particularly essential when it comes to the analysis of the geography of production. Depending on the labour process in an industry, the organizational cultures of the players involved, the nature of the areas in which activity is located and the market or macroeconomic circumstances surrounding individual sectors, a diversity of industrial geographies can be produced, with each offering different options of the spectrum between locational fixity and global mobility.

The emergence of new localized production complexes, to conclude this section, should be noted seriously. But this cannot become a basis for assuming, as two observers have done, that 'the mode of production, has in a sense, gone back to the future', with 'local economies ... already on the march' (Cooke and Imrie, 1989: 326). If localities are on the march, it is, if anything, as argued in the next section, to the tune of globalizing forces in the organization of production — a process in which local territorial integrity is far from guaranteed.

A reformulation: global networks

So far we have followed a fairly standard critique of the current literature on the resurgence of local economies. The literature has rather limited analytical power, most particularly because of a tendency to cling to a model which is locally based and which does not therefore recognize the importance of emerging global corporate networks. In this paper we still want to retain the notion of 'localization', but we want to relocate the account in two ways. First, we want to consider industrial districts and local complexes as the outgrowths of a world economy which is still rapidly internationalizing and which is still a world of global corporate power. Second, against an incipient economism in explaining the strengths of localized production, in order to provide a reformulation of the significance of local networking, we want to build on Marshall's work on 'industrial atmosphere', trying to analyse why such an atmosphere might still prove central in a world economy where transactions are increasingly indirect. In other words, we want to cross the new international political economy with the new economic sociology.

We take it that an important shift has occurred in the 1970s and 1980s and that this is a move from an international to a global economy. This global economy has many characteristics, of which four are particularly important. First, industries increasingly function on an integrated world scale, through the medium of global corporate networks. As a consequence, the control that multinational corporations (whether foreign or domestically owned) exert over employment, investment and trade has continued to grow in most developed economies through the 1980s and into the 1990s. Second, corporate power has continued to advance, so that the new global industries are increasingly oligopolistic, progressively cartelized. This intensifying concentration is best seen in the various merger booms around the world in the 1980s. For example, the EC saw a massive 25% increase in the number of industrial mergers across a wide range of sectors (Jacquemin *et al.*, 1989), leading to a growth in seller concentration levels, increasingly on an international rather than only national scale.

But, third, and importantly, today's global corporations have themselves become more decentralized through increased 'hollowing out', new forms of subcontracting, new types of joint ventures, strategic alliances and other new 'networked' forms of corporate organization. Thus corporations increasingly resemble 'flattened hierarchies'. However, there is little evidence to suggest that operational and organizational decentralization has resulted in a similar degree of devolution of power and control. 'Hollowing out', for example, has led to forward vertical integration by market leaders in order to secure strategic control over markets and distribution networks. Also, new developments in subcontracting have often led to preferred status being bestowed on fewer suppliers, linking them more tightly into corporate hierarchies and threatening the survival and growth of other suppliers. Strategic alliances, too, appear to be global partnerships between major oligopolies seeking to share markets and R & D costs. In other words, some form of centred control still exists, and in the hands of global corporations, thus suggesting that new developments like those noted above may well represent simply an extension and sharing of power beyond the boundaries of the individual firm, among key actors in a value-added network, rather than a genuine spread of authority to smaller or local players in corporate networks (Amin and Dietrich, 1991).

Further, it is by no means clear that these new developments are local phenomena or even have inevitable localized consequences. Increasing corporate integration may well be accompanied by increasing *geographical* integration, as more and more places are drawn into, or excluded from, the web of global corporate networks. Thus, against the benefit that the operational status of sites within these networks might be more complex and more autonomous than that of sites trapped in rigid (fordist) intra-corporate hierarchies, the fact remains that they are still locked into a global corporate web and therefore not restricted to local ties. In other words the sites might be relatively autonomous but they are not free agents.

Fourth, and finally, there is a new, more volatile balance of power between nation-states and corporations, which we might call 'short-run corporatism'. Nation-states' ability to intervene in the world economy has been weakened because states have been 'hollowed out' (Held, 1991), and because states are often internally divided. But corporations also find it difficult to impose a strategic direction on the world economy for reasons such as the unstable nature of business alliances and the contradictory interests embedded in individual industries (and even, sometimes, individual firms). The result is the increasing prominence of cross-national issue coalitions 'uniting fragments of the state, fragments of particular industries and even fragments of particular firms in a worldwide network' (Moran, 1991: 133).

The net result of these four developments has been the growth of increasingly integrated global production filieres orchestrated and coordinated by large corporations. But, because these filieres are more decentralized and less hierarchically governed, there are in fact a number of very considerable problems of integration and coordination. Three of these, each related to the other, stand out. The first problem is one of *representation*. Information has to be gathered and analysed about what is happening in these filieres. But the benefit of advances in global communications and information processing capacity to the growth of a globally oriented business press (Kynaston, 1985) and a whole industry of industrial research and analysis is a two-edged one. It has produced a massive increase in the quantity and even the quality of information, but the problem of how that information is interpreted and who has access to the interpretations remains; indeed, it is probably more pressing. There is *a growing interpretive task*. What can be said is that the 'stories' that are circulated about a global production filiere, and how they are scripted, constitute that filiere's understanding of itself. Indeed, new work in 'economic sociology' (e.g. Adler and Adler, 1984; Block, 1990; Zukin and Dimaggio, 1990) suggests that the only way of making sense of large and complex economic systems is through the formation of social cliques within which these stories circulate. This point relates to those of Giddens (1990; 1991) and Strange (1988; 1991), who have pointed out that what they call 'expert systems' or 'knowledge structures' have become a critical part of the global economy, and that these systems or structures are increasingly asymmetric.

The second problem is one of social *interaction*. Global production filieres are not just social structures, they are sociable structures. There is constant social interaction within them. Indeed, this is one of the ways in which they are able to be understood. Interaction promotes particular discourses and taps into particular knowledge structures. Perhaps one of the most misleading articles of the early 1980s was Offe and Wiesenthal's (1979) where they suggested that the capitalist class is a monological entity. Of course, capitalists have many advantages, but they are still dialogical beings. In particular, social interaction is still needed to gather information and to tap into particular knowledge structures, to make agreements and coalitions, and continually to cement relations of *trust*, of implicit contract (Marceau, 1989). New or heightened forms of corporate interaction, like joint ventures and strategic alliances, have made social interaction more rather than less central to many aspects of corporate life.

The third problem is one of tracking *innovation*. The problem is how to keep up levels of product and process innovation in a decentralized system (especially when the pressure to produce more products has increased) and, perhaps more importantly still, how to successfully market products in the early customized stages when they can succeed or fail and when a small critical mass of customers is needed. In turn, success depends critically on representation and interaction, since the stories that are told and who they are recounted to influence a product's chance of success.

Thus the world economy may have become more decentralized, but it is not necessarily becoming decentred. Centres are still needed, even in a world of indirect communication, for three reasons related to the problems above regarding representation, interaction and innovation.

Centres, in the first place, are needed to represent, that is to generate and disseminate discourses, collective beliefs, stories about what world production filieres are like. These discourses are constitutive of the direction in which industries and corporations can go — whether we are talking about new fashions in design or products or new management trends (like strategic alliances). They are the understandings industries make of themselves. Centres are also needed as points at which knowledge structures, many of which carry considerable social barriers to entry, can be tapped into. Often such centres can constitute a local knowledge structure which has 'gone global'. In other words, these are centres of *authority*.

Secondly, centres are needed to interact, that is to act as centres of sociability, so gathering information, establishing or maintaining coalitions and monitoring trust and implicit contracts. Thirdly, they are needed in order to develop, test and track innovations. Centres produce a discursive mass sufficient to help generate innovations; contact with numerous knowledgeable people identifies gaps in the market, new uses for technologies and so on. In the case of certain products (e.g. financial products), development occurs in close liaison with most who are the potential sources of reward. Centres are associated to provide sufficient mass in the early stages of innovation. Their social networks provide rapid reactions and an initial market. The success or failure of products, especially in the early stage, can depend upon the stories told about them. Finally, centres are still needed, to some at extent at least, to keep track of innovations; that is, as judges of their success.

These have to be *geographical* centres, that is, place-bound communities in which the agglomeration and interaction between firms, institutions and social groups acts to generate and reinforce that 'industrial atmosphere' which nurtures the knowledge, communication and innovation structures required for retaining competitive advantage in a given global production filiere. In other words, the localization of the functions of the 'head' contributes towards resolving problems outlined earlier facing global corporate networks.

To be sure, this form of localization is quite different from the older and more familiar habit of the vertically integrated, hierarchical firm of concentrating its strategic functions, its 'head', in headquarters located in major metropolitan cities. In contrast to this latter example of 'tight-encasement' of the 'head' within a closed corporate structure, neo-Marshallian nodes in global networks act, as it were, as a collective 'brain', as centres of excellence in a given industry, offering for collective consumption local contact networks, knowledge structures and a plethora of institutions underwriting individual entrepreneurship (see Peck, 1991; Todling, 1991; Tornqvist, 1991).

Such 'socialization' of the functions of the head, amounting to the 'valorization' of a local community, itself as an active factor serving to help local industry to maintain industrial supremacy, appears to be of particular relevance in industries characterized by knowledge-based competition, rapidly changing technological standards and volatile markets. All three are conditions of a greater spread of costs and risks between individual agents, as long as continuity in the flow of information, goods and services between firms, institutions and social groups is maintained. This is precisely the advantage which Marshallian nodes are able to offer — an industrial atmosphere and infrastructure which firms, small and large, isolated or interconnected, can dip into as and when required.

Two examples

Two examples follow to illustrate the significance of centred places in a global system. In the case of Santa Croce, a vertically disintegrated small-firm industrial district of the classical Marshallian type has evolved within the last 20 years, but is now experiencing a further change. In the case of the City of London, what was a Marshallian industrial district of considerable local integrity has evolved into an industrial complex which still has some Marshallian features but which now relies on global networking to cement these features in place.

These districts may appear dissimilar at first sight. For example, the City has a much larger employment base than Santa Croce. But we would argue that such dissimilarities are outweighed by some notable similarities. In particular, five similarities can be noted. First, they are in global industries with similar market conditions (product volatility, reduced product life-cycles, design intensity, flexibility of volume etc.). Second, in both districts, despite intensifying external linkage formation, many needs can be met locally. In Santa Croce this is the result of cumulative historical circumstance. In the City it is essentially the result of sheer scale. Third, both districts rely on strong knowledge structures. Fourth, both districts have strong traditions of 'thick' social interaction and 'collective consciousness', the result of distinctive institutional mixes. Fifth and finally, both districts are under threat. In the case of Santa Croce there is the threat of its incorporation back into global networks through vertical reintegration. In the case of the City, that threat is no longer relevant. The real threat now comes from an American-forced change from implicit to explicit contracts as the financial system becomes increasingly rule-based. In turn, this shift might increase transaction costs and reduce the need for specialized knowledge to the point where many crucial intermediaries are no longer necessary or viable. (However, increasing automation may well bring transaction costs down, thereby countering this tendency.)

Marshall in Tuscany: leather tanning in Santa Croce sull'Arno

Santa Croce is a small town in the lower Arno Valley, 40 kilometres east of Pisa, which specializes in the production of medium- to high-quality cured bovine leather for predominantly the 'fashion' end of the shoe and bag industries. There are only two other major leather tanning areas in Italy: Arzignano in the Veneto, which is dominated by a small number of large, vertically integrated and highly mechanized tanneries, orientated towards the furnishing and upholstery industry; and Solofra in the South (Campania), which specializes in less refined, non-bovine, cured leather for the clothing industry. The lower Arno Valley accounts for about 25% of the national employment in the leather and hide tanning industry.

In Santa Croce, an area no larger than 10 square kilometres, are clustered 300 artisan firms employing 4500 workers and 200 subcontractors employing 1700 workers. The real figures are probably much higher, as the latter capture only those firms officially registered with the Santa Croce Association of Leather Tanners and the Association of Subcontractors respectively. In 1986, the combined turnover of these 500 firms was £860 million (one-tenth of which was that of the subcontractors). On average, the area derives 15% of its sales revenue from exports, almost 80% of which are destined for the EC. Although the share of exports has been growing, the industry is still heavily dependent on the Italian market, particularly upon buyers in Tuscany, who account for over 40% of the domestic market.

Twenty years ago, Santa Croce was not a Marshallian industrial district. There were many fewer firms, production was more vertically integrated, the product was more standardized (albeit artisanal) and the balance of power was very much in favour of the older and larger tanneries. Today, Santa Croce is a highly successful 'flexibly specialized' small-firm industrial district, which derives its competitive strength from specializing in the seasonally based fashionwear niche of the industry. Typically, market conditions in this sector — e.g. product volatility, a very short product life-cycle, design intensity, flexibility of volume — demand an innovative excellence and organizational flexibility which Santa Croce has been able to develop and consolidate over the last two decades by building upon its early artisan strengths.

The boom in demand for Italian leather fashionwear in the 1970s and 1980s provided the occasion for area-wide specalization and growth in the output of cured leather. That such growth was to occur through a multiplication of independent small firms supported by a myriad of task-specialist subcontractors was perhaps more a result of specific local peculiarities than an outcome of the new market conditions. Opposed to the highly polluting effects of the tanning process — Santa Croce is one of those places in which you can recognize the Marshallian 'industrial atmosphere' by its smell — the local Communist administration was unsympathetic to factory expansion applications and also refused, until very recently, to redraw the Structure Plan to allow for more and better factory space. This, together with the strong tradition of self-employment and small-scale entrepreneurship in rural Tuscany, effectively led to a proliferation of independently owned firms scattered in small units all over Santa Croce. Two further encouragements to this process of fragmented entrepreneurship were, firstly, the preference of local rural savings banks to spread their portfolio of loans widely but thinly to a large number of applicants as a risk-minimization strategy, and secondly, the variety of fiscal and other incentives offered by the Italian state to firms with less than 15 employees.

This initial, and somewhat 'accidental', response to a situation of rapidly expanding demand was gradually turned into an organizational strength capable of responding with the minimum of effort and cost to new and rapidly changing market signals. The tanners — many of whom call themselves 'artists' — became more and more specialized, combining their innate 'designer' skills with the latest in chemical and organic treatment techniques to turn out leathers of different thickness, composition, coloration and design for a wide variety of markets. The advantage for buyers, of course, was the knowledge that any manner of product could be made at the drop of a hat in Santa Croce.

The small firms were also able to keep costs down without any loss of productive efficiency, via different mechanisms of cooperation. One example is the joint purchase of raw materials in order to minimize on price. Another is the pooling of resources to employ export consultants. The main device for cost flexibility, however, has been the consolidation of an elaborate system of putting-out between tanners and independent subcontractors (often ex-workers). The production cycle in leather tanning is composed of 15–20 phases, of which at least half are subcontracted to task-specialist firms (e.g. removal of hair and fat from the uncured skins, splitting the hide, flattening and drying). Constantly at work, and specializing in operations which are most easily mechanized, the subcontractors have been able to reduce drastically the cost of individual tasks at the same time as providing the tanners with the numerical flexibility demanded by their market. This articulate division of labour among and between locally based tanners and sub-contractors, combining simultaneously the advantages of complementarity between specialists and competition between the numerous firms operating in identical market niches, is perhaps the key factor of success.

But other factors have also played their part. One is area specialization. Santa Croce, like other industrial districts past and present, is a one-product town which offers the full range of agglomeration and external economies involved in local excellence along the entire filiere of activities associated with leather tanning. In the area there are the warehouses of major international traders of raw and semi-finished leather as well as the offices of independent import agents, brokers and customs specialists. There are the depots of the major multinational chemical giants as well as locally owned companies selling paints, dyes, chemicals and customer-specific treatment formulae to the tanners. There are at least three savings banks which have consistently provided easy and informal access to finance. There are several manufacturers of plant and machinery, tailor-made for the leather tanning industry, and there is a ready supply base for second-hand equipment and maintenance services. There are several scores of independent sales representatives, export agents and buyers of finished leather in the area. The local Association of Leather Tanners, the mayor's office, the bigger local entrepreneurs and the Pisa offices of the Ministry of Industry and Trade also act as collective agents to further local interests at national and international trade fairs. There are several international haulage companies and shipping agents capable of rapidly transporting goods to any part of the world. There is, at the end of the value-added chain, a company which makes glue from the fat extracted from the hides and skins. Finally, there is a water purification depot collectively funded by the leather tanners, the effluence of which is sold to a company which converts the non-toxic solids into fertilizer. All in an area of 10 square kilometres!

The entire community in Santa Croce, in one way or another, is associated with leather tanning. This provides new opportunities, through spin-off, along the value-added chain, which in deepening and refining the social division of labour guarantees the local supply of virtually all the ingredients necessary for entrepreneurial success in quality based and volatile markets. To use the language of neoclassical economics, over and above firm-specific and asset-specific advantages, there exists an area-wide asset which individual entrepreneurship draws upon. This 'valorization' of the milieu is a product of the progressive deepening of the social division of labour (vertical disintegration) at the local level. The area not only produces specialized skills and artisan capability, but also powerful external economies of agglomeration and a constant supply of industry-specific information, ideas, inputs, machinery and services — Marshall's 'industrial atmosphere'.

Thus far, the success of Santa Croce as a Marshallian industrial district has been ascribed to two broad sets of factors. One is the 'fortuitous' combination, since the early 1970s, of new market opportunities (the fashionwear sector) and a minimum set of inherited local capabilities (leather-tanning skills, a craft culture and so on). The second is the progressive vertical disintegration of the division of labour and its local containment. But there is also a third factor which has come to play a key role in safeguarding the success of the area. This is the institutionalization, at the local level, of individual sectional interests (e.g. the Association of Leather Tanners, the Association of Subcontractors, savings banks, the mayor's office, trade union branches etc.), as well as a sense of common purpose which draws upon Santa Croce's specialization in one industry and the intricate interdependences of a vertically disintegrated production system. Not only has this prevented the growth of rogue forms of individual profiteering which may destabilize the system of mutual interdependence, but it has also created a mechanism for collectivizing opportunities and costs as well as ensuring the rapid transmission of information and knowledge across the industrial district.

The 'collectivization of governance' has been of particular importance for the industrial district in recent years, as it has tried to cope with new pressures. By the mid-1980s, a honeymoon period of spectacular success for virtually all enterprises was coming to an end. This was the result of growing competition in international markets from fashionwear-oriented tanneries in Southeast Asia, a decline in demand from the Italian footwear industry, big price increases coupled with shortages in the availability of uncured skins and hides, and new costs attached to the introduction of environmental controls on effluence discharge. These are problems which have affected the entire community, problems which different interest groups have not been able to resolve individually. Resulting collective responses have ranged from joint funding by the tanners of an effluence treatment plant and multi-source funding (involving tanners, subcontractors, a local bank and the regional authorities) of an information service centre which offers advice on market trends, management skills and information technology, through to frequent and heated debates in the bar of the central piazza on new trends affecting the industry. How successful these efforts will be is not to be known. What matters, however, is that Santa Croce continues to possess a local institutional capability to respond collectively and swiftly to new market pressures and to steer the evolution of the industrial district in a particular direction.

This said, however, there is already some evidence to suggest that, into the 1990s, the organization of industry in Santa Croce will be 'post-Marshallian', that is, less locally confined and less vertically disintegrated. Increasingly, the trend is for tanners to import semi-finished leather, owing to difficulties in obtaining uncured hides and skins. If this practice becomes the norm, more than half of the production cycle will be eliminated from the area, to the detriment of locally based hide importers, subcontractors and chemical treatment firms. There is also a threat of 'forward' internationalization of the division of labour. A handful of companies — the oldest and the most powerful — have begun to open distribution outlets overseas as well as tanneries, usually through joint ventures, in countries either producing hides and skins or promising growth in the leather goods industries. They have also gone into the business of selling turnkey tanneries[3] for the east European countries — a development which stands to threaten Italian tanneries, including those in Santa Croce, if the finished leather ends up being imported by the domestic leather goods industry.

The risk, then, is that Santa Croce will come to perform only specific tasks in an internationally integrated value-added chain, thus threatening a shakeout of firms dependent upon tasks no longer performed locally. Through a narrowing of functional competencies, the area's industrial system will become less vertically disintegrated. Such a narrowing

3. These are factories in which all aspects of the plant, including technology and support services, are supplied by the seller, thus allowing the purchaser to produce the commodity at the 'turn of a key'.

runs the risk of threatening the institutional synergy and richness of activity which hitherto has secured the area's success as an industrial district. It is also possible that, with functional simplification and the offer of larger and better premises more recently by the local authority in its new Structure Plan, the larger tanners will seek to internalize individual production tasks more than before. Initial signals of such a development include the recommendation by the Associations of Leather Tanners and Subcontractors that transfers to the new industrial zone involve horizontal mergers, stricter loan scrutiny by banks of applications for business start-ups and the grouping of firms into business consortia in order to maximize on firm-level scale economies in such activities as purchasing and marketing.

If the twin processes of internationalization of the division of labour and vertical integration at the local level become the dominant trend, Santa Croce will lose its current integrity as a self-contained 'regional' economy. But, and this is the point, it will continue to remain a central node within the leather-tanning industry. Twenty years of Marshallian growth have made Santa Croce into a nerve centre of artisan ability, product and design innovation and commercial acumen within the international fashion-oriented leather goods filiere. This unrivalled expertise will guarantee its survival as a centre of design and commercial excellence, even if the activities of the 'hand' are reduced or internalized. The open question is whether, without the hand, the head will lose its might or successfully engineer a transition into other industrial ventures.

Marshall in London: the City and global financial services

Throughout much of the nineteenth century, and well into the twentieth century, the City of London[4] could have been characterized as a classical Marshallian industrial district. During this period it consisted of a network of small financial service firms, and a set of markets and market-clearing mechanisms, in close contact with and close proximity to one another, employing as many as 200,000 people. These firms and markets consistently minimized transaction costs through social and spatial propinquity, creating an industrial 'soup' rather than an industrial atmosphere, an upper-middle-class craft community of quite extraordinary contact intensity (Thrift and Leyshon, 1992).

This 'old City' was dynamized by the dictates of mercantile capital accumulation as described by Marx in the *Grundrisse*, that is, on the one hand its members strove towards endless expansion in the search for new products for old markets or new markets for old products so as to gain comparative advantage over rivals, and on the other, they tried to enlist non-economic power to regulate the system and to give monopolistic advantages to its members (usually relating to price competition). This regime of accumulation could survive precisely because of its mode of social regulation. Indeed, in a sense, it was the mode of regulation that generated the regime of accumulation. This statement is born out by the following observations.

First, the City was run on 'mesocorporatist' (Cawson, 1985; Moran, 1991) lines: that is, it was a largely self-regulating system of collective governance with the Bank of England acting to protect it from pluralist regulatory systems and politics. 'Representation and regulation were fused: the associations and institutions in City markets had authority because this was recognized by the Bank and because of this they were able to operate restrictive practices benefiting their members' (Moran, 1991: 63). Second, the City was tightly socially integrated. All its key workers were drawn from highly specific social backgrounds based on clearly drawn divides of class, gender and ethnicity, and this social specificity was reinforced throughout the workers' lives by various socializing processes at work and after work (the firms themselves, often partnerships, the markets, the livery companies, the Masonic lodges, the clubs and so on). This commonality of background

---

4. Clearly, over time, the City of London has consisted of a heterogeneous set of markets and firms. In this necessarily abbreviated account we have had to treat these markets and firms as rather more homogenous than they actually were, but the contact intensity between all these markets and firms is not in doubt (see, for example, Dunning and Morgan, 1971).

maximized face-to-face communication in a small area, generated a 'collective consciousness' or common 'gentlemanly' discourses, and also afforded quick assessments of character and, effectively, economic viability. Thus the old City was a trust-maximizing system and the knowledge structure was, in effect, the social structure. (That said, much of the City's economic dynamism and innovative capacity came from members of the City excluded from its mainstream — refugees, Jews, etc. — who were Weberian pariah capitalists in the strongest sense. It was those who were outside the system and therefore able to 'see' it who were most able to innovate (Chapman, 1984).)

Third, the City's markets were large and liquid, although in comparison with the later period they were based on comparatively few products and relatively slow rates of innovation. Again, to participate in these markets it was often important to know the right people, the right gentlemanly discourses and the right occupational languages. And fourth, the City was tightly concentrated in space, the result of the ability of the key mesocorporatist institutions like the Bank of England and the various markets to demand propinquity, the existence of various market-clearing mechanisms with quite restricted spatial ranges, and the peculiar importance of face-to-face transactions, an importance which can be interpreted as both a case and a result of the dense spatial arrangement (Pryke, 1991). (Indirect systems of communication like the telegraph and the telephone were a feature of the City from an early period, but these probably had the effect of strengthening rather than weakening social interaction.)

Thus, the old City of London was a protected, self-regulating, socially and culturally specific enclave able to wield very considerable social power (Ingham, 1984). It was, of course, tied in, in the strongest possible way, to the world economy of the time, but here was a case of the 'the local going global' (Thrift, 1987; 1990; Thrift and Leyshon, 1991; Pryke, 1991).

Since the late 1950s or early 1960s there has been a sea change in the way that the City of London has been able to go about its business, sufficient to be able to write of the emergence of a 'new City'. In particular, at least five major changes have threatened the City's integrity as an industrial district. The first change has been the emergence, through successive rounds of corporate restructuring, of larger and larger oligopolistic financial service firms, whether in the form of institutional investors (such as pension funds and insurance companies), securities houses or banks. Many of these firms are substantial multinational corporations that span the globe. The second change has been in the nature of international financial markets, which have altered in a number of ways. To begin with, they have become increasingly international in both space and time, often operating around the clock. They have also become, through a massive investment in telecommunications capacity, increasingly electronic, which means that more and more communication is taking place at a distance, with some markets becoming almost entirely decentralized. Most importantly, perhaps, many new fictitious capital markets are increasingly based on securitized products — not only bonds and securities but also various derivatives which demand new 'disintermediated' relationships between financial service firms.

The third change has been in the mode of regulation of the City. As Moran (1991) points out, neither the word 'deregulation' nor 'reregulation' adequately describes what has occurred. What has occurred has been the progressive accretion[5] of a more carefully codified, institutionalized and legalistic mesocorporatism, forced on the City by specific market circumstances, by a state succumbing to American pressure for change and by the diffusion of an American ideology of correct and incorrect practice. As Moran puts it, 'The story of the financial services revolution is the story of the rapid creation of new

5. Thus, the City's banking sector underwent its main period of deregulation in the 1960s, while the securities markets were not deregulated until the 1980s in the process known as 'Big Bang'. One of the chief reasons for taking the late 1950s or early 1960s as the dividing line between an old and a new City is, of course, that it was in this period that deregulation first started to bite, and with it came a progressive loss of social power.

institutions struggling for regulatory jurisdiction, the development of increasingly complex and unclear rules and the creation of growing numbers of regulators inside and outside firms, all quarrelling over the meaning of an expanding, contradictory and unclear body of jurisprudence' (Moran, 1991: 134). In other words, formal contracts have replaced implicit contracts.

A fourth change has occurred in the City's social structure. It has become more open in terms of class, gender and ethnicity — although no doubt it is still not as open as might be wished for (Rajan, 1988; 1991; Thrift and Leyshon, 1991). Socially specific selection of personnel of the old style has certainly been retained by a few small parts of the City, but these are increasingly sidelined. The City's social structure is now more mixed and more keyed into the knowledge structure in financial services. The fifth change has been in the City's rate of product innovation (which in turn has been facilitated by the investment in telecommunications technologies). Innovation in products is now much more rapid than before, but they are also more likely to fail (de Cecco, 1987). For example, of the 19 futures contracts launched by the London International Financial Futures Exchange, 8 have failed. Failure of products is a particular problem in financial services because liquidity and success are very tightly connected and because products and markets are so tightly interconnected.

In this process of multiple economic and social change, the City, like the financial services industry as a whole, has moved from a mercantile capitalist model to something more closely resembling an industrial model of production or, at least, an industrialized mercantile model of production. But, given the scope of the changes that have taken place and the undoubted loss of social power that they describe, one might well ask why the City of London has persisted at all. Why has something very like an industrial district, still relatively socially and spatially concentrated, persisted in and near to the confines of the old City of London, especially in the face of competition from other European financial centres?

The answer to this question can be related to the imperatives of the new global financial services production filiere. In its old incarnation, the City was the result of the local going global. In its new incarnation, the City is a result of the global going local. In particular, the City survives for three related reasons. First, it is a centre of representation. The City has become one of the chief points of surveillance and scripting of the global financial services filiere; London is where the stories are. Thus, much of the *world's* financial press (which has also rapidly globalized) operates from or near to the City; so does much research analysis, information processing and so on (Kynaston, 1985; Driver and Gillespie, 1991). The concentration of this activity in one place has the advantage that the City can watch and script the global financial services filiere industry by watching itself. Equally important, the City represents an important part of the knowledge structure of world financial services. Although in principle these collective assets could be decentralized, the fact is that it is still more convenient to have this massive body of knowledge, much of which is highly specialized, easily accessible in one place.

This point relates to the second reason for the city's persistence as an industrial district: its role as a centre of interaction. The City is a social centre of the global corporate networks of the financial service industry, with a large throughflow of workers from other countries, and people simply meeting. It is accepted in this role partly because of the concentration of headquarters and regional headquarters of financial services corporations in and around the City, partly because of a symbolic value which it explicitly plays up through 'trappings of trust' such as 'traditional' oak-lined meeting rooms (the City is still symbolically equated with financial services in a powerful way throughout the world), partly because of its accessibility in real and electronic space to markets and firms, and partly because of its knowledge base (which means that the wherewithal is always at hand to watch and script, buy and sell, borrow and lend). Thus the City is still a vital meeting place in which important deals can be made, issues marketed, coalitions formed, syndicates established and trust/implicit contracts cemented.

The third reason for the city's persistence is as a proving ground for product innovation. An important characteristic of financial service products is that product innovation and marketing are very closely tied together. A product only survives if a primary market can be established, and usually a secondary market too (not even counting the derivatives markets which are now so important). Thus a product cannot take off unless it is aggressively marketed, usually to a quite specific set of people in large investment institutions and companies. A place like the City allows new products to be evolved and tested quickly and efficiently, since there is always sufficient liquidity on tap and sufficient placement power available to assess an innovation's worth. More than this, the availability in one place of representatives of many investment institutions and companies means that products can be easily socialized and customized. Many derivatives products first started life as quite specific products for quite specific customers.[6]

Clearly these three reasons are closely interrelated. For example, product innovation depends on a 'thick' discursive atmosphere, on requisite knowledge structures and on availability of custom.

To conclude, the City is still an industrial district, but it is an industrial district which depends on fewer, larger firms, which still need a place from which and with which they can represent and analyse their world, a place where they can meet in order to add flesh and trust to electronically mediated personae, and a place which can be counted on to continue to invent new products and markets. Clearly, the City faces, and will continue to face, challenges from other financial centres, especially European financial centres such as Frankfurt and Paris (see Leyshon and Thrift, 1992). In part, this will depend on the social power it can still muster in what has become a complex geopolitical game, but in part it will also clearly continue to depend on its constitution as a local and global social space.

Conclusion

The argument of the first half of this paper is that there does not appear to be any inexorable trend towards the localization of production. This is not to deny that the trend towards vertical disintegration may have become more pronounced than in the past, nor is it to play down the significance that 'networking' may have in encouraging the resurgence along Marshallian lines of some regions as self-contained units of economic development. Against this, however, it has to be stressed that networking is also a global phenomenon, one which has come to coexist with, rather than replace, more orthodox forms of internationalization. Contemporary organizational change is very much a process of layering of new global corporate networks upon old international production hierarchies.

In this age of intensifying global hierarchies and global corporate networks, with both, as proposed in this paper, representing a reworked centralization of corporate command and control, it can only be a truism to propose that local economic prospects are becoming more dependent upon global corporate organizational forces. In such a context, it is difficult to think of localities as independent regional economies which participate freely, as the Marshallians would have it, in a global system integrated only by trade. But, then, what of the argument that an integral component of globalization is more localization of corporate activity, as companies turn decentralization and locational proximity into key conditions for flexibility and innovation? It cannot be denied that the growth of networking could lead to greater functional and operational decentralization down the corporate hierarchy as well as greater reliance on local external linkages for profitable production. However, and this is the point, the rediscovery of place is only occurring in a quite restricted set of

6. This is not, of course, to suggest that financial products always need to be provided locally, and it is certainly not to suggest that they will continue in localized fashion if they are successful.

localities, and efforts to encourage Marshallian growth in other areas through the formation of highly localized production systems are likely to fail. The examples of Santa Croce and the City of London illustrate only too clearly that the conditions for such growth are difficult to capture through even the most innovative policy measures, *unless certain basic structures are already in place*. These include a critical mass of know-how, skills and finance in rapidly evolving growth markets, a socio-cultural and institutional infrastructure capable of scripting and funding a common industrial agenda, and entrepreneurial traditions encouraging growth through vertical disintegration of the division of labour.

Collective intervention, both private and public, may be able to build upon and manipulate these basic structures, but it cannot generate them. Such an awareness of the limits of policy intervention is, in our view, important. Otherwise, typically Marshallian efforts to regenerate local economies through locally regulated ventures such as efforts to strengthen links between firms and between business and other local institutions (e.g. training colleges, development authorities etc.) run the risk of doing little more than legitimizing a false belief in the possibility of achieving solutions for what are global problems beyond local control.

Somewhat bleakly, then, we are forced to conclude that the majority of localities may need to abandon the illusion of the possibility of self-sustaining growth and accept the constraints laid down by the process of increasingly globally integrated industrial develop-ment and growth. Concretely, this may simply amount to pursuing those interregional and international linkages (trade, technology transfer, production etc.) which will be of most benefit to the locality in question. It may also involve — and on this point, the literature on industrial districts is helpful — upgrading the position of the locality within international corporate hierarchies and networks by improvements to a locality's skill, research, supply and infrastructure base in order to attract 'better quality' branch investments.

This, of course, is not much of a solution. On the other hand, it has to be stressed that as things are today even the neo-Marshallian nodes of global corporate networks are finding it difficult to retain their status. Furthermore, it has to be noted that the stakes for achieving the status of a node at the apex of an international filiere are truly high, and must be discounted for the vast majority of local areas which either lack or have lost the social and cultural infrastructure for innovation and transaction-rich competition. *Plus ça change* in the postfordist economy?

Ash Amin, Centre for Urban and Regional Development Studies, University of Newcastle upon Tyne, Newcastle upon Tyne NE1 7RU, and **Nigel Thrift**, Department of Geography, University of Bristol, Bristol BS8 1SS

References

Adler, P. and P. Adler (1984) (eds) *The social dynamics of financial markets*. JAI Press, Greenwich, CONN.

Amin, A. (1989) Flexible specialisation and small firms in Italy: myths and realities. *Antipode* 21.1, 13–34.

_____ and M. Dietrich (1991) From hierarchy to 'hierarchy': the dynamics of contemporary corporate restructuring in Europe. In A. Amin and M. Dietrich (eds), *Towards a new Europe?*, Edward Elgar, Aldershot.

_____ and K. Robins (1990) The re-emergence of regional economies? The mythical geography of flexible accumulation. *Environment and Planning D: Society and Space* 8.1, 7–34.

Bianchini, F. (1990) The 'Third Italy': model or myth? Mimeo. Centre for Urban Studies, University of Liverpool.

Block, F. (1990) *Post industrial possibilities*. University of California Press, Berkeley, CA.

Cawson, A. (1985) Varieties of corporatism: the importance of the meso-level of interest intermediation. In A. Cawson (ed.), *Organised interests and the state*, Sage, London.

Chapman, S. (1984) *The rise of merchant banking*. Allen and Unwin, London.

Cooke, P. and R. Imrie (1989) Little victories: local economic development in European regions. *Entrepreneurship and Regional Development* 1.4, 313—27.

de Cecco, M. (1987) *Money and innovation*. Blackwell, Oxford.

Driver, S. and A. Gillespie (1991) Spreading the word? Communications technologies and the geography of magazine print publishing. Newcastle PICT Working Paper I, Centre for Urban and Regional Development Studies, University of Newcastle.

Dunning, J.H. and K. Morgan (1971) *An economic study of the City of London*. Allen and Unwin, London.

Gertler, M. (1988) The limits to flexibility: comments on the post-Fordist vision of production and its geography. *Transactions of the Institute of British Geographers* 13, 419—32.

Giddens, A. (1990) *Consequences of modernity*. Polity Press, Cambridge.

____ (1991) *Modernity and self-identity*. Polity Press, Cambridge.

Glasmeier, A. (1991) Technological discontinuities and flexible production networks: the case of Switzerland and the world watch industry. Mimeo. Department of Geography, University of Texas at Austin.

Harrison, B. (1990) Industrial districts: old wine in new bottles? Working Paper 90-35, School of Urban and Public Affairs, Carnegie Mellon University, Pittsburgh.

Harvey, D. (1989) *The condition of postmodernity: an inquiry into the origins of cultural change*. Blackwell, Oxford.

Held, D. (1991) Democracy, the nation state and the global system. *Economy and Society* 20.12, 138—72.

Hirst, P. and J. Zeitlin (1989) Flexible specialisation and the competitive failure of UK manufacturing. *Political Quarterly* 60.3, 164—78.

____ and ____ (1991) Flexible specialisation vs. post-Fordism: theory, evidence and policy implications. *Economy and Society* 20.1, 1—56.

Ingham, G. (1984) *Capitalism divided*. Macmillan, London.

Jacquemin, A., P. Buiges and F. Ilzkovitz (1989) Horizontal mergers and competition policy in the European Community. *European Economy* 40 (May), CEC Directorate-General for Economic and Financial Affairs.

Kynaston, D. (1985) *The* Financial Times. *A centenary history*. Viking, London.

Leyshon, A. and N.J. Thrift (1992) European integration and the international financial system. *Environment and Planning A* 24, 49—81.

Marceau, J. (1989) *A family business? The making of an international business elite*. Cambridge University Press, Cambridge.

Moran, M. (1991) *The politics of the financial services revolution*. Macmillan, London.

Offe, K. and J. Wiesenthal (1979) Two logics of collective action. In M. Zeitlin (ed.), *Power and social theory*, JAI Press, Greenwich, CONN.

Peck, J. (1991) Labour and agglomeration: vertical disintegration, skill formation and flexibility in local labour markets. Mimeo. School of Geography, University of Manchester.

Piore, M. and C.F. Sabel (1984) *The second industrial divide*. Basic Books, New York.

Pryke, M. (1991) An international city going global. *Environment and Planning D: Society and Space* 9, 197—222.

Rajan, A. (1988) *Create or abdicate?* Witherley Press, London.

____ (1991) *Capital people*. Industrial Society, London.

Sabel, C.F. (1989) Flexible specialisation and the re-emergence of regional economies. In P. Hirst and J. Zeitlin (eds), *Reversing industrial decline? Industrial structure and policies in Britain and her competitors*, Berg, Oxford.

Sayer, A. (1989) Post-Fordism in question. *International Journal of Urban and Regional Research* 13.4, 666—95.

Scott, A.J. (1988) *New Industrial Spaces: flexible production organisation and regional development in North America and western Europe*. Pion, London.

____ (1991) The role of large producers in industrial districts: a case study of high-technology systems houses in southern California. UCLA Research Paper in Economic and Urban Geography, No. 2 (February).

Storper, M. (1991) Technology districts and international trade: the limits to globalisation in an age of flexible production. Mimeo. Graduate School of Urban Planning, University of California, Los Angeles.

_____ (1989) The transition to flexible specialisation in the US film industry: the division of labour, external economies and the crossing of industrial divides. *Cambridge Journal of Economics* 13.2, 273–305.

_____ and A.J. Scott (1989) The geographical foundations and social regulation of flexible production complexes. In J. Wolch and M. Dear (eds), *The power of geography: how territory shapes social life*, Unwin Hyman, Winchester, MA.

Strange, S. (1988) *States and markets*. Pinter, London.

_____ (1991) An eclectic approach. In C.N. Murphy and R. Tooze (eds), *The new international political economy*, Lynne Rienner, Boulder, CO.

Thrift, N.J. (1987) The fixers: the urban geography of international commercial capital. In J. Henderson and M. Castells (eds), *Global restructuring and territorial development*, Sage, London.

_____ (1989) The perils of transition models. *Environment and Planning D: Society and Space*, 7, 127–9.

_____ (1990) The perils of the international financial system. *Environment and Planning A* 22, 1135–7.

_____ (1990) Doing global regional geography. In R.J. Johnston and J.A. Hoekveld (eds), *Regional geography*, Routledge, London.

_____ and A. Leyshon (1991) In the wake of money. In L. Budd and C.S. Whimster (eds), *Global finance and urban living*, Routledge, London.

_____ and _____ (1992) *Making money*. Routledge, London.

Todling, F. (1991) The geography of innovation: transformation from Fordism towards post-Fordism? Mimeo. Institute for Urban and Regional Studies, University of Economics and Business Administration, Vienna.

Tornqvist, G. (1991) Swedish contact routes in the European urban landshape. Mimeo. Department of Social and Economic Geography, University of Lund.

Zukin, S. and P. Dimaggio (eds) (1990) *Structures of capital: the social organisation of the economy*. Cambridge University Press, Cambridge.

[8]

Sticky Places in Slippery Space:
A Typology of Industrial Districts*

Ann Markusen

Project on Regional and Industrial Economics, Rutgers University, New Brunswick, NJ 08901-1983

Abstract: As advances in transportation and information obliterate distance, cities and regions face a tougher time anchoring income-generating activities. In probing the conditions under which some manage to remain "sticky" places in "slippery" space, this paper rejects the "new industrial district," in either its Marshallian or more recent Italianate form, as the dominant paradigmatic solution. I identify three additional types of industrial districts, with quite disparate firm configurations, internal versus external orientations, and governance structures: a hub-and-spoke industrial district, revolving around one or more dominant, externally oriented firms; a satellite platform, an assemblage of unconnected branch plants embedded in external organization links; and the state-anchored district, focused on one or more public-sector institutions. The strengths and weaknesses of each are reviewed. The hub-and-spoke and satellite platform variants are argued to be more prominent in the United States than the other two. The findings suggest that the study of industrial districts requires a broader institutional approach and must encompass embeddedness across district boundaries. The research results suggest that a purely locally targeted development strategy will fail to achieve its goals.

Key words: industrial districts, regional growth.

The Puzzle of Stickiness in an Increasingly Slippery World

In a world of dramatically improved communications systems and corporations that are increasingly mobile internationally, it is puzzling why certain places are able to sustain their attractiveness to both capital and labor. Movement is, of course, costly and disruptive to both. David Harvey's (1982) work on capital's need for "spatial fix" and Storper and Walker's (1989) work on labor and reproduction suggest generic reasons why hypermobility cannot completely obliterate production ensembles in space. But neither account explains why certain places manage to anchor productive activity while others do not.

The problem is most acute in advanced capitalist countries, where wage levels and standards of living are substantially higher than in newly incorporated labor-rich and increasingly technically competent countries (Howes and Markusen 1993). Production space in these countries has become increasingly "slippery," as the ease to capital of moving plants grows and as new competing lines are set up in lower-cost regions elsewhere. Often the only alternative for the region of exit or any other aspirant appears to be matching local production conditions to those in the

* This research was supported by the National Science Foundation, Program in Geography and Regional Science, and by Rutgers University, though responsibility for the results remains the author's. The author would like to thank Clelio Campolina Diniz, Masatomi Funaba, Amy Glasmeier, Elyse Golob, Mia Gray, Bennett Harrison, Candace Howes, Andy Isserman, Mary Ellen Kelley, Yong Sook Lee, Marlen Llanes, Michael Oden, Sam Ock Park, Eric Parker, Mohamad Razavi, Masayuki Sasaki, and Frank Wilkinson for their insights from joint research and conversation on this subject, and Bill Beyers, Dan Knudsen, and three anonymous reviewers for comments on a previous version. Special thanks to Barbara Brunialti for research support.

competitor place, lowering wages and reproduction costs to the lower common denominator. Much of the stress on improving local "business climates" in a country like the United States in the past two decades is driven by the belief that localities have no other options.

Alarmed by the welfare implications of such a strategy, economists, geographers, and economic development planners have sought for more than a decade for alternative models of development in which existing activities are sustained or transformed in ways that maintain relatively high wage levels, social wages, and quality of life. They have done so largely with inductive methods, searching for the exceptions to the rule and examining the structure and operation of such "sticky places." One extensively researched formulation is that of the "flexibly specialized" or "new industrial district" (NID), based on the phenomenon of successful expansion of mature industries in the Emilio-Romagna region of Italy (Best 1990; Goodman and Bamford 1989; Piore and Sabel 1984; Scott 1988a, 1988b; Storper 1989). NIDs owe their stickiness to the role of small, innovative firms, embedded within a regionally cooperative system of industrial governance which enables them to adapt and flourish despite globalizing tendencies.

In this paper, I argue that there are at least three other types of industrial districts, or "sticky places," that have demonstrated resiliency in the postwar period in advanced industrialized countries. Stickiness connotes both ability to attract as well as to keep, like fly tape, and thus it applies to both new and established regions. Based on an inductive analysis of the more successful metropolitan regions in the United States, I show that structures and dynamic paths quite different from those captured in the NID formulation have enabled both relatively mature and up-and-coming regions to weather heightened capital mobility. Contrary to the emphasis on small firms in the NID formulation, these alternative models demonstrate the continued power of the state and/or multinational corporations under certain circumstances to shape and anchor industrial districts, providing the glue that makes it difficult for smaller firms to leave, encouraging them to stay and expand, and attracting newcomers into the region. These models exhibit greater propensities for networking across district lines, rather than within, and a much greater tendency to be exogenously driven and thus focused on external policy issues than do NIDs. From a welfare point of view, the four types perform quite differently with regard to income distribution, permissiveness toward labor organization, short-to-medium-term cyclicality, and longer-term vulnerability to secular change.

Identifying and Analyzing Sticky Places

The three alternative models of sticky places developed in this paper were constructed through a process of inductive inquiry similar to that used in researching NIDs. In the NID literature, intensive research on particular cases, sometimes comparing across several, has been used to identify causal forces and structural configurations. Piore and Sabel (1984, 1989) studied the Third Italy intensively in developing their notions of flexible specialization and industrial districts. In the United States, Christopherson and Storper's work on the film industry in Los Angeles (1986), Scott (1986) and Scott and Paul's work on Orange County (1990), and Saxenian's work on Silicon Valley (1990, 1991a, 1991b, 1994) enabled these authors to derive propositions about how secular changes in technology and markets enable and reward new forms of regional industrial organization. Vigorous debate on the accuracy and applicability of the NID formulation ensued, enlivening the economic geography literature for the better part of a decade (e.g., Amin and Robins 1990; Amin and Thrift 1992; Ettlinger 1992; Florida and Kenney 1990; Gertler

1988; Glasmeier 1988; Harrison 1992; Lovering 1990, 1991; Malecki 1987; Markusen 1991; Pollert 1988; Schoenberger 1988).

The research summarized here had its origins in a larger research project to determine the extent to which the NID model could explain the durability and flourishing of regional economies in the United States, Japan, Korea, and Brazil as adequately as it appeared to do so in the Third Italy. Empirical testing of the NID model has been surprisingly thin. Few attempts have been made to determine whether existing agglomerations are "flexibly specialized"—an exception is Feldman's (1993) remarkable study of U.S. industrial agglomeration—or to determine whether major industries are well characterized by this post-Fordist formulation (see Luria (1990) for an excellent investigation of the auto industry in this regard). No author has rigorously set out the features of new industrial districts in ways that permit easy assessment of their incidence and growth across space and time. The limits of the flexibly specialized new industrial district as an emergent paradigmatic form (a claim made by Scott (1988a, 1988b)) are best established by demonstrating that other industrial district profiles are both theoretically plausible and empirically demonstrable.

In each country studied in our larger project, it was clear that certain mature as well as newer agglomerations exhibited an ability to weather the leveling effect of accelerated world market integration and the global search for profitability, attributes that make space "slippery." But most of these enclaves did not match the features of the flexibly specialized industrial district of the NID literature. Just as deindustrializing regions are quite remarkably distinguishable from each other, as Massey and Meegan have deftly shown (Massey and Meegan 1982; Massey 1984), regions hosting rapid growth and/or escaping industrial decline exhibit distinctly different structures. Through inductive research, we were able to identify three alternative patterns.

Our method involved a two-stage process. First, we surveyed metropolitan growth since 1970 for each of the four countries, identifying the universe of those who posted growth rates significantly higher than the national average (tables showing each of these regional sets may be found in Markusen (1995)). We then chose a subset of each of these for further case study research, relying on both disaggregated data on industrial structure and expert opinion on industrial organization. For each country, we selected at least one case with apparent conformity to the NID formulation and three to five others whose industrial structure and organization appeared to be quite different. We used techniques pioneered in social science case study research (Yin 1984) and leaned heavily on interviews with business firms, trade associations, trade unions, and regional economy watchers, incorporating and adding to the literature on enterprise studies and corporate interviews as a method for studying regions (McNee 1960; Krumme 1969; Schoenberger 1985, 1991; Healey and Rawlinson 1993; Markusen 1994).

Conceptually, we inquired into the presence or absence of features specified in the NID formulation: firm size distribution, up- and downstream industrial linkages, degree of vertical disintegration, networks among district firms, districtwide governance structures, innovative capabilities, the organization of production. In addition, we explored a number of features not generally incorporated into NID studies (Park and Markusen 1994). First, we examined the role of the state at both the national and regional/local level as rule maker, as producer and consumer of goods and services, and as underwriter of innovation, with consequences for the distribution and anchoring of employment within and across regions (Christopherson 1993, 1994; Linge and Rich 1991; Markusen et al. 1991; Markusen and Park 1993; Saxenian 1995). Second, we scrutinized the role of large firms, especially those with internal and external market power, in industrial agglomerations (Amin and Robins 1990; Dicken 1992; Gereffi and Korzeniewicz 1994; Harrison 1994; Sayer 1989). Third, we examined the embeddedness of firms both within their districts and in non-local networks extending across national and international space (Granovetter 1985; Storper and Harrison 1991; Linge 1991; Markusen 1994). Fourth, since sources of profitability vary over the course of an industry's maturation and are linked to changing forms of competition, organizational structures, occupational characteristics, and locational tendencies (Markusen 1985), we investigated the longer-term developmental dynamic of major industries and their constituents present in the district, to determine their resiliency and/or vulnerability to longer-term atrophy. Fifth, we assessed the long-term dynamic potential of each region, including the likely trajectory and future competitiveness of its existing industrial ensemble and the ability of the latter to release locally anchored resources, human and physical, into new, unrelated specialized sectors. Finally, we searched for connections between district structure and operation and a number of social welfare metrics, including employment growth rates over time, cyclical stability, associated income and wealth distribution, trade union presence, and political diversity.

A bit more may be said about this final component of the research. Evaluation of the welfare implications of each type of sticky place is a complex task and rarely undertaken. Scholars of NID literature have generally written in a normatively favorable if implicit way about the virtues of NIDs as providers of good jobs and long-term stability and dynamism; this is especially palpable in the treatments of Piore and Sabel (1984), Best (1990), and Saxenian (1994). A sticky place is "better," in our normative view, if it (1) ensures average or better-than-average growth for a region as a whole over time; (2) insulates a region from the job loss and firm failures of short-to-intermediate-term business or political spending cycles; (3) provides relatively good jobs, ameliorates tendencies toward income duality, and prevents undue concentration of wealth and ownership; (4) fosters worker representa-

tion and participation in firm decision making; and (5) encourages participation and tolerates contestation in regional politics.

Our research findings enabled us to develop several schematic alternatives to NIDs. Like Storper and Harrison (1991), we opt for an expansive connotation of industrial district, which does not confine it to the most common usage, called here the Marshallian (or Italianate variant) district. Elsewhere, we offer the following definition: an industrial district is a sizable and spatially delimited area of trade-oriented economic activity which has a distinctive economic specialization, be it resource-related, manufacturing, or services (Park and Markusen 1994).

In what follows, I present four distinctive industrial spatial types: (1) the Marshallian NID, with its recent Italianate variety; (2) the hub-and-spoke district, where regional structure revolves around one or several major corporations in one or a few industries; (3) the satellite industrial platform, comprised chiefly of branch plants of absent multinational corporations—this type of district may either be comprised of high-tech branch plants or consist chiefly of low-wage, low-tax, publicly subsidized establishments; and (4) the state-centered district, a more eclectic category, where a major government tenant anchors the regional economy (a capital city, key military or research facility, public corporation). The hypothesized features of each are summarized in Table 1. Schematic visual models of each of the first three, showing relative firm size and interfirm connections, both inside and outside the district, are offered in Figure 1. Here, firm relationships within the region are depicted inside the circle versus those outside of it—suppliers to the left, customers to the right. A real-world district may be an amalgam of one or more types, and over time districts may mutate from one type to another. This conceptualization complements the geographic industrialization schema of Storper and Walker (1989, Fig. 3.1). While theirs is process-centered, the one offered here is region-centered, with a

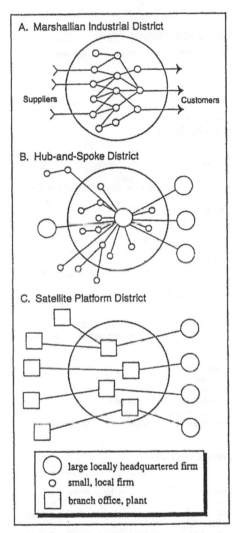

A. Marshallian Industrial District

Suppliers

Customers

B. Hub-and-Spoke District

C. Satellite Platform District

○ large locally headquartered firm

○ small, local firm

▢ branch office, plant

Figure 1. Firm size, connections, and local versus nonlocal embeddedness.

focus on firm size, interconnections, and internal versus external orientations.

Each spatial type is presented with a set of hypothesized traits, and the resilience and/or vulnerability of each to events in the changing global economy are noted. Districts which are sticky in one era may fail to cohere in the longer run—the glue may dry up, become brittle and lose its adhesive quality. Central to

the differences among sticky places and their ability to persist are presence (or absence) of distinctive and lopsided power relationships, sometimes within the district and sometimes between district entities and those residing elsewhere. Examples of each type can only be mentioned in passing here, but are the subject of complementary papers (e.g., Gray, Golob, and Markusen 1996; Golob et al. 1995; Markusen and Park 1993; Markusen and Sasaki 1994; Markusen 1994; Park and Markusen 1994).

The focus on rapidly growing industrial spaces helps us to develop an impressionistic sense of the relative contribution of each type to overall regional restructuring. In the United States, for instance, the fastest-growing industrial cities (as opposed to residentiary cities, where retirement communities account for the bulk of growth) include the 15 listed in Table 2, all of which added manufacturing employment at rates of 50 percent or more over the period 1970 to 1990, compared with a zero rate of growth nationally. These may be contrasted with the performance of the four major older industrial centers of New York, Boston, Chicago, and Los Angeles, at the bottom of the table. Very few of these fast-growing regions, I shall argue, can be characterized as NIDs, but many of them reproduce the conditions present in the other three models of "sticky places."

Marshallian and Italianate Industrial Districts

An extensive and recent literature on industrial districts focuses on the Marshallian industrial district and its more cooperative, embedded Italianate progeny. Since the characteristics hypothesized for these districts are relatively well known, I summarize them briefly here, with particular emphasis on those which may be contrasted to the district types presented below.

In his original formulation of the industrial district, Marshall envisioned a region where the business structure is comprised of small, locally owned firms

298   ECONOMIC GEOGRAPHY

Table 1
Hypothesized Features of New Industrial District Types

Marshallian industrial districts
- Business structure dominated by small, locally owned firms
- Scale economies relatively low
- Substantial intradistrict trade among buyers and suppliers
- Key investment decisions made locally
- Long-term contracts and commitments between local buyers and suppliers
- Low degrees of cooperation or linkage with firms external to the district
- Labor market internal to the district, highly flexible
- Workers committed to district, rather than to firms
- High rates of labor in-migration, lower levels of out-migration
- Evolution of unique local cultural identity, bonds
- Specialized sources of finance, technical expertise, business services available in district outside of firms
- Existence of "patient capital" within district
- Turmoil, but good long-term prospects for growth and employment

Italianate variant (in addition to the above)
- High incidence of exchanges of personnel between customers and suppliers
- High degree of cooperation among competitor firms to share risk, stabilize market, share innovation
- Disproportionate shares of workers engaged in design, innovation
- Strong trade associations that provide shared infrastructure—management, training, marketing, technical or financial help, i.e., mechanisms for risk sharing and stabilization
- Strong local government role in regulating and promoting core industries

Hub-and-spoke districts
- Business structure dominated by one or several large, vertically integrated firms surrounded by suppliers
- Core firms embedded nonlocally, with substantial links to suppliers and competitors outside of the district
- Scale economies relatively high
- Low rates of turnover of local business except in third tier
- Substantial intradistrict trade among dominant firms and suppliers
- Key investment decisions made locally, but spread out globally
- Long-term contracts and commitments between dominant firms and suppliers
- High degrees of cooperation, linkages with external firms both locally and externally
- Moderate incidence of exchanges of personnel between customers and suppliers
- Low degree of cooperation among large competitor firms to share risk, stabilize market, share innovation
- Labor market internal to the district, less flexible
- Disproportionate shares of blue-collar workers
- Workers committed to large firms first, then to district, then to small firms
- High rates of labor in-migration, but less out-migration
- Evolution of unique local cultural identity, bonds
- Specialized sources of finance, technical expertise, business services dominated by large firms
- Little "patient capital" within district outside of large firms
- Absence of trade associations that provide shared infrastructure—management, training, marketing, technical or financial help, i.e., mechanisms for risk sharing and stabilization
- Strong local government role in regulating and promoting core industries in local and provincial and national government
- High degree of public involvement in providing infrastructure
- Long-term prospects for growth dependent upon prospects for the industry and strategies of dominant firms

Satellite industrial platforms
- Business structure dominated by large, externally owned and headquartered firms
- Scale economies moderate to high
- Low to moderate rates of turnover of platform tenants
- Minimal intradistrict trade among buyers and suppliers
- Key investment decisions made externally
- Absence of long-term commitments to suppliers locally
- High degrees of cooperation, linkages with external firms, especially with parent company
- High incidence of exchanges of personnel between customers and suppliers externally but not locally
- Low degree of cooperation among competitor firms to share risk, stabilize market, share innovation
- Labor market external to the district, internal to vertically integrated firm
- Workers committed to firm rather than district

(continued)

Table 1

(continued)

- High rates of labor in-migration and out-migration at managerial, professional, technical levels; little at blue- and pink-collar levels
- Little evolution of unique local cultural identity, bonds
- Main sources of finance, technical expertise, business services provided externally, through firm or external purchase
- No "patient capital" within district
- No trade associations that provide shared infrastructure—management, training, marketing, technical, or financial help, i.e., mechanisms for risk sharing and stabilization
- Strong local government role in providing infrastructure, tax breaks, and other generic business inducements
- Growth jeopardized by intermediate-term portability of plants and activities elsewhere to similarly constructed platforms

State-anchored industrial districts

- Business structure dominated by one or several large, government institutions such as military bases, state or national capitals, large public universities, surrounded by suppliers and customers (including those regulated)
- Scale economies relatively high in public-sector activities
- Low rates of turnover of local business
- Substantial intradistrict trade among dominant institutions and suppliers, but not among others
- Key investment decisions made at various levels of government, some internal, some external
- Short-term contracts and commitments between dominant institutions and suppliers, customers
- High degrees of cooperation, linkages with external firms for externally headquartered supplier organizations
- Moderate incidence of exchanges of personnel between customers and suppliers
- Low degree of cooperation among local private-sector firms to share risk, stabilize market, share innovation
- Labor market internal if state capital, national if university or military facility or other federal offices for professional/technical and managerial workers
- Disproportionate shares of clerical and professional workers
- Workers committed to large institutions first, then to district, then to small firms
- High rates of labor in-migration, but less out-migration unless government is withdrawing or closing down
- Evolution of unique local cultural identity, bonds
- No specialized sources of finance, technical expertise, business services
- No "patient capital" within district
- Weak trade associations to share information about public-sector client
- Weak local government role in regulating and promoting core activities
- High degree of public involvement in providing infrastructure
- Long-term prospects for growth dependent on prospects for government facilities at core

that make investment and production decisions locally. Scale economies are relatively low, forestalling the rise of large firms. Within the district, substantial trade is transacted between buyers and sellers, often entailing long-term contracts or commitments. Although Marshall did not explicitly say so, linkages and/or cooperation with firms outside the district are assumed to be minimal. The Marshallian industrial district is depicted in the top portion of Figure 1, with many small firms buying and selling from each other for eventual export from the region. The arrows show necessary purchases of raw materials and business services from outside the region on the left and sales to

external markets on the right, in the form of exchange rather than cooperative relationships external to the region.

What makes the industrial district so special and vibrant, in Marshall's account, is the nature and quality of the local labor market, which is internal to the district and highly flexible. Individuals move from firm to firm, and owners as well as workers live in the same community, where they benefit from the fact that "the secrets of industry are in the air." Workers are committed to the district rather than to the firm. Labor out-migration is minimal, while in-migration occurs as growth permits. The district is seen as a relatively stable community,

Table 2

Selected U.S. Metropolitan Employment Growth Rates, 1970–1990

	Employment 1990 (in thousands)	Employment Change (%), 1970–90	Manufacturing Employment, 1990 (in thousands)	Manufacturing Change (%), 1970–90	Service Employment, 1990 (in thousands)	Service Change (%), 1970–90
Colorado Springs, Colo.	228	104	24	261	60	214
Austin, Tex.	471	178	50	249	131	253
Reno, Nev.	145	155	9	202	70	184
Tucson, Ariz.	316	123	28	199	101	219
Huntsville, Ala.	163	76	34	177	42	82
Orlando, Fla.	569	246	56	162	236	465
Albuquerque, N. Mex.	305	125	22	131	98	184
Melbourne/Titusville, Fla.	202	112	31	122	66	119
San Jose, Calif.	1,015	128	273	119	301	199
San Diego, Calif.	1,397	120	141	109	390	254
Anaheim-Santa Ana, Calif.	1,552	192	261	111	464	352
Raleigh-Durham, N.C.	513	123	66	94	145	175
Seattle, Wash.	1,339	114	227	73	362	206
Madison, Wis.	262	73	26	53	62	147
Elkhart-Goshan, Ind.	116	64	52	50	20	123
Los Angeles-Long Beach, Calif.	5,200	56	893	9	1,707	129
Boston-Lawrence-Salem, Mass.	1,672	30	340	−12	894	108
Chicago, Ill.	3,673	23	569	−33	1,128	101
New York, N.Y.	4,765	2	428	−51	1,704	50
United States	110,321	56	19,742	0	37,573	126

Source: U.S. Department of Commerce, Bureau of Economic Analysis (1970, 1990). Estimates of suppressed data were computed by Andrew Isserman and Oleg Smirnov, Regional Research Institute, West Virginia University, and compiled by Ann Markusen and Mia Gray.

which enables the evolution of strong local cultural identity and shared industrial expertise.

The Marshallian district also encompasses a relatively specialized set of services tailored to the unique products/ industries of the district. These services include technical expertise in certain product lines, machinery and marketing, and maintenance and repair services. They include local financial institutions offering so-called "patient capital," willing to take longer-term risks because they have both inside information and trust in the entrepreneurs of local firms.

All of these features are subsumable under the notion of agglomeration, which suggests that the stickiness of a place resides not in the individual locational calculus of firms or workers, but in the external economies available to each firm from its spatial conjunction with other firms and suppliers of services. In Marshall's formulation, it was not necessary

that any of these actors should be consciously cooperating with each other in order for the district to exist and operate as such. But in a more recent formulation, emerging from research on Italian industrial districts and extended to other venues in Europe and the United States, researchers have argued that concerted efforts to cooperate among district members and to build governance structures to improve districtwide competitiveness can improve prospects—that is, increase the stickiness of the district.

Features characterizing Italianate districts are articulated in intensive case studies on the Italian case (Piore and Sabel 1984; Bellandi 1989; Bull, Pitt, and Szarka 1991; Goodman 1989; Sforzi 1989). These have been reworked and adapted to American cases—Orange County (Scott 1986; Scott and Paul 1990) and Silicon Valley (Saxenian 1994)—though not without debate (Malecki 1987; Florida and Kenney 1990; Saxenian 1991a). The unifying notion is that firms (often with the help of regional governments and trade associations) consciously "network" to solve problems of cycles and overcapacity and to respond to new demands for flexibility (Amin and Thrift 1992). In the American version, rigidities in older industrial cities tend to encourage these agglomerations to root anew in relatively virgin locations (Markusen 1991; Scott 1988b; Storper and Walker 1989). Few cases have been identified outside of Europe or the United States, but good candidates for study are subdistricts such as the southern sector of Tokyo and Kangwan, a south side district in Seoul.

Unlike the passivity of Marshall's firms, Italianate districts exhibit frequent and intensive exchanges of personnel between customers and suppliers and cooperation among competitor firms to share risk, stabilize markets, and share innovation. Disproportionate shares of workers are engaged in design and innovative activities. Activist trade associations provide shared infrastructure—management, training, marketing, technical, or financial help—as well as providing forums to hammer out collective strategy. Local and regional governments may be central in regulating and promoting core industries. Trust among district members is central to their ability to cooperate and act collectively (Harrison 1992; Saxenian 1994), although critics argue that the power of large corporations to shape Italian industrial districts has been understated (see the discussion in Harrison 1994, Chap. 4).

In assessing the growth, stability, equity, and politics of Italianate industrial districts, the Italian variety must be distinguished from the Silicon Valley and Orange County cases, and each from their Marshallian predecessors. In terms of growth and stability, as long as agglomeration economies remain and are not replicated in other locales, both Marshallian and Italianate industrial districts retain good long-term prospects for growth and development. Although more standardized functions may be hived off and driven elsewhere by inflated regional

costs, innovation (so the theory goes) will ensure the revitalization of these "seedbeds of innovation." But other hypotheses have been advanced. Agglomerative specialization and success in one industry, especially when associated with some degree of market power and/or dominance over regional factor markets, can actually impede the development of other sectors, whose presence might diversify the economy and counteract maturation or instability in the original sector. Pittsburgh in the late nineteenth century and Detroit in the early decades of the twentieth century resembled Italianate districts and Silicon Valley, but the evolution of oligopoly and the crowding out of other sectors left both quite vulnerable to the inevitable maturation and decentralization of those industries (Chinitz 1960; Markusen 1985).

On the equity front, the high-tech Silicon Valleys and Orange Counties depart strikingly from the Italian industrial districts. Italian industrial districts are often the creatures of resilient cultures, organized politically on the basis of long-standing communities, unions, and the Italian communist party. Fundamental to their governance structures are strong leadership roles for unions and guarantees that most enterprises will be stabilized and nurtured, even during downturns. This has helped to stabilize incomes and assure relatively good income distributions within the districts. In the California cases, in contrast, district cooperation, where it exists, is purely between entrepreneurs and firms, who operate in a non-union environment and where there is little preexisting community to ameliorate vicious competition and failure in periods of instability. Income distribution tends to be highly dualized in such regions (Saxenian 1983; Harrison 1994). Furthermore, politics within such districts tends toward the conservative, laissez-faire end of the spectrum—Orange County is famous as the home of the John Birch society and Silicon Valley as a hotbed of free trade and anti-union business activism.

Despite the often extravagant claims of some of its protagonists, the "new industrial district" approach has much to offer and has deservedly captured the imagination of scholars and local economic development activists alike. But many of the faster-growing regions of the world turn out not to be primarily characterized by these same features. Furthermore, other structural forms may be associated with superior welfare and political cultures. It is to these other types of sticky places we now turn.

Hub-and-Spoke Industrial Districts

Another quite different type of industrial district is present in regions where a number of key firms and/or facilities act as anchors or hubs to the regional economy, with suppliers and related activities spread out around them like spokes of a wheel. Examples are Seattle and central New Jersey, United States; Toyota City, Japan; Ulsan and Pohang, South Korea; San Jose dos Campos in Brazil. A simple version of this form is depicted in the middle frame of Figure 1, where a single large firm (e.g., Boeing in Seattle or Toyota in Toyota City) buys from both local and external suppliers and sells chiefly to external customers, who may be large (e.g., the airlines, the military in the case of Boeing) or masses of individual consumers (Toyota). Intensive case studies of hub-and-spoke districts include Seattle (Gray, Golob, and Markusen 1996), central New Jersey (Fineberg et al. 1993), San Jose dos Campos and Campinas, Brazil (Diniz and Razavi 1994).

The dynamism in hub-and-spoke economies is associated with the position of these anchor organizations in their national and international markets. Other local firms tend to have subordinate relationships to them. If over time the anchors evoke a critical mass of agglomerated skilled labor and business services around them, they may set off a more

diversified developmental process where new firms form few connections to hub firms other than benefiting from the urbanization and agglomeration economies they have created.

Hub-and-spoke districts are thus dominated by one or several large, vertically integrated firms, in one or more sectors, surrounded by smaller and less powerful suppliers. Hub-and-spoke districts may exhibit either a strongly linked form, where smaller firms are quite dependent upon the large anchor firm or institution for either markets or supplies, or a weaker, more nucleated form, in which small firms enjoy the agglomerative externalities of the larger organization's presence without necessarily buying or selling to them. In some versions, the large player(s) may be oligopolists in a single industry, as with the Big Three auto corporations in Detroit or Toyota in Toyota City. Unrelated or loosely linked hubs in several industries may also coexist in a region. In Seattle, for instance, the economy is organized around Weyerhauser as the dominant resource-sector company, Boeing as the dominant industrial employer (commercial aircraft and military/spacecraft), Microsoft as the leading services firm, the Hutchinson Cancer Center as the progenitor of a series of biotechnology firms, and the Port of Seattle as the transportation hub. Core firms or institutions are embedded nonlocally, with substantial links to suppliers, competitors, and customers outside the district. Internal scale and scope economies are relatively high, and turnover of firms and personnel is relatively low except in third-tier suppliers or in major downturns in hub industries. Key investment decisions are made locally, but their consequences are spread out globally.

Hub-and-spoke districts may exhibit intradistrict cooperation, but it will generally be on the terms of the hub firm. Substantial intradistrict trade will take place among suppliers and hub firms, often embodied in long-term contracts and commitments. Cooperation may entail efforts to upgrade supplier quality, timeliness, and inventory control, and it may extend outside district boundaries to suppliers farther afield. Exchanges of personnel may take place, though not to the extent found in Italianate industrial districts. Markedly lacking is the coopera-tion among competitor firms to share risk, stabilize the market, and share innova-tion. Strategic alliances on the part of the larger firms are more apt to be forged with partners outside the region.

The labor market in hub-and-spoke districts is internal to both large hub firms and to the district, though it is less flexible than in the Italianate model. Workers' loyalties are to core firms first, then to the district, and only after that to small firms. If jobs open up in hub firms, workers will often abandon smaller employers to get onto the hub firms' payroll. This factor makes it tougher for smaller firms in some segments of the industry to survive. Hub firms attract new labor into the conurba-tion, however, which helps to counterbal-ance the power imbalance in the labor market.

Hub-and-spoke districts do evolve unique local cultures related to hub activities. Detroit is known as Motor City, and sports teams of many cities have been named after dominant sectors—the Oil-ers, the Steelers, the Brewers, the Pis-tons, the Millers (the old Minneapolis team). They develop considerable exper-tise in the labor pool in specialized industrial capabilities, and they engender specialized business service sectors tai-lored to their needs. Although these business services are focused on the large hub firms, some can become less depen-dent by extending their markets to other competitor firms in far-flung locales. An extensive discussion of how a small firm experiences its position in a hub-and-spoke economy is included in Markusen (1994).

Districts of this sort lack some of the more celebrated governance structures of the Italianate industrial districts. They often lack "patient capital," local venture capital specially tailored to start-ups in their industry. The largest returns to trade tend to be tied up as retained earnings in the major hub firms, who are happy to redeploy it wherever across the globe their strategic plans call for. The few trade associations that exist are relatively weak, often because top hub managers absent themselves from their deliberations and activities. Hub firms will concern themselves with state and local governmental activities that impinge upon their land use, tax, and regulatory situations and will try politically to ensure that area politicians represent the inter-ests of their firm and industries at the national and international levels. They may also be actively involved in issues that affect their work force and ability to do business—especially in improving area educational institutions and the provision of infrastructure.

In the long run, hub-and-spoke districts are quite dependent on their major industries and firms within them for their stickiness. Growth and stability can be jeopardized by intermediate-term porta-bility of plants and activities away from the region, or by the long-term decline of the industry, or by poor management of the principal firms. But stickiness also depends on the degree to which mature sectors can release local resources into new, unrelated sectors. A sobering histor-ical example of the vulnerability of hub-and-spoke districts is Detroit, where a turn-of-the-century Marshallian district (perhaps with some Italianate features) transformed itself into a hub-and-spoke district around the auto oligopoly by the 1930s. Here, to vastly oversimplify, De-troit's vitality was severely taxed by the oligopolistic rigidity of the locally head-quartered auto industry, combined with concerted investment on the part of the Japanese state and auto corporation in building a rival agglomeration around Toyota near Nagoya, Japan. Furthermore, tight oligoponistic control over the De-troit area's resources prevented the diver-sification of its economy (Chinitz 1960). A counter example is Seattle, where several unique features of Boeing as the undis-puted anchor to the regional economy (and the undisputed lead firm in the world aerospace industry) have contributed to (or at least not prevented) the region's diversification into other sectors—port-related activities, software, biotechnology—positioning it well to withstand retrenchment and global decentralization in the aircraft industry (Gray, Golob, and Markusen 1996).

Hub-and-spoke industrial districts may be characterized by relatively good income distributions. If so, this is due to both structural and institutional causes. Market power, often present in hub-and-spoke cases, results in relatively high returns to capital, a necessary though not sufficient condition for sharing of such returns with the work force in the form of higher wages. The presence of large anchor firms, nonprofit and public institutions may also reflect natural economies of scale, which are associated with large capital outlays and therefore high levels of labor productivity, available for distribution in wages. Securing this labor share is most often dependent on the presence of unions or the threat of their emergence. More vigorous political competition between probusiness and prolabor constituencies is apt to hold sway in such districts.

Satellite Platforms

Yet a third variant of rapidly growing industrial districts may be termed the satellite platform—a congregation of branch facilities of externally based multiplant firms. Often these are assembled at a distance from major conurbations by national governments or entrepreneurial provincial governments as a way of stimulating regional development in outlying areas and simultaneously lowering the cost of business for competitively squeezed firms bristling under relatively high urban wages, rents, and taxation. Tenants of satellite platforms may range from routine assembly functions to relatively sophisticated research, but they must be able to more or less "stand alone," detachable spatially from either up- or downstream operations within the same firm or from agglomerations of

competitors and external suppliers or customers (Glasmeier 1988).

Satellite platforms may be found in almost all countries, regardless of development. An outstanding high-end example in the United States is the internationally much-admired Research Triangle Park, a collection of unrelated research centers of major multinational corporations (Luger and Goldstein 1990), while a comparable low-end U.S. case is Elkhart, Indiana, where a number of auto-related branch plants have been attracted by relatively low-wage labor. In South Korea, Kumi constitutes a low-end textile and electronics platform, while Ansan operates as an odd collection of disparate industrial polluters grouped together (Park and Markusen 1994). In Japan, some of the better-performing technopoles, such as Oita and Kumamoto, fall into this category (Markusen and Sasaki 1994). In Brazil, a remarkable case is the state-sponsored expansion of Manaus as an import/export zone (Diniz and Borges Santos 1995).

In satellite platforms, business structure is dominated by large, externally situated firms that make key investment decisions. Scale economies within each facility are moderate to high, and rates of turnover of platform tenants are low to moderate. Minimal intradistrict trade or even conversation takes place among platform tenants. Orders and commitments to local suppliers are conspicuously absent. Since platforms generally host heterogeneous firms in terms of product if not industry and are remotely controlled, they do not operate as cooperative ventures among resident plants to share risk, stabilize the market, or engage in innovative partnerships. In this they differ from hub-and-spoke district, where the large multilocational firm or institution is locally based. This type of sticky place is presented in the lower portion of Figure 1; its most conspicuous feature is the absence of any connections or networks within the region and the predominance of links to the parent corporation and other branch plants elsewhere.

It is not as if branch operations, however, are not embedded in relationships external to the facility. They cooperate and communicate daily with the parent company. Personnel exchanges are common between branch operations and the headquarters firm, but not locally with other branch facilities. To buttress this nonplace embeddedness, the labor market within which each facility operates, at least in the high-end version and for management and some technical talent in the low-end version, cuts across district boundaries; it is internal to the vertically integrated firm, rather than to the district. This means that there will be high rates of labor migration in and out of the district at the managerial, professional, and technical levels. Often skilled professionals who originated from the region will be disproportionally represented. Only blue- and pink-collar labor will be hired locally, which may, however, not be inconsequential.

Over time, districts built around platforms may begin to host growth of suppliers, oriented toward platform tenants, and they may enjoy some increase in local entrepreneurship because the platform enhances the pool of skilled personnel resident in the region. But in cases studied to date, the incidence of such activity is small, and the aggregate growth of the region is still very much tied to the number of tenants that can be attracted and to the ability to retain them (Howes 1993).

A number of features of the satellite platform constrain its development into a better-articulated regional economy. First of all, the main sources of finance, technical expertise, and business services are external to the region, furnished through corporate headquarters. Satellite districts have little "patient capital" to draw upon, and because substantive activities are diverse, they lack industry-specific trade associations that would provide shared infrastructure and help with management, training, and marketing problems. These will only be partially compensated for by strong national or local government efforts and services offered by Chambers of Commerce and other associations of local fixed capital.

Satellite platforms' future growth is jeopardized by the intermediate-term portability of plants and activities elsewhere to similarly constructed platforms. Those concentrating on higher-end activities, where stability and amenities in the residential sphere are essential to drawing and keeping skilled personnel, will be less vulnerable in this regard, while purely low-cost districts will be more so, especially if fixed capital investment is low. Since individual plants and facilities are disparate and outward looking, satellite platforms do not engender the development of unique local cultural bonds or new identities, even though they may destroy preexisting ones. Thus they may be less sticky, especially if less skilled, than other types of district. Hosting communities face the challenge of trying to parlay resources assembled by such facilities into other diversifying and homegrown sectors. They do remain sticky, however, to the extent that large capital investments are made in the process of occupying them.

The record on income distribution in satellite platforms is mixed. In all countries studied, the entry of such platforms into previously depressed regions does contribute to higher overall per capita incomes (and perhaps a depression of those in regions of exit). Within the region, income distributional consequences depend on the nature of the industry and activity. Good blue-collar jobs in a depressed agricultural region will improve the income distribution. In technical branch plant platforms, the creation of a significant number of clerical and technician jobs may help to ameliorate the skewness introduced by operations that are top-heavy with managers and professionals. This seems to have occurred in the case of Research Triangle Park (Luger and Goldstein 1990). However, satellite platforms by their very nature artificially cordon off employment in some operations of a corporation from those in other regions, spreading income inequality out spatially. Somewhat better jobs for rural Japan or small-town Alabama placed on a satellite platform obscure the concentration of top-paid corporate jobs elsewhere and the deterioration in the income distribution in a Detroit or inner-city· Tokyo, especially for blue-collar workers.

The implications for the complexion of politics in satellite platform regions are also mixed. In some cases—Japan, for instance—the creation of such platforms under the technopolis strategy has co-opted militant, often antibusiness prefectural movements for environmental cleanup and an improved quality of life, redirecting their energies and local resources into speculative economic development activities. In other cases, new satellite platforms have helped break the stranglehold of traditionally dominant "good old boy" parties by introducing educated people and new immigrants into the region and contributing to more contested local politics.

State-anchored Districts

A fourth form of sticky place is what we call the state-anchored industrial district, where a public or nonprofit entity, be it a military base, a defense plant, a weapons lab, a university, a prison complex, or a concentration of government offices, is a key anchor tenant in the district. Here, the local business structure is dominated by the presence of such facilities, whose locational calculus and economic relationships are determined in the political realm, rather than by private-sector firms. This type of district is much more difficult to theorize, because contingencies particular to the type of activity involved color its operation and characteristics. It is apt to look much like the hub-and-spoke district in Figure 1, although a facility can operate with few connections to the regional economy, resembling the satellite platform case. · Nevertheless, some commonalities may be noted.

Before doing so, however, I shall simply cite examples of such districts. Many of the fastest growing industrial districts in the United States and elsewhere owe their

performance to the presence, new location, or expansion of state facilities. Military bases, military academies, and weapons labs, for instance, explain the phenomenal postwar growth of U.S. cities like Santa Fe, Albuquerque, San Diego, and Colorado Springs, while defense plants contributed dramatically to the growth of Los Angeles, Silicon Valley, and Seattle (Markusen et al. 1991). State universities and/or state capitals explain the prominence of cities like Madison, Ann Arbor, Sacramento, Austin, and Boulder among fastest growing U.S. cities. Denver owes much of its postwar growth to its hosting of the second largest concentration of federal government offices in the nation. In Japan and South Korea, the government research complexes at Tsukuba and Taejon, respectively, have fueled growth in their environs. In Brazil, Campinas owes much to its top-ranked university, while San Jose dos Campos's growth is based on the government-owned, military-oriented aerospace complex (Diniz and Razavi 1994).

In general, scale economies are relatively high in such complexes. Because state-owned or state-dependent facilities are so large, supplier sectors do grow up around them, dependent on the level of public expenditure. Short-term contracts and commitments do exist between state "customers" and their suppliers, subject to political change. In the case of state capitals and universities, high degrees of cooperation may exist between the customer and suppliers, and activity will be relatively immune from the threat of exodus. This is less true for national facilities, especially in times of fiscal stringency or redundancy of function (e.g., the current closing of military bases in the United States). In nationally funded facilities, decisions are made external to the district and may be more indifferent to regional development impacts.

When government contracting is involved, especially in areas like defense, the arcane and elaborate nature of the contracting process may encourage the development of long-term supply relationships, based on a fairly strong degree of trust and cooperation. However, these ties need not be localized; they may span thousands of miles between Los Angeles or Silicon Valley and Washington, D.C., for instance, or most of the length of a country like Korea, as between Changwon and Seoul (Golob et al. 1995; Markusen and Park 1993).

Labor markets will be tailored to the particular state activity hosted. For state capitals, the labor market will tend to be relatively local or regional. Personnel may cycle between state customers and local suppliers. For universities and national facilities, labor markets will operate externally for the higher-skilled occupations. In the case of military bases, blue-collar and unskilled positions will also be filled from a labor market national in scope. Workers' loyalties will be devoted to large state institutions and/or state-dependent facilities first, then to the district, then to firms.

Indigenous firms will play less of a role in these districts than in Marshallian or hub-and-spoke districts. Some may emerge out of specialized technology transfer (universities) or business service functions (lobbying). Firms will not tend to cooperate to stabilize markets or hedge against risk since they are not preoccupied with stabilizing demand in the same way that Marshallian districts with mature industries might. In general, trade associations will be relatively weak, and local government's role in regulating and promoting district activities will be minimal (consider, for instance, the District of Columbia's almost complete absence of power). Local fixed capital and government may adopt a sycophantic form of boosterism, designed to enhance the ability of the anchor facility to maintain or increase levels of external funding or protect it against closure.

In state-anchored industrial districts, long-term growth prospects depend on two factors: the prospects for the facility at the core of the region, and the extent to which the facility encourages growth within the region by spawning local suppliers, spinning off new businesses, or supplying labor or other factors of produc-

tion to the local economy. Often, the mammoth size of the facility—New Mexico's Los Alamos Laboratories, for instance, with an annual budget of $1.4 billion, mostly for personnel, or New London, Connecticut's Electric Boat submarine manufacturing facility, with its 20,000 workers—overwhelms any contribution, real or potential, that may be made through second effects. This means that local business and political energies tend to be focused on solidifying the facility's commitment and its level of funding. This must be pursued through politics at the relevant level and thus requires a relatively unique governance structure.

Politics in state-anchored industrial districts tend to be complex and tailored to the particularities of the form of government involvement. Military-industrial districts range from the remarkably conservative (Colorado Springs) to the remarkably liberal (New England). University towns and state capitals tend to be more liberal than other cities of similar size, even within their own states, while towns hosting military bases and prisons tend to line up on the conservative end of the spectrum.

Sticky Mixes

Although the presence of Marshallian industrial districts, even the Italianate version, can be confirmed in a number of American instances, the claims made for the paradigmatic ascendancy of this form of new industrial space (Scott's rubric) do not square with the experience of most rapidly growing agglomerations in industrialized and industrializing countries. In the United States, for instance, most rapidly growing industrial regions do not exhibit the characteristics of the Third Italy. Indeed, the lessons of the Italian industrial district experience are being adopted most fruitfully in the Industrial Midwest as a way of stemming deindustrializing and retaining jobs in small and medium-sized firms, not in explaining new industrial spaces. Even Silicon Valley, as we show elsewhere, is more a mix of industrial district types than a pure case of Italianate industrial district (Golob et al. 1995). In Japan, South Korea, and Brazil, it is difficult to find a single instance of a flexibly specialized industrial district outside of subareas of the major metropolis. Most rapidly growing metropolitan areas owe their performance to hub firms or industries, satellite platforms, and/or state anchors, or some combination thereof.

In the United States, the fast-growing industrial cities in Table 2 may be allocated to one or more of our industrial district types. Colorado Springs, Huntsville, Melbourne/Titusville, and San Diego, all military or space-dependent cities, belong in the fourth, government facility-anchored growth areas. Madison, Austin, and Albuquerque also belong in this category, the first two because they house both the state university and state capital, and Albuquerque because it hosts the state capital, state university, and various military-related facilities, including nearby Los Alamos and Sandia laboratories. Reno and Orlando's growth is primarily entertainment-related, although in recent years Reno has benefited from warehousing and related operations fleeing California's tax structure. Seattle, Los Angeles, and the latter's Anaheim/Santa Ana neighbor are hub-and-spoke districts organized around large defense and commercial corporations, with universities playing larger or smaller roles. Raleigh-Durham is a prototypical case of a successful high-tech satellite platform, while Elkhart-Goshen has flourished from low-wage, non-union capacity additions in aging industries.

The models of sticky places presented above are suggestive rather than definitive products of an inductive research method. Further application of these to an even broader set of regional economies will be necessary to determine how well each is constructed and how common its incidence is in real space. Comprehensive comparative work across a larger applied set could tell us much about district forms

and how they vary by type of industry and degree of maturity, national and regional rules and cultures, and firm and local economic development strategy.

Many localities, especially larger metropolitan areas, exhibit elements of all four models. Silicon Valley, for instance, hosts an industrial district in electronics (Saxenian 1994) but also revolves around several important hubs (Lockheed Space and Missiles, Hewlett Packard, Stanford University), as well as hosting large "platform" type branch plants of U.S., Japanese, Korean, and European companies (e.g., IBM, Oki, NTK Ceramics, Hyundai, Samsung). Furthermore, Silicon Valley is now and has been the fourth largest recipient of military spending contracts in the nation, a fact that shapes it defense electronics and communications sector (Saxenian 1985; Markusen et al. 1991; Golob et al. 1995).

An intriguing question is whether regions can maintain their stickiness by transforming themselves from one type of district to another. Historically, as I have pointed out, Detroit made the transition from a Marshallian district to a hub-and-spoke district. Localities that host satellite platforms may be able to encourage backward and forward linkages that transform them into more Marshallian or hub-and-spoke type districts; scholars are debating whether this is occurring around large Japanese auto transplants in the United States. A state-centered district might do the same. A hub-and-spoke district which loses its anchor tenant may be able to create a Marshallian district in its wake, as some are trying to do in the Los Angeles aerospace industry. Recruitment or incubation of a new hub could transform a Marshallian or state-centered district into a hub-and-spoke variant, which is what Colorado Springs has been doing with new organizational headquarters like the U.S. Olympics and the right-wing Christian Focus on the Family. More work could be done on the conditions that impede or facilitate these mutations.

The research reported here was methodologically confined to places doing better than average, simply because this ensured that they met the criterion of superior growth performance. However, many localities with stable or slowly declining growth patterns are struggling to be sticky places, and many are succeeding in stanching their losses by remaking their industrial structures. New England, for instance, began as early as the 1950s to transform itself into a diversified military-industrial complex, escaping the deeper displacement that occurred post-1970 in the Industrial Midwest (Markusen et al. 1991). Although New England has not as a region posted above-average long-term growth rates, even during the Reagan military buildup, it deserves study as a sticky place. Midwestern cities like Chicago, Milwaukee, and Cleveland with little comparative advantage in military-industrial sectors are trying to make themselves more sticky by anchoring and upgrading existing expertise in industries like metals, machining, and automobiles.

Our study was conducted at the metropolitan scale, equivalent more or less to a regional labor shed. However, industrial district features may characterize smaller agglomerations within metropolitan areas. Extension of these models to the subregional scale might require relaxing one or more assumptions and altering some hypotheses.

Research and Policy Implications

This exercise in distinguishing among types of sticky places illustrates the diversity in spatial form, industrial complexion and maturity, institutional configurations, and welfare outcomes found in contemporary regional economies. It cautions that the singular enthusiasm for flexibly specialized industrial districts, especially the high-tech American variant, is ill-founded on both growth/stability and equity grounds. In large part, the problem here lies in the limits of the research strategy used in the NID literature, which intensively studies particular localities extracted from their embeddedness in a larger global economy. It is useful to study why certain places appear to be different and/or more successful as a means of developing hypotheses regarding features that may contribute to such success. Once identified, these then need to be tested against a larger sample, one more representative of the universe of localities.

Furthermore, the study of industrial districts and networks within them has generally been confined to smaller firms in particular industries; their links to larger firms and to other firms and institutions outside the region have been ignored. As a result, conclusions have been drawn about the endogeneity of growth in such districts that, when viewed on a larger, more comprehensive canvas, are not warranted. Nor is the zero-sum nature of much of this growth acknowledged—that certain places grow at the expense of other places, that high-wage employment in some regions is linked to low-wage employment in others, and that only a few places can possibly aspire to become Silicon Valleys of the future.

In reality, sticky places are complex products of multiple forces: corporate strategies, industrial structures, profit cycles, state priorities, local and national politics. Their success cannot be studied by focusing only on local institutions and behaviors, because their companies (through corporate relationships, trade associations, trade, government contracts), workers (via migration and international unions), and other institutions (universities, government installations) are embedded in external relationships— both cooperative and competitive—that condition their commitment to the locality and their success there.

These reflections on research approach are applicable to economic development policy at both regional and national levels as well. At the regional level, economic developers would be well advised to assess their existing district structures accurately and design a strategy around them, rather than committing to a fashionable strategy of small-firm networking within the region. Improving cooperative relationships and building networks that reach outside of the region may prove more productive for some localities than concentrating on indigenous firms. Furthermore, our work on hub-and-spoke and satellite platform structures suggests that large firms can be significant contributors to regional development, albeit posing problems of dominance and vulnerability, and that recruitment of an external firm or plant may be a good strategy for a region at a particular developmental moment. Regions might also be well advised to target national-level policies shaping the competitive status of their industries and allocating public infrastructure and procurement contracts.

At the national level, a strategy to ameliorate regional competition and differential growth rates would (1) attempt to determine how many districts of each type the national economy might be expected to sustain, (2) develop a strategy for stabilizing existing districts and channeling new ones to deficit areas, (3) ban the use of public funds to subsidize competition among regions, and (4) monitor and if necessary alter national policies with substantial regional implications, such as devolution of powers and responsibility to subnational levels, new trade regimes (e.g., North American Free Trade Agreement, General Agreement on Tariffs and Trade), macroeconomic policy initiatives (e.g., deficit reduction and fiscal austerity versus stimulus), financial market structures, Third World development, international labor and human rights, international environmental standards, immigration restrictions, social safety nets, and infrastructure provisions. In the United States at present, only the third of these has any near-term possibility of being undertaken and then only as a result of considerable bipartisan clamor in Congress.

The prominence of hub-and-spoke and satellite platforms among U.S. sticky places suggests that economic development strategies built on cross-regional

alliances might be as important to localities as purely local networking approaches. Cross-regional networks might be forged to shore up progressive institutions under attack (labor, environmental and community development gains) and create better ones at national and international levels to curb the worst products of capitalist development—poverty, insecurity, income inequality, environmental degradation. While NID district builders struggle to create governance structures at the local level, multinational finance and industrial leaders have crafted a World Trade Organization that would be highly undemocratic and preempt many of the existing rights and safeguards that workers and communities have fought for and won. More sophisticated and pluralistic profiles of industrial districts and how they operate, both internally and externally, must be joined with more intensive study of multinational corporations and state institutions if a more powerful geographic contribution to progressive strategy is to emerge.

References

Amin, A., and Robins, K. 1990. The re-emergence of regional economies? The mythical geography of flexible accumulation. *Environment and Planning D: Society and Space* 8:7– 34.

Amin, A., and Thrift, N. 1992. Neo-Marshallian nodes in global networks. *International Journal of Urban and Regional Research* 16:571– 87.

Bellandi, M. 1989. The industrial district in Marshall. In *Small firms and industrial districts in Italy*, ed. E. Goodman and J. Bamford, 136– 52. London: Routledge.

Best, M. 1990. *The new competition: Institutions of industrial restructuring.* Cambridge: Harvard University Press.

Bull, A.; Pitt, M.; and Szarka, J. 1991. Small firms and industrial districts, structural explanations of small firm viability in three countries. *Entrepreneurship and Regional Development* 3:83– 99.

Chinitz, B. 1960. Contrasts in agglomeration: New York and Pittsburgh. *American Economic Association, Papers and Proceedings* 40:279–89.

Christopherson, S. 1993. Market rules and territorial outcomes: The case of the United States. *International Journal of Urban and Regional Research* 17:274– 88.

_____. 1994. Rules as resources in investment and location decisions. Paper presented at the Centenary Conference for Harold Innis, University of Toronto, September.

Christopherson, S., and Storper, M. 1986. The city as studio, the world as back lot: The impact of vertical disintegration on the location of the motion picture industry. *Environment and Planning D: Society and Space* 4:305–20.

Dicken, P. 1992. *Global shift: The internationalization of economic activity.* 2d ed. New York: Guilford.

Diniz, C. C. 1994. Polygonized development in Brazil: Neither decentralization nor continued polarization. *International Journal of Urban and Regional Research* 18:293– 314.

Diniz, C. C., and Borges Santos, F. 1995. Manaus: A satellite platform in the Amazon region. Working Paper, CEDEPLAR, Universidad Federal de Minas Gerais, Brazil, May.

Diniz, C. C., and Razavi, M. 1994. Emergence of new industrial districts in Brazil: Sao Jose dos Campos and Campinas cases. Working Paper, CEDEPLAR, Universidad Federal de Minas Gerais, Brazil, November.

Ettlinger, N. 1992. Modes of corporate organization and the geography of development. *Papers in Regional Science* 71:107– 26.

Feldman, M. 1993. Agglomeration and industrial restructuring. CPAD Working Paper 93-02, Graduate Program in Community Planning and Area Development, University of Rhode Island, June.

Fineberg, D.; Gilmore, R. W.; Krantz, J.; Lianes, M.; Miller, R.; Mann, U.; and Schmitt, B. 1993. The biopharmaceutical industry in New Jersey: Prescriptions for regional economic development. Report to the Princeton/Rutgers Research Corridor, Department of Urban Planning and Policy Development, Rutgers University.

Florida, R., and Kenney, M. 1990. Silicon Valley and Route 128 won't save us. *California Management Review* 33:68– 88.

Gereffi, G., and Korzeniewicz, M., eds. 1994. *Commodity chains and global capitalism.* Westport, Conn.: Greenwood.

Gertler, M. 1988. The limits to flexibility: Comments on the post-Fordist vision of production and its geography. *Transactions of the Institute of British Geographers* 13:419– 32.

Glasmeier, A. 1988. Factors governing the development of high tech industry agglomerations: A tale of three cities. *Regional Studies* 22:287– 301.

Golob, E.; Gray, M.; Markusen, A.; and Park, S. O. 1995. Valley of the heart's delight: Silicon Valley reconsidered. Working Paper, Project on Regional and Industrial Economics, Rutgers University, presented at the Regional Science Association annual meetings, Niagara Falls, Canada, November, 1994.

Goodman, E. 1989. Introduction: The political economy of the small firm in Italy. In *Small firms and industrial districts in Italy,* ed. E. Goodman and J. Bamford, 1–3. London: Routledge.

Goodman, E., and Bamford, J., eds. 1989. *Small firms and industrial districts in Italy.* London: Routledge.

Granovetter, M. 1985. Economic action and social structure: The problem of embeddedness. *American Journal of Sociology* 91:481– 510.

Gray, M.; Golob, E.; and Markusen, A. 1996. Big firms, long arms: A portrait of a "hub and spoke" industrial district in the Seattle region. *Regional Studies,* forthcoming.

Harrison, B. 1992. Industrial districts: Old wine in new bottles? *Regional Studies* 26:469–83.

_____. 1994. *Lean and mean: The changing landscape of corporate power in the age of flexibility.* New York: Basic Books.

Harvey, D. 1982. *The limits to capital.* London: Basil Blackwell.

Healey, M., and Rawlinson, M. 1993. Interviewing business owners and managers: A review of methods and techniques. *Geoforum* 24:339–55.

Howes, C. 1993. Constructing comparative disadvantage: Lessons from the U.S. auto industry. In *Trading industries, trading regions,* ed. H. Noponen, J. Graham, and A. Markusen, 45–91. New York: Guilford.

Howes, C., and Markusen, A. 1993. Trade, industry and economic development. In *Trading industries, trading regions, ed.* H. Noponen, J. Graham, and A. Markusen, 1–44. New York: Guilford.

Krumme, G. 1969. Toward a geography of enterprise. *Economic Geography* 45:30–40.

Linge, G. J. R. 1991. Just-in-time: More or less flexible? *Economic Geography* 67:316–32.

Linge, G. J. R., and Rich, D. C. 1991. The

state and industrial change. In *The state and the spatial management of industrial change*, ed. G. J. R. Linge and D. C. Rich, 1–21. London: Routledge.

Lovering, J. 1990. Fordism's unknown successor: A comment on Scott's "Theory of flexible accumulation and the re-emergence of regional economies." *International Journal of Urban and Regional Research* 14:159–74.

——. 1991. Theorising post-Fordism: Why contingency matters (A further response to Scott). *International Journal of Urban and Regional Research* 15:298–301.

Luger, M., and Goldstein, H. 1990. *Technology in the garden*. Chapel Hill: University of North Carolina Press.

Luria, D. 1990. Automation, markets and scale: Can "flexible niching" modernize U.S. manufacturing? *International Review of Applied Economics* 4:127–65.

McNee, R. 1960. Toward a more humanistic economic geography: The geography of enterprise. *Tijdschrift voor Economische en Social Geografie* 51:201–05.

Malecki, E. J. 1987. Comments on Scott's "High tech industry and territorial development: The rise of the Orange County complex, 1955–1984." *Urban Geography* 8:77–81.

Markusen, A. 1985. *Profit cycles, oligopoly and regional development*. Cambridge: MIT Press.

——. 1991. The military industrial divide: Cold war transformation of the economy and the rise of new industrial complexes. *Environment and Planning D: Society and Space* 9:391–416.

——. 1994. Studying regions by studying firms. *The Professional Geographer* 46:477–90.

——. 1995. The interaction of regional and industrial policies: Evidence from four countries. *Proceedings, World Bank's conference on development economics, 1994*. Washington, D.C.: World Bank.

Markusen, A.; Hall, P.; Campbell, S.; and Deitrick, S. 1991. *The rise of the Gunbelt*. New York: Oxford University Press.

Markusen, A., and Park, S. O. 1993. The state as industrial locator and district builder: The case of Changwon, South Korea. *Economic Geography* 69:157–81.

Markusen, A., and Sasaki, M. 1994. Satellite new industrial enclaves: A comparative study of United States and Japanese cases. Working Paper, Project on Regional and Industrial Economics, Rutgers University, New Brunswick, N.J., January.

Massey, D. 1984. *Spatial divisions of labor: Social structures and the geography of production*. New York: Methuen.

Massey, D., and Meegan, R. 1982. *The anatomy of job loss: The how, why and where of employment decline*. London: Methuen.

Park, S. O. and Markusen, A. 1994. Generalizing new industrial districts: A theoretical agenda and an application from a non-Western economy. *Environment and Planning A* 27:81–104.

Piore, M., and Sabel, C. 1984. *The second industrial divide: Possibilities for prosperity*. New York: Basic Books.

Pollert, A. 1988. Dismantling flexibility. *Capital and Class* 34:42–75.

Sabel, C. 1989. Flexible specialization and the reemergence of regional economies. In *Reversing industries decline*, ed. P. Hirst and J. Zeitlin, 17–70. New York: St. Martin's Press.

Saxenian, A. 1983. The urban contradictions of Silicon Valley. *International Journal of Urban and Regional Research* 17:236–57.

——. 1985. The genesis of Silicon Valley. In *Silicon landscapes*, ed. P. Hall and A. Markusen, 20–34. Boston: Allen and Unwin.

——. 1990. Regional networks and the resurgence of Silicon Valley. *California Management Review* 32:89–112.

——. 1991a. Silicon Valley and Route 128 won't save us: Response to Richard Florida and Martin Kenney. *California Management Review* 33:136–42.

——. 1991b. The origins and dynamics of production networks in Silicon Valley. *Research Policy* 20:423–37.

——. 1994. *Regional networks: Industrial adaptation in Silicon Valley and Route 128*. Cambridge: Harvard University Press.

Sayer, A. 1989. Post-Fordism in question. *International Journal of Urban and Regional Research* 13:666–95.

Schoenberger, E. 1985. Foreign manufacturing investment in the United States: Competitive strategies and international location. *Economic Geography* 61:241–59.

——. 1988. From Fordism to flexible accumulation: Technology, competitive strategies, and international location. *Environment and Planning D: Society and Space* 6:245–62.

——. 1991. The corporate interview as a

research method in economic geography. *The Professional Geographer* 44:180–89.

Scott, A. 1986. High tech industry and territorial development: The rise of the Orange County complex, 1955–1984. *Urban Geography* 7:3–45.

———. 1988a. Flexible production systems and regional development: The rise of new industrial space in North America and Western Europe. *International Journal of Urban and Regional Research* 12:171–86.

———. 1988b. *New industrial space.* London: Pion.

Scott, A., and Paul, A. 1990. Collective order and economic coordination in industrial agglomerations: The technopoles of Southern California. *Environment and Planning C: Government and Policy* 8:179–93.

Sforzi, F. 1989. The geography of industrial district in Italy. In *Small firms and industrial districts in Italy,* ed. E. Goodman and J. Bamford, 153–73. London: Routledge.

Storper, M. 1989. The transition to flexible specialization in industry: External economies, the division of labor and the crossing of industrial divides. *Cambridge Journal of Economics* 13:273–305.

Storper, M., and Harrison, B. 1991. Flexibility, hierarchy and regional development: The changing structure of industrial production systems and their forms of governance in the 1980s. *Research Policy* 20:407–22.

Storper, M., and Walker, R. 1989. *The capitalist imperative: Territory, technology and industrial growth.* New York: Basil Blackwell.

U.S. Department of Commerce, Bureau of Economic Analysis. 1973, 1990. *Country business patterns.* Washington, D.C.: Government Printing Office.

Yin, R. 1984. *Case study research: Design and methods.* Applied Social Research Methods Series 5. Beverly Hills, Calif.: Sage.

[9]

'Globalizing' regional development: a global production networks perspective

Neil M Coe*, Martin Hess*, Henry Wai-chung Yeung†, Peter Dicken* and Jeffrey Henderson‡

Recent literature concerning regional development has placed significant emphasis on local institutional structures and their capacity to 'hold down' the global. Conversely, work on inter-firm networks – such as the global commodity chain approach – has highlighted the significance of the organizational structures of global firms' production systems and their relation to industrial upgrading. In this paper, drawing upon a global production networks perspective, we conceptualize the connections between 'globalizing' processes, as embodied in the production networks of transnational corporations, and regional development in specific territorial formations. We delimit the 'strategic coupling' of the global production networks of firms and regional economies which ultimately drives regional development through the processes of value creation, enhancement and capture. In doing so, we stress the multi-scalarity of the forces and processes underlying regional development, and thus do not privilege one particular geographical scale. By way of illustration, we introduce an example drawn from recent research into global production networks in East Asia and Europe. The example profiles the investments of car manufacturer BMW in Eastern Bavaria, Germany and Rayong, Thailand, and considers their implications for regional development.

key words globalization global production networks regional development Asia Europe

*School of Environment and Development, University of Manchester, Oxford Road, Manchester M13 9PL
email: neil.coe@manchester.ac.uk
†Department of Geography, National University of Singapore
‡Manchester Business School, and the School of Environment and Development, University of Manchester

revised manuscript received 10 May 2004

Introduction and theoretical context

One of the many paradoxes of the processes of 'globalization' is the continued significance of 'regions', in the sense of sub-national spaces as foci of economic activity. Systemic processes of rapid technological change, enhanced capital mobility and neoliberally inspired inter-regional competition for investment have focused attention on the need for regional-level interventions among a broad community of academics and policymakers. Two recent strands of work attempt to tackle the links between globalization dynamics and notions of 'regional development'. One strand places particular emphasis on endogenous institutional structures and their capacity to 'hold down' global networks (for overviews see MacLeod 2001a; Scott 1998; Storper 1997). The other strand, focusing specifically on inter-firm networks and global commodity/value chains (GCCs/GVCs), considers the organizational structures of global firms' production systems and explores how particular regions 'slot into' these networks with varying impacts on industrial upgrading (see Gereffi and Kaplinsky 2001; Gereffi 1994 1996).

In their early formulations, both of these literatures could be criticized for their failure to effectively conceptualize regional economic development in an era of globalization. The new regionalism literature seemed overly pre-occupied with local transactions and institutional forms at the expense of the many extra-local connections within which regions are embedded, while the functional connections between seemingly desirable regional institutional configurations and actual levels of economic development were open to question (Amin and Thrift 1994). The GCC/GVC approaches, in turn, operated largely at the national scale, saying little about how particular sub-national spaces and their institutions are integrated into, and shaped by, transnational production systems (for recent critiques, see Henderson *et al.* 2002; Smith *et al.* 2002; Dicken *et al.* 2001).

Recent developments in these two fields, however, have begun to address these shortcomings and to move somewhat closer together. The 'new regionalism' literature, for example, places increased weight on the extra-local dynamics shaping economic growth within regions (both knowledge, capital and labour flows and also the wider institutional structures within which regions are embedded (e.g. Amin 2002; MacKinnon *et al.* 2002; Bunnell and Coe 2001; MacLeod 2001a; Lovering 1999). Conversely, a number of GCC and GVC studies explicitly explore how regional clusters and industrial districts are incorporated into global production systems, and consider their implications for local economic development and industrial upgrading (e.g. Bair and Gereffi 2001; Gereffi *et al.* 2001; Humphrey 2001; Sturgeon 2001; Humphrey and Schmitz 2000).

In this paper, we seek to make a primarily *conceptual* contribution to these converging research agendas. Drawing upon a global production networks (GPN) perspective (see Henderson *et al.* 2002), and deriving insights from both the new regionalist and GCC and GVC literatures, our approach focuses on the dynamic 'strategic coupling' of global production networks and regional assets, an interface mediated by a range of institutional activities across different geographical and organizational scales.[1] Our contention is that regional development ultimately will depend on the ability of this coupling to stimulate processes of value creation, enhancement and capture.

We regard regional development as a set of *relational processes* (see Amin 2002). It is also, by definition, an *interdependent* process (Massey 1984). The fortunes of regions are shaped not only by what is going on within them, but also through wider sets of relations of control and dependency, of competition and markets. These relations may be with other regions within the same national territory, but increasingly occur at the international scale. Hence, our conceptualization of a region is not as a tightly bounded space, but as a porous territorial formation whose notional boundaries are straddled by a broad range of network connections (Amin 2002; Allen *et al.* 1998).

The paper is organized into two main sections. First, we explore how the strategic coupling of global production networks and regional assets may (or may not, depending on the context) facilitate the processes of the creation, enhancement and capture of value upon which regional development ultimately depends. Second, in order to illustrate how our conceptual framework might be utilized empirically, we present an illustrative case study of the German car manufacturer BMW and its interactions with regional development processes in Eastern Bavaria, Germany and Rayong, Thailand.

'Globalizing' regional development: towards a re-conceptualization

In developing a broad conceptual framework for understanding regional development we need to pay analytical attention *both* to endogenous growth factors within specific regions *and* also to the strategic needs of trans-local actors coordinating global production networks (cf. Scott and Storper 2003). In our framework, regional development is conceptualized as *a dynamic outcome of the complex interaction between territorialized relational networks and global production networks within the context of changing regional governance structures*. In that sense, it resonates with Amin's topological/relational view (Amin 2002; see also Dicken 2004). We aim to specify the interactive complementarity and coupling effects between localized growth factors and the strategic needs of trans-local actors in propelling regional development. We argue that it is these *interactive effects* that contribute to regional development, not inherent regional advantages or rigid configurations of globalization processes. Despite certain path-dependent trajectories, regional development remains a highly contingent process that cannot be predicted *a priori*. This conceptualization, however, does not mean that regional institutions are unimportant. On the contrary. Often, such complementarity and coupling effects can be enhanced and exploited through particular sets and practices of 'regional' institutions. The

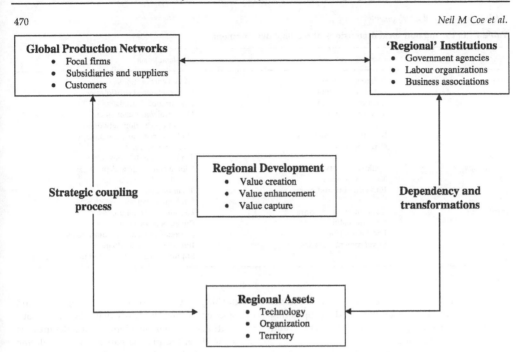

Figure 1   A framework for analysing regional development and global production networks

term 'regional' must be used with care here. We place it in scare quotes to indicate that, in reality, regional development is not just shaped by regionally specific institutions, but also by a variety of extra-local institutions (e.g. national, supra-national) that will impact on activities *within* a region. This 'scaling' of institutional influence is critical. In short, regional development at any particular historical moment requires the *necessary* co-presence of three inter-related sets of conditions:

1 the existence of economies of scale and scope within specific regions;
2 the possibility of localization economies within global production networks; and
3 the appropriate configurations of 'regional' institutions to 'hold down' global production networks and unleash regional potential.

We have summarized these conditions and their interactions in Figure 1.

Regional advantages, global production networks and economies of value.
Our analytical framework starts with the premise that endogenous factors are necessary, but in-

sufficient, to generate regional growth in an era in which competition is increasingly global. There is no doubt that, for development to take place, a region must benefit from economies of scale and scope derived from what Storper (1997, 26) terms the 'holy trinity' of technology–organization–territory. In Figure 1, we use the term 'regional assets' to describe this necessary precondition for regional development. In general, these assets can produce two types of economies. First, economies of *scale* can be achieved in certain regions through highly localized concentrations of specific knowledge, skills and expertise. This concentration of technological advantages embodied in and performed by social actors located in specific regions creates economies of scale in particular technologies that can be exploited through the agglomeration of firms that in turn provide employment and generate economic outputs within similar high tech industries. Second, economies of *scope* can exist if these regions are able to reap the intangible benefits of learning and the cooperative atmosphere embedded in these agglomerations. These are famously known as 'spillover' effects. A variety of different high value-added activities may be

Table I   Local and non-local dimensions of regional development

Dimensions	Local manifestations	Non-local forms
Firms	Indigenous SMEs Industrial clusters Intra-regional markets Venture capitalists	Global corporations Entrepreneurial subsidiaries Distant global markets Decentralized business and financial networks Global production networks
Labour	Skilled and unskilled workers Permanent migrants	Skilled experts and technologists Transient migrants Transnational business elites
Technology	Spillover effects Tacit knowledge Infrastructure and assets	Global standards and practices Intra-firm R&D activities Technological licensing Strategic alliances
Institutions	Conventions and norms Growth coalitions Local authorities Development agencies	Labour and trade unions Business associations National agencies and authorities Inter-institutional alliances Supranational and international organizations

located or developed in these regions because the tendencies towards learning and cooperation facilitate a broad spectrum of production and entrepreneurial activities.

We argue that economies of scale and scope embedded within specific regions are only advantageous to those regions – and bring about regional development – insofar as such region-specific economies can complement the *strategic needs* of trans-local actors situated within global production networks. As shown in Figure 1, when such a complementary effect exists between regions and global production networks, a coupling process will take place through which the relational advantages of regions interact with the strategic needs of actors in these global production networks. Regional development thus depends on such a coupling process that evolves over time in relation to the rapidly changing strategic needs of global production networks and the rather slow transformations in regional economies of scale and scope. Before we analyse such a coupling process, it is important to unpack what we mean by the *strategic needs* of actors in global production networks. We define *global production networks* as the globally organized nexus of interconnected functions and operations by firms and non-firm institutions through which goods and services are produced and distributed. Such networks not only integrate firms (and parts of firms) into structures which blur traditional organizational boundaries through the development of diverse forms of equity and non-equity

relationships, but also integrate regional and national economies in ways that have enormous implications for their developmental outcomes. At the same time, the precise nature and articulation of firm-centred production networks are deeply influenced by the concrete socio-political contexts within which they are embedded. The process is especially complex because while the latter are essentially territorially specific (primarily, though not exclusively, at the level of the nation-state and/ or the region), the production networks themselves are not. Global production networks 'cut through' national and regional boundaries in highly differentiated ways, influenced in part by regulatory and non-regulatory barriers and local socio-cultural conditions, to create structures that are 'discontinuously territorial' (see Henderson *et al.* 2002; Dicken and Malmberg 2001).

Put in these conceptual terms, it becomes clear that local actors in specific regions (e.g. labour and the state) and non-local actors in global production networks (e.g. TNCs and financial capital) are differentiated by their degree of *territorial embeddedness* which, in turn, will have very significant implications for regional development (see Table I). This distinction in territorial embeddedness is important because it shapes how value and power are distributed in their relational interactions, a point we develop in the next section (see also Hudson 2001).

As key local actors in regional development, the organizational strength and flexibility of labour is critical to the alignment of the region with the strategic needs of focal firms in global production networks. While labour has been internationalizing through inter-institutional alliances and international organizations (see Table I), the reality remains that in most cases labour is spatially entrapped in local labour markets (see Herod 2001; Peck 1996). To Castree *et al.*, workers face a particular kind of geographical dilemma because 'what might make sense for them at one geographical scale may have unfortunate consequences at other scales' (2004, 119). They thus recognize that

> it's not just that workers may be tempted to put local interests first, it's also that the very *nature* of local interests varies depending on the specifics of local industry, local standards of living, local living wages and so on. (Castree *et al.* 2004, 120; original emphasis)

In short, there is a *prima facie* case that economies of scale and scope in particular regions can be reaped more effectively by focal firms in global production networks through labour's spatial immobility and flexibility in skills. The local and the regional become the most important geographical scales through which labour interacts with the strategic needs of key actors in global production networks. Their interactive effects tend to favour trans-local actors embedded in these global production networks because, as pointed out some time ago by Massey (1984) among others, these global actors can engage in 'spatial switching' much more easily than workers themselves.

Similarly, the state and its development agencies are institutions that are strongly embedded locally in specific regions (see Table I again). This institutional dimension of regional development has been well theorized in the new regionalism literature (e.g. MacLeod 2001b). It is sufficient to say that the increasing devolution of political and economic authority from the nation state to local and regional institutions has led not only to the rise of growth coalitions within specific regions, but also to a higher degree of uneven regional development. The latter phenomenon occurs primarily because different regions have very different configurations of state institutions that in turn shape how these regions are articulated into global production networks. This situational power and role of the state (and labour) and its manifestations in local and regional institutions has very important implications for understanding the distributional aspects of regional development. In regions that

have strongly embedded local labour markets, we argue that focal firms in global production networks can better exploit economies of scale through technology- or expertise-specific production systems (e.g. in biotechnology or cultural industries). In regions with more flexible labour markets, economies of scope might be better achieved through the co-presence of a variety of different industries that reap the benefits of 'untraded interdependencies'. The role of state institutions is important here through their regulation of labour and its organizations. In some regions, state institutions may work with labour organizations and labour market intermediaries to increase the skill levels of labour and the flexibility of local labour markets (see Benner 2003; Peck 2000; Jones 1999). In other regions, the adversarial and confrontational relationship between the state and labour may significantly reduce the region's attractiveness to focal firms in global production networks (see Kelly 2002).

Before we move on to unravel the complexity behind the strategic considerations of focal firms in global production networks, it is useful to consider one significant category of non-local actors that impact significantly on local and regional development: financial capital institutions. While global production networks may not directly encapsulate financial capital in their network configuration, it is useful to distinguish three types of financial capital in relation to their differential territorial embeddedness: local venture capital, national banking institutions and globally decentralized financial networks (see Table I). From the perspective of global production networks, venture capital tends to be highly localized primarily because talents and expertise are often embodied in people within a particular region that are known to venture capitalists through interpersonal networks of relationships. Venture capital is important to regional development both in terms of its financing of high risk ventures that are more likely to be at the cutting edge of technological development and in terms of its financing of supporting industries that supply to global production networks. The nature and organization of local venture capital, however, is embedded within national banking systems.

In some countries, venture capital is much less active because of the close relationships between banks and industries (e.g. Germany and Japan). Regional developmental trajectories are highly dependent on the direction and influence of national banking institutions (see Dore 2000; Pauly and Reich 1997). In other countries (e.g. the US and the UK), banking institutions play much less significant roles *vis-à-vis* globally decentralized financial networks that are mediated through global financial centres (e.g. New York and London). Regional development in these countries is much less dependent on the presence of banking institutions and more on the articulation of those regions into global financial networks. For example, the availability of investment and equity funds has been critical to the continuous growth and development of Silicon Valley. Such funds emerge from a variety of financial networks that are decentralized in terms of their origin and composition (e.g. US pension funds vs Taiwanese private capital). We argue that the uneven access to these local and non-local forms of financial capital can both enhance the strategic importance of some regional economies to global production networks and diminish others. These different forms of capital also embody different territorial logics, with venture capital being mostly local in its orientation, and decentralized financial networks more global in nature (see also Clark *et al.* 2002).

Hence, spanning national boundaries and market areas, the strategic needs of *focal firms* – defined as dominant firms spearheading the global organization of production networks through their corporate and market power – in global production networks do not always and necessarily intersect with regional advantages. Global integration of activities within these production networks, for example, may not be beneficial to some regions because of the likelihood of greater external control of the regional economy. Indeed, many focal firms in global production networks may pursue different organizational configurations in order to reap economies of scale and scope in these networks. In general, *economies of scale* in global production networks can be achieved through globally integrated R&D, sourcing, production and marketing activities that take place only in specific locations. The smaller the number of firms engaging in each of these functions, the greater the economies of scale will be in a particular global production network. This is because each of these firms can specialize in the designated function, e.g. R&D or assembly operations. *Economies of scope* in global production networks, on the other hand, exist through differentiation in the functional activities of firms in the network such that a variety of firms

may be used for R&D, sourcing, production and marketing activities. These different firms often offer learning and knowledge possibilities that are not available if the function is performed by a single firm, as in the practice of global sourcing or R&D. As Nohria and Ghoshal (1997) argue, many leading global corporations are increasingly tapping into differentiated advantages among different subsidiaries and supplier networks.

Regional development and notions of value creation, enhancement and capture

How then is this complex organization of different actors in global production networks related to regional development? In Figure 1, this relationship can work through the creation, enhancement and capture of *value*. Here we use the term 'value' to refer to various forms of *economic rent* (Kaplinsky 1998) that can be realized through market as well as non-market transactions and exchanges. Alongside value creation through the labour process, for instance, value can take the form of technological rents by way of access to particular product or process technologies, or may be manifested as relational rents, based on inter-organizational links improving know-how transfer and collective learning. Other forms of rent identified by Kaplinsky may derive from organizational attributes, trade policy and branding. This conception of value as economic rent has two significant implications for analysing regional development. First, different forms of rent can be created and captured by local and non-local actors in global production networks such that some regions might be better in creating and retaining a particular form of rent (e.g. technological rents in Silicon Valley vs brand-name rents in London). A region needs neither to create nor to retain all forms of rent. Instead, a region that is endowed with certain configurations of labour, capital and state institutions might be better off by specializing and being competitive in one kind of economic rent – thus reaping economies of scale. A region with a highly competitive labour market, an active pool of venture capitalists and a pro-growth coalition of institutions will likely be engaged in the creation of value through new growth industries (e.g. biotechnology) that require rapid flows of knowledge embodied in the local workforce, high risk-taking financing and a stable institutional environment. On the other hand, a region burdened by a weakly organized and abundant supply of labour, the virtual absence of venture and banking capital and an unstable institutional structure may create value through performing highly labour-intensive work for focal firms in global production networks. Endowed with different configurations of assets, both regions perform very different roles in terms of value creation *vis-à-vis* global production networks.

Second, it should be noted that value takes on different forms in this spatialized network of flows. At the time when value is created in one region, it may take a particular form of, say, relational rent in regions embued with a Marshallian-style cooperative atmosphere. When this value is transferred through global production networks to other regions, it may take on other forms (e.g. technological and/or brand-name rents). This multiplicity in the forms of rent indicates that the analysis of value creation and capture in regional development must go beyond simply tracking market values of goods and services produced. More importantly, we need to unpack the different forms of rent that these values encapsulate.

The fact that a region is 'plugged' into global production networks, therefore, does not automatically guarantee its positive developmental outcome because local actors in this region may be creating value that does not maximize the region's economic potential. A region filled with cooperative atmosphere should be much more successful in creating relational rent, although in some cases cultural and institutional impediments may prevent such value from being created. Local actors in a region also may not be able to capture much of the value created in the region (cf. Amin and Thrift 1992). From the regional development perspective, the creation and retention of value within the region is imperative. For example, a region may have an advantage in the quantity of labour, but much of the value created in the utilization of this abundant pool of labour may be transferred out of the region through the repatriation of profits (realized value) and eventually the relocation of the production networks to other regions. At the other end of the value-creation spectrum, nevertheless, a region with substantial 'relational assets' (e.g. cooperative learning and venture capital formation) may be successful in creating value in team-based projects that require face-to-face interaction in spatially proximate clusters. However, such a value creation process may run out of steam when these highly localized conventions and norms in learning are so binding and constraining that they

hinder the development of alternative mode of learning, say, through decentralized and distanciated networks facilitated by greater mobility of actors and a series of other technologies of contact and translation (see Coe and Bunnell 2003; Amin 2002; Bunnell and Coe 2001).

Hence, regional assets can become an advantage for regional development *only if* they fit the strategic needs of global production networks. The process of 'fitting' regional assets with strategic needs of global production networks requires the presence of appropriate *institutional structures* that simultaneously promote regional advantages and enhance the region's articulation into global production networks (see Figure 1). Again, it is crucial to remember that our notion of 'regional' institutions includes not only regionally specific institutions, but also local arms of national/supranational bodies (e.g. a trade union's 'local' chapters), and extra local institutions that affect activities within the region without necessarily having a presence (e.g. a national tax authority). These regional institutions are important because they can provide the 'glue' that ties global capital and unleashes regional potential. Three dimensions of such institutional structures are crucial to regional development. The first dimension involves *the creation of value* through the efforts of regional institutions in attracting the location of value-added activities, e.g. training and educating the local workforce, promoting start-up firms and supplier networks, facilitating venture capital formation and encouraging entrepreneurial activities (see also Phelps and Raines 2003). Although it is often unclear whether such a process involves too much 'tying' the region to the value activities of particular focal firms or global production networks (e.g. Phelps *et al.* 1998), the efficacy of this relational coupling between the region and the focal firm hinges on the region's capacity to enhance and capture value from the process. However, in the absence of such a coupling process, the question of regional development remains a moot point since no value will be created, let alone enhanced and captured.

More importantly, the second and third dimensions refer to the capacity of regional institutions in *value enhancement* and *value capture*. Value enhancement essentially involves knowledge and technology transfer and industrial upgrading (from design and final production of commodities). The influence of regional institutions via government agencies, trade unions, employer associations and so on can be significant here. On the one hand, regional institutions may promote specific 'regional assets' (e.g. cooperative industrial relations) that are conducive to high value-added production activities because these activities incur high costs of fixed investment (i.e. sunk costs) and are difficult to be relocated within a short period of time. There is thus a mutually beneficial interaction between regional institutions and regional assets (see Figure 1). On the other hand, regional institutions can promote the value enhancement activities of focal firms in global production networks. This occurs when regional institutions are prepared to invest in developing the infrastructure and human resources required for value enhancement (e.g. highly stable power supply and skilled engineers for wafer fabrication). Over time, more value enhancement activities within global production networks may occur in these regions when focal firms are induced to bring in their core technologies and expertise. The development of sophisticated local supplier networks, for example, is important in enhancing the value activities of focal firms through 'reverse' transfer of local knowledge and experience (see Chew and Yeung 2001). In short, not all regional assets are complementary to the enhancement of value by focal firms in global production networks. The key issue is the appropriateness and complementarity of these assets, not their mere presence.

The third dimension of regional institutions in promoting regional development rests with their capacity to ensure *value capture*. It is one thing for value to be created and enhanced in some regions, but it may be quite another for it to be captured for the benefit of these regions. The issues of *power* and *control* are critical in the analysis of value capture and the distributional aspects of regional development. While the concept of power is complex in social thought (see Lukes 1986), we follow Allen (2003), who defines power not as a capacity or a repertoire of resources possessed by actors, but rather as *relational effects* of social interaction. To Allen,

> power as an outcome cannot and should not be 'read off' from a resource base, regardless of its size or scope ... It is, as suggested, a relational effect, not a property of someone or some 'thing'. (2003, 5)

Conceived as such, the role of regional institutions in negotiating these issues of power and control with focal firms in global production networks is linked to their development policies, ownership

patterns and corporate governance. Clearly, focal firms in global production networks have enormous corporate control of resources through their ability to collect and process information on a global basis. This information asymmetry may afford very strong bargaining positions to some focal firms when they interact with regional institutions (Dicken 1994 2003). The more a region is articulated into global production networks, the more likely it is able to reap the benefits of economies of scale and scope in these networks, but the less likely it is able to control its own fate. Put in terms of power and control, this non-local origin of regional development happens because

> the exercise of power in particular places may well originate beyond those places, at some other location, yet remains part of power's active presence. In other words, the power relations in place are affected by what happens elsewhere and the network of connections of which it is a part. (Allen 2003, 180–1)

This dimension of the external control or dependency of regional development has long preoccupied economic geographers (e.g. Massey 1978; Dicken 1976).

But equally, regional institutions may mobilize their region-specific assets to bargain with these focal firms such that their power relations are not necessarily one-way in favour of the latter. The bargaining position of these regional institutions is particularly high when their region-specific assets are highly complementary to the strategic needs of focal firms – these regional institutions become really powerful through their relational interaction with focal firms in selected global production networks. For example, focal firms that are under severe competitive cost pressures are more likely to allow for some forms of value to be captured in regions that offer not only significantly cheaper factors of production (labour rent), but also highly cooperative labour relations. In this case, these focal firms may choose to invest further to upgrade the local workforce that may better support future regional development through capturing skills and technological rents. Moreover, a region can achieve greater value capture if the reinvestment of retained earnings in localized subsidiaries is critical to a particular function of the global production network (e.g. new process technology) and the focal firm fails to secure further investments through globally decentralized financial networks (e.g. downturns in major stock markets). Such retention of value through reinvestment in local subsidiaries and/or suppliers may also be enhanced through the availability of venture capital formation or favourable support from national banking institutions.

In both scenarios, those local actors involved directly in these global production networks (e.g. workforce, suppliers, venture capitalists) are likely to benefit from the enhancement and retention of value through skill upgrading, technological innovation and new venture formation. For example, Simmie (2003) recently found that the most innovative firms in Europe tend to concentrate in a minority of key metropolitan regions, combining a strong local knowledge capital base with high levels of connectivity to similar regions elsewhere in the global economy. The likelihood of value capture in specific regions is therefore greatly enhanced by a cooperative set of state, labour and business institutions that offer region-specific assets to focal firms in global production networks. As such, the capacity of regions to capture value is a dynamic outcome of the complex bargaining process between regional institutions and focal firms in global production networks. The presence of region-specific assets is only relevant in this process if these assets are complementary to the strategic needs of trans-local actors embedded in global production networks.

What is new about 'globalizing' regional development?

The above re-conceptualization of regional development from the global production network perspective complements existing frameworks in at least two ways. First, it takes a dynamic approach to analysing regional development as a 'moving target'. While we recognize the path dependency in the evolution of regional assets (see Figure 1), our framework does allow for regions to break out of this trajectory of lock-in. This possibility occurs when specific regions are confronted with economic crises that do not necessarily originate from these regions. For example, a region may enjoy a relative advantage in a particular global industry (e.g. electronics) or segments of a global industry (e.g. integrated circuits). Even if the path dependency of this regional advantage has been set in motion, the region can still experience major problems of development when crises occur within the entire industry on the global scale. Such crises may be due to technological change that produces a

substitution effect (e.g. the development of super-conductors) or financial instability (e.g. over-capacity and over investment). Such a 'global' crisis in an industry may force the region in question to seek alternative development pathways that, if successful, will lead to the end of its path dependency. In this sense, our framework allows for a dynamic view of regional evolution without placing too much emphasis on endogenous structures that inhibit change and transformations.

Second, our framework is explicitly *comparative* because an analysis of the interactive complementarity and coupling effects requires us to examine how such effects materialize in one region but not another region. All too often in the new regionalism literature, we have been told how one region develops because of its endogenous growth factors. What is absent in this analytical approach is how other regions with similar growth factors either fail to develop or evolve through drastically different trajectories. It also ignores the complex interdependencies *between* regions that will shape regional development *within* regions. An explicitly comparative approach to regional development helps us appreciate better the critical mechanisms through which *some* regions gain developmental momentum whereas other regions miss the opportunity.

Global production networks and regional development: an illustrative example

Clearly, global production networks connect regions in a complex and highly variegated manner and, as a result, the developmental outcomes of these connections will differ in significant ways, depending on the focal firms' strategies, the institutional frameworks and sectoral/technological specifics. We will illustrate this for two regions in Europe and East Asia involved in the global production network of the German car manufacturer BMW Group (see Figure 2) in order to show the strategic coupling of a region's assets with the strategic needs of trans-regional actors. Given space constraints, this example is meant merely to be suggestive of how our framework might be mobilized empirically.[2]

In many ways, the case of BMW resembles the strategies, locational impacts and organizational characteristics of all major car manufacturers. However, while market entry modes and strategies, production and lean management systems, and the JIT-based clustering of suppliers around the main assembly plants are similar to those of

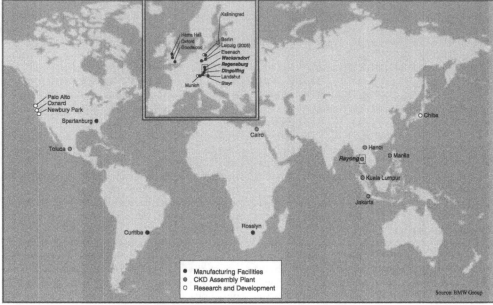

Figure 2 Global locations of BMW group

other firms in the sector, BMW has some specific characteristics that distinguish it from other car manufacturers. As a niche producer for the upmarket and luxury markets, the BMW Group has relatively low production volumes and thus the global production network strategy is to some extent different from that of the mass manufacturers. Also, the company has a very strong base in Bavaria and is approximately 47 per cent owned by members of the Quandt family. This shareholding structure has a decisive impact on inter-firm relationships within BMW's global production network.

BMW in Eastern Bavaria, Germany

Economic development in eastern Bavaria, for a long time a peripheral and economically weak region, has been transformed – not least by the arrival of BMW's production plants – since the late 1960s. Headquartered in Munich, the company was looking for new manufacturing sites to expand production and in 1967 took over the small car manufacturer Hans GLAS with its production facilities at Landshut and Dingolfing. The crucial regional asset that attracted BMW to Eastern Bavaria – apart from the proximity to Munich and

the takeover opportunity – was the availability of skilled labour and, increasingly important, a flexible workforce. Since unemployment in the region was high, the recruitment of people at competitive wages was much easier than through the job market in the prospering Munich area. Also, there was a willingness on the part of the workforce to accept flexible working hours and shift structures that allowed BMW to enhance capacity – for example, extending the machine time to 99 hours a week – and thus reduce unit costs. This was supported by cooperative workers' councils and regional labour unions, as well as government aid (through regional development programmes) to boost the weak regional economy. In the wake of BMW's 1982 decision to build another assembly plant in Eastern Bavaria near Regensburg, intensive and continuous negotiations between BMW and the national labour union IG Metall took place, illustrating the regional impact of decisionmaking by, and relational power of, actors inside and outside the region. Against the opposition of the IG Metall head office in Frankfurt, the regional branch of the union engaged in cooperation with Munich-based BMW to implement a work-shift model that has become a model for the German car industry. A recent example of this process is the negotiations between Volkswagen AG and IG Metall over a new tariff contract (cf. Pries 2002).

So far, BMW has invested some 7 billion Euro and now operates three plants and one supplier park in the area, directly employing about 35 000 people, whilst an estimated further 20 000 jobs have been created by local first-tier suppliers. This represents more than one tenth of all manufacturing employment in the region. The wages and salaries paid to the employees at the company's regional sites exceed 2 billion Euro per year and constitute a considerable source of consumer purchasing power within the region.

The development of BMW's supplier and innovation park in Wackersdorf near Regensburg in the 1990s has increased the integration of its operations in Eastern Bavaria with global production networks. After the German government was forced to abandon the building of a nuclear waste management facility at Wackersdorf, private companies were able to use the location as an industrial park, administered by the agency formerly responsible for setting up the nuclear site. Investors were attracted by the low prices of land available and about 500 million Euro in compensation payments to the region, paid by the federal and state governments, which accelerated infrastructure development. Again, this opportunity matched BMWs strategic needs at that time. The distribution of power between the relevant political and societal actors in Eastern Bavaria, however, has proven to be complex and has played out at different scales – analogous to the different scales of influence from the labour union's and workers' side. The federal and state governments have been under pressure to find alternatives to the planned nuclear facility and thus were obliged to regional political and civil actors, improving the bargaining power of the latter. BMW could reap the benefits from this political struggle and implement their plans in the region at comparatively lower costs.

The Just-In-Time system, introduced at BMW's plants in Regensburg and Dingolfing, increasingly required the co-location of major suppliers, while the modularization of production forced component manufacturers to integrate and coordinate their business. Initially, BMW started to produce convertibles at the new Wackersdorf plant, after it promised to create at least 1600 jobs in the region, but later changed the plans for this area. Using its

significant buying power, BMW persuaded global first tier suppliers like Lear Corp. and Modine (both from the US) to establish plants next to each other in the innovation park. This not only guaranteed the functioning of the production network, but created an innovative context, where suppliers (some of them competitors on the world markets) share tacit knowledge and continuously improve products and processes (cf. Hess 2001). That way, new innovative structures have been created through the BMW-induced arrival of foreign firms in the region, contributing not only to direct employment but also providing the environment for spillover effects that benefit the regional economy. Some of these global suppliers' branch plants have now become leading plants within their parent companies, setting benchmarks for other plants within the production network. The globalization of BMW's production network itself resulted in the establishment of a logistics centre at the same site in Wackersdorf, from which all of BMW's international parts and components distribution to its foreign plants in the US, South Africa, Russia and East Asia is organized, with a daily shipment of 2.5 million parts carried out by a third party logistics provider.

Eastern Bavaria's regional economy has, without doubt, benefited from globalizing processes linked to the region via BMW's production network. Apart from value creation in the form of both domestic and foreign investment, as well as direct and indirect employment, the skills and technology transfer from BMW and its foreign-owned suppliers to local companies ensures a noteworthy degree of value enhancement and capture, which is essential for the region's sustainable economic development. The strategic coupling process between regional assets and BMW's needs to develop its GPN has not only led to repeated rounds of capital and technological investments, but also provided opportunities for regional actors to capture different forms of rent, notably organizational and relational rents. The process of value creation and enhancement has been mediated through ongoing interaction and negotiation between firm, government and labour representatives at regional and trans-regional scales. However, as BMW is one of the leading economic actors within a quasi-hierarchical regional network, it has got considerable power vis-à-vis institutions and other firms, whereby quite a strong dependency on its commitment to the region remains.

BMW in Rayong Province, Thailand

BMW remained, until very recently, essentially a German-based company. Like other major car manufacturers, however, it has had to respond to globalizing forces by creating a geographically more extensive production network (Figure 2). In this context, BMW's entry into East Asia is potentially very significant both for the company itself and for the region, particularly Thailand. Since the 1960s, Thailand has become the centre of Southeast Asia's automotive industry, employing about 120 000 people in the sector. Motor vehicles have become the third biggest export category after computers and electrical circuits: 230 000 cars from a total output of 760 000 in 2003 were exported (*Bangkok Post* 2003). Due to a nationally implemented cluster policy (cf. Lecler 2002), most of the companies in this sector, including BMW, are located to the south of Bangkok, in the Rayong and Samutprakarn provinces of Thailand's Eastern Seaboard (see Figure 2). To date, there are almost two-dozen car manufacturers operating in the region, surrounded by more than 700 first-tier suppliers, 50 per cent of which are fully or partly foreign-owned. The Thai auto component industry is generating an annual turnover of about US$4bn. In this case, the nation state plays an important role in coupling the regional assets with the strategic needs of global companies and their networks, not least due to the weaknesses or lack of regional institutions. Since BMW is one, rather small player among a fairly large number of companies within the Thai car industry, its power in relation to regional and national institutions is limited. Hence the bargaining position of these institutions is considerable, supported by the fact that south-central Thailand has become first choice location for many foreign firms and thus creates opportunities towards the enhancement of regional assets.

Like many other car manufacturers, BMW has chosen Thailand's gulf area as its prime location for the Southeast-Asian region, because the country had no national car programme and hence this sector was fairly liberalized compared to other countries in the region, especially Malaysia and Indonesia (cf. Tucher 1999). In anticipation of a potentially large market after the completion of an Asian Free Trade Agreement (AFTA), the rationale for production in Asia was mainly to avoid current tariff and non-tariff trade barriers as well as to integrate the region into BMW's global production

network. Initial investments of 25 million Euro in the manufacturing facilities for the 3 series cars will be topped up by an additional 15 million Euro to install a new assembly line for the production of 7 series cars in 2003, to be sold in the domestic and regional markets. Currently, the Rayong plant of BMW employs about 250 people, assembling nearly 4000 cars annually from vehicle kits imported from Germany. These rather small figures suggest a negligible contribution to regional development in Thailand's Rayong province.

Indeed, while local content regulations existed until the year 2000, the bulk of value added parts are not manufactured in the region, but rather brought in from abroad. Since the market is not yet big enough, low production volumes do not allow for economies of scale by establishing a production site, and therefore completely knocked down (CKD) assembly with comparatively little regional value-added prevails. For a BMW car, about 40 per cent of value added is achieved through local content, but this means production and sourcing within ASEAN countries, therefore the value added within the Thai auto cluster is lower than the 40 per cent figure suggests. Furthermore, the regional assets in the South of Thailand do not necessarily match all the needs of car manufacturers like BMW, as the level of skills among the workforce and organizational sophistication have yet to reach the required standards. The fact that most of the suppliers are partly or wholly owned by foreign companies reflects the problems in upgrading the industrial base and transferring skills and technology to local companies. There have been considerable initiatives by the Thai government to adapt to the changing strategic needs of manufacturers like BMW and to participate more strongly in their value-added networks. Through governmental and quasi-governmental organizations, most notably the Board of Investment and the Thailand Automotive Institute, a series of attempts are being made to improve the vocational skills of the Thai workforce and to support domestic SMEs, in close collaboration with foreign manufacturers (see also Lauridsen 2003; Techakanont 2003). In addition, foreign assemblers train their workforce either in-house or put them on training courses within the parent company abroad, with BMW being no exception. However, know-how transfer to domestic suppliers is still rather limited.

The success of BMW's Thai venture depends, to a great extent, on the final implementation of supranational economic integration under an Asian free trade agreement. BMW has chosen the Rayong plant, its only wholly owned facility in Asia, to become an integrated production site once the institutional framework (AFTA) allows. Rayong will become a full production site whilst its other Southeast Asian locations will serve as BMW centres of excellence. In this way, economies of scale will be achieved through production specialization and exchange within Southeast Asia as a whole. Other companies, e.g. Toyota, follow similar strategies of regional complementation (cf. Yoshimatsu 2002), which provide an opportunity for the Thai auto cluster to upgrade and develop in the region. Once a critical mass is reached, BMW will not only be able to attract additional foreign suppliers to the region, but also will be more likely to invest more in upgrading and developing local suppliers which the company might use in the future. As it stands, BMW is currently bridging the gap between its strategic needs and the territorial assets in the form of local supply companies through a 'mediated' process of technology transfer and industrial upgrading.

For example, in order to secure local sourcing for side glass for its E46 cars assembled in Thailand, BMW approached a supplier they used elsewhere. However, the negotiations failed and since production volumes were too low to persuade its German glass supplier to locate in Rayong, BMW went to the Thai subsidiary of a Japanese glass manufacturer and arranged a technological cooperation process between the German supply company and the Japanese/Thai manufacturer to upgrade and technically release their products and processes. That way, the German supplier became the main facilitator of technology and know-how transfer, based on long-term relations and familiarity with BMW's technical and organizational standards, without BMW having to deal with double investments. A similar, triangular technology transfer arrangement exists between BMW, one of their European suppliers and a domestic, fully Thai-owned supplier. That way, technological rents are generated within the Thai auto cluster, although to date they are still rather confined to joint ventures and foreign-owned suppliers.

Summing up, in order to achieve the goal of upgrading, a number of adjustments are needed to facilitate a positive strategic coupling process between global production networks and regional assets. These include efforts by the Thai govern-ment to create regional institutions that help to transform and enhance the regional assets, especially in the field of education and vocational training, laid out in the current Automotive Industry Master Plan, and progress in supra-national negotiations to pave the way towards an integrated production system, with the global forces of multinational companies like BMW and supra-national economic policy arrangements being the main drivers for regional development in Rayong and Samutprakarn.

As this empirical example has illustrated, regional development is strongly linked to external influences in the form of both non-firm institutions and economic actors (see Figure 3). While the activation of endogenous resources to foster sustainable development is an important task for regional economic policy, it is not sufficient in itself but rather has to take into account the strategic coupling process between global production networks and regional assets. Figure 3 shows a strong intra-regional connectivity between actors for the region of Eastern Bavaria, both in terms of material flows and technological/organizational cooperation, supported by regional institutions. However, the main drivers of development are extra-local, based on BMW's production strategy and investment, while previous policy decisions and subsequent capital flows from the Bavarian and Federal Governments helped kick-start regional development. The BMW-linked production network in the Rayong/Samutprakarn area, on the other hand, currently shows comparatively few regional linkages. Most of the parts and components are imported from Germany via the Wackersdorf logistics centre, and investment as well as technology transfer to Thai suppliers has so far been rather modest. Additionally, the future of a prospering automotive cluster at Thailand's Eastern Seaboard depends to a large extent on supra-national free trade negotiations, making the international dimension of regional development ever more obvious.

Conclusion

In this paper we have proposed an integrated conceptual framework for 'globalizing' regional development that takes external or 'global' forces as well as regional assets into account. In short, our framework highlights the dynamic 'strategic coupling' of global production networks and regional assets, an interface that is crucially mediated by a

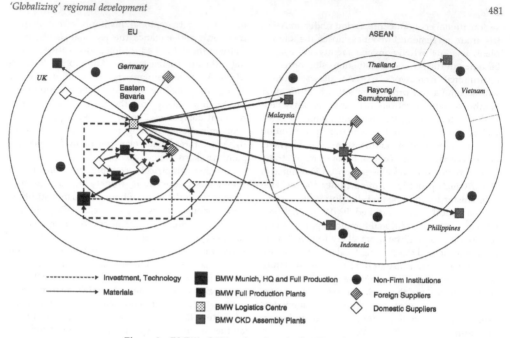

Figure 3   BMW's GPN and regions in the EU and ASEAN

range of institutional activities across different spatial scales. Our key argument is that regional development depends on the ability of this coupling mechanism to facilitate processes of value creation, enhancement and capture. Empirically, due to the constraints of space, we have only been able to offer a brief illustration of how our framework might be mobilized for a very simple global production network configuration consisting of one firm's activities across two regions. However, the BMW case clearly shows that the developmental impact of the coupling process is highly variable and contingent, and by no means automatically beneficial for the region.

Clearly, regional development does not take place on a level playing field. For the processes of value creation, enhancement and capture to benefit economic development in particular regions, the balance of power between the different actors involved is a crucial variable in determining the potential for value enhancement and, ultimately, value capture. Governance structures in different territorial contexts are variable and hence the possibilities for developmental policies to impact on a region's assets will differ as well. In many of

the newly industrialized countries, national politics sets the dominant framework for regional development, with regional institutions often weakly developed or completely missing. On the other hand, in countries with a more decentralized structure, regional institutions attempt to develop their bargaining power *vis-à-vis* focal firms in the context of nation-state governance structures and inter-regional competition. In every case, however, the exercise of power is a *multi-scalar* process, with varying combinations of actors cooperating or playing off one against the other. Knowledge about these territorially specific power configurations, therefore, is elementary for regional institutions to take appropriate measures for transforming a region's assets and to maximize their bargaining power and impact.

More specifically, what policy lessons might be drawn from our analysis for actors at the regional level? Four are particularly pertinent, although they can only be mentioned here because of space constraints. Firstly, policymakers clearly need to accumulate considerable stocks of knowledge not only about the various assets contained within their region, but also about how they relate to the needs of various global production networks, many of which will originate outside their territory. Secondly, such knowledge is clearly sectorally specific, and effective policy interventions will have to be designed for, and targeted at, particular segments of particular industries. Such flexibility would appear to be a necessary precondition for effective intervention in the contemporary era. Thirdly, 'out there' knowledge clearly needs to extend beyond the needs of specific production networks to incorporate a reflexive understanding of the multi-scalar institutional configurations in which all policymakers are situated. Some of these institutional relations will pull against a particular endeavour, whilst others will support it. An appreciation of this complexity is a crucial first step in appropriately harnessing (or resisting) extra-local connections. Finally, it is clear that policy interventions need to be underpinned by an awareness of the differences between value creation, enhancement and capture, so that strategic couplings that prioritize the latter two processes can be identified and supported. All of these are policy issues that require substantial attention in future work on regional development in a globalizing world.

Acknowledgements

This article draws on work conducted under the auspices of the Economic and Social Research Council project, *Making the Connections: Global Production Networks in Europe and East Asia* (Grant # R 000 238535). We are grateful to the ESRC for their support. Thanks also to Adam Tickell and three anonymous referees for their insightful comments on an earlier version of this paper.

Notes

1 Our notion of 'strategic coupling' here may appear to resonate quite closely with the coupling process between the mode of social regulation and regime of accumulation in the French regulationist theory of economic development (see Boyer 2000; Tickell and Peck 1992) and its refinement through Jessop's (1990 2001) strategic-relational approach to state theory. However, our use of the concept is not a direct import from the regulationist theory that focuses primarily on the coupling process of two meta-structures within particular national economies (i.e. the national scale). At the risk of developing a functionalist interpretation of regional development, we use the term to characterize the very complex scalar juxtapositions that drive

regional development processes. While we acknowledge the term is imperfect and may be perceived as a rather crude structural interpretation of regional development, it is used here for heuristic purposes and should be understood as the coupling process between regional economies and global production networks that is mediated through specific action and practices of key actors and institutions. In this sense, we hope to offer a middle-ground interpretation that is both structural and actor-centric.

2 A brief note on methodology. This illustrative case study is derived from the ESRC funded project *Making the Connections: Global Production Networks in Europe and East Asia* (Grant # R000238535) that ran from 2000 to 2003 and explored economic connections within and between Europe (including Central and Eastern Europe) and East Asia in three sectors: auto components, retailing and telecommunications. Framed within the GPN conceptual framework (Henderson *et al.* 2002), over 150 semi-structured corporate and institutional interviews were conducted in 12 countries in addition to an extensive programme of secondary data and documentary analysis. The aim of the research was both to 'map out' leading GPNs in these three sectors and to explore their developmental impacts in different localities.

References

Allen J 2003 *Lost geographies of power* Blackwell, Oxford

Allen J, Massey D and Cochrane A 1998 *Rethinking the region* Routledge, London

Amin A 2002 Spatialities of globalisation *Environment and Planning A* 34 385–99

Amin A and Thrift N 1992 Neo-Marshallian nodes in global networks *International Journal of Urban and Regional Research* 16 571–87

Amin A and Thrift N 1994 Living in the global in **Amin A and Thrift N** eds *Globalisation, institutions and regional development in Europe* Oxford University Press, Oxford 1–22

Bair J and Gereffi G 2001 Local clusters in global chains: the causes and consequences of export dynamism in Torreon's blue jeans industry *World Development* 29 1885–1903

Bangkok Post 2003 *Year-end economic review* Bangkok, Thailand

Benner C 2003 Labour flexibility and regional development: the role of labour market intermediaries *Regional Studies* 37 621–33

Boyer R 2000 The political in the era of globalization and finance: focus on some *Régulation* school research *International Journal of Urban and Regional Research* 24 274–322

Bunnell T G and Coe N M 2001 Spaces and scales of innovation *Progress in Human Geography* 25 569–89

Castree N, Coe N M, Ward K and Samers M 2004 *Spaces of work: global capitalism and geographies of labour* Sage, London

Chew Y-T and Yeung H W-C 2001 The SME advantage: adding local touch to foreign transnational corporations in Singapore *Regional Studies* 35 429–46

Clark G L, Mansfield D and Tickell A 2002 The German social market in the world of global finance: pension investment management and the limits of consensual decision-making *Transactions of the Institute of British Geographers* 27 91–110

Coe N M and Bunnell T 2003 'Spatializing' knowledge communities: towards a conceptualization of transnational innovation networks *Global Networks* 3 437–56

Dicken P 1976 The multiplant business enterprise and geographical space: some issues in the study of external control and regional development *Regional Studies* 10 401–12

Dicken P 1994 Global–local tensions: firms and states in the global space economy *Economic Geography* 70 101–19

Dicken P 2003 *Global shift* 4th edn Sage, London

Dicken P 2004 Geographers and 'globalization': (yet) another missed boat? *Transactions of the Institute of British Geographers* 29 5–26

Dicken P and Malmberg A 2001 Firms in territories: a relational perspective *Economic Geography* 77 345–63

Dicken P, Kelly P F, Olds K and Yeung H W-C 2001 Chains and networks, territories and scales: towards a relational framework for analysing the global economy *Global Networks* 1 89–112

Dore R 2000 *Stock market capitalism, welfare capitalism: Japan and Germany versus the Anglo-Saxons* Oxford University Press, Oxford

Gereffi G 1994 The organization of buyer-driven global commodity chains: how US retailers shape overseas production networks in **Gereffi G and Korzeniewicz M** eds *Commodity chains and global capitalism* Praeger, Westport CT 95–122

Gereffi G 1996 Global commodity chains: new forms of co-ordination and control among nations and firms in international industries *Competition and Change* 1 427–39

Gereffi G and Kaplinsky R eds 2001 The value of value chains: spreading the gains from globalization *IDS Bulletin* 32

Gereffi G, Humphrey J, Kaplinsky R and Sturgeon T 2001 Introduction: globalisation, value chains and development *IDS Bulletin* 32 1–8

Henderson J, Dicken P, Hess M, Coe N and Yeung H W-C 2002 Global production networks and the analysis of economic development *Review of International Political Economy* 9 436–64

Herod A 2001 Labor internationalism and the contradictions of globalization: or, why the local is sometimes still important in a global economy *Antipode* 33 407–26

Hess M 2001 Globalisierung und regionale Innovationsnetzwerke: Beispiele für 'Glokalisierung' in der deutschen Fahrzeugindustrie in **Schätzl L and Grotz R** eds *Regionale Innovationsnetzwerke im internationalen Vergleich* LIT, Münster 83–100

Hudson R 2001 *Producing places* Guilford, New York

Humphrey J 2001 *Opportunities for SMEs in developing countries to upgrade in a global economy* (http://www.ids.ac.uk/ids/global/pdfs/clusterchainsv4.pdf) Accessed 7 February 2003

Humphrey J and Schmitz H 2000 *Governance and upgrading: linking industrial cluster and global value-chain research* Institute of Development Studies, University of Sussex, IDS Working Paper 120

Jessop B 1990 *State theory: putting capitalist states in their place* Polity Press, Cambridge

Jessop B 2001 Institutional re(turns) and the strategic-relational approach *Environment and Planning A* 33 1213–35

Jones M 1999 *New institutional spaces: TECs and the remaking of economic governance* Jessica Kingsley, London

Kaplinsky R 1998 *Globalisation, industrialisation and sustainable growth: the pursuit of the nth rent* Institute of Development Studies, University of Sussex, IDS Discussion Paper 365

Kelly P F 2002 Spaces of labour control: comparative perspectives from Southeast Asia *Transactions of the Institute of British Geographers* 27 395–411

Lauridsen L 2003 The role of the state in linkage formation between TNCs and local Thai enterprises Paper presented at the workshop on 'Understanding FDI-assisted Economic Development' 22–25 May 2003 TIK-centre, University of Oslo

Lecler Y 2002 The cluster role in the development of the Thai car industry *International Journal of Urban and Regional Research* 26 799–814

Lovering J 1999 Theory led by policy: the inadequacies of the 'New Regionalism' *International Journal of Urban and Regional Research* 23 379–95

Lukes S ed 1986 *Power* Blackwell, Oxford

MacKinnon D, Cumbers A and Chapman K 2002 Learning, innovation and regional development: a critical appraisal of recent debates *Progress in Human Geography* 26 293–311

MacLeod G 2001a New regionalism reconsidered: globalization and the remaking of political economic space *International Journal of Urban and Regional Research* 25 804–29

MacLeod G 2001b Beyond soft institutionalism: accumulation, regulation, and their geographical fixes *Environment and Planning A* 33 1145–67

Massey D 1978 In what sense a regional problem? *Regional Studies* 13 233–43

Massey D 1984 *Spatial divisions of labour* Macmillan, London

Nohria N and Ghoshal S 1997 *The differentiated network: organizing multinational corporations for value creation* Jossey-Bass, San Francisco

Pauly L W and Reich S 1997 National structures and multinational corporate behavior: enduring differences in the age of globalization *International Organization* 51 1–30

484   *Neil M Coe et al.*

Peck J 1996 *Workplace: the social regulation of labor markets* Guilford, New York

Peck J 2000 *Workfare states* Guilford, New York

Phelps N and Raines P eds 2003 *The new competition for inward investment: companies, institutions and territorial development* Edward Elgar, Cheltenham

Phelps N, Lovering J and Morgan K 1998 Tying the firm to the region or tying the region to the firm: early observations on the case of LG in South Wales *European Urban and Regional Studies* 5 119–38

Pries L 2002 5000x5000: Ende gewerkschaftlicher Tarifpolitik oder innovativer betrieblich-tariflicher Sozialpakt? *Industrielle Beziehungen* 9 222–35

Scott A J 1988 *New industrial spaces* Pion, London

Scott A J and Storper M 2003 Regions, globalization, development *Regional Studies* 37 579–93

Simmie J 2003 Innovation and urban regions as national and international nodes for the transfer and sharing of knowledge *Regional Studies* 37 607–20

Smith A, Rainnie A, Dunford M, Hardy J, Hudson R and Sadler D 2002 Networks of value, commodities and regions: reworking divisions of labour in macro-regional economies *Progress in Human Geography* 26 41–63

Storper M 1997 *The regional world: territorial development in a global economy* The Guilford Press, New York

Sturgeon T 2001 How do we define value chains and production networks? *IDS Bulletin* 32 9–18

Techakanont K 2003 *Globalization strategy of assemblers and changes in inter-firm technology transfer in the Thai automobile industry* Working paper series vol 2003-23 The International Centre for the Study of East Asian Development, Kitakyushu

Tickell A and Peck J 1992 Accumulation, regulation and the geographies of post-Fordism: missing links in regulationist research *Progress in Human Geography* 16 190–218

Tucher M v 1999 *Die Rolle der Auslandsmontage in den internationalen Wertschöpfungsnetzwerken der Automobilhersteller* VVF, Munich

Yoshimatsu H 2002 Preferences, interests, and regional integration: the development of the ASEAN industrial cooperation agreement *Review of International Political Economy* 9 123–49

[10]

Theorizing Economic Geographies of Asia

Henry Wai-chung Yeung

Department of Geography, National University of Singapore,
1 Arts Link, Singapore 117570
geoywc@nus.edu.sg

George C. S. Lin

Department of Geography, University of Hong Kong,
Pokfulam Road, Hong Kong SAR
GCSLIN@hkucc.hku.hk

Abstract: Economic geographies of Asia are highly fascinating, not the least because Asia has increasingly emerged as a significant economic player in all spheres of global competition: production, consumption, and circulation. This dynamic mosaic of economic landscapes in Asia was further complicated during the 1997–1998 economic crisis and thereafter. While some aspects of these economic geographies of Asia have already received research attention, many complex economic geographic processes in Asia have been undertheorized in the literature. This agenda-setting article makes two critical observations. First, the theorization of dynamic economic changes in Asia needs to be more critical of economic geography theories developed elsewhere in the Anglo-American context. The Asian case may significantly challenge existing theories in economic geography. Second, certain geographic processes in Asia require fundamentally new approaches to theorization that may contribute to the development of broader theories in economic geography. The economic dynamism of Asia has provided a useful site for the development of theory and empirical understanding in contemporary economic geography. To support our arguments and observations, we discuss the situatedness and specificity of influential theories of economic geography and offer some constructive suggestions for an intellectual agenda for developing new theories in economic geography.

Key words: economic geography, Asia, theory, epistemology, intellectual agenda.

The articles in this special issue originated in a special session held at the 98th annual meeting of the Association of American Geographers, Los Angeles, 19–23 March 2002. We thank Fulong Wu for co-organizing the session and all presenters for their participation. The submitted papers went through a rigorous review process and different rounds of revisions. We are grateful to the authors, numerous reviewers, and editorial board members for their cooperation and hard work in putting this issue together. An earlier version of this introductory article was presented as a keynote address at the annual meeting of the Canadian Association of Geographers, Toronto, Canada, 29 May–1 June 2002. Henry Yeung thanks Philip Kelly and Glen Norcliffe for their kind invitation to attend this conference and the CAG Study Group on Social and Economic Change for funding his travel. David Edgington, the paper's discussant, provided useful comments. We also received helpful comments from David Angel, Tim Bunnell, and Adrian Smith. This article was subsequently presented to various audiences in the United Kingdom and we owe a great deal to insightful comments from the participants: Raymond Bryant, David Demeritt, Chris Hamnett, Keith Hoggart, and Geoff Wilson, of King's College London; Ash Amin, Emma Mawdsley, Joe Painter, and Janet Townsend, of the University of Durham; and Clive Barnett, Keith Bassett, Tony Hoare, Simon Naylor, Nigel Thrift, and Adam Tickell, of the University of Bristol. While we have incorporated their comments as much as possible into this version, none of these institutions and individuals should be responsible for any shortcomings of this article.

For a long time economic geographers have almost taken for granted that theories emerging from geographic studies of Silicon Valley or the City of London have been naturalized unequivocally as what may be termed "mainstream economic geography"—the influential core of Anglo-American economic geography.[1] One needs only to glance through recent major collections of economic geography published in English (e.g., Bryson, Henry, Keeble, and Martin 1999; Clark, Feldman, and Gertler 2000; Sheppard and Barnes 2000) to reinforce the point that an overwhelming majority of the chapters tend to address theoretical and empirical issues specific to only a handful of advanced industrialized economies (see Yeung 2002a). This heavy concentration of economic geography theories in relation to their sites of production and dissemination has certainly shaped the directions of research in economic geography in many other countries and/or regions, albeit each at a different pace of diffusion and adoption. Studies of economic geography of other localities have not only tended to follow the "templates" that have been institutionalized and legitimized by this mainstream economic geography, but also have earned the strange title as some kind of "regional geography." In this vein, geographic research on industrial locations in China and export-processing zones in Malaysia is often labeled as "Asian geography"; studies of the informal sector in Africa, as "African geography"; and investigations of gender relations in Latin American labor markets, as "Latin American geography." Potter (2001, 423; original italics) vividly described this bias in economic geography as follows:

> Those who work outside the Euro-North American orbit are excluded, or at best marginalized, from the specialisms which see themselves making up the *core* of the discipline of *Geography*. Quite simply, they are regarded as "ists" of the Latin American, Caribbean, African or Asian variety. If they endeavour to be comprehensive in their consideration of other regions of the globe, then they may qualify as the ultimate "ists": as full-blown "developmentalists"!

Such geographic specificity in constructing both leading theories in economic geography and the "other geographies" or "distant geographies" perhaps should not be surprising in light of the institutionalization of geography as an academic discipline (see Johnston 1997; Barnes 2000; Scott 2000). Few economic geographers have attempted to contextualize this specificity in the epistemology of economic geography and to offer suggestions for what may be done to redress it (see Yeung 2001a; Olds and Poon 2002; Smith 2002). In this article, we focus on a particular historical-geographic moment—the rise of Asia—and outline our vision for the development of theory in economic geography emanating from a rapidly growing number of geographers working on the dynamic economic transformations of Asia. We term this effort "theorizing economic geographies of Asia." The plurality of the term represents a deliberate attempt to theorize the diverse experiences and trajectories of economic transformations in Asia. There is thus no singular economic geography of Asia but, rather, multiple pathways and diversities. By the same token, there should be many models and theories of these transformations in economic geography.

---

[1] We use the term *economic geography* to refer to a discipline that deals with "the nature of world areas in their direct influence upon the production of commodities and the movement of goods" (Gotz, quoted in Barnes 2001c, 531) or what Scott (2000, 484) described as "the spatial and locational foundations of economic life." Economic geography supposedly has a wide spectrum of subjects, ranging from agrarian and pastoral economies to resource utilization and changes in land use. However, we have observed that major theoretical advancements in economic geography in recent decades have been overwhelmingly focused on the transformation of industrial economies in North America and Western Europe. We propose that economic geographers need to move forward along the lines of reconstructing a kind of *global economic geographies* that are broader in perspective and more inclusive in both sectoral and geographic terms.

This article has two interrelated aims. First, we offer a brief *critique* of leading theoretical perspectives in Anglo-American economic geography. We identify their inherent limits in relation to their analytical focus on historically and geographically specific industrial transformations. Second, we outline our *advocacy* for more theorization in future research on economic geographies of Asia. In steering the direction and content of this special issue, we were particularly driven by two concurrent trends—one intellectual and another empirical—that we believe will powerfully shape the future of economic geography. On the intellectual front, our efforts and that of the contributors to this special issue echo the recent institutional turn in economic geography from nationalistic economic geography to *global economic geographies*. Traditional economic geography has been concerned mainly with explaining patterns and processes within national space-economies. When such work is done within the Anglo-American countries, the subsequent models and theories are deemed universally true and applicable. More recently, however, an increasing number of economic geographers have begun to question seriously the situatedness of theories and knowledge of the global economy. This new kind of economic geography has become much more inclusive and open to ideas and opinions conceived outside a few dominant cores. This turn clearly supports Taylor's (1996) broader call for abandoning the "embedded statism" in the social sciences to open up to the new intellectual spaces of global economic geographies. It is interesting to note that this opening up in geography has been well recognized by scholars from other social scientific disciplines. For example, political theorist Martin Shaw (2000, 73–74; original italics) argued, in the context of geography's role in globalization debates, that

> [t]he disciplines of anthropology, geography and international relations have shown greater openness to *global* understanding than economics, politics and sociology, the histor-

ically defining fields of social science. Interestingly, the former are all fields in which historically the national-international nexus was formerly not just a methodological bias, but more or less *explicitly* constitutive. The openness of both social anthropology and geography to globalization debates follows their abandonment of nineteenth- and early twentieth-century nationalist and imperialist constructions of their subjects. These subjects underwent theoretical and ideological transformations earlier in the post-war period, which have prepared the way for the recognition of globalization.

In redressing the thematic (industrial) and geographic (Anglo-American) specificity in mainstream theories of economic geography, we aim to develop what Slater (1999, 67) called "reverse discourses" in order for non-Western work to "theorize back" at the West. These discourses should constitute "counterposed imaginations and visions emanating from different sites of experience and subjectivity." Similarly, Appadurai (1999, 237) argued for a conversation about and an imagination of research "to which scholars from other societies and traditions of inquiry could bring their own ideas about what counts as new knowledge and about what communities of judgment and accountability they might judge to be central in the pursuit of such knowledge."

This last point relates to the second concurrent trend in the *empirical realm* that has made the economic geographies of Asia highly fascinating. Asia has increasingly emerged as a significant economic player in different spheres of global competition: production, consumption, and circulation. This dynamic mosaic of economic landscapes in Asia was further complicated during the 1997–1998 economic crisis and afterward. Although some aspects of these economic geographies of Asia have already received research attention, many complex economic geographic processes in Asia have been undertheorized in the geographic literature, which leads to two important possibilities for future research. First, the theorization of dynamic economic changes in Asia needs to be more critical in adopting economic geography theories developed elsewhere in the Anglo-American context. As shown in the four articles in this special issue, the "Asian" case can significantly challenge existing theories in mainstream economic geography. Second, the economic dynamism and geographic processes in Asia require a fundamentally new approach to theorization that may contribute to the development of broader theories in economic geography. Through this process, we may witness the emergence of new kinds of theories that can account for differences and differentiation in global economic geographies—the distinguishing theme of economic geography reemphasized by Clark, Feldman, and Gertler (2000).

In the remainder of this article, we discuss the situatedness of mainstream theories of economic geography and show how Asia has been theorized in mainstream economic geography. We then examine how economic geographers may move from straightforward applications of "Western" theories in mainstream economic geography to the critical interrogation of these theories and the development of new theories through carefully grounded empirical research. We also offer some constructive suggestions for an intellectual agenda for developing new theories in economic geography.

Geographies of Economic Geography: The Situatedness of Theories

We have now been well told by historians of economic geography that dominant theories have always emerged from particular historical and geographic contexts (see Barnes 1996; Scott 2000). From locational models to spatial divisions of labor and from flexible specialization to local embeddedness, leading theories of economic geography have their peculiar histories and geographies. Their histories are very much outcomes of the conscious efforts of individual economic geographers in the context of creative tensions among different "paradigms" (see Barnes and Curry 1983; Sidaway 1997; Thrift

and Walling 2000; Barnes 2001b). In this and the next sections, we attempt to answer two related questions to explore further the situatedness of dominant theories in economic geography. First, why are economic geography theories, from the quantitative revolution and Marxism to flexible specialization and the recent "cultural turn," so dominant as if they were universal theories capable of explaining diverse economic geographic processes? Yet, why are they so little used in the economic geographic studies of other regions? Second, why have theoretical insights that have emerged from area studies and regional geography failed so far to capture the imaginations of mainstream economic geographers?

There is a noticeable gap between the obsession of some mainstream economic geographers with the *universalization* of their Western-based theories and the preoccupation of regional geographers with the task of meticulously sorting out the *geographic specificities* of particular countries or regions. We argue that this gap has been the consequence of historically specific circumstances, including the legacy of earlier colonialism, or what Hudson (1977, 12) referred to as the interests of European and American imperialism in world commerce and territorial acquisition (see also Barnes 2001c, 530); provincialism during and after the Vietnam War; linguistic and cultural barriers; and an intellectual environment that was dominated, until recently, by the Enlightenment and modernization school of thought. The persistence of this bifurcation in epistemology and methodology has led to the phenomenon of "the tragedy of the commons" in economic geography— theories that are derived from specific historical geographies become universalized among the former group of economic geographers, and descriptive specificities of regional geographies have little generality to offer to geographic studies in other countries and/or regions. We believe that such a tragedy of the commons has severely hindered the growth of a new kind of economic geography, known as *global economic geographies,* that must be built on *comparative* understandings of economic geographic processes emerging from *and* interconnecting different regions of the global economy.

In Table 1, we summarize several leading theoretical perspectives in mainstream economic geography that rose to prominence during the past two decades or so.

In constructing this table, we did not intend to "fit" different economic geography theories (and their proponents) into specific boxes. Rather, the table should be read as a heuristic device for the purpose of this article. Furthermore, we did not intend the table or the finality and "paradigmatic" nature of these theoretical themes to be complete and all-inclusive. We regret that

Table 1

Leading Theoretical Perspectives in Economic Geography
and Their Historical Geographies

Theoretical Perspectives	Period of Prominence	Key Authors	Geographic Specificities of Research
1. Location theory and the behavioral location model	1960s–1970s	Brian Berry Peter Haggett Peter Dicken F. E. I. Hamilton	United States United Kingdom
2. Spatial divisions of labor	1980s	Doreen Massey Gordon Clark	United Kingdom United States
3. Flexible specialization and new industrial spaces	1980s–1990s	Allen Scott Michael Storper David Harvey Richard Florida Meric Gertler Andrew Sayer	United States Italy Germany
4. Networks and embedded-ness	1990s	Nigel Thrift Peter Dicken Gernot Grabher Philip Cooke Ash Amin	United Kingdom Europe
5. Regional agglomerations and clusters	mid-1990s	Michael Storper Allen Scott Philip Cooke Kevin Morgan Anders Malmberg Peter Maskell Ray Hudson John Lovering	United States United Kingdom Europe
6. Regulation theory and governance	mid-1990s	Jamie Peck Adam Tickell Erik Swyngedouw	France United Kingdom Europe
7. Cultural economies	mid-1990s	Nigel Thrift Ash Amin Erica Schoenberger Linda McDowell Trevor Barnes J.-K. Gibson-Graham Roger Lee Jane Wills	United Kingdom United States and Canada (to a lesser extent)

certain key histories of (mainstream) economic geography have not been included in this table and the text. For example, one may notice that most theoretical perspectives in Table 1 deal with *industries*, rather than rural economies, natural resources and land use, financial markets, development processes, and so on. Yet largely for reasons suggested later, these theoretical perspectives are the most influential in contemporary economic geography as if they constitute the *core* of economic geography. Although a full critique of this industrial bias in contemporary economic geography is beyond the scope of this article, it is clear that even within Anglo-American economic geography, there is a significant marginalization of research into spheres of economic activities other than industries. This observation, we believe, has a lot to do with the historical and geographic specificity of these theoretical perspectives.

Bearing in mind these caveats, we point out that none of the major proponents of these theories of economic geography originated from outside the Anglo-American countries. Neither do most of them conduct their empirical research outside these advanced industrialized economies. This sweeping generalization points to the geographic specificities of these *leading* or *dominant* theories—they have been really leading and dominant among English-speaking economic geographers (see also Olds and Poon 2002). Location theory, for example, originates from what Barnes (2001a, 546) termed "epistemological theorizing," which assumes "that spatial economic phenomena could be expressed in an explicitly abstract, formal, and rationalist vocabulary and directly connected to the empirical world." This assumption allows for location theory to be universally generalizable from one geographic site to another. We should therefore expect it to be well applied in research on the economic geography of Asia. The reality, however, seems to work on the contrary. With the exception of G. William Skinner's (1964) influential work on marketing and social structure in rural China (see Cartier 2001 for a critique),

much of the research in economic geography in Asia during the 1960s and the 1970s remained descriptive and aligned with area studies and regional geography (Spencer 1954; Spate and Learmonth 1967; Ginsburg and Brush 1958; McGee 1967; Wheatley 1971; Murphey 1953; McGee and Yeung 1977).

Subsequent critiques of location theory and its variant in the behavioral locational model by such radical economic geographers as Massey (1973, 1984) and Walker and Storper (1981) led to the development of alternative theories on how to explain spatial economic phenomena. On the basis of their empirical studies of (de)industrialization in the United Kingdom and the United States, Clark (1981), Massey (1984), and Storper and Walker (1989) arrived at their respective theories of spatial divisions of labor and spatial switching by capital (see Table 1). These theories attempted to explain why (de)industrialization occurred in some but not all regions in the United Kingdom and the United States. The objective of the project was to specify the interdependent links between social processes of capitalist production and the spatial structures and distribution of industry, work, and classes. Although these theoretical perspectives on spatial divisions of labor generated much heated and exciting debates in subsequent studies of industrial restructuring and specific localities, most of these studies remained grounded in the *industrial* landscapes of the Anglo-American countries. Given their prominence in the mainstream economic geography of the 1980s, we would expect these perspectives to be universally applied to other research and empirical contexts. To the best of our knowledge, however, there has not been a significant diffusion of these theories of economic geography to geographic studies of other regions and countries that are concurrently experiencing dramatic processes of industrialization, economic restructuring, and rural development. This observation is certainly applicable to studies of the economic geography of Asia, although it is equally interesting to note that some Asian economies experienced unprecedented processes of industrialization during *exactly* the same periods—the 1970s and the 1980s. Mainstream economic geography thus fails to extend its analytical lens to examine geographic industrialization—let alone other aspects of economic transformations (e.g., rural changes and deprivation)—in other developing regions of the world economy. In the next section we outline what theoretical insights emerged from other social scientific studies of industrialization in Asia during these two decades.

Into the 1990s, mainstream economic geography certainly experienced a kind of intellectual renaissance through which a plethora of complementary theoretical perspectives were proposed—flexible specialization, networks and embeddedness, agglomerations and clusters, and regulation and governance. As is summarized in Table 1, these perspectives were concerned with why certain territorial ensembles—whether regions or new industrial spaces—emerged as the motors of growth in a particular country. It is no historical coincidence that during the late 1980s and the 1990s, several regions in the United States and some European economies became the leading engines of growth in the global economy. These theories vividly mirrored the historical and geographic specificities of the global space-economy. On the basis of his empirical analysis of the growth of high-tech industries in California, Italy, and France, Scott (1988) argued that a major shift was under way in contemporary capitalist industries—from mass production of the Fordist kind toward a post-Fordist form of flexible specialization and customization of production. These highly geographically specific observations led Scott (1988, 4; original italics) to conclude that "a series of *new industrial spaces* had come into existence and were beginning to form important alternative centers of capitalist accumulation based on a strong social division of labor, proliferations of small to medium-sized industrial establishments, and the marked reagglomeration of production." Although Scott's conclusions were not entirely new

vis-á-vis Piore and Sabel's (1984) earlier study of the Second Industrial Divide, his arguments for the rise of new industrial spaces did make a major impact on research in economic geography up to the mid-1990s (see a review in Yeung 1994).

New theoretical insights were also required to explain the geographic organization of production through firms and networks in these so-called new industrial spaces. Geographic agglomeration, proximity, processes of tacit knowledge and learning, and cooperative networks were conceptual categories proposed within this genre of theoretical and empirical research that has come to dominate much of Anglo-American economic geography since the late 1980s. More recent theoretical work on "relational assets," agglomeration economies, and institutional governance in the United Kingdom and the United States has reinforced the resurgence of the "regional world" of production as the dominant research theme in mainstream economic geography (see Yeung 2000). This resurgence, however, must again be situated in its peculiar historical and geographic contexts. Flexible production methods and agglomeration economies have been in existence for centuries, as found in craft industries and so on. The rise of these production methods and geographic economies to intellectual prominence within Anglo-American economic geography must have something to do with the "crisis of Fordism" during which an earlier wave of methods of mass production and economies of scale could no longer maintain a competitive edge with incumbent firms and corporations in advanced industrialized economies. This crisis, nevertheless, can also be understood from the historical perspective of the immense Japanese challenge to Anglo-American industrial might during the late 1970s and the 1980s (see the next section). These new theories of economic geography thus emerged as an unintended response to historical urgency—to explain the downfall of Fordist firms and industries and the rise of new propulsive industries (e.g., in Silicon Valley and elsewhere).

An important question remains: why did not these important and innovative theories in economic geography emerge from research on Fordism's competitors—Japan and the newly industrialized economies (NIEs) in Asia? Why did we not have economic geography theories that examined the crisis of Fordism and the rise of new industrial spaces *in relation* to the growing economic might of Japan and the NIEs? We return to this question in the next section.

The geographic context of these theories in economic geography is equally intriguing. Two observations are critical here. First, most theoretical work was based on empirical studies of a few selected regions in advanced industrialized economies in the United States (e.g., Silicon Valley and Route 128) and Europe (e.g., the M4 corridor and Cambridge, in the United Kingdom, the Third Italy, Baden-Württemberg in Germany, and the Scientific City in France). This spatial selectivity of empirical cases places an upper limit on the applicability of these theories even to different regions of the same country, let alone countries with contrasting forms of capitalism (Clegg and Redding 1990; Whitley 1999; Stark and Bruszt 2001). The geographic specificity of these theories explains why certain (propulsive) industries are privileged in their explanatory matrixes and why other important issues in economic geography are sidestepped. While flexible specialization may be crucial to understanding the economic transformation of the United States during the 1980s, one may surely argue that the transformation of agrarian economies under the auspices of neoliberal economic policies may be equally significant to the economic geographic understanding of many developing economies. We do not deny that a less-visible community of economic geographers, or those who have been branded as "development geographers" by the mainstream, have been studying the latter phenomenon. But their influence in economic geography remains limited precisely because mainstream economic geographers during the past two decades were narrowly focusing on a few industries

and regions in a handful of advanced economies. Second and relatedly, most leading proponents of these theories come from a few prestigious research departments in the United States and the United Kingdom. This geographic situatedness of authorship does not automatically invalidate the general applicability of their theories. But it does explain why certain theories emerge and become influential through more intensive interactions among like-minded scholars and research activities in these institutions. Their applicability to other geographic contexts needs to be interrogated and validated through carefully designed empirical research (see the articles in this special issue).

The situatedness of theories of economic geography is perhaps best illustrated by the recent "cultural turn" in economic geography toward a kind of "new economic geographies" that is much more reflexive and open in its nature and subject matter (Thrift and Olds 1996; Lee and Wills 1997; Yeung forthcoming). Indeed, Thrift and Olds (1996, 313) argued that we need to "make a space for new kinds of economic geography that can supplement or even replace the older forms of economic geography." In this process, Wills and Lee (1997, xvii) stated, we must appreciate how to "contextualize rather than to undermine the economic, by locating it within the cultural, social and political relations through which it takes on meaning and direction." According to Barnes (2001a, 551), this mode of "hermeneutic theorizing" differs significantly from the "epistemological theorizing" manifested in the quantitative revolution because it "[1] rejects fixed and final foundations . . . [2] promotes experimentation and engagement with radically different vocabularies, pressing them as far as they will go . . . [3] cultivates critical self-awareness of social and historical location and recognizes its influence on knowledge . . . [and] [4] is interested in keeping the conversation going." Precisely because of the inherent reflexivity and openness in the new economic geographies, it is very difficult to summarize diverse strands of theories and empirical findings. At the least, new economic geographers have refig-ured the economic through an excursion into the cultural and the political. As Barnes (1999, 17) noted, the basic explanatory cate-gories become "social power, cultural identity and institutional situatedness rather than economic ownership, universal defin-itions and individual agency." Several features of the new economic geographies include understanding the social embed-dedness of economic action, mapping shifting identities of economic actors, and exploring the role of context in explaining economic behavior (see a review in Barnes 2001a; Yeung 2001b; forthcoming).

To sum up our sympathetic critique, mainstream economic geography, which was developed in the Anglo-American countries, has experienced tremendous internal trans-formations and metamorphism during the past four decades. Its theoretical core has moved from universalizing location theory during the quantitative revolution to geographically specific theories of territorial development during the 1980s and the 1990s and, recently, to the more reflexive cultural turn that champions heightened sensitivity to the positionality of knowledge and theo-ries and the context in which these theo-ries emerge. This unprecedented intellec-tual movement in the epistemology of economic geography provides an exciting and important opportunity for us to recon-sider what theories of economic geography may be if we situate these theories in specific regions *beyond* the Anglo-American coun-tries (see the articles in this issue). More significant, we may have arrived at a time when new *kinds* of theories of economic geography are needed to account for the diversity of experience and transformations in the global economy. In this way, we may be able to construct genuine global economic geographies that are attuned to the histor-ical and geographic specificities of our theo-ries and yet are capable of producing a much broader and comparative understanding of dramatic economic transformations in the new millennium. Before we theorize the economic geographies of Asia, it is impor-tant for us to situate the region in the emerging global economic geographies.

Situating Asia in Global Economic Geographies

As a prelude to our advocacy, we argue that if we look seriously beyond North America and Western Europe, we can undoubtedly find innovative theoretical insights from social science studies of other regions. Although these theoretical insights are no less historically and geograph-ically specific than those championed in mainstream economic geography, few have really originated from the work of economic geographers or human geographers in general. Instead, those theoretical insights have emerged mostly from developmental studies, anthropology and sociology, and political economy. The situated nature of theoretical insights gathered from intensive studies of specific countries and/or regions should not be surprising if we take theories as hermeneutics or discursive formations that must be firmly grounded in material realities. These realities, however, differ from one historical moment to another and from one geographic setting to another. For example, whereas studies of Latin America have given rise to dependency theory, social science studies of Asia have similarly gener-ated many important theoretical insights, some of which have been followed up in the recent literature on economic geography: (1) the flying geese hypothesis (Hart-Landsberg and Burkett 1998; Edgington and Hayter 2000); (2) the new international divi-sion of labor (Frobel, Heinrichs, and Kreye 1980; Henderson 1989); (3) the devel-opmental state (Douglass 1994; Clark and Kim 1995; Brohman 1996; B.-G. Park 1998; Yeung 1999; Hsu and Cheng 2002); (4) social capital (Leung 1993; Hsing 1998; Yeung 1998c; Olds 2001a); and, more recently, (5) transnationalism (Mitchell 1995; Olds and Yeung 1999; Ley 1999; Hsu and Saxenian 2000; Yeoh and Chang 2001; Zhou and Tseng 2001; G. C. S. Lin 2002a; Ma and Cartier 2003).

It is important to note, however, that in stark contrast to such theories as spatial divisions of labor and flexible specialization, these theoretical interrogations that are grounded in Asia have not yet made a significant impact on the development of mainstream economic geography described in the previous section. Instead, mainstream economic geography produces the "right" kind of theories, emanating from specific cases in the Anglo-American countries that remain to be tested as universal principles that are equally applicable to other, more marginal, regions of the global economy. Economic geographers fail to heed Appadurai's (1999, 230) telling warning, in the context of area studies, that "the more marginal regions of the world are not simply producers of data for the theory mills of the North." How, then, has this highly unequal division of labor in research on economic geography emerged? We analyze this phenomenon in relation to three groups of geographers: (1) those who have engaged in mainstream enquiry, (2) those who have engaged in area studies, and (3) those who have interrogated mainstream theories on the basis of the Asian experience.

Historically, Asia—just like Africa, Russia, and Eastern Europe—has never really attracted serious attention in *mainstream economic geography*, despite the discipline's celebrated interest in spatial differentiation and uneven development. Even if it did, Asia was treated as "the others" in "the Far East," previously a market to be opened up by colonialism and now posing a challenge to the industrial might in Europe and North America (see Amsden 2001). This tendency toward what Said (1978) termed "Orientalism" is no less significant in economic geography than in the humanities and the other social sciences. The real difference, however, rests with the fact that Asia, along with Africa and Latin America, has been well studied by development specialists outside mainstream economic geography. Development is clearly central to the studies of economic geography and has always been one of the most exciting topics to university students of economic

geography in North America and Western Europe. Ironically, development geography has been largely constituted outside mainstream economic geography, in which overwhelming attention was devoted to industrial transformation in a few advanced economies and/or regions. This tension between development geography and economic geography has effectively marginalized, if not excluded, Asia on the research agenda of economic geography.

Meanwhile, the empirical landscape of Asia has undergone dramatic transformations since the 1970s, when Japan began to emerge as the leading competitor and alternative to the Anglo-American model of industrial capitalism. Ezra Vogel's (1979) influential book *Japan as Number One* was widely circulated in major intellectual and policy circles (that are often based in Washington, D.C., or the two Cambridges—one in England and the other in New England). Coupled with the emergence of Asian NIEs and the 1973 oil crisis, the rise of Japan triggered what was later conceptualized as flexibility, post-Fordism, and globalization. It is interesting that the social scientists who first realized the indispensable role of Japan and Asia in their theorization of global economic change came largely from area studies (Vogel 1989; Frank 1998), political science (Johnson 1982; Amsden 1989; Wade 1990), and economic sociology (Hamilton and Biggart 1988; Redding 1990).

More specifically, Japan was significantly featured in Piore and Sabel's (1984) *The Second Industrial Divide* and Womack, Jones, and Roos's (1990) *The Machines That Changed the World*. Both MIT (Cambridge, Mass.) products have fundamentally shaped the subsequent debates about America's and, by extension, the world's industrial future. During the same period (the 1980s and early 1990s), flexible specialization attracted substantial attention from economic geographers (see Table 1). As we noted earlier, much of this work was inspired by empirical studies in California, the Third Italy, and other European regions, with limited applicability to the Asian context (see Patchell

1993a, 1993b; Eng 1997). In comparison with other major social sciences and with a few exceptions (e.g., Dicken 2003; Florida and Kenney 1990; Angel 1994), mainstream economic geography has certainly missed the boat in exerting its intellectual influence on major policy debates in the United States and Europe about the imminent economic challenge from Asia and elsewhere (cf. Reich 1991; Tyson 1993; World Bank 1993).

Vogel's influential warning in *Japan as Number One*, nevertheless, was short lived. With the downturn of the Japanese economy since the early 1990s and the Asian financial crisis during 1997–1998, few people now take seriously "the Asian miracles" and the dawn of "the Asian century." In deconstructing "the myth of the Asian miracles" and the recent Asian economic crisis, mainstream economic geographers once again have failed to assume intellectual leadership in the broader social sciences. Curiously, it is the economists and their political science counterparts who have spearheaded the debates about the downfall of Japan, the Asian economic crisis (Krugman 1994, 1998; Radelet and Sachs 1998; Wade and Veneroso 1998), and the alleged rise of China as a threat (Vogel 1989; Goodhart and Xu 1996; Gertz 2000). Asking "Where have all the geographers gone?," Kelly, Olds, and Yeung (2001, x–xi) noted the absence of economic geographers in debates on the origins and impact of the Asian economic crisis. This lacuna is unfortunate because the crisis has much to offer to our understanding of the destabilization of a global financial architecture that is essentially built on the Bretton Woods institutions. The sheer scale and scope of the crisis also provided a unique case for reworking the fundamental nature and future of global capitalism and its powerful institutions (see Stiglitz and Yusuf 2001; Wade 2002).

If mainstream economic geography cannot contribute much to our understanding of the complex economic landscapes of Asia (other than making available situated theories for superficial testing and straightforward applications elsewhere), can we turn to *area studies specialists* (often known as development geographers), who may offer such an understanding from a more grounded perspective? Asia has long been studied by human and economic geographers who are interested in what cultural anthropologist Geertz (1973) termed a "thick description" of the land and the people outside Europe and North America. These geographers include indigenous scholars living in the regions and Western scholars who are interested in Asia. The former group has as sizable a population as their counterpart in Western Europe and North America. For instance, in the 1970s, over 6,000 professional geographers were working and teaching in socialist China, a size similar to if not greater than that in the United States (Pannell 1980, 176). Today, these Chinese professional geographers are all full-time researchers or university professors on the payroll of the state. Unfortunately, indigenous geographers in Asian countries have never been able to make any significant impact on mainstream enquiry in economic geography, largely because of the linguistic and cultural barriers as well as their different methodological traditions that have severely hindered meaningful scholarly exchanges with mainstream economic geographers. On the other hand, Western scholars who are interested in Asia have been in the minority, preoccupied with sorting out the facts about a mysterious region in the "Far East" for the occasional curiosity of Western academics and continuous strategic and economic interests.

With the exception of perhaps the flying geese model, studies of the economic geography of Japan have never occupied any prime position equivalent to the global significance of its national economy—not even in the flexible-specialization debate of the late 1980s and early 1990s.[2] Geographic studies of China and India, whose combined

---

[2] We regret that no article on Japan is included in this special issue, primarily because of insufficient submissions. See recent work by Peck and Miyamachi (1994), Aoyama (2000), and Patchell (2002).

population constitutes nearly half the people of the world, have not generated research published in English-language media that is anywhere near the global significance of their population (see also Potter 2001). A systematic search of articles published in the top 10 international journals in human geography from 1971 to 2000 found that only 66 articles, less than 1 percent of the total, dealt with the geography of China (G. C. S. Lin 2002b, 1813). Most of these studies were empirical and "had shallow roots, received little nourishment and predictably bore few and unappetizing fruits" (Leeming 1980, 218). It has only been recently that geographers who are interested in China have ventured to formulate some contextually sensitive theories (Fan, Ma, Pannell, and Tan forthcoming). The article by Yu Zhou and Tong Xin in this special issue thus significantly contributes to the critique of the literature on innovative regions that has primarily focused on localized endogenous factors in sustaining innovation and regional development. Through their intensive research in a high-tech cluster in Beijing, Zhou and Xin found important interactive and interdependent relations between local Chinese firms and global corporations in the joint development and commercialization of new information technologies. Their work complements the growing literature in economic geography on the critical importance of nonlocal and decentralized learning and innovations for understanding regional development (e.g., Bunnell and Coe 2001; MacKinnon, Cumbers, and Chapman 2002).

A relatively small group of geographers have managed to develop original theoretical insights from their grounded studies of the Asian experience. Through their work, some *grounded theories* have emerged that have proved to be influential in certain subfields of human geography and, to a lesser extent, economic geography. A prominent example of such grounded theories is Terry McGee's (1967, 1971) model of the Southeast Asian city (see also Armstrong and McGee 1985). In this morphological model of the internal structure of the city, McGee argued that different urban-economic activ-

ities have different spatial requirements and locational characteristics. For example, the informal sector tends to be located in the inner ring of the city. McGee's model has subsequently been well applied to the geographic study of other Third World cities. More crucially, it originated from empirical studies of such cities; it was not developed from studying the internal structures of advanced industrialized countries and then applied universally to Third World cities (akin to "epistemological theorizing" described earlier). Despite its generality in studies of urbanization and urban economic activities, McGee's model regrettably had only a limited impact on mainstream economic geography of the 1970s and 1980s, which was preoccupied with radical Marxism and post-Fordism.

More recently, geographic studies of *transnational business activities* and *transnationalism* represent a significant attempt to bring grounded theories of economic geographies of Asia back into mainstream economic geography (Leung 1993; Mitchell 1995; Yeung 1997; Zhou 1998; Hsu and Saxenian 2000; Olds 2001a). In particular, this body of literature on economic geography has managed to blend into its theoretical framework two important ingredients—a special blend that is well grounded in the Asian context. First, it has brought to its analytical forefront the conceptual lenses of *networks* and *embeddedness*. Although these conceptual categories did not originate from economic geographic studies of Asia, it is equally important to note that mainstream economic geographers did not develop them either. Indeed, these conceptual categories were first proposed by economic sociologists (Polanyi 1944; Granovetter 1985) and subsequently introduced into and appropriated by mainstream economic geography during the debate on flexible specialization (see Dicken and Thrift 1992). It is true that mainstream economic geography has further enhanced the theoretical sophistication of both conceptual categories through major debates on industrial districts (Asheim 2000), the spatial transfer of technologies (Gertler 2001), organizational change (Yeung 1994,

1998a; Schoenberger 1997), and institutionalism in urban and regional development (Amin 1999). But then it must be equally valid to argue that economic geographic studies of Asia during the 1990s significantly advanced the spatialization of these conceptual categories by theorizing the complex interactions among business networks, ethnicity/culture, embeddedness, and historical specificity. This effort to theorize the spatial rudiments of networks and embeddedness is no less significant than that of debates on industrial districts and so on (see Table 1).

Second, economic geographic studies of Asian diasporas and their worldwide webs in Europe and North America have made significant inroads into the debate on *globalization*. In many ways, this body of literature has contributed to enhancing economic geography's growing visibility in social scientific debates on globalization. Once again, geographers have not been well represented among leading scholars of globalization: one can literally think of only two works by geographers—Peter Dicken's (2003) *Global Shift* and David Harvey's (1989) *The Condition of Postmodernity*—that have represented different kinds of geographic takes on globalization and thereby have attracted different sorts of audiences. Economic geographers, however, have something significant to say about the geographic specificities of globalization in relation to the origins, processes, and outcomes of globalization (see Amin 1997; Yeung 1998b, 2002b; Kelly 1999; Peck and Yeung 2003).[3] Put in this perspective, economic geography research on Asian diasporas and their global networks

augments well the key mission of mainstream economic geography to ground globalization processes in specific territorial ensembles and formations. It helps not only to demystify the "faceless" representations of globalization by its ultra-supporters (e.g., Ohmae 1990), but also to make economic geographers aware of the highly uneven geographic outcomes of processes associated with globalization. To us, this geographic research on the globalization of Asia is worth as much intellectual capital as is other equally worthy research on globalization in economic geography on changing urban and regional governance (Brenner 1999) and the organizations of economic activities (Dicken, Kelly, Olds, and Yeung 2001).

Theories Wanted! An Intellectual Agenda for Economic Geographies of Asia

We are at a critical juncture in economic geography. There are unprecedented opportunities for mainstream economic geographers to give up our long-standing Euro-American-centric bias and develop theories that account for differences and differentiation in an era of accelerated globalization (cf. McGee 1991; Olds 2001b; Yeung 2001a). Although concerns about national security remain large, especially after the September 11, 2001, tragedy, earlier warped provincialism can no longer inhibit economic geographers who actively study the rapid transformation of regional economies in Asia. Major funding agencies, such as the National Science Foundation in the United States, the Economic and Social Research Council in the United Kingdom, and the Social Sciences and Humanities Research Council in Canada, have recently supported a growing number of research projects on Asia. Many leading universities in Anglo-America have actively recruited geographers to work on Asia. On the other side of the Pacific, most of the Asian economies have rearticulated themselves actively and openly to take part in the theater of global capital

---

[3] The recent special issue of *Economic Geography* (2002) on geographies of global economic change represents an important milestone in this endeavor to put geographers back on the intellectual table of globalization studies. The special issue originated from the 2001 Leir conference (http://www.clarku.edu/leir/index.shtml) funded by the U.S. National Science Foundation, to rethink how to build theories of geographies of global economic change and to unveil the theoretical and methodological obstacles to the process of theory building.

accumulation. Indigenous geographers in Asia have made special attempts to overcome linguistic and cultural barriers as they forge bilateral scholarly linkages and research collaborations with geographers in the English-speaking world. Intellectually, universalism as one of the defining features of the Enlightenment and modernization school of thought has given way to a more open-ended, plural, and contextually sensitive perspective on changing geographies in different world regions. Overall, the institutional setting that previously separated the regional geography of Asia from mainstream economic geography in the Anglo-American countries has undergone a profound transformation in a direction favorable to the development of *global economic geographies.*

Indeed, we are beginning to witness such a change in the direction of mainstream economic geography—more inclusive quality control in the academic production of knowledge and more intellectual activities organized outside the Anglo-American centers (see Barkema, 2001 for management studies). It is now incumbent on economic geographers who are interested in Asia to move from area studies to engage more actively with mainstream theoretical (re)constructions and interrogations. In this sense, there is a need for two intellectual movements. First, we must avoid uncritical applications of "Western" theories in mainstream economic geography as if these theories were universally true. We need to interrogate these theories critically through our detailed research on economic geographies of Asia. Through this process of critical engagement with mainstream theories, we will be able not only to contribute to economic geography through our refinement and reconstruction of these theories, but also to understand the economic landscapes of Asia from a grounded perspective. Second, we must turn away from doing what may be termed "Asian economic geography" because such a parochial approach to economic geographies of Asia will provide few significant theoretical insights that may be useful in other geographic contexts.

Rather, we must endeavor to develop *new* theories, grounded in Asia, that might better inform our understanding of the "economic" in economic geography at large. Commenting on management studies, Barkema (2001, 616) noted that "[i]f different management concepts, theories, and practices apply in different cultural and institutional settings, international research might lead to novel theory and evidence showing how." In this sense, there should not be a "mainstream" economic geography on the basis of geographic divides (the Anglo-American centers versus the rest of the world) or thematic divides (industrial geography versus rural or development geography).

Why is such an intellectual turn toward more-inclusive global economic geographies necessary? As outlined briefly in the introduction, we believe that two important concurrent trends warrant this turn. The first trend is inevitably related to *the globalization of knowledge and theories.* Economic geographers from major Anglo-American centers are increasingly reaching out to Asia so that Asia can be integrated into their theories and comparative analyses. As we argue later in the empirical realm, Asia is becoming too important to be ignored by economic geographers; it is, of course, also too important to be left to economic geographers only. This interest in Asia is exemplified by the recent work on flexibility, globalization, social capital, the cultural turn, the institutional turn, and the relational turn in economic geography. There is a growing interest among economic geographers in investigating territorial formations outside the Anglo-American contexts *not* as an anomaly or "other economic geography" from the perspective of Anglo-American economic geography, but instead as an original subject of inquiry in its own right. In this special issue, Smart and Lee critically engage one major theoretical strand in recent work in economic geography—regulation theory (see Table 1). Analyzing the vital role of real estate and property assets in Hong Kong's regime of accumulation and economic development during the past two decades, Smart and Lee argue that the Hong Kong case should not be interpreted as an anomaly that deviates from the developmental trajectory of Anglo-American capitalism (see also Kerr 2002 for an analysis of Japan). The distinctive features of Hong Kong's regime of accumulation mean that it is indeed possible to "examine them as forerunners of a possible future property-based mode of regulation that *might emerge in the West*, and thus to diagnose the potential problems and opportunities of such a path" (Smart and Lee 2003, 153; our emphasis). This "theorizing back" does not entail an unproblematical application of Western-centric theories per se. Instead, it uses conceptual apparatuses in these theories and empirical evidence in Asia to open up new directions for understanding the future of a variety of capitalisms that, no doubt, include the dominant Anglo-American genre.

This trend toward reaching out to Asia clearly does not represent a one-way flow in the globalization of knowledge and theories. Today, more economic geographers who are interested in Asia are themselves Asians who have received their academic training in the Anglo-American centers (e.g., both authors of this article and most contributors to this special issue). This two-way intellectual fertilization allows them to benefit from the best of both worlds, so to speak. On the one hand, their particular backgrounds and origins in Asia enable their work to be firmly grounded in the material realities of Asia. Their emic understanding of Asia is difficult to emulate by geographers from other regional origins. On the other hand, these geographers are well equipped with sophisticated theoretical ideas and rigorous methodological procedures to enhance their research on economic geographies of Asia. They are certainly capable of growing out of Asia in their theoretical work to make major contributions to global economic geographies. The articles in this special issue clearly exemplify this theoretical sophistication and methodological rigor among economic geographers with Asian origins. For example, in explaining the recent liberalization and globalization of the South Korean automobile industry, Park not only draws on theoretical ideas from the debate on the multiscalar processes of globalization, but makes an original contribution to the literature by highlighting the interscalar contestation between the national state and the local community in shaping the globalization processes of the industry. Although this national-local tension has been analyzed in the regional governance literature in economic geography (see Table 1), it has been mostly ignored in the geographic studies of the impact of globalization that have focused primarily on global-local tensions. Similarly, Poon and Thompson examine rigorously the concept of embeddedness in explaining the parent-subsidiary relationship in global corporations. By unpacking the nature of the embeddedness of subsidiaries in the global networks of their parent companies, they draw our attention away from localized embeddedness that has been well documented in the "new regionalism" literature (see Table 1) and contribute to our understanding of the spatial organization of economic activities among global corporations.

What in Asia attracts these economic geographers such that Asia becomes their key subject of inquiry? Our answer lies with the empirical trend toward the rapid and dramatic transformations in the economic landscapes of Asia in more recent decades. Asia has become a new site for theory development and empirical analysis in economic geography, as amply shown by the four excellent articles in this special issue. The importance of this new site does not merely rest with its internal transformations. More crucially, Asia's importance for economic geography is predicated on its potential to facilitate the production of new theoretical insights and, in Slater's (1999) words, "counter discourses" that allow economic geographers to "theorize back" at our situated knowledge emanating from Europe and North America (e.g., Park, Smart and Lee, and Zhou and Tong in this issue). As we noted earlier, Japan came to the forefront of social scientific inquiry during the 1980s because of its technological and economic prowess. Together with insights from other newly industrialized economies (except Hong Kong), research on Japan's rapid post–World War II economic development points to the role of the developmental state. From an economic geography perspective, this theorization allows economic geography to reconsider location theory and the development of industrial districts in novel ways that otherwise are unlikely to be achieved single-handedly through research on the Anglo-American countries (see Markusen and Park 1993; Park and Markusen 1995). It allows for the attainment of the "translocal" understanding and development advocated by Smith (2002). This movement in theorization entails more than just using different empirical contexts for theorization. Rather, it produces a new kind of theory that challenges, for example, the market-state dualism that is so ingrained in Anglo-American economic thinking, particularly among the neoliberals and "deregulationists" as labeled by Storper and Salais (1997, 246).

The rise of China since the late 1970s represents another critical juncture in the historiography of economic geography. For decades, the transformation of the Chinese space-economy under socialist authoritarianism has often been considered to be too unique or peculiar and thus incompatible with the international norms and theoretical templates. The peculiarity of the Chinese experience, plus the lack of necessary information for meaningful studies, had made it extremely difficult for economic geography theorists and China geographers to have fruitful communication (see also Liu and Lu 2002). In recent years, however, the Chinese space-economy has undergone profound structural and spatial transformations as the post-Mao regime changed its approach from rigid utopian socialism to market-oriented pragmatism and from self-isolation to active participation in globalization. A fascinating mosaic of plan and market, state and private sectors, central authoritarianism, and local corporatism has emerged to recontour the economic landscape (G. C. S. Lin 1997;

Marton 2000; Wei 2000). Given the fundamental importance of both the Chinese culture and its restructured socialist institutional setting to the transformation of the Chinese space-economy, incorporating the Chinese case into the development of theory in mainstream economic geography seems to be timely and appropriate (e.g., Zhou and Tong in this issue).

Recent institutional and economic processes in China have not only invalidated our received wisdom of the geography of industrialization and economic transition, but also present themselves as fertile grounds for the development of new theories. First, economic geographers have been accustomed to industrialization occurring virtually hand in hand with rapid urbanization and industrial activities located within urban areas. The core geographic argument for this trend toward urban-biased industrialization is related to the Marshallian notion of agglomeration economies and, more recently, to increasing returns to scale, as demonstrated in the endogenous growth models (Martin and Sunley 1998). This abstract theorization of industrialization and regional growth, however, ignores historical specificity and institutional rigidities that continue to exert strong effects in the case of China. A process of *rural industrialization* has come to characterize the postreform pattern of industrialization in China through which the labor force is expected to stay in rural areas and industrial activities are brought to their doorsteps (Marton 2000). The enormous contributions of township and village enterprises (TVEs) to China's gross domestic products (GDP) and employment are one such indicator of the pervasive extent of rural industrialization. Widely scattered all over the vast countryside, the TVEs generated over 30 percent of China's GDP and provided employment opportunities to 27 percent of the total rural labor force by the year 2000 (Editorial Board 2001, 4–5). This finding, of course, does not mean that urban-centered industrialization does not take place in China. But it does call for a reconceptualization of industrialization and urban-regional development in China not as a special case of geographic industrialization, but as an original subject of inquiry that may yield new theoretical insights into urbanization, industrialization, rural development, and agrarian change.

Second, China's transitional economy allows for new theories of economic transition and organizational change that are just making significant inroads into the major social sciences, such as sociology and economics. Sociologists like Victor Nee (1989), Andrew Walder (1995), Nan Lin (1995; N. Lin and Bian 1991), and Doug Guthrie (1997) have worked on China's postreform development and collectively developed what may be termed the market-transition theory (see Stark 1996 and Pickles and Smith 1998 for the case of Eastern Europe). This theory has certainly reinvigorated sociological studies of changing social structures and economic organization in transitional economies. In economics, Barry Naughton (1991), Thomas Rawski (1994), Alwyn Young (2000), and others have shown how conventional neoclassical economics has failed to provide a valid theoretical model for explaining China's economic development (see also Amsden 1991; Young 1995). Alternative economic models are therefore called for that account for China's unprecedented economic transformations. Although we have not yet observed a similar theoretical development and disciplinary impact of research on Asia in economic geography, we have certainly noticed some novel conceptualizations arising from recent work on China (Hsing 1998; Olds 2001a; Fan 2002; Zhou and Tong in this issue) and other economies in East and Southeast Asia (Kelly 2001a, 2001b; Hsu and Saxenian 2000; Coe and Kelly 2002; Park in this issue).

To conclude this extended introduction to the special issue, we believe that theorizing economic geographies of Asia is clearly an unfinished intellectual project. In fact, we go so far as to suggest that it simply marks the beginning of a new intellectual era for economic geography toward the development of global economic geographies. Economic geographies of Asia must not be a subject of theorization from the perspectives of mainstream Anglo-American economic geography. But equally, they are too important to be left to Asian economic geographers alone. Building on a growing body of economic geography research on Asia, what we aim to achieve through this special issue is a further and, we hope, significant step toward more genuine *theoretical dialogues* among economic geographers with different regional interests. This bold aim cannot be achieved without more theoretical work that is grounded in the material realities of Asia but that speaks to an audience that is well tuned into the transmission frequency of global economic geographies. The future of economic geography must be bright and exciting. In this sense, we fully concur with Barnes and Sheppard's (2000, 0, our emphasis) assessment: "There is a Chinese saying: 'May you live in interesting times.' Our argument is that they are here *now* in economic geography." Obviously, we—both ethnic Chinese writers—cannot agree more.

References

Amin, A. 1997. Placing globalization. *Theory, Culture and Society* 14:123–37.

———. 1999. An institutionalist perspective on regional economic development. *International Journal of Urban and Regional Research* 23:365–78.

Amsden, A. 1989. *Asia's next giant: South Korea and late industrialization.* New York: Oxford University Press.

———. 1991. Diffusion of development: The late industrializing model and greater Asia. *American Economic Review* 81:282–6.

———. 2001. *The rise of "the rest": Challenges to the West from late-industrializing economies.* New York: Oxford University Press.

Angel, D. P. 1994. *Restructuring for innovation: The remaking of the U.S. semiconductor industry.* New York: Guilford Press.

Aoyama, Y. 2000. Networks, *keiretsu* and locations of the Japanese electronics industry in Asia. *Environment and Planning A* 32:223–44.

Appadurai, A. 1999. Globalization and the research imagination. *International Social Science Journal* 51:229–38.

Armstrong, W., and McGee, T. G. 1985. *Theatres of accumulation: Studies in Asian and Latin American urbanization.* London: Methuen.

Asheim, B. T. 2000. Industrial districts: The contributions of Marshall and beyond. In *The Oxford handbook of economic geography*, ed. G. L. Clark, M. A. Feldman, and M. S. Gertler, 413–31. Oxford, U.K.: Oxford University Press.

Barkema, H. 2001. From the editors. *Academy of Management Journal* 44:615–17.

Barnes, T. J. 1996. *Logics of dislocation: Models, metaphors, and meanings of economic space.* New York: Guilford Press.

———. 1999. Industrial geography, institutional economics and Innis. In *The new industrial geography: Regions, regulation and institutions*, ed. T. J. Barnes and M. S. Gertler, 1–22. London: Routledge.

———. 2000. Inventing Anglo-American economic geography, 1889–1960. In *A companion to economic geography*, ed. E. Sheppard and T. J. Barnes, 11–26. Oxford, U.K.: Blackwell.

———. 2001a. Retheorizing economic geography: From the quantitative revolution to the "cultural turn." *Annals of the Association of American Geographers* 91:546–65.

———. 2001b. Lives lived and lives told: Biographies of the quantitative revolution. *Environment and Planning D: Society and Space* 19:409–29.

———. 2001c. In the beginning was economic geography–a science studies approach to disciplinary history. *Progress in Human Geography* 25:521–44.

Barnes, T. J., and Curry, M. 1983. Towards a contextualist approach to geographical knowledge. *Transactions of the Institute of British Geographers* 8:467–82.

Barnes, T. J., and Sheppard, E. 2000. Introduction: The art of economic geography. In *A companion to economic geography*, ed. E. Sheppard and T. J. Barnes, 1–8. Oxford, U.K.: Blackwell.

Brenner, N. 1999. Globalization as reterritorialisation: The European re-scaling of urban governance in the European Union. *Urban Studies* 36:431–51.

Brohman, J. 1996. Postwar development in the Asian NICs: Does the neoliberal model fit reality? *Economic Geography* 72:107–30.

Bryson, J.; Henry, N.; Keeble, D.; and Martin, R., eds. 1999. *The economic geography reader: Producing and consuming global capitalism.* Chichester, U.K.: John Wiley & Sons.

Bunnell, T. G., and Coe, N. M. 2001. Spaces and scales of innovation. *Progress in Human Geography* 25:569–90.

Cartier, C. 2001. *Globalizing South China.* Oxford, U.K.: Blackwell.

Clark, G. L. 1981. The employment relation and spatial division of labor: A hypothesis. *Annals of the Association of American Geographers* 71:412–24.

Clark, G. L.; Feldman, M. A.; and Gertler, M. S., eds. 2000. *The Oxford handbook of economic geography.* Oxford, U.K.: Oxford University Press.

Clark, G. L., and Kim, W. B., eds. 1995. *Asian NIEs in the global economy.* Baltimore, Md.: Johns Hopkins University Press.

Clegg, S. R., and Redding, S. G., eds. 1990. *Capitalism in contrasting cultures.* Berlin: de Gruyter.

Coe, N. M., and Kelly, P. F. 2002. Languages of labour: Representational strategies in Singapore's labour control regime. *Political Geography* 21:341–71.

Dicken, P. 2003. *Global shift: Transforming the world economy.* 4th ed. London: Sage.

Dicken, P.; Kelly, P.; Olds, K.; and Yeung, H. W.-c. 2001. Chains and networks, territories and scales: Towards an analytical framework for the global economy. *Global Networks* 1:89–112.

Dicken, P., and Thrift, N. 1992. The organization of production and the production of organization: Why business enterprises matter in the study of geographical industrialization. *Transactions, Institute of British Geographers* 17:279–91.

Douglass, M. 1994. The "developmental state" and the NIEs of Asia. *Environment and Planning A* 26:543–66.

Economic Geography 2002. Special issue on global economic change. 78(3).

Edgington, D. W., and Hayter, R. 2000. Foreign direct investment and the flying geese model: Japanese electronics firms in the Asia Pacific. *Environment and Planning A* 32:281–304.

Editorial Board. 2001. *Almanac of China's township and village enterprises (2001).* Beijing: China Agricultural Press.

Eng, I. 1997. Flexible production in late industrialization: The case of Hong Kong. *Economic Geography* 73:26–43.

Fan, C. C. 2002. The elite, the natives, and the outsiders: Migration and labor market segmentation in urban China. *Annals of the Association of American Geographers* 91:103–24.

Fan, C. C.; Ma, L. J. C.; Pannell, C. C.; and Tan, K. C. Forthcoming. China geography in North America. In *Geography in America at the dawn of the 21st century*, ed. G. L. Gaile and C. J. Willmott. New York: Oxford University Press.

Florida, R., and Kenney, M. 1990. *The breakthrough illusion*. New York: Basic Books.

Frank, A. G. 1998. *ReORIENT: Global economy in the Asian age*. Berkeley: University of California Press.

Frobel, F.; Heinrichs, J.; and Kreye, O. 1980. *The new international division of labour*. Cambridge, U.K.: Cambridge University Press.

Geertz, C. 1973. *The interpretation of cultures*. New York: Basic Books.

Gertler, M. S. 2001. Best practice? Geography, learning and the institutional limits to strong convergence. *Journal of Economic Geography* 1:5–26.

Gertz, B. 2000. *The China threat: How the People's Republic targets America*. Washington, D.C.: Regnery.

Ginsburg, N. S., and Brush, J. E. 1958. *The pattern of Asia*. Englewood Cliffs, N.J.: Prentice Hall.

Goodhart, C., and Xu, C., 1996. *The rise of China as an economic power*. Cambridge, Mass.: Institute for International Development, Harvard University.

Granovetter, M. 1985. Economic action, and social structure: The problem of embeddedness. *American Journal of Sociology* 91:481–510.

Guthrie, D. 1997. Between markets and politics: Organizational responses to reform in China. *American Journal of Sociology* 102:1258–304.

Hamilton, G. G., and Biggart, N. W. 1988. Market, culture, and authority: A comparative analysis of management and organization in the Far East. *American Journal of Sociology* 94:S52–S94.

Hart-Landsberg, M., and Burkett, P. 1998. Contradictions of capitalist industrialization in East Asia: A critique of "flying geese" theories of development. *Economic Geography* 74:87–110.

Harvey, D. 1989. *The condition of postmodernity: An enquiry into the origins of cultural change*. Oxford, U.K.: Basil Blackwell.

Henderson, J. 1989. *The globalisation of high technology production*. London: Routledge.

Hsing, Y.-t. 1998. *Making capitalism in China: The Taiwan connection*. New York: Oxford University Press.

Hsu, J.-Y., and Cheng, L.-L. 2002. Revisiting economic development in post-war Taiwan: The dynamic process of geographical industrialization. *Regional Studies* 36:897–908.

Hsu, J.-Y., and Saxenian, A. 2000. The limits of Guanxi capitalism: Transnational collaboration between Taiwan and the USA. *Environment and Planning A* 32:1991–2005.

Hudson, B. 1977. The new geography and the new imperialism: 1870–1918. *Antipode* 9:12–19.

Johnson, C. 1982. *MITI and the Japanese economic miracle*, Stanford, Calif.: Stanford University Press.

Johnston, R. J. 1997. *Geography and geographers: Anglo-American human geography since 1945*, 5th ed. London: Arnold.

Kelly, P. F. 1999. The geographies and politics of globalization. *Progress in Human Geography* 23:379–400.

———. 2001a. The political economy of local labor control in the Philippines. *Economic Geography* 77:1–22.

———. 2001b. Metaphors of meltdown: Political representations of economic space in the Asian financial crisis. *Environment and Planning D: Society and Space* 19:719–42.

Kelly, P. F.; Olds, K.; and Yeung, H. W.-c. 2001. Introduction: Geographical perspectives on the Asian economic crisis. *Geoforum* 32:vii–xiii.

Kerr, D. 2002. The "place" of land in Japan's postwar development, and the dynamic of the 1980s real-estate "bubble" and 1990s banking crisis. *Environment and Planning D: Society and Space* 20:345–74.

Krugman, P. 1994. The myth of Asia miracle. *Foreign Affairs* 73:62–78.

———. 1998. Asia: What went wrong? *Fortune* 137(4):32.

Lee, R., and Wills, J., eds. 1997. *Geographies of economies*. London: Arnold.

Leeming, F. 1980. On Chinese geography. *Progress in Human Geography* 4:218–37.

Leung, C.-k. 1993. Personal contacts, subcontracting linkages, and development in the Hong Kong–Zhujiang Delta region. *Annals of the Association of American Geographers* 83:272–302.

Ley, D. 1999. Myths and meaning of immigration and the metropolis. *Canadian Geographer* 43:2–19.

Lin, G. C. S. 1997. *Red capitalism in South China: Growth and development of the Pearl River Delta*. Vancouver: University of British Columbia Press.

———. 2002a. Transnationalism and the geography of sub-ethnicity in Hong Kong. *Urban Geography* 23:57–84.

———. 2002b. Changing discourses in China geography: A narrative evaluation. *Environment and Planning A* 34:1809–31.

Lin, N. 1995. Local market socialism: Local corporatism in action in rural China. *Theory and Society* 24:301–54.

Lin, N., and Bian, Y. 1991. Getting ahead in urban China. *American Journal of Sociology* 97:657–88.

Liu, W., and Lu, D. 2002. Rethinking the development of economic geography in mainland China. *Environment and Planning A* 34:2107–126.

Ma, L. J. C., and Cartier, C., eds. 2003. *The Chinese diaspora: Space, place, mobility and identity*, Boulder, Colo.: Rowman & Littlefield.

MacKinnon, D.; Cumbers, A.; and Chapman, K. 2002. Learning, innovation and regional development: A critical appraisal of recent debates. *Progress in Human Geography* 26:293–311.

Markusen, A., and Park, S. O. 1993. The state as industrial locator and district builder: The case of Changwon, South Korea. *Economic Geography* 69:157–81.

Martin, R., and Sunley, P. 1998. Slow convergence? The new endogenous growth theory and regional development. *Economic Geography* 74:201–27.

Marton, A. M. 2000. *China's spatial economic development: Regional transformation in the lower Yangzi Delta*. London: Routledge.

Massey, D. 1973. Towards a critique of industrial location theory. *Antipode* 5:33–9.

———. 1984. *Spatial division of labour: Social structures and the geography of production*. London: Macmillan.

McGee, T. G. 1967. *The southeast Asian city*. London: Bell.

———. 1971. *The urbanization process in the Third World: Explorations in search of a theory*. London: Bell.

———. 1991. Eurocentrism in geography— The case of Asian urbanization. *Canadian Geographer* 35:332–44.

McGee, T. G., and Yeung, Y.-m. 1977. *Hawkers in southeast Asian cities: Planning for the bazaar economy*. Ottawa: International Development Research Centre.

Mitchell, K. 1995. Flexible circulation in the Pacific Rim: Capitalism in cultural context. *Economic Geography* 71:364–82.

Murphey, R. 1953. *Shanghai: Key to modern China*. Cambridge, Mass.: Harvard University Press.

Naughton, B. 1991. Why has economic reform led to inflation? *American Economic Review* 81:207–11.

Nee, V. 1989. A theory of market transition: From redistribution to markets in state socialism. *American Sociological Review* 54:663–81.

Ohmae, K. 1990. *The borderless world: Power and strategy in the interlinked economy*. London: Collins.

Olds, K. 2001a. *Globalization and urban change: Capital, culture and Pacific Rim mega projects*. New York: Oxford University Press.

———. 2001b. Practices for "process geographies": A view from within and outside the periphery. *Environment and Planning D: Society and Space* 19:127–36.

Olds, K., and Poon, J. 2002. Theories and discourses of economic geography. *Environment and Planning A* 34:379–83.

Olds, K., and Yeung, H. W.-c. 1999. (Re)shaping "Chinese" business networks in a globalising era. *Environment and Planning D: Society and Space* 17:535–55.

Pannell, C. W. 1980. Geography. In *Science in contemporary China*, ed. L. A. Orleans, 167–87. Stanford, Calif.: Stanford University Press.

Park, B.-G. 1998. Where do tigers sleep at night? The state's role in housing policy in South Korea and Singapore. *Economic Geography* 74:272–88.

Park, S. O., and Markusen, A. 1995. Generalizing new industrial districts: A theoretical agenda and an application from a non-Western economy. *Environment and Planning A* 27:81–104.

Patchell, J. 1993a. From production systems to learning systems—lessons from Japan. *Environment and Planning A* 25:797–815.

———. 1993b. Composing robot production systems—Japan as a flexible manufacturing system. *Environment and Planning A* 25:923–44.

———. 2002. Linking production and consumption. The coevolution of interaction systems in the Japanese house industry. *Annals of the Association of American Geographers* 92:284–301.

Peck, J. A., and Miyamachi, Y. 1994. Regulating Japan? Regulation theory versus the Japanese experience. *Environment and Planning D: Society and Space* 12:639–74.

Peck, J. A., and Yeung, H. W.-c., eds. 2003. *Remaking the global economy: Economic-geographical perspectives*. London: Sage.

Pickles, J., and Smith, A., eds. 1998. *Theorising transition: The political economy of post-Communist transformations*. London: Routledge.

Piore, M. J., and Sabel, C. F. 1984. *The second industrial divide: Possibilities for prosperity*. New York: Basic Books.

Polanyi, K. 1944. *The great transformation.* New York: Holt, Rinehart.

Potter, R. 2001. Geography and development: "Core and periphery"? *Area* 33:422–7.

Radelet, S., and Sachs, J. D. 1998. The East Asian financial crisis: Diagnosis, remedies, prospects. *Brookings Papers on Economic Activity* 1:1–90.

Rawski, T. G. 1994. Chinese industrial reform: Accomplishments, prospects, and implications. *American Economic Review* 84:271–5.

Redding, S. G. 1990. *The spirit of Chinese capitalism.* Berlin: de Gruyter.

Reich, R. B. 1991. *The work of nations: Preparing ourselves for 21st century capitalism.* New York: Vintage Books.

Said, E. W. 1978. *Orientalism.* London: Routledge.

Schoenberger, E. 1997. *The cultural crisis of the firm.* Oxford, U.K.: Basil Blackwell.

Scott, A. J. 1988. *New industrial spaces: Flexible production, organisation and regional development in North America and Western Europe.* London: Pion.

———. 2000. Economic geography: The great half-century. *Cambridge Journal of Economics* 24:483–504.

Shaw, M. 2000. *Theory of the global state: Globality as unfinished revolution.* Cambridge, U.K.: Cambridge University Press.

Sheppard, E., and Barnes, T. J., eds. 2000. *A companion to economic geography.* Oxford, U.K.: Blackwell.

Sidaway, J. D. 1997. The production of British geography. *Transactions of the Institute of British Geographers* 22:488–504.

Skinner, G. W. 1964. Marketing and social structure in rural China (Part I). *Journal of Asian Studies* 24:3–44.

Slater, D. 1999. Situating geopolitical representations: Inside/outside and the power of imperial interventions. In *Human Geography Today,* ed. D. Massey, J. Allen, and P. Sarre, 62–84. Cambridge, U.K.: Polity Press.

Smith, A. 2002. Trans-locals, critical area studies and geography's others, or why "development" should not be geography's organising framework: A response to Potter. *Area* 34:210–3.

Spate, O. H. K. and Learmonth, A. T. A. 1967. *India and Pakistan: A general and regional geography.* London: Methuen.

Spencer, J. E. 1954. *Asia, East by South: A cultural geography.* New York: John Wiley & Sons.

Stark, D. 1996. Recombinant property in East European capitalism. *American Journal of Sociology* 101:993–1027.

Stark, D., and Bruszt, L. 2001. One way or multiple paths: For a comparative sociology of East European capitalism. *American Journal of Sociology* 106:1129–37.

Stiglitz, J. E., and Yusuf, S., eds. 2001. *Rethinking the East Asian miracle.* Washington, D.C.: World Bank.

Storper, M., and Salais, R. 1997. *Worlds of production: The action frameworks of the economy.* Cambridge, Mass.: Harvard University Press.

Storper, M., and Walker, M. 1989. *The capitalist imperative: Territory, technology and industrial growth.* Oxford, U.K.: Basil Blackwell.

Taylor, P. J. 1996. Embedded statism and the social sciences: Opening up to new spaces. *Environment and Planning A* 28:1917–28.

Thrift, N., and Olds, K. 1996. Refiguring the economic in economic geography. *Progress in Human Geography* 20:311–37.

Thrift, N., and Walling, D. 2000. Geography in the United Kingdom 1996–2000. *Geographical Journal* 166:96–124.

Tyson, L. D'A. 1993. *Who's bashing whom? Trade conflicts in high-technology industries.* Washington, D.C.: Institute for International Economics.

Vogel, E. F. 1979. *Japan as number one: Lessons for America.* Cambridge, Mass.: Harvard University Press.

———. 1989. *One step ahead in China: Guangdong under reform.* Cambridge, Mass.: Harvard University Press.

Wade, R. 1990. *Governing the market: Economic theory and the role of government in East Asian industrialization.* Princeton, N.J.: Princeton University Press.

———. 2002. U.S. hegemony and the World Bank: The fight over people and ideas. *Review of International Political Economy* 9:215–43.

Wade, R., and Veneroso, F. 1998. The Asian crisis: The high debt model versus the Wall Street-Treasury-IMF complex. *New Left Review* 228:3–23.

Walder, A. G. 1995. Local governments as industrial firms: An organizational analysis of China's transitional economy. *American Journal of Sociology* 101:263–301.

Walker, R., and Storper, M. 1981. Capital and industrial location. *Progress in Human Geography* 5:473–509.

Wei, Y. D. 2000. *Regional development in China: States, globalization, and inequality.* London: Routledge.

Wheatley, P. 1971. *The pivot of the four quarters: A preliminary enquiry into the origins and character of the ancient Chinese city.* Chicago: Aldine.

Whitley, R. 1999. *Divergent capitalisms: The social structuring and change of business systems.* New York: Oxford University Press.

Wills, J., and Lee, R. 1997. Introduction. In *Geographies of economies,* ed. R. Lee and J. Wills, xv–xviii. London: Arnold.

Womack, J. P.; Jones, D. T.; and Roos, D. 1990. *The machines that changed the world.* New York: Rawson Associates.

World Bank 1993. *The East Asian miracle.* Oxford, U.K.: Oxford University Press.

Yeoh, B. S. A., and Chang, T. C. 2001. Globalising Singapore: Debating transnational flows in the city. *Urban Studies* 38:1025–44.

Yeung, H. W.-c. 1994. Critical reviews of geographical perspectives on business organisations and the organisation of production: Towards a network approach. *Progress in Human Geography* 18:460–90.

———. 1997. Business networks and transnational corporations: A study of Hong Kong firms in the ASEAN region. *Economic Geography* 73:1–25.

———. 1998a. The social-spatial constitution of business organisations: A geographical perspective. *Organization* 5:101–28.

———. 1998b. Capital, state and space: Contesting the borderless world. *Transactions of the Institute of British Geographers* 23:291–309.

———. 1998c. *Transnational corporations and business networks: Hong Kong firms in the ASEAN region.* London: Routledge.

———. 1999. Regulating investment abroad? The political economy of the regionalisation of Singaporean firms. *Antipode* 31:245–73.

———. 2000. Organising "the firm" in industrial geography, I: Networks, institutions and regional development. *Progress in Human Geography* 24:301–15.

———. 2001a. Redressing the geographical bias in social science knowledge. *Environment and Planning A* 33:2–9.

———. 2001b. Regulating "the firm" and sociocultural practices in industrial geography, II. *Progress in Human Geography* 25:293–302.

———. 2002a. Doing what kind of economic geography? *Journal of Economic Geography* 2:250–2.

———. 2002b. The limits to globalization theory: A geographical perspective on global economic change. *Economic Geography* 78:285–305.

———. Forthcoming. Practicing new economic geographies: A methodological examination. *Annals of the Association of American Geographers* 93(2).

Young, A. 1995. The tyranny of numbers— Confronting the statistical realities of the East Asian growth experience. *Quarterly Journal of Economics* 110:641–80.

———. 2000. The razor's edge: Distortions and incremental reform in the People's Republic of China. *Quarterly Journal of Economics* 115:1091–1135.

Zhou, Y. 1998. Beyond ethnic enclaves: Location strategies of Chinese producer service firms in Los Angeles. *Economic Geography* 74:228–51.

Zhou, Y., and Tseng, Y.-F. 2001. Regrounding the "Ungrounded Empires": Localization as the geographical catalyst for transnationalism. *Global Networks* 1:131–54.

Part III
Firms, Workers and Places

Part III
Firms, Workers, and Places

[11]

Sunk costs: a framework for economic geography

Gordon L Clark* and Neil Wrigley**

A framework for research on corporate strategy and restructuring in economic geography which focuses explicitly upon the role of sunk costs is proposed. The management of sunk costs in contrasting 'domains of competition' is discussed, and fifteen analytical propositions about the economic and spatial logic of sunk costs are outlined. Claims are advanced that the logic of sunk costs can accommodate both spatial fixity and spatial plasticity and can help bridge the separate, apparently distinctive, notions of restructuring and post-Fordism. Illustrative examples are drawn from both the manufacturing and service sectors. The regional effects of alternative corporate strategies with respect to sunk costs and an assessment of the increasing significance of new financial instruments designed to liberate firms from the history and geography of production are outlined. The conclusion sets out a comparative evaluation of sunk costs and transaction costs as heuristic frameworks in economic geography.

key words   sunk costs   corporate strategy   market entry and exit

*Halford Mackinder Professor of Geography, School of Geography, University of Oxford, Mansfield Road, Oxford OX1 3TB; **Professor, Department of Geography, University of Southampton, Southampton SO17 1BJ

revised manuscript received 5 December 1994

Introduction

In March 1993, the giant US retail company Sears Roebuck announced a previous year's loss of $2.56 billion and plans radically to restructure the corporation.[1] These plans included laying off 50 000 employees, terminating the mail-order catalogue business and closing many retail outlets. The restructuring of Sears will have drastic consequences for the welfare of communities whose businesses have specialized in the provision of services, like the printing of catalogues, to the company. Likewise, the company's plans drastically to down-size their network of retail outlets will affect smaller stores located in poorer suburbs of large metropolitan areas and stores located in declining rural areas of the United States. Whereas Sears' network of stores was once an asset for the corporation, combining market position and a distinctive product range with advantageous locations of individual stores, their network is now a liability. Even a decade ago, many

of these stores could have been sold for their property value. Now, many of these same stores have no real value; they will be abandoned.

In August 1993, the major UK financial institution (previously building society) Abbey National announced the sale of its Cornerstone estate agency chain of 347 outlets for £8 million. The chain had been purchased in 1987 for more than £200 million in an attempt to develop a corporate strategy in which the estate agencies became a distribution network through which a suite of financial products associated with the process of home ownership could be sold at the time of house purchase (Beaverstock et al. 1992). In one of the most humiliating and expensive episodes of collective corporate misjudgement in recent British economic history, Abbey National's losses were mirrored by many other UK building societies/financial institutions. The Prudential insurance company, for example, recovered only £30 million of its outlay of £334 million on the sale of its expensively created network of estate agencies in 1991. Its network had, like that of Sears, become an ever-increasing liability with the potential to damage long-term corporate viability on a scale sufficiently grave for the company to be willing to accept, in preference, the huge costs of market exit.

The Sears, Abbey National and Prudential cases are in some senses merely yet more examples of the patterns and consequences of corporate restructuring which have formed such a significant part of economic geography since the late 1970s. Yet focused as they are on issues of market exit and non-recoverable investment, they serve to direct attention to a topic which has to a large extent been neglected in economic geography and regional analysis: the role of *sunk costs* in corporate decision-making and corporate restructuring. In this paper we propose a framework for research on corporate restructuring in economic geography which focuses explicitly upon the role of sunk costs, where sunk costs are defined as those costs of a firm 'which are irrevocably committed to a particular use, and therefore are not recoverable in case of exit' (Mata 1991, 52).[2] We suggest that the logic of sunk costs offers not only a valuable analytical framework for understanding the spatial patterns of restructuring but also serves as a vitally important bridging concept, linking firm-specific case studies with more abstract notions of spatial fixity and plasticity (cf. Tomaney 1994).

Our presentation begins with a review of some problematic issues in contemporary theoretical debate in economic geography. We advance claims that sunk costs can provide a useful analytical device through which to understand the significance of flexibility, can accommodate notions of both spatial fixity and spatial plasticity and can provide a logic which can bridge the separate, apparently distinctive, conceptions of restructuring and post-Fordism. To this end we provide a description of the nature and characteristics of sunk costs, discuss the management of such costs in two contrasting 'domains of competition' which we term 'structure-dependent' and 'structure-focused',[3] and attempt to interrelate types of sunk costs, characteristics of sunk costs, and market structures. Fifteen analytical propositions that we believe can be made about the economic and spatial logic of sunk costs are identified and debated, before moving on to an examination of how a sunk costs framework can help economic geographers understand why spatial plasticity will be a more significant phenomenon over

the coming decade. The paper concludes with a comparison of the value of sunk costs versus transactions costs frameworks as heuristic devices in economic geography.

Theory and economic geography

Making sense of the vast literature related to the issues of restructuring and post-Fordism is a difficult and complex task. Not only have new explanatory frameworks come and gone at an increasing rate, the powers of critique and evaluation have taken on greater importance compared to the construction of core theoretical claims. The exchange between Gertler (1988, 1989) and Schoenberger (1989) on the meaning and status of flexibility, that between Graham (1990, 1992) and Peet (1992) on the status and stability of concepts such as class and capital in economic geography, and the response of Dicken and Thrift (1992) to Walker (1989) on corporate geography are indicative of this trend towards critical evaluation. Not only are the theoretical skills of economic geographers more developed than in previous generations, the increasing importance of social theory in economic analysis, combined with a realization of the problematic status of theory *qua* theory when separated from close-grained analyses of localities and events, have together created waves of theoretical discourse.

The threads and themes of research evident in the literature reflect this preoccupation with theoretical debate. In the main, those working on industrial restructuring focus either upon the geographical and employment consequences of corporate decision-making (reflecting the importance of Bluestone and Harrison's (1982) work) or the logic of corporate decision-making itself, including reference to the regulation of corporate decision-making (e.g. Clark 1993b; Wrigley 1992b, 1995a; and Christopherson 1993). Here, case studies are the accepted method of analysis. To the extent that these kinds of studies are explicit about their theoretical vantage points, the literature emphasizes the historical and geographical specificity of decision-making, matching in substance a predisposition towards specificity in the corporate-strategy literature (cf. Collis 1991). Implied by specificity is a concern with sunk costs.[4] In this sense, one of the most important research questions to be resolved has to do with the relative importance of spatial fixity compared to spatial plasticity (the status of sunk costs) now and in the near future (cf. Harvey 1982). Rather than attaching great theoretical importance to the end-point of restructuring (reflecting, perhaps, a putative new regime of accumulation), the literature tends to suggest that case studies are indicative of the range of paths of adjustment and experimentation set within more general crises (economic and ideological) of economy and society (see Gibson and Graham 1992).

By comparison, the literature on post-Fordism seems at once abstract and focused upon the end-point of global restructuring. There are, of course, many versions of post-Fordism and some overlap between studies of restructuring and post-Fordism. For example, Florida and Kenney's (1992) paper on restructuring in the US steel industry emphasizes both the spatial fixity of experimentation, reflecting embedded regional and organizational institutions, and their belief that Japanese production systems are the blueprint of a new, post-Fordist industrial world order. In that sense, the interplay between restructuring and post-Fordism is the object of analysis. More generally, however, the literature on post-Fordism presumes that the patterns and processes of restructuring are of less consequence compared to the putative hegemony of the emerging new industrial order. Presumably, if corporate strategy is so affected by sunk costs, experiments not consistent with the economic and organizational imperatives of flexibility sketched by Piore and Sabel (1984) will fail. It often appears as if an evolutionary metaphor justifies indifference to counter-examples (see Alchian (1950) for an early application of the evolutionary metaphor to dominance through economic competition).[5] When combined with the implicit equilibrium logic of the French regulation school, where crisis is a precondition for the destruction of the old order and the emergence and ultimate stability of the new order (see Storper and Scott 1992), arguments for a new industrial order presume a degree of coherence and all-inclusive logic at odds with scepticism about theory (if not theories).

This is the context in which we propose a sunk costs framework for economic geography. In subsequent sections, we show how the framework can help explain the persistence of spatial fixity, why space (the inherited configuration of production) is not simply annihilated by the imperatives of economic competition and why history and geography matter in formulating corporate strategy. An appreciation of the logic of sunk costs can then help in understanding the distinctive spatial patterns of restructuring, how restructuring is designed and implemented. At the same time, we also show how sunk costs can be a useful analytical device through which to understand the significance of flexibility (of capital and strategy) and other corporate strategies. In this sense, the logic of sunk costs can accommodate both spatial fixity and spatial plasticity. It is a logic which can bridge the separate, apparently distinctive conceptions of restructuring and post-Fordism.

In making this argument about the value of a sunk costs framework for economic geography we are conscious of the problematic status of sweeping claims which appear essentialist in appearance or intent. So it is important to be quite explicit about the basis of our argument. How can a sunk costs framework be justified? First, it can be justified on empirical grounds. The management of sunk costs is a very important part of the restructuring process, a claim which may be substantiated across a broad range of sectors and countries. The need efficiently to manage sunk costs has prompted firms to seek ways to liberate themselves from those costs. Flexibility is a common strategic reaction to the actual and potential burdens of sunk costs. Clearly, these empirical arguments are open to dispute and counter-interpretations. They are subject to the same arguments made against post-Fordists, that is, our interpretation of recent events is not exclusive of other rival interpretations. Even so, our framework has one advantage over other related frameworks in that it does not dispute the claimed significance of flexibility or spatial fixity. Rather, it explains how both can coexist at different times and places, in different industries and economies.

In the next section, it will become apparent that the sunk costs framework depends upon certain basic assumptions about the nature of firms' costs and prices. Unfortunately, few case studies of restructuring are explicit about the underlying theoretical logic of costs and prices. Most analysts are content to reference Massey (1984) believing, somehow, that this absolves them from detailing their assumptions about costs and prices. On the other side, those writing about post-Fordism are more explicit about the economic logic of their models though there is a tendency to emphasize the macroeconomic consequences of their arguments rather than their microeconomic foundations. We would suggest that, one way or another, both sets of literature depend upon similar assumptions about costs and prices. This should not be so surprising considering that Marglin (1984) has shown that neoclassical, Keynesian and Marxist theorists share common microeconomic roots in classical economic theory. While there is disagreement about the practical significance of marginal pricing and about how to specify and estimate production functions amongst such theorists, Marglin argues that their basic analytical tools are much the same. They differ significantly, however, on the lessons to be drawn from those models. Are there also significant normative differences between those who focus on restructuring and those who focus on post-Fordism?

These two reasons for a bridging framework will not satisfy sceptics of theory *qua* theory. Perhaps we can never assuage their doubts. Still, accepting that the truth of any one claim about the world is problematic, accepting there cannot be universal agreement about the terms and conditions of truth and accepting the plurality of worlds and ways of world-making does not deny the utility of theoretical frameworks. We just have to accept the inevitable incompleteness of frameworks (Rorty 1992). Following Goodman (1978), we believe that theory-making is a process of construction rather than deconstruction.[6] In this respect, theory-making is doubly contingent; contingent upon the predicament in which we find ourselves and contingent upon the theoretical context within which we work. As best, theory-making is what Nussbaum (1990) would describe as a process of persuasion and argument. At worst, advocates of rival worlds are at war with one another for hegemony. In any event, without theoretical frameworks we are all at the mercy of those who control the identification and representation of events, their causes and consequences (Clark *et al.* 1992). Theory is a blunt instrument used to fashion worlds out of the materials we inherit and co-opt from others (Siebers 1992).

Sunk costs and market structure

Sunk costs

Textbook treatments of production costs normally distinguish between fixed costs and variable costs. Fixed costs are then divided into those costs that would disappear if production were to cease and those 'sunk' costs that would remain even if output was zero (Carlton and Perloff 1990). In general, therefore, sunk costs can be defined as those costs of production that do not vary with output (unlike variable costs) and do not vary directly with scale (unlike fixed costs). Sunk costs normally have no

market value. By definition, they cannot be recovered by selling a part of the plant, capital or equipment to another competitor. Empirically, Mata's (1991) definition of sunk costs is

$$\text{SUNK} = \text{KR}(1 - \text{n ANDEP}) (1 - \text{RESEL})$$

where KR is the initial capital investment, n is the expected number of years that an entrant will be in the market, ANDEP is the annualized depreciated proportion of KR at the time of exit and RESEL is the recoverable portion of that investment.

Three types of sunk costs can usefully be identified. The first can be termed *set-up* sunk costs (initial capital investment); the second, *accumulated* sunk costs (or what Kessides (1991, 29) refers to as 'normal costs of doing business'); and the third, *exit* sunk costs. To illustrate the difference between these types, we may note that while labour costs are normally assumed to be either variable costs (wages) or fixed costs (minimum level of staffing required for a plant to operate), the training of labour for a new production facility could be thought to be a set-up sunk cost, the seniority of labour an accumulated sunk cost and the pension entitlements of labour an exit sunk cost.

In addition to identifying the various types of sunk costs – set-up, accumulated and exit – it is also useful to consider the important characteristics of sunk costs. For purposes of discussion, we highlight four such characteristics: recoverability, transferability, longevity and recurrent financial need. *Recoverability* refers to the likelihood of being able to sell, or in some way retrieve, the market value or some discounted value of the initial investment. A necessary condition for a cost to be sunk is that potential recoverability will be negligible. *Transferability* refers to the potential of shifting the burden of a sunk cost to another agent. A transfer strategy may include public and private issuing agents, and it may also involve bankruptcy. Transferability involves, in essence, shifting the burden so that the financial capacity of the parent corporation is not harmed over the long run. *Longevity* refers to the time horizon over which a sunk cost has use-value. It may be measured against the operating life of a plant or a piece of equipment, or against a firm's operating revenue. Finally, *recurrent financial need* refers to the timing of financial obligations associated with sunk costs. Some sunk costs have low recurrent financial need, just a one-time initial investment. Other kinds of sunk costs require recurrent, non-recoverable funding.

Taking both the types and characteristics of sunk costs into account provides us with a more sensitive understanding of the nature and scope of sunk costs. For example, compare the sunk costs in labour training with labour pensions. Neither may be recovered – in the US at least, skilled labour cannot be sold between firms and firms' pension commitments are administered by third-party trust funds whose conduct is regulated by the government. Labour training can be transferred between firms' production processes, depending upon the quality of training and the overlap between processes, the longevity of skills and the flexibility of work practices and labour contracts. Notwithstanding the longevity of many skills (embodied in human life) relative to the life of capital equipment – and given the recurrent needs for retraining – there may come a point in time where the use-value of such sunk costs is practically exhausted. But unless workers' pension entitlements and inherited environmental obligations can be transferred to another party, the longevity and recurrent financial needs of those obligations may be so significant that keeping the inherited configuration of production intact is more desirable than restructuring. This assumes, of course, that firms remain in control of sufficient market to continue to finance those obligations. Once the regime of prices changes against incumbent firms, corporate strategies may face extraordinary financial pressure from within and without. In the long run, large-scale restructuring may be the only option (Clark 1993a).

Domains of competition

Studies of industry (entry and exit) strategy have noted that firm size is an important variable in determining the significance of sunk costs. For instance, Sleuwaegen and Dehandschutter (1991, 117) show that the minimum efficient scale of an industry (which necessarily involves a significant sunk costs component) can be 'a major entry-impeding factor'.[7] While we accept that the scale of production is important, we also suggest here that firm size cannot be reduced simply to scale. The size of a firm is related to the type of firm and to the range and nature of corporate strategies which it is able to adopt in particular market structures. Elsewhere, Clark (1994) has considered this linkage between corporate strategy and market structure within the context of what he terms 'domains of competition'. In particular he contrasts 'structure-dependent' and 'structure-focused' domains of com-

petition. The *structure-dependent domain* is viewed as a world of small firms in which their strategic options are narrowly circumscribed by the rules of market competition. In this domain, inter-firm price competition dominates and corporate strategy is all about realizing a profit in circumstances which cannot be controlled. By contrast, the *structure-focused domain* is viewed as one in which firms aim at transcending the normal boundaries of competition, whether by cunning, innovation or restructuring. It is a world of large firms, financial capital and hierarchical control of the processes of profit-making. It is our view that, at least initially, the significance and management of sunk costs can best be understood through the lens provided by these contrasting domains of competition.

We begin our analysis by focusing upon small, closely held firms whose goals are simply to minimize costs and maximize profits. Here, at least, there is no need to consider the behavioural consequences of the different interests of owners and managers; principals and agents are one (see Baird and Picker 1991). It is also assumed that these firms operate in a structure-dependent world; the structure or organization of the industry dominates the formation of firms' competitive strategies. This means that the structure of competition is beyond the capacity of individual agents, the inherited configuration of production is a profound constraint upon their strategic options and the market embodies the rules of the game – the set of incentives and disincentives by which all firms must function. Let us also assume that, for a given market selling-price of a common product, each firm arranges its level of output so as to minimize costs and maximize profits.[8] We then assume that there are three types of costs: variable costs, fixed costs and sunk costs. For the moment, let us imagine that variable costs and fixed costs dominate total costs. So if firms share the same production function (the same methods of production) and if the output of firms increases in response to an autonomous increase in market prices, variable costs will increase but fixed costs need not. Thus we can imagine all firms will share in decreasing average costs of production and increasing rates of profit until the time when variable costs and/or fixed costs begin to increase with output. But what if each firm in the industry has a distinctive inherited configuration of production, the result of its particular history and geography? In effect, imagine a heterogeneous mix of capital (production methods and techniques) and imagine that there are significant, non-recoverable set-up costs (initial capital investment) involved in shifting from one method to another. The existence of firm-specific sunk costs will result in a differentiated industry cost structure and differential firm performance over a range of market prices.

To the extent to which there is a distinctive history and geography of the industry, firm-specific performance will result in well-defined spatial patterns of economic activity. Over the long term, however, given the structure of market incentives and disincentives, one method of production (hence one place of production) may come to dominate other methods over a given range of market prices. If firms located outside the core of the industry can afford to restructure their inherited configuration of production, if they can afford to absorb new set-up costs and abandonment costs, they will invest in the dominant method of production. If they can afford to invest in methods of production which are competitively effective (flexible?) over a wider range of market prices (even over a broader range of products) they will do so up to the constraints set by the structure of competition within the industry. If they cannot, the evolutionary metaphor holds that the imperatives of competition will force lower performing firms and places out of the industry.[9]

The structure-dependent world of inter-firm, intra-industry competition provides a first view of the nature and characteristics of sunk costs. Let us now shift the focus of analysis from small, closely held firms to large, management-controlled corporations. We will assume that ownership (shareholding) is so diffuse that agents effectively run the corporation.[10] For the sake of argument, we will also assume that the industry is dominated by a small group of corporations which compete with one another on the basis of price, quality and product differentiation. Their world, unlike the world of small, closely held firms, is oligopsonistic. Making a profit in this context is a more complex problem than in the previous case. At the enterprise level, costs and prices still drive managers' decisions but higher costs per unit produced (compared to contracting-out production) need not be indicative of a lower rate of profit. Higher costs could be associated with higher quality and higher prices, just as higher costs could be associated with a particular product style and market niche. At this level, maximizing profits is, in part, a problem of balancing unit costs against desired market prices given increasing returns to scale and the threat of entry

into the industry of potential competitors. At the corporation level, making a profit depends upon managing company assets efficiently; that is, managing the various businesses and markets that make up the company in ways that add more value than the sum of its parts.

Unlike the representative small, closely held firm, the strategies of representative large management-controlled corporations are not dependent upon market structure. These firms have the resources and market position to be able to formulate and implement strategies aimed at changing their place in the market, even strategies aimed at changing the long-term configuration of the market. What is more, these firms have the resources and expertise (and managerial ambition) to treat the inherited configuration of production of their separate business enterprises as a portfolio of malleable assets and liabilities. For the most part, the competitive strategies of such firms are focused upon structure: the structure of market opportunities and constraints, and the internal structure of their various business activities. By comparison, small, closely held firms confronting sunk costs in the harsh world of price competition face a two-sided dilemma. If operating outside the core of the industry, the set-up costs involved in moving towards the most efficient mode of production over a competitive range of market prices may be so high that they are, in effect, a barrier to long-term competitiveness. On the other hand, those firms located at the core of the industry, unable effectively to control either the market or the actions of others and dominated by marginal pricing, may reasonably suppose that any accumulated sunk costs would threaten their long-term viability. In the domain of structure-dependent firms, the inherited configuration of production may be valuable for deterring entry to the market but is more likely an impossible burden on long-term competitiveness.

For large, management-controlled corporations, the inherited configuration of production can be valuable in the following contexts: where set-up sunk costs deter entry into the market; where accumulated sunk costs represent firm-specific knowledge about production efficiency and market demand; and where, in cases of very large sunk costs, exit costs are so significant that they become an irreversible long-term commitment to the industry. By this logic, spatial fixity is a virtue and a persistent attribute of this type of firm. Of course, it is just as possible that large sunk costs may, in the end, threaten the very future of the corporation. This is what has happened to Sears Roebuck and to many other large manufacturing and service-related firms. Clearly, sunk costs may lead to bankruptcy and encourage strategies designed fundamentally to alter the inherited debt structure and ownership of the firm. It should not be surprising, therefore, that management-led buyouts of enterprises within large conglomerates occur more often in firms and industries characterized by large sunk costs. To the extent they are better able to manage sunk costs, the existence of those costs are an opportunity for owner-managers to extract hidden wealth while discounting variable and fixed costs. Here, flexibility is less important than the virtues of being in place with a distinctive, non-replicable configuration of production.

These general observations about the nature of sunk costs and their relation to market structure serve two purposes. First, they emphasize the close connection which has to be made between market structure and firms' competitive strategy. The decision to accept and maintain sunk costs reflects firms' policies of competitive strategy and the scope of such policies given the structure of competition. Secondly, these observations also emphasize the contingency of sunk costs; that is, the fact that sunk costs can have real advantages for some firms but threaten the existence of other types of firms. To illustrate these two points and provide a worked-through example which bears directly on regional economic analysis we now turn to an example based upon a manufacturing firm which operates in the structure-focused domain. This example is not intended to be exhaustive of all possible examples, nor is it meant to limit subsequent analysis of sunk costs; rather it is illustrative of current corporate practice.

Fifteen propositions about sunk costs

On 26 June 1992, *The Wall Street Journal* reported that BMW were to begin production of Series-3 autos at a new plant in South Carolina. Although the US auto market is seen by industry analysts as increasingly 'treacherous' because of intense competition between US and Japanese auto-makers, a combination of circumstances prompted the new venture. Unfavourable currency exchange rates, the North American Free Trade Agreement and lack of sufficient export-orientated capacity at BMW's European plants left the company without any

reasonable alternative to a new US plant. The site is in a Right-to-Work state, close to other German and French parts and tyre manufacturers. According to the *Journal*, the new plant will be very sophisticated being able to produce a variety of models efficiently at a variety of scales. It will marry flexible techniques of production with flexible modes of work organization. While we could explain the selection of this site by comparing it against the landscape of US labour relations (Clark 1989), just as we could invoke the related literature concerning flexible accumulation and spatial agglomeration (Storper and Scott 1992), here we emphasize the significance of sunk costs in BMW's decision to '[structure] the deal to distance itself from as much risk as it [could]'.

This analysis is embedded in a more general goal; in this section we draw together and identify fifteen basic propositions about sunk costs which inform our analysis. The BMW example and others are used to illustrate these propositions. We identify those propositions which are important across a variety of competitive domains, though it should be acknowledged that the BMW example is probably most relevant to firms in the structure-focused domain. Note also that these propositions are not meant to be exhaustive and no claims are made about the analytical separability or exclusiveness of our statements. Indeed, there is a certain cumulative logic (from entry to exit) inherent in the narrative order in which the first ten propositions are presented. These propositions merely underpin our belief that the management of sunk costs across a variety of competitive domains is a vital component in any explanation of the spatial patterns of restructuring. It should also become apparent that firms like BMW are very conscious of sunk costs in managing their entry into new markets. Entry and exit strategies and the spatial character of sunk costs are closely intertwined.

Proposition 1: sunk costs need not be fixed and fixed costs need not be sunk. As Baumol and Willig (1981, 407) suggest, the distinction between sunk and fixed costs is more than a 'mere terminological quibble'. Although fixed costs are not reduced by decreases in output, they can be eliminated by the total cessation of production and a portion of capital invested can be salvaged (sold or re-deployed). Sunk costs, on the other hand, cannot be eliminated by the cessation of production. This has important consequences for the spatial patterns of industry investment and restructuring. Baumol and Willig

(1981) note, for example, that the airline industry is one in which relatively low sunk costs attach to the operation of particular routes; hence a carrier can reverse its entry into a particular route market relatively cheaply. As a result, the fixed costs associated with a route might exceed its sunk costs. By contrast, new assembly plants in the automobile industry involve huge set-up costs, accumulated sunk costs and exit costs. Cessation of production, the fate of Volkswagen's (VW) assembly plant at Westmoreland (Pennsylvania), may involve huge losses unless those sunk costs can be transferred to a third party.

Proposition 2: for most companies, there are significant set-up costs involved in entering a market, many of which have limited salvage value. As part of its commentary, *The Wall Street Journal* (26 June 1992) noted that BMW had been able to negotiate with South Carolina to have the state 'commit to paying much of the up-front costs'. These costs included purchase of the plant site, provision of physical infrastructure like road access and airport facilities and worker training programmes. If paid for by BMW, it is doubtful that any of these set-up sunk costs could be salvaged by the company. For the state all these investments may have a general benefit to the local economy even if the particular company does not remain in business. The level of benefit depends, of course, on the specificity and durability of sunk costs. As is illustrated by VW's Westmoreland plant, the economic landscape is littered with unused plant and equipment – a burden to state and local governments – because of the spatial and functional specificity of investment. Not surprisingly, BMW's initial investment over the first three years was reported to be modest. But there are countervailing factors to be considered in any decision about the scale of initial investment.

Proposition 3: the need to sink costs can be a significant barrier to entry.[11] As we have noted, firms are neither omniscient nor omnipotent (Clark 1993b). Thus, market-entry decisions are made under uncertainty and conjectural interdependence: entry involves 'conjectures by potential entrants about the quasi rents they can earn by committing resources to a market' (Caves and Porter 1977, 242). If incumbents are structure-focused oligopolists, the quasi-rents conjectured by a potential entrant will depend upon the reaction of incumbents. Here, though, it is also the case that 'those reactions

depend on the incumbents' prior provisions for the contingency of entry, themselves based on conjectures about what will deter entry and how any actual entrants will behave' (*ibid.*, 242). In general, potential entrants and incumbents face different incremental costs and risks. Incumbents have *already* incurred such costs and are *already* exposed to the risks of market participation. However, the potential entrant, as Baumol and Willig (1981) demonstrate, must take a corresponding amount of liquid capital and deliberately turn it into a frozen asset. Therefore, the *incremental* costs of entry include the full amount of sunk costs which are bygone to incumbents. The potential entrant also faces the threat of non-recoverable entry costs as a result of retaliatory strategic or tactical responses by incumbents. Incumbents may overinvest because they may have an interest in raising the entry barrier that sunk costs represent.[12]

Proposition 4: sunk costs as entry-barrier investments increase the risks faced by incumbents and their susceptibility to accidents of history and geography. It is tempting to suppose that sunk costs are a burden only to new entrants to a market. In fact, the specificity and durability of sunk costs may be a burden on incumbents facing the threat of new entrants. The history and geography of the US auto industry could be identified as important factors inhibiting and channelling the strategic response of US manufacturers to new competitors (Clark 1986). Here, history and geography represent the specificity of the domestic industry and its firms (even their inherited spatial configuration of production). By contrast, new entrants like BMW have a chance to learn from the history and geography of an industry and invest in a more efficient spatial configuration. BMW's plans to maximize internal plant-wide flexibility and to out-source from networks of local parts and components suppliers is believed by many to be the most efficient model of the organization of production (see Florida and Kenney 1992). Only BMW (or a similarly 'placed' new entrant with sufficient resources) can take advantage of the history and geography of incumbents' sunk costs. According to this logic, the configuration of others' sunk costs are a benefit to new entrants and a risk to incumbents.

Proposition 5: significant sunk costs can, in certain circumstances, be used as a means to overcome barriers to entry. It should be noted that the need to sink costs will not *always* act as a barrier to entry. There are circumstances discussed by Baumol and Willig (1981, 419) when sunk costs can be used by the potential market entrant in a sufficiently aggressive manner to overcome other barriers to entry. This occurs when an entrant deliberately incurs substantial sunk costs, perhaps substituting them for variable costs to a greater degree than would otherwise be profitable over the short term. In Baumol and Willig's terms, 'the entrant, in effect, chooses to burn his bridges so that he is left with far less to lose by remaining in the field'. In this way, sunk costs can be used to signal to incumbents that the new entrant will be difficult to dislodge. On the other hand, an entry strategy that was too cautious could signal to incumbents that the new competitor may be an easy target of retaliation. Oddly, because BMW's entry strategy focuses upon minimizing the sunk costs of exit should retaliation be impossible to withstand, it invites retaliation because of the risk minimizing structure of the deal. It can be seen, therefore, that for both incumbents and potential entrants, sunk costs play a vital role in the process of competition for market and dominance.

Proposition 6: sunk costs imply commitment-intensive choices. Following Ghemawat (1991), commitment can be defined as the tendency of corporations to persist with particular competitive strategies at lower than average returns. Even if there are other opportunities which offer higher returns, the existence of sunk costs that are specific and durable lock firms in to distinctive paths of accumulation at particular places of production (cf. Storper 1992). Recognition of the irrevocable commitment embodied in actions which involve sunk costs demands, in turn, consideration of the future consequences of current actions. Once the persistence of corporate strategies and the role of sunk costs in corporate commitment are accepted, history and geography matter: past actions in particular places limit present options but not the scope of future consequences of current actions. In a world where the past is the future, where the basic structure of market competition is given, commitment is essential for firms to maintain competitive position. On the other hand, where the future is uncertain a structure-focused strategy is necessary if firms are to overcome the limits on strategic response imposed by sunk costs. There is, then, an uneasy tension between commitment and expected competition. In BMW's case, that tension has been resolved (for the short term)

by limiting sunk costs in order to maximize the range of strategic options to competition.

Proposition 7: the existence of sunk costs suggests the possibility of strong first-mover advantages. Sunk costs are often treated as simply the inherited stock of capital. The implication is that once an entry decision has been made, the firm is saddled with the capital it has brought into the market. But this ignores the fact that, with success, firms tend to add to and specialize their initial capital base, in effect exchanging industry-specific capital for firm-specific capital. Recognizing the dangers with increasing specificity, why should firms add to their sunk costs? Why should they increase the proportion of total costs which are sunk costs? There are a variety of explanations, most of which can be attributed to the informational advantages of being a first-mover in the industry rather than being an outsider contemplating entry into the industry (see Krouse 1990). Being in the industry allows for the accumulation of knowledge which cannot be found on the outside – knowledge of production techniques, market opportunities and the options of other incumbents – which tends to reinforce prior choices. Indeed, incumbents have very strong incentives to make their sunk costs work for them in ways that are not directly replicable by other competitors or potential rivals. The irony is that just as incumbents cannot match BMW's new configuration of production, BMW cannot access incumbents' special knowledge of existing techniques of production.

Proposition 8: the potential transferability, longevity and frequency of financial commitment needed to sustain sunk costs are vital secondary characteristics of sunk costs. By making a deal with the State of South Carolina, BMW has sought to both reduce its liability for the set-up sunk costs associated with the new plant, forestall its responsibility for maintaining those sunk costs and, if necessary, transfer the risks of forced exit to the State. The firm's initial investment has been limited to $400 million over three years. By contrast, the State's set-up costs were $130 million and carried a continuing long-term obligation to fund the maintenance of those investments. In this sense, firms have learnt that entry and exit strategies related to sunk costs necessarily involve secondary strategies of risk management, the significance of which is determined by the transferability, longevity and frequency of financial commitment inherent in the choice of strategy and capital stock. It is clearly the case that for many incumbents in the US auto industry, exit sunk costs are unavoidable even if bankruptcy were to be the final solution. For instance, the longevity and recurrent financial obligations of hazardous waste dumps cannot ordinarily be transferred to other parties or even abandoned without incurring significant financial penalties. Managing the secondary characteristics of sunk costs is so significant because the long-term value of companies is now being calculated with respect to these kinds of costs.

Proposition 9: sunk costs are hardly ever without merit. At issue for all companies, large and small, structure-focused and structure-dependent (and in between), is how best to manage sunk costs given the circumstances of each firm and the advantages and disadvantages of being committed to a certain path of accumulation. We have stressed throughout this paper the differences between structure-dependent and structure-focused firms, emphasizing the threat that accumulated sunk costs pose for structure-dependent firms in their strategic relationships with other similar firms. It is easy enough to understand how and why sunk costs can be of strategic value to firms in the structure-focused competitive domain. It is less clear in the structure-dependent domain. We have noted, however, that one attribute of the structure-dependent domain is its contestability: high rates of entry and exit. But such high rates are possible only if set-up sunk costs are small in relation to other costs and if those costs can be recouped by sale or re-use. Inevitably, in these circumstances, set-up sunk costs are industry-specific, not firm-specific. But it is also clear that if structure-dependent firms are to prosper and thereby shift into more protected markets, industry-specific set-up sunk costs must be transformed into firm-specific sunk costs. The efficient and innovative utilization of set-up sunk costs is a necessary condition for any strategic shift into a different domain of competition. Sunk costs are hardly ever without merit; their value, though, depends on the nature of competition and the circumstances of competitors (see Clark and Wrigley 1995).

Proposition 10: the implications of sunk costs for firms' exit strategies are relatively unexplored. Understanding the nature and characteristics of sunk costs is of vital importance in understanding the design and implementation of exit strategies. However, a considerable amount of the work in industrial economics related to sunk costs has been devoted to analysing firms' entry strategies, not exit strategies. While there is obviously significant interaction between entry and exit strategies, knowing the former is not sufficient to know the latter as 'the exit decision differs from the entry decision in the way the firm treats sunk costs' (Bresnahan and Reiss 1993, 5). By contrast, much of the work in economic geography devoted to corporate restructuring is actually about the interaction between sunk costs – the specificity and durability of the inherited configuration of production – and spatial fixity. Indeed, it might reasonably be argued that the notion of sunk costs is literally the logic of spatial persistence. In this sense, the significance of sunk costs is not limited to questions of corporate strategy. For economic geography and industrial economics, sunk costs (even in their absence) are an essential lens through which to understand the patterns and processes of restructuring.

The previous ten propositions mix basic economic claims with history and geography. In doing so, they provide a means of matching or imitating the underlying logic behind the decision-making of BMW and similar companies. These ten propositions could be interpreted, though, as having described a set of structural (price-driven) behavioural imperatives – imperatives which, if ignored, would threaten the survival of the firm. While plainly important, we do not mean to imply that the sunk cost logic has such independent causal status. We need, in fact, to go further and to make more explicit the strategic interaction between sunk costs and geography. As a result, we next identify a further five propositions that bear directly on the status of geography in corporate strategies of efficiently managing sunk costs.

Proposition 11: sunk costs result in 'zones of inaction' or 'zones of persistence' where firms neither enter nor exit a market, hence a particular level of demand can support several different market structures. Market entry and exit are, by definition, forward-looking decisions. It is clearly the case, therefore, that incumbents may delay exit from a market because of the uncertain costs of re-entry. In addition, it can be anticipated that even major declines in the level of demand in a market will not necessarily trigger exit. Indeed, there is likely to be a considerable range in the levels of demand which will be tolerated before firms will accept the sunk costs implications of

214

Gordon L Clark and Neil Wrigley

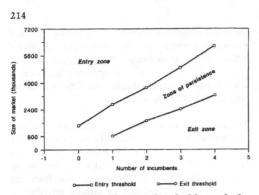

Figure 1. Entry and exit thresholds and the 'zone of persistence'
Source: adapted from Bresnahan and Reiss (1993)

market exit. What this implies, therefore, is that it is impossible to read from a particular level of demand to a given market structure. Rather, sunk costs will result in one level of demand supporting several different market structures and in the existence of considerable 'zones of inaction or persistence'.

In this context, Bresnahan and Reiss (1993) have developed the (somewhat simplified) diagram shown in Figure 1 from an empirical study of location-specific fixed and sunk costs. The upper line connects estimated market-entry thresholds and the lower line exit thresholds. Between the thresholds, fixed and sunk costs preclude entry or exit. By fixing a market size on the horizontal axis it is possible, therefore, to read off the number of market structures consistent with that level of demand and to make sense of apparent contradictions (e.g. why some smaller markets can apparently support far more incumbents than larger markets). More generally, Bresnahan and Reiss (1993) interpret the difference between entry and exit thresholds as measuring the importance of sunk costs. They suggest that sunk costs impede exit from declining markets and they attribute the lumpiness of entry and exit observed in concentrated industries to differences in market dynamics in small and large markets.

Proposition 12: both agglomerated locations and dispersed networks of sources provide ways of managing sunk costs. Ways can be found to minimize set-up and exit sunk costs, perhaps utilizing programmes of local and state development agencies. This is not sufficient, however, to determine the location of a

new market entrant like BMW. By this criterion, there remain many possible sites to locate production. Companies like BMW have also to consider the management of accumulated sunk costs. Recognizing this factor, not all localities are equally attractive. For instance, the company could have located in a large agglomeration, thereby utilizing specialized producer services from firms already located in the region. In effect, given the relatively small size of the company, it could have chosen a location in a large centre specializing in auto-industry services so as to externalize the sunk costs of growth. By so obtaining specialist services on contract through existing firms, it would be neither responsible for the provision of those services nor for the long-term maintenance of favoured suppliers. This kind of strategic decision-making matches in form, if not substance, the claims about the virtues of urban agglomerations made by those who favour models of flexible accumulation (see Storper and Scott 1992) and more conventional models of urban growth (see Krugman 1992).

Whilst locating close to suppliers is one way of externalizing accumulated sunk costs, it is entirely possible that the most efficient management of sunk costs presupposes dispersed networks of suppliers. Instead of locating close to industry-specific suppliers of services, trading upon the actions and investment of previous firms in the industry, some firms could manage sunk costs by sourcing services from firms located at a significant distance from the principal firm. Why would a firm prefer its sources to be at a distant location from the firm? We can think of two reasons. First, the principal firm may use a variety of sources, using its market power to force down the price of individual service contracts. By sourcing services from the immediate urban agglomeration, firms like BMW may simply encourage service providers to collude on prices thereby forcing the firm to bear more of its own sunk costs. Service providers have their own market strategies. Secondly, there are some services that can be best provided at specialized distant locations rather than internally to the firm. For example, if provided internally, financial services may be sunk costs (set-up and accumulated) to the firm, whereas for firms that specialize in providing those services to the industry the risks of capitalization are spread across a large number of clients. In this case, both the service consumer (BMW) and the service provider have an interest in keeping service providers at a distance from the client.

Proposition 13: market competition may encourage new entrants to locate in peripheral regions which allow for the mutual development of firm-specific knowledge. Again, it is important to stress that long-term competitiveness depends, in part, upon the accumulation of firm-specific expertise, not simply industry-specific expertise. While this kind of strategy entails considerable risks, including the possibility of profoundly negative accidents of history and geography (path dependence which ends up at a dead end), it is equally true that average industry performance is not sufficient for long-term growth. So, for a new market entrant without established networks (sites) of suppliers and producers, the inherited industrial landscape offers few immediate opportunities. Even the richest sites (those with long histories of industry specialization) are not necessarily attractive locations for new entrants. In this respect a small city in South Carolina, far from the traditional sites of auto production, may be a very attractive proposition. The fact that this city has already attracted German and European parts producers which have ties with the headquarters of the firm may also be an advantage. At least the existence of these independent firms promises an opportunity to develop firm-specific knowledge in a way that allows for the externalization of accumulated sunk costs *and* for mutual, reinforcing development.

Proposition 14: the spatial centralization of management may be an efficient short-term solution for managing sunk costs but a long-term disaster. It is common to discuss sunk costs in terms of physical assets and liabilities; for example, the particular configuration of a firm's plant and equipment. In small firms, operating in the structure-dependent domain, the efficient management (minimization) of sunk costs is an essential ingredient for success. Thus, centralized control is vital. Lack of centralized control may allow sections of the firm to accumulate sunk costs which threaten the long-term survival (and wealth) of owners (or owner-managers). But in circumstances of growth, where firms move through structure-dependent strategies to the intermediate domain of competition which Clark (1993b) has termed 'structure-limited', it is equally possible that centralized control reduces the capacity of owners to make informed judgements about the value of accumulated sunk costs. Indeed, it is entirely possible that centralized managers are unable to appreciate the distinctive attributes of local situations. In

these circumstances, over the long term centralized managers become merely accountants of others' actions rather than being strategic agents in their own right. Thus, the corporate bureaucracy becomes an accumulated sunk cost. This is a danger that faces BMW (America).

Proposition 15: tendencies towards the spatial differentiation of sunk costs are mediated by the value of spatial homogeneity. There is an argument in the literature to the effect that there is value to be had in firm-specific capital. To the extent that there are local industrial complexes of related activities, it may be also argued that there are shared, strong incentives to develop firm- and region-specific capital. By their combination, local complexes may sustain the paths of accumulation of individual firms. Whilst an attractive proposition for many reasons, the prospect of sunk costs limits the extent to which spatial differentiation is a desirable attribute for any one firm (and even region). In the end, at the point of exit from a market or industry, the value of the firm is, in part, determined by the value of its capital. The resale value of capital is determined by its durability and specificity; the former factor ensures extended resale value while the latter factor limits resale value. The incentives towards spatial differentiation (distinctive paths of accumulation) are conditioned by the incentive towards spatial homogeneity (the resale value of capital). Otherwise, if spatial differentiation overwhelms spatial homogeneity, all capital for all local firms becomes sunk costs – a barrier to exit. In the event of firm failure, realizing that a path of accumulation has become a dead end, the well-being of the entire region would be at risk.

Sunk costs and the year 2000

With respect to our sunk costs framework, what can be said about the future? Can the putative framework provide insights about the year 2000 (a near-term future dependent upon current investment choices)? To answer these questions requires, as a first step, making a set of assumptions about the nature of agents' rationality and foresight. What is remarkable about the last two decades is the extent to which corporate strategists have learnt from their experience. With respect to new competition in US industry, corporate strategies have evolved from accommodation and consolidation around core markets to strategies of spatial capital switching and ultimately to *in situ* restructuring. In doing so, corporations have come to value their history and geography although the inherited configuration of production has also been targeted for massive restructuring. Restructuring, experimentation and joint partnerships have all been closely tied to the nature of firms' sunk costs (see Clark 1993a; and Florida and Kenney 1992). But firms' strategists have also learnt something else: competitive stability is a chimera even in those industries protected by historically significant set-up sunk costs. Strategists have come to expect recurrent rounds of competition and restructuring. Around the year 2000, competitive success will depend upon the ability of firms efficiently to manage their sunk costs in relation to the volatile world of global competition.

Implied by this scenario are four analytical presumptions about the interests and capacities of economic agents (cf. Casetti 1981). First, we presume that economic agents plan the best use of their resources (including investment) with respect to the past and expectations of the future. Secondly, economic agents are, of course, located in distinct competitive domains; their strategic responses are very much related to the circumstances and constraints of those domains. Thirdly, economic agents are neither omnipotent nor omniscient but they are self-conscious and other-regarding; they are able to appreciate the limits of their knowledge and the competence of others. In a world of apparently increasing uncertainty, corporate managers are seeking ways of managing sunk costs to their advantage while minimizing the risks of being overburdened by them if circumstances turn against them. Our fourth presumption is that ruin is to be avoided at all costs. Closure and bankruptcy is more than just the simple re-deployment of resources to a higher end (the unexamined presumption of standard neoclassical theory); we presume that ruin carries with it high risks for owner-managers (realizing their equity) and managers (their reputations and henceforth their potential for re-employment) (see Rose-Ackerman 1991).

With respect to the year 2000, closely held firms in the structure-dependent competitive domain must minimize costs. This is a condition for the success of any long-term strategy which has as its objective a location in a more stable market niche. Assuming that our representative firm is already established, it has an interest in increasing the set-up sunk costs for any new firm entering the market.

Given the considerable risks of taking on sunk costs, however, our firm would prefer that higher set-up sunk costs are imposed on new entrants in the form of prospective regulation. Without the protection of higher set-up sunk costs, firms in this domain must be sure that any accumulated sunk costs do not impede their flexibility in responding to changes in the market. This could mean remaining relatively undercapitalized whilst maximizing the rate of labour exploitation, or investing in plant and equipment that can be used efficiently over a range of potential products and prices, joined with labour practices which can be re-defined as required by owner-managers. As for exit sunk costs, the closely held firm faces a dilemma. To realize the wealth of the business for consumption or investment in another arena requires exit costs to be as small as possible. Moreover, to maximize their long-term wealth, owner-managers must try to avoid any recurrent financial obligations attached to exit sunk costs. However, the firm knows that such costs and recurrent obligations might well deter new entrants into the market.

In effect, if the primary goal of owner-managers is protection of the short-term realizable ('cash-out') value of their firm, there may never be a long term. Owner-managers may choose to locate production in regions which protect them (incumbents) from new competitors, which allow for high rates of labour exploitation thereby limiting the rate of accumulation of sunk costs and which also allow for the abandonment or transfer of exit sunk costs to the public sector. If some regions deliberately, or by default, maintain a policy regime which is relatively attractive for owner-managers whose primary interest is the short-term realization of their wealth, those regions will become havens for a peculiar kind of flexibility. By the year 2000, those regions may well be thought successful in terms of their rates of new firm formation (and destruction). But rates of labour exploitation and capital switching between firms in different sectors would also be very high. Furthermore, rates of private investment may remain low but the public-sector burden of abandoned exit sunk costs may be very significant indeed. In effect, the public sector may become impoverished in the interests of maintaining the wealth of the private sector.

What if owner-managers are more focused upon long-term wealth rather than on protecting their short-term investment? Here, the previously described short-term 'cash-out' strategy may come to be recognized by economic agents as self-defeating. If owner-managers do not accumulate sunk costs that add value to their market position whatever the rate of labour exploitation, there is a risk that their initial investment will be devalued by competitors able to capitalize over the long term on their particular market positions (history and geography). Switching capital between short-term opportunities may become the only opinion left to short-term, wealth-orientated owner-managers. Elsewhere, Clark (1994) details the ways in which investment in particular settings can add value by virtue of the localization of expertise, knowledge of the market and the exploitation of higher value-added market niches. As local paths of capital accumulation become self-reinforcing, they can also become dangerously narrow – on one side promising higher returns, on the other side threatening ruin if the path becomes a dead end. If successful, history and geography become realizable wealth. If unsuccessful, history and geography become accumulated sunk costs which may translate into an exit sunk cost burden with long-term financial obligations.

Recognizing the risks entailed in sunk costs, firms focused upon long-term growth and a place in the intermediate 'structure-limited' competitive domain may join together with other complementary firms to share the burden of those costs. To the extent to which some regions facilitate joint ventures, while also protecting incumbent firms from the rapacious actions of short-term wealth-orientated owner-managed firms located elsewhere, those regions will become havens for cooperative growth firms.[13] By the year 2000, those regions could be characterized by high rates of investment (relative to other regions which began with similar but non-cooperative firms), high rates of product innovation and, perhaps, lower rates of labour exploitation. But there is a danger in this scenario, well-appreciated by firms themselves. If history and geography are an initial market advantage for a particular group of firms that same location could, over time, become a threat to their individual and collective survival. They may invest in the wrong kind of capital and the wrong kind of technology. History and geography may conspire to isolate them from the cutting edge of the market. Other histories and geographies may be integrated into the market, radically altering the comparative advantage of incumbent firms. In that case, the wealth of whole (integrated) regions is at risk. Recognizing these

risks is one thing; having sufficient resources to avoid the risks is quite another.

The preceding analysis utilizes the sunk costs framework to make some fairly elementary points about the timing and logic behind the choice of corporate strategy and the likely regional consequences that flow from such strategies. It tells a story which bridges case studies of restructuring and the post-Fordist vision of a 'flexible' future. But another story could be told, one which begins in a different domain of competition, about structure-focused, management-controlled firms that have the resources to accommodate the risks of sunk costs. This story is more empirical and of the moment than an exercise in logic and generality. It depends upon our knowledge of recent innovations in financial services, particularly the arbitraging of risks associated with capital investment by financial institutions. The story is premised upon an assertion: that one of the profound lessons of the last two decades has been the realization that plant and equipment can, in practice, be just as easily viewed as sunk costs as they are normally viewed as assets. The risk is that as the global economy evolves through to the year 2000, firms will find that their accumulated plant and equipment are not appropriate to their needs. Not only may firms find their plant and equipment relatively inefficient, they may also find that they have no positive market value, only future financial obligations. So how are these risks to be accommodated?

Imagine a major international airline (cf. Baumol *et al.* 1988). To be competitive it must have the most efficient and the most modern aircraft. New Boeing 747s cost a fortune. In present circumstances few international airlines can afford to purchase them outright and some countries face severe short-term balance-of-payments problems when their airlines purchase new 747s. Not only is uncertainty about future gasoline prices a major consideration, the style, comfort and amenities of aircraft are important if the airline is to attract an increasing share of business passengers. But if an airline was to exercise its option to buy a new Boeing 747, there is a real prospect that the aircraft will be less than competitive on these criteria within five or six years, a time frame much shorter than the time it would take to recover its purchase price (say ten years). While there may always be a market for a standard Boeing 747 whatever its age (up to say 25 years), the more distinctive and differentiated the amenities of its purchased aircraft (necessary for enhanced competitiveness on major international routes) the more likely the corporation will be unable to realize its initial investment if forced to sell the aircraft before ten years. Moreover, the airline runs the risk of having to accept the non-recoverable expenses of retrofitting (standardizing) the airplane before being able to sell it on the open market. To purchase a new Boeing 747 is to accept all three kinds of sunk costs: set-up costs (for new entrants, the maintenance facilities), accumulated sunk costs (the costs of differentiation) and exit costs (the costs of disposing of the aircraft).

The solution is to lease the new airplane. But someone (a company) has to buy it to lease it to the airline. Few companies can afford to carry the full purchase price especially if the airline requires more than one airplane. Given the risks associated with buying the plane (the same risks the airline was unwilling to accept) no single company can afford to carry the purchase price of airplanes outfitted to the airline's particular specifications. So, instead of a single company purchasing the plane and leasing it to the airline, the solution is for a consortium of finance companies to purchase the plane and lease it to the airline for ten years. However, since the airline is reluctant to lease the plane for more than five years and needs to upgrade its fleet on a regular basis, the lease agreement is likely to allow for replacement of the airplane with another new Boeing 747 at the end of the fifth year. There are, of course, extra transaction costs involved in this arrangement which must be borne by the airline. The rollover clause means, however, that by the tenth year the consortium could own two Boeing 747s. In this respect, the consortium faces the same risk identified by the airline: an unknown future market for specialized but older Boeing 747s. In this case, however, the risk will be dispersed amongst the consortium partners; being one small risk amongst many other risks paying an agreed 15 per cent return per year per partner. To guard against the ten-year risk being held against the consortium partners and to guard against any liability if the plane crashes, each partner may buy shares in a special-purpose holding company located in the Cayman Islands. If it is unable to sell both airplanes after ten years, the holding company will simply be liquidated; in this way the costs of bankruptcy proceedings in the US or elsewhere will be avoided by the consortium partners.

Actually, the consortium partners will aim to protect the value of their assets so as to maximize

the resale price of the airplanes. all partners will want to share in the potential 'cash throw-off' that comes with disposing of an airplane. To do so, however, brings them up against the classic principal-agent problem: the principal (the holding company) has different interests from the agent (the airline).[14] Having been freed from the burden of financing the purchase of their new airplane, the airline will aim to maximize its use of the airplane and minimize its running costs in order to maximize its profits. Recognizing the different interests of the parties, the holding company might respond by setting stringent use and maintenance standards as a way of protecting the long-term value of its asset. But there are many opportunities for the airline to subvert standards. Happily, the airline's reluctance to take on the accumulated sunk costs of higher-quality maintenance facilities can provide both parties with an agreeable solution. The consortium partners will bring more specialized companies into the holding company, including the jet engine manufacturer and the audiovisual electronics manufacturer. In this way, the consortium partners will actually have sold parts of the airplane back to manufacturers and made them responsible (in conjunction with the airline) for ensuring the highest quality maintenance of the asset.

This and other similar deals in the industry are priced (the yearly leasing charge) in accordance with the holding partners' assessment of the anticipated performance of the airline. Thus, an airline with a reputation for high-quality performance would pay a lower-risk premium than an airline with a poorer reputation. Likewise, an airline which, for competitive purposes, requires a distinctive and unusual Boeing 747 will pay a higher-risk premium (against the expected disposal price of the airplane) than an airline willing to accept a standard aircraft. Similarly, there are competitive risks to be accounted for by the airline in making these deals. All these risks (and their risk premiums) are offset against the expected revenue of the stylized airplane in particular markets *and* the expected volatility of global markets at the conclusion of the life of the deal.

The implications of this study, involving as it does separation of the ownership of capital from its use, are many-fold. Here, we want to emphasize three points. First, the goal of plasticizing, making malleable, the inherited configuration of production which the story implies is, in effect, a dual goal of increasing the rate of capital turnover and spatial turnover. The goal of management is their liberation from the history and geography of production. Secondly, this goal matches, in effect, the perceived long-term volatility of global markets and prices. It is as if corporate managers have taken the lessons of restructuring during the last decade and have planned to be completely unencumbered for another new round of global economic restructuring in the near future. Whereas liberation from capital was shown to be an expensive proposition over the last decade, by the year 2000 corporate managers hope that liberation will not be their problem but that of whoever finally holds the ownership of redundant capital. Whether or not this is actually a plausible goal, given the specificity of markets and the advantages of differentiated capital for competitive strategy, remains to be seen (cf. Ghemawat 1991). What if many firms write these kinds of deals? What if these deals are close in time and space and all come due about the year 2000? Our third point is that in the near future there may be enormous volatility as redundant capital swamps capital markets and product markets. In their haste to rid themselves of the potential burdens of sunk costs, corporate strategists may be simply adding more productive capacity in the near future (at much reduced prices) to market segments (most likely in the structure-dependent domain) already characterized by significant competition.

Conclusions

Whether focused upon case studies of restructuring or upon the logic of post-Fordism, it is clear from the literature that many writers believe western countries are caught in economic crisis. There are reasons to doubt the generality of this crisis (is it one crisis or a set of different but overlapping crises?), just as there are reasons to be cautious about the proximate and long-term causes of the crisis (how important was the Bretton Woods agreement and its subsequent collapse for global monetary stability?).[15] Still there can be no doubt that, compared to the economic performance of southeast Asia and the Asian newly industrialized countries (NICs), the past decade has seen many western industrialized countries stagnate and some decline in real terms (Clark and Kim 1995).[16] Given the volatility of global currency markets and the unknown consequences of new supra-regional trading blocs, just two facets of the political economy of international trade, there is great

unease about the short-term economic prospects of regions let alone their long-term prospects.

In this context, it is important for economic geography to develop tools of analysis amenable and applicable to the emerging global economy. The standard price-theoretic models of location and market competition presume that prices (commodity prices, factor prices, wages, etc.) determine the shape and evolution of the landscape (see Lloyd and Dicken 1977). Over the last decade, many economic geographers have moved away from the narrowness implied by this approach by adding a concern for the spatial differentiation and historical roots of particular places as opposed to ideal landscapes. Even so, for some analysts there remains an unresolved tension between the imperatives of price theory and the texture of localities (see Scott 1988). And yet, for those working at the interface between analysis and policy, price-theoretic models have many advantages, not least of which is the generality of the approach. According to Easterbrook (1992), price-theoretic models provide general rules that facilitate public decision-making, especially the efficiency (timing, costs of litigation, etc.) of judicial decision-making. By his analysis, price theory as *per se* rules enable judges to go beyond the specifics of cases and eschew disputes of interpretation about the causes and consequences of those cases. By extension, the interpretive efficiency of theory seems to be at the heart of Harvey's (1989) dismay about the consequences of post-modernism.

Notwithstanding the fifteen propositions about the nature of sunk costs, our ability to generate price-theoretic *per se* rules for (academic and judicial) decision-making seems remote. One reason to be sceptical of such rules applied to specific cases is to be found in the very practice of applying them. Standards, conventions and methods of linking rules with circumstances are subject to as much dispute and disagreement as are competing interpretations of those cases.[17] But it is also true that our assumptions about the rationality of economic agents and their capacity for self-consciousness and other-regarding cuts against the power of general rules. If it is known that accumulating sunk costs in the structure-dependent domain is a real threat to firms' long-term strategic flexibility, we should expect those firms 'to seek whatever means they can to free themselves from it [the competitive process] or, at least, to reduce the harshness of its regime' (Baumol and Ordover 1992, 87). Once we recognize the scope of self-consciousness and other-regarding, then we should also recognize that price-theoretic models of behaviour are not determinate in the manner assumed by economic structuralists; corporate strategy (agency) is essential for realizing the specific manifestations of general economic imperatives. By virtue of the market opportunities available to economic agents who are similarly aware of the indeterminacy of structure, the gap between *per se* rules and specific cases is part of the very fabric of competition. Opportunism and games of strategic interaction are at the heart of this world.

What then is the status of a sunk cost approach compared to other, recent developments in economic theory? One response to this question would be to suggest that the sunk cost framework is an heuristic device, perhaps similar to Williamson's (1985) transaction cost economics. In summary terms, the transaction cost framework supposes that

transactions (which differ in their attributes) are aligned with governance structures (which differ in their costs and competencies) in a discriminating (mainly, transaction cost economising) way. (Williamson 1992, 148)

By analogy it could be said that firms economize on their sunk costs, managing those costs with respect to their relative value (relative to potential competitors). Both frameworks presume (transaction and sunk) costs are inevitable (even in their absence), both presume that the proper focus of analysis is the firm and its particular circumstances and both presume there are a range of options open to firms in how they respond to those costs; indeed, neither approach proscribes the limits of strategy. In our case, however, we would note that the range of options available to firms in managing their sunk costs is closely related to the competitive domain they occupy or would wish to occupy.

Notwithstanding their shared analytical roots in classical price theory there are, however, some significant differences between these two frameworks which may resist accommodation. Most importantly, the sunk costs framework has a clearer and more contained empirical agenda. The three types of sunk costs (set-up, accumulated and exit) coupled with their four general characteristics (recoverability, transferability, longevity and recurrent financial need) provide a ready recipe for empirical analysis across very different settings. The transaction costs framework, while pregnant with implications for empirical study, seems uncomfortably unbounded. Even if the whole world is governed by transaction costs there are few guidelines

in the literature that can help distinguish between particular transaction costs and their relative status compared to other kinds of costs. Another important difference is the fact that sunk costs are closely allied with other related costs of production, allowing us to see how and why the management of sunk costs is related to the management of variable and fixed costs. This allows us to bridge explanations of spatial structure that emphasize variable costs (wages, for example) with explanations which stress the long-term flexibility of production.

Acknowledgements

A version of this paper was presented at the Annual Meeting of the Association of American Geographers, Atlanta, Georgia, 7–9 April 1993. Thanks to those who participated in the special session on corporate strategy and spatial structure for comments and advice. Kathie Gibson, Kevin O'Connor, Rod Francis and Erica Schoenberger also provided useful comments on a previous draft, and William Baumol provided encouraging comments on the overall project. None of the above should be held responsible for the final product.

Notes

1. See *Business Week* (15 March 1993). About a month later *The Wall Street Journal* (9 April 1993) reported that Sears plans to spend $4 billion to 'transform its struggling general merchandise retailer into a moderately priced department store chain'.

2. This definition, like Baumol *et al.*'s (1988), stresses the non-recoverable nature of sunk costs even if production were to cease and the production unit were closed. Most studies of sunk costs focus upon the interaction between entry and exit strategies, arguing that sunk costs are a significant barrier to entry in some industries and hence a means of guarding against competition. There are very few studies of the role of sunk costs in determining the patterns of restructuring (but see Wrigley 1992a, 1995b; and Clark 1994).

3. These two competitive domains (and a third, intermediate domain) are described and analysed in more detail by Clark (1994).

4. An early empirical demonstration of this point can be found in Caves and Porter (1976). While they do not mention sunk costs explicitly, their work is devoted to explicating the effect that durable and specific assets (what they term as 'DSAs') have on firms' exit strategies and their persistence in an industry at low levels of profitability. Generally,

they note that DSAs 'may be specific to the particular business or productive activity, to the company employing them, to the physical location, or to any combination of these' (p. 40).

5. There is considerable interest in the rhetoric of theory-making in the social sciences, including geography. For instance, Barnes (1992) provides a critical assessment of the role and significance of metaphor in economic geography. Here, we do not mean to imply criticism of the use of metaphor. In any narrative analysis of the economy, metaphor will have an important role for representing complex ideas. For now, at least, by identifying the metaphorical device used to represent the dynamics of competition we are simply noting the underlying reference point for the analysis.

6. Putnam (1992) describes Goodman as an irrealist. By Goodman's account, as there are many worlds, there are many ways of describing the world. The plausibility of worlds cannot be adjudicated (ordered in terms of their value) by reference to a 'uniquely true description of reality' (p. 110). Not all worlds need be incompatible but some may be so.

7. In a related vein, Kessides (1991, 29) notes 'the act of entry requires the conversion of liquid assets into frozen physical capital, only part of which may be recoverable through disinvestment. For the incumbent, however, these commitments have either already been made (initial capital investment) or they constitute a normal cost of doing business (advertising). Thus, the entrant's incremental cost includes the full amount of the sunk costs which are largely bygone to the incumbent'.

8. This is a conventional, simplifying assumption we make regarding firms' competitive decision-making. It reflects our belief about the significance of market incentives in structuring decision-making. Of course, in a less constrained world, firms may also treat their selling price as a variable price set in accordance with their optimal level of output. The consequences of such actions are noted in Hay and Morris (1991).

9. There are, however, more complex stories which could be told about evolution and competition. Rather than imaging a world in which it is competition for single-species dominance, firms and places could co-evolve in a reciprocal manner, even in a parasitic manner whereby lower performing firms adjust to new competition by sheltering under the minor opportunities provided by dominant firms. See Thompson (1982) for a detailed analysis of the logic and character of coevolution.

10. Elsewhere, Clark (1994) expands the argument to include instances where ownership rivals management for control of large corporations.

11. A barrier to market entry can be defined as *any factor that decreases the likelihood, scope, or speed of entry when firms in the market are exercising market power* and it should be noted that it is this definition which is currently utilized by the US Department of Justice (Rill 1990). Within the context of this definition it is accepted that entry barriers may arise from *costs that are sunk (non-recoverable) if entry fails*.

12. Examples of such entry-discouraging investment noted by Caves and Porter (1977) include excess production capacity (making credible the threat of price warfare against the entrant); product differentiation by advertising, product-oriented R and D, proliferation of brands, etc. to reduce the cross-elasticity of demand between the products of the incumbent and the entrant; and upward shifting of the production cost function. However, investment in the raising of entry barriers is not without risks for the incumbent. It may increase the probability of ruinous loss by increasing the fixity of costs or lowering the average salvage value of the firm's assets.

13. We do not mean to imply that regions ought to protect their firms from the rapacious actions of other firms located elsewhere in the economy (national or global). Whatever the moral significance of these actions, we simply observe that such regional strategies are common and part of the never-ceasing competition for employment investment between jurisdictions. Like Baumol and Ordover (1992, 87), we believe that '[t]he competitive process is one that is not calculated to be pleasant to the firms [or communities] that are subject to its pressures'.

14. Pratt and Zeckhauser (1985, 2–3) describe the problem quite succinctly: '[t]he challenge in the agency relationship arises whenever – which is almost always – the principal cannot perfectly and costlessly monitor the agent's action and information'. They argue that this is a 'pervasive' problem of corporate governance, one that has had profound implications for the organization of business.

15. One of the problems of discussing the crisis (or crises) of western economies is establishing a stable and common reference point. If there are crises, what is the implied or explicit point (some time in the past) of comparison? For most, if not all commentators, an often-cited reference point has been the immediate postwar international trading system known generally as the Bretton Woods agreement. But recent research on the Bretton Woods system has raised doubts about its timing and its coherence and stability. Reporting on the proceedings of a conference on the system, Bordo (1992/3, 8), suggests that the 'heyday of Bretton Woods was from 1959 to 1967'. If this was the point of stability and coherence against which to

compare the present world, it was a very narrow point indeed.

16. See Goldman Sachs (1993) for a detailed comparison of the economic performance of the EU nations, the US and Japan.

17. See Clark (1993b, ch. 7) on Easterbrook's theory of statutory interpretation. With respect to US plant-closing legislation, it is shown that Easterbrook's standards of 'reasonable' and 'competent' allow for a wider variety of interpretive practice that he would seem to accept.

References

Alchian A 1950 Uncertainty, evolution and economic theory *Journal of Political Economy* 58 211–21

Baird D and Picker R 1991 A simple noncooperative bargaining model of corporate reorganisations *Journal of Legal Studies* 20 311–50

Barnes T 1992 Reading the texts of economic geography in **Barnes T and Duncan J** eds *Writing worlds: discourse, text and metaphor in the representation of the landscape* Routledge, London

Baumol W and Ordover P 1992 Antitrust: source of dynamic and static inefficiencies in **Jorde T and Teece D** eds *Antitrust, innovation, and competitiveness* Oxford University Press, New York 82–97

Baumol W Panzor J and Willig R 1988 *Contestable markets and the theory of industry structure* 2nd edn Harcourt Brace Jovanovich, New York

Baumol W and Willig R 1981 Fixed costs, sunk costs, entry barriers, and sustainability of monopoly *Quarterly Journal of Economics* 95 405–31

Beaverstock J Leyshon A Rutherford T Thrift N and Williams P 1992 Moving houses: the geographical reorganization of the estate agency industry in England and Wales *Transactions of the Institute of British Geographers* NS 17 166–82

Bluestone B and Harrison B 1982 *The deindustrialization of America* Basic Books, New York

Bordo M 1992/3 The Bretton Woods international monetary system: a lesson for today. *NBER Reporter* winter 1992/3 National Bureau of Economic Research, Cambridge, MA

Bresnahan T F and Reiss P C 1993 Measuring the importance of sunk costs Working paper, Stanford Centre for Economic Policy Research, Stanford University

Carlton D and Perloff J 1990 *Modern industrial organisation.* Scott, Foresman and Co, Glenview, IL

Casetti E 1981 A catastrophe model of regional dynamics *Annals of the Association of American Geographers* 71 572–9

Caves R and Porter M 1976 Barriers to exit in **Masson R and Qualls P** eds *Essays on industrial organisation in honor of Joe S Bain* Ballinger, Cambridge, MA 36–69

Caves R and Porter M 1977 From entry barriers to mobility barriers: conjectual decisions and contrived deterrence to new competition *Quarterly Journal of Economics* 91 241–61

Christopherson S 1993 Market rules and territorial outcomes: the case of the United States *International Journal of Urban and Regional Research* 17 274–88

Clark G L 1986 The crisis of the midwest auto industry in **Scott A and Storper M** eds *Production, work, territory: the geographical anatomy of industrial capitalism* Allen and Unwin, London

Clark G L 1989 *Unions and communities under siege: American communities and the crisis of organized labour* Cambridge University Press, Cambridge

Clark G L 1993a Costs and prices, corporate competitive strategies and regions *Environment and Planning A* 25 5–26

Clark G L 1993b *Pensions and corporate restructuring in American industry: a crisis of regulation* Johns Hopkins University Press, Baltimore

Clark G L 1994 Strategy and structure: corporate restructuring and the scope and characteristics of sunk costs *Environment and Planning A* 25 9–32

Clark G L and Kim W B eds 1995 *Asian NIEs in the global economy* Johns Hopkins University Press, Baltimore

Clark G L McKay J Missen G and Webber M 1992 Objections to economic restructuring and the strategies of coercion: analytical evaluation of policies and practices in Australia and the United States *Economic Geography* 68 43–59

Clark G L and Wrigley N 1995 The rudiments of corporate geography: sunk costs and the spatial structure of production Paper presented at the Annual Conference of the Institute of British Geographers, Newcastle, 5 January 1995

Collis D 1991 Corporate strategy: a research agenda (mimeo) Harvard Business School, Boston

Dicken P and Thrift N 1992 The organisation of production and the production of organization: why business enterprises matter in the study of geographical industrialization *Transactions of the Institute of British Geographers* NS 17 279–91

Easterbrook F 1992 Ignorance and antitrust in **Jorde T and Teece D** eds *Antitrust, innovation, and competitiveness* Oxford University Press, New York 119–36

Florida R and Kenney M 1992 Restructuring in place: Japanese investment, production organisation, and the geography of steel *Economic Geography* 68 146–73

Gertler M 1988 The limits to flexibility: comments on the post-Fordist version of productivity and its geography *Transactions of the Institute of British Geographers* NS 13 419–32

Gertler M 1989 Resurrecting flexibility? A reply to Schoenberger *Transactions of the Institute of British Geographers* NS 14 109–12

Ghemawat P 1991 *Commitment: the dynamics of strategy* Free Press, New York

Gibson K and Graham J 1992 Rethinking class in industrial geography: creating a space for an alternative politics of class *Economic Geography* 68 109–27

Goldman Sachs 1993 *Portfolio strategy: world investment highlights* Goldman Sachs International Limited, London

Goodman N 1978 *Ways of worldmaking* Hackett Publishing Co, Indianapolis

Graham J 1990 Theory and essentialism in marxist geography *Antipode* 22 53–66

Graham J 1992 Anti-essentialism and over-determination: a response to Dick Peet *Antipode* 24 141–56

Harvey D 1982 *The limits to capital* University of Chicago Press, Chicago

Harvey D 1989 *The condition of postmodernity* Blackwell, Oxford ch. 17

Hay D A and Morris D J 1991 *Industrial economics and organisation* Oxford University Press, Oxford ch. 1

Kessides I N 1991 Entry and market contestability: the evidence from the United States in **Geroski P A and Schwalbach J** eds *Entry and market contestability: an international comparison* Blackwell, Oxford 23–48

Krouse C 1990 *Theory of industry economics* Blackwell, Cambridge, MA ch. 10

Krugman P 1992 First nature, second nature, and metropolitan location (mimeo) Department of Economics, MIT, Cambridge, MA

Lloyd P and Dicken P 1977 *Location in space* 2nd edn Harper & Row, New York

Marglin S 1984 *Growth, distribution, and prices* Harvard University Press, Cambridge, MA

Massey D 1984 *Spatial divisions of labour: social structures and the geography of production* Macmillan, London

Mata J 1991 Sunk costs and entry by small and large plants in **Geroski P A and Schwalbach J** eds *Entry and market contestability: an international comparison* Blackwell, Oxford 49–62

Nussbaum M 1990 *Love's knowledge: essays on philosophy and literature* Oxford University Press, New York

Peet D 1992 Some critical questions for anti-essentialism *Antipode* 24 113–30

Piore M and Sabel C 1984 *The second industrial divide* Basic Books, New York

Pratt J and Zeckhauser R 1985 Overview in **Pratt J and Zeckhauser R** eds *Principals and agents: the structure of business* Harvard Business School, Boston 1–35

Putnam H 1992 *Renewing philosophy* Harvard University Press, Cambridge, MA

Rill J J 1990 Sixty minutes with Honorable James F Rill, Assistant Attorney General, Antitrust Division, US Department of Justice *Antitrust Law Journal* 5(9) 45

Rorty R 1992 The pragmatist's progress in **Collini S** ed. *Interpretation and over-interpretation* Cambridge University Press, Cambridge 89–108

Rose-Ackerman S 1991 Risk taking and ruin: bankruptcy and investment choice *Journal of Legal Studies* 20 277–310

Schoenberger E 1989 Thinking about flexibility: a response to Gertler *Transactions of the Institute of British Geographers* NS 14 98–108

Scott A 1988 *Metropolis: from the division of labor to urban form* University of California Press, Berkeley and Los Angeles ch. 11

Siebers T 1992 *Morals and stories* Columbia University Press, New York

Sleuwaegen L and Dehandschutter W 1991 Entry and exit in Belgian manufacturing in **Geroski P A and Schwalbach J** eds *Entry and market contestability: an international comparison* Blackwell, Oxford 111–20

Storper M 1992 The limits to globalisation: technology districts and international trade *Economic Geography* 68 60–93

Storper M and Scott A eds 1992 *Pathways to industrialization and regional development* Routledge, New York and London

Thompson J N 1982 *Interaction and coevolution* Wiley, New York

Tomaney J 1994 Alternative approaches to restructuring in traditional industrial regions: the case of the Maritime sector *Regional Studies* 28 543–50

Walker R 1989 A requiem for corporate geography: new directions in industrial organisation, the production of place and uneven development *Geografiska Annaler B* 71 43–68

Williamson O 1985 *The economic institutions of capitalism* Free Press, New York

Williamson O 1992 Antitrust lenses and the uses of transaction cost economics reasoning in **Jorde T and Teece D** eds *Antitrust, innovation, and competitiveness* Oxford University Press, New York

Wrigley N 1992a Sunk capital, the property crisis, and the restructuring of British food retailing *Environment and Planning A* 1521–7

Wrigley N 1992b Antitrust regulation and the restructuring of grocery retailing in Britain and the USA *Environment and Planning A* 24 727–49

Wrigley N 1995a Retailing and the arbitrage economy: market structures, regulatory frameworks, investment regimes and spatial outcomes in **Barnes T and Gertler M** eds *Regions, institutions and technology* MacGill-Queens University Press, Toronto

Wrigley N 1995b Sunk costs and corporate restructuring: British food retailing and the property crisis in **Wrigley N and Lowe M S** eds *Retailing, consumption and capital: towards a new retail geography* Longman, London

[12]

Firms in Territories: A Relational Perspective*

Peter Dicken

*School of Geography, University of Manchester,
Manchester M13 9PL, U.K.
p.dicken@man.ac.uk*

Anders Malmberg

*Department of Social and Economic Geography,
Uppsala University, S-751 20 Uppsala, Sweden
anders.malmberg@kultgeog.uu.se*

Abstract: The role of space and place in shaping the transformation of firms and industries and the impact of such transformations on the wider processes of territorial development at local, regional, national, and global scales are basic research issues in economic geography. Such analyses tend to be compartmentalized, focusing on a specific economic activity or on a specific territory, rather than on the relationships between them. It is difficult simultaneously to conceptualize economic activities (including such phenomena as firms, industries, and other types of systems of networked economic activity), on the one hand, and territorially defined economies, on the other. In this paper, we address the interconnections between economic activities and territories through an exploration of the mutually constitutive relationships between firms and territories: the firm-territory nexus. The focus of our analysis is the nexus of three major dimensions—firms, industrial systems, and territories—embedded in turn in the overall macro dimension of governance systems.

Key Words: firms, territory, industries, regions, clusters, economic development.

Much economic geography research asks questions about the role of space and place in shaping the transformation of economic activity systems and about the impact of such transformations on the wider processes of territorial development at local, regional, national, and global scales. Typically, however, such analyses tend to be compartmentalized, focusing either on a specific activity system or on a

* This paper began its life in a series of discussions during the spring of 1999, when the authors were both fellows at the Swedish Collegium for Advanced Study in the Social Sciences (SCASSS) in Uppsala. We would like to thank the directors of SCASSS, as well as our co-fellows, for providing such a stimulating intellectual environment. In addition to valuable referee comments, we acknowledge constructive suggestions from Dominic Power, Bjørn Asheim, and Michael Taylor.

specific territory, rather than on the relationships between them. It is extraordinarily difficult to conceptualize economic activity systems and territorially defined economies simultaneously.

The starting point for analysis of an economic activity system may be the *individual business organization (the firm)*—which can be anything from a small-scale, family-owned business to a giant transnational corporation (TNC) operating multiple lines of business in various parts of the world. Alternatively, the starting point may be some notion of a *system of similar and related firms*, such as the industry, the production chain, or the cluster. In either case, territory may enter the picture as geographic distance, as the spatial distribution of location factors, or as a notion of territorial milieu. Thus, firms may be deeply rooted in one particular place or they may be more or less mobile, while systems of firms may be agglomerated (i.e., co-located within one narrowly defined geographic area) or more globally dispersed.

In the analysis of a *territorial economy*, on the other hand, the point of departure is some bounded segment of geographic space, such that the territorial economy comprises all the economic activities taking place within its geographic boundaries. Such territorial economies may be based on a few very large economic operations or on a large number of small activities. The activities in the territory may comprise a set of strongly interrelated economic activities, known as a localized cluster, or they may be made up of firms operating in a number of nonrelated segments of the economy. Furthermore, the firms in a territory may be "inward-looking," in the sense that they source a large share of their inputs from within the region and/or find the outlet for their goods or services within the same territory. More often, though, firms will be linked, to a greater or lesser extent, to the outside world, with the majority of their suppliers and customers being located in other places.[1] Finally, the firms in a specific territory may be largely locally owned and operated or they may be nonlocal, in the sense that they are owned by, and integrated with, larger business organizations whose "home bases" lie outside. The latter may be active in many different territories—in the extreme case all around the world. Any given territory, therefore, has a distinctive firm ecology.

In this paper, we address this issue of the interconnections between economic activities and territories through an exploration of one aspect: the mutually constitutive relationships between firms and territories, or the *firm-territory nexus*.

[1] As a general rule, of course, the degree of inwardness will be positively correlated with the (economic) size of the territory: the larger the territorial economy, the more "self-sufficient" it will tend to be.

Conceptualizing Firm-Territory Relationships

Although the firm-territory relationship constitutes a long-standing economic geography focus, it is, to use Markusen's (1999) terminology, a "fuzzy concept." Such relationships need clearer articulation and understanding. There has been a resurgence of interest by a diverse academic and policy community in the region as a focus of economic activity and as a key component of national economic competitiveness. Within what is now a vast literature (new and old) on the regional economy, little attention has been paid to the precise nature of that relationship. There are, of course, studies that focus, for example, on small firms as key players in regional economies (usually seen as positive in their effects) or on the role of branch plants of multilocational firms, both domestic and, especially, foreign (usually seen as negative in their effects). But much of this literature tends to restate rather tired conventional positions, so that the relationships between firms and territories are weakly conceptualized.

We need, therefore, a better understanding of how firms are being organized and reorganized; how internal and external power structures are configured and reconfigured; how business strategies are developed and implemented, as part of the dynamics of the wider industrial systems of which firms are a part; and how each of these dimensions are "territorialized." This involves recognizing the nature of firms not only as legally bounded entities and owners of proprietary assets (both tangible and intangible) but also as institutions with permeable and highly blurred boundaries—in other words, conceptualizing them as "networks within networks" or "systems within systems." In turn, this exercise also requires us to conceptualize more clearly what is meant by "territory," a term that we, along with Storper and Walker (1989, 182), prefer to "region," simply because it allows us to adopt an approach which is not confined to the subnational scale. In this sense, our approach is consistent with that of Amin (1998), who conceives of territories in terms of multiscalar interconnections.

At the same time, we need to be sensitive to the *particularities of specific territories*. Part of the nexus of relationships between firms within a specific territory and that territory itself is the specific ways in which the particularities of a firm (with all its attributes and "histories") enmesh with the particularities of that territory (with all *its* attributes and histories). However, in focusing on the "firm-territory nexus" per se there is a real danger of decontextualizing both firms and territories—of seeing them as entities separate from the broader structures of which they are a part. Such separation is, indeed, one of the major causes of misunderstanding. In Figure 1 we attempt to capture (in overly simplistic terms) the major dimensions of this highly complex set of relationships.

In this schema, the central focus of our analysis is identified as the nexus of three major dimensions—firms, industrial systems, and territories—embedded in turn in the overall macro dimension of governance systems. The institutions and processes of *governance*—the sets of institutions, rules, and conventions that form the regulatory context of industrial systems, firms, and territories—pervade all aspects of the firm-territory nexus.[2] Such governance elements operate at a range of geographic scales. The national scale, at least in the past, was held to be preeminent, although systems of governance have evolved at multiple scales, both above and below the national scale. They also operate at a range of *organizational* scales, including that of the firm itself. The kinds of governance systems that impact upon both firms and territories are

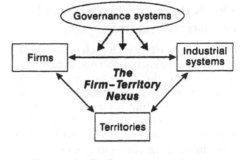
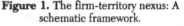

Figure 1. The firm-territory nexus: A schematic framework.

diverse in their institutional form, their specific function, and in their geographic span. The governance systems that envelop and regulate firms and territories may be general to the economy as a whole or specific to particular segments of the system (e.g., individual industries).

The paper is structured as follows. First, we discuss the notion of industrial systems: the particular constellation of processes and institutions in a specific segment of similar and related economic activity. The industrial systems within which all firms are embedded fundamentally influence firm behavior and, by extension, the firm-territory nexus. The interconnected nature of industrial system structures and dynamics—their essentially networked form—ensures that the firm cannot simply be depicted as "an island of co-ordination within a sea of market relations."[3] Second, we move on to explore the ways in which firms are encapsulated within, but are also the primary agents of, industrial systems. In doing so, we emphasize that firms are to be seen as "networks within networks" rather than as hierarchies with clear-cut boundaries in relation to the rest of the world. Finally, we focus specifically on the way firms and industrial systems relate to territorial processes.

---

[2] The "system" is, of course, more than merely economic. Indeed, one of the major characteristics of current research is its emphasis on the inherently sociocultural nature of the economic (Thrift and Olds 1996).

[3] Coase (1937) used the "island" metaphor in his seminal discussion of the nature of the firm.

348   Economic Geography

The Nature and Dynamics of Industrial Systems

Firms interacting with each other can be said to form industrial systems, as can firms acting within the same broader business environment. Thus, in principle, there are two possible starting points for identifying industrial systems. Either one can start with the *contact network* across which firms interact (i.e., firms interacting in various ways form a system) or from the *arena* (i.e., firms active within the same market, regulatory, or institutional framework form a system). A systemic approach to industrial transformation may thus imply focusing on interactions and relations between firms as well as on the relations between firms and their wider social surroundings. The relational type of system is sometimes referred to as functional, while the arena-based type is referred to as territorial. In the following, we will argue that it is essential to our understanding of the firm-territory nexus to clarify how these two aspects of "systemic-ness" relate—and should be related—to each other.

A large number of "systems concepts" exist which partly complement and partly compete with each other. Interfirm relations binding functional systems together can be of different types, such as transactions (linking customers and suppliers); competition (linking firms operating in the same product or factor market); technological collaboration (linking firms taking part in joint R&D projects or strategic alliances); knowledge spillovers (linking firms between which knowledge flows from one to the other, with or without the intention of the parties involved). "Arena-based," or territorial, systems, on the other hand, stem from geographic proximity (linking neighboring firms which will almost automatically monitor each other and are likely to interact, if nothing else, in the local labor market and in relation to local authorities); a shared regulatory and institutional framework (linking firms operating in the same local milieu, since they are rather uniformly affected by for-

mal and informal rule systems, including attitudes and "business climate").

Faced with the richness and variety of systemic notions, we make two observations. The first is that existing concepts or models differ in the way they incorporate the spatial or territorial dimension, in the definition of the system, and in the analysis of its working. Some concepts, such as industrial networks (Håkansson 1989), commodity chains (Gereffi 1994), or industry clusters (Porter 1990), are primarily functional. This means that the system is defined by various types of manifest relations—notably business transactions, collaboration, and competition—between the actors/firms who make up the system. Within this overall category of relational/functional systems, individual approaches vary in the degree to which they emphasize the spatial dimension of these relations. In general, however, they are relatively weak in specifying the role of space and territory. On the other hand, approaches taking their point of departure in territorially defined systems—such as industry agglomerations (Malmberg and Maskell 1997), industrial districts (Becattini 1990), innovative milieus (Maillat 1998), national innovations systems (Lundvall 1992), or learning regions (Asheim 1996)—elaborate more carefully the spatial aspects of the system, even if they are sometimes suspiciously vague about defining the system's particular territorial scale.

Our second observation is that the various systems approaches differ in the degree to which they focus on capturing the complexity of production organization and the logic of learning and innovation. Some approaches clearly aim at coming to grips with what Schumpeter referred to as "the circular flow of economic life" as it runs "on in the same channels year after year" (Schumpeter 1934 [1959]). These approaches focus on the way the division of labor between firms unfolds differently in different industries, how production chains are organized, and how relations between trading partners are managed. Other approaches are less interested in the material and production-oriented aspects of economic life and focus more on how innovations occur and learning processes unfold. Of course, these two types of systems—production and innovation—are in reality intertwined. Much learning and innovation takes place as a result of interactions generated by the needs of the "circular flow" of production systems.

"Productionist" Systems

The concept of the *industry* is probably the more basic; nevertheless, compared with the extensive discussions on the concept of the firm, considerably less interest has been devoted to the concept of the industry (notable exceptions include Walker (1988) and Storper and Walker (1989)). An industry is usually defined on the basis of certain similarities in production technology, product characteristics, and service content or on combinations of these (see Walker (1988) or Maskell et al. (1998) for a detailed discussion). The relationships between firms in an industry are horizontal: they produce similar outputs and are mainly related through competition. Such horizontal relations may take on many forms, from head-on antagonistic price and/or quality competition, through more or less formalized agreements on joint developmental work, to temporary collusion. The general definition of an industry is usually based on competition in product markets.

Most systems approaches, however, emphasize vertical interrelatedness. Thus, a second way of conceptualizing the systemic nature of production is through the notion of the *chain* (also variously termed value chain, production chain, and commodity chain, or *filière*; see Dicken 1994; Raikes, Jensen, and Ponte 2000). Disregarding the more or less subtle differences between concepts, the vertical dimension of an industrial system may be conceived of as a sequence of activities or functions that are combined in order to produce a given output (i.e., a good or service, or a combination of both). There is a long tradition in economic analysis of describing the structure of production in terms of a division of labor held functionally together by networks of input-output linkages (Scott 1995). The individual functions performed in such a system—logistics, production operations, research and development, marketing and sales, and after-sales services—are thus complementary to each other and coordinated. Coordination takes place either through the mechanism of market transactions, intrafirm planning, and control procedures or through some sort of network arrangement.

The most fully worked-out version of this conceptualization is Gereffi's *global commodity chain* (Gereffi 1994, 1995, 1996; Gereffi and Korzeniewicz 1994).[4] Gereffi has concentrated overwhelmingly on the *governance* structure of such chains, arguing that two dominant types of governance have evolved to coordinate transnational production systems: *producer-driven* and *buyer-driven*.[5] Storper and Harrison (1991), like Gereffi, also focus on the governance dimension of production systems. In their conceptualization, "a real production system is a function of an input-output system in the context of a governance structure," where governance structures refers to the degree of hierarchy and leadership (or their opposites, collaboration and cooperation) in coordinating the input-output system (1991, 412).

"Innovationist" Systems

Systems notions that focus explicitly on innovation and learning take only marginal,

---

[4] For a critical evaluation, see Dicken et al. (2001); Raikes, Jensen, and Ponte (2000).

[5] Gereffi regards producer-driven chains as being characteristic of manufacturing industries dominated by large TNCs (e.g., autos, computers, aircraft, electrical machinery), whereas buyer-driven chains are most commonly found in industries dominated by large retailers, brand-name merchandisers, and trading companies (e.g., garments, footwear, toys, housewares).

if any, interest in the actual organization of production. Instead, they focus mainly or exclusively on the way varied skills and competencies are combined through various interactions to result in new knowledge and innovation. The latter is often defined in Schumpeterian terms as the introduction of new goods, the introduction of new production methods, the opening of new markets, the conquest of new sources of supply, and the creation of new organizational systems.

In a *competence bloc,* as defined by Eliasson (2000), innovations occur when various competencies are brought together within the framework of a system comprising different types of actors, including competent *customers* demanding new and sophisticated solutions, *innovators* who create such solutions, *entrepreneurs* who identify innovations and create businesses based on them, *venture capitalists* who supply finance to innovators and entrepreneurs, and *industrialists* who produce and market the new product/service/method on a large scale. For Perroux (1955), Schumpeterian activities occurred in *pôles de croissance* (growth poles), which he saw as clusters of linked activities in nongeographic (i.e., economic) topological space (see Lasuén 1972). For Porter (1990), innovations occur within the framework of *industry clusters.* These consist of producers of primary goods, suppliers (of specialty inputs, machinery, and associated services), customers, and related industries. The mechanisms creating dynamism and innovation in such a cluster are captured in Porter's much-quoted diamond model and partly overlap with those identified by Eliasson and Perroux. Thus, the presence of sophisticated customers plays a key role in Porter's model as well. Other important factors include competent suppliers, appropriate factor conditions (including the interesting and novel concept of selective factor disadvantage), and rivalry between the core firms of the cluster.

Our argument here, in sum, is that, when approaching the firm-territory nexus, an important first step is to realize that firms have to be understood as parts of broader industrial systems. For analytical convenience such systems are defined in functional, rather than territorial, terms, although, of course, all systems are simultaneously both functional *and* territorial. Focusing on systems means, in a way, downplaying the significance and role of firms, in the same way as sociological analyses of the dynamics of social groups and networks tend to downplay the physical and mental attributes of individuals. A systemic approach forces us to concentrate on *structures* at the macro or meso scale. However, the primary *agent* within industrial systems is the firm, "the crucible within which both macro- and micro-forces meet and are played out" (Taylor 1984, 8). Although we cannot understand how firms behave without a clear understanding of their structural position, neither can we fully understand the workings of industrial systems without an understanding of firms.

Firms within Industrial Systems: Firms as Networks within Networks

Business firms are both primary agents of and, at the same time, encapsulated within the kinds of industrial system discussed above. Having set out the systemic context within which firms operate, we now turn to a focus on the firm itself. There are a plethora of definitions for "the firm" in the literature, to the extent that, as Taylor (1999) points out, no generally accepted definition exists. In the context of this paper, it is not necessary to indulge in a discussion of alternative theoretical conceptualizations of the firm (for a recent review, see Maskell 2000). What we do need, however, is a framework that delimits the boundaries of firms, the rationale for firm action, and firm organization. For our purposes, therefore, Whitley's (2000, 65–66) definition is a sufficient point of departure: "The significance of firms in capitalist economies lies in their combination of *financial control over resources* with *employment.* As ownership-based units of decision-making and control, they are clearly central collective actors in the mobilization, allocation, and use of assets, especially human labour power" (emphasis added).

Firm Boundaries

Virtually all conceptualizations of the firm tend to assume that firms are *clearly bounded* institutions: that where a firm's "territory" ends and others' begin—what is "inside" the firm and what is "outside"—is essentially given and unproblematic. There are two major issues here. The first relates to the extent to which firms' boundaries are changing. There is abundant empirical evidence to show that, through a whole variety of processes, the boundary between the functions performed within a firm ("in-house") and those which are externalized to other firms ("out-sourced") is in a continuous state of flux, especially among large firms. What is evident is both a complex reconfiguring of organizational boundaries (including the proliferation of various forms of interfirm collaboration) and a continuing extension of geographic boundaries, particularly among firms transnationalizing their operations. Although there may be disagreement about the extent of such boundary extensions, the question is essentially empirical.

The second, and more fundamental, issue is: do firms actually have boundaries at all? In one sense—the legal definition of the firm that defines the locus of corporate responsibility—the answer is obviously "yes." As Markusen (1999, 878) rightly observes, the boundaries of a firm are, ultimately, "not at all fuzzy—they are written down in asset, cost, and revenue statements that owners and managers, whether private or public, scrutinize carefully every quarter . . . as organizations and institutions, firms are clearly bounded."[6] In a

broader sense, however, the answer to the boundary question is more ambiguous. Badaracco (1991), one of the few writers to address this broader question, responds:

> In one sense, the answer is not really. Nations, states, and backyards have boundaries, the skin is more or less the boundary of the body, but no one ever sees or touches or steps across the boundaries of a firm. The phrase "the boundaries of the firm". . . simply conveys the common-sense notion that government bodies, competition, and markets are somehow "outside" a firm, while managers, employees, equipment, and inventory are more or less "inside" it. (Badaracco 1991, 293)

Instead of conceptualizing the firm as having sharply defined boundaries or, conversely, of not having boundaries at all, Badaracco (1991, 314) portrays it as "a dense network at the center of a web of relationships." These relationships are defined by "ownership, hierarchical control, centralized power, managerial discretion, social bonds of membership, loyalty, and shared purpose, and formal, legal contractual arrangements." The central domain of the firm blends slowly into its surrounding environment as ownership, hierarchy, control, power, social bonds, classic contracting, and other boundary-defining devices diminish in significance or are shared with other organizations. Ultimately, "the gradual attenuation of these relationships reaches a point at which the firm exercises neither power nor influence. Here, the genuinely external environment of the firm begins" (Badaracco 1991, 314).

Rationality and Order

Another basic tendency of many conceptualizations of the firm is to see it as "an ordered, autonomous and ultimately rational economic subject that operates according to a central logic and manifests pre-

[6] Markusen's reference to quarterly scrutiny of firms' financial performance reflects an essentially Anglo-American perspective. The same practice does not necessarily apply in other sociocultural contexts.

dictable dynamics" (O'Neill and Gibson-Graham 1999, 12). The prevailing view still projects an undersocialized conceptualization of the firm. In particular, there is a tendency to underplay the contested nature of processes of decision making within firms and to accept their (self-)representation as designing and implementing strategies through logically structured organizational forms. However, in the real world, firms are not like that at all. Schoenberger (1997, 113), for example, demonstrates that corporations do not necessarily "act in their own best interests, even when they have access to good information about what to do." She attributes such "deviant" behavior to specificities of corporate culture. In a similar vein, Thrift and Olds (1996, 319–20) point to the need to revise this view of business organizations and to recognize what they call "the disorganization of organization." They neither see firms as being enclosed "shells through which transactions with the outside world [take] place to a greater or lesser degree" nor as "characterized by preset goals which they [work] towards." Rather, they emphasize the intensely practical and ad hoc character of most organizations, seeing them as improvising, tentative, and temporary entities, always in action and "on the move." Thus, "organizations are talked into being at each moment by actors who often have to fashion informal solutions to formal goals" (Thrift and Olds 1996, 319–20).

Governance and Power: Struggle over Strategy and Organization

Even in the archetypal hierarchical business organization, in which the governance system is essentially top-down, it is by no means the case that each component part simply operates as a passive recipient of decisions handed down from on high. On the contrary, intrafirm relationships within all firms are highly contested processes—the manifestation of internal governance systems, formal and informal power structures, and bargaining relationships—and these have profound implications for

national, regional, and local economies (Halford and Savage 1997). As Cawson and his colleagues observe: "firms are themselves systems of power with constituent groups (e.g., of engineers, managers, workers, R&D staff) challenging each other's power . . . [and] in which different kinds of interests within the firm try to pursue their own . . . strategies" (Cawson et al. 1990, 8, 27).[7]

This representation of firms as diverging from the uniformities of the "standard" business organizational models has implications for our understanding of firm-territory relationships (Dicken and Thrift 1992). At this point we examine the changing nature of firms' *organizational architectures*, avoiding, for the time being, their intrinsic relationships with geographic territory. Our basic position is that firms are essentially *networks within networks*. In addition to arguing, as many writers do, that networks are external structures within which a firm is embedded, we argue that firms are *themselves* networks, configured and organized in particular forms of power structure and governance (Dicken et al. 2001). We thus reject the concept of the network as a new and different organizational form between hierarchies and markets, the common position in much of the current literature (e.g., Powell 1990). Ours is emphatically not a reductionist position; on the contrary, it helps us avoid the common practice of seeing the network organi-

---

[7] Dicken and Miyamachi's (1998) study of the Japanese *sogo shosha* in Europe provides some evidence of this internal contestation and of the dissonance between the formal organizational structures represented in firms' organizational charts and the reality of actual decision processes. More recently, O'Neill and Gibson-Graham (1999) have shown how the apparently logical and rational restructuring decisions of a major Australian transnational corporation (BHP) were anything but that. Rather they were the outcome of highly fragmented, often contradictory, and certainly contested discourses and logics embodied within key actors within the firm.

zation as something novel. The challenge instead is to unravel networks' specific processes and structures. From this perspective, then, we can conceive of production systems (and territories) as consisting of a *spectrum* of organizational forms and types of governance, from the "temporary coalition" (Taylor 1999) to the centralized, hierarchical corporation.

Changing Organizational Architectures?

The business-organizational literature is replete with discussions of evolving organizational structures. As such, it is a rich source of material for understanding the changing organizational architecture of firms and, therefore, for illuminating our understanding of firm-territory relationships. A common thread woven through much of this literature is the assertion that the traditional hierarchical form of top-down control is being replaced by flatter architectures in which the relationships between the component elements of the organization are based on governance systems characterized more by reciprocity than by authority. In this scenario, hierarchies are being replaced by *heterarchies* (Hedlund 1986), or even, some would argue, by *virtual organizations*. But whatever its precise form, whether hierarchically organized or more flexibly networked, the firm (defined in its broadest sense to incorporate its entire set of transactions) has to be *coordinated*.

Miles and his co-authors (1999) provide an interesting perspective on the supposed evolution of dominant organizational forms, tracing a path from the hierarchy, through the network, to the cellular form. The underlying parameters of this progression include a shift in the firm's key asset—from capital goods, through information, to knowledge—and a shift in the firm's key capability—from specialization and segmentation, through flexibility and responsiveness, to design creativity. In their view, "the new organizational form for a new economic era" (1999, 162) will be a cellular

organization. Although this may be the way firms should—or indeed will—evolve, it is misleading to extrapolate from this to the world at large. Bartlett and Ghoshal (1989) have made similar predictions of organizational transformation toward more integrated network forms of organization within transnational corporations. They argue that the perceived inadequacies of existing organizational architectures for firms operating over extensive geographic space will inevitably lead to the adoption of what they term (rather confusingly) the "transnational solution." For us, however, the value of Bartlett and Ghoshal's work lies less in its prescriptive/predictive element than in its empirically grounded identification of a variety of organizational architectures (see Dicken, Forsgren, and Malmberg 1994).[8]

Indeed, the point we would emphasize is not the possible development of a particular organizational form at some point in the future but, rather, the *diversity* of organizational forms existing at the present. Firms organized on hierarchical principles not only still exist, but they may even be in a majority. The empirical evidence tends to suggest that the newer organizational forms are confined to a limited number of innovative firms in certain sectors. In an organizational sense, the real world is incredibly messy. Firms come in all shapes, sizes, and forms of governance. Their internal architecture reflects not only the external constraints and opportunities they have to face in going about their business but also a strong element of *path dependency*. Such path dependency consists of two major components. One is what Heenan and Perlmutter (1979) called the firm's *strategic disposition*, derived from the cultural and administrative heritage of accepted practices built up over the course of the firm's history. "This administrative

[8] Harzing (2000) provides some empirical validation of Bartlett and Ghoshal's basic organizational typology, using data from 166 subsidiaries of 37 companies headquartered in 9 countries.

heritage represent[s] both a major asset and a powerful impediment in the change process" (Bartlett and Ghoshal 1989, 35), a characteristic clearly illustrated by Schoenberger (1997). The second component derives from a firm's *geographic embeddedness,* an issue to which we will return in the next section of the paper.

The "messiness" of firm architectures is compounded by the volatility of firms' external networks, most notably in the proliferation of collaborative ventures and in the continued pervasiveness of mergers and acquisitions.[9] Both of these tendencies impact on firms' organizational structures, though often in hard-to-read ways. Collaborative arrangements, themselves of great diversity, create organizational pressures on firms because they are both hybrid and not fully internalized. The participants retain their basic identities and, in many cases, continue to compete fiercely with one another. Mergers and acquisitions, on the other hand, involve the internalization of formerly independent organizations (with their own path dependencies) into a single organization. Not surprisingly, the empirical evidence shows that creating a new structure to incorporate formerly separate ones is fraught with difficulty and often unsuccessful. The process of incorporating both external collaborative ventures and mergers and acquisitions into a firm's organizational architecture is made even more difficult when the participants differ culturally. Yet, one of the most marked features of the contemporary economic landscape is that both domestic and cross-border collaborations and mergers continue to proliferate.

[9] An increasing number of mergers and acquisitions occur across national boundaries as part of the process of transnationalization. The United Nations estimates that the value of completed cross-border deals (involving more than a 10 percent equity share) rose from less than $100 billion in 1987 to $720 billion in 1999 (UNCTAD 2000).

Firms, Systems, and Territories

The conventional view of territory sees it as a *bounded* portion of space "occupied by a person, group, local economy or state. . . . Sometimes territory is used as equivalent to such spatial concepts as place or region, conveying the sense of a clustering or concentration of people or activities" (Agnew 2000, 824). At one level, therefore, territories can be regarded as "containers" of sets of physical, social, cultural, political, and economic attributes. Contrary to much received wisdom that conceptualizes territories as a series of nested geographic scales, rather like a set of Russian dolls (see, e.g., Scott 1998), they are, in fact, arrayed along a *continuum of scales,* each individual scale being contingent and socially constructed.[10] Of course, certain territorial scales tend to have particular significance at particular historical junctures, the most obvious example being the national state. In recent years, as noted in the introduction to this paper, the subnational or "regional" scale has become the center of attention. Our argument, developed in the preceding section of the paper, is that we need to conceive of the dynamics of firms not in some formulaic manner that depicts a clear and unambiguous trajectory toward a dominant organizational form but, rather, as constituting a tapestry of continuing variety. This must be taken into account in any attempt to understand how firms impact on territories; at the same time, the very existence of territorial differentiation within which firms are intrinsically embedded helps to explain the variety of firms themselves.

In the following, we concentrate on three specific aspects of the firm-territory nexus: the firm as a complex spatial structure, localized clusters of firms as territor-

[10] There is a large literature on the issue of geographic scale. See, in particular, Brenner 1998, 2001; Cox 1998; Kelly 1997, 1999; Marston 2000; Swyngedouw 1997.

ial systems, and governance tensions between firms and bounded territories.

Firms as Complex Spatial and Territorial Structures

Firms, like all other forms of social organization, are fundamentally and intrinsically *spatial* and *territorial*. They are spatial in the sense that they are responsive to geographic distance and to spatial variations in the availability of necessary resources and business opportunities. Such spatiality may have—indeed, most often has—a territorial manifestation. Hence, firms arc territorial as well as spatial, in the sense that the "surface" from which firms originate and on which they operate is most commonly made up of a tessellated structure of territorial entities arrayed along a continuum of variable and overlapping scales. Firms themselves have *territorial extent:* the roughly bounded area over which they conduct their operations (e.g., their market area, their labor catchment area, their supply area). For some functions of the firm the territory may be intensely local, for others it may approach the global. Such firm territories, however, are usually temporally volatile, spatially discontinuous, and not clearly bounded. Indeed, firms in competition interpenetrate each other's territories in highly complex and contested ways.

More particularly from the perspective of this paper, firms are territorial in that they derive some of their characteristics from, and also directly influence, the characteristics of specific territories and places. We have explored the first of these two firm-place/territory relationships elsewhere (Dicken 2000; Dicken, Forsgren, and Malmberg 1994; Dicken and Thrift 1992) and argued for a rejection of the "placeless" firm scenario so popular in much academic literature. Where firms come from—their home country, even their *local* origin—appears to remain significant in influencing how firms behave and, therefore, in how they use and respond to territorial variations, including

how the diverse places in which TNCs operate may themselves exert an influence on a firm's behavior.

The business-organizational literature has tended to have a naive view of the spatial character of firms and of the ways in which firms relate to territory. Mostly, spatiality has either been assumed (and, therefore, ignored) or conceptualized simplistically and dichotomously in terms of "domestic" and "foreign" operations or headquarters and subsidiary roles. The "local" dimension of firms' activities is rarely, if ever, specified (Dicken 1994), and invariably refers to the national scale or even to such supra-national "regional" groupings as the European Union or the North American Free Trade Agreement. The connections between the organizational architectures of firms, discussed earlier, and their geographic forms have been poorly conceptualized. In one sense, we do not appear to have moved very far beyond Hymer's (1972) simplistic projection of the hierarchical-organizational structure of the large TNC on to geographic space to create a crude international division of labor.

We are gradually beginning to understand that, in fact, the relationships between different parts of firms located in different places and occupying different positions within the firm's organizational architecture are not only far more complex than this but also that such complexity is deeply embedded within the firm's spatiality and territoriality. Schoenberger (1999) argues that different "places" within the firm, organizationally and geographically, develop their own identities, ways of doing things, and ways of thinking over time, the reason being that they "live in different places and must confront and respond to the particularities of these places across a whole range of practices and issues" (1999, 211). She goes on to suggest that "in effect, the large firm is internally regionalized and what goes on in its regions is important." Thus, the firm's dominant culture necessarily contains multiple subcultures. Some of these may "revolve around functions and cut across places . . . but some will have real geographical locations . . . the interesting locus of study and of transformative processes is not only where 'the firm' (conceived as unitary agent) meets the world (competitors, markets, suppliers), but also internally as competing subcultures strive for validation and expression" (Schoenberger 1999, 211).

Schoenberger herself provides some case study evidence to support her claims. But there is also a growing literature within international management that focuses on the changing relationships between different parts of the firm, notably between headquarters and subsidiaries (see, e.g., Birkinshaw 1996, 2000; Birkinshaw and Hood 1998; Ivarsson 1996; and Taggart 1996, 1997). Two related aspects of this literature are especially relevant to the current debate on firm-territory relationships: the nature of the internal power and bargaining relationships between headquarters and subsidiary units and between subsidiary units themselves, and the processes whereby subsidiary units are able to carve out distinctive and influential roles within the firm. Much attention has been focused on the influence of such variables as the size of subsidiary units or the role of dynamic and powerful individual subsidiary managers and management groups on these two dimensions, and these are, no doubt, important in intrafirm bargaining processes. But there is a growing body of evidence to suggest that the particularities of a unit's *territorial embeddedness* are centrally important to the nature and influence of subsidiary units within firm structures.

A subsidiary's level of competence determines the strength of the subsidiary within the corporate network and influences its power to make strategic and operational decisions. The genesis of such competence lies in the location- or territory-specific advantages that emanate from the uneven distribution of resources, knowledge, and technological or organizational capabilities among territories. In this respect, the competence of a firm's subsidiary unit is driven (at least partly) by

environmental factors derived from the dynamics of the location in which it is situated. The competencies of a corporate unit are created over extended periods as a firm interacts with its surrounding environment. Birkinshaw's (1996) findings into how transnational subsidiary units gained world product mandates provide indirect support for this.

Localized Industrial Clusters as Territorial Systems

Despite the alleged homogenizing effects of globalization, individual territories—nations, and individual cities or regions within nations—continue to exhibit dramatic differences in terms of specialization, competitiveness, and industrial dynamics. Successful industries in a territory often retain their leading edge over extended periods of time, despite attempts by others to imitate their success. Competitiveness and, still more important, innovative capacity, have a markedly territorial element (Maskell et al. 1998).

There are various reasons for this. While people in some professions do indeed form and interact in global networks, labor markets are still predominantly local; some technologies move across the globe, while others remain spatially "sticky" (i.e., they are kept "secret" or are retained in certain individuals) and are therefore relatively immobile (Malmberg and Sölvell 1997; Markusen 1996). Network relations between buyers and suppliers, within which new products and technologies are developed and where production flows are perfected, are traditionally stronger if they are local rather than international (Gertler 1995). Within a given territory, people speak the same language and tend to trust each other (though, of course, "trust" is a somewhat complex and contested concept; see Malmberg and Maskell (2001)). Furthermore, local networks often include links between firms and universities, research institutes, and public authorities. Even the most modern forms of communication technology are inferior to face-to-face contacts between people when it comes to building trustful relations and to communicating noncodified types of information. In addition, personal travel is both costly and time-consuming (Malmberg and Maskell 1997).

Thus, the more codified the knowledge involved, the more easily it can be transferred across territorial boundaries. The more tacit the knowledge, the more attached it is to the local milieu. Knowledge and competencies which are embodied in people or are "in the air"—for example, in the routines of organizations (or local milieus)—can often neither be articulated, nor moved, from one place to another.

Porter (1990) argues that certain *territorial* circumstances determine the innovative, and thus competitive, strength of a given industry. All the four determinants in Porter's "diamond model" relate to the existence of a system of interdependent firms or industries embedded in a specific territory. Specialized factors of production are seen as being formed historically in interactions between firms and institutions. Firms needing products with specific characteristics by and large raise sophisticated demand, and to meet such demand on the domestic market implies the coexistence of buyers and sellers in the same territory. The notions of related and supporting industries and of interfirm rivalry presuppose the presence, in the territory, of more than one firm in a particular industry. In a similar vein, Maskell and Malmberg (1999) talk about localized capabilities to denote those infrastructures, resources, institutions, and competencies that make certain territories especially apt when it comes to promoting the development of certain types of economic activity. Likewise, Storper (1997) argues that a territory's or region's *relational assets* help to determine both *where* multilocational firms place their individual functions and operations and *how well* they perform (i.e., their level of competence).

Thus, a territory's competitive firms and industries are not spread evenly through the economy but are connected in *clusters of industries* related by links of various kinds. Successful firms and industries in a territory are often linked through vertical (buyer/supplier) or horizontal (common customers or technology) relationships. In this sense, the localized cluster is the nexus where industrial and territorial systems amalgamate. In the localized cluster, the firm is embedded both functionally and territorially. Clusters may differ in terms of their dynamism, degree of maturity, spatial extension, and local embeddedness. Not all spatial agglomerations of similar or related industrial activity are "Hollywoods" or "Silicon Valleys." The most decisive factor concerns how powerful the cluster is in terms of its long-term innovative capacity. Thus, when addressing the issue of how firms are related to territory, the dynamism of the local cluster is a major determining force. Cluster characteristics are assumed to affect the average level of competence in all business units located in the cluster (including foreign subsidiaries), and it will also affect the formal mandate given to local subsidiaries by the firm's headquarters.

We may expect that the more dynamic a local cluster is, the more resources and capabilities the subsidiary will build up, and a stronger de facto mandate will emerge. And the stronger the mandate a corporate unit possesses, the more autonomy it has to make its own strategic decisions (Malmberg and Sölvell 2000). The potential negative effects of foreign ownership identified in the literature, such as cutbacks, closures, and drainage of resources and competencies (Firn 1975; Dicken 1998), are also less likely to materialize—that is, the risks associated with distance to corporate headquarters are less pronounced.

Governance Tensions between Firms and Bounded Territories

In our discussion of firms as complex spatial structures and of localized clusters of firms as territorial systems, we have treated territory as a "natural" outcome of firm-organizational processes operating in geographic space. In so doing, we have deliberately ignored territories as *bounded systems of governance*, an omission we must now repair, albeit briefly. One of the major differentiating dimensions of geographic space, of course, is its demarcation into multiscalar spaces of governance, encompassing the "local state" through to the national, as well as such supranational jurisdictions as the European Union. Such governed spaces are one of the major ways in which the resources on which firms depend are "packaged." Despite the increasing permeability of their boundaries, states continue to act as significant "containers" of distinctive sociocultural practices, regulatory institutions, and processes, and of constellations of factors of production and other firm-relevant assets.

Even though the spatial/territorial structures of firms and industrial systems rarely coincide with political-territorial units, it is the very existence of such governed spaces that stimulates multilocational firms to seek out the most favorable package, frequently engaging in what has been termed "regulatory arbitrage" (Dicken 1992). Their existence also stimulates states/localities to strive to attract mobile investment, in a process of competitive bidding (Dicken 1998; UNCTAD 1995, Chap. 3). The outcome of this process depends on the relative bargaining power of firms and "states." Although the degrees of freedom of multilocational—especially transnational—firms to move into and out of territories at will is often exaggerated, the potential for such mobility is obviously there. This creates a profound problem for states and localities whose objectives are, at least in part, to attract and retain as much productive investment as possible within their boundaries. By definition, however, the flexible "territories" of multilocational/transnational firms overlap, and interpenetrate, the relatively fixed territories of states, both national and local. This tends to create an asymmetry of bargaining

power between firms and states, although, as Stopford and Strange (1991, 216) argue, it is important to distinguish between a general shift in bargaining power from states to firms and the specific bargain that may be struck under particular conditions. In general, however, the tensions between firms and states have increased as processes of firm-organizational restructuring have intensified and become geographically more extensive.

In general, the relative bargaining power of the "states" is primarily a function of their size, their control over and access to locationally specific resources and capabilities, and their political status. Other things being equal, large states will tend to have more bargaining power vis-à-vis multilocational firms than small states, but this may be tempered by a smaller state's possession of an especially sought-after resource (including, of course, human knowledge–based resources). However, political status is especially important for territories below the national scale. Territories within highly centralized national political systems invariably have far less autonomy (for example, in financial and fiscal affairs) than those embedded in federal systems. This means that the bargaining power of subnational governance units may be highly dependent upon the locational uniqueness of their bounded assets. For such subnational territories in general, their economic prospects may be influenced as much, if not more, by national policies as by local actions. Their abilities to bargain with multilocational firms will be constrained by national policies. In a global context, the key power interactions are played out between nation-states and transnational corporations (Dicken 1994).

Conclusion

In this paper, we have argued that the relationships between firms and territories are complex. Insofar as firms are networks embedded in the broader networked structure of industrial systems, the firm-territory nexus is not just an issue of how individual firms are related to the territory within which they happen to be located. It also involves the more complex issue of how firms are embedded in systems relating them to other firms and institutions, and how such systems, in turn, are territorialized.

Webs of intra- and interorganizational relationships are woven across geographic space in ways which not only connect organizations, and parts of organizations, together but which also connect dispersed *territories* together through networked flows coordinated primarily by firms. In one sense, therefore, the economies of territories reflect the ways in which they are "inserted" into the organizational *spaces* of firms—either directly, as the geographic locus of particular functions, or indirectly, through customer-supplier relationships with other (local) firms. Multilocational firms, especially those operating across national boundaries (TNCs), have the potential to manipulate geographic space and to use territory as an intrinsic part of their competitive strategies. Thus, the ability to "control" space and to use the resources (in the broadest sense) of specific territories are diagnostic characteristics of multilocational (especially transnational) firms, although the nature and effectiveness of such control varies from firm to firm. In this paper, we have proposed a perspective on these issues that has significant implications for how economic geography research might approach the firm-territory nexus, in order to escape the pitfalls that follow from analyzing firms and territories separately.

First, we have argued that adopting a systemic approach is essential to an understanding of the dynamics of modern economies. Thus, the relationship between firms and territories can never be properly specified unless it is mediated through the notion that each individual economic activity is bound up in a web of relationships with other activities. Rather than understanding industrial transformation through the restructuring of the operations of indi-

vidual firms, the point of departure for analyses in economic geography should be some conceptualization of an industrial system. One benefit of adopting a systemic approach is that it allows us to transcend the division of the economy into a number of basically unnatural dichotomies, which so often tend to characterize—and weaken—analyses in the field. Such "false dichotomies" typically include divisions between, for example, small firms and large firms, manufacturing and services, high-tech and low-tech activities, and domestic and foreign firms. When notions of industrial systems are brought to the fore, such distinctions become less significant. Within any single industrial system, one may find activities belonging on both sides of false dichotomies. The analysis of how industrial systems develop and transform will typically display interlinkages and dependencies between very dissimilar types of activities.

We have also argued that perhaps the most fundamental aspect of industrial systems is their role in the generation of learning and innovation. While "productionist" system analyses certainly have value in many contexts, when it comes to approaching the long-term economic viability of a specific territory, it seems more important to focus on the industrial system as the context for knowledge production. In terms of research agendas in economic geography, it follows that we need to develop methods for analyzing interactions and relationships, particularly those involving the exchange of information and knowledge—across industrial systems, for example—in order to assess the extent to which such interactions and exchanges are carried out within bounded territories.

Second, we have emphasized the intrinsically networked character of firms themselves, as well as their unclear boundaries, multiple—and often contested—goals and rationalities, and their varying organizational architectures. The research implication of this way of defining firms is, again, the need to focus on interrelationships and contact patterns when analyzing firm behavior. Particularly in the case of multi-locational firms, we believe that economic geography has much to gain from abandoning stereotyped positions in which the behavior and development of a particular business unit can be simply inferred from its formal organizational position. Following Schoenberger (1999), we have argued that business strategy and development are the outcome of complex actions and interactions that unfold over time as different parts of the firm, organizationally and territorially, develop their own identities, ways of thinking, and ways of doing things. A future research challenge is to try to understand, more precisely, how this process of creating identities and routines at the level of the business unit works. Which interactions and interrelations are most decisive in this context?

Third, such questions are particularly important when we focus specifically on territory. The relationships between different parts of firms located in different territories and occupying different positions within the firm's organizational architecture are deeply embedded within the firm's spatiality and territoriality. From the point of view of firm-territory relationships, the nature of the internal power and bargaining relationships between headquarters and subsidiary units and between subsidiary units themselves are important, as are the processes whereby subsidiary units are able to carve out distinctive and influential roles within the firm.

We have argued that certain *territorial* circumstances help to determine the innovative, and thus competitive, strength of a given industry. Regardless of whether we conceptualize these circumstances in terms of Porterian "diamond dynamics," as localized capabilities (Maskell et al. 1998), or as relational assets (Storper 1997), the implication is that we expect a territory's competitive firms and industries not to spread evenly through the economy but to be connected in *localized clusters*. The localized cluster is the amalgamation of industry, firm, and territorial systems. In the local-ized cluster, the firm is embedded both functionally and territorially.

Therefore, when addressing the issue of how firms are related to territory, the dynamism of the local cluster is a major determining force. Cluster characteristics affect the overall competitiveness of business units located within them, including local subsidiaries of foreign firms. We may expect that the more dynamic a local cluster, the more resources and capabilities the firms will build up. From this it follows that analyses of territorial industrial development should focus on the ways in which localized clusters gain dynamism and on the actual competencies and mandates of the business activities in the territory. In that sense, therefore, the formal position of local units within multilocational, notably transnational, corporations becomes a secondary issue.

In such an analysis, however, we should not forget that the spatial/territorial structures of firms and systems of firms rarely, if ever, coincide with political-territorial units. Such a disjuncture is a source of considerable tension between "states"—at a whole range of geographic scales—and firms, as well as between such "states" themselves. Intense institutional competition has developed between governed spaces, which, among other things, stimulates multilocational firms to seek out favorable resource "packages" and to engage in "regulatory arbitrage" (Dicken 1992). The relative bargaining power of firms and "states" is important in this context. Although most firms never, or rarely ever, move into and out of territories, the potential for such mobility is there, and it does create a problem for states and localities whose objectives are, at least in part, to attract and retain as much productive investment as possible within their boundaries.

In sum, the firm-territory nexus is an intriguing, complex phenomenon, the study of which should be a major focus of research in economic geography. In proposing a relational perspective on this nexus, we hope to have contributed to the formation of a conceptual point of departure for its further analysis.

References

Agnew, J. A. 2000. Territory. In *The dictionary of human geography*, 4th ed., ed. R. J. Johnston, D. Gregory, G. Pratt, and M. Watts, 824. Oxford: Blackwell.

Amin, A. 1998. Globalisation and regional development: A relational perspective. *Competition and Change* 3:145–65.

Asheim, B. T. 1996. Industrial districts as "learning regions": A condition for prosperity? *European Planning Studies* 4:379–400.

Badaracco, J. 1991. The boundaries of the firm. In *Socio-economics: Towards a new synthesis*, ed. A. Etzioni and P. R. Lawrence, Chap. 17. Armonk, N. Y.: M. E. Sharpe.

Bartlett, C. A., and Choshal, S. 1080. *Managing across borders: The transnational solution.* Boston: Harvard Business School Press.

Becattini, G. 1990. The Marshallian industrial district as a socio-economic notion. In *Industrial districts and inter-firm cooperation in Italy*, ed. F. Pyke, G. Becattini, and W. Sengenberger, 37–51. Geneva: International Institute for Labour Studies.

Birkinshaw, J. 1996. How multinational subsidiary mandates are gained and lost. *Journal of International Business Studies* 27:467–96.

———. 2000. *Entrepreneurship in the global firm.* London: Sage.

Birkinshaw, J., and Hood, N. 1998. *Multinational corporate evolution and subsidiary development.* London: Macmillan.

Brenner, N. 1998. Between fixity and motion: Accumulation, territorial organization and the historical geography of spatial scales. *Environment and Planning D: Society and Space* 16:459–81.

———. 2001 forthcoming. The limits to scale? Methodological reflections on scalar structuration. *Progress in Human Geography* 25.

Cawson, A.; Morgan, R.; Webber, D.; Holmes, P.; and Stevens, A. 1990. *Hostile brothers: Competition and closure in the European electronics industry.* Oxford: Clarendon Press.

Coase, R. H. 1937. The nature of the firm. *Economica* 4:386–405.

Cox, K. 1998. Spaces of engagement, spaces of dependence and the politics of scale. *Political Geography* 17:1–23.

Dicken, P. 1992. International production in a volatile regulatory environment: The influence of national regulatory policies on the spatial strategies of transnational corporations. *Geoforum* 23:303–16.

———. 1994. Global-local tensions: Firms and states in the global space-economy. *Economic Geography* 70:101–28.

———. 1998. *Global shift: Transforming the world economy.* 3d ed. London: Paul Chapman.

———. 2000. Places and flows: Situating international investment. In *The Oxford handbook of economic geography*, ed. G. L. Clark, M. P. Feldman, and M. S. Gertler, 275–91. Oxford: Oxford University Press.

Dicken, P.; Forsgren, M.; and Malmberg, A. 1994. The local embeddedness of transnational corporations. In *Globalisation, institutions and regional development in Europe*, ed. A. Amin and N. Thrift, 23–45. Oxford: Oxford University Press.

Dicken, P.; Kelly, P. F.; Olds, K.; and Yeung, H. W.-C. 2001. Chains and networks, territories and scales: Towards a relational framework for analysing the global economy. *Global Networks* 1:89–112.

Dicken, P., and Miyamachi, Y. 1998. "From noodles to satellites": The changing geography of the Japanese *sogo shosha.* *Transactions of the Institute of British Geographers* 23:55–78.

Dicken, P., and Thrift, N. 1992. The organization of production and the production of organization: Why business enterprises matter in the study of geographical industrialization. *Transactions of the Institute of British Geographers* 17:279–91.

Eliasson, G. 2000. Industrial policy, competence blocks and the role of science in economic development. *Journal of Evolutionary Economics* 10:217–41.

Firn, J. R. 1975. External control and regional development: The case of Scotland. *Environment and Planning A* 7:383–414.

Gereffi, G. 1994. The organization of buyer-driven global commodity chains: How US retailers shape overseas production networks. In *Commodity chains and global capitalism*, ed. G. Gereffi and M. Korzeniewicz, 95–122. Westport, Conn. : Praeger.

———. 1995. Global production systems and Third World development. In *Global change, regional response: The new international context of development*, ed. B. Stallings, 100–142. New York: Cambridge University Press.

————. 1996. Commodity chains and regional divisions of labor in East Asia. *Journal of Asian Business* 12:75–112.

Gereffi, G., and Korzeniewicz, M., eds. 1994. *Commodity chains and global capitalism.* Westport, Conn.: Praeger.

Gertler, M. S. 1995. "Being there": Proximity, organization, and culture in the development and adoption of advanced manufacturing technologies. *Economic Geography* 71:1–26.

Håkansson, H. 1989. *Corporate technological behavior—co-operation and networks.* London: Routledge.

Halford, S., and Savage, M. 1997. Rethinking restructuring: Embodiment, agency and identity in organizational change. In *Geographies of economies,* ed. R. Lee and J. Wills, 108–17. London: Arnold.

Harzing, A.-W. 2000. An empirical analysis and extension of the Bartlett and Ghoshal typology of multinational companies. *Journal of International Business Studies* 31:101–20.

Hedlund, G. 1986. The hypermodern MNC—A heterarchy? *Human Resource Management* 25:9–35.

Heenan, D. A., and Perlmutter, H. 1979. *Multinational organizational development: A social architectural perspective.* Reading, Mass.: Addison-Wesley.

Hymer, S. 1972. The multinational corporation and the law of uneven development. In *Economics and world order,* ed. J. N. Bhagwati, 113–40. London: Macmillan.

Ivarsson, I. 1996. Integrated international production: A survey of foreign transnational corporations in Sweden. Department of Geography, University of Göteborg, Series B, 90.

Kelly, P. F. 1997. Globalization, power and the politics of scale in the Philippines. *Geoforum* 28:151–71.

————. 1999. The geographies and politics of globalization. *Progress in Human Geography* 23:379–400.

Lasuén, J. 1972. On growth poles. In *Growth centers in regional economic development,* ed. N. M. Hansen, 20–49. New York: Free Press.

Lundvall, B.-Å., ed. 1992. *National systems of innovation: Towards a theory of innovation and interactive learning.* London: Pinter.

Maillat, D. 1998. From the industrial district to the innovative milieu: Contribution to an analysis of territorialised productive organisations. *Recherches Economiques de Louvain* 64:111–29.

Malmberg, A., and Maskell, P. 1997. Towards an explanation of regional specialization and industry agglomeration. *European Planning Studies* 5:25–41.

————. 2001. The elusive concept of localization economies—Towards a knowledge-based theory of spatial clustering. Paper prepared for the "Industrial Clusters" Revisited: Innovative Places or Uncharted Spaces? session, AAG Annual Conference, New York, 27 February–3 March.

Malmberg, A., and Sölvell, Ö. 1997. Localized innovation processes and sustainable competitive advantage of firms: A conceptual model. In *Interdependent and uneven development: Global-local perspectives,* ed. M. Taylor and S. Conti, 67–85. Aldershot, U.K.: Avebury.

————. 2000. Does foreign ownership matter? Subsidiary impact on local clusters. Paper prepared for the Marcus Wallenberg Symposium on Critical Perspectives on Internationalisation, 10–11 January, Uppsala, Sweden.

Markusen, A. 1996. Sticky places in slippery space: A typology of industrial districts. *Economic Geography* 72:293–313.

————. 1999. Fuzzy concepts, scanty evidence, policy distance: The case for rigour and policy relevance in critical regional studies. *Regional Studies* 9:869–84.

Marston, S. 2000. The social construction of scale. *Progress in Human Geography* 24:219–42.

Maskell, P. 2000. Why are not all firms equally well suited for application within the conversation of economic geography. Paper presented at workshop on "The Firm in Economic Geography," 9–10 March, University of Portsmouth, U.K.

Maskell, P.; Eskelinen, H.; Hannibalsson, I.; Malmberg, A.; and Vatne E. 1998. *Competitiveness, localised learning and regional development—Specialisation and prosperity in small open economies.* London: Routledge.

Maskell, P., and Malmberg, A. 1999. Localised learning and industrial competitiveness. *Cambridge Journal of Economics* 23:167–85.

Miles, R.; Snow, C. C.; Mathews, J. A.; and Miles, G. 1999. Cellular-network organizations. In *Twenty-first century economics: Perspectives of socioeconomics for a changing world,* ed. W. E. Halal and K. B. Taylor, 155–74. New York: St. Martin's Press.

O'Neill, P., and Gibson-Graham, J. K. 1999. Enterprise discourse and executive talk:

Stories that destabilise the company. *Transactions of the Institute of British Geographers* 24:11–22.

Perroux, F. 1955. Note sur la notion de 'pôle de croissance.' *Économie Appliqué* 8:25–49.

Porter, M. E. 1990. *The competitive advantage of nations.* London: Macmillan.

Powell, W. W. 1990. Neither market nor hierarchy: Network forms of organization. *Research in Organizational Behaviour* 12:295–336.

Raikes, P.; Jensen, M. F.; and Ponte, E. 2000. Global commodity chain analysis and the French *filière* approach: Comparison and critique. *Economy and Society* 29:390–417.

Schoenberger, E. 1997. *The cultural crisis of the firm.* Oxford: Blackwell.

———. 1999. The firm in the region and the region in the firm. In *The new industrial geography: Regions, regulation and institutions,* ed. T. Barnes and M. Gertler, 205–24. London: Routledge.

Schumpeter, J. [1959] 1934. *The theory of economic development: An inquiry into profits, capital, credit, interest, and the business cycle.* Cambridge: Harvard University Press.

Scott, A. J. 1995. The geographic foundations of industrial performance. *Competition and Change* 1:51–66.

———. 1998. *Regions and the world economy: The coming shape of global production, competition, and political order.* Oxford: Oxford University Press.

Stopford, J. M., and Strange, S. 1991. *Rival states, rival firms: Competition for world market shares.* Cambridge: Cambridge University Press.

Storper, M. 1997. *The regional world: Territorial development in a global economy.* New York: Guilford.

Storper, M., and Harrison, B. 1991. Flexibility, hierarchy and regional development: The changing structures of production systems and their forms of governance in the 1990s. *Research Policy* 21:407–22.

Storper, M., and Walker, R. 1989. *The capitalist imperative: Territory, technology and industrial growth.* Oxford: Blackwell.

Swyngedouw, E. 1997. Neither global nor local: "Glocalization" and the politics of scale. In *Spaces of globalization,* ed. K. Cox, 137–66. New York: Guilford.

Taggart, J. H. 1996. Multinational manufacturing subsidiaries in Scotland: Strategic role and economic impact. *International Business Review* 5:447–68.

———. 1997. Autonomy and procedural justice: A framework for evaluating subsidiary strategy. *Journal of International Business Studies* 28:51–76.

Taylor, M. 1984. Industrial geography and the business organization. In *The geography of Australian corporate power,* ed. M. Taylor, 1–12. Sydney: Croom Helm.

———. 1999. The small firm as a temporary coalition. *Entrepreneurship and Regional Development* 11:1–19.

Thrift, N., and Olds, K. 1996. Refiguring the economic in economic geography. *Progress in Human Geography* 20:311–37.

United Nations Conference on Trade And Development (UNCTAD). 1995. *World investment report 1995: Transnational corporations and competitiveness.* New York: United Nations.

———. 2000. *World investment report 2000: Cross-border mergers and acquisitions and development.* New York: United Nations.

Walker, R. 1988. The geographical organization of production systems. *Environment and Planning D: Society and Space* 6:377–408.

Whitley, R. 2000. *Divergent capitalisms: The social structuring and change of business systems.* Oxford: Oxford University Press.

[13]

Labor and Agglomeration: Control and Flexibility in Local Labor Markets*

Jamie Peck

School of Geography, Manchester University, Manchester M13 9PL, UK

Abstract: Through a critique of Allen Scott's recent work on the economics of labor market agglomeration, the paper develops an approach to labor and agglomeration in which labor control has a central role. Scott's work provides new insights into the processes of labor market agglomeration but understates the importance of labor control in particular and social relations in general. The social character of labor is therefore denied. Consequently Scott does not adequately confront the contradictory relationship between labor and agglomeration. I identify a variety of labor control strategies and also suggest that the strategy deployed reflects, first, the particular "coupling" of the labor process and the supply of labor, and second, the ways in which struggles around workplace discipline and labor reproduction are socially regulated at the local level. Setting labor markets and labor control strategies in their local social context throws light on the question of why firms in some situations find it necessary to relocate in order to restructure their employment relationships, while others are able to effect *in situ* adjustments. Following from this, further work is needed on the relationship between processes of industrial restructuring and the political dynamics of local labor markets.

Key words: labor control, agglomeration, labor markets, social regulation, industrial restructuring, new industrial spaces, flexibility, skills, social reproduction, transactions costs, vertical disintegration.

Of all the major problems that form part of the urban question in contemporary capitalism, the mechanisms of local labor market adjustment and their effects on intraurban [industrial] location are assuredly among the most perplexing and underresearched. (Scott 1988b, 120)

Our knowledge of processes of local labor market restructuring is doubtless partial and underdeveloped. Yet recent debates around the emergence of new industrial spaces under the ostensibly ascendant regime of flexible accumulation have pushed these issues into the foreground of empirical and theoretical research (Scott 1986, 1988a, 1988b, 1988c, 1992; Scott and Storper 1990; Storper and

* Prepared as part of the European Science Foundation's scientific program on Regional and Urban Restructuring in Europe. Helpful comments were received from Ash Amin, Graham Haughton, Ray Hudson, Nigel Thrift, Adam Tickell, and Ian Winter.

Scott 1989, 1990). For Allen Scott in particular, labor market processes have an important role in generating those renewed tendencies for industrial agglomeration believed to underpin the formation of new industrial spaces. As the social division of labor deepens under flexible accumulation, he argues, labor market forces and the logic of intensified interfirm transacting combine to unleash powerful centripetal tendencies (Scott 1988c, 31). Scott's argument, although persuasive, remains by his own admission a partial one, based as it is upon a rather narrow conception of labor market processes. This paper will critically examine Scott's analysis of local labor markets before going on to suggest some preliminary theoretical pointers toward a reevaluation of labor and agglomeration. Here, I shall emphasize processes of labor control, the operation of which underscores the fundamentally *contradictory* relationship between labor and agglomeration.

After briefly summarizing Scott's account of labor, agglomeration, and new industrial spaces, I critically evaluate several of the implications of his analysis, namely the definition of labor market flexibility, the role of labor segmentation, and the constitution of new industrial spaces. The third section of the paper proposes a conception of production, labor, and agglomeration in which the imperatives of labor control are afforded a decisive theoretical role. By conceiving the labor market as a system of power relations (linked to the imperatives of labor control), I establish connections between industrial restructuring and labor market restructuring that place labor and agglomeration in a new light. In particular, I argue that industrial organization and structure are strongly related, in the first instance, to the need to maintain (or extend) control within the labor process and, in the second instance, to the structure of the local labor market and the nature of segmentation therein. These links between industrial organization and the operation of local labor markets have implications for capitalist restructuring strategies. In particular, they have a bearing on firms' choices between employment restructuring in place (i.e., through *in situ* adjustments) and restructuring through space (i.e., through physical relocation).

Scott's Account

In his books *Metropolis* and *New Industrial Spaces,* Allen Scott seeks to specify the linkages between industrialization, urban structure, and the formation of territorial production complexes. He assigns a "strong analytical privilege to the functioning of the production apparatus and its expressive effects in the division of labor" (Scott 1988b, 2). While much of his analysis is pitched at the meso scale, at the level of the production complex, the lineage of Scott's argument is very much rooted in an institutionalist theory of the

firm (Scott 1988b, 29–33). Following Coase (1937) and Williamson (1975, 1985), Scott pursues a transaction costs approach in which the boundaries of the firm and the configuration of the production system are understood to be determined by the relative costs of internal (intrafirm) and external (market) transactions. The virtue of this approach for Scott (1988c, 24) is that

> it allows us to seize production in general (a confusing assemblage of labor processes, technologies, physical stocks, and so on) as a coordinated system of internal hierarchies and external markets. . . . [Moreover] since there is not in reality a sharp break between internal (hierarchical) and external (market) relations, but an irregular continuum extending over a variety of intermediate forms (joint ventures, partnerships, quasi-vertical integration, and so on), we can see production as a complex but rationally comprehensible organizational structure rooted in the polarities of the firm and the market.

Scott deploys a transaction costs approach to establish the incipient agglomeration tendencies seen as inherent in contemporary capitalist industrialization. His argument proceeds in four stages.

First, a series of forces in contemporary capitalism—which Scott captures within the rubric of the transition from Fordism to flexible accumulation—are engendering a need for *enhanced flexibility* in production systems (Scott and Cooke 1988, 241–42). The fragmentation of consumer markets and heightened levels of business uncertainty are two key factors behind this shift (Scott 1988b, 35–37). Second, this flexibility is stimulating a process of *dynamic vertical disintegration* in the production system, as firms seek to enhance their flexibility and responsiveness by externalizing many of the functions previously performed within the firm. This amounts to more than a "breaking up" of established production chains, because flexible production systems are seen as expansionist and innovation rich: as the system expands, new and independent forms of specialist production emerge (Scott 1988c, 27–28). Third, in this *deepened social division of labor,* individual producers become locked into "networks of extremely malleable external linkages and labor market relations" (Scott 1988a, 174). The redrawing of the boundaries of individual production units, with the concomitant downsizing of plant sizes and narrowing of organizational specialisms, brings about a heightened *interdependence* in the production system as firms become embedded within complex webs of interorganizational transactions. Fourth, the twin requirements of minimizing external (interfirm) transaction costs and of establishing an appropriate set of labor market relations bring about a marked *agglomeration* of economic activity. New industrial spaces, then, are seen to coalesce around "dense networks of transactional interrelations" (Scott 1988c, 31) and are associated with the establishment of new labor market norms, based upon the principles of numerical and functional flexibility (Scott 1988a, 177; 1992, 266).

Scott conceives new industrial spaces as highly integrated territorial production complexes, the logic of which is a collective one and therefore represents more than the simple outcome of aggregated individual behaviors (Scott 1988b, 42; 1988c, 29). Of particular concern here is the collective logic associated with labor market operation in new industrial spaces. The search for flexibility in production is paralleled in Scott's account with a search for flexibility in the labor market. Storper and Scott (1990, 575) identify three forms of labor market flexibility under flexible accumulation. First, there is an attempt to *individualize* the employment relation, moving away from (institutionalized and therefore comparatively rigid) collective bargaining and negotiation systems in key areas such as wage setting. Second, firms are seeking to achieve enhanced *internal flexibility* through labor process changes such as multiskilling and reduced job demarcation. Third, *external flexibility* is being sought through strategies (such as the deployment of part-time and temporary workers) that enable rapid quantita-

tive adjustments to the labor intake to be made in accordance with fluctuating production needs.

These strategies for enhancing the flexibility of firms' labor relations are generating increased polarization and segmentation in the labor market. More often than not, they are associated with the diminution of labor market security and with the establishment of regressive work norms (Scott 1988a; Storper and Scott 1990). Exceptionally high levels of labor turnover (even among upper-tier workers), low rates of union density, deinstitutionalization of worker protection, and the presence of a large contingent labor pool of politically marginalized workers (such as immigrants, women, and agricultural laborers) are all characteristic of flexible labor markets in the new industrial spaces (Scott 1984, 1988c; Storper and Scott 1989, 1990). These labor markets are consequently seen to combine high levels of flexibility with deep segmentation.

In the labor market, too, significant agglomeration economies can be obtained. "All such labor market fluidity," Scott (1988a, 177) argues, "is enhanced as the size of the local pool of jobs and workers increases." These agglomeration economies follow from five interrelated labor market conditions (Scott 1988b, 120–38; 1988c, 33–38). First, labor turnover rates tend to correlate positively with the overall size of the local labor market. Given the strong random component in job separations and accessions, these flows are more likely in large labor markets than in smaller ones to converge toward equilibrium. This enhanced "predictability" in labor market flows allows firms to fine-tune their labor utilization to changing production needs: large, volatile labor markets provide a ready supply of workers who can be hired as and when they are needed. Second, workers are better able to accommodate to unstable job tenures (and hence less likely to out-migrate) in large and turbulent labor markets where the supply of alternative jobs is plentiful, enabling workers to move from job to job with increased ease and rapidity.

Third, information and search costs tend to fall as the size of the local labor market increases. High rates of job changing and job filling place a premium on efficient job search mechanisms. The greater the size of the labor market, the lower the marginal cost of search activities (such as newspaper advertising for firms or visits to a labor exchange for workers). Fourth, workers accustomed to secondary labor market conditions tend to be drawn into large, volatile labor markets, where their chances of finding (and refinding) work are highest. In turn, employers of such labor (and those with labor-intensive production systems in general) will be induced to gravitate toward the spatial core of their preferred labor supply. A recurrent relationship consequently exists between the tendencies for firms to cluster around appropriate labor supplies and for workers' residences to cluster around major centers of employment opportunities. Fifth, processes of local socialization will tend to reinforce agglomeration tendencies, as workers become acclimatized to particular work rhythms and as they develop appropriate labor market coping strategies. Over time, patterns and processes of labor market behavior thus become socially embedded and, to a certain extent, self-perpetuating.

In these ways, labor market agglomeration tendencies become mutually reinforcing and self-sustaining. At this point in Scott's analysis, labor market forces interact with those associated with the restructuring of organizational systems. Economic change and agglomeration exist in a reciprocal relationship:

> Thus, on the one side, the social division of labor provokes spatial agglomeration as a way of lowering external transaction costs; on the other side, agglomeration encourages further social division of labor and in-migration of new producers precisely because it lowers these costs; and so the cycle of action and reaction continues until its inner energy is exhausted. . . . With the rise of any industrial agglomeration, local labor markets are set in motion, and they

too help boost processes of spatial concentration and growth. (Scott 1988c, 33)

For Scott, the most theoretically significant conjuncture at which this "inner energy" of cumulative causation breaks down is the transition between regimes of accumulation. Each regime of accumulation is held to have a distinctive geographical logic: the shift to flexible accumulation is associated with the emergence of a new form of uneven development. The locational logic of flexible accumulation consists, according to Scott, of a twin tendency for, first, the active *evasion* of Fordist labor pools (with their politicized working class, institutionalized labor processes, and high cost structures) and, second, the selective spatial *reagglomeration* of production in locations insulated (either socially or, more often, geographically) from the core regions of Fordist industrialization (1988c, 11–15). Thus, empirical examples of new industrial spaces range from enclaves within the old manufacturing regions, through suburban extensions of major metropolitan areas, to craft communities; from established central business districts to rural communities with an agricultural heritage (Scott and Cooke 1988; Storper and Scott 1990). Significantly, innovative production and labor market norms established in new industrial spaces are, through a kind of "backwash effect," imported back into the devalorized old industrial regions as the new regime of accumulation becomes established (Scott 1988c, 108).

Toward a Critique

Scott's work on new industrial spaces is clearly innovative and challenging. Not surprisingly, it has provoked considerable critical interest. The purpose here is not to develop a comprehensive critique of Scott's work, a task under way elsewhere (see Sayer 1989; Amin and Robins 1990; Gertler 1988, 1992; Lovering 1990, 1991; Henry 1992; Tickell and Peck 1992) and which has begun to yield a response (Scott 1991a, 1991b, 1992). My objective is to critically interrogate Scott's specific assertions concerning labor and agglomeration. The bases for these assertions are examined in three areas: the controversial issue of increasing labor market flexibility, the question of changing forms of labor segmentation, and the problem of the geographic logic of the new industrial spaces.

New Flexibility?

A central tenet of Scott's argument is that increasing flexibility in the production system is related to increasing flexibility in labor markets. Trends such as the rise in the relative and absolute size of contingent work forces (such as those employed on a temporary, part-time, or contract basis) and the shift toward decentralized pay bargaining are indeed indicative of increased labor market flexibility (Organization for Economic Cooperation and Development 1986a, 1989; Williams 1986; Hakim 1987; Boyer 1988; Laflamme et al. 1989). The question of how to *interpret* such changes, in either theoretical or political terms, is, however, intensely problematic (Organization for Economic Co-operation and Development 1986b; Standing 1986; Boyer 1987; Harrison and Bluestone 1987; Tarling 1987; Hyman 1988; Pollert 1988; Hudson 1989; Rodgers and Rodgers 1989; Rosenberg 1989b; Howell 1992). Several interpretative difficulties seem immediately apparent. First, cyclical, temporary, and ephemeral phenomena must be distinguished from longer-term, structural changes. Second, the search for flexibility must be understood in terms of the historical and geographic evolution of particular labor market and production systems. Third, it is necessary to establish the extent to which such changes relate to (and subsequently further modify) the balance of power in the labor market. Fourth, instances of and strategies for labor market flexibility must be interpreted in light of the restructuring of the state's regulatory systems and with regard to contemporary political discourses.

Hudson (1989), in his study of labor market restructuring in old industrial regions, argues that many of the labor market phenomena being interpreted as evidence of rising flexibility are in fact manifestations of a quite different set of processes. He finds the erosion of the power of organized labor, the imposition of hostile neoliberal regulatory systems, the presence of a permanent pool of long-term unemployed workers, and the need to restructure established (Fordist and pre-Fordist) industries to be the most persuasive explanations for rising labor market "flexibility." Hudson concludes that such flexibility reflects not so much the emergence of a new regime of flexible accumulation as a "reworking of existing accumulation strategies as capital takes advantage of its greatly enhanced strength vis-à-vis labor on the market" (1989, 25). Particular care, then, must be taken in the interpretation of labor market flexibility, especially when links to processes of production restructuring are being asserted.

To be certain, there is no straightforward connection between industrial structure and labor market structure, let alone between flexibility in production and flexibility in labor markets (Wilkinson 1983; Peck 1989). Indeed, such "surface" manifestations of flexibility and fragmentation as the vertical disintegration of production or the breaking up of collective labor market structures are often associated with powerful tendencies toward the further *concentration* of social capital and of oligopolistic control (Michon 1987; Sayer 1989; Amin and Robins 1990). Disintegration and fragmentation on the surface often reflect underlying processes of integration and centralization. This is certainly true in the labor market context, where many of the so-called flexibility strategies—"individualized" employment relations, plant-level pay bargaining, incentives-based contracts—are established means of deepening control over the labor process in ways of which Frederick Taylor might have been proud.

If the links between flexible production and flexible labor markets are not direct, the two may not be causally related in the present historical conjuncture. Far from being a recent phenomenon associated with the rise of flexible accumulation, flexible labor markets have a long history. Indeed, *different forms* of flexibility have been associated with different historical phases in the process of industrialization (see Gordon, Edwards, and Reich 1982). There is a need, then, to clarify, first, which aspects of contemporary flexible labor markets are residuals of earlier phases of capitalist accumulation; second, which are the product of contingent conditions such as those suggested by Hudson (1989); and most importantly, third, which are necessary features of flexible accumulation. Inevitably, these three processes will overlap and interact: part of the "see-sawing" logic of the historical geography of capitalist industrialization is the exploitation of economically disenfranchized labor pools created in earlier periods (Webber 1982; Massey 1984; Smith 1984). Firms also opportunistically seize upon changing external conditions, such as the tendency for cyclical recession to ebb the power of trade unions, in their efforts to restructure the labor process (Braverman 1974; Wood 1982, 1989). The point at issue here is not the manner in which flexible accumulation can initially "take hold" in a particular labor market (for example, the form in which it engages with the disorganized immigrant labor pools of inner Los Angeles), but how it is subsequently sustained and, most importantly, reproduced.

Despite its theoretical significance, very little work has explored the reproduction of flexible labor markets. Given that such labor markets are quite possibly no more than manifestations of relatively short-term phenomena (such as high levels of unemployment, the imposition of regressive employment policies, capitalist "plundering" strategies), the question of their sustainability as economic structures is doubly important to address. There may

indeed be significant obstacles to sustainable labor market flexibility. Drawing upon the French experience, Michon (1987, 42–43) has argued that structural limits hamper the development of external flexibility through such mechanisms as subcontracting and homeworking:

> [Once] personnel management and/or the supply of labour and/or the training and supervision of the workforce are involved, the externalisation of activities runs up against some particular difficulties . . . linked to the control of the labour force. . . . [The] need to control the workforce seriously restricts the use of externally supplied manpower. More generally, the desire to retain control over all manufacturing processes, and to avoid any sort of dependence on other individual capitals considerably restricts the scope for implementing an extreme division of labour. In other words, the processes in the division of labour seem to be doubly dependent, first, on the processes of capitalist competition, and second on the need to control the workforce, and thus on the processes that determine the "rapport salarial" [employment relation].

While external flexibility favors some forms of labor control (for example, its role in breaking up collective labor structures), it simultaneously undermines others (for example, direct supervision or on-the-job, firm-specific skill formation).

The contradictions of flexibility strategies are not confined to external flexibility; internal flexibility too is problematic. Through strategies such as multiskilling, the broadening of job categories, and the formation of flexible work teams, firms are seeking to enhance the qualitative or functional flexibility of their core work forces (Atkinson 1985). Such restructuring of firms' internal labor markets may, however, have undesired effects (see Saxenian 1983; Gertler 1988; Sayer 1989). First, the formation of multiskilled core work forces, with bundles of specialist skills crucial to the production process, can afford this group of workers substantial bargaining power vis-à-vis management. Second, multiskilling may also increase the "tradeability" of core workers' skills on the external labor market, with the result that labor turnover increases and firms become reluctant to reinvest in training. A further implication of this second condition is that wage inflation across the local labor market may ensue, which in turn may stimulate yet higher rates of mobility among skilled workers.

Flexible labor markets will begin to break down if appropriate mechanisms of labor control and reproduction are not set in place. The technopoles of Southern California, indeed, are already displaying signs of "market failure" within their training systems (Scott and Paul 1990), while wage inflation became a serious problem in British growth centers such as Cambridge (Crang and Martin 1991). The internal economies of scale, which are associated with training and the long-term and risk-laden nature of skill investments, suggest that the process of skill formation will always be problematic in labor markets composed of networks of small interdependent firms, particularly if the degree of interfirm competition is high (Peck and Haughton 1991). Critical for the effective operation of any system of skill formation is the establishment of an appropriate system of social regulation.[1] Indeed, the existence of such an institutional system is a measure of maturity in a local labor market (Scott 1988b, 182). Few such fully coherent systems are currently in place in either Europe or North America (Harrison and Bluestone 1988;

Boyer 1988; Rosenberg 1989c; Storper and Scott 1990).

The current vogue for "deregulation" in policy-making circles is unlikely to provide any solutions in this regard. In Britain, for example, attempts to create a "market-driven" training system have ended in predictable failure: the retrenchment of state funding for training, the removal of mechanisms for the control of labor poaching, and the dissolution of effective planning frameworks have resulted in a skills crisis of unprecedented proportions (Finegold and Soskice 1988; Peck 1991). Such neoliberal approaches to skill formation are consequently unlikely to sustain labor market development in the medium term. Other institutional experiments are, however, under way. The ongoing process of reform in education and training systems in many of the advanced industrial countries is an uncertain and multidirectional one, conditioned as it is by political struggles in and around the state (Green 1992). Such struggles might assume increasing importance as nation-states begin to concentrate on supply-side policies, in the context of globalizing economic forces and the erosion of national macroeconomic integrity (Jessop 1991; Peck and Tickell 1992). What is clear, though, is that we remain a long way from any sort of institutional "solution" to the problem of skill formation.

There are serious structural flaws in the process of labor reproduction within flexible labor markets. Certainly, the model of the flexible labor market which is based upon institutional deregulation and high levels of external flexibility seems particularly crisis-prone. Unless workable political and economic mechanisms for the reproduction of flexible labor markets can be established, these economic structures may prove to be no more than transitory economic phenomena, a combined product of short-term "plundering" strategies on the part of capital and political opportunism on the part of neoliberal nation-states. This has wider implications for the flexibility de-

---

[1] Social regulation is defined here in regulationist terms as the complex of state forms, habits and norms, social mores, and laws which together play a role in underwriting the process of accumulation. In regulation theory, the mode of social regulation (MSR) is regarded as a temporary institutional "fix" that contains in the medium term the inherent crisis tendencies in capitalist development. Not functionally determined, the MSR is shaped through theoretically indeterminate political struggles, particularly—in the original theory at least—in and around the nation-state (see Tickell and Peck 1992).

bate, in which greater care needs to be taken in distinguishing between reproducible social structures and (possibly nonsustainable) experimental strategies (see Jessop 1992; Leborgne and Lipietz 1992; Peck and Tickell 1992).

New Segmentation?

The phenomenon of labor market segmentation is certainly not new (Doeringer and Piore 1971; Reich, Gordon, and Edwards 1973). Also long established are the strategies of restructuring and redefining labor segmentation in accordance with changing economic conditions (Sengenberger 1981; Villa 1986; Nohara and Silvestre 1987). The early dualist distinction between stable primary and volatile secondary labor markets was rediscovered in the 1980s, when concepts such as the "flexible firm" and core-periphery labor markets were popularized (Atkinson 1985, 1987). More sophisticated work has highlighted the finely grained character of empirical instances of segmentation and the multicausal nature of segmentation processes (Labour Studies Group 1985; Villa 1986). The diverse causal roots of segmentation have now been traced to technological requirements, to divergent industry structures, to imperatives of managerial control, to processes of skill formation, to the activities of the state and of trade unions, and to the dynamics of the reproduction sphere (see Peck 1989; Rosenberg 1989a).

New industrial spaces tend to be characterized by deepening labor market segmentation and by income polarization (Storper and Scott 1990). The search for enhanced flexibility induces firms to restructure primary sector labor markets and to export certain labor functions to the expanding secondary sector. Thus, in accounts of new industrial spaces, the processes of labor market flexibility and labor market segmentation evolve side by side. In fact, there is a great deal of conceptual slippage between these two sets of processes, despite their quite different intellectual origins. While labor

market segmentation theory is now associated with a position critical of economic orthodoxy (Wilkinson 1981, 1983; Tarling 1987), much of the discourse of labor market flexibility has its roots in supply-side economics, with its emphasis on the logical priority (and real-world desirability) of competitive labor markets, freed of "imperfections" (such as trade unions and welfare support) and operating in accordance with the laws of supply and demand (Pollert 1988; Standing 1989). For the supply-side economist, then, segmentation, with its associated discontinuities in labor mobility and internalized "shelters" from labor market competition, is anathema to flexibility. This approach, which has dominated policymaking in much of Europe and North America in the 1980s, is to deregulate the labor market, to remove barriers to the mobility of labor and to the adjustability of wage rates.

Contrary to the canons of supply-side economics, deregulated labor markets (to the extent that it is possible to create such conditions) tend to be associated *not* with convergence upon equilibrium but with a disturbing trend toward job and wage polarization (see Standing 1986; Harrison and Bluestone 1988; Rosenberg 1989c; Howell 1992). So are labor markets becoming more flexible or are they being resegmented, or both? Present evidence is inconclusive and, again, the central issue becomes one of interpretation. In Atkinson's (1987) core-periphery model, different types of flexibility strategies are associated with different segments of the labor market: qualitative flexibility in the primary sector and quantitative flexibility in the secondary sector. For Wilkinson (1988), segmented labor markets can also be flexible labor markets; indeed, given the state of dynamic interdependence existing between the primary and secondary sectors, flexibility is a part of the very rationale of segmentation. Developing this view, Rosenberg (1989a, 392) regards the search for flexibility as part on an ongoing process of labor market restructuring: "As the search for flexibility proceeds and labor markets become re-structured, the boundaries of labor market segmentation change. The labor markets do not necessarily become less segmented. And even if they become less segmented, that does not necessarily mean that they are more flexible."

In analyses of labor market restructuring, it is not at all helpful to conflate different notions of flexibility. It is particularly important to make a distinction between the *competitive flexibilities* associated with deregulation, individualized employment relations, and sharpened competition (and relate to Leborgne and Lipietz's (1988) Californian model of labor market regulation) and the *structured flexibilities* associated with regulated and collectivized labor markets (and relate to Leborgne and Lipietz's (1988) Saturnian model). Each of these have different rationales, different causes, and different impacts. Flexible labor markets, then, do not have a single, universal "logic," but a variety of "logics." Their nature tends to vary from one place to another, along with the local regulatory and social context (Peck 1993). These differences, and their causes, must be unravelled with great care. It is far from certain that they are rooted in a "common causal dynamic," the defining theoretical feature of the new industrial spaces (Scott 1988b; Storper 1990; cf. Henry 1992).

One way of overcoming this impasse is again to focus less on the short-term mechanisms of restructuring than on the longer-term processes of labor market *reproduction*, less on strategies than on *structures*. Two points are worthy of mention here. First, the means by which highly polarized labor markets are *politically* reproducible in the medium to long term have yet to be specified (Mahon 1987; Brunhes 1989; Storper and Scott 1990; Tickell and Peck 1992; cf. Davis 1986; Jessop et al. 1988). Widening socioeconomic inequalities and political tensions are predictable consequences. As Saxenian (1983, 256) demonstrated in her analysis of the urban contradictions of industrialization in Silicon Valley, these

problems may be of sufficient magnitude to put a brake on local economic growth and even to trigger restructuring in the productive sphere:

> [It] became increasingly difficult to accommodate and reproduce both segments of [the semiconductor industry's] dichotomized workforce within the same metropolitan area. Inflation of housing prices, transportation congestion, labour shortages and the no-growth movement are all manifestations of the limitations of the local spatial structure for accommodating the industry's bifurcated workforce. These urban contradictions eventually caused the industry itself to restructure, thus preserving Silicon Valley as a site for headquarters, high level research and prototype production activities.

Second, the intrinsic problems of *skill formation* in flexibly externalized labor markets discussed in the previous section imply that the enlargement of the secondary sector may not be a sustainable option. As Doeringer and Piore (1971) emphasized, one of the main causes of segmentation is the differentiated nature of skills and training: primary sector employers create internal labor markets in order to reduce employee turnover following costly investments in training, while lower and more irregular skill needs in the secondary sector enable employers to forego training investments and recruit directly from the external labor market. The important implication of this differentiated process of skill formation is that the generalized enskilling of the labor force anticipated in some visions of post-Fordism (Piore and Sabel 1984; Applebaum and Shettkat 1990; cf. Graham 1992) may be *incompatible* with deepening segmentation, the structural growth of the secondary sector, and the emergence of a small firms-oriented economy.

Although there are some early signs of modes of skill formation that may be compatible with small firms-based economies (Perulli 1990) and with the higher echelons of the secondary labor market (Peck and Haughton 1991), all such systems are invariably fragile. Skill formation and its accompanying system of social regulation, indeed, seems destined to be one of the decisive factors in determining whether economies take the "high road" or the "low road" from Fordism (Mahon 1987; Leborgne and Lipietz 1988, 1990; Brunhes 1989; Rojot 1989; Boyer 1990; Michon 1990; Sengenberger and Pyke 1990; Green 1992). "Low road" approaches, based on competitive flexibility, already seem to be faltering: unregulated competition breeds short-termism and a reluctance to invest in either skills or technology, "resulting in a vicious, downward-spiralling cycle" (Sengenberger and Pyke 1990, 10). "High road" approaches based on structured flexibility, on the other hand, certainly seem to be more sustainable, with their high standards of social protection for workers and collectivized economic systems, but in practice such approaches remain comparatively rare (Leborgne and Lipietz 1988; Brusco 1990).

One of the most testing issues facing the nascent industrial spaces (and for that matter facing those who are attempting to theorize them) is the establishment of effective forms of "local governance" (Storper and Harrison 1991) or, more broadly, "modes of local social regulation" (Peck and Tickell 1992). These relate to broad-ranging issues of local social reproduction (such as the socialization of the rising generation or the establishment of sustainable household divisions of labor) as well as to the more specific concerns of social regulation in labor markets and in interfirm relations (such as measures to ensure that firms can safeguard their skills investments or to contain to socially acceptable levels the rate of exploitation of labor). One of the most basic requirements of such a regulatory system is that it should be capable of sustaining an environment in which risk-laden, long-term investments in innovation, technologies, and skills can be made. Nowhere is this more important than in the area of skill formation, where unregulated competition leads to underinvestment. If institutions and networks appropriate for

the reproduction of labor skills—specifically, to *socialize* the process of skill formation—are not set in place, then economic growth will break down. There is consequently a possibility that industrial disintegration and labor market segmentation, upon which the logic of flexible production systems is predicated in Scott's analysis, will ultimately undermine the nascent growth model of flexible accumulation itself.

New Industrial Spaces?

A characteristic of flexible production is, for Scott (1988c, 14), that it is "always some distance—socially or geographically—from the major foci of Fordist industrialization." Thus in some instances flexible production will be found outside the old manufacturing regions, in others within it: "new" industrial spaces are sometimes located within old industrial spaces. So in some cases capital seeks out a *tabula rasa*, where flexible employment relations can be created "with a minimum of local obstruction" (Scott 1988a, 178), whereas in others it is apparently possible to reconstruct these employment relations *in situ*. Scott's account of the geographic formation of flexible employment relations is therefore ambiguous.

Capital's periodic need to reconstruct the employment relation induces the abandonment of regions inculcated with socialized labor processes deemed inappropriate for continued accumulation (Webber 1982; Storper and Walker 1983; Smith 1984). This imparts an important dynamic to the process of uneven development. Webber (1982) has suggested that this process of spatial switching is a necessary one: each regime of accumulation will create new, geographically defined, core-periphery relations. From this perspective, the process of spatial restructuring Scott describes for the new regime of flexible accumulation seems partial, almost hesitant, in that some colonization of "new" regions co-exists with a great deal of staying put. Does this mean that the process of accumulation can be revived *in situ?*

Certainly, possibilities for reconstructing employment relations *within* as opposed to *between* local labor markets do exist, for example by recruiting from different segments of the labor market or from the rising generation (Morgan and Sayer 1985; Barnes, Hayter, and Grass 1990; Peck 1992). It is just as important to understand why capital does *not* move, why local labor markets do *not* decline—the processes of *in situ* restructuring—as it is to understand the processes of relocation and restructuring through space. Indeed, this is all part of the same question. Presumably, firms do not consider relocation until the possibilities for *in situ* adjustment have been exhausted. Unfortunately, geographers are sometimes guilty of focusing primarily or exclusively on spatial restructuring processes and relocation strategies and of ignoring *in situ* ones. Scott implies as much in his comment that we know little about the processes of local labor market adjustment (1988b, 120). In calling also for more work on *in situ* employment restructuring, Barnes, Hayter, and Grass (1990, 146) note that because geographers have tended to emphasize "the geographical solution, they have not always unpacked the internal dynamics of particular segments of local labour markets in a given place and time."

The existence of such *in situ* restructuring strategies, by which capital is able to reconstruct its local labor market relations, poses some problems for conventional conceptions of industrial restructuring. Put simply: what determines whether restructuring capital will opt for an *in situ* or a "spatial" strategy? Harvey (1989a, 187) has maintained that the existence "side by side within the same space" of alternative labor strategies enables capitalists "to choose at will between them." But where does this leave the process of spatial restructuring? Why do firms relocate? Clearly, *in situ* employment restructuring has limits, such as those imposed by the continued existence of inappropri-

ate local regulatory frameworks (see Moulaert 1987; Heckscher 1988).

One of the key tasks for theorists of flexible accumulation, and indeed of capitalist restructuring in general, must be to determine the conditions under which these "choices" between restructuring strategies are being made and to specify likely outcomes in different situations. Given the body of knowledge already accumulated on processes of spatial restructuring (Bluestone and Harrison 1982; Massey 1984; Clark, Gertler, and Whiteman 1986; Scott and Storper 1986; Storper and Walker 1989), priority here must be placed upon mechanisms of local labor market adjustment and their relationships with the process of industrial restructuring.

Production, Labor, and Agglomeration

Labor Control and the Labor Market

I have argued that several issues remain problematic or unresolved in Scott's approach to labor and agglomeration. Some of these difficulties—the problem of skill formation in flexible labor markets, the role of *in situ* versus spatial restructuring strategies—derive from Scott's somewhat mechanistic approach to labor market analysis. Transaction costs analysis shares with mainstream neoclassical economics a vulnerability to criticisms of methodological individualism (see Marsden 1986). Labor markets are seen to be populated by utility maximizers and configured by cost considerations. As Henry (1992) points out, Scott sees the firm as an exchange mechanism rather than as a production organization, driven among other things by the imperative to maintain labor control. Segmentation, power relations, and imperatives of labor control are afforded no more than a subsidiary role in an approach that places logical priority upon the market-related phenomena of transaction costs.

Scott maintains that his approach is "complementary to, and not in opposition with" those which emphasize the "tense force-field of capitalist-worker relations" (1988c, 24), but such considerations play no more than a minor role in his subsequent analysis. Scott's approach is less unbendingly orthodox (cf. Lovering 1990) than "selectively eclectic," a combination of institutionalist economics, a particular reading of regulation theory and "post-Weberian" economic geography (Scott and Storper 1990, 5–6). Aside from the epistemological problems of such an eclectic approach (see Fincher 1983), Scott's emphasis on transaction costs represents an implicit subordination of forces rooted in the capital-labor relation. This may have serious implications for his argument, for the transaction-costs "market" upon which the burden of explanation is ultimately placed is itself socially constituted. Correspondingly, transactions "costs" are socially constructed, the malleable outcome of social struggles, rather than the mathematical derivative of some set of ostensibly apolitical economic processes.

The role of social relations needs to be integrated into analyses of labor and agglomeration, rather than "added on" in a post hoc fashion. Thus, I argue that labor markets should be regarded, first and foremost, as political constructions imbued with profoundly asymmetrical power relations (see Giddens 1980; Offe 1985; Rueschemeyer 1986; Solow 1990) and only secondarily as economic systems governed by transaction costs. Asymmetries of power within the labor market, many of which are rooted in the labor process, are principal determinants of the structure and segmentation of labor markets (Kreckel 1980; Peck 1989). In addition to the "primary" asymmetry between capital and labor, a series of "secondary" asymmetries reflect the uneven distribution of labor market power—first, among employers (for example, between monopoly and competitive firms, between those in stable and unstable product markets) and second, among the work force (for example, between unionized and non-unionized sectors, between different age,

gender, and ethnic groups). The distribution of primary and secondary sector jobs is shaped by these secondary asymmetries within the work force: politically marginalized groups such as blacks, women, and young people must bear the brunt of low pay, poor working conditions, and unemployment, which are associated with the secondary sector. Some ascribe capital a conscious role in the exploitation of these divides within the working class (Reich, Gordon, and Edwards 1973; Gordon, Edwards, and Reich 1982).

These profound discontinuities in the labor market are legitimated through the attribution of *skilled status*. In this sense, skill should not be seen simply as a "resource" that is rewarded in accordance with human capital theory, but as an *ideological* construct reflecting the distribution of power in the labor market (Allen 1977; Cockburn 1983). A common characteristic of "secondary workers" is that they are denied access to skilled status, even though the "technical" requirements of their job may warrant this (Craig et al. 1982). Accordingly, the process of skill formation represents considerably more than the simple activity of training workers for jobs: it is one of the principal mechanisms by which inequalities in pay and power are produced, reproduced, and legitimated. Not surprisingly, struggles over skill have been among the most bitter in the history of industrial relations (More 1980), and many recent struggles have been triggered by attempts to redefine skills in some way (Mahon 1987).

Gender and ethnicity are particularly important dimensions of this social construction of skills. The continuing segregation of the labor market along the lines of gender and ethnicity feeds, and is fed by, the social processes through which the "value" of skills is determined (see Horrell, Rubery, and Burchell 1990). It is no coincidence, then, that flexibility for white, male workers tends to be about multiskilling, responsible autonomy, and task redefinition, whereas for many women and black workers it means lower pay, irregular employment, and harder work. Nielsen (1991) has argued, for example, that flexibility must be regarded as a gendered concept. New forms of labor flexibility are associated with new forms of labor control, new forms of labor exploitation. Again, these processes are themselves socially constructed, and as a result will tend to vary in character and intensity from one place to another (see, for example, Harrison and Bluestone 1987; Capecchi 1989; Hadjimichalis and Papamichos 1990; Crang and Martin 1991). Stated rather differently, variations in local forms of labor segmentation, in local norms of labor control, in local patterns of informal working, and so on, represent a geographically variable prior set of possibilities for the establishment of new labor market relations. Even within Scott's (1988c) sample of new industrial spaces, the social relations of the labor market vary considerably, reflecting the diverse economic histories and social conditions of these areas.

Contemporary flexibility strategies seem to be bringing about a widening of both primary and secondary asymmetries in the labor market: increased work flexibility is often associated with the reassertion of managerial dominance (Pollert 1988), while in terms of secondary asymmetries, "the burden of adjustment [is placed] on the shoulders of the weakest" (Meulders and Wilkin 1987, 4). Flexibility in the labor market is then as much a political as an economic phenomenon: there will be losers as well as winners under a new flexible regime (Kern and Schumann 1987). This point sometimes seems to be lost on academic advocates (for this is the role they are increasingly fulfilling) of the new flexibility. There is a need, then, to challenge aspects of the emergent politics of post-Fordism (Graham 1992).

Such a view of the labor market as a system of power relations brings into question those interpretations of flexibility as a "new deal" for labor (for example, Piore and Sabel 1984). Pollert may indeed be closer to the truth when she asserts that the search for labor market flexibility

represents an "ideological offensive which celebrates pliability and casualization, *and makes them seem inevitable*" (1988, 72, original emphasis). Power relations in the labor market profoundly affect the course of economic development. These forces simply cannot be held in suspension while transaction costs are calculated, but should be afforded a central role at the very outset. Through such a reorientation it should be possible to add a new dimension to Scott's analysis of labor and agglomeration. By placing the question of labor control at the center of the analysis, for example, some of the *contradictions* of agglomeration become more readily apparent. The massing of workers at one location facilitates the process of trade union organization (Lane 1988, Harvey 1989b); agglomeration can consequently undermine labor control, unless appropriate countermeasures (for example, the construction of internal labor markets, the monopolization of a segment of the labor supply) can be set in place.

Labor control strategies vary both historically and geographically. In *Contested Terrain*, Edwards (1979) developed a periodization of labor control strategies (from simple to technical to bureaucratic), which Gordon, Edwards, and Reich (1982) were later to integrate into a more broad-ranging treatment of historical shifts in the social structures of accumulation in the United States. This approach is useful in that it links, albeit in a rather schematic way, different forms of labor control with different segments of the labor supply. A relationship is consequently established between the labor process and the labor market (although for Edwards and his colleagues, the former was seen to shape the latter). Building to a certain extent upon these insights, Burawoy (1985) deepened the approach to labor control by taking into consideration the role of the state and of the (related) conditions surrounding the reproduction of labor. His explicitly politicized treatment of the labor process centers on the notion of "factory regimes," the overall political form of production that brings together the labor process and those "*political apparatuses* which reproduce [the] relations of the labour process through the regulation of struggles" (Burawoy 1985, 122, original emphasis).

Burawoy demonstrates how these political apparatuses are shaped and how they subsequently shape the labor process under different historical and geographic conditions, though he stops short of considering the theoretical implications of this spatially uneven development in the politics of production. Alan Warde has taken up this point, arguing that "the nature of the *local* labour market and the impact of the *local* political system often have immediate consequences for factory discipline" (1989, 51, original emphasis). As Warde's case study of Lancaster revealed, conditions of labor market dependency (following from its domination by a handful of large employers and the paucity of alternative employment opportunities) played an important role in explaining the exceptional quiescence of the local working class and in explaining the labor control strategies adopted by local firms. Even in this case, labor control strategies were not defined in accordance with the free will of management, but were formed within the parameters set by the local political and social milieu. In this sense, the *contested terrain* of labor control has a significant territorial, as well as historical, dimension.

Firms' labor strategies are consequently defined within a complex political arena. Among the factors that shape labor control strategies are the role of the state in the reproduction of labor, the relationship between internal and external labor markets, the dynamics of local political movements, and the structure of household divisions of labor. The outcomes of these social struggles are by no means predetermined. Reducing these complex and locally articulated processes to a set of ostensibly universal rules based on transactions costs denies the social nature of labor itself. *Labor strategies are not calculated, they are struggled over.* Moreover, these struggles—mobilizing as they

do particular political resources and capacities, constrained as they are by different sets of political and economic parameters, regulated as they are in contrasting ways—hold the key to explaining what is "local" about the local labor market. Consideration of how labor markets are regulated in locally specific ways does not have a place in Scott's rather economistic account, but this must surely play a part in shaping new geographies of labor. As we will suggest below, this may also exercise an influence on industrial organization.

Vertical Disintegration and Labor Control

For Scott, industry structure (measured in terms of degrees of integration/disintegration) is determined by transaction costs. These, I have argued, are in fact socially constructed, reflecting among other things the balance of power in the labor market. Where issues such as labor control are considered in Scott's account, they seem to be just one of the factors in the transaction cost equation (an equation which itself addresses only the unidimensional question of internalization versus externalization). The issue of labor control is much more complex and multidimensional than this. I see labor control as referring to the interrelated processes of, first, securing an appropriate labor supply; second, maintaining control within the labor process; and third, effectively reproducing this set of social relations. None of these factors is integrated into Scott's account.

Imperatives of labor control significantly influence industry structure, including industrial integration/disintegration (see Friedman 1977; Gordon, Edwards, and Reich 1982). Employers have used subcontracting relations, for example, to extend control over the labor process; they move jobs from large, unionized factories to small, non-unionized workshops, thereby redefining the social relations of work (Holmes 1986; Peck and Lloyd 1989). Industry structures (and by implication, industrial restructuring) consequently reflect existing possibilities for establishing different forms of labor control *within* a local labor market. The chain of logic here is as follows:

- A given production chain or industry comprises a series of interrelated labor processes.
- Each of these labor processes tends to be associated with qualitatively different labor supplies (although the chosen labor supply is not determined functionally by the "technical" characteristics of the labor process).
- Each of these "couplings" between a particular labor process and a particular labor supply is associated with a different system of *labor control*.
- Industry structure (including the degree of integration/disintegration) will be configured in such a way as to secure, maintain, and develop appropriate forms of labor control.
- Industry structure is consequently a reflection of both the technical configuration of its constituent labor processes *and* the structure of the local labor market (representing as it does a profile of different labor supplies, a set of institutionalized procedures for accessing those labor supplies, and a particular system of social regulation).
- A recursive relationship therefore exists between industry structure (with its set of labor processes) and the local labor market (with its segmented and compartmentalized labor supply, regulatory norms, etc.), as these structures will be configured in such a way as to secure the most appropriate forms of labor control available within the local labor market.

Because industry structure is influenced by both labor process *and* labor market factors, it is shaped by particular *local* conjunctures. Strategies such as industrial homeworking, for example, require the prior existence of an appropriate set of labor process relations (such as a technically divisible labor process or relatively weak labor organization) *and* an appropriate set of labor market relations (such as the existence of a pool of home-based laborers whose ability to participate in the "external" wage labor market is restricted or the close physical proximity of this pool of workers to the employing organization). The form and scale of industrial homeworking reflect the historical and geographical conjuncture of these processes (Peck 1992).

Massey (1984) developed a similar analysis to explain how multiplant firms were "stretching out" production hierarchies over space in order to tap into appropriate labor supplies. In this way, the social division of labor was seen to imply the formation of spatial divisions of labor. Although Massey concentrated on the spatial restructuring of production, a parallel set of propositions can be developed for the processes of *in situ* restructuring. The search for a new labor control "fix" (following, say, technical change in the labor process or a shift in the local labor supply) can trigger either relocation—a *spatial* strategy for engaging with a new labor supply—or changes in an industry's local labor market relations—an *in situ* strategy for engaging with a new labor supply. In this sense, labor control considerations bear on the costs of restructuring in place versus restructuring through space.

Restructuring: *In Situ* and Through Space

In Scott's analysis, firms deploying flexible labor strategies will tend to agglomerate in order to reap scale economies in the labor market. In transaction costs terms, these are defined as economies in information and search costs and in the costs associated with job accession and separation. The force of Scott's argument is, however, undermined by his failure to consider labor control; his analysis neglects the decisive role of skill formation and the contradictory nature of urban labor market locations.

Investments in skill formation are costly, risky, and are only realized in the medium term. Predictably, firms making such investments will take steps to ensure that they, not their competitors (for labor), realize the benefits. Dual labor market theorists see this as a principal motivation for constructing internal labor markets within the primary sector, in that internalized labor markets impede labor turnover and capture the benefits of skill formation within the enterprise. The formation of internal labor markets is also costly: premium wages must be paid in order to retain workers, and the stable employment conditions that are created may engender a labor process that is vulnerable to institutionalization, inertia, and collective labor organization. Firms with internalized labor markets are largely "insulated" from external labor market conditions and so are theoretically free to locate in a variety of spatial contexts. Some firms, however, may seek further guarantees of immunity from interfirm labor competition by opting for relatively isolated locations. These primary sector firms can locate in such environments because they can, in effect, reproduce their own labor supply: the process of skill formation is internalized.

In contrast, firms utilizing labor from the secondary sector of the labor market, where skill levels are generally lower and skill demands tend not to be enterprise-specific, must depend upon the haphazard process of external skill formation (i.e., "buying in" skills from the external labor market). Often such firms will "poach" labor skills formed either in the primary sector (when, for example, they recruit redundant workers), or in the domestic sphere (when, for example, they hire workers for their physical strength, basic mental capabilities, the ability to drive a car or operate a sewing machine). Inevitably, this process of externalized skill formation is unreliable and results in a low degree of control over labor supply. An example of

such a precarious skill formation strategy is the clothing industry's exploitation of immigrant labor pools in major cities (Sassen 1988; Peck 1992).

Firms dependent upon externalized skill formation will be subject to powerful agglomeration tendencies. The process of clustering in large urban labor markets permits the socialization of the costs of skill formation. In effect, these firms seek to become free-riders on the social infrastructure of the city. Skills formed in other spheres (such as homes, schools, and other workplaces) and possessed by marginal workers (such as women, immigrants, young people, and redundant workers) are plundered by this group of secondary firms able to survive themselves only on the margins of the labor market. Such "parasitic" strategies are tenable, of course, only when the host organism is sufficiently large.

That agglomeration tendencies are inherent in the process of skill formation supports some of Scott's arguments. This, however, is only a partial account: labor market agglomeration also produces contradictory tendencies. While users of flexible labor will tend to agglomerate in order to socialize the costs of labor reproduction, in so doing they inadvertently initiate a set of countervailing forces. Agglomeration raises the level of interfirm competition for labor, which may trigger local wage inflation and will almost certainly lead to labor recruitment and retention difficulties for firms operating on the margins of the labor market. In this way, the agglomeration of secondary sector employers in urban labor markets begins to undermine the utility of these locations themselves. The purchase of such firms on secondary labor supplies is inevitably tenuous; with heightening labor competition, even this tenuous grip may be lost. Through over-exploitation—to return to the metaphor—the parasites may eventually destroy the host organism. Although ultimately self-destructive, such recourse to plundering strategies is commonplace where capitalists are unable to place a floor under the process of labor market competition (Harvey 1989b, 134).

How can secondary sector firms respond to this contradictory situation? Suburbanization is rarely an option, as such firms typically depend upon urban locations in a variety of ways (such as for externalized skill formation or, because these are typically small firms, for subcontracting). Such firms are, in Cox and Mair's (1988) terms, "locally dependent." As a result, they must evolve *in situ* coping strategies by reorganizing the labor process or by tapping into different labor supplies within the urban labor market.[2] These strategies, which effectively entail adjustments in firms' local labor market relations, often involve yet further plundering of secondary labor supplies. Strategies such as homeworking and casualization allow such firms, perhaps only temporarily, to establish a "new" coupling between the (restructured) labor process and a segment of the urban labor supply. (Significantly, this reactive coping strategy might be mistaken by advocates of post-Fordism as evidence of rising flexibility!) The most favored of such strategies will be those allowing firms to establish some degree of monopoly control over their labor supply. The preference, therefore, will be for labor supplies in which the degree of interfirm labor competition is low (such as women bound to the home by domestic responsibilities, or immigrant workers with poor language skills). In these ways, secondary sector employers embark on a destructive, spiraling descent into yet more regressive labor practices. The collective urban problem of skill reproduction is intensified, not solved.

Urban labor markets are consequently profoundly contradictory places: although

they may offer possibilities for socializing the costs of skill formation (along with those other scale economies Scott identifies), the process of agglomeration also triggers contradictory and ultimately self-destructive plundering activities. While major cities are the sites of secondary labor, such urban labor markets are precarious. Because the basic requirements for the reproduction of the labor supply are not met, these labor markets are *systemically* crisis-prone. The challenge for the flexible labor markets of the new industrial spaces is therefore to overcome these self-destructive tendencies and to establish effective mechanisms for labor reproduction (see Mahon 1987; Perulli 1990; Sengenberger and Pyke 1990; Peck 1993). This, in essence, is a problem of social regulation. In the absence of appropriate regulatory structures, flexible labor markets may turn out to be no more than temporary phenomena, a manifestation of decline rather than a symbol of growth.

Work on the restructuring of urban labor markets must recognize that these are complex social as well as economic structures, that they are sites of competition and conflict among firms and among other potential users of labor, that they are the product of interacting geographies of paid and unpaid work and of production and servicing work, and that their tensions and contradictions pose immense problems of social regulation. Labor market analysis therefore needs to be integrated within a broadly conceived urban political economy. Elements of such an approach have been established (for example, Cox and Mair 1988; Harvey 1989b; Davis 1990; Sassen 1991), though many questions remain unresolved. Given that urban sites are those in which agglomeration tendencies in the labor market play themselves out, universalist claims about new relationships between production, labor, and agglomeration are perhaps premature.

In order to understand the mechanisms of local labor market reproduction, further research is required at the interface of

[2] More broadly, these conditions of local dependence may have implications for the politics of local economic development, as Cox and Mair (1988) have suggested. As such, they constitute one of the specifically *local* influences on the process of social regulation in the labor market.

industrial and labor market restructuring. In particular, there is a need to probe the other side of the coin of spatial restructuring (and strategies involving relocation): *in situ* restructuring through which firms adjust their labor process and local labor market relations. The focus of attention might usefully be placed on the formation of distinctive local labor regimes and the means by which struggles in and around these are socially regulated. What is theoretically significant here is the *reproduction* of local labor relations. Spectacular bursts of economic growth, such as those seen in many of the new industrial spaces during the 1980s, may prove to be unsustainable. There are many aspects to this question of sustainable growth (see Harrison 1992; Peck and Tickell 1992). Labor relations, I have argued, are central to this question of sustainability.

Conclusion

Through a critique of Allen Scott's recent work on the economics of labor market agglomeration, I have argued that closer attention must be paid to processes of labor control and to the reproduction of local labor markets. Scott has offered a stimulating analysis of this issue, but his approach remains a partial one. The richness and complexity of the social relations of employment cannot be reduced to calculations of transactions costs. The labor process and labor market are much more than cost structures; they are arenas of political power and conflicting class, gender, and ethnic forces. Firms' labor adjustment strategies are not calculated on the back of the proverbial envelope; they are formulated within the confines of ongoing imperatives of labor control.

For firms, the question of the "cost" of labor is inextricably tied up with the problem of labor control. Decisions concerning where to locate the boundaries of the firm and how far to internalize or externalize labor are influenced by labor control factors as well as cost calculations. This is particularly important in flexible

production complexes, where the progressive externalization of labor is likely to be associated with increasing problems of labor control and reproduction. The questions of if and how these problems will become manifest are essentially regulatory questions; they refer to the local social regulatory context, to local institutions, and to the political dynamics of the local labor market. The strategies for labor flexibility that are feasible in one local context may be untenable in others. The highly competitive environments associated with neoliberalism are, however, not effective bases on which to build sustainable flexible employment systems. The struggle over the redefinition of labor control strategies and institutions is not yet lost, nor has it been won.

A multitude of labor control "fixes"— defined here as the social relations surrounding the coupling of a particular labor process with a particular labor supply and its subsequent reproduction—are possible. These may or may not involve restructuring through space. While we have developed a good understanding of the processes of spatial restructuring, we know much less about the processes of *in situ* labor adjustment, about the relationship between industrial restructuring and the changing dynamics of local labor markets. This gap in our knowledge concerning the transformation of labor relations *in place* needs to be filled if the complex relationship between labor and agglomeration is to be adequately understood.

The relationship between labor and agglomeration is a contradictory one. Agglomeration tendencies cannot be assumed to follow, universally and unproblematically, from changes in the organization of labor markets. These labor market changes, moreover, are themselves far from straightforward and unilinear: the shift to flexible labor may or may not prove to be sustainable in the medium term. This question of sustainability warrants further consideration in local labor market research. For as Warde (1989, 61) has observed, "different modes of institutionalisation [of labor reproduction have] differing effects upon industrial discipline. . . . The way in which the reproduction of labour power is organised locally is an important determinant of workplace discipline, a key element in the means used to regulate struggles around the relations of domination in any workplace." This might provide one way of responding to Scott's (1988b, 120) understandable observation that local labor market processes remain "perplexing and underresearched." In turn, a clearer understanding of the operation of urban labor markets provides a different—and challenging—way of cutting into the question of contemporary capitalist restructuring.

References

Allen, V. 1977. The differentiation of the working class. In *Class and class structure*, ed. A. Hunt, pp. 61–79. London: Lawrence and Wishart.

Amin, A., and Robins, K. 1990. The re-emergence of regional economies? The mythical geography of flexible accumulation. *Environment and Planning D: Society and Space* 8:7–34.

Applebaum, E., and Shettkat, R. 1990. The impacts of structural and technological change: An overview. In *Labour market adjustments to structural change and technological progress*, ed. E. Applebaum and R. Shettkat, 3–14. New York: Praeger.

Atkinson, J. 1985. *Flexibility, uncertainty and manpower management*. Institute for Manpower Studies Report 89. Brighton: University of Sussex.

———. 1987. Flexibility or fragmentation? The United Kingdom labour market in the eighties. *Labour and Society* 12:87–105.

Barnes, T.; Hayter, R.; and Grass, E. 1990. MacMillan Bloedel: Corporate restructuring and employment change. In *The corporate firm in a changing world economy*, ed. M. de Smidt and E. Wever, 145–65. London: Routledge.

Bluestone, B., and Harrison, B. 1982. *The deindustrialization of America: Plant closings, community abandonment and the dismantling of basic industry*. New York: Basic Books.

Boyer, R. 1987. Labour flexibilities: Many forms, uncertain effects. *Labour and Society* 12(1):107–27.

———, ed. 1988. *The search for labour market flexibility: The European economies in transition*. Oxford: Clarendon Press.

———. 1990. The impact of the single market on labour and employment: A discussion of macro-economic approaches in the light of research in labour economics. *Labour and Society* 15:109–42.

Braverman, H. 1974. *Labor and monopoly capital: The degradation of work in the twentieth century*. New York: Monthly Review Press.

Brunhes, B. 1989. Labour flexibility in enterprises: A comparison of firms in four European countries. In *Labour market flexibility: Trends in enterprises*. Paris: Organization for Economic Co-operation and Development.

Brusco, S. 1990. Small firms and the provision of real services. Paper presented to the International Conference on Industrial Districts and Local Economic Regeneration, International Institute for Labour Studies, Geneva, October.

Burawoy, M. 1985. *The politics of production: Factory regimes under capitalism and socialism*. London: Verso.

Capecchi, V. 1989. The informal economy and the development of flexible specialization in Emilia-Romagna. In *The informal economy: Studies in advanced and less developed countries*, ed. A. Portes, M. Castells, and L. Benton, 189–215. Baltimore: Johns Hopkins University Press.

Clark, G. L.; Gertler, M. S.; and Whiteman, J. 1986. *Regional dynamics: Studies in adjustment theory*. Boston: Allen and Unwin.

Coase, R. H. 1937. The nature of the firm. *Economica* 4:386–405.

Cockburn, C. 1983. *Brothers: Male dominance and technological change*. London: Pluto.

Cox, A., and Mair, A. 1988. Locality and community in the politics of local economic development. *Annals of the Association of American Geographers* 78(2):307–25.

Craig, C.; Rubery, J.; Tarling, R.; and Wilkinson, F. 1982. *Labour market structure, industrial organisation and low pay*. Department of Applied Economics Occasional Paper 54. Cambridge: University of Cambridge.

Crang, P., and Martin, R. L. 1991. Mrs Thatcher's vision of the "new Britain" and other sides of the "Cambridge phenomenon." *Environment and Planning D: Society and Space* 9:91–116.

Davis, M. 1986. *Prisoners of the American dream.* London: Verso.

———. 1990. *City of quartz.* London: Verso.

Doeringer, P., and Piore, M. J. 1971. *Internal labor markets and manpower analysis.* Lexington, Mass.: D. C. Heath.

Edwards, R. C. 1979. *Contested terrain: The transformation of the workplace in the twentieth century.* New York: Basic Books.

Fincher, R. 1983. The inconsistency of eclecticism. *Environment and Planning A* 15:607–22.

Finegold, D., and Soskice, D. 1988. The failure of training in Britain: Analysis and prescription. *Oxford Review of Economic Policy* 4:21–53.

Friedman, A. L. 1977. *Industry and labour: Struggle at work and monopoly capitalism.* London: Macmillan.

Gertler, M. S. 1988. The limits to flexibility: Comments on the post-Fordist vision of production and its geography. *Transactions of the Institute of British Geographers* 13:419–32.

———. 1992. Flexibility revisited: Districts, nation-states, and the forces of production. *Transactions of the Institute of British Geographers* 17(3):259–78.

Giddens, A. 1980. *The class structure of the advanced societies.* 2d ed. London: Hutchinson.

Gordon, D. M.; Edwards, R.; and Reich, M. 1982. *Segmented work, divided workers: The historical transformation of labor in the United States.* Cambridge: Cambridge University Press.

Graham, J. 1992. Post-Fordism as politics: The political consequences of narratives on the left. *Environment and Planning D: Society and Space* 10(4):393–410.

Green, F. 1992. On the political economy of skill in the advanced industrial nations. *Review of Political Economy* 4(4):413–35.

Hadjimichalis, C., and Papamichos, N. 1990. "Local" development in southern Europe: Towards a new mythology. *Antipode* 22:181–210.

Hakim, C. 1987. Trends in the flexible workforce. *Employment Gazette* 95 (11):549–61.

Harrison, B. 1992. Industrial districts: Old wine in new bottles. *Regional Studies* 26(5):469–85.

Harrison, B., and Bluestone, B. 1987. *The dark side of labour market flexibility: Falling wages and growing income inequality in the United States.* World Employment Programme Working Paper. Geneva: International Labour Organisation.

———. 1988. *The great U-turn: Corporate restructuring and the polarizing of America.* New York: Basic Books.

Harvey, D. 1989a. *The condition of postmodernity.* Oxford: Basil Blackwell.

———. 1989b. *The urban experience.* Oxford: Basil Blackwell.

Heckscher, C. C. 1988. *The new unionism.* New York: Basic Books.

Henry, N. 1992. The new industrial spaces: Locational logic of a new production era? *International Journal of Urban and Regional Research* 16(3):375–96.

Holmes, J. 1986. The organization and locational structure of production subcontracting. In *Production, work, territory: The geographical anatomy of industrial capitalism,* ed. A. J. Scott and M. Storper, 80–106. Boston: Allen and Unwin.

Horrell, S.; Rubery, J.; and Burchell, B. 1990. Gender and skills. *Work, Employment and Society* 4:147–68.

Howell, C. 1992. The dilemmas of post-Fordism: Socialists, flexibility, and labor market deregulation in France. *Politics and Society* 20(1):71–99.

Hudson, R. 1989. Labour-market changes and new forms of work in old industrial regions: Maybe flexibility for some but not flexible accumulation. *Environment and Planning D: Society and Space* 7:5–30.

Hyman, R. 1988. Flexible specialization: Miracle or myth? In *New technology and industrial relations,* ed. R. Hyman and W. Streeck, 48–60. Oxford: Basil Blackwell.

Jessop, B. 1991. Structural competitiveness and strategic capacities: Implications for the state in the 1990s. Department of Sociology, Lancaster University. Mimeo.

———. 1992. Fordism and post-Fordism: A critical reformulation. In *Pathways to industrialization and regional development,* ed. M. Storper and A. J. Scott. London: Routledge.

Jessop, B.; Bonnett, K.; Bromley, S.; and Ling, T. 1988. *Thatcherism: A tale of two nations.* Cambridge: Polity Press.

Kern, H., and Schumann, M. 1987. Limits of the division of labour. *Economic and Industrial Democracy* 8:151–70.

Kreckel, R. 1980. Unequal opportunity structure and labour market segmentation. *Sociology* 14:525–49.

Labour Studies Group. 1985. Economic, social and political factors in the operation of the labour market. In *New approaches to economic life: Restructuring, unemployment and the social division of labour,* ed. B. Roberts, R. Finnegan, and D. Gallie, 105–23. Manchester: Manchester University Press.

Laflamme, G.; Murray, G.; Belanger, J.; and Ferland G., eds. 1989. *Flexibility and labour markets in Canada and the United States.* Geneva: International Labour Organisation.

Lane, T. 1988. The unions: Caught on the ebb tide. In *Uneven re-development: Cities and regions in transition,* ed. D. Massey and J. Allen, 188–97. London: Hodder and Stoughton.

Leborgne, D., and Lipietz, A. 1988. New technologies, new modes of regulation: Some spatial implications. *Environment and Planning D: Society and Space* 6:262–80.

———. 1990. How to avoid a two-tier Europe. *Labour and Society* 15:177–99.

———. 1992. Conceptual fallacies and open questions on post-Fordism. In *Pathways to industrialization and regional development,* ed. M. Storper and A. J. Scott, 332–48. London: Routledge.

Lovering, J. 1990. Fordism's unknown successor: A comment on Scott's theory of flexible accumulation and the re-emergence of regional economies. *International Journal of Urban and Regional Research* 14:159–74.

———. 1991. Theorizing postfordism: Why contingency matters. A further response to Scott. *International Journal of Urban and Regional Research* 15:298–301.

Mahon, R. 1987. From Fordism to ?: New technology, labour markets and unions. *Economic and Industrial Democracy* 8:5–60.

Marsden, D. 1986. *The end of economic man? Custom and competition in labour markets.* Brighton: Wheatsheaf Books.

Massey, D. 1984. *Spatial divisions of labour: Social structures and the geography of production.* Basingstoke: Macmillan.

Meulders, D., and Wilkin, L. 1987. Labour market flexibility: Critical introduction to an analysis of a concept. *Labour and Society* 12:3–17.

Michon, F. 1987. Segmentation, employment structures and productive structures. In *Flexibility in labour markets,* ed. R. Tarling, 23–55. London: Academic Press.

———. 1990. The "European Social Community," a common model and its national variations? Segmentation effects, societal effects. *Labour and Society* 15:215–36.

More, C. 1980. *Skill and the English working class.* London: Croom Helm.

Morgan, K., and Sayer, A. 1985. A "modern" industry in a "mature" region: The remaking of management-labour relations. *International Journal of Urban and Regional Research* 9:383–404.

Moulaert, F. 1987. An institutional revisit to the Storper-Walker theory of labour. *International Journal of Urban and Regional Research* 11:309–30.

Nielsen, L. D. 1991. Flexibility, gender and local labour markets—some examples from Denmark. *International Journal of Urban and Regional Research* 15:42–54.

Nohara, H., and Silvestre, J.-J. 1987. Industrial structures, employment trends and the economic crisis: The case of France and Japan in the 1970s. In *Flexibility in labour markets,* ed. R. Tarling, 147–76. London: Academic Press.

Organization for Economic Co-operation and Development (OECD). 1986a. *Flexibility in the labour market: The current debate.* Paris: OECD.

———. 1986b. *Labour market flexibility: Report of a high-level group of experts to the secretary general.* Paris: OECD.

———. 1989. *Labour market flexibility: Trends in enterprises.* Paris: OECD.

Offe, C. 1985. *Disorganized capitalism: Contemporary transformations of work and politics.* Cambridge: Polity Press.

Peck, J. A. 1989. Labour market segmentation theory. *Labour and Industry* 21:119–44.

———. 1991. The politics of training in Britain: Contradictions of the TEC initiative. *Capital and Class* 44:23–34.

———. 1992. "Invisible threads": Homeworking, labour-market relations, and industrial restructuring in the Australian clothing trade. *Environment and Planning D: Society and Space,* forthcoming.

———. 1993. Regulating labour: The social regulation and reproduction of local labour markets. In *Holding down the global: Global networks and local institutions,* ed. A. Amin and N. J. Thrift. Forthcoming.

Peck, J. A., and Haughton, G. F. 1991. Youth training and the local reconstruction of skill: Evidence from the engineering industry of North West England, 1981–88. *Environment and Planning A* 23:813–32.

Peck, J. A., and Lloyd, P. E. 1989. Conceptualizing processes of skill change: A local labour market approach. In *Labour, environment and industrial change*, ed. G. J. R. Linge and G. A. van der Knaap, 107–27. London: Croom Helm.

Peck, J. A., and Tickell, A. 1992. Local modes of social regulation? Regulation theory, Thatcherism and uneven development. *Geoforum* 23:347–63.

Perulli, P. 1990. Industrial flexibility and small firm districts: The Italian case. *Economic and Industrial Democracy* 11:337–53.

Piore, M. J., and Sabel, C. 1984. *The second industrial divide: Possibilities for prosperity.* New York: Basic Books.

Pollert, A. 1988. Dismantling flexibility. *Capital and Class* 34:42–75.

Reich, M.; Gordon, D. M.; and Edwards, R. C. 1973. A theory of labor market segmentation. *American Economic Review* 63:359–65.

Rodgers, G., and Rodgers, J., eds. 1989. *Precarious jobs in labour market regulation: The growth of atypical employment in western Europe.* Geneva: International Labour Organisation.

Rojot, J. 1989. National experiences in labour market flexibility. *Labour market flexibility: Trends in enterprises.* Paris: OECD.

Rosenberg, S. 1989a. From segmentation to flexibility. *Labour and Society* 14:363–407.

———. 1989b. Labor market restructuring in Europe and the United States: The search for flexibility. In *The state and the labor market*, ed. S. Rosenberg, 3–16. New York: Plenum Press.

———. 1989c. The state and the labor market: An evaluation. In *The state and the labor market*, ed. S. Rosenberg, 235–49. New York: Plenum Press.

Rueschemeyer, D. 1986. *Power and the division of labour.* Cambridge: Polity Press.

Sassen, S. 1988. *The mobility of capital and labor: International investment and labor flow.* New York: Cambridge University Press.

———. 1991. *The global city: New York, London, Tokyo.* Princeton: Princeton University Press.

Saxenian, A. 1983. The urban contradictions of Silicon Valley: Regional growth and the restructuring of the semiconductor industry. *International Journal of Urban and Regional Research* 7:237–62.

Sayer, A. 1989. Postfordism in question. *International Journal of Urban and Regional Research* 13:666–96.

Scott, A. J. 1984. Territorial reproduction and transformation in a local labour market: The animated film workers of Los Angeles. *Environment and Planning D: Society and Space* 2:277–307.

———. 1986. Industrial organization and location: Division of labor, the firm and spatial process. *Economic Geography* 62: 215–31.

———. 1988a. Flexible production systems and regional development: The rise of new industrial spaces in North America and western Europe. *International Journal of Urban and Regional Research* 12:171–86.

———. 1988b. *Metropolis: From the division of labor to urban form.* Berkeley: University of California Press.

———. 1988c. *New industrial spaces: Flexible production organization and regional development in North America and western Europe.* London: Pion.

———. 1991a. Flexible production systems: Analytical tasks and theoretical horizons—a reply to Lovering. *International Journal of Urban and Regional Research* 15:130–34.

———. 1991b. A further rejoinder to Lovering. *International Journal of Urban and Regional Research* 15:231.

———. 1992. The role of large producers in industrial districts: A case study of high technology systems in Southern California. *Regional Studies* 26(3):265–76.

Scott, A. J., and Cooke, P. 1988. The new geography and sociology of production. *Environment and Planning D: Society and Space* 6:241–44.

Scott, A. J., and Paul, A. 1990. Collective order and economic coordination in industrial agglomerations: The technopoles of Southern California. *Environment and Planning C: Government and Policy* 8:179–93.

Scott, A. J., and Storper, M., eds. 1986. *Production, work, territory: The geographical anatomy of industrial capitalism.* London: Allen and Unwin.

———. 1990. *Regional development reconsidered.* Lewis Center for Regional Policy Studies Working Paper 1. Los Angeles: University of California.

Sengenberger, W. 1981. Labour market segmentation and the business cycle. *The dynamics of labour market segmentation*, ed. F. Wilkinson, 235–49. London: Academic Press.

Sengenberger, W., and Pyke, F. 1990. Small firm industrial districts and local economic development: Research and policy issues. *Labour and Society* 16:1–24.

Smith, N. 1984. *Uneven development: Nature, capital and the production of space.* Oxford: Basil Blackwell.

Solow, R. M. 1990. *The labor market as a social institution.* Cambridge, Mass.: Basil Blackwell.

Standing, G. 1986. *Unemployment and labour market flexibility: The United Kingdom.* Geneva: International Labour Organisation.

————. 1989. Labour flexibility in Western European labour markets. In *Flexibility and labour markets in Canada and the United States,* ed. G. Laflamme, G. Murray, J. Belanger, and G. Ferland, 37–60. Geneva: International Labour Organisation.

Storper, M. 1990. Responses to Amin and Robins: Michael Storper replies. In *Industrial districts and inter-firm cooperation in Italy,* ed. F. Pyke, G. Becattini, and W. Sengenberger. Geneva: International Labour Organisation.

Storper, M., and Harrison, B. 1991. Flexibility, hierarchy and regional development: The changing structure of industrial production systems and their forms of governance in the 1990s. *Research Policy* 20:407–22.

Storper, M., and Scott, A. J. 1989. The geographical foundations and social relations of flexible production complexes. In *The power of geography: How territory shapes social life,* ed. J. Wolch and M. Dear, 21–40. Boston: Unwin Hyman.

————. 1990. Work organisation and local labour markets in an era of flexible production. *International Labour Review* 129:573–91.

Storper, M., and Walker, R. 1983. The theory of labour and the theory of location. *International Journal of Urban and Regional Research* 7:1–41.

————. 1989. *The capitalist imperative: Territory, technology and industrial growth.* Oxford: Basil Blackwell.

Tarling, R., ed. 1987. *Flexibility in the labour market.* London: Academic Press.

Tickell, A., and Peck, J. A. 1992. Accumulation, regulation and the geographies of post-Fordism: Missing links in regulationist research. *Progress in Human Geography* 16:190–218.

Villa, P. 1986. *The structuring of labour markets: A comparative analysis of the steel and construction industries in Italy.* Oxford: Clarendon Press.

Warde, A. 1989. Industrial discipline: Factory regime and politics in Lancaster. *Work, Employment and Society* 3:49–63.

Webber, M. J. 1982. Agglomeration and the regional question. *Antipode* 14:1–11.

Wilkinson, F., ed. 1981. *The dynamics of labour market segmentation.* London: Academic Press.

————. 1983. Productive systems. *Cambridge Journal of Economics* 7:413–29.

————. 1988. Deregulation, structured labour markets and unemployment. In *Unemployment: Theory, policy, structure,* ed. P. J. Pedersen and R. Lund, 167–85. Berlin: Walter de Gruyter.

Williams, F. 1986. The changing labour markets. *European Trends* 2:60–66.

Williamson, O. E. 1975. *Markets and hierarchies: Analysis and antitrust implications.* New York: Free Press.

————. 1985. *The economic institutions of capitalism.* New York: Free Press.

Wood, S., ed. 1982. *The degradation of work?* London: Hutchinson.

————, ed. 1989. *The transformation of work? Skill, flexibility and the labour process.* London: Unwin Hyman.

[14]

The Economic Geography of Talent

Richard Florida

Heinz School of Public Policy and Management, Carnegie Mellon University

The distribution of talent, or human capital, is an important factor in economic geography. This article examines the economic geography of talent, exploring the factors that attract talent and its effects on high-technology industry and regional incomes. *Talent* is defined as individuals with high levels of human capital, measured as the percentage of the population with a bachelor's degree and above. This article advances the hypothesis that talent is attracted by diversity, or what are referred to as low barriers to entry for human capital. To get at this, it introduces a new measure of diversity, referred to as the diversity index, measured as the proportion of gay households in a region. It also introduces a new measure of cultural and nightlife amenities, the coolness index, as well as employing conventional measures of amenities, high-technology industry, and regional income. Statistical research supported by the findings of interviews and focus groups is used to probe these issues. The findings confirm the hypothesis and shed light on both the factors associated with the economic geography of talent and its effects on regional development. The economic geography of talent is highly concentrated. Talent is associated with the diversity index. Furthermore, the economic geography of talent is strongly associated with high-technology industry location. Talent and high-technology industry work independently and together to generate higher regional incomes. In short, talent is a key intermediate variable in attracting high technology industries and generating higher regional incomes. *Key Words: amenities, diversity, human capital, high-technology industry, talent.*

> What is important for growth is integration not into an economy with a large number of people, but rather into one with a large amount of human capital.
>
> —(Romer 1990, S98)

The distribution of talent, or human capital, is an important factor in economic geography. Geographers have paid considerable attention to the geography of labor, suggesting that key factors in the location decisions of firms include labor costs and labor quality. Jacobs (1961) long ago called attention to the role of cities in attracting and mobilizing talented and creative people. Ullman (1958) also recognized the role of talent or human capital in his classic work on regional development and the geography of concentration. Lucas (1988) has argued that the driving force behind the growth and development of cities and regions is the productivity gains associated with the clustering of talented people or human capital. Research by Glaeser (1998, 1999, 2000) and others (Glaeser, Sheinkman, and Sheifer 1995; Glendon 1998; Simon 1998) provides empirical evidence of the association between human capital or talent and regional economic growth (see Mathur 1999 for a review). Florida (2002a, 2002b; Florida and Gates 2001) argues that regional innovation and economic growth are associated with regional openness to creativity and diversity.

There has been less research on the factors that attract talent and shape its economic geography. For the most part, geographers and social scientists have viewed the economic geography of talent as a function of employment opportunities and financial incentives. A growing stream of research suggests that amenities, entertainment, and lifestyle considerations are important elements of the ability of cities to attract both firms and people (Glaeser, Kolko, and Saiz 2001; Lloyd 2001; Lloyd and Clark 2001; Florida 2002a, 2002b).

This article explores the economic geography of talent, focusing in particular on the factors that attract human capital or talent. It advances the main hypothesis that the economic geography of talent is associated with diversity or openness—what I refer to as *low barriers to entry for human capital*. It also explores the effect of the economic geography of talent on high-technology industry and regional incomes, suggesting that concentrations of talent are associated with both.

To shed light on these issues, this article summarizes the results of both qualitative and quantitative research on the factors associated with the economic geography of talent and its effects on high-technology industry location and regional income. As a proxy for human capital, it measures talent as percentage of the population with a bachelor's degree and uses two supplementary measures: percentage of total persons employed that are scientists and engineers, and similarly, percentage that are professional and technical workers. The article introduces a new measure of diversity: the diversity index, based on the proportion of coupled gay households in a region's population. Another new measure, the coolness index, is introduced to account for cultural and nightlife amenities.

The findings of the research shed considerable light both on the factors associated with the economic geography of talent and on the effects of that geography on regional development. The economic geography of talent is highly concentrated at the regional level. Talent is associated with the diversity index, confirming the hypothesis that talent is attracted to places with low entry barriers for human capital. In contrast to much of the recent literature on amenities and city growth, this study finds that talent is more attracted to diversity than to measures of climate, recreational, and cultural amenities. Furthermore, talent is strongly associated with high-technology industry location. Talent and high-technology industry work independently and together to generate higher regional incomes. In short, talent is a key intermediate variable in attracting high-technology industries and generating higher regional incomes.

Concepts and Theory

The literature on the roles of employment, labor, and human capital in geography is vast (see Mathur 1999 and Hanson 2000 for reviews of aspects of this literature). Below I offer a brief review of this literature, focusing on the following: talent and regional growth; the location of talent; and the role of diversity.

Talent and Regional Growth

Jacobs (1961, 1969) called attention to the central role played by people in the generation and organization of economic activity in cities. In her view, cities play a crucial role in economic development, through the generation and mobilization of new knowledge. The scale of cities and their diversity of inhabitants create the interactions that generate new ideas. In other words, the diversity of economic actors within cities and their high level of interaction promote the creation and development of new products and new technology. Ullman (1958) also noted the role played by human capital or talent in the process of regional development and the geography of concentration. Andersson (1985) and Desrochers (2001) noted that the ability to incubate and nurture creativity and to attract creative people is a central factor in regional development. The *new growth theory* associated with Romer (1990) formally highlights

the connection between knowledge, human capital, and economic growth.

Building upon these insights, Lucas (1988) essentially argued that cities function to collect and organize human capital, giving rise to strong external economies, which he refers to as external human capital. These economies increase productivity and spur growth:

> If we postulate the usual list of forces, cities should fly apart. The theory of production contains nothing that holds a city together. A city is simply a collection of factors of production—capital people and land—and land is always far cheaper outside cities than inside. Why don't capital and people move outside, combining themselves with cheaper land and thereby increasing profits? Of course people like to live near shopping and shops need to be located near their customers, but circular considerations of this kind explain shopping centers, not cities. . . . It seems to me that the "force" we need . . . to account for the central role of cities in economic life is of exactly the same character as external human capital. What can people be paying in Manhattan or downtown Chicago rent FOR, if not to be near other people? (Lucas 1988, 38–39)

Empirical studies support the human capital-regional growth connection. Eaton and Eckstein (1997) and Black and Henderson (1999) have suggested that given spillovers in the accumulation of human capital, workers are more productive when they locate around others with high levels of human capital. Other empirical studies have found that human capital is strongly associated with urban and regional growth. Rauch (1993) found that both wages and housing rents were higher in cities with higher average education levels. Glaeser, Sheinkman, and Sheifer (1995) found a strong relationship between human capital and city growth. They found that cities that begin with more educated populations exhibit higher rates of population growth as time goes on. Simon and Nardinelli (1996) examined the connection between human capital and city growth in the United States and Great Britain, finding that the level of human capital in 1880 predicted city growth in subsequent decades. Glaeser (2000) found that access to common pools of labor or talent is what underpins the tendency of firms to cluster together in regional agglomerations, rather than interfirm linkages. Simon (1998) and Glendon (1998) found a strong relationship between the average level of human capital and regional employment growth over a considerable time frame. Florida (2002a) found a positive relationship between technological creativity (measured as regional innovation and high-technology industry) and cultural creativity (measured by a "bohemian index," the regional share of artists, musicians, and cultural producers). Florida and Gates (2001) found a positive relationship between regional concentrations of high-technology industry and several measures of diversity, including the percent of the population that is foreign-born, the percent that is gay, and a composite diversity measure. Florida (2002b) argued that regional economic outcomes are tied to the underlying conditions that facilitate creativity and diversity.

The Location of Talent

The literature suggests that places attract human capital or talent through two interrelated mechanisms. The traditional view offered by economists is that places attract people by matching them to jobs and economic opportunity. More recent research suggests that places attract people by providing a range of lifestyle amenities (see Gottlieb 1995). This is particularly true of highly educated, high human-capital individuals who possess resources, are economically mobile, and can exercise considerable choice in their location. Lloyd and Clark (2001; Lloyd 2001) argue that amenities are a key component of modern cities, referring to this lifestyle-oriented city as an "Entertainment Machine." Kotkin (2000) argues that high-technology industries and workers are attracted to a range of lifestyle amenities. Glaeser, Kolko, and Saiz (2001, 48) found a significant relationship between amenities and city growth. They suggest not only that high human-capital workers increase productivity, but that high human-capital areas are pleasant places to live in, concluding that "If cities are to remain strong, they must attract workers on the basis of quality of life as well as on the basis of higher wages." In a review of the literature, Glaeser (1999) notes that cities attract people as well as firms through the interplay of both market and nonmarket forces at work in cities.

The Role of Diversity

A central argument of this article is that diversity plays an important role in attracting talent or human capital. Urban and regional economists have long argued that diversity is important to regional economic performance. In the main, the term "diversity" is used to refer to the diversity of firms or regional industrial structures. In a major review of the field, Quigley (1998) suggests that regional economies benefit from the location of a diverse set of firms and industries.

The argument advanced here is different. It suggests that diversity plays a key role in the attraction and retention of the kinds of talent required to support high-technology industry and generate regional growth. Jacobs (1961) called attention to the role of diversity and immigration

in powering innovation and city growth. Following Jacobs, Desrochers (2001) notes the relationships between diversity, creativity, and regional innovation. Zachary (2000) argues that openness to immigration is a key factor in innovation and economic growth. He notes that the United States' competitiveness in high-technology fields is directly linked to its openness to outsiders, while the relative stagnation of Japan and Germany is tied to "closedness" and relative homogeneity. In an empirical study of Silicon Valley, Saxenian (1999) found that roughly one-quarter of new business formations had a Chinese- or Indian-born founder and that roughly one-third of the region's scientists and engineers were foreign-born. Florida and Gates (2001) found a positive relationship between high-technology industry concentration and diversity.

This article suggests that diversity—or low entry barriers for talent—increase a region's ability to compete for talent. At any given time, regions, like firms, compete with one another for talent. To support high-technology industries or a wide range of economic activity in general, regions compete for a variety of talent across a wide variety of fields and disciplines. Regions that are open to diversity are thus able to attract a wider range of talent by nationality, race, ethnicity, and sexual orientation than are those that are relatively closed. Simply put, regions that are open and possess low barriers to entry for human capital gain distinct economic advantage in the competition for talent or human capital and, in turn, in their ability to generate and attract high-technology industries and increase their incomes. Figure 1 outlines the structure of these relationships.

Research and Methods

This article reports on an empirical analysis of the economic geography of talent, the factors that attract

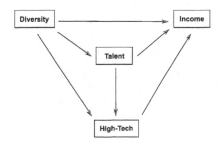

Figure 1. Structure of relationship between diversity, talent, and high technology.

talent, and talent's effects on high-technology industry location and growth, and regional income. Qualitative research, including interviews and focus groups, was initially conducted to better understand the structure and mechanics of these relationships and to generate testable hypothesis. Unstructured, open-ended interviews were conducted with more than 100 people who were making or had recently made location decisions. Structured focus groups were conducted to further assess the factors involved in personal location decisions. The original focus groups were conducted in March 1999 in Pittsburgh, with the assistance of a professional focus group organization. The author and the research team worked together with the focus group organization to screen focus group participants and develop the instrument. The focus group instrument probed respondents who were in the process of making location decisions or had recently made such decisions about the key factors that mattered to them in the choice of particular locations. It also probed respondents about the key economic, cultural, and lifestyle factors that affected their choices of particular locations in which to live and work. Four structured groups were conducted involving graduating undergraduate students in technical fields, graduating undergraduate students in nontechnical fields, graduating graduate students in business and technical fields, and professionals who had recently made location decisions. The focus groups took place over the course of a week and were conducted in a specialized facility with a one-way mirror for observation. Florida (1999) summarizes the results of the focus groups. Subsequent field research and personal interviews were conducted with individuals making location decisions in various cities and regions across the United States. The qualitative research was exploratory in nature and designed to shed light on and help structure the quantitative research, which was confirmatory in nature and approach.

Data, Variables, and Methods

Statistical analysis examined the geography of talent, the factors associated with that observed geography, and the effect of talent on characteristics of regional economies. It included descriptive statistics, correlation or bivariate analysis, multivariate regression analysis, and path analysis. Table 1 provides descriptive statistics for the various measures used in this research.

Talent Index

The basic talent index is a measure of highly educated people, defined as those with a bachelor's degree and

Table 1. Descriptive Statistics

Variable	Obs	Mean	Standard Deviation	Minimum	Maximum
Diversity index	50	1.32	0.87	0.19	5.39
Tech-Pole Index	50	1.40	1.88	0.06	8.24
Talent index	50	0.24	0.05	0.14	0.42
Coolness index	43	6.35	1.51	1.00	10.00
Median house-value ($000)	48	84.65	30.60	51.39	186.20
Cultural amenities	50	1,804.76	1,458.98	482.00	9,375.56
Recreational amenities	50	2,275.82	727.94	933.00	4,390.00
Climate	50	579.91	116.79	293.00	903.00
Per-capita income ($)	50	24,350.10	3,264.02	19,412.92	34,751.28
Per-capita income change ($)	50	2,881.09	982.89	297.38	4,682.39

above. This index is normalized on a percentage basis or per thousand people and based on the 1990 decennial census Public-Use Microdata Sample (U.S. Bureau of the Census 1993, 1995). Two additional measures of talent are also used: professional and technical workers, and scientists and engineers. Both of these are normalized on a percentage basis or per thousand people and based on the 1990 decennial census Public-Use Microdata Sample.

Amenity Measures

Several measures of amenities are used. These are based on traditional indicators of climate, cultural, and recreational amenities adapted from the 1989 *Places Rated Almanac* (Boyer and Savageau 1989).

Coolness Index

This measure is adapted from the so-called coolness factor used by *POV Magazine* (December–January 1999). The measure is based on the percentage of population ages 22–29 (with points added for diversity), nightlife (number of bars, nightclubs, and the like per capita) and culture (number of art galleries and museums per capita).

Diversity Index

The research employs a unique measure of openness or diversity—the diversity index, which is also known as the gay index. It is a measure of the fraction of the population that is gay (see Black et al. 2000 for a discussion of this measure). The gay index is a good proxy for diversity, defined as lower barriers to entry for human capital. The reason for this is that the gay population is a segment of the population that has long faced discrimination and ostracism. The presence of a relatively large gay population thus functions as a signal indicator of a region that is very open to various other groups. The diversity or gay index is based on data from the 1990 decennial census (5-percent sample), identifying households in which a householder and an unmarried partner were both of the same sex (in this case, male). Approximately 0.01 percent of the population was composed of gay, coupled men. The index is basically a location quotient that measures the number of gay households compared to the national population of gay households divided by the population in the city compared to the total national population.

Median House-Value

Median house-value is used to examine the effects of talent on housing costs. Furthermore, since Rosen (1974), researchers have argued that amenities are at least partially capitalized in land rents. This measure is also adapted from the 1990 decennial census.

Tech-Pole Index

The analysis examines the effect of talent on the location of high-technology industry. The measure of high-technology concentration is based on the Milken Institute's Tech-Pole Index. The Tech-Pole Index is a composite measure based on the percent of national high-technology real output multiplied by the high-technology real-output location quotient for each metropolitan statistical area (MSA) (see DeVol et al 1999).

Regional Income

The research also examines the effect of talent on regional income. Two measures of income are used: per-capita income level and absolute income change. Income level is for 1997, and income change covers the period from 1991 to 1997. These data are from the U.S. Bureau of Economic Analysis.

Statistical and Econometric Analysis

Both bivariate and multivariate analyses are conducted to examine the factors associated with the economic geography of talent and the effect of that geography (controlling for other factors) on high-technology industry location and regional income. Path analysis is used to better understand the structure of relationships

among these variables. Path analysis can help to discern the path of relationships in a model with multiple competing paths of causality. It should be pointed out that path analysis does not prove the direction of causality, but can provide support for a certain path of causality.

The analysis is based on the fifty largest metropolitan regions, each with populations of 700,000 and above in 1990. For most regions, the MSA is employed as the unit of analysis. MSAs that are part of a consolidated metropolitan statistical area (CMSA) are combined into their CMSA as a single unit of analysis. MSA-level variables are weighted by their proportion of the CMSA and then summed at the CMSA level. The CMSA is used as the unit of analysis for the five largest regions: San Francisco,

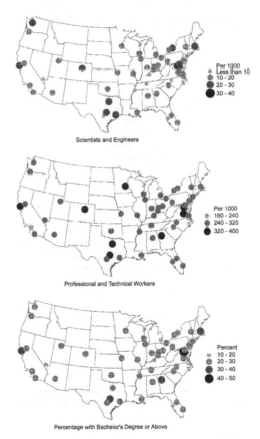

Figure 2. The geography of talent. *Source:* U.S. Bureau of the Census 1993, 1995.

Table 2. Correlation Analysis Results

	Diversity Index	Tech-Pole Index	Talent Index	Coolness Index	Median House-Value	Culture	Recreation	Climate	Income	Income Change
Diversity index	1									
Tech-Pole Index (98)	0.7677***	1								
Talent index	0.7181***	0.723***	1							
Coolness index	0.3769**	0.4285***	0.4687***	1						
Median house-value	0.4464	0.5064***	0.5384***	0.3552**	1					
Culture	0.2886**	0.4933***	0.4298***	0.5693***	0.4446***	1				
Recreation	0.1568	0.1587	−0.0482	0.2464	0.3983***	0.2494*	1			
Climate	0.4466***	0.4641***	0.2198	0.1458	0.432***	0.2049	0.2907**	1		
Income	0.4983***	0.6014***	0.5882***	0.4167	0.3597**	0.5209***	0.0977	0.2171	1	
Income change	0.1991	0.3205**	0.2916**	0.2368	−0.1263	0.1817	−0.1865	−0.1192	0.5165***	1

* Significant at the 0.10 level; ** Significant at the 0.05 level; *** Significant at the 0.01 level.

Los Angeles, Miami-Fort Lauderdale, New York, and Dallas-Fort Worth.

Findings

The findings of the research are presented in three sections. The first provides a descriptive overview of the economic geography of talent. The second examines the factors that attract talent and shape that geography. The third turns to the effect of talent on high-technology industry location and regional incomes.

The Economic Geography of Talent

The economic geography of talent is uneven, as Figure 2 shows. Roughly 42 percent of the population of the top-ranked region, Washington, DC, had a bachelor's degree or above in 1990. Washington, DC was followed by Boston, San Francisco, Austin, Atlanta, and Seattle, in all of which more than 30 percent of the population held a bachelor's degree or above. However, in more than thirty of the top fifty regions, less than 25 percent of the population had a bachelor's degree or above in 1990. Just 14 percent of the population of the region ranked fiftieth, Las Vegas, had a bachelor's degree or above. Similar patterns hold for scientists and engineers and professional and technical workers. Table 2 presents the results of a correlation analysis. Figure 3 shows maps for cultural amenities, the coolness index, and the diversity/gay index. The graphs in Figure 4 plot the relationships between talent, amenities, the coolness index, and the diversity index.

Amenities. The results of the correlation analysis indicate that talented individuals appear to be attracted

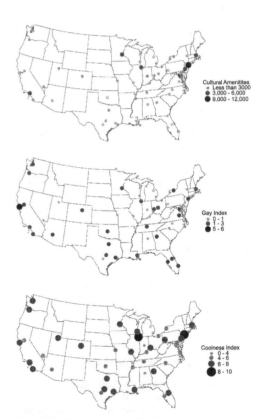

Figure 3. Quality of place. *Sources: Money Magazine* (http://pathfinder.com/money/bestplaces/); U.S. Bureau of the Census 1993, 1995; *POV Magazine.*

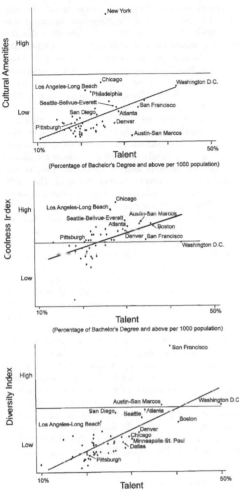

Figure 4. Talent versus amenities, coolness, and diversity. *Sources:* U.S. Bureau of the Census 1993, 1995; *POV Magazine.*

the findings of the interviews and focus groups, which indicate that high human-capital individuals exhibit a strong preference for cultural amenities. The correlations between talent and measures for both recreational amenities and climate are weak and mixed.

It is important to interpret these results with the following caveat in mind. The participants in the focus groups and interviews drew a sharp distinction between active outdoor recreation and spectator sports, such as professional baseball and football. The focus groups and interviews clearly indicate that talented individuals are attracted to places with high levels of active outdoor recreation. Here, it is important to note that the recreation measure is biased toward spectator sports. Since no reliable measures for such active outdoor recreation could be identified for the sample MSAs, the statistical research is unable to address the direct effect of active outdoor recreation.

Coolness. The correlation coefficient between the coolness measure and the talent index is 0.469. This finding is in line with the interview and focus group results, which indicate that highly educated, talented people— particularly younger workers who are active and those in knowledge-industry labor markets—are attracted to energetic and vibrant places. The focus group and interview subjects strongly emphasized the importance of visual and audio cues such as outdoor dining, active outdoor recreation, a thriving music scene, active nightlife, and bustling street scene as important attractants.

Median House-Value. Median house-value is positively associated with talent, the correlation being 0.538. The focus groups and interviews suggest that high human-capital individuals are willing to pay more for higher levels of lifestyle and amenities. Indeed, median house-value is correlated with coolness (0.355), the diversity index (0.446), and cultural amenities (0.445). This stands in some contrast to conventional wisdom on the subject, which suggests that lower costs of living (reflected in lower median house-values) may comprise an advantage in attracting people to a location.

Diversity. Talent is strongly associated with the diversity index. The correlation coefficient is 0.718, making it the highest correlation coefficient among this group of measures. This is also reflected in the scatterplot for talent and diversity (Figure 4). These results reflect the findings of the focus groups and interviews, which found that talented people are attracted to locations that have a high degree of demographic diversity and are distinguished by a high degree of openness and relatively

more by cultural amenities than by recreational amenities or climate. The correlation coefficient for the basic talent index and cultural amenities is positive and significant (0.429; see Table 2). The same is true for professional and technical workers, but not for scientists and engineers, where the correlation coefficient is negative and insignificant. These relationships are in line with low barriers to entry. The diversity index can be thought of as a leading indicator of these characteristics. Places that are open to and supportive of a gay population, our proxy measure for diversity, are likely to be open and supportive of other groups. Simply put, the diversity index reflects an environment that is characterized by low entry barriers to human capital.

Multivariate Analysis

Multivariate regressions were used to further probe the factors associated with the economic geography of talent. Several models were run to gauge the effects of amenity measures (climate, culture, and recreation), coolness, and diversity on the location of talent. The results of the various models generated R-squared values between 0.65 and 0.75, suggesting a robust relationship (see Table 3).

The most consistent finding is for diversity. The coefficient for the diversity index is consistently positive and highly significant (at the 0.001 level) in all permutations of the model. These include both basic models and more complex ones where it is included alongside an array of other variables. This suggests that diversity (measured by the gay index) is strongly associated with the location of talent. The interviews and focus group findings are in line with this result. The focus groups and interview participants report that diversity is particularly important in the location decisions of high human-capital individuals. Talented people making location decisions report a clear preference for places with a high degree of demographic diversity. The findings for the diversity index suggest that talented people prefer locations where anyone from any background, race, ethnicity, gender, or sexual orientation can easily plug in. In formal terms, this preference for diversity can be interpreted as reflective of places with low barriers to entry for human capital.

The coolness measure is also associated with the location of talent. While it sometimes has significance in models where it is run alongside the diversity index, it is typically significant in models that do not include that index. The focus group and interview findings also suggest that high human-capital individuals, particularly younger ones, are drawn to places with vibrant music scenes, street-level culture, active nightlife, and other signifiers of being "cool."

The results for the amenity measures suggest that these cultural factors are not associated with the location of talent. The coefficients for cultural amenities are positive but never significant. The coefficients for climate are typically negative and are significant (at the 0.10 level) in only one permutation of the model. The coefficients for recreational amenities are negative and significant. These findings suggest that talent is not necessarily drawn to warmer climates, greater recreational amenities, or cultural amenities. These findings, in part, can be attributed to shortcomings with existing measures of amenities. For example, available measures of culture and recreation take into account only certain types of amenities. The interview and focus group findings suggest that talented people are drawn to cultural and recreational amenities that are more broad-based, open, and participative, such as active outdoor recreation or a vibrant music scene, which these measures do not reflect.

In addition, the focus group and interview findings suggest that these nonmarket or lifestyle factors work in concert with economic opportunity in shaping the economic geography of talent. Clearly, people need to make a living and thus require jobs and gainful employment. Furthermore, the field research results indicate that high human-capital people have many employment options and change jobs relatively frequently, and thus they strongly favor locations that possess thick labor markets (see Florida 2002b). Simply put, high-paying, challeng-

Table 3. Regression Model Findings: Talent, Diversity, and Amenities

| | Dependent Variable: Talent (BA and above) | | | | | | | |
| | Model 1 | | Model 2 | | Model 3 | | Model 4 | |
Variables	Coef.	P-value	Coef.	P-value	Coef.	P-value	Coef.	P-value
Diversity index	0.033	0.00***	0.036	0.00***	0.029	0.00***	0.033	0.00***
Coolness index			0.008	0.01***	0.008	0.01**	0.005	−0.17
Median house-value	0.000	−0.14			0.001	0.04**	0.001	0.02**
Culture							0.000	−0.2
Recreation					0.000	0.00***	0.000	0.00***
Climate							0.000	−0.13
Observations	48		43		42		42	
R-squared	0.58		0.58		0.72		0.75	

* Significant at 0.1; ** Significant at 0.05; *** Significant at 0.01.

ing employment is a necessary but insufficient condition to attract talent. Because high human-capital individuals are mobile and have many options, all of these conditions—particularly diversity—must be in place to attract them.

Talent and High-Technology Industry

I now turn to the relationship between talent and high-technology industry. A number of trends are readily apparent. Talent is quite closely correlated with high-technology industry, as measured by the tech-pole index—a coefficient of 0.723 (see Table 2). High-technology industry is positively correlated with cultural amenities (0.493), climate (0.464), coolness (0.429), and median house-value (0.506), but not with recreational amenities. But high-technology industry is even more closely correlated with the diversity index—a correlation coefficient of 0.768. Figure 5 provides scatterplots of high-technology industry and talent, and high-technology industry and diversity.

Multivariate regressions and path analysis were used to further probe the relationships between talent, diversity, and high-technology industry (see Table 4). The ad-

justed R-squared values for these models range from 0.64 to 0.68, which suggests a considerable relationship among these variables. High-technology industry is associated with talent and diversity in virtually all versions and permutations of the model. In the basic structure of the model, where talent and diversity are included as the only independent variables, both are positive and significant. The adjusted R-squared for this model is 0.635. Interestingly, while high-technology industry is associated with diversity and talent, it does not appear to be associated with amenity variables or coolness. The coefficients for these variables are insignificant in most permutations of the model.

The results of the field research support these statistical findings. The interviews suggest that the availability of talent is an increasingly important location factor for these firms. They indicate that firms in knowledge-based industries are less concerned with traditional factors, such as land costs, labor costs, tax rates, or government incentives. Such firms report that they orient their location decisions to attract and retain talent. Places with large available talent-pools reduce the costs associated with search and recruitment of talent. This is particularly important in highly competitive and highly innovative industries where speed to market is a critical success factor.

Path analysis was used to further explore the path of causality among these variables. Figure 6 provides a schematic depiction of the key variables in the path analysis. A number of paths are of note. First, talent is strongly associated with high-technology industry: the direct effect of talent on high-technology industry location is 0.42. Second, diversity is associated both with talent and high-technology industry: the direct effect of diversity on talent is 0.59. Diversity also works indirectly on high-technology industry via its effect on talent. This indirect effect is 0.25.[1] In addition, diversity has a direct effect on high-technology industry of 0.35. When combined, the total effect of diversity on high-technology industry is 0.60. Third, the path analysis suggests that the effects of other variables, such as coolness or other amenity measures, are weak and frequently negative (not shown in Figure 6). For example, coolness has a weak positive effect (0.15) on talent but a negative effect (−0.024) on high-technology industry. Cultural amenities have a weak positive effect on both talent (0.14) and high-technology industry (0.16). Recreational amenities have a weak negative effect (−0.34) on talent and a weak positive effect (0.05) on high-technology industry. Climate has a weak negative effect (−0.17) on talent and a small positive effect (0.20) on high-technology industry.

Taken as a whole, the findings suggest the following relationship between diversity, talent, and high technology. Talent is attracted to regions with low entry barriers as

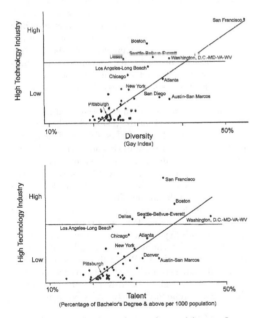

Figure 5. High-technology industry, talent, and diversity. *Sources:* U.S. Bureau of the Census 1993, 1995; DeVol 1999.

Table 4. Regression Model Findings: Talent and High-Technology Industry

	Dependent Variable: Tech-Pole Index							
	Model 1		Model 2		Model 3		Model 4	
Variables	Coef.	P-value	Coef.	P-value	Coef.	P-value	Coef.	P-value
Diversity index	1.1070	0.000***	0.857	0.009***	1.0816	0.000***	1.3074	0.000***
Talent index	13.8415	0.006***	13.2809	0.027**	11.7028	0.028**		
Coolness index			−0.0225	0.881				
Median house-value					0.0074	0.256	0.0055	0.431
Culture			0.0002	0.104			0.0003	0.012**
Recreation			0.00002	0.947			−0.0002	0.517
Climate			0.0025	0.133			−2.2359	0.015**
R-squared	0.6502		0.7281		0.6720		0.6958	
Adjusted R-squared	0.6354		0.6828		0.6497		0.6596	
Observations	50		43		48		48	

* Significant at 0.1; ** Significant at 0.05; *** Significant at 0.01.

measured by the diversity index. In turn, high-technology industries are attracted to places with high levels of talent.

Talent and Regional Income. A large and influential body of research notes the close relationship between human capital and income. This work has focused on the direct effects of human capital on income at the regional level (Simon 1998). The research presented here builds upon this line of work by examining the effects of human capital or talent on income while controlling for the effects of high-technology industry, diversity, and other factors. The analysis employs two income measures: (1) per-capita income and (2) absolute change in per-capita income from 1991 to 1997.

Per-Capita Income Level. There is substantial variation in per-capita income among the top 50 MSAs. The top-ranked MSAs are San Francisco and New York, with per-capita income levels exceeding U.S.$30,000. But thirty-six of the top fifty MSAs have per-capita incomes

below $25,000, and eight of these have per-capita income levels below $20,000.

Talent is positively correlated with per-capita income, a finding that is in line with the literature (see Table 2). The correlation coefficient between talent and per-capita income level (1997) is 0.588. More interesting, however, is the strong positive correlation between income and the diversity index (0.498). This suggests that places that are open and supportive of diversity will not only attract talent, but tend to have higher income levels as well. Based on this, one can theorize that low entry barriers to talent (represented by the diversity index) translate into higher regional incomes. Income is also positively correlated with cultural amenities, coolness, and median house-values, as well as high-technology industry.

Multivariate regression models were used to further investigate the nature of the relationships between income, talent, and other factors (see Table 5). The adjusted R-squared values for these models are 0.57 and 0.65 respectively, suggesting a reasonably positive and robust relationship. The talent coefficient is positively and significantly associated with per-capita income level in all permutations of the model. The coefficient for cultural amenities is also positively and significantly associated with per capita income. Per-capita income level is also associated with high-technology industry. This suggests that talent and technology work together in creating regional income effects. While this analysis does not address the chicken-or-the-egg question of what comes first—talent or high-technology jobs—it does suggest that talent is an important factor in its own right.

Income Change. It is also useful to examine the relationship between talent and income change. As Table 2

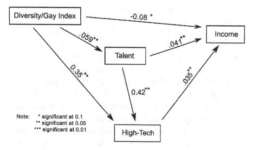

Note: * significant at 0.1
       ** significant at 0.05
      *** significant at 0.01

Figure 6. Path-analysis results.

Table 5. Regression Model Findings: Talent and Income Change

Variables	Dependent Variable: Income				Dependent Variable: Income Change	
	Model 1		Model 2		Model	
	Coef.	P-value	Coef.	P-value	Coef.	P-value
Diversity index	−640.33	0.244			−201.5561	0.3780
Tech-Pole Index	911.88	0.001***	549.60	0.048**	177.3669	0.1100
Talent index	27629.77	0.005***	24958.17	0.028**	8782.428	0.0310**
Coolness index			−236.73	0.348		
Median house-value	−9.59	0.399	−22.36	0.101	−14.3988	0.0040***
Culture			0.91	0.001***		
Recreation			0.93	0.758		
Climate			−22.36	0.101		
R-squared	0.6028		0.7114		0.2909	
Adjusted R-squared	0.5659		0.6520		0.225	
Observations	48		42		48	

* Significant at 0.1; ** Significant at 0.05; *** Significant at 0.01.

shows, the correlation coefficient for absolute income change (1991–1997) and talent (1990) is 0.337. That is, the level of talent in 1990 predicts the absolute dollar change in income between 1991 and 1997. The results of the regression analysis suggest that this relationship is robust (see Table 5). The dependent variable in the model is absolute change in income (1991–1997), and the independent variables are talent, diversity, high-technology industry, and median house-value. The adjusted R-squared value for the model is 0.225. Talent is the only variable in the model that is positively and significantly associated with income change.

Path analysis was used to further probe the structure of relationships among these variables (see Figure 6). Here several findings are of note. First, talent has a direct effect on income (0.41) as well as a direct effect on high-technology location (0.42). This is greater than the direct effect of high-technology industry on income (0.35). The estimated total effect of talent on income is 0.56. Furthermore, while diversity has no direct effect on income, it has a substantial indirect effect. This analysis indicates that diversity works indirectly on income through two additional paths. Working through high-technology industry, the indirect effect of diversity on income is 0.12. Working indirectly through talent and then high-technology industry, the indirect effect of diversity on income is 0.24. The estimated total effect of diversity on income is 0.37.

Taken in combination with the results of the field research, the statistical findings suggest the following set of relationships among these variables. Talent is associated with diversity, as diverse and open environments attract high-human capital individuals. Diversity is directly as-

sociated with talent and also with concentrations of high-technology industry. High-technology industry is attracted to places with high levels of human capital and high levels of diversity. Talent and high-technology industry work independently and in concert to generate higher regional incomes. Talent is thus a key intermediate variable in attracting high-technology industries and generating higher regional incomes.

Summary and Discussion

This article set out to examine the economic geography of talent and to explore the factors that shape that geography and its effects on the location of high-technology industry and regional income. Talent was defined as individuals with high levels of human capital, measured as the percentage of the population with a bachelor's degree or above. It advanced the hypothesis that talent is attracted by diversity, defined as low barriers to entry for human capital. To get at this, it introduced a new measure of diversity, the gay index, as a proxy for these low entry barriers. This article also used a new measure of cultural or nightlife amenities, the coolness index, as well as conventional measures of amenities, high-technology industry, and regional income. Statistical research supported by the findings of interviews and focus groups was used to probe the factors associated with the economic geography of talent and its effects on regional development.

The findings of the research confirm the hypothesis and shed light on both the factors associated with the economic geography of talent and its effects on regional development. The economic geography of talent is highly concentrated by region. Talent is associated with high levels of diversity. Talent is more closely associated with diversity than with conventional measures of climate, cultural, and recreational amenities. Taken together, the findings suggest that talent is not only associated with economic opportunity, as conventional theory allows, but is drawn to places with low entry barriers for human capital. It turns out that low entry barriers of this sort are not just important to the location of talent but are also directly associated with concentrations of high-technology industry.

Furthermore, the research indicates that the economic geography of talent exerts considerable effects on the location of high-technology industries and regional incomes. Talent is strongly associated with high-technology industry location. These findings support the human capital-growth connections noted by Lucas (1988), Glaeser (1998, 1999, 2000) and Simon (1998) and suggest that human capital works both directly and indirectly through high-technology industry to affect regional income. In short, talent is a key intermediate variable in attracting high-technology industries and generating higher regional incomes.

The findings have a number of implications for regional development. Taken together with the work of Jacobs (1961, 1969) and Lucas (1988) and the empirical findings of Glaeser (1998, 1999, 2000), and others (Glaeser, Sheinkman, and Sheifer 1995; Glendon 1998; Simon 1998), they suggest that talent, or human capital, is a driving factor in regional development. Going beyond this literature, however, they further suggest that talent is not just an endowment or stock that is in place in a given region, but that certain regional conditions are required to attract talent. In other words, talent does not simply show up in a region; rather, certain regional factors appear to play a role in creating an environment or habitat that can attract and retain talent or human capital. Paramount among these factors, the findings suggest, is openness to diversity or low barriers to entry for talent. This, in turn, suggests that a more efficacious approach to regional development may be to emphasize policies and programs to attract human capital, as opposed to conventional approaches that focus on the attraction of firms and the formation of industrial clusters. Regions may have much to gain by investing in a "people climate" as a complement to their more traditional "business climate" strategies (see Florida 2002b). It also appears that diversity has a significant impact on a region's ability to attract talent and to generate high-technology industries. Thus, regions would appear to have much to gain by introducing measures to support and enhance diversity. This suggests that diversity is more than just a social goal—it may have direct economic benefits as well.

This article has tried to illustrate the importance of the relationship between talent, diversity, and regional development. It is clear that diversity helps to attract talent, and that talent is in turn related to high-technology industry and regional growth. More research is encouraged to delineate the precise nature of causality among these factors.

Acknowledgments

The research was supported in part by the Alfred P. Sloan Foundation, Richard King Mellon Foundation, and the Heinz Endowments. Gary Gates collaborated on aspects of this research. Elizabeth Currid, Brian Knudsen, Sam Youl Lee, and Ji Woong Yoon provided research assistance. Ashish Arora, Gordon Clark, Meric Gertler, Mark Kamlet, John Paul Jones, Kevin Stolarick, and the anonymous reviewers provided comments on aspects of this work.

Note

1. This indirect effect is calculated by multiplying the effect of diversity on talent (0.59) and the effect of talent on high-technology industry (0.42).

References

Andersson, Å. E. 1985. Creativity and regional development. *Papers of the Regional Science Association* 56:5–20.

Black, D., G. Gates, S. Sanders, and L. Taylor. 2000. Demographics of the gay and lesbian population in the United States: Evidence from available systematic data sources. *Demography* 73 (2): 139–54.

Black, D., and V. Henderson. 1998. A theory of urban growth. *Journal of Political Economy* 107 (2): 252–84.

Boyer, R., and D. Savageau. 1989. *Places rated almanac: Your guide to finding the best places to live in North America.* New York: Prentice-Hall Travel.

Desrochers, P. 2001. Diversity, human creativity, and technological innovation. *Growth and Change* 32 (Summer): 369–394.

DeVol, R. C., P. Wong, J. Catapano, and G. Robitshek. 1999. America's high-tech economy: Growth, development, and risks for metropolitan areas. Report of the Milken Institute, Santa Monica, CA. http://www.milkeninstitute.org/pub14/pub14_research.html (last accessed 29 October 2002).

Eaton, J., and Z. Eckstein. 1997. Cities and growth: Theory and evidence from France and Japan. *Regional Science and Urban Economics* 27 (4–5): 443–74.

Florida, R. 1999. Competing in the age of talent. Report to the R. K. Mellon Foundation, Pittsburgh, PA. http://www.heinz.cmu.edu/~florida (last accessed 21 August 2002).

———. 2002a. Bohemia and economic geography. *Journal of Economic Geography* 2:55–71.

———. 2002b. *The rise of the creative class: And how it's transforming work, leisure, and everyday life*. New York: Basic Books.

Florida, R., and G. Gates. 2001.*Technology and tolerance: The importance of diversity to high-tech growth*. Washington, DC: Brookings Institution, Center for Urban and Metropolitan Policy.

Glaeser, E. L. 1998. Are cities dying? *Journal of Economic Perspectives* 12:139–60.

———. 1999. *The future of urban research: Nonmarket interactions*. Washington, DC: Brookings Institution.

———. 2000. The new economics of urban and regional growth. In *The Oxford handbook of economic geography*, ed. Gordon Clark, Meric Gertler, and Maryann Feldman, 83–98. Oxford: Oxford University Press.

Glaeser, E. L., J. Kolko, and A. Saiz. 2001.Consumer city. *Journal of Economic Geography* 1:27–50.

Glaeser, E. L., J. A. Sheinkman, and A. Sheifer. 1995. Economic growth in a cross-section of cities. *Journal of Monetary Economics* 36:117–43.

Glendon, Spencer. 1998. *Urban life cycles*. Working paper. Cambridge, MA: Harvard University.

Gottlieb, Paul D. 1995. Residential amenities, firm location and economic development. *Urban Studies* 32:1413–36.

Hanson, G. H. 2000. Firms, workers, and the geographic concentration of economic activity. In *The Oxford handbook of economic geography*, ed. Gordon Clark, Meric Gertler, and Maryann Feldman, 477–94. Oxford: Oxford University Press.

Jacobs, J. 1961. *The death and life of great American cities*. New York: Random House.

———. 1969. *The economy of cities*. New York: Random House.

Kotkin, J. 2000. *The new geography*. New York: Random House.

Lloyd, R. 2001. Digital bohemia: New media enterprises in Chicago's Wicker Park. Paper presented at the annual meeting of the American Sociological Association, Anaheim, CA, August.

Lloyd, R., and T. N. Clark. 2001. The city as entertainment machine. In *Research in urban sociology*, vol. 6, *Critical perspectives on urban redevelopment*, ed. Kevin Fox Gatham, 357–78. Oxford: JAI/ Elsevier.

Lucas, R. E., Jr. 1988. On the mechanics of economic development. *Journal of Monetary Economics* 22:1–42.

Mathur, V. K. 1999. Human capital-based strategy for regional economic development. *Economic Development Quarterly* 13 (3): 203–16.

Money Magazine, http://pathfinder.com/money/bestplaces (last accessed 29 October 2002).

POV Magazine, December–January 1999.

Quigley, J. 1998. Urban diversity and economic growth. *Journal of Economic Perspectives* 12 (2): 127–38.

Rauch, J. E. 1993. Productivity gains from geographic concentrations of human capital: Evidence from cities. *Journal of Urban Economics* 34:380–400.

Romer, P. M. 1990. Endogenous technological change. *Journal of Political Economy* 98 (5): S71–S102.

Rosen, Sherwin. 1974. Hedonic prices and implicit markets: Product differentiation in pure competition. *Journal of Political Economy* 82:4–55.

Saxenian, A. 1999. *Silicon Valley's new immigrant entrepreneurs*. Berkeley: Public Policy Institute of California.

Simon, C. 1998. Human capital and metropolitan employment growth. *Journal of Urban Economics* 43:223–43.

Simon, C., and C. Nardinelli. 1996. The talk of the town: Human capital, information and the growth of English cities, 1861–1961. *Explorations in Economic History* 33 (3): 384–413.

Ullman, E. L. 1958. Regional development and the geography of concentration. *Papers and Proceedings of the Regional Science Association* 4:179–98.

U.S. Bureau of Economic Analysis. Regional accounts data, local area personal income. http://www.bea.gov/bea/regional/reis/ (last accessed 21 August 2002).

U.S. Bureau of the Census. 1993. Census of population and housing, 1990: Public use microdata sample: 1/10,000 of the census [Computer file]. Washington, DC: U.S. Dept. of Commerce, Bureau of the Census.

U.S. Bureau of the Census. 1995. Census of population and housing, 1990: Public use microdata sample: 5-Percent sample [Computer file]. 3rd release. Washington, DC: U.S. Dept. of Commerce, Bureau of the Census.

Zachary, P. G. 2000. *The global me: New cosmopolitans and the competitive edge—Picking globalism's winners and losers*. New York: Perseus Books.

Correspondence: Software Industry Center, Heinz School of Public Policy and Management, Carnegie Mellon University, Pittsburgh, PA 15213, e-mail: florida@cmu.edu.

Part IV
Culture, Technology and the
Geographies of Knowledge

[15]

Telecommunications and the Changing Geographies of Knowledge Transmission in the Late 20th Century

Barney Warf

[Paper first received. March 1994; in final form, July 1994]

Summary. Recent innovations in telecommunications and computing, enhanced by a global wave of deregulation and the emergence of post-Fordist production regimes, have unleashed profound transformations of various service sectors in the global economy. This paper first reviews the geographical repercussions of the explosion of information services, including the birth of electronic funds transfer systems, the growth of global cities and the dispersal of back offices to low-wage sites across the globe. Secondly, it explores the political economy and spatiality of the largest of these systems, the Internet. Thirdly, it summarises how the global division of labour has recently engendered the birth of 'new information spaces', places whose recent growth is contingent upon the introduction of telecommunications, citing as examples Singapore, Hungary and the Dominican Republic.

The late 20th century has witnessed an explosion of producer services on an historic scale, which forms a fundamental part of the much-heralded transition from Fordism to post-Fordism (Coffey and Bailly, 1991; Wood, 1991). Central to this transformation has been a wave of growth in financial and business services linked at the global level by telecommunications. The emergence of a global service economy has profoundly altered markets for, and flows of, information and capital, simultaneously initiating new experiences of space and time, generating a new round of what Harvey (1989, 1990) calls time–space convergence. More epistemologically, Poster (1990) notes that electronic systems change not only what we know, but how we know it.

The rapid escalation in the supply and demand of information services has been propelled by a convergence of several factors, including dramatic cost declines in information-processing technologies induced by the microelectronics revolution, national and worldwide deregulation of many service industries, including the Uruguay Round of GATT negotiations (which put services on the agenda for the first time), and the persistent vertical disintegration that constitutes a fundamental part of the emergence of post-Fordist production regimes around the world (Goddard and Gillespie, 1986; Garnham, 1990; Hepworth, 1990). The growth of traditional financial and business services, and the emergence of new ones, has ushered in a profound—indeed, an historic—transformation of the ways in which information is collected, processed and circulates, forming

Barney Warf is in the Department of Geography, Florida State University, Tallahassee. FL 32306-2050. USA.

what Castells (1989) labels the 'informational mode of production'.

This paper constitutes an ambitious overview of the development, spatial dynamics and economic consequences of international telecommunications in the late 20th century as they arise from and contribute to the expansion of a global service economy. It opens with a broad perspective of recent changes in trade in producer services, particularly international finance, as the propelling force behind a large and rapidly expanding telecommunications infrastructure. Secondly, it explores the political economy of one of the largest and most renowned electronic systems, the Internet. Thirdly, it dwells upon the spatial dimensions of the mode of information, including the flowering of a select group of global cities, offshore banking centres, and the globalisation of clerical functions. Fourthly, it traces the emergence of what is called here 'new informational spaces', nations and regions reliant upon information services at the core of their economic development strategies. The conclusion summarises several themes that arise persistently in this discussion.

The Global Service Economy and Telecommunications Infrastructure

There can be little doubt that trade in services has expanded rapidly on an international basis (Kakabadse, 1987), comprising roughly one-quarter of total international trade. Internationally, the US is a net exporter of services (but runs major trade deficits in manufactured goods), which is one reason why services employment has expanded domestically. Indeed, it could be said that as the US has lost its comparative advantage in manufacturing, it has gained a new one in financial and business services (Noyelle and Dutka, 1988; Walter, 1988). The data on global services trade are poor, but some estimates are that services comprise roughly one-third of total US exports, including tourism, fees and royalties, sales of business services and profits from bank loans.

From the perspective of contemporary social theory, services may be viewed within the context of the enormous series of changes undergone by late 20th-century capitalism. In retrospect, the signs of this transformation are not difficult to see: the collapse of the Bretton-Woods agreement in 1971 and the subsequent shift to floating exchange rates; the oil crises of 1974 and 1979, which unleashed $375b of petrodollars between 1974 and 1981 (Wachtel, 1987), and the resulting recession and stagflation in the West; the explosive growth of Third World debt, including a secondary debt market and debt–equity swaps (Corbridge. 1984), the growth of Japan as the world's premier centre of financial capital (Vogel, 1986); the explosion of the Euromarket (Pecchioli, 1983; Walter, 1988); the steady deterioration in the competitive position of industrial nations, particularly the US and the UK, and the concomitant rise of Japan. Germany and the newly industrialising nations, particularly in east Asia; the transformation of the US under the Reagan administration into the world's largest debtor, the emergence of flexible production technologies (e.g. just-in-time inventory systems) and computerisation of the workplace; the steady growth of multinational corporations and their ability to shift vast resources across national boundaries; the global wave of deregulation and privatisation that lay at the heart of Thatcherite and Reaganite post-Keynesian policy; and finally, the integration of national financial markets through telecommunications systems. In the 1990s, one might add the collapse of the Soviet bloc and the steady integration of those nations into the world economy. This series of changes has been variably labelled an 'accumulation crisis' in the transition from state monopoly to global capitalism (Graham *et al.*, 1988), or the end of one Kondratieff long wave and the beginning of another (Marshall, 1987). What is abundantly clear from these observations is the emergence of a new global division of labour, in which services play a fundamental role.

The increasing reliance of financial and business services as well as numerous multinational manufacturing firms upon telecommunications to relay massive volumes of information through international networks has made electronic data collection and transmission capabilities a fundamental part of regional and national attempts to generate a comparative advantage (Gillespie and Williams, 1988). The rapid deployment of such technologies reflects a conjunction of factors, including: the increasingly information-intensive nature of commodity production in general (necessitating ever larger volumes of technical data and related inputs on financing, design and engineering, marketing and so forth); the spatial separation of production activities in different nations through globalised sub-contracting networks; decreases in price and the elastic demand for communications; the birth of new electronic information services (e.g. on-line databases, teletext and electronic mail); and the high levels of uncertainty that accompany the international markets of the late 20th century, to which the analysis of large volumes of data is a strategic response (Moss, 1987b; Akwule, 1992). The computer networks that have made such systems technologically and commercially feasible offer users scale and scope economies, allowing spatially isolated establishments to share centralised information resources such as research, marketing and advertising, and management (Hepworth, 1986, 1990). Inevitably, such systems have profound spatial repercussions, reducing uncertainty for firms and lowering the marginal cost of existing plants, especially when they are separated from one another and their headquarters over long distances, as is increasingly the case.

Central to the explosion of information services has been the deployment of new telecommunications systems and their merger with computerised database management (Nicol, 1985). This phenomenon can be seen in no small part as an aftershock of the microelectronics revolution and the concomitant switch from analogue to digital information formats: the digital format suffers less degradation over time and space, is much more compatible with the binary constraints of computers, and allows greater privacy (Akwule, 1992). As data have been converted from analogue to digital forms, computer services have merged with telecommunications. When the cost of computing capacity dropped rapidly, communications became the largest bottleneck for information-intensive firms such as banks, securities brokers and insurance companies. Numerous corporations, especially in financial services, invested in new communications technologies such as microwave and fibre optics. To meet the growing demand for high-volume telecommunications, telephone companies upgraded their copper-cable systems to include fibre-optics lines, which allow large quantities of data to be transmitted rapidly, securely and virtually error-free. By the early 1990s, the US fibre-optic network was already well in place (Fig. 1). In response to the growing demand for international digital data flows beginning in the 1970s, the United Nation's International Telecommunications Union introduced Integrated Service Digital Network (ISDN) to harmonise technological constraints to data flow among its members (Akwule, 1992). ISDN has since become the standard model of telecommunications in Europe, North America and elsewhere.

The international expansion of telecommunications networks has raised several predicaments for state policy at the global and local levels. This topic is particularly important because, as we shall see, state policy both affects and is affected by the telecommunications industry. At the international level, issues of transborder data flow, intellectual property rights, copyright laws, etc., which have remained beyond the purview of traditional trade agreements, have become central to GATT and its successor, the International Trade Organization. At the national level, the lifting of state controls in telecommunications had significant impacts on the profitability, industrial organisation and spatial structure of information services. In the US, for example, telecommunications underwent a profound reorganisation following the dissolution of ATT's monopoly in 1984, leading to secular declines in the price

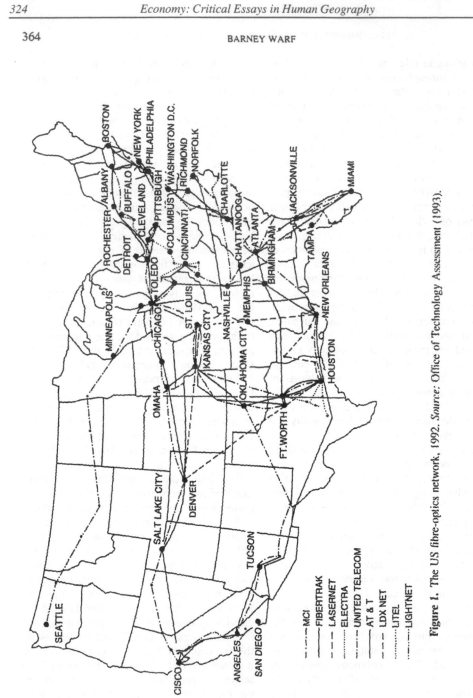

Figure 1. The US fibre-optics network, 1992. *Source: Office of Technology Assessment* (1993).

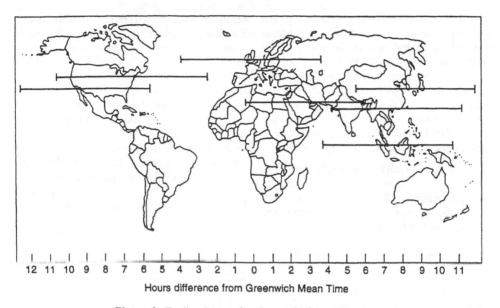

Figure 2. Trading hours of major world financial centres.

of long-distance telephone calls. Likewise, the Thatcher government privatised British Telecom, and even the Japanese began the deregulation of Nippon Telegraph and Telephone.

Telecommunications allowed not only new volumes of inter-regional trade in data services, but also in capital services. Banks and securities firms have been at the forefront of the construction of extensive leased telephone networks, giving rise to electronic funds transfer systems that have come to form the nerve centre of the international financial economy, allowing banks to move capital around a moment's notice, arbitraging interest rate differentials, taking advantage of favourable exchange rates, and avoiding political unrest (Langdale, 1985, 1989; Warf, 1989). Citicorp, for example, erected its Global Telecommunications Network to allow it to trade $200bn daily in foreign exchange markets around the world. Such networks give banks an ability to move money—by some estimates, more than $1.5 trillion daily (*Insight*, 1988)—around the globe at stupendous rates. Subject to the process of

digitisation, information and capital become two sides of the same coin. In the securities markets, global telecommunications systems have also facilitated the emergence of the 24-hour trading day, linking stock markets through the computerised trading of stocks. Reuters and the Chicago Mercantile Exchange announced the formation of Globex, an automated commodities trading system, while in 1993 the New York stock exchange began the move to a 24-hour day automated trading system. As Figure 2 indicates, the world's major financial centres are easily connected even with an 8-hour trading day. The volatility of stock markets has increased markedly as hair-trigger computer trading programmes allow fortunes to be made (and lost) by staying microseconds ahead of (or behind) other markets, as exemplified by the famous crashes of 19 October 1987. It is vital to note that heightened volatility, or the ability to switch vast quantities of funds over enormous distances, is fundamental to these capital markets: speculation is no fun when there are no wild swings in prices (Strange, 1986).

Within the context of an expanding and ever more integrated global communications network, a central role in the formation of local competitive advantage has been attained by teleports, which are essentially office parks equipped with satellite earth stations and usually linked to local fibre-optics lines (Lipman *et al.*, 1986; Hanneman, 1987a, 1987b and 1987c). The World Teleport Association defines a teleport as:

> An access facility to a satellite or other long-haul telecommunications medium, incorporating a distribution network serving the greater regional community and associated with, including, or within a comprehensive real estate or other economic development. (Hanneman, 1987a, p. 15)

Just as ports facilitate the transshipment of cargo and airports are necessary for the movement of people, so too do teleports serve as vital information transmission facilities in the age of global capital. Because telecommunications exhibit high fixed costs and low marginal costs, teleports offer significant economies of scale to small users unable to afford private systems (Stephens, 1987; Burstyn, 1986). Teleports apparently offer a continually declining average cost curve for the provision of telecommunications services. Such a cost curve raises important issues of pricing and regulation, including the tendency of industries with such cost structures to form natural monopolies. Government regulation is thus necessary to minimise inefficiencies, and the pricing of telecommunications services becomes complex (i.e. marginal revenues do not equal marginal costs, as in non-monopolistic, non-regulated sectors) (Rohlfs, 1974; Saunders *et al.*, 1983; Guldmann, 1990).

In the late 1980s there were 54 teleports in the world, including 36 in the US (Hanneman, 1987a). Most of these are concentrated in the industrialised world, particularly in cities in which data-intensive financial and business services play a major economic role. In Europe, London's new teleport in the Docklands will ensure that city's status as the centre of the Euromarket for the near future; Hamburg, Cologne, Amsterdam and Rotterdam are extending telematic control across Europe.

Tokyo is currently building the world's largest teleport. In the 1980s, the Japanese government initiated a series of high-technology 'technopolises' that form part of a long-term 'teletopia' plan to encourage decentralisation of firms out of the Tokyo region to other parts of the nation (Rimmer, 1991). In 1993 the city initiated the Tokyo Teleport on 98 ha of reclaimed land in Tokyo harbour (Tokyo Metropolitan Government Planning Department, 1993). The teleport's 'intelligent buildings' (those designed to accommodate fibre optics and advanced computational capacity), particularly its Telecom Centre, are designed to accommodate ISDN requirements. Wide Area Networks (WANs) provide local telecommunications services via microwave channels, as do Value Added Networks on fibre-optic routes. The site was originally projected to expand to 340 ha, including office, waterfront and recreational functions, and employ 100 000 people, but may be scaled back in the light of the recent recessionary climate there.

The world's first teleport is named, simply, The Teleport, located on Staten Island, New York, a project jointly operated by Merrill Lynch and the Port Authority of New York and New Jersey. Built in 1981, The Teleport consists of an 11-acre office site and 16 satellite earth stations, and is connected to 170 miles of fibre-optic cables throughout the New York region, which are, in turn, connected to the expanding national fibre-optic network. Japanese firms have taken a particularly strong interest in The Teleport, comprising 18 of its 21 tenants. For example, Recruit USA, a financial services firm, uses it to sell excess computer capacity between New York and Tokyo, taking advantage of differential day and night rates for supercomputers in each city by transmitting data via satellite and retrieving the results almost instantaneously (Warf, 1989).

In addition to the US, European and

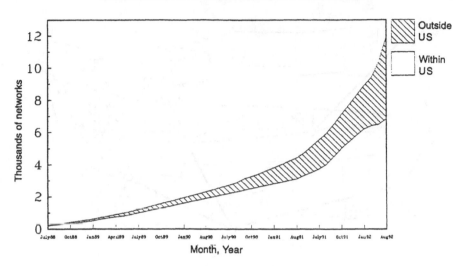

Figure 3. Growth of Internet. *Source:* Broad (1993).

Japanese teleports, some Third World nations have invested in them in order to secure a niche in the global information services economy. Jamaica, for example, built one at Montego Bay to attract American 'back office' functions there (Wilson, 1991). Other examples include Hong Kong, Singapore, Bahrain and Lagos, Nigeria (Warf, 1989).

The Internet: Political Economy and Spatiality of the Information Highway

Of all the telecommunications systems that have emerged since the 1970s, none has received more public adulation than the Internet. The unfortunate tendency in the popular media to engage in technocratic utopianism, including hyperbole about the birth of cyberspace and virtual reality, has obscured the very real effects of the Internet. The Internet is the largest electronic network on planet, connecting an estimated 20m people in 40 countries (Broad, 1993). Further, the Internet has grown at rapid rates, doubling in networks and users every year (Figure 3); by mid 1992, it connected more than 12 000 individual networks worldwide. Originating as a series of public networks, it now includes a variety of private systems of access,

in the US including services such as Prodigy, CompuServe or America On-Line (Lewis, 1994), which allow any individual with a microcomputer and modem to 'plug in', generating a variety of 'virtual communities'. By 1994, such services connected almost 5m people in the US alone (Lewis, 1994).

The origins of the Internet can be traced back to 1969, when the US Department of Defense founded ARPANET, a series of electronically connected computers whose transmission lines were designed to withstand a nuclear onslaught (Schiller, 1993). Indeed, the very durability and high quality of much of today's network owes its existence to its military origins. In 1984, ARPANET was expanded and opened to the scientific community when it was taken over by the National Science Foundation, becoming NSFNET, which linked five supercomputers around the US (Figure 4). The Internet, which emerged upon a global scale via its integration with existing telephone, fibre-optic and satellite systems, was made possible by the technological innovation of packet switching, in which individual messages may be decomposed, the constituent parts transmitted by various channels (i.e. fibre optics, telephone lines, satellite), and

Economy: Critical Essays in Human Geography

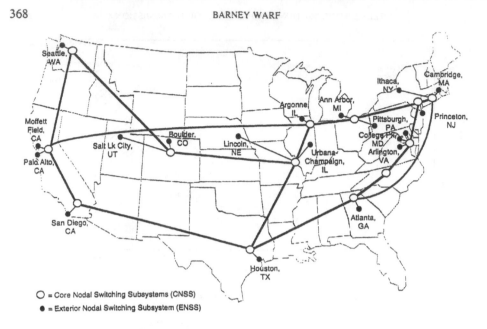

Figure 4. Distribution of NSFNET Backbone Service. *Source:* Office of Technology Assessment (1993).

then reassembled, seamlessly and instanta-neously, at the destination. In the 1990s such systems have received new scrutiny as cen-tral elements in the Clinton administration's emphasis on 'information superhighways'.

The Internet has become the world's single most important mechanism for the transmission of scientific and academic knowledge. Roughly one-half of all of its traffic is electronic mail, while the remainder consists of scientific documents, data, bibli-ographies, electronic journals and bulletin boards (Broad, 1993). Newer additions in-clude electronic versions of newspapers, such as the *Chicago Tribune* and *San Jose Mercury News*, as well as an electronic li-brary, the World Wide Web. In contrast to the relatively slow and bureaucratically mon-itored systems of knowledge production and transmission found in most of the world, the Internet and related systems permit a thor-oughly unfiltered, non-hierarchical flow of information best noted for its lack of over-lords. Indeed, the Internet has spawned its own unregulated counterculture of 'hackers' (Mungo and Clough, 1993). However, the

system finds itself facing the continuous threat of commercialisation as cyberspace is progressively encroached upon by corpora-tions, giving rise, for example, to new forms of electronic shopping and 'junk mail' (Weis, 1992). The combination of popular, scientific and commercial uses has led to an enormous surge in demand for Internet capacities, so much so that they frequently generate 'traffic jams on the information highway' as the transmission circuits become overloaded (Markoff, 1993).

Despite the mythology of equal access for everyone, there are also vast discrepancies in access to the Internet at the global level (Schiller, 1993; Cooke and Lehrer, 1993). As measured by the number of access nodes in each country, it is evident that the greatest Internet access remains in the most economi-cally developed parts of the world, notably North America, Europe and Japan (Figure 5). The hegemony of the US is particularly nota-ble given that 90 per cent of Internet traffic is destined for or originates in that nation. Most of Africa, the Middle East and Asia (with the exceptions of India, Thailand and Malaysia), in contrast, have little or no access. There is, clearly, a reflection here of the long-standing bifurcation between the First and Third Worlds. To this extent, it is apparent than the geography of the Internet reflects previous rounds of capital accumulation—i.e. it exhibits a spatiality largely preconditioned by the legacy of colonialism.

There remains a further dimension to be explored here, however, the bifurcation between the superpowers following World War II. As Buchner (1988) noted, Marxist regimes favoured investments in television rather than telephone systems: televisions, allowing only a one-way flow of information (i.e. government propaganda), are far more conducive to centralised control that are telephones, which allow multiple parties to circumvent government lines of communication. Because access to the Internet relies heavily upon existing telephone networks, this policy has hampered the emerging post-Soviet 'Glasnet'. Superimposed on top of the landscapes of colonialism, therefore, is the landscape of the Cold War.

A rather curious yet revealing byproduct of the Internet's expansion concerns the international transmission of computer viruses, programmes written deliberately to interfere with the operations of other software systems. Although viruses are not new to users of computers, the rapid growth of electronic systems in the 1980s has markedly accelerated their capacity to travel internationally, indicating both the extent and speed with which knowledge circulates through such networks as well as the vulnerability of these systems to unwanted intrusions. In 1992, for example, the Michelangelo virus disrupted software systems of users ranging from South African pharmacists to the San Francisco police department. More ominous is the 'Bulgarian virus machine' (Mungo and Clough, 1993). In the 1980s, Bulgaria was the designated computer producer for the Soviet bloc, and Sofia University produced large numbers of skilled engineers and programmers to serve it. As communism collapsed in the late 1980s, many of such

individuals, including bored young men who comprise the vast bulk of hackers, took to writing viruses and releasing them on international networks, including those of the UN. Simultaneously, Sofia University began to export its anti-virus software on the world market. Although some of the worst excesses of Bulgarian hackers have been curtailed, some indications are that they are being joined by Russian, Thai and other counterparts.

Geographical Consequences of the Mode of Information

As might be expected, the emergence of a global economy hinging upon producer services and telecommunications systems has led to new rounds of uneven development and spatial inequality. Three aspects of this phenomenon are worth noting here, including the growth of world cities, the expansion of offshore banking centres and the globalisation of back offices.

World Cities

The most readily evident geographical repercussions of this process have been the growth of 'world cities', notably London, New York and Tokyo (Moss, 1987a; Sassen, 1991), each of which seems to be more closely attuned to the rhythms of the international economy than the nation-state in which it is located. In each metropolitan area, a large agglomeration of banks and ancillary firms generates pools of well-paying administrative and white-collar professional jobs; in each, the incomes of a wealthy stratum of traders and professionals have sent real estate prices soaring, unleashing rounds of gentrification and a corresponding impoverishment for disadvantaged populations. While such predicaments are not new historically—Amsterdam was the Wall Street of the 17th century (Rodriguez and Feagin, 1986)—the magnitude and rapidity of change that global telecommunications have unleashed in such cities is without precedent.

London, for example, boomed under the

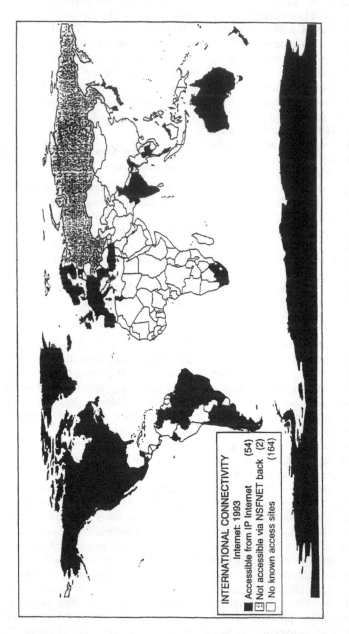

Figure 5. The geography of access to the Internet, 1993. *Source:* unpublished data from the Internet Supervisory Oversight Committee.

impetus of the Euromarket in the 1980s, and has become detached from the rest of Britain (Thrift, 1987; Budd and Whimster, 1992). Long the centre of banking for the British Empire, and more recently the capital of the unregulated Euromarket, London seems to have severed its moorings to the rest of the UK and drifted off into the hyperspaces of global finance. State regulation in the City—always loose when compared to New York or Tokyo—was further diminished by the 'Big Bang' of 1986. Accordingly, the City's landscape has been reshaped by the growth of offices, most notably Canary Wharf and the Docklands. Still the premier financial centre of Europe, and one of the world's major centres of foreign banking, publishing and advertising, London finds its status challenged by the growth of Continental financial centres such as Amsterdam, Paris and Frankfurt.

Similarly, New York rebounded from the crisis of the mid 1970s with a massive influx of petrodollars and new investment funds (i.e. pension and mutual funds) that sustained a prolonged bull market on Wall Street in the 1980s (Scanlon, 1989; Mollenkopf and Castells, 1992; Shefter, 1993). Today, 20 per cent of New York's banking employment is in foreign-owned firms, notably Japanese giants such as Dai Ichi Kangyo. Driven by the entrance of foreign firms and increasing international linkages, trade on the New York stock exchange exploded from 12m shares per day in the 1970s to 150m in the early 1990s (Warf, 1991). New York also boasts of being the communications centre of the world, including one-half million jobs that involve the collection, production, processing, transmission or consumption of information in one capacity or another (Warf, 1991). This complex, including 60 of the largest advertising and legal services firms in the US, is fuelled by more word-processing systems than in all of Europe combined. The demand for space in such a context has driven an enormous surge of office construction, housing 60 headquarters of US Fortune 500 firms. Currently, 20 per cent of New York's office space is foreign-owned, testi-

mony to the need of large foreign financial firms to establish a presence there.

Tokyo, the epicentre of the gargantuan Japanese financial market, is likely the world's largest centre of capital accumulation, with one-third of the world's stocks by volume and 12 of its largest banks by assets (Masai, 1989). The Tokyo region accounts for 25 per cent of Japan's population, but a disproportionate share of its economic activity, including 60 per cent of the nation's headquarters, 65 per cent of its stock transactions, 89 per cent of its foreign corporations, and 65 per cent of its foreign banks (Cybriwsky, 1991). Tokyo's growth is clearly tied to its international linkages to the world economy, particularly in finance, a reflection of Japan's growth as a major world economic power (Masai, 1989; Cybriwsky, 1991). In the 1980s, Japan's status in the global financial markets was unparalleled as the world's largest creditor nation (Vogel, 1986; *Far Eastern Economic Review*, 1987). Tokyo's role as a centre of information-intensive activities includes a state-of-the-art telecommunications infrastructure, including the CAPTAIN (Character and Pattern Telephone Access Information Network) system (Nakamura and White, 1988).

Offshore Banking

A second geographical manifestation of the new, hypermobile capital markets has been the growth of offshore banking, financial services outside the regulation of their national authorities. Traditionally, 'offshore' was synonymous with the Euromarket, which arose in the 1960s as trade in US dollars outside the US. Given the collapse of Bretton Woods and the instability of world financial markets, the Euromarket has since expanded to include other currencies as well as other parts of the world. The recent growth of offshore banking centres reflects the broader shift from traditional banking services (loans and deposits) to lucrative, fee-based non-traditional functions, including debt repackaging foreign exchange transactions and cash management (Walter, 1989).

Today, the growth of offshore banking has occurred in response to favourable tax laws in hitherto marginal places that have attempted to take advantage of the world's uneven topography of regulation. As the technological barriers to capital have declined, the importance of political ones has thus risen concomitantly. Several distinct clusters of offshore banking may be noted, including, in the Caribbean, the Bahamas and Cayman Islands; in Europe, Switzerland, Luxembourg and Liechtenstein; in the Middle East, Cyprus and Bahrain; in southeast Asia, Singapore and Hong Kong; and in the Pacific Ocean, Vanuatu, Nauru and Western Samoa. Roberts (1994, p. 92) notes that such places "are all part of a worldwide network of essentially marginal places which have come to assume a crucial position in the global circuits of fungible, fast-moving, furtive money and fictitious capital." Given the extreme mobility of finance capital and its increasing separation from the geography of employment, offshore banking can be expected to yield relatively little for the nations in which it occurs; Roberts (1994), for example, illustrates the case of the Cayman Islands, now the world's fifth-largest banking centre in terms of gross assets, where 538 foreign banks employ only 1000 people (less than two apiece). She also notes that such centres are often places in which 'hot money' from illegal drug sales or undeclared businesses may be laundered.

Offshore markets have also penetrated the global stock market, where telecommunications may threaten the agglomerative advantages of world cities even as they reinforce them. For example, the National Associated Automated Dealers Quotation System (NASDAQ) has emerged as the world's fourth-largest stock market; unlike the New York, London, or Tokyo exchanges, NASDAQ lacks a trading floor, connecting half a million traders worldwide through telephone and fibre-optic lines. Similarly, Paris, Belgium, Spain, Vancouver and Toronto all recently abolished their trading floors in favour of screen-based trading.

Global Back Offices

A third manifestation of telecommunications in the world service economy concerns the globalisation of clerical services, in particular back offices. Back offices perform many routinised clerical functions such as data entry of office records, telephone books or library catalogues, stock transfers, processing of payroll or billing information, bank cheques, insurance claims, magazine subscriptions and airline frequent-flyer coupons. These tasks involve unskilled or semi-skilled labour, primarily women, and frequently operate on a 24-hour-per-day basis (Moss and Dunau, 1986). By the mid 1980s, with the conversion of office systems from analogue to digital form largely complete, many firms began to integrate their computer systems with telecommunications.

Historically, back offices have located adjacent to headquarters activities in downtown areas to ensure close management supervision and rapid turnaround of information. However, under the impetus of rising central-city rents and shortages of sufficiently qualified (i.e. computer-literate) labour, many service firms began to uncouple their headquarters and back office functions, moving the latter out of the downtown to cheaper locations on the urban periphery. Most back office relocations, therefore, have been to suburbs (Moss and Dunau, 1986; Nelson, 1986). Recently, given the increasing locational flexibility afforded by satellites and a growing web of inter-urban fibre-optics systems, back offices have begun to relocate on a much broader, continental scale. Under the impetus of new telecommunications systems, many clerical tasks have become increasingly footloose and susceptible to spatial variations in production costs. For example, several firms fled New York City in the 1980s: American Express moved its back offices to Salt Lake City, UT, and Phoenix, AZ; Citicorp shifted its Mastercard and Visa divisions to Tampa, FL, and Sioux Falls, SD, and moved its data processing functions to Las Vegas, NV, Buffalo, NY, Hagerstown, MD, and Santa Monica, CA; Citibank moved its cash management services to New Castle, DE; Chase Manhattan housed its credit card operations in Wilmington; Hertz relocated its data entry division to Oklahoma City; Avis went to Tulsa. Dean Witter moved its data processing facilities to Dallas, TX; Metropolitan Life repositioned its back offices to Greenville, SC, Scranton, PA, and Wichita, KS; Deloitte Haskins Sells relocated its back offices to Nashville, TN; and Eastern Airlines chose Miami, FL.

Internationally, this trend has taken the form of the offshore office (Wilson, 1991). The primary motivation for offshore relocation is low labour costs, although other considerations include worker productivity, skills, turnover and benefits. Offshore offices are established not to serve foreign markets, but to generate cost savings for US firms by tapping cheap Third World labour pools. Notably, many firms with offshore back offices are in industries facing strong competitive pressures to enhance productivity, including insurance, publishing and airlines. Offshore back office operations remained insignificant until the 1980s, when advances in telecommunications such as trans-oceanic fibre-optics lines made possible greater locational flexibility just when the demand for clerical and information processing services grew rapidly (Warf, 1993). Several New York-based life insurance companies, for example, have erected back office facilities in Ireland, with the active encouragement of the Irish government (Lohr, 1988). Often situated near Shannon Airport, they move documents in by Federal Express and the final product back via satellite or the TAT-8 fibre-optics line that connected New York and London in 1989 (Figure 6). Despite the fact that back offices have been there only a few years, Irish development officials already fret, with good reason, about potential competition from Greece and Portugal. Likewise, the Caribbean has become a particularly important locus for American back offices, partly due to the Caribbean Basin Initiative instituted by the Reagan administration and the guaranteed access to the US market that it provides. Most back offices in the Caribbean have chosen Anglophonic nations, particularly Jamaica and Barbados. American Airlines has paved the way in the Caribbean through its subsidiary Caribbean Data Services (CDS), which began when a data processing centre moved from Tulsa to Barbados in 1981. In 1987, CDS opened a second office near Santo Domingo, Dominican Republic, where wages are one-half as high as Barbados (Warf, forthcoming). Thus, the same flexibility that allowed back offices to move out of the US can be used against the nations to which they relocate.

New Information Spaces

The emergence of global digital networks has generated growth in a number of unanticipated places. These are definitely not the new industrial spaces celebrated in the literature on post-Fordist production complexes (Scott, 1988), but constitute new 'information spaces' reflective of the related, yet distinct, mode of information. Three examples—Singapore, Hungary and the Dominican Republic—illustrate the ways in which contemporary telecommunications generate repercussions in the least expected of places.

Singapore

Known best perhaps as a member of the East Asian newly industrialised countries (NICs), Singapore today illustrates what may be the most advanced telecommunications infrastructure in the world, creating an 'intelligent island' with high-speed leased circuits, a dense telephone and fibre-optic network, household teleboxes for electronic mail and ubiquitous remote computer access (Dicken, 1987; Corey, 1991). Singapore's government has led the way in this programme through its National Computer Board and Telecommunications Authority. This transformation has occurred as part of a sustained shift in the island's role from unskilled, low-wage assembly functions to exporter of high value-added business services and as the financial hub of south-east Asia, a process hastened by the flight of capital from Hong Kong

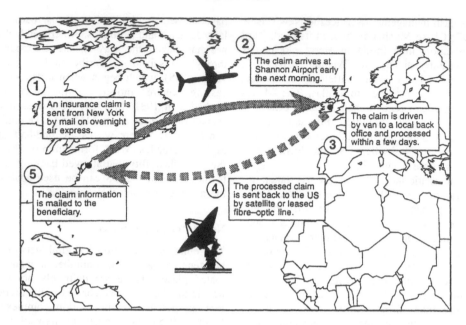

Figure 6. Mechanics of back office relocation to Ireland.

(Jussawalla and Cheach, 1983). Exports of services have now become Singapore's largest industry in terms of employment and foreign revenues. Reuters, for example, uses Singapore as its news hub in south-east Asia. In part, this transformation reflects the island's relatively high wages and fears of competition from its larger neighbours. Today, more than one-third of Singapore's labour force is engaged in skilled, white-collar employment. In addition, Singapore uses its telecommunications network for advanced Electronic Data Interchange (EDI) services to facilitate maritime shipping, in congruence with its status as the world's largest port.

Hungary

Before the collapse of the Soviet Union, Hungary suffered many of the same telecommunications problems as other underdeveloped nations: outdated technology, unsatisfied demand and few advanced services. Today, largely due to deregulation and foreign investment, the Hungarian telecommunications system is the most advanced in the former Soviet bloc, subsuming 10 per cent of the nation's total investment capital. The leader in this process has been the postal service, Magyar Posta, and its successor, the Hungarian Telecommunications Company (Matav), which introduced innovative pricing based on market, not political criteria, fees for telephone connections, time-differentiated and distance-sensitive pricing and bond financing. Concomitantly, an administrative reorganisation decentralised control of the firm, breaking the inefficient stranglehold of the bureaucratic, Communist *apparatchik* (Whitlock and Nyevrikel, 1992). The birth of the new Hungarian telecommunications network was invaluable to the nation's emerging financial system, centred in Budapest, which has expanded beyond simple loans and stocks to include database management and stock transfers (Tardos, 1991). Thus, in this respect, Hungary serves as a model for other nations making the transition from state socialism to market economies.

The Dominican Republic

In the 1980s, the Dominican government introduced a policy designed to develop non-traditional exports, particularly tourism and information services, as part of a strategy to reduce the country's reliance upon agricultural exports. For a small, relatively impoverished nation, the nation possesses a well-endowed information services infrastructure (Warf, forthcoming). The national telephone company Codetel (Compania Dominicana de Telecommunicaciones), for example, has provided the Dominican Republic with near-universal telephone access, high-speed data transmission services on fibre-optics lines, digital switching equipment, cellular telephones and microwave service to all neighbouring nations except Cuba. Codetel also sells a variety of high value-added services such as electronic mail and data-bases, telex, remote terminals, facsimile services, Spanish–English translations and leased lines. This infrastructure has made the Dominican Republic the most advanced nation in telecommunications in Latin America and has attracted numerous foreign firms. IBM-Santo Domingo, for example, engages in a complex, worldwide system of sub-contracting with its subsidiaries, purchasing, for example, printers from Argentina, disk drives from Brazil, CPUs from the US or Brazil and software, written in Canada, the US and Denmark, through its distributor in Mexico. A similar firm is Infotel, which performs a variety of computer-related functions for both domestic and international clients, including compilation of telephone directories, photo-composition, data conversion, computerised, on-line sale of advertising images, desktop publishing and map digitising. Infotel serves a variety of domestic and foreign clients, including Dominican utilities and municipal governments, the GTE telephone-operating companies, the US Geological Survey and the Spanish telephone network. Another service attracted to the Dominican Republic is back offices. American Airlines, and its subsidiary Caribbean Data Services, processes medical and dental insurance claims, credit card applications, retail sales inventories, market surveys and name and address listings at a Free Trade Zone near the capital.

Concluding Comments

What lessons can be drawn from these observations about the emergence of a globalised service economy and the telecommunications networks that underpin it? As part of the broad sea-change from Fordist production regimes to the globalised world of flexible accumulation, about which so much has already been said, it is clear that capital—as data or cash, electrons or investments—in the context of global services has acquired a qualitatively increased level of fluidity, a mobility enhanced by the worldwide wave of deregulation unleashed in the 1980s and the introduction of telecommunication networks. Such systems give banks, securities, insurance firms and back offices markedly greater freedom over their locational choices. In dramatically reducing the circulation time of capital, telecommunications have linked far-flung places together through networks in which billions of dollars move instantaneously across the globe, creating a geography without transport costs. There can be no doubt that this process has real consequences for places, as attested by the current status of cities such as London, New York, Tokyo and Singapore and the Cayman Islands. Generally, such processes tend to concentrate skilled, high value-added services, e.g. in global cities, while dispersing unskilled, low value-added services such as back offices to Third World locations.

In short, it is vital to note that, contrary to early, simplistic expectations that telecommunications would 'eliminate space', rendering geography meaningless through the effortless conquest of distance, such systems in fact produce new rounds of unevenness, forming new geographies that are imposed upon the relics of the past. Telecommunications simultaneously reflect and transform the topologies of capitalism, creating and rapidly recreating nested hierarchies of spaces technically articulated in the architecture of computer networks. Indeed, far from eliminating variations among places, such systems permit the exploitation of differences between areas with renewed ferocity. As Swyngedouw (1989) noted, the emergence of hyperspaces does not entail the obliteration of local uniqueness, only its reconfiguration. That the geography engendered by this process was unforeseen a decade ago hardly needs restating; that the future will hold an equally unexpected, even bizarre, set of outcomes is equally likely.

References

AKWULE, R. (1992) *Global Telecommunications: The Technology, Administration, and Policies.* Boston: Focal Press.

BLAZAR, W. (1985) Telecommunications: harnessing it for development, *Economic Development Commentary*, 9, pp. 8–11.

BROAD, W. (1993) Doing science on the network: a long way from Gutenberg, *New York Times*, 18 May, B5.

BUCHNER, B. (1988) Social control and the diffusion of modern telecommunications technologies: a cross-national study, *American Sociological Review*, 53, pp. 446–453.

BUDD, L. and WHIMSTER, S. (Eds) (1992) *Global Finance and Urban Living: A Study of Metropolitan Change.* London: Pergamon.

BURSTYN, H. (1986) Teleports: at the crossroads. *High Technology*, 6(5), pp. 28–31.

CASTELLS, M. (1989) *The Informational City.* Oxford: Basil Blackwell.

COFFEY, W. and BAILLY, A. (1991) Producer services and flexible production: an exploratory analysis, *Growth and Change*, 22, pp. 95–117.

COOKE, K. and LEHRER, D. (1993) The Internet: the whole world is talking, *The Nation.* 257, pp. 60–63.

CORBRIDGE S. (1984) Crisis, what crisis? Monetarism. Brandt II and the geopolitics of debt, *Political Geography Quarterly*, 3, pp. 331–345.

COREY, K. (1991) The role of information technology in the planning and development of Singapore, in: B. BRUNN and T. LEINBACH (Eds) *Collapsing Space and Time*, pp. 217–231. London: HarperCollins.

CYBRIWSKY, R. (1991) *Tokyo: The Changing Profile of an Urban Giant.* Boston: G.K. Hall and Co.

DICKEN, P. (1987) A tale of two NICs: Hong Kong and Singapore at the crossroads. *Geoforum.* 18, pp. 151–164.

DICKEN, P. (1992) *Global Shift: The Internationalization of Economic Activity.* 2nd edn. New York: Guilford Press.

FAR EASTERN ECONOMIC REVIEW (1987) Japan banking and finance, 9 April, pp. 47–110.

GARNHAM, N. (1990) *Capitalism and Communication: Global Culture and the Economics of Information.* Beverly Hills: Sage.

GILLESPIE, A. and WILLIAMS, H. (1988) Telecommunications and the reconstruction of comparative advantage, *Environment and Planning A.* 20, pp. 1311–1321.

GODDARD, J. and GILLESPIE. A. (1986) *Advanced Telecommunications and Regional Development.* Newcastle-upon-Tyne: Centre for Urban and Regional Development Studies.

GRAHAM, J. ET AL.(1988) Restructuring in U.S. manufacturing: the decline of monopoly capitalism, *Annals of the Association of American Geographers*, 78, pp. 473–490.

GULDMANN, J. (1990) Economies of scale and density in local telephone networks, *Regional Science and Urban Economics.* 20, pp. 521–533.

HALL, P. and PRESTON, P. (1988) *The Carrier Wave: New Information Technology and the Geography of Innovation, 1846–2003.* London: Unwin Hyman.

HANNEMAN, G. (1987a) The development of teleports, *Satellite Communications.* March, pp. 14–22.

HANNEMAN, G. (1987b) Teleport business, *Satellite Communications*, April, pp. 23–26.

HANNEMAN, G. (1987c) Teleports: the global outlook, *Satellite Communications.* May, pp. 29–33.

HARVEY, D. (1989) *The Condition of Postmodernity.* Oxford: Basil Blackwell.

HARVEY, D. (1990) Between space and time: reflections on the geographical imagination, *Annals of the Association of American Geographers*, 80, pp. 418–434.

HEPWORTH, M. (1986) The geography of technological change in the information economy, *Regional Studies.* 20, pp. 407–424.

HEPWORTH, M. (1990) *Geography of the Information Economy.* London: Guildford Press.

Insight. (1988) Juggling trillions on a wire: is electronic money safe? 15 February, pp. 38–40.

JUSSAWALLA, M. and CHEAH, C. (1983) Towards an information economy: the case of Singapore, *Information Economics and Policy*, 1. pp. 161–176.

KAKABADSE, M. (1987) *International Trade in Services: Prospects for Liberalisation in the 1990s.* London: Croom Helm.

LANGDALE, J. (1985) Electronic funds transfer and the internationalisation of the banking and finance industry, *Geoforum*, 16, pp. 1–13.

LANGDALE. J. (1989) The geography of international business telecommunications: the role of leased networks, *Annals of the Association of American Geographers*, 79, pp. 501–522.

LEWIS, P. (1994) A boom for on-line services, *New York Times*, 12 July, C1.

LIPMAN, A., SUGARMAN, A. and CUSHMAN, R. (1986) *Teleports and the Intelligent City*. Homewood, IL: Dow Jones.

LOHR, S. (1988) The growth of the global office, *New York Times*, 18 October, D1.

MARKOFF, J. (1993) Traffic jams already on the information highway, *New York Times*, 3 November, p. 1, C7.

MARSHALL, M. (1987) *Long Waves of Regional Development*. London: Macmillan.

MASAI, Y. (1989) Greater Tokyo as a global city, in: R. KNIGHT and G. GAPPERT (Eds) *Cities in a Global Society*. Newbury Park, CA: Sage.

MOLLENKOPF, J. and CASTELLS, M. (Eds) (1992) *Dual City: Restructuring New York*. New York: Russell Sage Foundation.

MOSS, M. (1987a) Telecommunications, world cities and urban policy, *Urban Studies*, 24, pp. 534–546.

MOSS, M. (1987b) Telecommunications and international financial centres, in: J. BROTCHIE, P. HALL and P. NEWTON (Eds) *The Spatial Impact of Technological Change*. London: Croom Helm.

MOSS, M. and DUNAU, A. (1986) Offices, information technology, and locational trends, in: J. BLACK, K. ROARK and L. SCHWARTZ (Eds) *The Changing Office Workplace*, pp. 171–182. Washington, DC: Urban Land Institute.

MUNGO, P. and CLOUGH, B. (1993) *Approaching Zero: The Extraordinary Underworld of Hackers, Phreakers, Virus Writers, and Keyboard Criminals*. New York: Random House.

NAKAMURA, H. and WHITE, J. (1988) Tokyo, in: M. DOGAN and J. KASARDA (Eds) *The Metropolitan Era, Volume 2: Mega-Cities*. Newbury Park, CA: Sage.

NELSON, K. (1986) Labor demand, labor supply and the suburbanization of low-wage office work, in: A. SCOTT and M. STORPER (Eds) *Production, Work, Territory*. Boston: Allen Unwin.

NICOL, L. (1985) Communications technology: economic and spatial impacts, in: M. CASTELLS (Ed.) *High Technology, Space, and Society*, pp. 191–209. Beverly Hills, CA: Sage.

NOYELLE, T. and DUTKA, A. (1988) *International Trade in Business Services*. Cambridge, MA: Ballinger.

OFFICE OF TECHNOLOGY ASSESSMENT (1993) *Automation of America's Offices*. Washington, DC: US Government Printing Office.

PECCHIOLI, R. (1983) *The Internationalization of Banking: The Policy Issues*. Paris: OECD.

POSTER, M. (1990) *The Mode of Information: Poststructuralism and Social Context*. Chicago: University of Chicago Press.

QUINN, J., BARUCH, J. and PAQUETTE, P. (1987) Technology in services. *Scientific American*, 257(6), pp. 50–58.

RIMMER, P. (1991) Exporting cities to the western Pacific Rim: the art of the Japanese package, in: J. BROTCHIE, M. BATTY, P. HALL and P. NEWTON (Eds) *Cities of the 21st Century*. Melbourne: Longman Cheshire.

ROBERTS, S. (1994) Fictitious capital, fictitious spaces: the geography of offshore financial flows, in: S. CORBRIDGE, R. MARTIN and N. THRIFT (Eds) *Money Power Space*. Oxford: Basil Blackwell.

RODRIGUEZ, N. and FEAGIN, J. (1986) Urban specialization in the world-system, *Urban Affairs Quarterly*, 22, pp. 187–219.

ROHLFS, J. (1974) A theory of interdependent demand for a communications service, *Bell Journal of Economics and Management Science*, 5, pp. 13–37.

SASSEN, S. (1991) *The Global City: New York, London, Tokyo*. Princeton, NJ: Princeton University Press.

SAUNDERS, R., WARFORD, J. and WELLENIUS, B. (1983) *Telecommunications and Economic Development*. Baltimore: Johns Hopkins University Press.

SCANLON, R. (1989) New York City as global capital in the 1980s, in: R. KNIGHT and G. GAPPERT (Eds) *Cities in a Global Society*. Newbury Park, CA: Sage.

SCHILLER, H. (1993) 'The information highway': public way or private road? *The Nation*, 257, pp. 64–65.

SCOTT, A.J. (1988) *New Industrial Spaces*. London: Pion.

SHEFTER, M. (1993) *Capital of the American Century: The National and International Influence of New York City*. New York: Russell Sage Foundation.

STEPHENS, G. (1987) What can business get from teleports? *Satellite Communications*, March, pp. 18–19.

STRANGE, S. (1986) *Casino Capitalism*. Oxford: Basil Blackwell.

SWYNGEDOUW, E. (1989) The heart of the place: the resurrection of locality in an age of hyperspace, *Geografiska Annaler*, 71, pp. 31–42.

TARDOS, A. (1991) Problems of the financial information system in Hungary, *Acta Oeconomica*, 43, pp. 149–166.

THRIFT, N. (1987) The fixers: the urban geography of international commercial capital, in: J. HENDERSON and M. CASTELLS (Eds) *Global Restructuring and Territorial Development*. Beverly Hills: Sage Publications.

TOKYO METROPOLITAN GOVERNMENT PLANNING DEPARTMENT (1993) *Tokyo Teleport*. Tokyo: Tokyo Metropolitan Government Information Centre.

VOGEL, E. (1986) Pax Nipponica? *Foreign Affairs*. 64, pp. 752–767.

WACHTEL, H. (1987) Currency without a country: the global funny money game. *The Nation*. 26 December, 245, pp. 784–790.

WALKER, R. (1985) Is there a service economy? The changing capitalist division of labor, *Science and Society*, Spring, pp. 42–83.

WALTER, I. (1988) *Global Competition in Financial Services: Market Structure, Protection, and Trade Liberalization*. Cambridge, MA: Ballinger.

WALTER, I. (1989) *Secret Money*. London: Unwin Hyman.

WARF, B. (1989) Telecommunications and the globalization of financial services, *Professional Geographer*, 41, pp. 257–271.

WARF, B. (1991) The internationalization of New York services, in: P. DANIELS (Ed.) *Services and Metropolitan Development: International Perspectives*, pp. 245–264. London: Routledge.

WARF, B. (1993) Back office dispersal: implications for urban development. *Economic Development Commentary*. 16, pp. 11–16.

WARF, B. (forthcoming) Information services in the Dominican Republic. *Yearbook of the Association of Latin American Geographers*.

WEIS, A. (1992) Commercialization of the Internet. *Electronic Networking*. 2(3), pp. 7–16.

WHITLOCK, E. and NYEVRIKEL, E. (1992) The evolution of Hungarian telecommunications. *Telecommunications Policy*. pp. 249–258.

WILSON, M. (1991) *Offshore relocation of producer services: the Irish back office*. Paper presented at the *Annual Meeting of the Association of American Geographers*. Miami.

WOOD, P. (1991) Flexible accumulation and the rise of business services. *Transactions of the Institute of British Geographers*. 16, pp. 160–172.

[16]

The cultural economy: geography and the creative field

Allen J. Scott
UCLA, LOS ANGELES, USA

Introduction

The cultural economy comprises all those sectors in modern capitalism that cater to consumer demands for amusement, ornamentation, self-affirmation, social display and so on. These sectors comprise various craft, fashion, media, entertainment and service industries with outputs like jewelry, perfume, clothing, films, recorded music or tourist services. Such outputs have high symbolic value relative to utilitarian purpose (see Bourdieu, 1971; Lash and Urry, 1994).

Major portions of the modern cultural economy are concentrated in global cities like Los Angeles, New York, Paris, Milan or Tokyo (Hall, 1998; Scott, 1997). Significant geographic fragments of the cultural economy can also be found in other locales, ranging from Las Vegas or the French Côte d'Azur to the neo-artisanal industrial districts of north-east and central Italy. One common characteristic of such places is that their participation in cultural–economic activities is based upon dense networks of producers combined with a dependence on complex local labor markets. My argument in what follows is that these geographic underpinnings are decisive for understanding processes of creativity and innovation in the cultural economy.

Art, science, culture

There is an extended literature suggesting that culture – even in Williams's (1982) rather abstract sense of the 'informing spirit' of a whole way of life – can best be comprehended as a social phenomenon rather than as the expression of some transcendent personalized impulse. Culture, in short, is an immanent construct whose character can only be seized in terms of the wider systems of human relationships with which it is intertwined (Bordwell et al., 1985; Crane, 1992; Negus, 1998; Wolff, 1981).

This argument may be advanced not just for the case of culture as embodied in the everyday artifacts that constitute the materiality of social life, but also as it is expressed in such domains of human activity as art or science. To be sure, there are powerful versions of philosophy that arrogate to themselves special authority to issue warrants for aesthetic or scientific practices, but this view is increasingly in retreat in the light of scholarly work showing how these practices connect even at their most intimate moments of genesis with concrete social conditions. Writers on the sociology of art and culture, such as Becker (1974, 1976, 1982), Bourdieu (1983), Crane (1992) or White and White (1965), and on the sociology of knowledge, such as Barnes (1974), Barnes et al. (1996), Latour and Woolgar (1979), Mannheim (1952) or Mulkay (1972) have built up an imposing array of arguments and evidence in favor of this sociological approach.

Four main points need to be made.

1. What can be identified as viable (i.e. inter-subjectively meaningful) topics for art works or scientific projects are given out of conditions of practical and political life.
2. Artistic or scientific work is always molded by the context in which it occurs. One of the more significant variables at play here is the division of labor in cultural production, even in such apparently elusive cases as a painter's studio or a scientific laboratory.
3. Art and science depend on inter-personal norms, methods, languages and so on, in order to achieve communicability. Both practitioners and audiences will therefore have had to undergo some degree of common socialization if they are to connect together.
4. The social profile of consumers of art and science (alternatively, the market) invariably plays a role in how producers conceive and present their finished work. This is all the more so where intermediaries (such as agents, editors or gallery owners) exert an influence on cultural production (Hennion, 1989; White and White, 1965).

These remarks, if correct, begin the task of situating cultural production firmly within the domain of the social. They serve, too, as a prelude to the proposition that the cultural economy in capitalism is just another way of producing human culture, though it does not produce just any culture; on the contrary, the outputs of the modern cultural economy bear a determinate relationship to their social (capitalistic) conditions of production.

In none of this is there any necessary denial of the talents or dispositions of the individual cultural worker. The point is not that these qualities are always submerged in the anonymous apparatus of commodified production, but that they are mobilized and channeled by the manner in which the apparatus works, including the ways in which specialized but complementary workers come together in the tasks of cultural production (see DiMaggio, 1977; DiMaggio and Hirsch, 1976). This statement applies as much to sectors like film or music where (some) workers' identities are overtly inscribed on the final product, as to sectors like furniture or clothing where workers' identities tend to be relatively hidden from the consumer. It applies because modern cultural–economic systems almost always take the form of complex networks of workers *within* firms, linked together by tightly wrought networks of transactions *between* firms, in which many different hands are brought to bear on products as they go through the process of conception, fabrication and final embellishment. The attributes of these networks, and hence of final outputs, are subject to strong economic pressures, which in cultural-products sectors often assume one of two mutually exclusive types. One type derives from efforts to economize on costs by standardizing production processes; the other seeks to ward off competitive threats by means of constant product differentiation.

The cultural content of cultural products, then, needs to be treated as endogenous to the system of production, but also as an authentic vehicle of aesthetic and semiotic expression (no matter how well or badly achieved in any given instance).

Cultural communities

Many types of cultural production, whether in the commodity form or not, are rooted in communities of workers anchored to particular places. Examples range from traditional craft communities, such as the brahmin painters of Nathdwara in north-west India as described by Maduro (1975), through the artistic and intellectual circles of 19th- and 20th-century Paris (see Bourdieu, 1977; Hall, 1998; Menger, 1993), to the actors, directors, writers and so on, that make up the film colony of contemporary Hollywood (Scott, 1984; Storper and Christopherson, 1987). Place-based communities such as these are not just foci of cultural labor in the narrow sense, but are also vortexes of social reproduction in which critical cultural competencies are generated and circulated. They are, too, magnets for talented individuals from other places who migrate to these centers in search of professional fulfillment (Denisoff and Bridges, 1982; Menger, 1983, 1993).

These examples hint at one of the representative features of such communities, namely, that they are less miscellaneous jumbles of individuals following many different and disconnected pursuits than they are collectivities whose members become caught up in mutually complementary and socially coordinated careers (see Montgomery and Robinson, 1993). A major factor binding such collectivities together is the conventions that virtually always come into being in any human community that has subsisted over a period of time. As such, communities are the repositories of an accumulated cultural capital that is one (though only one) of the defining factors of the creative field.

Cultural capital in this sense is further sustained by the institutional infra-structures with which most such communities are endowed. Specialized schools, training establishments, apprenticeship programs and so on are a recurrent feature of well-established communities of specialized workers and provide a ready supply of appropriately socialized neophytes. Workers' organizations, such as unions or professional associations, contribute further to the maintenance of local standards of cultural and economic performance. Other, more idiosyncratic institutions, ranging from museums or associations devoted to keeping alive memories of past accomplishments, to annual festivals focused on celebrating the achievements of the immediate present also reinforce communal cultural frameworks.

This kind of overarching order is in large degree what Marshall (1919) meant by the term 'industrial atmosphere', a term that he applied to the specialized social assets of the manufacturing districts of 19th-century England. Atmosphere in this sense represents an externality, a common set of resources facilitating the adaptation of workers to employment routines and providing a platform for creative and innovative activity; it eases the tasks of intra-community communication, and represents the common ground on which localized groups of firms or workers come together in the solution of workaday problems. Clearly, though, it can also sometimes operate in a negative fashion by making certain options invisible and encouraging premature termination of the search for solutions.

Each community of cultural workers represents a unique and complex case in which inventiveness on the one hand and conventional or institutional constraint on the other give rise to many varied outcomes. In the traditional brahmin painter community described by Maduro (1975) a restrictive tradition keeps creativity and innovation within tight bounds. In the case of the country-music performers who congregate in large numbers in and around Nashville, Tennessee, adherence to a strongly demarcated musical genre is tempered by occasional bravura demonstrations of originality that periodically stretch its boundaries (see Peterson and DiMaggio, 1975; Ryan and Peterson, 1982). In contemporary Hollywood, a new wave of creativity and innovation has been unleashed by the digital effects revolution which has prompted a wholesale transformation of the art of film-making.

Place, community and the cultural economy are thus often closely interconnected. For this reason, many kinds of outputs are indelibly associated in the mind of the discriminating consumer with particular locales. Theater in London, Parisian *haute couture* or furniture from Italy all illustrate this reputation effect. As Molotch (1996) has argued, place endows products with a guarantee, somewhat akin to the accumulated symbolic value of the fashion designer's label (Bourdieu and Delsaut, 1975), and thus is the source of location-specific monopoly rents.

Production system and milieu

Communities of skilled and socialized cultural workers are one thing; mobilizing them into patterns of productive labor is another. It is only when we introduce the much more expansive notion of the production system and its milieu that we really start to grapple centrally with the logic of the creative field. In this regard, we need to go well beyond 'gatekeeping' models of cultural production (Hirsch, 1969, 1972). These models describe the cultural economy as a filtering process through which some products pass while others are rejected along the way; but they are silent on the central question of how the cultural economy itself begets the essential features of final outputs.

Let us begin with a brief allusion to the work of Hennion (1981, 1983, 1989) and Kealy (1979) on the internal operations of recording studios in the popular music industry, though we might equally as well have started off with such analogous sites of cultural labor as the fashion designer's workshop or a work-team creating a multimedia game.

In his 1989 article, Hennion likens the production of recorded music to the execution of scientific experiments in that the recording studio and the laboratory represent organized social milieux where groups of workers seek by trial and error to obtain results that can then be publicized. What is never made public, however (except by the inquisitive anthropologist), is how these results are influenced by the purely social and often quite messy internal order of the laboratory or studio.

In the recording studio, the interactions between the composer/arranger, the performers, the producer, the sound engineer and other critical individuals constitute an inherently collective sphere of artistic experimentation, and even the efforts of the performers themselves do not necessarily always comprise the most decisive ingredient of what is realized on the final recording (Kealy, 1979). The finished products that emerge from the cultural economy are the results of a collaborative labor process that involves many specialized operations by many different individuals (see Frith, 1992; Negus, 1996; Ryan and Peterson, 1982). Even the stars who occupy the pinnacle positions on the work ladder of the cultural economy are in important ways an endogenous expression of its logic.

There is no special reason, however, why we should call a halt to our investigations at the outer walls of the recording studio. The studio is only one element, a sort of microcosm, of a much more extensive domain of activities in the cultural economy, and hence of the creative field. This domain almost always takes the form of an agglomerated production system and its immediate geographic milieu. Two remarks are apposite.

First, the cultural economy is typically a site of dense inter-firm transactions, as illustrated by the modern film industry with its propensity to horizontal and vertical disintegration (Scott, 1999a; Storper and Christopherson, 1987). These transactions involve both traded and untraded interdependencies. In the cultural economy they are often extremely unstable, finely grained, frequent and mediated by face-to-face contact, which means also that they tend to absorb significant resources of time and energy. Interrelated firms thus commonly put a high premium on mutual proximity to one another.

Second, as production proceeds, workers need to be assembled daily at work sites, and labor markets are constantly recalibrated by means of job search and recruitment activities. Spatial clustering of the participants in the economy again has beneficial consequences since it means that firms and workers occupy mutually accessible locations and that local labor markets function as sets of quasi-pooled resources (see Scott, 1998a).

Inter-firm transacting and local labor market processes in the cultural economy thus encourage agglomeration, and this tendency is greatly boosted by the increasing returns effects (including 'atmosphere') that usually emerge. Agglomeration and related increasing returns effects not only enhance system *efficiency* but also *creativity*, and perhaps nowhere more so than in the case of cultural-products complexes. It is within such systems of spatial relations that the creative field emerges in its definitive form.

One of the claims of the literature on learning and change in industrial systems is that many improvements in product and process configurations flow steadily from the multiple, small, unrecorded, day-by-day transactions that occur in any production complex (Lundvall and Johnson, 1994; Von Hippel, 1988). Transactions like these are particularly common in agglomerations where production is deeply disintegrated. Transactions that involve negotiations over the design specifications of products and services as they pass along the input–output chain appear to be a remarkably rich source of creative and innovative energy. In a study of ceramic tile production in Sassuolo, Italy, Russo (1985) showed that a stream of small-scale but cumulatively important improvements in production practices could be traced back to detailed interactions between specialized firms as they engaged in business with one another.

In cultural-products agglomerations, these types of interactions are usually well developed (O'Connor, 1991). They are apt to be characterized by close collaboration between the different parties involved (e.g. in the planning of a television show, or in the design and fabrication of a piece of jewelry), so that the consequences in terms of learning will tend to be all the more evident. The gains in know-how and the beneficial on-the-job adjustments that occur in this manner, refer not only to concrete practices and techniques, but also to the emotive content of products. Workers caught up in this sort of activity are apt to emerge with an altered awareness of the imaginative possibilities that lie within their work (see Csikszentmihalyi, 1990).

The instability of markets for cultural products means that these interactions are often subject to rapid rotation, thus sharpening the associated learning effects. With sufficiently large pools of specialized firms and workers, the number of different combinatorial variations in the structure of (inter-firm) production teams is effectively unlimited, and in industries where there is an incessant search for fashion or novelty effects in final products, this system flexibility is critical.

Perhaps the most dramatic instance of this phenomenon can be found in the popular music industry where recording companies maintain an ever-changing flow of releases in the effort to win a place on the hit-parade. Accordingly, frequent access to a large variety of relevant skills is paramount. Los Angeles and New York, the main concentrations of the recording industry in America, have a disproportionate capacity for producing hit singles, and this is evident even after controlling for such complicating factors as the presence of music-industry majors in both places (Scott, 1999b).

These remarks evoke the notion of the learning region as identified by theorists such as Cooke and Morgan (1998) or Storper (1996). The notion is generally brought up in relation to technology-intensive production, though it can be usefully extended to include the cultural economy as well. Thus, it refers here not only to agglomerations of technologically dynamic firms but also to production places where qualities such as cultural insight and imagination are stimulated by the action and needs of the local economy. At the level of the region as a whole, these qualities can be described as reflections of a many-sided economic and geographic system of production, above and beyond the sum of individual creative efforts. This comment is illustrated by modern film-industry agglomerations where scores if not hundreds of different firms typically come together on particular projects. In these circumstances, any film – even a *film d'auteur* – is actually a huge collective venture, and the final product is always a cultural and an economic artifact at one and the same time. From this perspective we can finally make sense of the celebrated but otherwise cryptic afterthought that Malraux (1946) appended to his study of the cinema to the effect that film-making is not only an art but also, intrinsically, an industry.

Temporality

Creativity and innovation in the cultural economy are marked not only by strong spatial patterns, but also by a robust temporal logic. A pioneering attack on this question was mounted by Peterson and Berger (1975) in their study of diversity in the recorded music industry in the USA. Peterson and Berger defined diversity in terms of the number of different recording companies with titles listed on *Billboard*'s Hot 100 music charts. Their study of data from these charts over the 1950s and 1960s concluded that the diversity of hit records is subject to cyclical oscillations over time. These oscillations appear to depend on the competitive interplay between major recording companies and small independents in final markets. Peterson and Berger observed that when majors dominate the market, product diversity diminishes; when independents are in the ascendant, diversity increases. These phases succeed one another as follows:

1. When the diversity of hit records is at a low level due to majors' control of the market, new niches appear beyond the margins of the current mainstream.
2. Many of these niches are explored by risk-taking independents, and as some of them find popular favor the diversity of entries on the hit parade increases.
3. Some of these niches become commercially lucrative, and the majors will proceed to take them over, thus creating a new mainstream with eventually diminished diversity of products.
4. New margins and exploratory possibilities for risk-taking independents appear; and so on.

This approach has been corroborated by analysts such as Alexander (1996), Christianen (1995) and Gronow (1983). In addition, Burnett (1990, 1992, 1993), Lopes (1992) and Scott (1999b) have pointed out that since the end of the 1970s an independent trend toward increasing diversity in Hot 100 hits has become intertwined with – and is possibly beginning to override – this cyclical process. In all likelihood, this circumstance is due to the growing tendency for majors to absorb independent recording companies as affiliates in the quest for broader market penetration. This is in line with a shift to increasing product diversity within the cultural economy at large, and it represents – up to a point – a move away from the kind of massification that Frankfurt School theorists, such as Adorno (1991), thought was to be the fate of popular culture.

Agglomerations of producers usually follow developmental trajectories shaped by the interpenetrating logic of all their component elements. Accordingly, and in view of their intricate structural make-up, cultural-products agglomerations are evidently subject to the system-wide branching and lock-in processes described by Arthur (1990) and David (1985). Once any agglomeration shifts in a given direction, further evolutionary possibilities are pre-empted in the sense that only derivative branches can now be followed while others become closed off. Depending on the past history of the system, these branches may lead on to further evolutionary possibilities, or to stagnation.

Creativity and innovation are very much at stake as this process works its course. One of the most pervasive forms of lock-in in the types of systems under scrutiny here is the situation exemplified by Maduro's brahmin painters where the codes and styles of cultural performance become so conventionalized that possibilities for change are negligible. It might be argued that much of the French cinema today, with its overarching structure of protective institutions is, paradoxically, in danger of approaching lock-in (Scott, 1999a). Similarly, in the immediate post-war years, the film industry of Hollywood was relatively locked in to a studio system of production, and this greatly impeded it from responding to the competitive pressures now coming from the burgeoning television industry (Maltby, 1981). Under the stimulus of this external threat, the industry painfully restructured over the 1950s and 1960s, above all by pursuing strategies of vertical disintegration. In parallel with these changes, new rounds of creativity and innovation were unleashed in the industry, in both business practices and in the cultural content of films, culminating in the Hollywood of today with its myriad small adaptable firms moving rapidly from project to project and its wholesale embrace of new digital technologies.

These temporal logics are deeply etched onto the geography of the creative field. As they unfold, particular places become dynamically integrated into the system of cultural production as aspects of their identity are assimilated into final products, and as these products in turn define and redefine their places of origin. This relationship can be relatively open-ended or can converge toward lock-in, leading in the one case to continual shifts in the images projected by product and place, and in the other to a mutually reinforcing stability.

Consider the symbiosis between the film industry, qua production apparatus, and Hollywood, the place. The film industry of Hollywood draws on a web of local cultural assets in the form of street scenes, natural landscapes, ways of life and so on. These assets play a crucial role in imparting to the products of the industry their distinctive look and mental associations. But these products also create new images (real or imagined) of Hollywood/Southern California and imbue earlier images with fresh meanings. These are then assimilated back into the region's fund of cultural assets where they become available as inputs to new rounds of production. This dynamic flow of images is also frequently tapped by other cultural-products industries in Southern California, such as music recording, television show production, clothing and so on (Scott, 1996). Equally, industries like high fashion in Paris or country music in Nashville draw upon associations rooted in place while simultaneously helping to make those same places as cultural constructs. 'Nashville', indeed, is virtually entirely a creature of its cultural-products industry (see Peterson, 1975). Las Vegas, another prominent instance of the same phenomenon, is literally a place and a cultural-products industry at one and the same time.

Conclusion

I have sought in this brief review to demonstrate that creativity and innovation in the modern cultural economy can be understood as social phenomena rooted in the production system and its geographic milieu, i.e. the creative field. Three main issues have been investigated in this regard, namely, (a) the formation of cultural communities, (b) the organization of the cultural economy and its propensity to agglomeration and (c) the temporal logic of cultural production complexes. Once again, this claim about the immanence of creativity and innovation does not in any way depreciate the role of the individual as a repository of specific aptitudes and ' imaginative capacities. Indeed, the individual as the bearer of these endowments is indispensable to the whole process. Creativity and innovation, however, are also imbricated in spatial and temporal fields of social action in the sense that they come into being in organized production environments where the talents and abilities of different individuals assume an interdependent character directed to economic ends.

When the Frankfurt School theorists wrote their pessimistic prognostications about the looming eternal sameness and alienating effects of commercialized culture, they were making inductions from then current trends which they no doubt correctly apprehended. These trends were evident in the deepening corporate control of culture, and its expression in Fordist or proto-Fordist production methods. However, the world has not quite turned out as the members of the Frankfurt School anticipated. For one thing, in the light of increasing evidence that *all* culture grows out of concrete social situations, one might plausibly argue that while large segments of the capitalist cultural economy will always produce dross of one sort and another, there is no reason in principle why other segments cannot function at the leading edges of cultural progress and experimentation (see Featherstone, 1995; Frith, 1996; Garnham, 1987; Negus, 1998; Rowe, 1995). For another, there has been a turn to increasing diversity in many different sectors of the cultural economy, a turn that is likely to be intensified as new cultural-products agglomerations make their appearance in the global economy (Scott, 1998b).

References

Adorno, T.W. (1991) *The Culture Industry: Selected Essays on Mass Culture.* London: Routledge.
Alexander, P.J. (1996) 'Entropy and Popular Culture: Product Diversity in the Popular Music Recording Industry', *American Sociological Review* 61: 171–4.
Arthur, W.B. (1990) 'Silicon Valley Locational Clusters: When do Increasing Returns Imply Monopoly?', *Mathematical Social Sciences* 19: 235–51.
Barnes, B. (1974) *Scientific Knowledge and Sociological Theory.* London: Routledge & Kegan Paul.
Barnes, B., D. Bloor and J. Henry (1996) *Scientific Knowledge: A Sociological Analysis.* London: Athlone.
Becker, H.S. (1974) 'Art as Collective Action', *American Sociological Review* 39: 767–76.
Becker, H.S. (1976) 'Art Worlds and Social Types', *American Behavioral Scientist* 19: 703–18.

Becker, H.S. (1982) *Art Worlds*. Berkeley and Los Angeles: University of California Press.

Bordwell, D., J. Staiger and K. Thompson (1985) *The Classical Hollywood Cinema: Film Style and Modes of Production to 1960*. New York: Columbia University Press.

Bourdieu, P. (1971) 'Le Marché des biens symboliques', *L'Année sociologique* 22: 49–126.

Bourdieu, P. (1977) 'La Production de la croyance: Contribution à une économie des biens symboliques', *Actes de la recherche en sciences sociales* 13: 3–44.

Bourdieu, P. (1983) 'The Field of Cultural Production, or: the Economic World Reversed', *Poetics* 12: 311–56.

Bourdieu, P. and Y. Delsaut (1975) 'Le Couturier et sa griffe: Contribution à une théorie de la magie', *Actes de la recherche en science sociales* 11: 7–36.

Burnett, R. (1990) *Concentration and Diversity in the International Phonogram Industry*. Gothenburg Studies in Journalism and Mass Communication 1. Gothenburg, Sweden: University of Gothenburg.

Burnett, R. (1992) 'The Implications of Ownership Changes on Concentration and Diversity in the Phonogram Industry', *Communication Research* 19: 749–69.

Burnett, R. (1993) 'The Popular Music Industry in Transition', *Popular Music and Society* 17: 87–114.

Christianen, M. (1995) 'Cycles in Symbol Production? A New Model to Explain Concentration, Diversity and Innovation in the Music Industry', *Popular Music* 14: 55–93.

Cooke, P. and K. Morgan (1998) *The Associational Economy: Firms, Regions, and Innovation*. Oxford: Oxford University Press.

Crane, D. (1992) *The Production of Culture: Media and the Urban Arts*. Newbury Park, CA: Sage.

Csikszentmihalyi, M. (1990) 'The Domain of Creativity', pp. 190–212 in M.A. Runco and R.S. Albert (eds) *Theories of Creativity*. Newbury Park, CA: Sage.

David, P. (1985) 'Clio and the Economics of QWERTY', *American Economic Review* 75: 332–7.

Denisoff, R.S. and J. Bridges (1982) 'Popular Music: Who are the Recording Artists?', *Journal of Communication* 32: 132–42.

DiMaggio, P. (1977) 'Market Structure, the Creative Process, and Popular Culture: Toward an Organizational Reinterpretation of Mass-Culture Theory', *Journal of Popular Culture* 11: 436–52.

DiMaggio, P. and P.M. Hirsch (1976) 'Production Organizations in the Arts', *American Behavioral Scientist* 19: 735–52.

Featherstone, M. (1995) *Undoing Culture: Globalization, Postmodernism and Identity*. London: Sage.

Frith, S. (1992) 'The Industrialization of Popular Music', pp. 49–74 in J. Lull (ed.) *Popular Music and Communication*. Newbury Park, CA: Sage.

Frith, S. (1996) *Performing Rites: On the Value of Popular Music*. Cambridge, MA: Harvard University Press.

Garnham, N. (1987) 'Concepts of Culture: Public Policy and the Cultural Industries', *Cultural Studies* 1: 23–37.

Gronow, P. (1983) 'The Record Industry: The Growth of a Mass Medium', *Popular Music* 3: 53–75.

Hall, P. (1998) *Cities in Civilization*. New York: Pantheon.

Hennion, A. (1981) *Les Professionels du disque: Une sociologie des variétés*. Paris: Éditions A.M. Métailié.

Hennion, A. (1983) 'The Production of Success: An Anti-Musicology of the Pop Song', *Popular Music* 3: 159–93.

Hennion, A. (1989) 'An Intermediary between Production and Consumption: The Producer of Popular Music', *Science, Technology and Human Values* 14: 400–23.

Hirsch, P.M. (1969) *The Structure of the Popular Music Industry*. Ann Arbor: Institute for Social Research, University of Michigan.

Hirsch, P.M. (1972) 'Processing Fads and Fashions: An Organization-Set Analysis of Cultural Industry Systems', *American Journal of Sociology* 77: 639–59.

Kealy, E.R. (1979) 'From Craft to Art: The Case of Sound Mixers and Popular Music', *Sociology of Work and Occupations* 6: 3–29.

Lash, S. and J. Urry (1994) *Economies of Signs and Space*. London: Sage.

Latour, B. and S. Woolgar (1979) *Laboratory Life: The Social Construction of Scientific Facts*. Beverly Hills, CA: Sage.

Lopes, P.D. (1992) 'Innovation and Diversity in the Popular Music Industry, 1969 to 1990', *American Sociological Review* 57: 56–71.

Lundvall, B.A. and B. Johnson (1994) 'The Learning Economy', *Journal of Industrial Studies* 1: 23–42.

Maduro, R. (1975) *Artistic Creativity in a Brahmin Painter Community*. Research Monograph 14. Berkeley: Center for South and Southeast Asia Studies, University of California.

Malraux, A. (1946) *Esquisse d'une Psychologie du Cinéma*. Paris: Gallimard.

Maltby, R. (1981) 'The Political Economy of Hollywood: The Studio System', pp. 42–58 in P. Davies and B. Neve (eds) *Cinema, Politics, and Society in America*. Manchester: Manchester University Press.

Mannheim, K. (1952) *Essays in the Sociology of Knowledge*. Henley-on-Thames: Routledge & Kegan Paul.

Marshall, A. (1919) *Principles of Economics*. London: Macmillan.

Menger, P.M. (1983) *Le Paradoxe du Musicien: Le Compositeur, le Mélomane et l'État dans la Société Contemporaine*. Paris: Flammarion.

Menger, P.M. (1993) 'L'Hégémonie Parisienne: Économie et Politique de la Gravitation Artistique', *Annales: Économies, Sociétés, Civilisations* 6: 1565–1600.

Molotch, H. (1996) 'LA as Design Product: How Art Works in a Regional Economy', pp. 225–75 in A.J. Scott and E.W. Soja (eds) *The City: Los Angeles and Urban Theory at the End of the Twentieth Century*. Berkeley and Los Angeles: University of California Press.

Montgomery, S.S. and M.D. Robinson (1993) 'Visual Artists in New York: What's Special about Person and Place?', *Journal of Cultural Economics* 17: 17–39.

Mulkay, M.J. (1972) *The Social Process of Innovation: A Study in the Sociology of Science*. London: Macmillan.

Negus, K. (1996) *Popular Music in Theory: An Introduction*. Hanover and London: Wesleyan University Press.

Negus, K. (1998) 'Cultural Production and the Corporation: Musical Genres and the Strategic Management of Creativity in the US Recording Industry', *Media, Culture and Society* 20: 359–79.

O'Connor, K. (1991) 'Creativity and Metropolitan Development: A Study of Media and Advertising in Australia', *The Australian Journal of Regional Studies* 6: 1–14.

Peterson, R.A. (1975) 'Single-Industry Firm to Conglomerate Synergistics: Alternative Strategies for Selling Insurance and Country Music', pp. 341–58 in J.F. Blumstein and B. Walter (eds) *Growing Metropolis: Aspects of Development in Nashville*. Nashville, TN: Vanderbilt University Press.

Peterson, R.A. and D.G. Berger (1975) 'Cycles in Symbol Production: The Case of Popular Music', *American Sociological Review* 40: 158–73.

Peterson, R.A. and P. DiMaggio (1975) 'From Region to Class, the Changing Locus of Country Music: A Test of the Massification Hypothesis', *Social Forces* 53: 497–506.

Rowe, D. (1995) *Popular Cultures: Rock Music, Sport and the Politics of Pleasure.* London: Sage.

Russo, M. (1985) 'Technical Change and the Industrial District: The Role of Inter-Firm Relations in the Growth and Transformation of Ceramic Tile Production in Italy', *Research Policy* 14: 329–43.

Ryan, J. and R.A. Peterson (1982) 'The Product Image: The Fate of Creativity in Country Music Songwriting', in J.S. Ettema and D.C. Whitney (eds) *Individuals in Mass Media Organizations: Creativity and Constraint.* Beverly Hills, CA: Sage.

Scott, A.J. (1984) 'Territorial Reproduction and Transformation in a Local Labor Market: The Animated Film Workers of Los Angeles', *Environment and Planning D: Society and Space* 2: 277–307.

Scott, A.J. (1996) 'The Craft, Fashion, and Cultural-Products Industries of Los Angeles: Competitive Dynamics and Policy Dilemmas in a Multisectoral Image-Producing Complex', *Annals of the Association of American Geographers* 86: 306–23.

Scott, A.J. (1997) 'The Cultural Economy of Cities', *International Journal of Urban and Regional Research* 21: 323–39.

Scott, A.J. (1998a) 'Multimedia and Digital Effects: An Emerging Local Labor Market', *Monthly Labor Review* (March): 30–8.

Scott, A.J. (1998b) *Regions and the World Economy: The Coming Shape of Global Production, Competition, and Political Order.* Oxford: Oxford University Press.

Scott, A.J. (1999a) 'French Cinema: Economy, Policy and Place in the Making of a Cultural-Products Industry', *Theory, Culture & Society.*

Scott, A.J. (1999b) 'The US Recorded Music Industry: On the Relations between Organization, Location, and Creativity in the Cultural Economy', *Environment and Planning A.*

Storper, M. (1996) 'Innovation as Collective Action: Conventions, Products, Technologies', *Industrial and Corporate Change* 5: 761–90.

Storper, M. and S. Christopherson (1987) 'Flexible Specialization and Regional Industrial Agglomerations: The Case of the US Motion Picture Industry', *Annals of the Association of American Geographers* 77: 260–82.

Von Hippel, E. (1988) *The Sources of Innovation.* New York: Oxford University Press.

White, H.C. and C.A White (1965) *Canvases and Careers: Institutional Change in the French Painting World.* New York: Wiley.

Williams, R. (1982) *The Sociology of Culture.* New York: Schocken Books.

Wolff, J. (1981) *The Social Production of Art.* New York: St Martin's Press.

[17]

"Being There": Proximity, Organization, and Culture in the Development and Adoption of Advanced Manufacturing Technologies*

Meric S. Gertler

Department of Geography, University of Toronto, Toronto, Ontario M5S 1A1, Canada

Abstract: Recent work on innovation and technology implementation suggests the importance of closeness between collaborating parties for the successful development and adoption of new technologies. "Closeness" is used here both in the literal sense, as allowing more frequent, effective, often unplanned interaction, and more broadly, to encompass common language, modes of communication, customs, conventions, and social norms. Such relationships are said to be particularly important in the case of production process innovations. These theoretical ideas are subjected to empirical scrutiny through a postal survey and set of interviews with users of advanced manufacturing technologies in Southern Ontario. Given the premise that intensive interaction and collaboration between users and producers of advanced process technologies is necessary for successful technology implementation to occur, then the chronically underdeveloped state of the Ontario advanced machinery sector ought to create major difficulties for manufacturers there trying to implement leading-edge processes. The analysis indicates that "closeness" between user and producer, defined physically, organizationally, and "culturally," is important for the successful implementation of these advanced technologies. The paper suggests a set of circumstances in which proximity of the user to the producer is most important. In doing so, it offers an interpretation of "culture" that goes beyond common language, codes of communication, and norms to incorporate shared workplace practices and training regimes. The paper concludes that industrial policies based on free trade in industrial machinery will be inadequate, on their own, to ensure that firms in mature regions make effective use of advanced process technologies.

Key words: user-producer interaction, flexible machinery, industrial networks, training.

During the past several years, a growing number of geographers and other social scientists have chronicled the apparent rise of post-Fordist economic systems (Scott and Storper 1987; Schoenberger 1988; Harvey 1989; Storper and Walker 1989; Boyer 1990). These systems are said to employ a flexible approach to production, reflected in employment relations, the organization of work within firms, and the broader social division of labor (Cooke and Morgan 1991). To some, the heart of this transformation lies in the

* This research was funded by the Social Sciences and Humanities Research Council of Canada, whose support is gratefully acknowledged. The author also wishes to acknowledge the extremely valuable research assistance provided by Jill Watson, Jonathan Hack, and Michael Skelly. Helpful comments contributing to this project have been received from John Britton, Susan Christopherson, Amy Glasmeier, Bennett Harrison, Dan Knudsen, Tod Rutherford, Anno Saxenian, Erica Schoenberger, Phil Shapira, Guy Steed, and three anonymous reviewers. The author would also like to acknowledge the useful feedback received during presentations of earlier drafts of this paper from seminar participants at McMaster University, Ohio State University, the TARP Working Group of the Ontario Federation of Labour, Wilfrid Laurier University, and the Atlanta Meetings of the Association of American Geographers (April 1993).

rise of a new set of forces of production (Walker 1994). In particular, they point to a new set of flexible process technologies whose programmable properties offer producers prospects of great versatility, limited downtime, unparalleled precision, and superior quality. The same technologies are said to hold the potential to unleash the creative potential of workers and to compel manufacturers to establish a new regime of cooperation on the shopfloor (Florida 1991).

Despite the popularity of such arguments, their unqualified acceptance has not been universal. A critical literature has arisen which, among other things, questions the pervasiveness of such practices, especially in locations outside the "paradigmatic" flexible production regions (Gertler 1988, 1992; Sayer 1989; Pudup 1992). It has come to light that, for example, while rates of adoption of flexible technologies such as computerized numerical control (CNC) are reasonably high among manufacturers in countries like the United States, Great Britain, and Canada, many firms in these countries have experienced great difficulty in trying to implement such technologies effectively (Jaikumar 1986; Kelley and Brooks 1988; Turnbull 1989; Oakey and O'Farrell 1992; Beatty 1987; Meurer, Sobel, and Wolfe 1987).

Furthermore, there is an apparent regularity to the geography of technology adoption difficulty that is highly suggestive of its roots. Many of these implementation difficulties seem to arise in older, mature industrial regions, where manufacturing firms are far removed from the major production sites of the new flexible production technologies (Gertler 1993). Increasingly, the leading producers of these process technologies are to be found in countries like Germany, Japan, and Italy, while once-dominant American machinery producers have seen their market shares drop significantly, both at home and abroad (Graham 1993).

At the same time, it must be noted that those flexible production regions that have received the most attention in recent years are located in the same countries that are home to the leading producers of advanced machinery. In fact, case studies suggest that the production and use of such advanced machinery frequently occurs within the very same region, whether it be textiles in Germany's Baden-Württemberg (Piore and Sabel 1984; Sabel et al. 1987), leather goods, ceramic tiles, and knitwear in Emilia-Romagna (Brusco 1986; Russo 1985), or semiconductors in Japan (Stowsky 1987). The same literature suggests that the spatial coincidence of machine production and use in the recently successful industrial districts is not merely coincidental. These regularities hint strongly at the importance of spatial context in determining the degree of success in technology implementation. It is precisely this spatial context and the interaction between users and producers of advanced machinery that constitutes the central subject for the empirical analysis presented in this paper.

Machine Production and Use from an Interactive Perspective

Although geographers have not devoted a great deal of attention to the spatial context of machinery production and use, the relationship between machine users and producers has been the subject of theoretical and empirical work, first by economic historians (Rosenberg 1976, 1982a, 1982b), and more recently by students of industrial organization and technological change (Lundvall 1985, 1988; Porter 1990). Based largely on case studies, this literature has suggested that "closeness" between the users and producers of advanced machinery is important for a variety of reasons, laid out briefly below.[1]

[1] The discussion here draws particularly from Lundvall's (1985, 1988) work, since he provides the most systematic treatment of these issues. For a more thorough review of this work, see Gertler (1993). Throughout this paper, the term "user" refers to a firm that has

The argument begins with the general insight that capital goods differ in important ways from the other kinds of inputs purchased by manufacturers. Their function, when combined with labor, is of obvious central importance to the success of a manufacturer's operations. Furthermore, they tend to be long-lived (by definition), and hence the firm will have to rely on such assets for a long period of time. Add to this the consideration that frequently—particularly when the firm is purchasing recently developed, leading-edge technology—a large degree of uncertainty surrounds the future use qualities of these capital goods. Because of these properties, the wise firm is less inclined by choice to purchase advanced capital goods through a simple, discrete, "off-the-shelf" market transaction, but is more inclined toward engaging in what might be described as an *organized market* transaction—that is, one in which there is extensive interaction and communication between the prospective user firm and the producer of the machinery.

This mode of purchase is said to offer a number of benefits to prospective machinery users and producers alike, so that complex production equipment is not only more likely to be *adopted* successfully when there is close and frequent interaction between producer and user, but is also likely to be *produced* more successfully as well. For users, this interactive mode of technology acquisition allows them to gather as much information as possible about the properties of the machinery under consideration, to gauge the reliability and trustworthiness of the producer. Furthermore, it may allow the user to make its technological needs more readily and clearly known to the producer, creating the conditions under which the useful customization of

the product to the user's particular intended application is more likely. To allow customization to occur, however, users must reveal to an outside firm certain proprietary details concerning their products or production processes, and they may be unwilling to do so unless they have been able to build up a sufficient level of trust with machine producers, resulting from a process of close interaction over an extended period of time.

At the same time, this kind of interaction is also important and beneficial for machinery producers. Research on capital goods innovation has demonstrated that prospective users—particularly demanding and technologically sophisticated customers—represent a vital source of creative stimulus for producers, who are more likely to develop important innovations when compelled to meet their customers' needs (von Hippel 1988; Lotz 1990; Teubal, Yinnon, and Zuscovitch 1991). These innovations not only help producers themselves compete more successfully, but also bring obvious benefits to the users who are fortunate enough to enjoy a close relationship with producers, characterized by this process of mutual learning.

Other benefits for users may arise in later stages of the machinery acquisition process. The interactive perspective sees this process as one of considerable duration, consisting of three distinct stages. Beyond the benefits arising in the preinstallation phase of design and production described above, the second phase (installation and start-up) is likely to proceed more smoothly when producer and user enjoy a close relationship, since (1) the producer will more likely be on hand to assist in the installation and break-in of the new machine or system, providing useful on-site training and assistance, and (2) the start-up process is likely to be easier (i.e., requiring less adaptative behavior by the user) when the machine or system has been tailored to meet the user's precise specifications. Similarly, the third phase—of "normal" operation of the

_____
acquired and implemented some form of advanced manufacturing technology in its production process. The term "producer" denotes the manufacturer of advanced equipment, machinery, or integrated systems.

new process postinstallation—will also be more successful (i.e., fewer breakdowns and more likely to meet users' expectations concerning productivity, quality, and functional capabilities) if a close relationship with the producer facilitates the adjustments and debugging that are bound to be necessary as the experience of regular use brings to light inevitable operational problems.

Clearly, these kinds of benefits, and the closeness between users and producers which is said to facilitate the technology acquisition process, are most likely to be important when the technology involved is expensive, complex, and rapidly developing. Presumably, when the process technology in question is less expensive, or represents more mature or familiar technology, an interactive mode of acquisition will be less important. For cheaper, well-established, or "tried and true" technologies, an off-the-shelf mode of purchase should suffice.

A number of significant implications flow directly from this literature. First, these arguments suggest that a large measure of the success enjoyed by manufacturers in the "canonical" industrial districts of Europe and Asia may be attributed to the close and constructive relationship they are able to maintain with nearby producers of innovative machinery. Second, viewed from this perspective, it should come as no surprise that machinery producers in these same regions have become highly successful competitors in international markets in recent years. Third, and perhaps most significantly for the present analysis, this literature suggests (although it does not make this point explicitly—see Gertler 1993) that many of the problems arising when manufacturers in the mature industrial regions of the United States, Britain, or Canada attempt to implement new, technologically advanced process technologies stem from the greater difficulty these users have in developing and maintaining a "close" relationship with advanced machine producers, as the latter are currently (and increasingly) more

likely to be located at a considerable distance from would-be users. Forced instead to acquire complex technologies using a mode approaching that of an off-the-shelf transaction, such users are much more likely to encounter the kinds of problems documented by previous case studies.[2]

Despite the compelling nature of these arguments, it is possible that certain countervailing forces might intervene to reduce or qualify the attenuating influence of simple physical distance between user and producer. Three, in particular, merit discussion here. First, it is possible that, despite a preference for direct, face-to-face interaction, users and producers might be able to communicate quite adequately for many purposes by using modern and increasingly effective telecommunications media (especially the fax and telephone). And for those functions that cannot be provided over the wire, rapid air transportation of technical personnel (or key parts) might serve as a reasonable compromise when users are far away from their advanced machinery producers. In addition, it should be recognized that many European and Asian producers contract with distributors, sales representatives, and maintenance firms in North America to perform on-site service functions on their behalf. Such intermediaries might be able to compensate for the large distances intervening between North American users and overseas producers.

Second, the case has frequently been made that large, multilocational (including multinational) firms serve as highly effective, distance-transcending vectors of (intrafirm) technology transfer.[3] According to

---

[2] For an early recognition of the off-the-shelf manner in which most independent Canadian firms are forced to acquire their production technology, as well as some of the inadequacies of this mode of technology acquisition, see Chapter 6 in Britton and Gilmour (1978).

[3] See the excellent discussion of this issue provided in Chapter 12 of Dicken (1992).

this view, production regions that might be viewed as peripheral (with respect to the location of leading machinery producers) may nevertheless be characterized by the presence of advanced machinery that is being used effectively by the local branch plant operations of such large, multisite firms. The close organizational ties between such branch plants and their foreign head offices (or sister plants elsewhere) may allow the branch operations to benefit from the considerable expertise and experience that exist within the larger firm. If the parent firm has developed its own advanced process technologies, at its head office or other production or research sites, these are likely to be transferred quite effectively to branch plant users at distant sites. According to this argument, then, what matters more than simple physical distance is what one might call *organizational distance*.[4]

Third, one might offer another interpretation of the difficulties encountered by the users of advanced machinery in "peripheral" locations which has more to do with differences in cultures, institutions, and the legacy of past industrial practices than with the problems caused by the intervening distance between users and producers. For example, at the most basic level, Lundvall (1988) argues the importance of a common culture and language shared by users and producers, to facilitate the transmission of highly encoded information concerning users' needs and the capabilities and proper operation of complex and rapidly changing process technologies (see also Storper 1992). Others point to differences in training cultures and attitudes toward technology as the crucial issues (Stowsky 1987; Gordon 1989), implicating the distinct set of practices and attitudes peculiar

to Anglo-American firms as the true source of technology implementation problems.

According to this latter view, the typical American, British, or Canadian firm regards technology as something *embodied* entirely within the physical properties and design of machinery and production systems themselves. This stands in sharp contrast to the approach more typical of European and Japanese manufacturers, who not only appreciate the necessity of social interaction for effective machine production and use, but also regard the technological capabilities of a production process as being produced through the interaction between machines and skilled workers who have built up a wealth of knowledge and problem solving abilities through many years of training and learning by doing.

The consequence of this difference is that Anglo-American users of advanced machinery, who espouse what Block (1990, 152) refers to in the macroeconomic context as an "intravenous" model of capital investment, typically expect to be able to extract the full capabilities of such technologies merely by installing them correctly and "flipping the switch." In contrast to their European and Asian counterparts, and in response to the national institutions shaping relations in their labor market, they tend systematically to undervalue the importance of training and to maintain shorter-term relations with their workers (instead making extensive use of external labor markets). A further consequence is that an advanced machine designed and built in, for example, Germany will be considerably more difficult to implement successfully in a North American user plant than in a German user plant because the "culture" of industrial practices peculiar to Germany (high skill levels of factory workers, stability of employment relation, cooperative decision making on the shopfloor, strong emphasis on training) have been incorporated into the design of the German-made machine. According to this view, then, physical distance is really just

---

[4] Lundvall (1988) offers similar speculations about the possible importance of spatial-organizational structure, but provides no concrete evidence to support or refute his observations.

6   ECONOMIC GEOGRAPHY

a proxy for *cultural distance*, where "culture" refers to a set of dominant workplace practices shaped in large part by legislative definitions of employment relations and the nature of the (public and private) industrial training system. Furthermore, this approach would seem to ascribe continuing importance to nation-states and the economic and social institutions created by them (see Gertler 1992).[5]

Research Questions and Methodology

The empirical study presented in this paper seeks to investigate the relationship between users and producers of advanced process technologies in order to shed more light on the source of users' difficulties in implementing complex new machinery. This study is motivated by several concerns. First, many of the contentions made in the literature reviewed above remain just that—that is, intelligent arguments based on theoretical speculation or limited empirical observation. Second, these questions have not yet been examined from an explicitly geographic perspective, though it is clear that the central relationships are inherently spatial.

Specifically, the study seeks to answer the following questions: (1) Is a high-quality (i.e., interaction-intensive) relationship between user and producer

---

[5] Storper (1992) provides a useful discussion of the importance of shared culture within territorial production complexes characterized by what he calls product-based technological learning. He also discusses the concept of user-producer interaction in more general terms, noting its relevance not only to users and producers of capital goods, but also between users and producers of components, materials, information, and final products. However, Storper's conception of "culture" depends far less on state regulatory frameworks, emphasizing instead the traditions, norms, and practices developing from close interaction between spatially clustered economic agents over an extended period of time.

necessary to support the successful implementation of advanced manufacturing technology? (2) Must users and producers be "close" to one another for such a high-quality relationship to develop? (3) Under what circumstances is "closeness" likely to be most important? (4) How close is "close," and is this to be understood only in terms of physical distance, or are organizational or "cultural" conceptions of distance more important?

I have sought to answer these questions through an investigation of advanced machinery implementation among manufacturers in the province of Ontario. Given the relatively underdeveloped state and small size of the Canadian advanced machinery sector (Science Council of Canada 1991), additional questions that drive this particular study are: in what ways do Ontario manufacturers suffer from the incomplete development of an indigenous advanced machine-building capability (as manifest in the kinds of implementation problems discussed above)? To what extent, and in what ways, do the typically long distances between Ontario users and (foreign) machinery producers contribute to the implementation difficulties experienced by users in Ontario? Previous work has shown that, despite relatively high rates of new technology adoption, Ontario manufacturers' productivity performance has lagged significantly behind that of their counterparts in other industrialized countries (Gertler 1993). In addressing such questions, it is hoped that this study might shed further light on the geographic roots of firms' competitive failures and difficulties.

The empirical analysis presented here is drawn from two modes of inquiry: (1) a postal questionnaire survey of users of advanced manufacturing technologies in Ontario manufacturing industries, and (2) follow-up plant visits and personal interviews with a subset of the firms that completed the mailed survey. It should be noted that the research described in this paper, emphasizing the experiences and characteristics of users, represents the first phase of a two-part study. In a subsequent phase, the "demand side" view presented here will be supplemented by a more systematic analysis of "supply side" considerations—that is, the characteristics, organization, and strategic behavior of the machinery producers. Nevertheless, the analysis offered in this paper does strive to shed light on the *relationship* between these two interacting partners, albeit as viewed from the perspective of the users.

The survey, mailed in the summer of 1991, targeted Ontario establishments in four sectors: transportation products (automotive parts and aerospace), electrical and electronic products, fabricated metal products, and plastic and rubber products. These sectors were selected on the basis of previous surveys of advanced technology adoption, which had identified plants in these industries as among the most frequent users of complex new process technologies (Statistics Canada 1989; Ontario 1989). The initial sample was drawn from the manufacturer's data base maintained by the Ontario Ministry of Industry, Trade and Technology. Entries in this data base provided basic information on names, addresses, and telephone numbers of manufacturing establishments, as well as ownership (location), major products (by four-digit SIC code), and employment size. Prior to mailing, plants were contacted by telephone to determine if they did in fact use any advanced manufacturing technologies (see Table 1) and, if so, were they willing to participate in the study. The sample of establishments for mailing was stratified by both plant size (employment) and ownership (Canadian or foreign-owned) to permit investigation of arguments on the importance of organizational and cultural influences.

The resulting framework sought a balanced sample in six different cells—three size categories (small, medium, large) by two ownership categories (domestic, foreign). Questionnaires were normally addressed to plant managers, production managers, or chief engineers, as these

Table 1

Advanced Manufacturing Process Technologies

Design and Engineering
 Computer-Aided Design (CAD) and/or
  Computer-Aided Engineering (CAE)
 Computer-Aided Design (CAD) and/or
  Computer-Aided Manufacturing (CAM)
 Digital data representation
Fabrication and Assembly
 Flexible Manufacturing Cells (FMC)
 Flexible Manufacturing Systems (FMS)
 NC/CNC machines
 Material working lasers
 Robots
 Pick and place robots
Automated Material Handling
 Automated Storage and Retrieval Systems
  (AS/RS)
 Automated Guided Vehicle Systems (AGVS)
Communications and Control
 Technical data network
 Factory network
 Programmable Logic Controller (PLC)
 Computers used for control on the factory floor
Manufacturing Information Systems
 Materials Requirements Planning (MRP)
 Manufacturing Resource Planning (MRP II)
Integration and Control
 Computer-Integrated Manufacturing (CIM)
 Supervisory Control and Data Acquisition
  (SCADA)
 Artificial Intelligence (AI)
 Expert systems

Source: Statistics Canada (1989).

individuals had been identified through telephone contact as being the most closely involved in the process of technology acquisition and implementation.[6] Several weeks after the initial mailing, respondents were contacted again by telephone if they had not returned their questionnaire. A total of 408 surveys were mailed, with 170 usable returns, for a response rate of roughly 42 percent.

The survey instrument was designed to solicit information for up to three of each plant's most advanced process technolo-

---

[6] In the case of smaller plants, surveys were more frequently directed to company owners, as they often tended to fulfill the role of production manager/engineer in such settings.

8   ECONOMIC GEOGRAPHY

gies in use. Consequently, the final data base contained information describing 407 distinct technology implementation cases (see Table 2). Although a list and glossary of some 22 types of advanced process technologies was appended to the questionnaire as a guide from which to choose (see Table 1), respondents were invited to discuss other advanced technologies not explicitly mentioned on the list. Such "write-ins" were normally treated as legitimate instances of advanced technology implementation, except in a handful of cases where I determined through a follow-up telephone call that the technology was more appropriately regarded as conventional. The questionnaire sought information on the type of advanced process technology used, where it was produced, whether or not the design was customized to suit the user's particular needs, and how it was acquired (i.e., directly from producer, indirectly via a distributor, produced in-house, produced in a sister plant owned by the same firm). It also collected information on the extent and nature of difficulties encountered during both the installation and operation phases of technology implementation, on the perceived importance of spatial proximity between user and producer, on the frequency of interaction between the user and producer of the machinery or system, on the type of interaction (site visit,

telephone, fax, other), and on whether some kind of collaborative relationship existed between machinery producer and user.

In order to gauge the effect of other possible influences on technology implementation success, the survey gathered further information on internal characteristics of the plant, including the occupational mix of employment, types of training provided to shopfloor workers, the extent to which decisions determining technological change were made in a consultative fashion by plant management, and the extent to which other external sources of technical assistance (besides machinery producers or distributors) were utilized by users. Finally, information on growth in sales, employment, and exports over the prior three-year period, on the average size of production run, and on the proportion of sales going to local markets was collected.

The end result was a data base providing a fairly comprehensive picture of each plant's experiences in technology implementation, set within the context of its more general internal characteristics and market relations. However, it was recognized that a postal survey would be limited in the extent to which it could collect information of a more qualitative nature on plants' technology implementation experiences and their relationships

Table 2

Distribution of Survey Returns

| | Plant Size | | | | | | | |
| | Small | | Medium | | Large | | All Sizes | |
Ownership	Cases[a]	Plants	Cases	Plants	Cases	Plants	Cases	Plants
Canadian	60	27	76	32	74	29	210	88
Foreign	49	24	93	39	53	18	197	82
Total	109	51	169	71	127	47	407	170

Note: For transportation and electrical/electronic products, small is 1–99 employees, medium is 100–249 employees, and large is 250+ employees. For fabricated metal products and plastic/rubber products, the size categories are, respectively, 1–49, 50–149, and 150+. The latter set of categories was necessary to accommodate plant size distributions much more heavily skewed toward the small end of the scale in these two sectors.
[a] Refers to the number of technology implementation cases. Since the questionnaire solicited information on up to three discrete implementation experiences per plant, the total number of cases reported exceeds the total number of plants.

with machinery producers. Hence, plant visits and personal interviews were conducted during the summer of 1992 with 30 of the survey respondents to allow more detailed probing of the kinds of relationships addressed in the questionnaire, as well as to provide a better understanding of the properties of the process technologies under consideration. The sample for the interviews was similarly structured to include establishments in all six size/ownership categories.

In the following discussion, information from the postal survey is presented in tabular format. As will become apparent, however, its interpretation is frequently assisted by use of qualitative information gleaned from the interviews with production managers or chief engineers. Although the initial intent was to structure the analysis of survey results by the four industrial sectors, this strategy was ultimately not pursued. Upon gaining a deeper appreciation of the kinds of process technologies being implemented in each of the sectors, it became clear that many of the same types of machinery and equipment were common across all four product groups. Hence, computer-controlled metal-cutting and forming technologies were evident not only in the fabricated metal products sector, but also in transportation products. Technologies for the assembly of electronic components

and products were also widely used in the transportation sector. Advanced machinery for the production of plastic products (e.g., robot-assisted injection molding machines) was commonly being used by firms in the electronic products sector (such as a large producer of telephone handsets), as well as in the automotive parts industry. Furthermore, initial analysis of survey returns indicated that users' reported degree of difficulty in the installation and operation of such equipment did not vary significantly by industry sector. In the discussion that follows, therefore, the analysis is not disaggregated by product sector.

User-Producer Interaction and the Implementation of Complex Process Technologies in Ontario Manufacturing Industries

Does Distance Matter?

Perhaps the most direct way to determine the significance of intervening distance between users and producers is simply to ask users. The results of such an inquiry are presented in Table 3, which assesses the importance attached by users to having their major machinery producers located at varying distances from them. There is an obvious and consistent pattern to the responses. Just under 40

Table 3

Importance to Users of Physical Proximity to Major Machinery Producers (Frequency and Percentage)

Degree of Importance	Geographic Region				
	Same City (within 20 km)	Same Region (within 75 km)	Same Province	Same Country	Same Continent
Very important	28	30	36	41	79
	17.2%	19.2%	23.4%	26.8%	52.3%
Somewhat important	34	51	50	63	37
	20.9%	32.7%	32.5%	41.2%	24.5%
Not important	101	75	68	49	35
	62.0%	48.1%	44.2%	32.0%	23.2%
Total users	163	156	154	153	151

percent of plants judged a very close physical link (20 kilometers or less) to be somewhat or very important. This figure rises steadily, however, as the spatial scale increases. More than half answered somewhat or very important when asked to comment on the importance of colocation of user and producer within the same region (defined for this purpose as within a 75 kilometer radius). By the time the national scale is reached, the figure jumps to exceed two-thirds of all plants (hinting at the possible importance of effects related to international border considerations).

At the continental scale, more than half of all firms judged colocation to be very important, and three-quarters answered very or somewhat important. The general conclusion to be drawn from this analysis is that having the producer of one's major machinery on the same continent is quite important to most users of advanced process technologies and, for many users, even shorter distances between them and producers are preferred. However, the reasons underlying this sentiment are not immediately clear.

To get some sense of these reasons, respondents were asked to explain the answers they gave to the prior question. This was also the subject of considerable discussion in the personal interviews. Table 4 provides a summary content analysis of the explanatory responses from

the questionnaire. For plants indicating that closeness to producers *was* important, the overwhelming consideration appeared to be the ability to obtain good service or spare parts quickly in the event of breakdowns. Interviews revealed a widely held view that service and spare part delivery was more likely to be reliable, timely, and effective when provided directly by the original producer of the machinery. Distributors and sales representatives were often seen as distinctly inferior substitute sources of service, repairs, modifications, and technical information. According to one manager of a medium-sized automotive parts plant east of Toronto, the comparison between a distributor/sales representative and the original producer when such service is required is analogous to the choice between "the guy who sold you your car and a properly trained mechanic when your automobile needs service or repairs. I know who I'd prefer to have working on my car."

A production manager at a small auto parts plant in southwestern Ontario reported that his firm had tried to deal with the Kentucky-based North American distributor of a German robotics producer, but was very unsatisfied with the quality and level of service offered. It chose instead to deal directly with the head office in Germany, a more expensive but also more effective strategy. Ultimately,

Table 4

Reasons Cited by Users for the Importance/Unimportance of Physical Proximity to Advanced Machinery Producers

Why Closeness to Machinery Producer *is* Important	Frequency of Mention	Why Closeness to Machinery Producer is *not* Important	Frequency of Mention
1. Service/spare parts	40	1. Local machinery reps are satisfactory	10
2. Communication	6	2. In-house or intrafirm expertise	9
3. Border hassles	5	3. Will go anywhere for the right technology	4
4. Time zones	3	4. Receive good service from distant sources	3
5. Travel costs	1	5. Standard technology requires little communication	2
Total	55	Total	28

the Canadian plant cemented its ties to the German firm by actually hiring one of their "best" field service personnel to work for them!

Other considerations underlying the importance of close contact between user and producer relate to physical issues such as different time zones (which may reduce or entirely eliminate the number of hours of the day during which both the user plant and the producer plant are operating—a problem when technical advice is required immediately) or the simple cost of transportation (for parts, service personnel, or those providing training in machinery use). Complications arising from intervening international borders were also identified as problematic. These might include everything from duties or tariff payments on imported machinery and parts to difficulty of entry by foreign service personnel.[7] Finally, communications problems were also singled out by user firms as a source of difficulty when producers were not "close by." These included both the problems arising when the producers' first language was not English, as well as the general problems associated with communicating complex technical information over long distances. While technical terms are notoriously difficult to translate from one language to another, many problems are

not easily described or solved over the phone or fax, even when the communicating parties are both native English speakers.[8]

These comments are helpful in suggesting why continental colocation of user and producer might be especially important, since overseas interaction is likely to be more difficult for all of the physical, logistical, and cultural reasons noted above. It is also apparent that for many users location of their producer on the same continent may not be good enough to suit their needs. Illuminating as these observations may be, they still beg further questions, such as: What kinds of user plants will receive "good" service, either from local or distant sources? What intervening impacts might size, ownership, or other firm and plant characteristics have on this relationship?

At the same time, Table 4 indicates that, in certain circumstances, closeness between user and producer is deemed unimportant. This appears to be the case when user establishments receive a high level of service, either from local reps of distant producers or directly from the producer (despite long intervening distances). Alternatively, other user firms appear to compensate for their "peripheral" location relative to machine producers by developing and exploiting strong technical expertise within the firm itself.

Finally, some users indicated that the nature of the technology being applied might be an important intervening consideration. Those more standardized forms of advanced technology in wider use might

---

[7] Present trade laws in Canada allow imported machinery to enter the country duty free if the importing firm can successfully convince federal customs officials that there is no domestic producer for a given piece of equipment. Interviews indicated that this process could often be difficult and time-consuming, despite the underdeveloped state of the Canadian machinery industry. As for entry of service personnel working for the machinery producer, while the Canada–United States Free Trade Agreement (FTA) was supposed to facilitate the flow of such personnel between the two countries, many user firms reported continued delays and difficulties. And, of course, the FTA has no bearing on such movements between Canada and countries in Europe and Asia.

[8] One medium-sized manufacturer of plastic products described how a service person for a machinery producer in Ohio failed repeatedly to solve the user's technical problem despite repeated contact via telephone and fax. In frustration, the producer finally sent "a box of parts" by courier for the user to try. This strategy too was unsuccessful. After the user had incurred considerable expense and delay, the service person finally paid a site visit to the user's plant, whereupon the problem was solved "in about five minutes."

require little or no communication with the producer to enable effective use, particularly if local technical colleges or service firms can offer assistance. At the other end of the spectrum, if the technology is truly leading-edge, some firms say they will "go anywhere" to obtain it, no matter how far away the producer may be. Of course, despite this willingness to transcend long distances to secure the "right" technology, it may still be the case that implementation difficulties will arise, for the reasons noted above. In such cases, when a large part of the firm's competitive strategy depends on access to innovative process technologies, presumably the plant would be willing to persevere through the implementation problems, since it feels it "has no choice." The question arising from this is: At what costs might such a strategy be pursued, in a relatively "peripheral" location? How does this affect the user firm's competitive position relative to its more "favorably" located machine-using competitors abroad?

When Does Distance Matter Most?

One of the questions driving this research seeks to uncover the conditions under which closeness between user and producer is likely to be most important. The empirical study addresses this issue in a variety of ways. First, one would expect, a priori, that closeness would be more important to smaller user plants, since they are more likely to lack the financial or organizational resources required to overcome the attenuating influence of distance between themselves and machinery producers. This expectation is confirmed in Table 5, which shows how the importance of closeness varies by size of establishment.

Here one sees two notable patterns. First, as in Table 4, the importance of closeness generally increases with spatial scale. However, there appear to be two notable departures from this pattern. For plants in the smallest size category, a marginally higher proportion indicate that it is very important for their major machinery producers to be within the same immediate vicinity than the proportion indicating importance within the same region. Furthermore, the mid-sized plants appear to be indifferent between the intermediate spatial scales of region, province, and country. The second general pattern evident in this table is that, for the most part, small and medium-sized plants attach greater importance to closeness to machine producer at every spatial scale examined. The largest plants seem to be the least concerned with the need for spatial proximity, particularly at subnational scales. At the national level, the percentage of large plants answering "very important" approaches that of the other two size groups, and at the continental scale this figure is even marginally

Table 5

Percentage of Plants Saying Physical Proximity to Machinery Producers is "Very Important," by Plant Size

User Plant Size[a]	Geographic Region				
	Same City	Same Region	Same Province	Same Country	Same Continent
Small	21.9	18.0	27.1	28.8	59.3
Medium	19.6	26.9	26.9	26.9	48.0
Large	11.9	11.9	14.3	24.4	48.8
All plants	17.2	19.2	23.4	26.8	52.3
	$p = .0010$ $\chi^2 = 18.55$				$p = .0005$ $\chi^2 = 20.08$

[a] See Table 2 footnote for size definitions.

higher than that of the medium-sized group (though both are substantially lower than the proportion of small plants deeming location on the same continent as their producer to be very important).

These results appear to confirm the idea that larger establishments can exploit their greater spatial reach to overcome the problems associated with long distances from machine user to producer. However, it seems that this reach (as far as acquisition of advanced machinery is concerned) is more accurately described as *continental* than *global*. Furthermore, interviews revealed an alternative interpretation of this pattern that has less to do with the resources supporting spatial reach and more to do with the plant's attractiveness to potential machinery producers. Simply put, larger plants constitute more significant and important customers for producers and hence are considerably more likely to receive good service, no matter where they are located. There appear to be two distinct dimensions to this. First, larger plants are more likely to spend larger sums on a given producer's machinery, both in any single purchase and over time. Second, it is apparent that, from the producer's perspective, large plants often represent more prestigious customers capable of lending credibility in the marketplace to the supplier. Installations in large plants are more likely to serve as "showcase" applications of a machinery producer's best product, holding the potential to help generate future sales to other user firms.

Of course, plant size may be less important than firm size. If the plant is merely one of many owned by a parent company, this fact will add to the single plant's significance in the eyes of the producer firm, securing "red carpet" treatment from producers for even comparatively small installations. To explore this idea further, we investigated how the importance of closeness to machine producer might vary by plant type. The results (Table 6) confirm the suspicion that organizational status is indeed important. At all spatial scales, single-plant

establishments (which, by virtue of the sample design, are by definition Canadian-owned) are by far the most likely of the three plant types to judge closeness to machinery producer as very important.[9]

It is also apparent that, within the multiplant category, ownership type (foreign versus domestic) does not appear to make much of a difference to the importance of closeness between user and producer. Canadian multiplant users are somewhat more likely to judge colocation in the same city as very important and considerably less likely to judge colocation within the same province as important, compared to foreign multiplant users. But at the national and continental scales the differences are small indeed, and even reveal Canadian multiplant establishments to feel slightly less bound to the same country or continent when shopping for machinery than are their foreign counterparts.

The Influence of "Cultural" Distance

Our earlier review of the literature drew attention to the possible importance of cultural commonality between user and producer as a facilitating influence on the technology implementation process. This was alleged to result from both a shared code of communication as well as a common legacy of industrial practices and institutions. To investigate these arguments, we asked responding user plants to gauge the degree of difficulty they have

[9] It should also be noted that, by definition, all foreign-owned plants sampled in this study fall into the category of multiplant since, if their ownership resides outside of Canada, and they operate production facilities in Ontario, they must also operate plants in other locations outside of the country. In many cases they also happened to operate multiple plants even within Ontario. In order to examine foreign-owned single-plant establishments in this study, the sample would have had to include single-plant establishments in other countries—clearly beyond the spatial reach, organizational scope (and budget) of this project.

14   ECONOMIC GEOGRAPHY

Table 6
Percentage of Plants Reporting Physical Proximity to Machinery Producers as "Very Important," by Ownership-Organization Type

User Plant Type	Geographic Region				
	Same City	Same Region	Same Province	Same Country	Same Continent
Foreign multiplant	13.8	15.2	23.4	26.3	50.7
Canadian multiplant	18.2	16.7	14.3	23.8	47.5
Canadian single-plant	23.1	31.4	34.3	31.4	61.1
All plants	17.2	19.2	23.4	26.8	52.3

had in operating their advanced machinery, and attempted to relate this to user plant ownership and the geographic origin of the process technology in question.

The results (Table 7) are very revealing. When the advanced machinery is produced in Canada, it appears that Canadian users are about as likely to describe their experience as unproblematic ("not difficult" or "on the easy side") as are foreign-owned users (59 versus 61 percent of plants, respectively). Hence, in both Canadian and foreign-owned plants, only about 40 percent of users acknowledge some difficulty in operating complex machinery obtained from Canadian producers, and Canadian plants are substantially more likely to situate their experience within the easiest category than are foreign-owned plants (17.4 versus 11.1 percent).

However, when the origin shifts to overseas (i.e., outside North America) machinery producers, the 60/40 easy/difficult split for Canadian users is almost reversed: now 57 percent of Canadian users report difficulties, and the share increase is especially notable in the "very difficult" category (from 13 to almost 24 percent). At the same time, foreign-owned users indicate only a modest increase in difficulty, with 57 percent of their experiences being rated as unproblematic. In other words, when the source of advanced machinery is overseas, foreign-owned plants are substantially less likely to encounter difficulty in operation than are Canadian-owned establishments. This is also evident when examining the "not difficult" row for foreign-owned plants: as one reads across the table from Canadian to overseas technology, the proportion of experiences falling into this easiest category increases dramatically (from 11.1 to 28.6 percent). It is worth pointing out here that the analysis in Table 7 effectively controls for plant size, since the sample was constructed to have

Table 7
Degree of Difficulty Operating Machinery, as Reported by Users (Percentage)

	Canadian-Produced Machinery		Overseas-Produced Machinery[a]	
	Canadian-Owned Plant	Foreign-Owned Plant	Canadian-Owned Plant	Foreign-Owned Plant
Not difficult	17.4	11.1	9.5	28.6
On easy side	41.3	50.0	33.3	28.6
On difficult side	28.3	33.3	33.3	33.3
Very difficult	13.0	5.5	23.8	9.5
	$p = .9351$ $F = 0.0067$		$p = .0407$ $F = 4.3292$	

[a] Denotes location outside North America.

similar size distributions of users in both the Canadian and foreign-owned categories.[10]

These results appear to lend credence to the idea that the degree of cultural commonality exerts an important influence on the. path of the technology implementation experience. This idea also came through very strongly in the personal interviews with users, as will be made evident below. Furthermore, it was a phenomenon that characterized both large and small plants and single- and multiplant firms alike. When this area was probed in conversation with production managers and engineers, lack of common language was usually the first consideration mentioned. However, many also indicated that the problem was deeper than one of language. After all, as was commonly noted by interviewees, most European and Japanese machine-producing firms know that their sales, engineering, and service personnel have to be reasonably proficient at speaking English if they wish to make a foreign sale.

Some of the more subtle forces at work became evident when those Canadian user-plant personnel interviewed described the often difficult implementation processes they had experienced. Interestingly, comments pertained to the entire technology acquisition process, from design and prepurchase negotiations to installation and training to postinstallation troubleshooting and support. Plant personnel made frequent references to different business cultures, noting how, for example, when dealing directly with European machine producers, it was necessary to establish a relationship over time, with repeated personal interaction, in order to obtain the kind of product and service they were seeking. Even after persevering, many commented on an apparent chauvinism on the part of

(especially northern) European machine builders, who were resistant to altering their designs to suit North American users' requests because they felt simply that "they [the producers] knew best" and that North American users, if only they were more technologically sophisticated, would see that they were right.

The case of a large automotive parts plant, part of a larger Canadian firm, illustrates these characteristics quite vividly. The plant had decided to acquire a new, multimachine production system to cope with rising demand. At the insistence of the firm's senior management, who believed steadfastly in the superiority of northern European machinery, the plant engineers selected a German supplier for the job. There was very little customizing of the machinery to the Canadian plant's needs, and most of the preinstallation interaction involved a small number of the Canadian personnel being sent to Germany to receive training in the producer's plant (at the user's expense).

Once the installation was complete, the plant experienced problems almost immediately, and a long and difficult break-in period ensued. The relationship between the Canadian user and the German producer became strained and adversarial as conflict arose over the true source of the operational problems. From the user's perspective, the system had not been properly installed and adjusted to suit their needs and the postinstallation support was considered to be woefully inadequate and excessively expensive (costing about $800 per day). However, according to the user firm, the producer was adamant in claiming that the problem lay with the customer, whose management, engineers, maintenance staff, and shopfloor workers (in the producer's estimation) lacked the requisite technical abilities to operate and maintain the machinery properly. The producer was reported to have said, "We have no problems running exactly the same machinery in Germany. The problem must be yours." Ultimately, relations became

---

[10] It should also be mentioned that similar results were obtained when the analysis was repeated using difficulty of *installation*, rather than operation.

strained to the point where the vendor's on-site personnel were asked to leave the user's plant, and the user, in apparent disgust, simply ceased all contact with the producer, preferring to struggle on by themselves.

This story is fairly typical of experiences at other Ontario plants where users have attempted to implement northern European technology while lacking the necessary "cultural" affinity. It is also noteworthy because the user plant was large, part of a multiplant firm, and, by North American standards, regarded as highly sophisticated from a technological point of view.

"Being There": Site Visits and Alternatives

It is clear from my earlier literature review that one of the alleged advantages of closeness between users and producers of advanced technologies is the greater ease of interaction and communication before, during, and after installation of machinery. For maximum effect, much of this interaction would need to be achieved through direct, face-to-face contact.

Indeed, our interviews revealed an emphatic and widespread agreement that site visits between user and producer—ideally throughout the entire acquisition process—were absolutely crucial for ultimate success when the technology being implemented was new, complex, and expensive. These visits could involve travel in both directions (user visiting

producer, producer visiting user) and were usually regarded as being especially important during the installation and initial operation phases of technology acquisition. Beyond providing an opportunity for the effective exchange of information, they were also seen as the most useful medium by which to deliver training to the user's technical and operative personnel.

Many users complained of the difficulties that arose when their contact with producers was strictly remote. In many such cases, "training" might be provided in the form of printed manuals, which were frequently viewed with utter disdain and regarded as "next to useless." In other cases, the producer might send the user a videotape in which the operation of the machinery was demonstrated. These too were regarded as poor substitutes for "being there."

These observations raise an important question—namely, what kinds of plants do (and do not) receive site visits? And what kinds of producers are more likely to provide this service to users? These questions are investigated in Tables 8 and 9, which provide information from the questionnaire on the proportion of plants receiving site visits, disaggregated by size, type, and geographic source of technology, for both the installation and operation phases of technology acquisition.

From Table 8, four major patterns are evident. First, the probability of a user plant receiving a site visit (at least, as reported by the user) is notably greater

Table 8

Percentage of Plants that Received Site Visits, by Plant Size

	Site Visit during Installation			Site Visit during Operation		
Source Country	1–174 Employees[a]	> 174 Employees	All Plants	1–174 Employees[a]	>174 Employees	All Plants
Canada	90.0	94.4	91.9	76.0	88.2	79.8
United States	68.5	88.8	78.5	59.2	77.8	68.5
Overseas	71.4	86.4	79.1	61.9	71.4	64.3

Note: Figures pertain only to technologies purchased directly from the producer ($N = 238$).
[a] Binary size classes are used here for presentational clarity. Tables based on a three-way size classification show similar results.

Table 9
Percentage of Plants that Received Site Visits, by Establishment Type

	Site Visit during Installation				Site Visit during Operation			
Source Country	Foreign Multi	Canadian Multi	Canadian Single	All Plants	Foreign Multi	Canadian Multi	Canadian Single	All Plants
Canada	89.5	96.7	88.9	91.9	75.7	76.7	94.4	79.8
United States	82.0	96.7	65.2	78.5	68.0	71.4	65.2	68.5
Overseas	81.8	78.6	71.4	79.1	71.4	57.1	57.1	64.3

Note: Figures pertain only to technologies purchased directly from the producer (N = 238).

during the installation phase than during the operation phase. Hence, it appears that Ontario users receive more attention in the earlier stages of technology acquisition than in the latter stages. Although this pattern holds for all source locations, the declining frequency of site visit reports as one moves from installation to operation seems to be greatest when the machinery in question originates overseas (from nearly 80 percent during installation to roughly 64 percent during operation). In this sense, then, distance does appear to "matter."

Second, the likelihood of receiving a site visit from the machinery producer — in either the installation or operation phase — declines as the distance from the machinery source increases. Third, while this "distance decay" effect is experienced by both large and small plants, the attenuating influence of distance seems to be especially strong for smaller plants. Fourth, for both phases, small plants are less likely to receive site visits than are large plants, no matter where the machinery producer is located. However, Canadian vendors appear to be more even-handed in their treatment of small and large plants. For example, while almost 95 percent of all large plants received a site visit from a Canadian vendor during installation, so too did 90 percent of the smaller plants. Compare this to the treatment accorded users by American vendors: almost 89 percent of large plants received visits, while the figure for small plants was only about 69 percent.

Many of the same patterns found in Table 8 apply to Table 9, where variation by plant type is explored. In addition, we see that the distance decay in the likelihood of receiving a site visit is especially marked in the case of Canadian plants (in both single and multiplant firms). Indeed, foreign-owned plants experience *almost no decline in the likelihood of receiving a site visit* as the distance from the source increases. Because of the rather stark differences between the experiences of foreign (by definition, multiplant) and Canadian multiplant establishments (especially during the operation phase), it is probably inaccurate to infer that the rather special treatment received by foreign-owned plants is a function of firm size. Instead, this pattern may be one more manifestation of the "cultural affinity" arguments made earlier, or may also provide evidence of intrafirm, though international, transfer of production technologies and assistance to foreign-owned plants located in Ontario.

The patterns evident in Tables 8 and 9 suggest that many plants, particularly those that are small and those that are Canadian-owned single-establishment operations, are less likely to receive site visits from machinery producers than are other types of user plants. Furthermore, this discrepancy appears to widen as the source of machinery becomes more distant, and this "distance decay" effect is especially marked during the operation phase of technology acquisition. The implication here is that those plants that receive lower levels of personal attention from producers are forced to resort to other means of communication to try to obtain the assistance they require.

Some evidence of this is provided in Table 10, which shows the proportion of plants (by type) ranking fax or telephone as the most important means of communication with machinery producers during the installation and operation phases. Consistent with Tables 8 and 9, it is evident that these indirect means of communication are less important during the installation phase than during the operation phase, for all types of plant. However, during the installation phase, a notably higher proportion of Canadian single-plant establishments rank fax/phone as their most important means of communication than is apparent for the other two plant types. A surprisingly high 58 percent of plants in this group list fax or phone as their most important communications medium, compared with figures below 40 percent for the other two groups.

For the operation phase, the Canadian single-plant establishments still show a greater reliance on fax or phone than is evident for their multiplant counterparts. However, the proportion of foreign multiplant users reporting fax or phone as their most important medium is also surprisingly high. This somewhat puzzling outcome may be explained by returning to the themes of organizational and cultural distance discussed earlier. From our interviews, it seems clear that smaller firms and those of the single-plant variety

resort to indirect means of communication because they have little in the way of alternatives. They are too small or insignificant to warrant a site visit from the producer, and hence must rely on phone calls and faxes. Alternatively, since the cost of site visits was determined in general to be borne, either directly or indirectly, by the customer, these kinds of plants may be less able to afford the considerable cost associated with a site visit and instead resort to cheaper methods. As noted earlier, this means of communication does not always provide effective results.

At the other end of the spectrum, when foreign-owned plants turn to telephone and fax use, this mode of communication is likely to prove more effective for them, for two reasons. First, if the machinery in question was developed within the same firm (though at another branch location), even though communication is between two geographically remote parties, the parties are employed by the same organization and are therefore more likely to share common firm-specific knowledge and codes of communication. Second, if the machinery was externally sourced (in both a geographic and organizational sense), chances are that the two communicating parties still share a common cultural background, which acts to facilitate communication for the reasons described earlier. Hence, in either of these situations, communication via phone or fax is likely to be significantly more effective than in those cases where the communicating parties share neither orga-

Table 10

Percentage of Plants that Ranked Fax and Phone as Important Means of Communication with Machinery Producers

Importance	During Installation			During Operation		
	Canadian Single	Canadian Multi	Foreign Multi	Canadian Single	Canadian Multi	Foreign Multi
First	58.1	36.6	39.6	70.4	63.5	69.7
Second	38.7	55.9	55.4	25.4	30.6	26.8
Third	1.6	5.4	4.3	1.4	2.4	2.8

nizational nor cultural affinity. Though an obvious point, it is also worth pointing out that a series of phone calls or faxes is still far cheaper than the cost of sending technical personnel on a long-distance trip, and even large firms will pay attention to such cost differentials.

Vendor-Provided Training: Who Provides It, and Who Gets It?

While these patterns concerning site visits and use of alternative communications means are quite revealing, the simple presence or absence of site visits by the machinery producer is still a rather coarse measure of the effectiveness of communication between user and producer. As noted above, our interviews confirmed that one of the most important potential benefits of a site visit is the opportunity for the producer to provide explicit, hands-on training for the employees of the user plant. This would suggest that it is important to have an understanding of the incidence of training by user type and producer location. It is also important to examine the effectiveness of this training provided by vendors (producers).

Table 11 provides an analysis of training incidence as revealed in the questionnaire returns, broken down by user plant type and size and producer location. The general pattern, consistent with earlier results, is that Canadian single-plant establishments and small plants are the least likely to receive vendor-provided training. As for the vendors themselves, it appears that U.S. producers are the least likely to provide training, although they do seem to serve the needs of the foreign-owned and larger plants better than the rest of their market. They are particularly unlikely to provide training to Canadian single-plant establishments and to those with fewer than one hundred employees. On the whole, it is the overseas vendors who appear to be serving their users' needs best, in terms of the proportion of users receiving training from them. Presumably this may be explained by the recognized unwillingness of their North American customers to purchase complex machinery made overseas unless the vendor is willing to provide some kind of training support.

Discussion of this issue during the interviews revealed that many users have been unhappy with the quality of the training support they have received from distant vendors—again, particularly from those located in Europe. The often-expressed general sentiment was that vendors were merely "going through the motions," in order to fulfill contractual obligations made at the time of sale. And, consistent with the patterns noted above,

Table 11
Percentage of Machinery Producers Providing Training

| Producer Location | User Establishment Type | | | User Plant Size[a] | | | All Plants |
	Foreign Multi	Canadian Multi	Canadian Single	Small	Medium	Large	
Canada	9/11[b]	9/9	6/8	8/10	7/9	9/9	24/28
	81.8%	100%	75.0%	80.0%	77.8%	100%	85.7%
United States	11/13	11/15	4/8	5/10	14/17	7/9	26/36
	84.6%	73.3%	50.0%	50.0%	82.4%	77.8%	72.2%
Overseas	11/13	8/8	1/1	7/7	7/7	6/8	20/22
	84.6%	100%	100%	100%	100%	75.0%	90.9%
All sources	31/37	28/32	11/17	20/27	28/33	22/26	70/26
	83.8%	87.5%	64.7%	74.1%	84.8%	84.6%	81.4%

[a] See Table 2 footnote for definitions of plant size categories.
[b] Ratios express actual number of cases receiving training, over the total number in each category.

Economy: Critical Essays in Human Geography

this dissatisfaction was particularly strongly (though not exclusively) felt among the smaller, independent Canadian-owned plants.

The extent of this dissatisfaction is evident in Table 12, which shows satisfaction levels by machinery source and user plant characteristics. The very low satisfaction rating for training provided by overseas vendors is striking. For all plants, only 50 percent of users were satisfied or very satisfied with the quality of this training. This figure drops even lower for Canadian-owned users (33.3 percent) and for small and medium-sized plants (around 43 percent). Hence, although Table 11 showed that overseas-sourced machinery is the product most likely to be accompanied by some form of training (no matter what the characteristics of the user), Table 12 indicates that this type of training is apparently not very satisfactory.

Within the same table, there is further evidence of a possible "cultural affinity" effect, in that Canadian machinery producers seem to receive the highest overall satisfaction ratings for their training, particularly from Canadian-owned users. At the same time, foreign-owned users are almost twice as likely as Canadian users to claim satisfaction with training provided by overseas machinery producers.[11] Interestingly, Canadian users also give a high satisfaction rating to vendor-provided training from U.S. producers, suggesting that the commonality of language and work practices may facilitate the transmission of technical knowledge to engineers and shopfloor workers in the Canadian user plants. And, following the same pattern, foreign-owned users are notably less likely to express satisfaction with training provided by a Canadian machinery vendor than are Canadian user firms. Surprisingly, however,

foreign-owned plants (which include a sizable number of American-owned establishments) also express satisfaction even *less* frequently for training provided by U.S. vendors, suggesting that they may represent a more sophisticated or demanding category of user that is generally harder to please.

Implications for Regional-Industrial Policy

We preceded our discussion of empirical findings by posing four research questions, to which we now return. First, the results reported here would appear to lend strong support to the idea that a high-quality (interaction-intensive) relationship between the user and producer of advanced capital goods is of major importance in facilitating implementation by the user. For many user establishments, interaction with machinery producers was not very intensive, and this often contributed to major implementation difficulties. Second, "closeness" between producer and user does indeed seem to facilitate the formation and maintenance of a high-quality, interaction-intensive relationship in which there is a more open flow of information. In cases where large distances intervened between producers and users, problems frequently arose, and producers' local distributors or sales representatives were generally regarded by users as distinctly "second best" alternatives to the producers themselves.

Third, lack of "closeness" appears to be especially onerous for smaller enterprises and for domestically owned single-plant establishments. Larger and multiplant establishments encounter fewer difficulties in dealing with machinery producers over long distances, because of their greater financial and other capacities to overcome the costs and difficulties of doing business on a long-distance basis, because of the potential for intrafirm transfers of technical assistance between plants to assist in the implementation

---

[11] Note again that, since the sample achieved similar plant-size distributions for the domestic and foreign-owned categories, the analysis presented in Table 12 controls for a possible plant size effect.

Table 12

Percentage of Plants Claiming to Be Either Satisfied or Very Satisfied with the Quality of Vendor-Provided Training

Producer Location	User Plant Ownership		User Plant Size[a]			
	Canadian-Owned	Foreign-Owned	Small	Medium	Large	All Plants
Canada	80.0	50.0	50.0	75.0	77.8	70.8
United States	80.0	45.5	80.0	57.1	71.4	65.4
Overseas	33.3	63.6	42.9	42.9	66.7	50.0

Note: These figures apply only to technologies purchased directly from the producer (*N* = 238).
[a] See Table 2 footnote for definitions of plant size categories.

process, or because their size and profile within the industry garner them high-quality service from even distant producers of advanced machinery and production systems.

Fourth, even large and multiplant establishments indicate overwhelmingly that, given the choice, they would rather do business with a producer of advanced machinery that is located on the same continent as they are. In fact, many users made clear that their first preference would be to have such producers located very close by, but this is often not possible as the local machinery sector is not sufficiently well developed, or the user plant is forced to trade off this consideration against many other locational influences. The research shows that "closeness" is to be understood in an organizational and cultural sense, as well as in the more traditional physical sense of the term. Considerations here run from the relatively banal (problems with time zones, border crossings) to the interesting (communications problems related to language differences, technical complexity of subject matter) to more fundamental concerns (problems arising from contextual differences in workplace practices, training cultures, and conceptions of "technology").

Of particular interest here are the findings concerning the difficulties arising from a mismatch in work cultures between machinery users and producers. Earlier, I argued that this stemmed from rather different understandings of the concept of technology and, hence, rather different strategies for attempting to maximize productivity in production operations. The research reported here suggests that what shows up in Canada or the United States as a "training problem" may in part be symptomatic of this "cultural" gulf between machinery users and the (overseas) firms producing their process technologies. Where advanced machinery is largely imported from countries whose industrial culture is based to a much greater extent upon a *socially determined* notion of technology (compared to the "embodied" conception dominant in the Anglo-American economies), implementation problems may be expected. Of course, the fact that machine producers are not only culturally, but frequently also physically distant from North American users only serves to exacerbate what one should expect a priori to be rather significant implementation difficulties. Presumably, if these two parties were closer to one another, then easier, cheaper, and more effective interaction would help to overcome these problems. Or, by improving the likelihood of effective customization of machinery by producers to suit users' needs, a closer physical link might reduce the likelihood that such problems would arise in the first place.

This argument holds a number of important implications for training policies to promote more competitive manufacturing. First, it suggests that, on its own, more training, whether provided by employers or by the state, will only provide a partial solution to a much more fundamental problem. In other words, this strategy addresses the symptom of the problem, rather than the root of the problem itself. In the absence of strategies to address the cultural and physical gap between machinery users and (largely) distant producers, training needs will not only remain chronically high, but will also be profoundly difficult to satisfy.

To state this argument in a somewhat different way, training-based strategies will be largely ineffective unless they are accompanied by two further changes. First, the above reasoning would suggest that a greater emphasis on training must be accompanied by a sea change in Anglo-American manufacturers' understanding of technology: they must acknowledge that productivity can best be generated from production processes in which skilled employees work closely with advanced machinery over extended periods of time. This amounts to espousing an employment relation based on stability, trust, and the free exchange of information between workers and managers. Only within the context of such a relationship can workers build up the kinds of skills and experience necessary to engage in effective learning by doing.[12]

Changing such attitudes is by no means easy. However, perhaps the most direct way to effect such change is to alter the regulatory and institutional environment that shapes workplace relations, recognizing the key role that such institutional features have played in, for example, the

German economy (Sorge and Streeck 1988). By enshrining stability of tenure, such a regulatory climate has provided a powerful incentive for employers (and governments) to invest in workplace training. Further, this has induced firms to pursue competitive strategies based on product innovation, quality, and the efficiency and innovativeness of the productive process, instead of trying to compete on the basis of cheap wages (Marshall 1992). Such strategies have been achieved largely through the use of internal, rather than external, labor markets.

The second change required to make training more effective would be to take steps to strengthen the relationship between Canadian (or U.S. or British) users of advanced process technologies and the firms producing these technologies. When it comes to the question of machinery sourcing, governments in North America have traditionally relied on the doctrine of comparative advantage and free trade, based on the rationale that countries like Canada, with a rather poorly developed domestic machinery industry, will maximize their welfare by importing machinery from more efficient producers abroad. If the arguments presented above have any validity, they would appear to cast some doubt on the wisdom of this strategy, which, after all, assumes that an advanced machine can be used equally effectively no matter how "close" or "distant" the user is from the producer. Instead of relying solely on the importation of advanced machinery, a number of options are open to policymakers.

First, efforts might be made to support the further domestic development of advanced machinery production, especially in those sectors in which the user country has a critical mass of demand and at least the rudiments of a domestic machinery-building capability already. This strategy has been pursued successfully elsewhere in countries like Japan and, more recently, South Korea (Amsden 1989). Of course, within the regulatory framework of the Canada-United States Free Trade Agreement (FTA) or its North

[12] This stands in sharp contrast to the prevailing North American tradition. For example, Noble (1984) has documented convincingly that the dominant design principles incorporated into American machine tool technology were consciously selected (by the U.S. military) for their capacity to control and limit workers' discretion on the shopfloor—hardly the model of a stable, open environment conducive to learning by doing.

American successor, NAFTA, Canadian policies that favor domestic firms may be subject to increasing international scrutiny and would undoubtedly attract responses in the form of countervailing trade actions by its trade bloc partners. Consequently, whatever assistance is offered to domestic machinery-building firms should also, under the principle of "national treatment," be made available to foreign-owned firms wishing to produce advanced machinery in Canada.

In fact, foreign direct investment by advanced machinery producers could be actively encouraged as a second way to enhance the "closeness" of domestic users of advanced manufacturing processes to the producers of such technologies. By encouraging such firms to locate production facilities (with design capabilities) in Canada—or, as a second-best solution, within the northern United States—policymakers would markedly improve the potential for mutual learning and productive interaction to occur between domestic users and foreign-owned producers. The success of this strategy would also rely on the kind of efforts described above to reduce the gap in industrial practices and governing institutions, so that users and producers are brought closer together not just in a physical sense, but in a cultural sense as well. The potential for mutual benefit inherent in this kind of convergence could be considerable.[13]

---

[13] In a complementary phase of this study, I conducted interviews with producers of advanced manufacturing technologies based in Europe who currently sell their machinery to North American users. From these discussions it is clear that such producers are beginning to recognize the difficulties associated with serving the North American market directly from Europe, or even indirectly through North American distributors and representatives. As a consequence, many of these producers (particularly the larger ones) are now taking steps to create some form of production capability in North America. A more complete

Conclusion: Implications for Theoretical Debates

This paper began by acknowledging the ongoing debate within economic geography and the other social sciences regarding the end of the Fordist era and the nature of its replacement. A central issue in this debate concerns the extent to which "flexible" production processes have already diffused throughout the industrial world, as well as their *potential* to become more widely adopted, particularly in more mature industrial regions. The findings reported in this paper shed some interesting light on these questions, since the machinery and systems in question here constitute the very heart of these flexible technologies.

First, it should be clear from the analysis presented here that the spread of such production methods to the industrial "diaspora" will be anything but unproblematic. Predictions of the "diffusion" of flexible production methods to mature industrial regions may be premature and overly optimistic, underestimating the difficulties involved in making this transition. As I have noted elsewhere (Gertler 1993), this "diffusionist" perspective is guilty of conceiving of the problem of manufacturing renewal in mature industrial regions as one of insufficient rates of uptake (or "adoption") of process technologies whose productive attributes are assumed to be wholly contained within a set of inanimate objects. If instead one espouses a more Marxian conception of capital as a social relationship,[14] one should *expect* difficulties (or what I have

---

discussion of these findings will be the subject of forthcoming papers.

[14] As Block (1990, 121) notes, Marx's conception of capital sprang from a critique of the traditional classical conception of capital as simply an inanimate "thing": "Marx insisted on the contrary that capital is a social relationship. For Marx, rather than being in itself a piece of capital, the machine is inserted into a set of social arrangements that endow it with certain properties."

called "implementation pathologies") to arise if the social relations surrounding the use of such technologies in their regions of origin are not substantially present in the region of implementation.

Second, the findings in this paper not only help to interpret the failings of firms in mature industrial regions, but also hold important implications for our understanding of the "success stories"—the canonical industrial districts of Europe, Asia, California, and elsewhere. A common interpretation of the success of such regions emphasizes the beneficial effects of transaction cost reductions stemming from the physical proximity of interacting firms in these industrial agglomerations (Scott 1988). However, this paper's findings strongly suggest that physical proximity alone does not constitute a full explanation of the success of users (and producers) in such districts. Nor are common language, shared codes of communication, trust, or "embeddedness" (Harrison 1992), on their own, sufficient to explain the success with which users are able to implement advanced process technologies. Underemphasized in this literature is the importance of work practices and training cultures (and the broader regulatory and institutional framework) that are shared by both users and producers. In the absence of this commonality, it is difficult to imagine the advanced machinery users in these districts attaining the levels of success they have been alleged to enjoy.

Third, the similarity between the arguments made in this paper and those offered by Rosenberg (1976, 1982a, 1982b) in his analysis of user-producer relations in British and American industrial clusters in earlier historical periods suggests that this thesis may have broader applicability to capitalist industrialization in general. Undoubtedly, many conditions have changed during the many years since the first British and American industrial revolutions. However, the effects of the more important of these changes may cancel out each other, as they work in different directions. On the one hand, tremendous improvements in transportation and communications technologies imply that more effec-

tive long-distance interaction between users and producers is now possible. On the other hand, the quantum leap in complexity of production technologies since the early 1800s (and, in particular, the recent transition to *electronic control* of mechanical systems—what Florida (1991) has dubbed "mechatronics") suggests that "closeness" between user and producer would be more important in the contemporary era than it was one hundred and fifty or two hundred years ago. Given the increasingly global nature of production systems and corporate organization, there is more than a little irony in this insight.

References

Amsden, A. H. 1989. *Asia's next giant: South Korea and late industrialization*. New York: Oxford University Press.

Beatty, C. A. 1987. *The implementation of technological change*. Research and Current Issues Series Report No. 49. Kingston: Industrial Relations Centre, Queen's University.

Block, F. 1990. *Postindustrial possibilities: A critique of economic discourse*. Berkeley: University of California Press.

Boyer, R. 1990. *The regulation school: A critical introduction*. New York: Columbia University Press.

Britton, J. N. H., and Gilmour, J. M. 1978. *The weakest link: A technological perspective on Canadian industrial underdevelopment*. Background Study 43. Ottawa: Science Council of Canada.

Brusco, S. 1986. Small firms and industrial districts: The experience of Italy. In *New firms and regional development in Europe*, ed. D. Keeble and E. Wever, 184–202. London: Croom Helm.

Cooke, P., and Morgan, K. 1991. *The network paradigm: New departures in corporate and regional development*. Regional Industrial Research Report No. 8. Cardiff: University of Wales.

Dicken, P. 1992. *Global shift: The internationalization of economic activity*. 2d ed. New York: Guilford Press.

Florida, R. 1991. The new industrial revolution. *Futures* 23:559–76.

Gertler, M. S. 1988. The limits to flexibility: Comments on the post-Fordist vision of production and its geography. *Transactions of the Institute of British Geographers* n.s. 13:419–32.

———. 1992. Flexibility revisited: Districts, nation-states, and the forces of production. *Transactions of the Institute of British Geographers* n.s. 17:259–78.

———. 1993. Implementing advanced manufacturing technologies in mature industrial regions: Towards a social model of technology production. *Regional Studies* 27:665–80.

Gordon, R. 1989. Beyond entrepreneurialism and hierarchy: The changing social and spatial organization of innovation. Paper presented at the Third International Workshop on Innovation, Technological Change and Spatial Impacts, Selwyn College, Cambridge, UK, 3–5 September.

Graham, J. 1993. Firm and state strategy in a multipolar world: The changing geography of machine tool production and trade. In *Trading Industries, trading regions*, ed. H. Noponen, J. Graham, and A. Markusen, 140–74. New York: Guilford Press.

Harrison, B. 1992. Industrial districts: Old wine in new bottles? *Regional Studies* 26:469–83.

Harvey, D. 1989. *The condition of postmodernity*. Oxford: Basil Blackwell.

Jaikumar, R. 1986. Postindustrial manufacturing. *Harvard Business Review* 64:69–76.

Kelley, M. R., and Brooks, H. 1988. *The state of computerized automation in U.S. manufacturing*. Project Report, Center for Business and Government, Kennedy School of Government, Harvard University.

Lotz, P. 1990. User-producer interaction in the Danish medical equipment industry. In *management of technology II*, ed. T. M. Khalil and B. Bayraktar, 129–39. Proceedings of the Second International Conference on Management of Technology, 28 February–2 March. Norcross, Ga.: Industrial Engineering and Management Press.

Lundvall, B-A. 1985. *Product innovation and user-producer interaction*. Industrial Development Research Series, Report No. 31. Aalborg: Aalborg University Press.

———. 1988. Innovation as an interactive process: From user-producer interaction to the national system of innovation. In *Technical change and economic theory*, ed. G. Dosi, C. Freeman, R. Nelson, G. Silverberg, and L. Soete, 349–69. London: Frances Pinter.

Marshall, R. 1992. Unions and competitiveness. In *Empowering workers in the global economy: A labour agenda for the 1990s*,

75–98. Background and Conference Proceedings, 22–23 October 1991. Toronto: United Steelworkers of America.

Meurer, S.; Sobel, D.; and Wolfe, D. 1987. *Challenging technology's myths: A report on the impact of technological change on secondary manufacturing in metropolitan Toronto*. Toronto: prepared for the Labour Council of Metropolitan Toronto.

Noble, D. F. 1984. *Forces of production: A social history of industrial automation*. New York: Knopf.

Oakey, R. P., and O'Farrell, P. N. 1992. The regional extent of computer numerically controlled (CNC) machine tool adoption and post adoption success in small British mechanical engineering firms. *Regional Studies* 26:163–75.

Ontario. 1989. *A comparison of Canadian and U.S. technology adoption rates*. Toronto: Ministry of Industry, Trade and Technology.

Piore, M. J., and Sabel, C. 1984. *The second industrial divide*. New York: Basic Books.

Porter, M. E. 1990. The competitive advantage of nations. *Harvard Business Review* 68:73–93.

Pudup, M. B. 1992. Industrialization after (de)industrialization: A review essay. *Urban Geography* 13:187 200.

Rosenberg, N. 1976. *Perspectives on technology*. Cambridge: Cambridge University Press.

———. 1982a. *Inside the black box*. Cambridge: Cambridge University Press.

———. 1982b. Technological progress and economic growth. In *Technical change, employment and investment*, ed. N. Rosenberg and L. Jörberg, 7–27. Lund: Department of Economic History, University of Lund.

Russo, M. 1985. Technical change and the industrial district: The role of interfirm relations in the growth and transformation of ceramic tile production in Italy. *Research Policy* 14:329–43.

Sabel, C. F.; Herrigel, G.; Kazis, R.; and Deeg, R. 1987. How to keep mature industries innovative. *Technology Review* (April): 27–35.

Sayer, A. 1989. Postfordism in question. *International Journal of Urban and Regional Research* 13:666–95.

Schoenberger, E. 1988. From Fordism to flexible accumulation: Technology, competitive strategies and international location. *Environment and Planning D: Society and Space* 6:245–62.

Science Council of Canada. 1991. *Canada's non-electrical machinery sector*. Sectoral Technology Report No. 14, July. Ottawa.

Scott, A. J. 1988. *New industrial spaces*. London: Pion.

Scott, A. J., and Storper, M. 1987. High technology industry and regional development: A theoretical critique and reconstruction. *International Social Science Journal* 112:215–32.

Sorge, A., and Streeck, W. 1988. Industrial relations and technical change: The case for an extended perspective. In *New technology and Industrial Relations*, ed. R. Hyman and W. Streeck. Oxford: Basil Blackwell.

Statistics Canada. 1989. *Survey of manufacturing technologies—1989: Statistical tables*. Ottawa: Science, Technology and Capital Stock Division, September, catalogue ST–89–10.

Storper, M. 1992. The limits to globalization: Technology districts and international trade. *Economic Geography* 68:60–93.

Storper, M., and Walker, R. 1989. *The capitalist imperative: Territory, technology,* *and industrial growth*. Oxford: Basil Blackwell.

Stowsky, J. 1987. The weakest link: Semiconductor production equipment, linkages, and the limits to international trade. Working Paper 27, Berkeley Roundtable on the International Economy, University of California at Berkeley.

Teubal, M.; Yinnon, T.; and Zuscovitch, E. 1991. Networks and market creation. *Research Policy* 20:381–92.

Turnbull, P. J. 1989. Buyer supplier relations in the UK automotive industry. Paper presented at the conference "A Flexible Future," Employment Research Unit, Cardiff Business School, University of Wales College of Cardiff, 19–20 September.

von Hippel, E. 1988. *The sources of innovation*. New York: Oxford University Press.

Walker, R. A. 1994. Regulation and flexible specialization: Challengers to Marx and Schumpeter? In *Spatial practices: Markets, politics and community life*, ed. H. Liggett and D. Perry. Beverly Hills: Sage, forthcoming.

[18]

Towards a Knowledge-based Theory of the Geographical Cluster

PETER MASKELL

(Center for Economic and Business Research and Department of Industrial
Economics and Strategy, Copenhagen Business School, Howitzvej 60,
DK-2000 Frederiksberg C, Denmark. Email: maskell@cbs.dk)

*Co-located firms within related industries enhance the ability to create knowledge by
variation and a deepened division of labour. The interdependent development between
economic activities and local institutions make the cluster attractive to some industries
and hostile to others. The very reasons why cognitive distance might be small within
the cluster tend to make cognitive distance great between clusters and make interfirm
co-operation across bodies of knowledge more costly. The additional value created when
clustering may justify the additional cost.*

1. Introduction

One of the most significant consequences of the present process of
globalization is the way in which it continues to turn inputs, previously crucial
to the competitiveness of firms, into ubiquities.[1] Ubiquities are inputs equally
available to all firms at more or less the same cost almost regardless of location
(Weber, 1909). A large domestic market is, for instance, no longer an
unquestioned advantage when global transport costs are becoming negligible;
when the loyalty of customers toward national suppliers is dwindling; and
when most trade barriers have eroded. Domestic suppliers of the most
efficient production machinery are, similarly, no longer a solid competitive
advantage, when the sales and marketing strategies of the suppliers reach
across borders, and their equipment becomes available world-wide at

---

[1] Globalization has increasingly been associated with the 'unbundling' of the previous relationship
between sovereignty, territoriality and state power (Ruggie, 1993) and, as a consequence, steadily
weakening nation states (see Maskell, 2000), but it is arguably the many economic consequences of
ubiquitification that has contributed most in making globalization the favourite business buzzword at the
turn of the 20th century.

Towards a Knowledge-based Theory of the Geographical Cluster

essentially the same cost. The omnipresence of organizational designs of proven value, furthermore, makes a long industrial track record less valuable. So when input becomes ubiquitous, all competing firms are, in a sense, placed on an equal footing. What everyone has cannot constitute a competitive advantage.[2]

Firms cope with this situation in various ways. Some invest heavily in order to increase productivity, while others outsource, leaving the old industrial areas in a slowly more and more desolate and jobless state. 'Automate, emigrate or evaporate', as the saying goes. Other firms, in contrast, confront the new competitive situation by sharpening their abilities to learn and create knowledge a little faster than their competitors.

The creation of knowledge is usually seen as a process that requires dedicated investments either as pre-competitive research and education through universities, etc., or at the level of the individual firm through R&D activities. At least as important is, however, the investment in incremental 'low-tech' learning and innovation (Laestadius, 1996; Maskell, 1998) that takes place when firms, also in fairly traditional industries, create strongly corroborated knowledge while handling and developing mundane day-to-day operations like resource management, logistics, production organization, personnel, marketing, sales, distribution, industrial relations, etc. (Malerba, 1992). The possessors might know little or nothing of the origin of the knowledge or how they have come to know it, but 'it's here' and 'it works' (Baumard, 1996; Spender, 1996).

However, scholars and policymakers have increasingly come to suspect that the specific spatial arrangement of economic activities might also *in itself* somehow influence the creation of knowledge and, consequentially, economic growth (OECD, 1999).

Broadly, we may recognize two major categories of agglomeration economies (Estall and Buchanan, 1961). First are those that accrue from the geographical propinquity of industries and services in general, usually referred to as 'urbanization economies' (Hoover, 1970). The second category is usually referred to as 'locational economies' and embraces those economies that arise from the geographical agglomeration of related economic activities. It is the second category of geographical agglomerations or 'clusters' that in particular have been selected in recent years by scholars from a number of different disciplines as *the* territorial configuration most likely to enhance learning processes.[3]

---

[2] The role of ubiquities in changing the competitive environment is discussed in more detail in Maskell *et al.* (1998) and in Maskell and Malmberg (1999).

[3] The terms 'geographical agglomeration' or 'cluster' are used almost synonymously in the literature

Towards a Knowledge-based Theory of the Geographical Cluster

Some justification for this choice has been found in empirical studies showing, for instance, how 'innovative activity, as measured by patent data, and the location of high-tech industries is . . . highly concentrated' (Breschi, 1995), and how the agglomeration of firms within one or a few interrelated industries in Italian industrial districts gave rise to superior performance and some of the highest regional income levels in Europe (Bellandi, 1989).[4] Today Silicon Valley and Hollywood are probably the world's best-known examples of successful, non-random, market-led clusters.[5]

Presumably, clusters of related firms have been contributing to economic growth for quite a while, but the contemporary turn towards a knowledge-based economy (Carter, 1994) in many parts of the world has certainly sharpened our interest in understanding the nature of this process.[6]

The existing literature provides two types of understanding of the phenomenon. One source of insight is to be found in ideographic, historical work on how clusters have originated and developed into fruition, occasionally accompanied by accounts of subsequent descents (Malmberg and Maskell, 2001). Another attempts to specify conceptually the mechanisms that provide advantages to be reaped by firms located in a cluster. The present article is concerned with the latter aspect. It suggests a way of structuring our perception regarding how the cluster might partake in knowledge creation. In dealing with this issue, it moves mainly within the world of concepts,

together with 'industrial agglomeration' or 'localization', while the term 'industrial district', initially used by Marshall (1890) for the result of locational economies, is now often applied when wishing explicitly to emphasize the values and norms shared by co-localized firms (see, for instance, Brusco, 1982).

[4] Nevertheless, it must be admitted that while the cluster discourse is characterized by an overabundance of valuable case studies, the lack of solid empirical evidence across cases, sectors and countries is still profound. The present article can be seen as an attempt to provide a renewed platform for subsequent empirical investigations.

[5] In order to exclude 'random' agglomerations, the number of co-localized firms must be larger than if no locational economies are present. Ellison and Glaeser (1994) note that if firms in an industry choose locations by throwing darts on a map, only six darts must be thrown at a map of the US before it is most likely that two will hit the same state (see also Malmberg and Maskell, 1997).

[6] The growing interest has occasioned a number of distinct schools of thought to develop, including the GREMI approach (Maillat, 1991, 1998; Camagni, 1995; Ratti *et al.*, 1998), the many largely Marshallian studies of the Italian industrial districts (Brusco, 1986, 1999; Brusco and Righi, 1989; Beccatini, 1990; Garofoli, 1992a,b, 1993; Dei Ottati, 1994a,b; Bellandi, 1996; Gottardi, 1996; Belussi, 1999a,b), the French 'proximité' tradition (Blanc and Sierra, 1999; Kirat and Lung, 1999), an econometric type of cluster analysis (Swann *et al.*, 1998), different 'systemic' analyses (Markusen *et al.*, 1986; Malecki, 1991; Saxenian, 1994), some of which have focused explicitly on the geography of innovation (Feldman, 1994; Stenberg, 1999; Breschi, 2000), as well as the cherished approach applied by Porter (1990). Until we have found the 'serene and luminous region of truth where all may meet and expatriate in common', it is impossible within the frame of a single article to take into consideration the diversity in these and other schools of thought and to pay due respect to even important distinctions and points made in this vaste literature. I apologize for this when allowing myself a certain degree of simplification in the following sections.

raising a set of questions regarding the way in which economic performance is related to space in general, and to the role of localized learning in particular.

The aim of the article is thus to investigate the nature of the cluster when knowledge creation becomes key. It does not necessarily assert that learning and innovation takes place in the cluster only, or deny that a good portion of all firms is happily located outside the cluster. Neither does the focus on the cluster exclude the fact that circumstances, events and decisions in distant parts of the world heavily influence many firms today. The article merely presupposes that the cluster play a role in knowledge creation that is by and large sufficiently important to affect what is going on in the world to warrant analysis.

The article is structured along the following lines. Section 2 looks briefly into previous cost-based accounts of how firms might benefit when being part of a cluster. It is suggested that such approaches often fall short when addressing the more fundamental question of the cluster: the existence of many co-localized firms in related industries rather than a single, but larger entity, carrying the same tasks. It is proposed that the reason for the existence of the cluster can be found in the enhanced knowledge creation that takes place along its horizontal and vertical dimensions. In Section 3 the learning advantages stemming from the intrinsic variation between co-localized firms with similar capabilities is discussed, while Section 4 deals with the division of labour and the interaction taking place among firms along the cluster's vertical dimension. The various factors contributing to the growth of the cluster are sketched out in Section 5 before mowing to the more detailed discussion on the boundaries of the cluster. In Section 6 it is suggested that the boundaries can be defined by the interdependence between certain kinds of economic activities on the one hand and their appropriate institutional framework on the other. An institutional endowment favourable towards one kind of economic activity can be hostile to others. The very reasons for why cognitive distance might be small within the cluster will, it is asserted, make the cognitive distance between clusters very great. When access to dissimilar bodies of knowledge is required in product innovation, too much clustering becomes perhaps a burden and further clustering ceases. The final section points to areas where future research is needed to expand and elaborate on the theory of the cluster.

2. *Existence of the Cluster*

At least since Alfred Marshall's initial reflections on localized industries and the industrial district were published in his *Principles of Economics* in 1890

Towards a Knowledge-based Theory of the Geographical Cluster

scholars from a range of different fields have regularly concerned themselves with the issue.[7] The bulk of the studies in most of the 20th century were, however, mainly ideographic, and the reasons why firms cluster were assumed or implied rather than carefully investigated and specified. It was almost as if the benefits associated with the cluster were considered self-evident enough to require little discussion (Feser, 1999). When an explanation was offered it was usually based on a model where the balance between centripetal and centrifugal forces determined the locational pattern of firms. The dispersing forces normally included the costs of congestion, or the bidding-up of prices for land and labour. The concentrating forces were, in contrast, often identified as the cost advantages in transportation or when sharing an environment made particularly agreeable by, for instance, a dedicated infrastructure, a pool of notably skilled labour, an educational systems of distinctive relevance, etc.

This model largely disappeared as the swelling interest in clusters towards the end of the 20th century occasioned a number of novel research propositions to unfold.[8] Instead, the main emphasis shifted towards explanations more or less explicitly based on transactions costs, including search and information costs, bargaining and decision costs, as well as policing and enforcement costs (Babbage, 1832; Dahlman, 1979).[9] As Coase pointed out:

> In order to carry out a market transaction, it is necessary to discover who it is that one wishes to deal with, to inform people that one wishes to deal and on what terms, to conduct negotiations leading up to a bargain, to draw up a contract, to undertake the inspection needed to make sure that the terms of the contract are being observed, and so on. These operations are often extremely costly, sufficiently costly at any rate to prevent many transactions that would be carried out in a world in which the pricing system worked without cost. (Coase, 1960, p. 15)

Much in this spirit, some of the recent cluster studies have emphasized how the local activity will rise and the economic growth rate increase when the co-localization of firms benefits from the information easily available on potential partners in the vicinity and, perhaps more importantly, by the ease of conducting business with such local firms. The reason for the latter is found

[7] Since the outstanding contributions by Marshall (1890, 1919), major works have been published by Weber (1909), Hoover (1948), Perroux (1950), Hirschman (1958), Ullman (1958), Jacobs (1961), Chinitz (1961), Greenhut (1970) and Pred (1976, 1977).

[8] Accounts of this literature can be found in Harrison (1992), Norton (1992), Storper (1995), Baptista (1998), Bianchi (1998) and Yeung (2000).

[9] Other costs of using the market include the cost of establishing the appropriate incentive arrangements (Foss, 1993).

Towards a Knowledge-based Theory of the Geographical Cluster

in the behavioural constraints imposed on co-localized firms by the knowledge of the unattractive consequences of misbehaving. In a cluster it will immediately be noticed if a firm attempts to overutilize asymmetrical information; or pass defective or substandard goods as first class; or create hold-ups in order to benefit at the expense of others in the local milieu. Information about such misbehaviour will be passed on to everyone, who in future will tend to take their business elsewhere. Worse still, by becoming a local outcast the firm is deprived of the flow of knowledge, including its tacit parts, which can prove very difficult to substitute. Co-localized firms will, therefore, it is asserted, often benefit from the emergence of a general climate of understanding and trust[10] that helps (i) to reduce malfeasance, (ii) to induce the volunteering of reliable information, (iii) to cause agreements to be honoured, (iv) to place negotiators on the same wavelength, and (v) to ease the sharing of tacit knowledge.

The cluster thus exists, it is often implied, because the co-location of firms cuts the cost of identifying, accessing or exchanging products, services or, not least, knowledge between firms.[11]

However, it is not always realized that such costs might be eliminated altogether by joining the different activities and placing them under one common authority or ownership. When it comes to reducing transaction costs only, the single firm is superior to all market configurations imaginable—even to the high-trust cluster. The benefits of substituting interfirm interaction with the managerial authority of a single firm is, incidentally, one of the most significant reasons identified in the management literature for the birth and rise of the successful multinational enterprise, as Teece, among others, has observed:

> Internal trading changes the incentives of the parties and enables the firm to bring managerial control devices to bear on the transaction, thereby attenuating costly haggling and disruptions and other manifestations of non-cooperative behaviour. Exchange can then proceed at lower cost and with higher returns to the participants.   (Teece, 1980, p. 232)

The joining together of co-localized firms in related industries under one

---

[10] Trust is in most of this literature defined along the lines suggested by Glaeser *et al.* (1999) as the commitment of resources to an activity where the outcome depends upon the co-operative behaviour of others.

[11] An overview of the broad literature emphasizing knowledge exchange is given in Malmberg (1996, 1997). Two major journals have recently devoted special theme issues to research along these lines. *European Urban and Regional Studies*, Vol. 61(1), January, 1999, brought out an issue on 'Localised Learning and Regional Economic Development', while the *Cambridge Journal of Economics*, Vol. 23(2), March, 1999, published on ' Learning, Proximity and Industrial Development'.

Towards a Knowledge-based Theory of the Geographical Cluster

common ownership will, in addition to possible scale economies,[12] both help to align incentives and to diminish transaction costs.[13] It seems to follow that no theory attempting to explain the existence of the cluster can be based only on the reduction of transport, information and transaction costs.

In order to get a grip on the problem at hand we need to start by recognizing how the continued formation and survival of the cluster attest that the total economic effect of curtailed information and transaction costs as well as of scale advantages are *inferior* to the locational economies available when *being separate firms*.[14]

But what *are* then the advantages of N co-localized firms of size S undertaking related activities that are not transferable to a single firm of size S × N doing the same? This is arguable the single most important question for understanding the existence of the cluster, yet largely ignored in discussions on the subject.

In order to structure the discussion that follows Richardson's (1972) now classical dichotomy can be helpful when distinguishing between the horizontal dimension of the cluster, consisting of firms with similar capabilities that carry out similar activities, and the vertical dimension composed of firms with dissimilar but complementary capabilities that carry out complementary activities.[15] Richardson explains:

> Now it is quite clear that similarity and complementarity . . . are quite distinct; clutch linings are complementary to clutches and to cars but, in that they are best made by firms with a capability in asbestos fabrication, they are similar to drain-pipes and heat-proof suits. Similarly, the production of porcelain insulators is complementary to that of electrical switch-gear but similar to other ceramic manufacture. And while the activity of retailing toothbrushes is complementary to their manufacture, it is similar to the activity of retailing soap.   (Richardson, 1972, p. 889)

[12] Economies of scale might be defined as those that result when the increased size of a single operating unit reduces the unit cost of production or distribution.

[13] Babbage (1835), for instance, observed how flour could be purchased cheaper on the market than if the government produced it themselves. Nevertheless the latter course of action was preferred rather than carrying the costs of verifying each sack of flour purchased. Information asymmetries give rise to monitoring costs that make authority more efficient than market governance.

[14] The advantages of proximate specialized suppliers and customers in the cluster is in principle equally available to one big firm as to, say, 20 smaller doing similar things, just as most of the advantages in relation to the skills developed in the local labour market might be just as big or small for 20 co-localized firms of a given size as for a single firm, 20 times bigger.

[15] Activities are defined broadly by Richardson (1972, p. 888) as 'related to the discovery and estimation of future wants, to research, development and design, to the execution and co-ordination of processes of physical transformation, the marketing of goods and so on'.

Complementarity signals scope for fruitful exchange while similarity in activities spells contest and market encounter. The firms in the vertical dimension of the cluster will, accordingly, often be business partners and collaborators. The horizontal dimension will, on the contrary, consist mainly of rivals and competitors. Both dimensions contain features that might contribute in explaining the existence of the cluster, and both will in turn be looked into below.

3. *The Horizontal Dimension of the Cluster*

Marshall (1890) long ago hinted at an explanation for the existence of the cluster along the horizontal dimension of the cluster.[16] Marshall's reflection concerns the advantages of variation that are caused by the parallel performance of similar tasks. It is based on the conjecture that firms (i.e. owners, managers and employees) have different perceptive powers, divergent insights and dissimilar attitudes. Their different valuation of the information at hand results from an idiosyncratic and at least partly tacit way by which the information is initially assembled and interpreted (Casson, 1982). Consequently, firms develop a variety of solutions as an intricate part of their daily operations when holding dissimilar beliefs about their chances of success if using one of several possible approaches to similar problems (von Hayek, 1937).

Even when trying hard it would be extremely difficult, and often impossible, for a single, multidivisional firm to replicate internally the process of parallel experimentation and testing of a variety of approaches that take place among a group of independent firms doing similar things in the cluster. For as Loasby points out:

> Competing visions between firms are necessary features of an evolutionary or experimental economy. But competing visions within firms, unless very carefully managed, and limited in scope, cause trouble. (Loasby, 2000, p. 11)

Co-localized firms undertaking similar activities find themselves in a situation where every difference in the solutions chosen, however small, can

---

[16] For some reason or another Marshall's explanation never really entered the discussion of the cluster before Brian Loasby (1999, 2000) recently reintroduced it. A crude and unsubstantiated hypothesis could be that those who has been occupied with clusters have focused their attention on Book IV in Marshall's *Principles of Economics* while those who also cared to read Book V did so as part of a different agenda and never felt inclined to become engaged in the cluster discourse.

Towards a Knowledge-based Theory of the Geographical Cluster

be observed and compared. While it might be easy for firms to blame the inadequate local factor market when confronted with the superior performance of competitors located far away, it is less so when the premium producer lies down the street. The sharing of common conditions, opportunities and threats make the strengths and weaknesses of each individual firm apparent to the management, the owners, the employees and everyone else in the cluster who cares to take an interest. Co-location, furthermore, provides firms with an arsenal of instruments to obtain and understand even the most subtle, elusive and complex information of possible relevance developed along the horizontal dimension of the cluster.

It is by watching, discussing and comparing dissimilar solutions—often emerging from everyday practices—that firms along the horizontally dimension of the cluster become increasingly engaged in the process of learning and continuous improvement, on which their survival depends. Harrison C. White saw this very clearly in his account for the essence of competition:[17]

> Markets are self-reproducing social structures among specific cliques of firms and other actors who evolve roles from observing each other's behavior. I argue that the key fact is that producers watch each within a market. Within weeks after Roger Bannister broke the four-minute mile, others were doing so because they defined realities and rewards by watching what other 'producers' did, not by guessing and speculating on what the crowds wanted or the judges said. Markets are not defined by a set of buyers, as some of our habits of speech suggest, nor are the producers obsessed with speculations on an amorphous demand. I insist that what a firm does in a market is to watch the competition in terms of observables. (White, 1981, p. 518)

If the firms operating along the horizontal dimension of the cluster were to be spread thinly throughout a large city among many unrelated businesses, their ability to monitor and subsequently learn from each other's mistakes and successes would be severely restricted. In the focused and transparent environment of the cluster, successful experiments can more easily be distinguished from the less successful by knowledgeable local observers. Sharing a communal social culture—including collective beliefs, values, conventions and language—often significantly assists them in this process. Promising avenues identified by one firm become available to others. Even when carefully guarded or protected by a patent, enough information often

---

[17] White's proposition can be found in several later works and his idea is at the core of Porter's (1990) concept of rivalry.

leaks out to set local competitors on the track and enable them to 'invent around' the protection (Maskell, 2001). Firms along the horizontal dimension of the cluster are constantly given the opportunity to imitate the proven or foreseeable success of others while adding some ideas of their own.

The resulting enhanced knowledge creation following from the ongoing sequence of variation, monitoring, comparison, selection and imitation of identified superiour solutions is in essence why N similar firms of size S are not equal to one firm of size $N \times S$ doing the same.

The advantages suggested stem from the specific forms of knowledge creation available to the individual firm when pursuing self-defined object-ives, but not to the division of a larger entity where instructions are received and actions restrained by some procedure or limitation imposed from above.

It might be worth emphasizing an essentially Darwinian feature of the process of variation: as long as the firms share a common language and certain codes that ease their interpretation of local events no trust is required as a prerequisite for learning. The sequence of variation, monitoring, comparison, selection and imitation can take place without any close contact or even an arm's-length interaction between the firms. While suppliers and customers simply *need* to interact with each other in order to do business, competitors don't. Most relationships in the cluster will therefore be along the vertical dimension.[18] This is not the same as implying that the firms in the horizontal dimension of the cluster never co-operate by helping each other in over-coming technical problems, by lending materials and swapping surplus capacity or by exchanging information. In fact, they may interact regularly, even intimately so, in order to forward some particular scheme (Allen, 1983). On the other hand, they might just as well hate each other intensely, never exchanging anything useful.

The proposition put forward here simply suggests that the cluster exists because of locational economies *that are independent of the internal degree of interaction* at least in principle. The sole requirement is that *many firms under-taking similar activities* are placed in circumstances by co-locating where they can monitor each other constantly, closely and almost without effort or cost.

Other arguments for the existence of the cluster can be found along the vertical dimension of the cluster and we shall turn to these next.

4. *The Vertical Dimension of the Cluster*

The vertical dimension of the cluster consists of firms linked through

---

[18] This theoretical point has been supported by empirical findings (Håkanson, 1987).

——————— *Towards a Knowledge-based Theory of the Geographical Cluster*

input/output relations.[19] Specialized suppliers and critical customers become attracted to the cluster, once established, by the particular opportunities available. The vertical dimension of the cluster might, however, also be developed by task partitioning, which tends to evolve spontaneously when economic agents are free to pursue their own advantage, as pointed out by Adam Smith more than 250 years ago:

> In a tribe of hunters and shepherds a particular person makes bows and arrows, with more readiness and dexterity than any other. He frequently exchanges them for cattle or for venison with his companions; and he finds at last that he can in this manner get more cattle and venison than if he himself went to the field to catch them. From a regard to his own interest therefore, the making of bows and arrows grows to be his chief business, and he becomes a sort of armourer.   (Smith, 1979, p. 119)

Some firms will thus gradually move from the horizontal to the vertical dimension of the cluster by concentrating on some particular process, where they believe they possess or might develop certain lucrative capabilities, dissimilar to others. Such distinct capabilities, once developed, will gradually be improved through a continuing process of learning-by-doing. As the cluster's vertical dimension develops and firms become more specialized, they often find solutions to problems otherwise overlooked and bypassed, even when specializing in performing some particularly trivial tasks. An extended division of labour is therefore often closely associated with an acceleration of the growth of knowledge in the cluster.

The steady deepening of the division of labour is limited not only by the extent of the market,[20] but also by information asymmetries and the costs of co-ordination. Knowledge dispersed needs to be reassembled in order to be useful and firms need to co-operate in matching their related plans in advance since '. . . the one that make the heads of the pins must be certain of the cooperation of the one who makes the points if he does not want to run the risk of producing pin heads in vain' (List, 1841, p. 150).

In addition, firms hold asymmetrical knowledge about products and market opportunities. These asymmetries arise as an unavoidable consequence of the way in which knowledge is produced. Interfirm learning is, therefore, always subject to both thresholds, before the knowledge bases of divided firms

---

[19] The product innovation literature has firmly established that firms learn from each other when interacting. See, for instance, Rosenberg (1972), Freeman (1982, 1991), Kline and Rosenberg (1986), Håkansson (1987), Hagedoorn and Schakenraad (1992) and OECD (1992).

[20] See Young (1928), Stigler (1951) and Smith (1979).

have diverged sufficiently for interaction to imply learning, after which the cognitive distance becomes too great for firms to bridge, and where learning will consequentially cease.

Firms in the cluster might have some advantages on both accounts compared to outsiders. The spatially defined community that often emerges when related firms co-locate makes it easier for them to co-ordinate and to bridge communication gaps resulting from heterogeneous knowledge endowments (Eliasson, 1996), and to understand motives and desires that in other circumstances would remain opaque. By reducing the costs of co-ordination and by overcoming problems of asymmetrical information, the process of clustering tilts the balance in favour of further specialization so that a higher level of knowledge creation might be obtained. The main advantages are not the ease of intra-cluster interaction as such, as our manner of speech sometimes seems to suggest, but the deepening of the knowledge base that it enables.

The analysis so far thus suggests a reason for the existence of the cluster along the vertical dimension supplementing the one offered in the previous section on the horizontal dimension. When creating an appropriate vertical differentiation, new economic activities become possible, knowledge creation is advanced, and the resulting extension of the internal market helps make the process self-reinforcing (Young, 1928).

It follows from the concept of variation dealt with in the previous section that if all firms in the cluster hold complementary capabilities, while no two firms hold similar capabilities, then all learning through variation and monitoring must necessarily cease. A continued division of labour among firms in the cluster might thus only be expedient for the overall knowledge creation up to a certain point. Beyond that, the benefits might be offset by the corresponding reduction in knowledge creation as variation is diminished and fewer possible avenues of progress are tried out in parallel. Only by a steady increase in the number of firms in the cluster would it be possible to create knowledge simultaneously by variation and by the division of labour.

5. *The Growth of the Cluster*

To the extent that incumbent firms in the cluster are able to reap the benefits of enhanced learning along its horizontal or vertical dimensions, a non-random improvement in performance is to be expected.

In addition to the expansion of incumbents the cluster might grow by an increase in the number of firms through three different processes. First, already existing firms located elsewhere might be tempted to relocate all or

Towards a Knowledge-based Theory of the Geographical Cluster

a part of their activities to the cluster because of the real or imagined advantages of getting better access to the local knowledge base or to the suppliers or customers already present. As the Finnish CEO of Nokia-Mobira, J. U. Nieminen, once stoically noticed when commenting on these cluster-based advantages:[21]

> When an inventor in Silicon Valley opens his garage door to show off his latest idea, he has 50 per cent of the world market in front of him. When an inventor in Finland opens his garage door, he faces three feet of snow. (van Tulder, 1988, p. 169)

Second, a dominant position will also attract entrepreneurs with ambitions to start firms in the particular industry. This is why many of the most talented wannabes within the film industry tend to end up in Hollywood and many of world's best specialists in information and communication technology at some stage find themselves in Silicon Valley. Immigrating individuals and firms can over time have quite spectacular effects by the way they fuel the growth of the cluster.

Third and finally, new firms come into being in the cluster by spin-offs; smaller or larger groups of former employees recognize a potentially profitable business opportunity and decide to exploit it by becoming entrepreneurs themselves (Belussi, 1999b).[22]

By starting activities close to what is already going on in the cluster, all new spin-offs—newcomer or local independent entrepreneur alike—can safely skip the burdensome and costly process of gathering a lot of circumstantial knowledge about the business environment otherwise crucial. When it works for the neighbour why shouldn't it also work for me? New start-ups are thus given for free the advantages of a business environment tailored to their specific needs, even in situations when they might still be unaware of what these needs might be or how they may best be accommodated.

The availability of a suitable business environment is, of course, important not only for starts-ups but also for incumbents. By their everyday practices

---

[21] Nokia, however, stayed in Finland and has by now become the world's leading developer and manufacturer of mobile or cellular phones.

[22] Dalum (1995) shows how most of the many firms currently active in the communication cluster in Northern Jutland (Denmark) can be traced back to one initial firm producing off-shore radio equipment. Similar genealogical accounts for the emergence of many Canadian clusters have been established by a variety of local organizations. The general tendency for new firms to have their main activity within similar or complementary industries to the ones already operating in the area has been shown empirically for Denmark by Maskell (1992). If all incumbents were placed in a region by industry matrix with 2496 (12 × 208) cells, most cells would be empty but only 12% of the new firms established throughout a 20-year period would be located in an empty cell.

both simultaneously rely on and contribute to the further development of this particular environment. It is to this issue that we shall now turn.

6. *The Boundaries of the Cluster*

The processes of knowledge creation along the horizontal and vertical dimensions of the cluster are rooted in the day-to-day operations of the firms but influenced by a complex set of institutions developed over time.[23] Some of these institutions are of a general nature, equally applicable and useful for promoting the economic activity in all clusters, or at least in a large number of clusters, almost regardless of the particular activities carried out by the firms located there. The emergence of general formal constraints, communal regimes of appropriation and a common climate of understanding and trust, discussed above, belongs to this category.[24]

Other institutions have, however, a definite scope and will differ from one cluster to the next.[25] It is reasonable to assume that the cluster's particular set of institutions has emerged as a response to the special requirements of the activities performed in the cluster.[26] There is thus a fundamental *interdependence* between the economic structure and the institutions of the cluster as they have developed over time.[27]

It has been suggested that while the cluster's particular set of activities affects what is done within and among the firms in the cluster and therefore *what is learnt*, it is the institutions in the cluster that define how things are done and consequently *how learning takes place* (Lundvall and Maskell, 2000).

Just as the set of firms undertaking similar and complementary activities differ between clusters, so do institutions. Different activities each have their own mode of learning that gives rise to different institutional outcomes. The

[23] See Cannan (1912). We might follow North (1994, p. 360) in defining institutions as 'humanly devised constraints that structure human interaction. They are made up of formal constraints (e.g., rules, laws constitutions), informal constraints (e.g., norms of behaviour, conventions, self-imposed codes of conduct), and their enforcement characteristics . . .' while explicitly adding Smith's (1997) economic (knowledge) infrastructures that result from conscious policy decisions and investment programmes and include special programmes in local schools and universities, government-supported technical institutions and training centres, specialized apprenticeship programmes, etc.

[24] It might be argued that the specific way by which trust is obtained will make it differ from cluster to cluster and that very few 'general institutions' can therefore be expected to be found in practice.

[25] This is in line with much of the innovation systems literature (Lundvall, 1992; Nelson, 1993).

[26] On the national level recent research has proved the existence of such a correlation between patterns of specialization in production and trade, on the one hand, and the knowledge base, on the other (Archibugi and Pianta, 1992).

[27] Some argue that the differences emanate mainly from structural characteristics (Breschi and Malerba, 1997), while others look at how institutional specificities affect the location of certain industries (Guerrieri and Tylecote, 1997). See also Gertler (1995a, 1996, 1998) and Maskell and Törnqvist (1999).

Towards a Knowledge-based Theory of the Geographical Cluster

resulting institutions in turn assist the firms of the cluster when facing the challenges and opportunities presented by changes in the outside world.

Resent research has established the close interaction between structure and institutions when investigating industrial failure in places with a particularly favourable factor endowment. Eskelinen and Kautonen (1997), for instance, demonstrate how Finland, with its bounteous supply of high-quality timber resources, high educational and training standards, and a long track record of world-class designers, has been loosing out in wooden furniture production to countries with obviously inferior resources on some or all of these counts. The wooden furniture industry is generally characterized by very flexible, small-batch production, modest capital intensity, skilled or semi-skilled labour, integrated design, frequent contact with many different and shifting groups of customers, few long-term contracts, and periods of high activity alternating with inactive spells of uncertain length (Lorenzen, 1998; Maskell, 1998).

However, the relevant Finnish institutions of the wood-processing industry were defined *not* by the wooden furniture industry, but by the far larger user of wood as a primary input: the manufacturing of paper and pulp. This industry, in contrast, is characterized by long production runs, extremely high capital intensity, few highly skilled employees, many low or semi-skilled skilled workers, no design, a rather stable set of very big customers, long-term contracts, and very structured systems of production and maintenance.

The dominance of paper and pulp has lead to certain highly significant idiosyncrasies in business behaviour in the timber market. As a result, an institutional environment is created that is highly supportive of the paper and pulp industry, but distorting to the wooden furniture industry (Kautonen, 1996). Important institutional features hostile to the wooden furniture industry are: no distinction between quality classes of relevance to furniture production; fixed standard pricing practices for roundwood; large volume discounts on purchase of wood combined with long-term wood contracts and upstream vertical integration; emphasis in governmental policy and managerial ethos on technology, scale economies and process innovation rather than on market presence, design and product innovation; output markets seen as fixed once and for all; hierarchical labour relations supported by labour market agreements structured in ways that penalize small firms.

The institutions developed and refined to accommodate the needs of firms in the paper and pulp industry created so unfavourable a business climate for the wooden furniture industry that it stagnated and declined, while the same industry blossomed in the neighbouring country of Denmark. To the uninformed observer, Danish firms faced a considerable handicap as their insufficient local timber supply forced them to import most of the wood needed from countries such as Russia, Sweden, Poland and—notably—also Finland. However, this cost disadvantage counted for very little compared to the advantages of *not* being burdened with an unfavourable institutional endowment created by a dominant industry like the Finnish paper and pulp industry (Lorenzen, 1998).

The lesson that can be learned by the fate of the Finnish wooden furniture industry is that the more helpful an institutional endowment becomes for one type of activity the less suitable it can be for others. The significance of an appropriate *fit* between industry and institution also suggests why certain types of activity are never found in the same cluster. A cluster producing fashion wear or financial services will simultaneously develop (dissimilar) institutions that most likely will turn out to be alien to the production of ships, coal or cars.

The restrained ability to 'stretch' an institutional endowment to serve different kinds of economic activities equally well might also help to explain why new clusters emerge; when knowledge grows and economic activities begin to diverge, requirements also start to diverge and new clusters are likely to be established with institutions of their own.

The boundaries of the cluster might therefore be defined by the *fit* between the economic activities carried out by the related firms of the cluster on the one hand and the particular institutional endowment developed over time to assist these activities on the other.[28] The expansion into new activities along the vertical dimension of the cluster ceases to be feasible when the fit begins to weaken.

This framework might also account for the dispersing forces at work when the additional value created from spanning across distant bodies of knowledge must justify the additional transaction costs involved. Firms heavily engaged in interfirm innovation across usually unrelated activities and bodies of knowledge might, perhaps, be better off by not being *too* embedded in a particular cluster in order to avoid be facing by an even greater cognitive distance to potential partners when interaction is required.

Furthermore, the framework might provide an explanation for the demise of clusters as exiting value chains at some point become fragile and new ones are being moulded. As the new vertical dimension is gradually developed, the required institutional adjustments will almost inevitably meet resistance from old incumbents struggling to survive. If some compromise is not found, the resulting tension can easily lead to steady decline.

---

[28] If no such mechanism restricted the cluster's institutional endowment to a *certain kind of related industries only*, we would ultimately expect to end up with a single and rather large cluster containing all economic activity.

Towards a Knowledge-based Theory of the Geographical Cluster

7. Final Comments

The core of the argument presented in this article is that any economic theory of the cluster must address certain basic questions in order to be satisfactory.

First, such a theory must at the very least contain an explanation for the *existence of the cluster*. The theory must specify the process or processes that impel related firms to assemble and stay together at one place and—by doing so—make them thrive. More specifically, the theory must provide an explanation for the advantages that *many* related and co-localized firms might accrue but which are not available to a hypothetical *single* firm carrying out precisely the same activities, even if at the same location, using the same suppliers, customers and workforce.

It is suggested that the cluster exists because of the enhanced knowledge creation stemming from the variation developed along the horizontal dimension of the cluster, supported by the reduced costs of co-ordinating dispersed knowledge, of overcoming problems of asymmetrical information and aligning incentives, as well as of easing the actual transactions taking place along the vertical dimension.

Second, a theory of the cluster must include an explanation for the *growth of the cluster*. It must identify how new firms emerge and add to the strength of the cluster.

It is argued above that the cluster, once established, acts as a selection device, attracting particular kinds of economic activity comparable with the incumbents and reducing the ambiguity and costs facing local entrepreneurs when keeping close to the activities already present.

This selection device carries with it a set of constraints that might hamper future prosperity when external changes make readjustments necessary.

Third, the theory of the cluster must be able to identify *the boundaries of the cluster* by specifying why the clustering of some economic activities precludes the integration of others.

The reason forwarded in this article is based on the idea of a closely interdependence or *fit* between the specific economic activity of a cluster and the particular institutional endowment developed. A growing mismatch leads to decreasing returns. Negative feedback loops start to develop.

Further work might reveal how some of the specific suggestions made in this article when attempting to flesh out a theory of the cluster are ill conceived, or that other issues than the three identified should be included. One such possible candidate will answer questions regarding the *external* fit between the characteristics of the cluster on the one hand and its broader environment on the other. The last decade's many research publications frequently perceive the cluster as the basically random outcome of present or

Towards a Knowledge-based Theory of the Geographical Cluster

historical processes. Relatively few have so far taken care to ponder the factors that might have made certain environments more or less suited for the emergence of the cluster.[29] The evidence available so far does not constitute the foundation for any general *ex ante* statements about the suitability of a given economic environment to sustain the growth of a non-random, market-led group of co-localized firms doing similar things. The number of case studies produced during the last decade does, however, suggest that an effort to develop such statements might lead to interesting results.

There are other important aspects that require further consideration in subsequent research. The question of the *internal organization* will, for instance, be concerned with the ways that different configurations within the cluster might influence its knowledge-creating abilities. The theory of the cluster might also be asked to further specify the reasons for the decline of the formerly successful cluster.

Maybe, over time, new research will also make us able to tell whether the possible mismatch between a slowly adjusting institutional endowment and the highly dynamic requirements of many contemporary industries is the primary reason why innovative firms also survive and prosper without being supported by the many proposed advantages of the cluster. The theory of the cluster will not be complete before we more fully understand the successful solitary firm.

Acknowledgements

When writing this paper I have benefited from discussions with Kirsten Foss, Mark Lorenzen, Brian Loasby, Michael Storper and other members of the DRUID research network. Anders Malmberg, Gabi Dei Ottati, Meric Gertler, Paivi Oinas, Edward J. Malecki, Ash Amin and the participants at the World Conference on Economic Geography, Singapore, 4–7 December 2000, have provided valuable comments to earlier drafts. I thank the Nordic Center for Spatial Development (NORDREGIO) for supporting financially the research on which this article is based. The usual disclaimers apply.

References

Allen, R. C. (1983), 'Collective Invention,' *Journal of Economic Behaviour and Organization*, 4, 1–24.

Archibugi, D. and M. Pianta (1992), *The Technological Specialization of Advanced Countries*. Kluwer: Dordrecht.

[29] For an exception see, for instance, Gertler (1993, 1995b).

Towards a Knowledge-based Theory of the Geographical Cluster

Babbage, C. (1832), *On the Economy of Machinery and Manufactures*. Charles Knight: London.

Baptista, R. (1998), 'Clusters, Innovation and Growth: A Survey of the Literature,' in P. G. M. Swann, M. Prevezer and D. Stout (eds), *The Dynamics of Industrial Clustering. International Comparisons in Computing and Biotechnology*. Oxford University Press: Oxford, pp. 13–51.

Baumard, P. (1996), 'Organizations in the Fog: An Investigation into the Dynamics of Knowledge,' in B. Moingeon and A. Edmondson (eds), *Organizational Learning and Competitive Advantage*. Sage: London.

Becattini, G. (1990), 'The Marshallian Industrial Districts as a Socio-economic Notion,' in F. Pyke, G. Becattini and W. Sengenberger (eds), *Industrial Districts and Inter-firm Co-operation in Italy*. International Institute for Labour Studies: Geneva, pp. 37–51.

Bellandi, M. (1989), 'The Industrial District in Marshall,' in E. Goodman and J. Bamford (eds), *Small Firms and Industrial Districts in Italy*. Routledge: London, pp. 136–152.

Bellandi, M. (1996), 'Innovation and Change in the Marshallian Industrial District,' *European Planning Studies*, 4, 357–368.

Belussi, F. (1999a), 'Path-dependency vs. Industrial Dynamics: An Analysis of Two Heterogeneous Districts,' *Human Systems Management*, 18, 161–174.

Belussi, F. (1999b), 'Policies for the Development of Knowledge-intensive Local Production Systems,' *Cambridge Journal of Economics*, 23, 729–747.

Bianchi, G. (1998), 'Requiem for the Third Italy? Rise and Fall of a Too Successful Concept,' *Entrepreneurship and Regional Development*, 10, 93–116.

Blanc, H. and C. Sierra (1999), 'The Internationalisation of R&D by Multinationals: A Trade-off between External and Internal Proximity,' *Cambridge Journal of Economics*, 23, 187–206.

Breschi, S. (1995), 'Identifying Regional Patterns of Innovation Using Patent Data,' paper presented at the workshop on Regional Innovation Systems, Regional Networks and Regional Policy, organized by the STEP group at Lysebu Conference Centre (Oslo), Norway.

Breschi, S. (2000), 'The Geography of Innovation: A Cross-sector Analysis,' *Regional Studies*, 34, 213–230.

Breschi, S. and F. Malerba (1997), 'Sectoral Innovation Systems,' in C. Edquist (ed.), *Systems of Innovation: Technologies, Institutions and Organizations*. Pinter: London.

Brusco, S. (1982), 'The Emilian Model: Productive Decentralisation and Social Integration,' *Cambridge Journal of Economics*, 6, 167–184.

Brusco, S. (1986), 'Small Firms and Industrial Districts: The Experience of Italy,' in D. Keeble and E. Wever (eds), *New Firms and Regional Development in Europe*. Croom Helm: London, pp. 184–202.

Brusco, S. (1990), 'The Idea of the Industrial District: Its Genesis,' in F. Pyke, G. Becattini and W. Sengenberger (eds), *Industrial Districts and Inter-firm Co-operation in Italy*. International Institute for Labour Studies: Geneva, pp. 10–19.

Brusco, S. (1999), 'The Rules of the Game in Industrial Districts,' in A. Grandori (ed.), *Interfirm Networks. Organization and Industrial Competitiveness*. Routledge: London, pp. 17–40.

Brusco, S. and E. Righi (1989), 'Local Government, Industrial Policy and Social Consensus: The Case of Modena,' *Economy and Society*, 18, 405–424.

Camagni, R. P. (1995), 'The Concept of "Innovative Milieu" and its Relevance for Public Policies in European Lagging Regions,' *Papers in Regional Science*, 74, 317–340.

Cannan, E. (1912), *The History of Local Rates in England in Relation to the Proper Distribution of the Burden of Taxation*. P. S. King & Son, Orchard House: London.

Carter, A. P. (1994), 'Measuring the Performance of a Knowledge-based Economy,' Working Paper no. 337, Departments of Economics, Brandeis University.

Casson, M. (1982), *The Entrepreneur. An Economic Theory*. Martin Robertson: Oxford.

Towards a Knowledge-based Theory of the Geographical Cluster

Chinitz, B. (1961), 'Contrasts in Agglomeration: New York and Pittsburgh,' *American Economic Review*, LI, 279–289.

Coase, R. H. (1960), 'The Problem of Social Cost,' *Journal of Law and Economics*, 3, 1–44.

Dahlman, C. J. (1979), 'The Problem of Externality,' *Journal of Law and Economics*, 22, 141–162.

Dalum, B. (1995), 'Local and Global Linkages. The Radiocommunications Cluster in Northern Denmark,' *Journal of Industry Studies*, 2, 89–109.

Dei Ottati, G. (1994a), 'Co-operation and Competition in the Industrial District as an Organisational Model,' *European Planning Studies*, 2, 463–483.

Dei Ottati, G. (1994b), 'Case study I: Prato and its Evolution in a European context,' in R. Leonardi and R. Y. Nanetti (eds), *Regional Development in a Modern European Economy: The Case of Tuscany*. Pinter: London, pp. 116–144.

Dei Ottati, G. (1996), 'Trust, Interlinking Transactions and Credit in the Industrial Districts,' *Cambridge Journal of Economics*, 18, 529–546.

Eliasson, G. (1996), *Firm Objectives, Controls and Organization. The Use of Information and the Transfer of Knowledge within the Firm*. Kluwer: Dordrecht.

Ellison, G. and E. L. Glaeser (1994), 'Geographical Concentration in the US Manufacturing Industries. A Dartboard Approach,' Working Paper no. 4840, National Bureau of Economic Research (NBER), Cambridge, MA.

Eskelinen, H. and M. Kautonen (1997), 'In the Shadow of the Dominant Cluster—The Case of Furniture Industry in Finland,' in H. Eskelinen (ed.), *Regional Specialisation and Local Environment—Learning and Competitiveness*. NordREFO Report 1979:3, Stockholm, pp. 171–192.

Estall, R. C. and R. O. Buchanan (1961), *Industrial Activity and Economic Geography*. Hutchinson: London.

Feser, E. J. (1999), 'Old and New Theories of Industrial Clusters,' in M. Steiner (ed.), *Clusters and Regional Specialisation. On Geography, Technology and Networks*. Pion: London, pp. 19–40.

Feldmann, M. P. (1994), *The Geography of Innovation*. Kluwer: Dordrecht.

Foss, N. J. (1993), 'More on Knight and the Theory of the Firm,' *Managerial and Decision Economics*, 14, 269–276.

Freeman, C. (1982), *The Economics of Industrial Innovation*. Pinter: London.

Freeman, C. (1991), 'Networks of Innovators: A Synthesis of Research Issues,' *Research Policy*, 20(5), 5–24.

Garofoli, G. (1992a), 'Industrial Districts: Structure and Transformation,' in G. Garofoli (ed.), *Endogenous Development and Southern Europe*. Avebury: Aldershot, pp. 49–60.

Garofoli, G. (1992b), 'The Italian Model of Spatial Development in the 1970s and 1980s,' in G. Benko and M. Dunford (eds), *Industrial Change and Regional Development*. Bellhaven Press: London, pp. 85–101.

Garofoli, G. (1993), 'Economic Development, Organization of Production and Territory,' *Revue d'Économie Industrielle*, 64, 22–37.

Gertler, M. S. (1993), 'Implementing Advanced Manufacturing Technologies in Mature Industrial Regions: Towards a Social Model of Technology Production,' *Regional Studies*, 27, 665–680.

Gertler, M. S. (1995a), '"Being There": Proximity, Organization, and Culture in the Development and Adaptation of Advanced Manufacturing Techologies,' *Economic Geography*, 71, 1–26.

Gertler, M. S. (1995b), 'Manufacturing Culture: The Spatial Construction of Capital,' presented at the annual conference of the Institute of British Geographers (mimeo).

Gertler, M. S. (1996), 'Worlds Apart: The Changing Market Geography of German Machinery Industry,' *Small Business Economics*, 8, 87–106.

Gertler, M. S. (1998), 'The Invention of Regional Culture,' in R. Lee and J. Wills (eds). *Geographies of Economies*. Hodder Headline: London, pp. 47–58.

Towards a Knowledge-based Theory of the Geographical Cluster

Glaeser, E. L., C. L. Laibson, J. A. Scheinkman and C. L. Soutter (1999), 'What Is Social Capital? The Determinants of Trust and Trustworthiness,' Working paper no. 7216, NBER, Cambridge, MA.

Gottardi, G. (1996), 'Technology Strategies, Innovation without R&D and the Creation of Knowledge within Industrial Districts,' *Journal of Industry Studies*, 3(2), 119–134.

Greenhut, M. L. (1970), *A Theory of the Firm in Economic Space*. Appleton-Century-Crofts: New York.

Guerrieri, P. and A. Tylecote (1997), 'Interindustry Differences in Technical Change and National Patterns of Technological Accumulation,' in C. Edquist (ed.), *Systems of Innovation: Technologies, Institutions and Organizations*. Pinter: London.

Hagedoorn, J. and J. Schakenraad (1992), 'Leading Companies and Networks of Strategic Alliances in Information Technologies,' *Research Policy*, 21, 163–181.

Håkansson, H. (ed.) (1987), *Industrial Technology Development—A Network Approach*. Croom Helm: London.

Harrison, B. (1992), 'Industrial Districts. Old Wine in New Bottles?,' *Regional Studies*, 26, 469–483.

Hirschman, A. O. (1958), *The Strategy of Economic Development*. Yale University Press: Clinton, MA.

Hoover, E. M. (1948), *The Location of Economic Activity*. McGraw-Hill: New York.

Hoover, E. M. (1970), *An Introduction to Regional Economics*. Alfred A. Knopf: New York.

Jacobs, J. (1961), *The Death and Life of Great American Cities*. Random House: New York.

Kautonen, M. (1996), 'Emerging Innovative Networks and Milieux: the case of the Furniture Industry in the Lahti Region of Finland,' *European Planning Studies*, 4, 439–456.

Kirat, T. and Y. Lung (1999), 'Innovation and Proximity: Territories as Loci of Collective Learning Processes,' *European Urban and Regional Studies*, 6, 27–38.

Kline, S. J. and N. Rosenberg (1986), 'An Overview of Innovation,' in R. Landau and N. Rosenberg (eds), *The Positive Sum Strategy: Harnessing Technology for Economic Growth*. National Academy Press: Washington, DC, pp. 275–306.

Laestadius, S. (1996), 'Technology Level, Knowledge Formation, and Industrial Competence within Paper Manufacturing,' Working paper, Department of Industrial Economics and Management, Kungliga Tekniska Högskolan.

List, F. (1841), *The National System of Political Economy*. Longmans, Green: London.

Loasby, B. J. (1999), 'Industrial Districts as Knowledge Communities,' in M. Bellet and C. L'Harmet (eds), *Industry, Space and Competition. The Contribution of Economists of the Past*. Edward Elgar: Cheltenham, pp. 70–85.

Loasby, B. J. (2000), 'Organisations as Interpretative Systems,' paper presented at the DRUID Summer Conference, Rebild, Denmark (www.business.auc.dk/druid).

Lorenzen, M. (ed.) (1998), *Specialisation and Localised Learning. Six Studies on the European Furniture Industry*. Copenhagen Business School Press: Copenhagen.

Lundvall, B.-Å. (ed.) (1992), *National systems of innovation: Towards a Theory of Innovation and Interactive Learning*. Pinter: London.

Lundvall, B.-Å. and P. Maskell (2000), 'Nation States and Economic Development—From National Systems of Production to National Systems of Knowledge Creation and Learning,' in G. L. Clark, M. P. Feldmann and M. S. Gertler (eds), *The Oxford Handbook of Economic Geography*. Oxford University Press: Oxford, pp. 353–372.

Maillat, D. (1991), 'Local Dynamism, Milieu and Innovative Enterprises,' in J. Brotchie, M. Batty, P. Hall and P. Newton (eds), *Cities of the 21st Century*. Longman: London.

Maillat, D. (1998), 'Innovative Milieux and New Generations of Regional Policies,' *Entrepreneurship and Regional Development*, 10, 1–16.

Malecki, E. J. (1991), *Technology and Economic Development: The Dynamics of Local, Regional and National Change*. Longman: Harlow.

Towards a Knowledge-based Theory of the Geographical Cluster

Malerba, F. (1992), 'Learning by Firms and Incremental Technical Change,' *Economic Journal*, 102, 845–860.

Malmberg, A. (1996), 'Industrial Geography. Agglomerations and Local Milieu,' *Progress in Human Geography*, 20, 392–403.

Malmberg, A. (1997), 'Industrial Geography: Location and Learning,' *Progress in Human Geography*, 21, 573–582.

Malmberg, A. and P. Maskell (1997), 'Towards an Explanation of Industry Agglomeration and Regional Specialization,' *European Planning Studies*, 5, 25–41.

Malmberg, A. and P. Maskell (2001), 'The Elusive Concept of Localization Economies. Towards a Knowledge-based Theory of Spatial Clustering,' *Environment and Planning A*, forthcoming.

Markusen, A., P. Hall and A. K. Glasmeier (1986), *High Tech America*. Allen and Unwin: Boston, MA.

Marshall, A. (1890), *Principles of Economics*. Macmillan: London.

Marshall, A. (1919), *Industry and Trade. A Study of Industrial Technique and Business Organization, and of their Influences on the Condition of Various Classes and Nations*. Macmillan: London.

Maskell, P. (1992), *Nyetableringer i industrien og industristrukturens udvikling [New Firm Formation and Industrial Restructuring in Denmark]*. Copenhagen Business School Press: Copenhagen.

Maskell, P. (1998), 'Successful Low-tech Industries in High-cost Environments: The Case of the Danish Furniture Industry,' *European Urban and Regional Studies*, 5, 99–118.

Maskell, P. (2000), 'Social capital and competitiveness,' in S. Baron, J. Field and T. Schuller (eds), *Social Capital. Critical Perspectives*. Oxford University Press: Oxford, pp. 111–123.

Maskell, P. (2001), 'Knowledge Creation and Diffusion in Geographic Clusters: Regional Development Implications,' in D. Felsenstein, R. McQuaid, D. McCann and D. Shefer (eds), *Public Investment and Regional Economic Development*. Edward Elgar: London, forthcoming.

Maskell, P. and A. Malmberg (1999), 'Localised Learning and Industrial Competitiveness,' *Cambridge Journal of Economics*, 23, 167–186.

Maskell, P. and G. Törnqvist (1999), *Building a Cross-border Learning Region. The Emergence of the Northern European Øresund Region*. Copenhagen Business School Press: Copenhagen.

Maskell, P., H. Eskelinen, I. Hannibalsson, A. Malmberg and E. Vatne (1998), *Competitiveness, Localised Learning and Regional Development. Specialisation and Prosperity in Small Open Economies*. Routledge: London.

Nelson, R. R. (ed.) (1993), *National Innovation Systems. A Comparative Analysis*. Oxford University Press: New York.

North, D. C. (1994), 'Economic Performance through Time,' *American Economic Review*, 84, 359–368.

Norton, R. D. (1992), 'Agglomeration and Competitiveness: From Marshall to Chinitz,' *Urban Studies*, 29, 155–170.

OECD (1992), *Industrial Policy in the OECD Countries*. Organisation for Economic Co-operation and Development: Paris.

OECD (1999), *Boosting Innovation: The Cluster Approach*. Organisation for Economic Co-operation and Development: Paris.

Perroux, F. (1950), 'Economic Space, Theory and Applications,' *Quarterly Journal of Economics*, LXIV, 89–104.

Porter, M. E. (1990), *The Competitive Advantages of Nations*. Macmillan: London.

Pred, A. (1976), 'The Interurban Transmission of Growth in Advanced Economies: Empirical Findings versus Regional-planning Assumptions,' *Regional Studies*, 10, 151–171.

Pred, A. (1977), *City Systems in Advanced Economies. Past Growth, Present Processes and Future Development Options*. Hutchinson: London.

Ratti, R., A. Bramanti and R. Gordon (eds) (1998), *The Dynamics of Innovative Regions. The GREMI Approach*. Ashgate: Aldershot.

Towards a Knowledge-based Theory of the Geographical Cluster

Richardson, G. B. (1972), 'The Organisation of Industry,' *Economic Journal*, **82**, 883–896.

Rosenberg, N. (1972), *Technology and American Economic Growth*. Sharpe: White Plains, NY.

Ruggie, J. G. (1993), 'Territoriality and Beyond: Problematizing Modernity in International Relations,' *International Organizaton*, **47**, 139–175.

Saxenian, A. (1994), *Regional Advantage. Culture and competition in Silicon Valley and Route 128*. Harvard University Press: Cambridge, MA.

Smith, A. (1979), *An Inquiry into the Nature and Causes of the Wealth of Nations* [1776]. W. Strahan and T. Cadell: London.

Smith, K. (1997), 'Economic Infrastructures and Innovation Systems,' in C. Edquist (ed.), *Systems of Innovation: Institutions, Organisations and Dynamics*. Pinter: London.

Spender, J.-C. (1996), 'Competitive Advantage from Tacit Knowledge? Unpacking the Concept and its Strategic Implications,' in B. Moingeon and A. Edmondson (eds), *Organizational Learning and Competitive Advantage*. Sage: London.

Stenberg, R. (1999), 'Innovative Linkages and Proximity: Empirical Results from Recent Surveys of Small and Medium Sized Firms in German Regions,' *Regional Studies*, **33**, 529–540.

Stigler, G. J. (1951), 'The Division of Labor is Limited by the Extent of the Market,' *Journal of Political Economy*, **LIX**, 185–193.

Storper, M. (1995), 'The Resurgence of Regional Economies, Ten Years Later: The Region as a Nexus of Untraded Interdependencies,' *European Urban and Regional Studies*, **3**, 191–221.

Swann, P. G. M., M. Prevezer and D. Stout (1998), *The Dynamics of Industrial Clustering. International Comparisons in Computing and Biotechnology*. Oxford University Press: Oxford.

Teece, D. J. (1980), 'Economies of Scope and the Scope of the Enterprise,' *Journal of Economic Behavior and Organization*, **3**, 223–247.

Ullman, E. L. (1958), 'Regional Development and the Geography of Concentration,' *Papers of the Regional Science Association*, **IV**, 179–198.

van Tulder, R. (1988), 'Small European Countries in the International Telecommunications Struggle,' in C. Freeman and B.-Å. Lundvall (eds), *Small Countries Facing the Technological Revolutions*. Pinter: London.

von Hayek, F. A. (1937), 'Economics and Knowledge,' *Economica*, **IV**(N.S.) (13), 33–54.

Weber, A. (1909), *Über den standort der industrien*. J. C. B. Mohr: Tübingen.

White, H. C. (1981), ;Where Do Markets Come From?,' *American Journal of Sociology*, **87**, 517–547.

Yeung, H. W. (2000), 'Organising "the Firm" in Industrial Geography I: Networks, Institutions and Regional Development,' *Progress in Human Geography*, **24**.

Young, A. (1928), 'Increasing Returns and Economic Progress,' *Economic Journal*, **38**, 527–542.

[19]

THE EVOLUTION OF TECHNOLOGIES IN TIME AND SPACE: FROM NATIONAL AND REGIONAL TO SPATIAL INNOVATION SYSTEMS

PÄIVI OINAS
Department of Economics, Erasmus University, Rotterdam, the Netherlands, oinas@few.eur.nl

EDWARD J. MALECKI
Department of Geography, The Ohio State University, Columbus, malecki@geog.ufl.edu

Complementing existing approaches on national innovation systems (NISs) and regional innovation systems (RISs), the proposed spatial innovation systems (SISs) approach incorporates a focus on the path-dependent evolution of specific technologies as components of technological systems and the intermingling of their technological paths among various locations through time. SISs utilize spatial divisions of labor among several specialized RISs, possibly in more than one NIS. The SIS concept emphasizes the external relations of actors as key elements that transcend all existing systems of innovation. The integrating role of these relations remains inadequately understood to date. This poses a challenge for future research.

This article aims to understand technological development from a perspective that both integrates and transcends contemporary discussions about national innovation systems (NISs) and regional innovation systems (RISs). It approaches technological development as path-dependent processes at the level of specific technologies that evolve in time and space. These technologies are components, or subsystems, of broader technological systems, which makes them interdependent. Furthermore, technological development is spatially bound; technological paths are shaped by the social relations involved in their production as well as consumption (in processes of adoption, adaptation, and rejection) and the interplay between them. This prompts us to pay attention to the role of many RISs (and possibly NISs) in shaping the various components of technological systems: to look at the historical coevolution of interdependent technological paths. Their evolution is inseparable from the

The authors wish to thank Stephan Schüller and two anonymous referees for perceptive comments. Päivi Oinas is grateful for the financial support of the J&A Wihuri Foundation, Finland, and the Dutch Central Board for the Retail Trade.

socioeconomic circumstances in the places in which they take place, as well as the broader competence endowments in their surrounding regions and nations. Technological frontiers create their specific "time geographies" as they evolve so as to take advantage of such circumstances. This makes us observe both the simultaneous evolution of technological paths in many RSIs or NISs and their occasional movements in space. Conjointly, these are viewed as forming spatial innovation systems (SIS), which consist of "overlapping and interlinked national, regional and sectoral systems of innovation which all are manifested in different configurations in space" (Oinas and Malecki 1999, 10). Although portions of this argument have been made previously, the notion of an SIS has not been elaborated on in much detail. This article aims to make some progress in this regard.

Much of the thinking on innovation systems in economic geography and regional science is centered on localities or regions. Different places are viewed as manifesting systems—industrial, technological, sociocultural, or otherwise. What we wish to suggest in our approach, in contrast, is that innovation systems are worked out differently in space; they exhibit different spatial configurations. They may originate in one place, but often they are spread beyond local, regional, and even national borders. Technological evolution occurs through the interplay between elements of national, subnational, and transnational innovation systems that produce flows of innovation and are to different degrees able to keep up with state-of-the-art practices in different technological frontiers. Central in the SIS approach are (1) the external relations of actors and (2) the variability of the relative weights of different places or regions as center points of particular technological paths in time. With these emphases, the SIS approach offers a complement to much of the literature on localized learning that emerged toward the end of the 1990s and assumed that proximate relationships are most conducive for learning and innovation (see Oinas 1999, 2000). This assumption has largely prevailed even though it has been observed that production or innovation systems are not necessarily delimited to localities or regions (see, e.g., Storper 1996, 787; Storper 1997, 71; Amin and Cohendet 2000). This issue seems to be drawing more attention in most recent scholarship, however (see Bunnell and Coe 2001).

The SIS approach also complements the earlier literature that paid abundant attention to industrial districts, new industrial spaces, and other specialized industrial agglomerations. While this literature highlighted the specialization of those regions, the SIS approach pays attention to the possibility of various types of regions being part of SISs, whether diverse or specialized. As a related matter, regions whose economies are associated mainly with technologically mature products and processes may also serve a role in SISs in addition to technologically more advanced ones.

The problem is that innovation systems are complex entities, and it is difficult to find clear patterns that would structure our observations on the relative importance of the local versus translocal elements in them. What this article aims to do, therefore, is to open up this complexity for further exploration. While the suggested SIS approach could be applied to innovation in various types of economic activities (such as organizational, financial, and design activities), the discussion in this article is delimited to technological innovations.

The article proceeds as follows. We first discuss briefly why the prevalent literatures on NISs and RISs do not provide a full understanding of how technological innovation evolves. We then outline the SIS approach. In subsequent sections thereafter, we discuss key elements of SISs: technological paths, types of RISs involved, proximate and distant relations between actors, and firms and individuals as connectors in SISs. We conclude by reflecting on the main argument of the article and by outlining major challenges related to further theoretical and empirical research in the SIS framework.

LIMITATIONS OF THE NIS AND RIS APPROACHES

The SIS view is a complement to the existing concepts of NISs and RISs. The NIS and RIS approaches largely center on the conditions for innovative activity in a territory—nation or region—at a particular point in time. We propose instead that it helps to put these discussions in a broader perspective by providing an approach to look at the intermingling of technological trajectories among various locations through time.

The NIS approach (Freeman 1987, 1995; Nelson 1993; Edquist 1997) generally focuses on institutional characteristics of innovation systems at the national scale and privileges those at the expense of other scales. The effect of an NIS is seen in the accumulation of specific types and levels of competences in a country. Besides the private sector, this body of research recognizes the involvement of the public sector in innovation, both directly (via universities and government laboratories) and indirectly (by creating incentive structures, education and training systems, and promoting exports through fiscal, monetary, and trade policy packages) (Patel and Pavitt 1994; Nelson 1993). Other factors also can be seen as influencing the emergence of distinct NISs, such as national culture and its effect on policy (Roobeek 1990), business management systems (Hampden-Turner and Trompenaars 1993; Hickson 1993), and financial systems, which configure the relative roles of subsidies, loans, shares, and other prevailing national financial arrangements (Christensen 1992; Guinet 1995).

There is a parallel stream in the NIS literature that focuses on networks and interaction. Indeed, Lundvall's (1992) approach to interfirm networks and interactive learning (see also Gelsing 1992) may be seen as suggesting a focus on a smaller scale, geographically (subnational spaces) or otherwise (e.g., development blocks; Edquist and Hommen 1999) (i.e., to the concrete contexts of the actual interactions where learning and innovation actually occur; Acs, de la Mothe, and Paquet 1996). In line with this observation, the NIS (or NSI) approach has been criticized by, for example, Kumaresan and Miyazaki (1999), whose concern is that

> while the concept of NSI is rich and has a strong foundation, it is too rich, too macro and broad—covering all aspects from institutional set up, interfirm relationships, organization of R&D, educational and training systems, natural resource endowments, financing mechanisms to even culture. Moreover, it is unable to deal with the diversity of industrial situations in one country. In other words it is difficult to analyze NSI without going through in-depth studies at the meso-level. At the micro-level, much of the work on dynamic capabilities has focused on the issue of corporate competencies. In order to analyze dynamic capabilities at the national level, we need to accumulate studies in meso-systems, focusing on the internal dynamics of network evolution. (P. 564)

The meso level has been also highlighted in other recent research, which is attempting to focus on a scale below the national (macro level) and above that of the firm (the micro level) (Braczyk, Cooke, and Heidenreich 1998; de la Mothe and Paquet 1998a). To Foss (1996), the meso level is crucially where nonproprietary and intangible higher order industrial capabilities are developed and maintained by the interactions among firms (cf. also Nooteboom 1999b, 2000). This implies, centrally, that the development of the capabilities crucial for innovation, as well as innovation itself, is a relation-specific process. We adopt the meso level of analysis as most appropriate for a focus on the relations and flows within a spatial innovation system.

Regions within countries share some of the aspects of the entire nation, but they also have different possibilities to "go their own ways" and ultimately end up diverging from a national average (in terms of, e.g., the nature of education and training systems, science and technology capabilities, industrial structure, interactions within the innovation system, and propensities to absorb from abroad; cf. Archibugi and Michie 1997, 127-28). Indeed, within countries, specific regions tend to bring about a large share of the outcomes which, in the NIS framework, would be regarded as the accomplishments of national systems of innovation (Ohmae 1995; Oinas and Malecki 1999; Scott 1998; Storper 1997, 218). Accordingly, an increasing awareness has grown among those sensitive to spatial issues that regions might be an appropriate scale for carrying out analysis on systems of innovations. A focus on regions does not lead to the denial of the importance of the NIS as a key context and facilitator of the smaller scale innovation systems.

Those smaller scale systems are variously called clusters, territorial production complexes, productive systems, territorial systems, milieus, and local systems (see, e.g., Acs, de la Mothe, and Paquet 1996; Asheim and Dunford 1997; Cooke 1996; de la Mothe and Paquet 1998a, 1998b; Enright 1996; Feser 1998; Porter 1998; Rosenfeld 1997; Steiner 1998), but they can be seen as belonging under the broad umbrella of RISs. Three features of regional and local systems stand out as important: (1) the collectivity that somehow encompasses—indeed defines—a region in its entirety, (2) the emphasis put on the soft aspects of economic activity, and increasingly, (3) extralocal connections. It is this third feature that has not received due attention in the literature on RISs and needs to be focused on more centrally. The SIS framework seeks to provide a remedy in this regard.

SPATIAL INNOVATION SYSTEMS

A technology is an industry-specific, time-specific, and place-specific way of doing things. In clusters of economic activity, developments in several industries become integrated and coordinated through strong links (as, e.g., in industrial clusters producing electronic appliances that involve producers in several industries such as metals, plastics, telecommunications, and electronics). These clusters of interrelated and thus coevolving industry-specific technologies form technological systems (cf. Carlsson and Stankiewicz 1991; Carlsson 1994). *Technological system* refers to sets of technologies in use in specific interlinked industries. Technological systems may be local, regional, or multinational, depending on the nature and extent of the networks involved.

From the standpoint of the dynamics of these technological systems, the evolution of the various technologies in technological systems can be seen as forming technological paths. This notion relates closely to Dosi's (1982) "technological trajectory." Both notions, of course, are metaphorical, but they have slightly different connotations. Dosi defined a technological trajectory as "the pattern of 'normal' problem solving activity (i.e., of 'progress') on the ground of a technological paradigm," where a technological paradigm is a " 'model' and a 'pattern' of solution of *selected* technological problems, based on *selected* principles derived from natural sciences and on *selected* material technologies" (p. 152). Technological paradigms, in his view, embody "strong prescriptions on the *directions* of technical change to pursue and those to neglect" (p. 152). Thus, paradigms form cognitive limits for actors involved in them. While they give direction to activities, they also delimit the options that might actually be available. In addition, institutionalized structures of relations around technological trajectories add inertia to them. Dosi's trajectory, then, appears to be reminiscent of the use of the term in ballistics: a technology develops in the direction to which it is set under initial conditions until, for any reason, a paradigm changes. We regard the metaphor of a "path" more appropriate, yet we share Dosi's idea that broader paradigms give direction to them: technological paths do not move to random directions. Accordingly, the evolution of technologies can actually sometimes be described as trajectories, due to the relatively stable direction in which they seem to be moving, sometimes for relatively long periods of time. Like Dosi, we emphasize that neither trajectories nor paradigms stay unchanged. Paradigms change and new trajectories are set in motion. Thus, technological evolution involves alternating periods of progress along a trajectory (and within a paradigm) as well as periods of change, resulting in settling on a new trajectory based on a new paradigm. Yet, Dosi (1982, 158) seems to suggest that a technology progresses along a trajectory, attaining incremental innovations, until the paradigm changes, due to a radical innovation, and a new trajectory is set in motion. We draw less of a sharp distinction between incremental and radical inno-vations (cf. Tidd, Bessant, and Pavitt 1997), which allows for the idea of a more evolving technological path. Our emphasis is on the possibility of continuous adjustments. For example, in the case of emerging technologies and in entirely new technological systems, no clear directions of trajectories can be seen but rather an apparently randomly winding path, until the new developments settle onto some-thing that could be called a trajectory. How steady and long lasting such trajectories are depends on the nature of the technologies and on the competitive environment.

The details related to technological change are being revealed in ongoing re-search, particularly where empirical situations are analyzed with evolutionary con-ceptual frameworks. Yet, there is a relatively broadly shared understanding of the evolution of technologies as having certain, if unpredictable, life-cycle characteris-tics (Nelson 1996). Technological development includes stages during which ideas emerge for new products and processes and subsequently standards and dominant designs evolve. We draw in the following on Tushman, Anderson, and O'Reilly's (1997) account on technology cycles and Nooteboom's (1999a, 2000) account on cycles of discovery. Both of these accounts, albeit with some differences, highlight subsequent periods characterized by

- the emergence of variation through technological discontinuity (novel combinations),
- consolidation (following a fermentation period including design competition),
- selection of dominant design and generalization of its application, and
- retention with incremental changes in the dominant design (Tushman, Anderson, and O'Reilly 1997) as well as differentiation as a result of applications in new contexts (Nooteboom 1999a, 2000).

These cyclical processes in the evolution of technologies keep technological tra-jectories or paths moving in one direction for a period of time, but relatively smaller adjustments in that direction are made in periods of retention and differentiation. More significantly, "turns" in a technological path are made during technological discontinuities as major technological discoveries are made (or as novel combina-tions are brought about).

What is described above refers to the progress that is made in a technological system and that takes place by advancing knowledge at the level of specific technol-ogies (or components, subsystems of technological systems). These components have their own technology cycles, but their development is influenced by develop-ments in other parts of the technological system. As one technology changes, adjustments have to be made in the rest that belong to the same technological sys-tem. Different subtrajectories have their own frontiers, which give them new direc-tion. Different frontiers may compete with each other even within the same techno-logical system.

In addition, technological frontiers are developed at different levels of techno-logical sophistication, as older and newer technologies are often developed

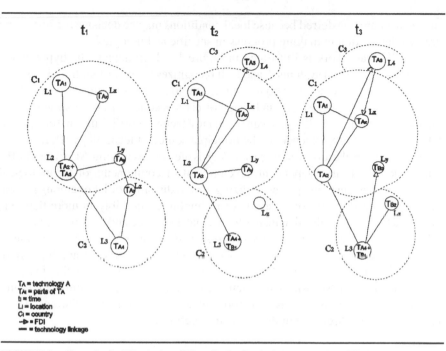

TA = technology A
TA₁ = parts of TA
t₁ = time
L₁ = location
C₁ = country
→ = FDI
— = technology linkage

FIGURE 1. **Hypothetical Evolution of Two Technological Systems (A and B) in the Spatial Innovation Systems Framework**

simultaneously to serve the needs of different customer groups. For example, in mobile telephony, the older NMT (*nordisk mobiltelefon*) technology is continuously improved for mobile phones used in remote areas even though most research and development (R&D) effort in Europe is being put into global system for mobile communications (GSM) applications and, increasingly, third-generation technologies. That is, parts of a technology system may make progress by the exploitation of existing technologies at different levels of advancement and incremental improvements in those, whereas the activities in frontiers are aimed at exploration: the search for novel combinations. These parts may be spatially and organizationally separate so as to receive support of suitable sets of actors, capabilities, and institutional environments in different RISs.

Accordingly, no innovation system is located in one place only. This is why it is not enough to focus on particular RISs in trying to understand technological change. Instead, the development of a technological system takes place via the coterminous evolution of its various components in space and time. It is supported by an interlinked set of social relations in a number of RISs of different levels of socioeconomic development, (semi-)integrated by the requirements of a technological system, resulting in a distinct spatial division of labor in that system. Technological systems are not autonomous of the place-specific RISs where they originate or are transferred because local conditions may be decisive for sustaining creative interaction in making progress in specific technologies.

The SIS framework is illustrated in Figure 1, which depicts the hypothetical paths of technological systems A and B. It recognizes the role of multinational corporations (MNCs) as actors who transfer technologies through international flows (e.g., the links from C_1 to C_2) via foreign direct investment (FDI) or strategic alliances. In addition, a small region can originate a technology (T_{B1} in L_3 in time period t_2), decreasing its dependence on the principal source of technology (L_2 in C_1).

Summing up the discussion so far, key issues in discussing SISs are (1) the simultaneous and interdependent development of components of technological systems possibly in many places, utilizing spatial divisions of labor among several RISs specialized in different aspects of technologies, possibly in more than one NIS, and (2) the "travels" that technologies make in space and over time as knowledge flows take place along with the progress made in the frontiers of those components. The key elements in the complex spatial innovation systems are the technological paths themselves, the RISs that participate in creating the technologies or parts of them, the actors whose interaction locally and over space ultimately brings technologies about, as well as their (proximate or more distant) relations. These elements will be discussed in the following sections.

TECHNOLOGICAL PATHS IN TIME AND SPACE

In line with the above, we portray technological evolution at the level of specific technologies that coevolve as part of a wider technological system along their specific paths. The directions that technological paths take are influenced, but not entirely determined, by each technology's frontier. A frontier is advanced by actors within sets of social relations, for example, in one (or several) RIS. It is in the idea of a frontier that it is brought about by unique knowledge and skills: exactly the same frontier (or a part of it) cannot be in two places at the same time. Thus, the collective action of those sets of interdependent actors at a technology's frontier is subject to the basic constraints identified by time geography: movement takes time, and the same actor cannot be in more than one place at the same time (Hägerstrand 1970). They are subject to situated interdependence (Jackson and Thrift 1996, 214) with others working with a specific technology. In other words, they are locally dependent on the RISs, which are able to support a particular level of and progress within a specific technology. As a result, the advancement of each particular technology has its own time- and space-specific developmental path behind (and ahead) of it. For many innovations, technological development proceeds simultaneously but focused on different specializations, in several places. Lasers, for example, are the result of research efforts in Germany, Japan, and the United States (Grupp 2000). In aircraft, innovation is the fruit of a complex web of producers in many places, as Frenken (2000) showed by tracing 863 aircraft models. Each firm, in its own location(s), has its unique design specialization (Frenken and Leydesdorff 2000).

The cyclical patterns of technological development referred to above also need to be seen as having spatial patterns that are part of their evolution: technological frontiers may change places. Spatial discontinuities, or shifts, in technological trajectories or paths do not necessarily (or even often) happen because the sets of social relations advancing them change places but rather due to reasons related to the dynamics of technological progress, as discussed above. What is different about the time geographies of technological paths compared to the more customary idea of individuals' time geographies is that technologies may change form and multiply. Even if a frontier moves ahead in space, the path that it leaves behind does not remain entirely unchanged, that is, there is a period of retention (Tushman, Anderson, and O'Reilly 1997). Besides, new developments are set in motion as adapters of the technologies created by a frontier remain along the path and may be successful in further developing the technology—giving rise to new paths, which either go their own way or start competing with the frontier, that is, there is a period of differentiation (Nooteboom 2000). As a product moves from R&D to production, firms (and places) that specialize in economies of scale play a more important role, as firms in Singapore and Taiwan do in semiconductors (O'hUallachain 1997).

Spatial discontinuities may relate to specific phases in technological cycles. Technological frontiers are in operation in the RISs that are involved in creating novel combinations and that are able to build local structures around emerging dominant designs and exploit them commercially. If those structures become too rigid in time, they cannot change as new variation emerges, beginning a new cycle of technological change. At this time, the actors at the frontier of technology may move to areas that fulfill their locational specifications (cf. Storper and Walker 1989). Alternatively, new technological frontiers may emerge in new areas as a result of the previous dominant design having been applied in a new context, possibly in a new region, where it becomes differentiated and may give rise to a new subtrajectory and maybe later to another novel combination beginning a new cycle with different set of actors involved.

In sum, technologies have their specific, path-dependent time geographies: technologies emerge somewhere, in a place—or sometimes similar technological solutions are invented in more than one place simultaneously (shown, e.g., by patent applications for similar technical solutions of different origins being received one after another by patent authorities)—and the further development of those technologies may take place in a new context and in a new place, where possibly new qualities are added to them. Technological development is the result of the intermingling of such technological paths, overlapping in content and possibly also in space. Each path is part of an industry- or product-specific technological system and epitomizes its developmental phases. An example is found in the hard disk drive industry (Christensen 1997), in which customers in different markets place priority on different types of performance (e.g., size, weight, speed, capacity). The industry has evolved to meet new needs, often through the emergence of new firms, taking advantage of skills and networks in new locations, such as Singapore and

Penang (Malaysia). The production skills in these new locations helped to shape the trajectories of several producers as new products were introduced (McKendrick, Doner, and Haggard 2000).

TYPES OF REGIONAL INNOVATION SYSTEMS

So far, we have aimed at giving a dynamic, albeit metaphorical, account of technological development as paths that the frontiers of specific technologies create as they evolve in time and as they travel and make connections in space. This section discusses the different kinds of places: the different, interlinked RISs that are involved in producing those paths. The RIS literature usually fails to provide distinctions between types of RISs, which may be top-down and poorly integrated regionally (regionalized NISs) or bottom-up, with considerable regional networking (territorially integrated innovation systems; Asheim and Cooke 1999; Hassink 2000). For our purposes, such distinctions are important, as SISs consist of various kinds of activities with different levels of sophistication organized in space (within and between different RISs) according to a division of labor that is specific to each SIS (cf. Figure 1). Our typology of RISs evolves out of a discussion on the relative technological advancement of regions and on the relative specialization versus diversity of their economic activities.

We start with a basic tenet of evolutionary accounts on technological development: that innovation requires diversity (Nelson and Winter 1982). In spatial analysis, the need for diversity has been documented in recent research showing that diverse locales (i.e., locales with relatively large numbers of different industries) are more important for promoting innovative firm behavior (Feldman and Audretsch 1999; Harrison, Kelley, and Gant 1996; Quigley 1998) than specialized ones (of, e.g., the industrial district type, as often assumed in the course of the 1990s). Small firms in particular benefit from regional industrial diversity (Kelley and Helper 1999) because they cannot create it internally.

These are important findings for the spatial analysis of technological change. The SIS perspective, however, prompts us to raise three additional issues.

Diversity and actual (innovative) relationships. As pointed out above, a meso-level approach to innovation pays attention to relations between actors. In regard to that, simple claims made at the level of numbers of industry sectors (whichever ISIC digit level) miss the point about the critical nature of relations within and between industries for innovation. This is the case even in regions with a broad diversity of industries. The mere presence of a variety of industries in a region obviously does not make a region a "territorially integrated innovation system" (Asheim and Cooke 1999); it does not reveal the basis of the relations between firms in any of those industries. Rather, this basis has to be seen in the potential relatedness between firms' knowledge and capabilities that may trigger their engagement in innovative interaction (Oinas and van Gils 2001).

Diversified and specialized regions complement each other (via external relations). While diverse regions may be more conducive for regional innovation relative to specialized ones, the SIS approach helps to point out that diversity as food for innovation is not always locally available in the right form and thus needs to be complemented by interaction with more distant actors—actors that can bring in specialized expertise based on their participation in another RIS. Within the SIS framework, there is no reason to think, ceteris paribus, that diversity originating locally is drastically different in terms of its potential input into innovation as compared to diversity originating elsewhere.

In line with our discussion of technological paths above, we regard it as one of the key functions of technological frontiers to search for diversity: to direct and redirect technological paths to regions where they can find suitable diversity to support innovation. When actors in one region do not provide enough diversity for innovativeness, a technological frontier either puts effort into making regional actors effectively connected to sources of diversity elsewhere, or the technology gradually loses its edge and the frontier moves to another region (yet, as discussed above, it may be another frontier, with a different set of actors and their specific social relations).

This does not exclude the possibility that even narrowly specialized regions may have a role in the evolution of technologies, by creating leading-edge specialized knowledge that supports a larger innovation system. Specialized regions do not operate in isolation but receive impulses for renewal and innovation from interaction with other innovative actors who are part of the same system even if they are not located in the same place. Thus, even a narrowly specialized region may be a substantial contributor if the part of the technology that it creates happens to be crucial at some point in time (Frenken 2000; Frenken and Leydesdorff 2000). This may be the result of the (possibly slow and incremental) evolution of specialized knowledge through local adjustments in a region that leads to a strong (possibly leading-edge) expertise in a narrow area of knowledge. It is possible that such a small region will not stay central for a long time, and the technology may or may not create spillover effects in its regional environment, but it may still be relevant for the historical evolution of the technological path.

Within regions, thus, each sector has its specific connections to extraregional partners, which enhances the innovative potential of those sectors' actors. In the case of diversified regions, external relations are likely to add to the total innovative potential of the region's actors by helping to sustain continuously higher and more diversified technological capabilities. In the case of specialized regions with a more narrow range of economic activities, external relations compensate for the lack of regional diversity.

Diversity versus specialization and technological advancement in SISs. What is important in the SIS framework, accordingly, is the variety of regions involved in whole innovation systems. Yet, regions differ not only in terms of their relative specialization but also in terms of their relative technological sophistication. The RISs that are involved in SISs range from genuine innovator regions (which may tend to be more diversified on the average) to regions that merely imitate or adopt innovations. Yet, each has a function in the SIS. We make a simple distinction in the following between three types of regions in SISs in terms of their ability to bring about innovation: adopters, adapters, and genuine innovators (Oinas and Malecki 1999) (these types still leave outside the large swath of the world which is technologically excluded; Sachs 2000).

1. Genuine innovators. These are the RISs in which genuinely novel combinations ("new to the world" innovations) take place and best practices emerge, in specific technologies. Sometimes all stages of innovation cycles (Nooteboom 1999a, 2000; Tushman, Anderson and O'Reilly 1997) may be carried out in them. Or, as innovations diffuse from them through imitation, they may host the actors that pick up problem signs or signals of new opportunities from actors in other regions exploiting existing, yet maturing, technologies (incremental innovation) and engage in exploration, to hit yet another novel combination (radical innovation), which might begin a new cycle of innovation. These regions also maintain competitive and/or collaborative relations with other leading-edge regions, which further propels their innovativeness. This involves close monitoring of what is going on in other key RISs in a particular technology. Many technologies evolve as products incorporating the knowledge contributions of firms and people in several places. Computers and peripherals, for example, are frequently the result of flows back and forth between Silicon Valley in California (prominently) and key Asian locations, such as Penang in Malaysia and Singapore (Gourevitch, Bohn, and McKendrick 2000; McKendrick, Doner, and Haggard 2000).
2. Adapters. While the main emphasis and interest of the scientific community has been in the regions that host actors creating best practices, innovation is not absent from less-than-best-practice regions. These regions do it by providing an environment for steady improvements and incremental innovations, possibly leading gradually to high quality. This takes place in RISs that are able to adopt new innovations from external sources relatively early and gradually improve them. The ability to learn from innovative firms in other places (i.e., imitating) is considered the best route for developing and maintaining innovative capability of this sort (Kim 1997; Mody, Suri, and Tatikonda 1995). Examples of regions include the newly industrializing countries of Southeast Asia, where incremental innovations are becoming common (Kim 1997; Leonard-Barton 1995; Singh 1995). Bangalore in India (Fromhold-Eisebith 1999), parts of Mexico, and the Zhong'guancun area of Beijing, China (Wang 1999), typify this environment. These areas attract a great deal of foreign direct investment, based on their productive workers, but they have not yet attained the perception from the outside as generating a steady flow of more fundamental innovations. Hobday (1994, 1995), Porter et al. (1996), and Roessner et al. (1996) include most of East Asia, including Singapore, as not yet at the stage at which local ability for innovation matches that originating from outside, except in production.
3. Adopters. RISs into which innovations diffuse relatively slowly (latecomers) are regional "imitator systems." They are characterized by actors employing an adopter strategy: they are able to import and use technological solutions (in end products, intermediaries, machinery, or appliances) from external, technologically more advanced sources. Via adopting technologies as users and through learning by imitating, they are able to adopt the production of mature products. Actors in such imitator sys-

tems are not capable of significantly improving those products. Yet, they form parts of innovation systems due to their specialization in more routine parts of production, or even just assembly (McKendrick, Doner, and Haggard 2000).

These different regions may maintain their roles in a rather static manner or they may upgrade their capabilities and gradually improve. This means that it is difficult to make clear distinctions in the real world, as many regions may host actors at varying levels of technological sophistication and, especially in diverse regions, possibly belonging to different technological systems. In a dynamic analysis, this may sometimes reveal to be a sign of a relative regional decline, sometimes of a relative upgradation of the overall regional capabilities, and sometimes even a sign of a regional structure that is fit for hosting the various actors involved in the whole innovation cycle (i.e., both those who are specialized in exploration and those who are specialized in exploitation. When progress is made in a region's innovativeness, the basis of its knowledge system changes over time, incorporating and diffusing successively more external technology (Bell and Albu 1999). This is how Kim (1999) described the process by which Korea built technological capability at the national scale. Korea went through three stages: (1) duplicative imitation of mature technology, (2) creative imitation of intermediate technology, and (3) innovation or emerging technology. Each stage required changes in Korea's national system of innovation.

Innovation systems do not operate in isolation but are the dynamic parts of production systems that are geared around getting the right goods for the right markets. Storper and Salais's (1997; Storper 1997, 116-26) typology of "worlds of production" is used in the following to elaborate briefly on the kinds of production systems innovation systems participate in.

In the "industrial world," generic and standardized products are produced for a market with undifferentiated demand. This can be done endlessly once the required skills have been learned. Actors belonging to industrial worlds are likely to be found in adopter RISs. Only through external shocks (drop in demand) may regions of this type start looking around for more sophisticated technologies to adopt themselves. As parts of innovation systems, they are able to adapt to new standards or requirements demanded by those more actively involved in innovating. Industrial worlds tend to be in lower labor cost regions or countries and mature industrial regions.

In the "market world," products are in many ways standardized (they consist of parts made according to standardized specifications), but they are produced for dedicated customers. Market worlds tend to be specialized production regions with large numbers of firms in an industry. Market worlds are likely to be in operation in adapter regions. Producers may be of the adapter type because they may engage in incremental innovation while adjusting their production to the needs of customers.

The "interpersonal world" produces for dedicated customers with specialized needs. For this purpose, specialized capabilities are needed as well. This world is found in technology and industrial districts, and it is the central locus of lead-ing-edge innovation. This world is obviously a genuine innovator RIS. This is the kind of system in which it is usually assumed that proximity among actors is required due to the need for frequent interpersonal communication and shared understandings to support it.

The "world of intellectual resources" uses scientific methods for developing new generic products for specialized purposes. The intellectual worlds are designed to specialize in the phase of particular innovation cycles that explore new knowledge, and they may comprise those parts of the interpersonal world that are geared toward innovation (e.g., R&D projects). As part of innovation processes, relations in this world are, indeed, maintained with actors in an interpersonal world. Storper (1997, 124-25) seemed to assume that interaction in innovation takes place in the districts, or RISs, of the interpersonal world. Productive activities may also take place over long distance as external transactions may happen over long dis-tances in predictable, formal, contractual governance regimes (Storper 1997, 124-25). This world may be the progressive core of a genuine innovator system, and it often works in close connection with the interpersonal world in the same or a closely connected location.

These worlds may be connected to each other through concrete production rela-tions. They may also become connected via forming key nodes of cycles of innova-tion over time: what is first discovered in the intellectual and interpersonal worlds is transferred to other places after or during a period of consolidation by actors in them (foreign direct investment, licensing, etc.) or imitated by actors in the indus-trial or market worlds. Sometimes these worlds may operate in the same place, as suggested above, and sometimes they are separated by space so that each type of activity is located in regions (RISs) where they find the best fit with other actors in the local environment. A spatial division of labor reflects the relative advantages of local environments for activities before and after the emergence of a dominant design (Utterback and Afuah 2000) or as technological clusters as opposed to oper-ational clusters (McKendrick, Doner, and Haggard 2000). Younger technologies are characterized by a wider, more open-minded perspective, based on many links to sources of knowledge. By the time production begins, the number of partners and suppliers is reduced to reflect the standardization of production.

The above considerations on the sectoral specialization versus diversity of regions and the relative maturity versus advancement of their technologies are brought together in Table 1, which outlines a typology of RISs involved in different types of SISs. Regions that host genuine innovators may be diversified or special-ized, but with specialization may come an inability to connect to other industries or shift to new technological regimes as times change (i.e., to sustain innovativeness in the region). Adapter regions may acquire a high level of competence, enhanced by diversity, which enables greater technological sophistication. Adopter regions exhibit innovativeness only in production, and many are unable to exhibit any innovativeness because of specialization in assembly with few local suppliers.

TABLE 1.   A Typology of Regional Innovation Systems

Characterization of Region	Sectoral Diversity	Sectoral Specialization
Genuine innovators (best practice places)	"Stars" (e.g., Silicon Valley, Cambridge, U.K.)	"Shooting stars" (e.g., Detroit, U.S., eighteenth-century Glasgow)
Adapters (relatively high levels of diverse competences)	"Living room lamps" (e.g., Hsinchu, Taiwan)	"Spotlights" (e.g., Bangalore, India)
Adapters (production-oriented competences)	"Chandeliers" (e.g., Bangkok, Thailand)	"Candles" (e.g., Dongguan, China)

As these ideas are at an experimental stage, the names of the various types of RISs are obviously playful. Among other things, we do not aim to approximate real scales (from stars to candles). The real-world examples we provide are also only suggestive as no in-depth analyses of the places are carried out.

- "Stars" are the suns for their surrounding planets: with the leading-edge innovations that they keep pushing to the market, they generate the energy that keeps other places going, either via imitation or enhanced innovativeness in other stars. They are kept strong by the multiple links among diverse industries as, for example, in the Silicon Valley where the venture capitalists that keep the electronics industry alive also finance start-up biotechnology companies. Actors in their key industries also monitor developments and maintain close links with other centers of excellence on the world scale.

- "Shooting stars" live as long as they are able to live on the strength of an innovation or a set of interrelated innovations, such as those related to technological and organizational innovations in automobile production in Detroit from 1910 to 1960 and in shipbuilding in Liverpool during the eighteenth century.

- "Living room lamp" regions host actors with relatively high levels of competences in a number of different sectors, each of which maintain close links with nonlocal sources of innovation. They may also be locally connected so as to collaborate in improving local production conditions; local cross-sectoral connections may also give rise to occasional technological improvements. It is possible that these regions become "rising stars" and later give rise to genuine innovations. Korea, but perhaps particularly the Seoul region, also fits this description.

- "Spotlights" get the stimuli to engage in mainly incremental innovation through their strong external connections. Through the high competences, they are able to respond to relatively advanced R&D-related improvements, for example, delegated by headquarters staff or in collaboration with main contractors, such as Nike's developed partners in Taiwan and Korea (Donaghu and Barff 1990).

- "Chandeliers" are regions where many sectors are colocated but where those sectors are not strongly linked to each other. Rather, they maintain relatively stronger links to their respective external customers, main contractors, and other sources of knowledge. Thus, chandeliers consist of several islands of locally isolated industrial activity. Their colocation may be supported, for example, by strong government support and consequently improved production environment (involving infrastructure, finance, education, etc.).

- "Candles" stay alive as long as their relatively simple production-oriented competences are utilized and supported by externally based customers, main contractors, or corporate structures. They may become efficient masters in certain production lines. Occasional incremental innovations in production activities may occur, but this is most likely to happen via imitation or knowledge transfer within corporate networks than through the initiative of local actors.

LOCAL AND DISTANT CONNECTIONS

It has been postulated in the literature on RISs and localized learning that the creation of noncosmopolitan (Storper 1997) or unique (Maskell 1999) knowledge through learning takes place more easily within proximate relations (e.g., Maskell and Malmberg 1999a, 1999b; Asheim and Cooke 1999). Yet, as the importance of links to nonregional networks is also a recurrent finding in recent research on industrial districts and technology districts (Amin and Thrift 1992; Tödtling and Kaufmann 1999; Maillat 1995; Mueller and Loveridge 1995; Storper 1993), it seems increasingly clear that the connections of regional actors to extraregional actors stand as momentous in technological progression. Connections to other networks in other regions provide access to a diversity of ideas and bases for comparison with local practices that are not internally generated (Amin and Thrift 1992, 1993; Camagni 1995; Maillat 1995; Tödtling 1995). An interesting example is seen in the immigrant communities from around the world that converge in and benefit Silicon Valley, partly by maintaining their previous connections (Saxenian 1999). External connections help actors within a regional system to stay in tune with what happens in the market, what happens among other producers (both competitors and collaborators), customers, scientists, regulators, support agencies, and other sources of technological knowledge and help them form fruitful relations with these agents.

It may be the case that the content of learning in nonlocal networks differs from the kind of learning that occurs in local relations (Oinas 2000). Overall, however, we do not seem to understand the nature and relative significance of proximate and distant connections in innovative activity very well to date. It is often assumed that only codifiable and hence nonculture-dependent, cosmopolitan-scientific, or professional languages can be communicated over longer distances (Storper 1997, 114). Noncosmopolitan knowledge is usually believed to involve a considerable tacit component that makes it glued to concrete local relationships. Yet, as Storper (1997) pointed out, "Noncosmopolitan knowledge is not necessarily associated with proximity or localization. The two are theoretically distinct: noncosmopolitan knowledge can be 'localized' in a restricted technological, organizational, or professional 'space', that is, in certain interpretative networks that transcend local geographical space" (p. 71). Due to the complexity of this issue, there is no pretence of exhausting it here. Let us point out that even if innovation systems were considered as localized (e.g., Asheim and Cooke 1999), it would not mean that they operate in total isolation. Being localized, then, must mean that either most relationships or key relationships in production systems or worlds of production take place in proximate relationships. What the SIS framework suggests, in addition, is that there is the possibility that neither most nor the key relationships are necessarily proximate.

In local and regional innovative systems, two sets of effects operate simultaneously (Camagni 1995; cf. Malmberg and Maskell 1997): proximity effects, such as reductions in costs because of quicker circulation of information, face-to-face contacts, and lower costs of collecting information or sharing knowledge, and socialization effects, related to collective learning, cooperation, and socialization of risks. These two processes are collective but not necessarily (explicitly) cooperative (meaning concrete, goal-oriented interaction whether in the form of supplier-customer relationship, joint R&D, or informal collaboration); they spread beyond bilateral interfirm relationships. In nonlocal relations, the proximity effect is missing and leaves only socialization effects and the possible forms that they may take over space.

Shared rationalities, or common frameworks of action (Storper 1997, 45), must be seen central in the socialization effect. Such frameworks of action, which are specific to the different worlds of production, are formed by conventions (Storper and Salais 1997, 15-17), which bring about coordination among actors (Storper 1997, 42-43). They "include taken-for-granted mutually coherent expectations, routines, and practices, which are sometimes manifested as formal institutions and rules but often not" (Storper 1997, 38). This implies that conventions are also key carriers of collectively shared tacit knowledge related to the functioning of the relevant innovation system. Following Blanc and Sierra (1999), the more precise content of Camagni's socialization effect can be interpreted to have four aspects: (1) organizational proximity (including formal relationships with suppliers), (2) relational proximity (which includes noneconomic relationships), (3) institutional proximity (especially of local informal institutions), and (4) temporal proximity (a shared vision of the future). Geographical proximity does not guarantee the other proximities, but those can partially substitute for geographical proximity (for a related discussion in terms of competence relatedness, see Oinas 1999; Oinas and van Gils 2001). There are complex trade-offs between the various proximities. When the potential involved in each type of proximity is actualized, it is manifested in shared context-specific conventions, coordinating both local and nonlocal relations.

The issues related to the question of how and to what degree conventional relations based on the various proximities are maintained over space remains largely unanswered to date. Exactly how knowledge grows and is shared in an agglomeration is beginning to be teased out in detailed studies (Henry, Pinch, and Russell 1996; Pinch and Henry 1999; Porter 1998). More work needs to be done on nonlocal relations and especially on the distant transferability or exchangeability of knowledge that involves a considerable share of tacitness.

TYPES OF CONNECTORS

The actors that create and maintain the relations that are emphasized in the SIS approach are centrally individuals (entrepreneurs, managers, employees, individuals in governmental or semigovernmental bodies, researchers, etc.) with their interpersonal networks (face-to-face, virtual, or a combination of these) and firms (multilocational/multinational) and their networks of various sorts: (advanced) customers; universities; research institutions; support organizations (such as chambers of commerce, knowledge centers, government bodies, and consultants). The reasons for the success of some places and the lack of success of others appear to be two interrelated things: first, interfirm differences in the degree to which active, extroverted behavior takes place and, second, the technical culture created within intensively connected communities of professionals, much of which is summed up by the characteristics of technologically successful regions (Malecki 1997; Sweeney 1991, 1999). This section discusses the nature of the actors creating those connections (cf. Oinas and Malecki 1999; Bunnell and Coe 2001). Innovation involving both local and distant relations often center around networks of these actors.

FIRMS AND THEIR NETWORKS

Kelley and Brooks (1992) distinguished between firms with primarily active and social external linkages and those with passive and asocial linkages (see also Amendola and Bruno 1990; Estimé, Drilhon, and Julien 1993). Indeed, the role of active, extroverted firms needs to be acknowledged in their role of making connections (Malecki and Poehling 1999; Patchell, Hayter, and Rees 1999). It is via the multilocational networks of facilities, alliances, and other linkages that such extroverted corporations and small and medium-sized enterprises alike are able to make SISs cut through possibly several RISs. These extroverted, active firms utilize written sources for acquiring information, interact with sales representatives, participate in trade shows, contact with vendors, and create close relationships with special-order customers for sharing of technical information (Malecki and Poehling 1999).

Via extroverted behavior, even small firms compensate for their size limitation in the adoption of new technology (Julien 1995; Rothwell 1992). Oerlemans, Meeus, and Boekema (1998) found that access to external resources increases innovation in small firms over those using only internal resources. Firms used four distinct types of external information: public knowledge infrastructure, private knowledge infrastructure, production column, and intermediaries. The most significant is the production column, comprised of buyers, suppliers, and other firms, reinforcing the view that links with customers, or producer-user connections, are the most beneficial. However, networks alone are not as effective as the combination of internal technical ability and effort with external networks (MacPherson 1997). The most likely firms to be active in seeking out external information are those with in-house R&D activity (Tsipouri 1991; Keeble et al. 1998; MacPherson 1992), which increases their absorptive capacity (Cohen and Levinthal 1989).

The wider networks of active, extroverted firms tend to encompass both more connections within the region and outside it. Extroverted firms are also more likely to aim at competition in international markets (MacPherson 1995). Externally oriented firms are able to overcome the constraints related to a peripheral location (Alderman 1999; Vaessen and Wever 1993). Vaessen and Keeble (1995) found that growth-oriented firms do more R&D and have more external programs for worker training regardless of their regional environment. Localized technological knowledge is highest where both the receptivity to nonlocal information and regional network connectivity provide access to and absorption of external information, combining it with internal competence (Antonelli 1999; MacPherson 1997). The combination of a critical mass and diversity of firms together with a set of fast-growing firms at technological frontiers appears to be the key to success at the level of a region (Chesbrough 1999).

In sum, the capability to innovate successfully at the firm level appears to be strongly conditioned by the ability to accumulate specific knowledge internally and to access sources of knowledge via external relations. The ideal case may exist when the firm's external networks can learn from strong local knowledge infrastructures, as well as maintain links to global networks of best practice in technologies, products, and services. Jacobs and de Man (1996) suggested that firms' strategies toward local and nonlocal clusters have different effects on which activities should be located in which locations. Local clusters allow greater cooperation and intensive user-producer interaction. Nonlocal clusters open possibilities to work with other clients and suppliers, and to tap—if not to become fully integrated—into different knowledge networks.

Multilocationality/multinationality is a form of extroversion. There is a growing tendency for companies to seek extraregional connections by using several home bases, including R&D and sophisticated production. External knowledge is most easily obtained by MNCs, with corporate facilities in various locations exploiting the relative advantages of their locations (Ferdows 1997), which may be seen as types of RISs. But such external knowledge must be internalized. To integrate knowledge residing in distant locations, firms must become locals in those places (Blanc and Sierra 1999; Cohendet et al. 1999; Gassmann and von Zedtwitz 1999; Reger 1999). This is evident in the five competencies that Amin and Cohendet (1999) suggested are now critical for globalized firms: (1) integrate the firm internally, (2) exploit advantages of proximity at many locations, (3) integrate fragmented pieces of localized learning, (4) invest continually in access to knowledge, and (5) focus on a small number of core competencies. This suggests three aspects to the information-age organization's structure: decentralization, information practices that promote both an awareness of external information and information-sharing within the organization, and a network structure for the outsourcing of noncore activities (Mendelson and Pillai 1999). For smaller firms, it is more difficult to be all things at once, but an effort to make external connections seems to be a minimal requirement.

Technology-based firms are particularly inclined to diversify their technology sources (Granstrand 1998) even though dispersed corporate networks do not necessarily have the result of diversifying firms' technological capabilities (Zander 1999). There is actually only scarce empirical evidence of projects that integrate knowledge across related technologies within MNCs internationally, yet, to some degree, it does happen (Zander 1998, 19).

INDIVIDUALS AND THEIR NETWORKS

The role of individual entrepreneurial initiative is obviously central in creating and transferring innovations, whether based on imitation and adaptation of technological solutions elsewhere (e.g., Fujimoto 1998, 23), differentiation (Nooteboom 2000), or novel combinations (e.g., palm-size devices and other hybrids of mobile telephones and portable computers). Competent and mobile individuals are equally an important group of connectors (e.g., Eliasson 1998). For instance, in a comparison of twelve U.S. semiconductor regions, Almeida and Kogut (1999) found the high level of intraregional mobility of engineers in Northern California unique.

Individuals seldom innovate alone, however. Interpersonal networks are increasingly seen as a powerful force in learning and maintaining (technological) capabilities. Their role can also be highlighted in making connections within technological systems. Recent research describes innovation networks as technological communities (Powell, Koput, and Smith-Doerr 1996; Rycroft and Kash 1999), or communities of practice (e.g., Aldrich 1998; Brown and Duguid 1994; Lave and Wenger 1991; Wenger and Snyder 2000). Through intensive relations, members of technological communities share common ways of thinking about work-related issues (the collaborative project, perceiving the problems to be solved, getting about solving problems, etc.), which enables the sharing of tacit knowledge. Moreover, they often share similarities in their educational backgrounds and features of lifestyle (in many cases including a highly international orientation accompanied by frequent traveling), which facilitates the process of learning to communicate meanings in a long-standing collaborative situation. While such technological communities may be locally or regionally based, they need not be. The literature on communities of practice usually refers to collective practice-based learning within business organizations. Amin and Cohendet (2000) observed that such communities of practice may also operate across space in multilocational firms. In addition, there is no reason to think that communities of practice would be limited to organizations only: tightly knit networks also consist of communities of professionals who may have an intimate understanding of each others work, whether or not they are (physically) located in the same (local) community (Oinas 2001). These communities also function effectively as connectors between firms and locations but often within technological systems.

CONCLUSIONS AND CHALLENGES FOR FUTURE RESEARCH

The notion of SISs has been used to refer to the organization of technological systems in space as well as their evolution in time. The SIS approach shares the view of the emerging meso-level analyses of technological evolution in that it regards as central both the concrete interactions through which innovations emerge and diffuse and the broader societal (techno-economico-cultural) context (cf. Green et al. 1999). It is distinct, however, in the sense that we emphasize centrally the spatial dimensions so as to pay attention to the evolution of technological trajectories in space. SISs are also seen as distinct from NISs because they do not necessarily reside within national boundaries. In regard to RISs, the SIS approach depicts that the capabilities and results of several RISs might be included in one SIS, simultaneously and/or over time. Accordingly, what we call the SIS refers to those (parts of) region-specific innovation systems that are relevant for the development of particular technological systems, involving the various interconnections of subsystems over space. In other words, "spatial innovation systems consist of overlapping and interlinked national, regional and sectoral systems of innovation which all are manifested in different configurations in space" (Oinas and Malecki 1999, 10). Thus, the SIS approach aims to highlight the "complex and evolving integration at different levels of local, national and global forces" (Archibugi and Michie 1997, 122). It seems that this complexity is increasingly recognized but that we are still at the stage where many basic concepts need to be searched and developed for pinning it down (e.g., Howells and Roberts 2000) and for finding patterns in that complexity. It is the aim of the SIS approach to provide some building blocks for analyzing the complex processes around innovation.

It is especially the connections between regional systems that remain relatively little understood. We know that local as well as nonlocal sources of innovative activity are decisive for innovations to occur and evolve, but we are just beginning to understand "the details related to the cofunctioning of proximity versus distance effects in various sorts of innovation" (Oinas and Malecki 1999, 25; cf. Blanc and Sierra 1999; Bunnell and Coe 2001; Gertler 1995; Hudson 1999; Oinas 1999, 2000; Oinas and Virkkala 1997).

This article was aimed at proposing a broad framework for analyzing SISs. In so doing, we have not penetrated into the details of actual technological systems and their evolution in time and space. We conclude by outlining several interrelated challenges that remain to be tackled in continued work on identifying and analyzing SISs.

1. An important issue is understanding dynamics: seeing the need for technological systems to evolve as firms and that their interactive patterns change, that is, as products, strategies, resource bases, and information bases change (Ebers and Grandori 1997; Galli and Teubal 1997). The evolutionary trajectories of firms must be matched rather closely by the evolution of their networks and the broader institutional environments. Yet, it is very difficult, for example, for regional support organizations to keep up with general trends as well as the varied and specific needs for firms for support (Braczyk and Heidenreich 1998; Cooke 1998).

2. We do not know very much about how successful firms build their local and extralocal networks of contacts. As Cantwell and Piscitello (2000) noted, "We need to know more . . .about changes in the exact geographical composition of technological activity in each industry" (p. 45). Does it matter whether local relations or linkages to other regions are the first to be built, as long as the firm can survive until the appropriate network is assembled? Does a region's success depend on a specific degree of globalness in its firms' networks? Or are the local relations really relatively most important? We have some hints about these matters. For example, the necessary progression by a firm from a technological focus to a market focus (Roberts 1990) typically coincides with a shift in linkages from local to national and international markets (Autio 1994; Christensen 1991; Christensen and Lindmark 1993). However, it is not yet clear which kinds of processes or activities of innovation are dependent on proximity (i.e., constrained by the need to establish close personal relations at close distance in specific institutional and conventional set-ups) and which are those that can be carried out over long distances. To start finding out, we assume that the appropriate units of analysis are the interactions related to specific technologies and models of products, which are typically organized within product families (Sanderson and Uzumeri 1995).

3. What remains to be further explored in the specific interactions within innovation systems are their "soft" sides. The degree to which the embeddedness of the relevant actors in their possibly different local institutional environments—involving their specific cultural conventions—affects their external relations is a key question. Local practitioners may remain tied to traditional factors as the basis for local development, and this may impede their ability to interact effectively with external actors. Alternatively, their embeddedness in local social relations involving strong interactions within professional communities with specific business cultures may provide the basis for finding useful complementarities with externally emerging technology, knowledge, and business cultures (cf. Malecki 2000; Wong 1998). Central is the question of the transferability of tacit knowledge as part of the operation of various communities of practice over space (Oinas 2001). There is very little empirical evidence that we can draw on concerning the travel of tacit knowledge over space, yet we should be reminded that the distinction between codified and tacit knowledge is not fixed in a "spiral of knowledge" (Krogh, Ichijo, and Nonaka 2000; Nonaka and Konno 1998). Complex and changing combinations of codified and tacit knowledge are likely to be found in innovative interactions in different spheres of activity in technological systems.

There is indeed a need to gain deeper understanding of the types of networks firms and individuals and firms create for different strategic purposes. While implementation networks (and the regional environments that support them) are highly important for firms to succeed in their existing competitive contexts, learning networks are more relevant for the competitive success of firms in the long run (Oinas and Packalén 1998). Minimally, differentiating between types of network relations will be helpful in understanding the types of connections actors create between RISs within SISs and the kinds of knowledge exchanges that are involved in them.

4. There is a need to incorporate conceptual insight into comprehensive empirical studies. Yet, technological systems are not identifiable with simple means. They involve knowledge systems, innovative capability, knowledge transfer, and so on—largely intangible objects that are difficult to define and investigate (Smith 1995, 86). The collective nature of technological development often has no formal manifestation but involves informal, invisible practices. This renders research difficult; data are not readily available. " 'Problem-solving' networks are what really define (technological) systems, not buyer-supplier links. Such relationships can only be identified and analyzed through primary data collection (via interviews, plant visits, etc.), which also needs to be oriented toward analyzing infrastructure and institutional arrangements" (Braunerhjelm and Carlsson 1999, 290). We are beginning to see the results of research along these lines in a few sectors, such as aircraft (Eriksson 1995; Frenken 2000) and hard disk drives (Gourevich, Bohn, and McKendrick 2000), but we do not know if these are special cases or the tip of a generally applicable iceberg.

REFERENCES

Acs, Z., J. de la Mothe, and G. Paquet. 1996. Local systems of innovation: In search of an enabling strategy. In *The implications of knowledge-based growth for micro-economic policies*, edited by P. Howitt, 339-59. Calgary, Canada: University of Calgary Press.

Alderman, N. 1999. Local product development trajectories: Engineering establishments in three contrasting regions. In *Making connections: Technological learning and regional economic change*, edited by E. J. Malecki and P. Oinas, 79-107. Aldershot, UK: Avebury.

Aldrich, H. 1999. *Organizations evolving*. London: Sage.

Almeida, P., and B. Kogut. 1999. Localization of knowledge and the mobility of engineers in regional networks. *Management Science* 45: 905-17.

Amendola, M., and S. Bruno. 1990. The behaviour of the innovative firm: Relations to the environment. *Research Policy* 19: 419-33.

Amin, A., and P. Cohendet. 1999. Learning and adaptation in decentralised business networks. *Environment and Planning D: Society and Space* 17: 87-104.

———. 2000. Organisational learning and governance through embedded practices. *Journal of Management and Governance* 4: 93-116.

Amin, A. and N. Thrift. 1992. Neo-Marshallian nodes in global networks. *International Journal of Urban and Regional Research* 16: 571-87.

———. 1993. Globalization, institutional thickness and local prospects. *Revue d'Economie Régionale et Urbaine* 3: 405-27.

Antonelli, C. 1999. *The microdynamics of technological change*. London: Routledge.

Archibugi, D., and J. Michie. 1997. Technological globalisation or national systems of innovation. *Futures* 29: 121-37.

Asheim, B. T., and P. Cooke. 1999. Local learning and interactive innovation networks in a global economy. In *Making connections: Technological learning and regional economic change*, edited by E. J. Malecki and P. Oinas, 145-78. Aldershot, UK: Ashgate.

Asheim, B., and M. Dunford. 1997. Regional futures. *Regional Studies* 31: 445-55.

Autio, E. 1994. New, technology-based firms as agents of R&D and innovation: An empirical study. *Technovation* 14: 259-73.

Bell, M., and M. Albu. 1999. Knowledge systems and technological dynamism in industrial clusters in developing countries. *World Development* 27: 1715-34.

Blanc, H., and C. Sierra. 1999. The internationalisation of R&D by multinationals: A trade-off between external and internal proximity. *Cambridge Journal of Economics* 23: 187-206.

Braczyk, H.-J., P. Cooke, and M. Heidenreich, eds. 1998. *Regional innovation systems: The role of governances in a globalized world.* London: UCL Press.

Braczyk, H.-J., and M. Heidenreich. 1998. Regional governance structures in a globalized world. In *Regional innovation systems: The role of governances in a globalized world,* edited by H.-J. Braczyk, P. Cooke, and M. Heidenreich, 414-40. London: UCL Press.

Braunerhjelm, P., and B. Carlsson. 1999. Industry clusters in Ohio and Sweden, 1975-1995. *Small Business Economics* 12: 279-93.

Brown, J. S., and P. Duguid. 1994. Organizational learning and communitites-of-practice: Toward a unified view of working, learning and innovation. In *New thinking in organizational behaviour,* edited by H. Tsoukas, 165-87. Oxford, UK: Butterworth-Heinemann.

Bunnell, T. G., and N. M. Coe. 2001. Spaces and scales of innovation. *Progress in Human Geography* 25. Forthcoming.

Camagni, R. 1995. Global network and local milieu: Towards a theory of economic space. In *The industrial enterprise and its environment: Spatial perspectives,* edited by S. Conti, E. J. Malecki, and P. Oinas, 195-214. Aldershot, UK: Avebury.

Cantwell, J., and L. Piscitello. 2000. Accumulating technological competence: Its changing impact on corporate diversification and internationalization. *Industrial and Corporate Change* 9: 21-51.

Carlsson, B. 1994. Technological systems and economic performance. In *The handbook of industrial innovation,* edited by M. Dodgson and R. Rothwell, 13-24. Aldershot, UK: Edward Elgar.

Carlsson, B., and R. Stankiewicz. 1991. On the nature, function and composition of technological systems. *Journal of Evolutionary Economics* 1: 93-118.

Chesbrough, H. 1999. The organizational impact of technological change: A comparative theory of national institutional factors. *Industrial and Corporate Change* 8: 447-85.

Christensen, C. M. 1997. *The innovator's dilemma: When new technologies cause great firms to fail.* Boston: Harvard Business School Press.

Christensen, J. L. 1992. The role of finance in national systems of innovation. In *National systems of innovation: Towards a theory of innovation and interactive learning,* edited by B.-Å. Lundvall, 146-68. London: Pinter.

Christensen, P. R. 1991. The small and medium-sized exporters' squeeze: Empirical evidence and model reflections. *Entrepreneurship and Regional Development* 3: 49-65.

Christensen, P. R., and L. Lindmark. 1993. Location and internationalization of small firms. In *Visions and strategies in European integration,* edited by L. Lundqvist and L. O. Persson, 131-51. Berlin: Springer-Verlag.

Cohen, W. M., and D. A. Levinthal. 1989. Innovation and learning: The two faces of R&D. *Economic Journal* 99: 569-96.

Cohendet, P., F. Kern, B. Mehmanpazir, and F. Munier. 1999. Knowledge coordination, competence creation and integrated networks in globalised firms. *Cambridge Journal of Economics* 23: 225-41.

Cooke, P. 1996. The new wave of regional innovation networks: Analysis, characteristics and strategy. *Small Business Economics* 8: 159-71.

———. 1998. Introduction: Origins of the concept. In *Regional innovation systems: The role of governances in a globalized world,* edited by H.-J. Braczyk, P. Cooke, and M. Heidenreich, 2-25. London: UCL Press.

de la Mothe, J., and G. Paquet. 1998a. Local and regional systems of innovation as learning socio-economies. In *Local and regional systems of innovation,* edited by J. de la Mothe and G. Paquet, 1-16. Dordrecht, the Netherlands: Kluwer.

de la Mothe, J., and G. Paquet. 1998b. National innovation systems, "real economies" and instituted processes. *Small Business Economics* 11: 101-11.

Donaghu, M. T., and R. Barff. 1990. Nike just did it: International subcontracting and flexibility in ath-
   letic footwear production. *Regional Studies* 24: 537-52.
Dosi, G. 1982. Technological paradigms and technological trajectories. *Research Policy* 11: 147-62.
Ebers, M., and A. Grandori. 1997. The forms, costs, and development dynamics of inter-organizational
   networking. In *The formation of inter-organizational networks*, edited by M. Ebers, 265-86. Oxford,
   UK: Oxford University Press.
Edquist, C. 1997. Systems of innovation approaches—Their emergence and characteristics. In *Systems
   of innovation: Technologies, institutions and organizations*, edited by C. Edquist, 1-35. London:
   Pinter.
Edquist, C., and L. Hommen. 1999. Systems of innovation: Theory and policy for the demand side. *Tech-
   nology in Society* 21: 63-79.
Eliasson, G. 1998. Competence blocs and industrial policy in the knowledge-based economy. *STI
   Review* 22: 209-41.
Enright, M. J. 1996. Regional clusters and economic development: A research agenda. In *Business networks: Prospects for regional development*, edited by U. H. Staber, N. V. Schaefer, and B. Sharma,
   190-213. Berlin: de Gruyter.
Eriksson, S. 1995. *Global shift in the aircraft industry: A study of airframe manufacturing with special
   reference to the Asian NIEs*, Gothenburg, Sweden: University of Gothenburg, Department of Human
   and Economic Geography.
Estimé, M.-F., G. Drilhon, and P.-A. Julien. 1993. *Small and medium-sized enterprises: Technology and
   competitiveness*. Paris: OECD.
Feldman, M. P., and D. B. Audretsch. 1999. Innovation in cities: Science-based diversity, specialization
   and localized competition. *European Economic Review* 43: 409-29.
Ferdows, K. 1997. Making the most of foreign factories. *Harvard Business Review* 75: 73-88.
Feser, E. J. 1998. Old and new theories of industry clusters. In *Clusters and regional specialisation: On
   geography, technology and networks*, edited by M. Steiner, 18-40. London: Pion.
Foss, N. J. 1996. Higher-order industrial capabilities and competitive advantage. *Journal of Industry
   Studies* 3: 1-20.
Freeman, C. 1987. *Technology policy and economic performance*. London: Pinter.
———. 1995. The "national system of innovation" in historical perspective. *Cambridge Journal of Eco-
   nomics* 19: 5-24.
Frenken, K. 2000. A complexity approach to innovation networks. The case of the aircraft industry
   (1909-1997). *Research Policy* 29: 257-72.
Frenken, K., and L. Leydesdorff. 2000. Scaling trajectories in civil aircraft. *Research Policy* 29: 331-48.
Fromhold-Eisebith, M. 1999. Bangalore: A network model for innovation-oriented regional development in NICs? In *Making connections: Technological learning and regional economic change*,
   edited by E. J. Malecki and P. Oinas, 231-60. Aldershot, UK: Ashgate.
Fujimoto, T. 1998. Reinterpreting the resource-based capability view of the firm: A case of the develop-
   ment-production systems of the Japanese auto-makers. In *The dynamic firm: The role of technology,
   strategy, organization, and regions*, edited by A. D. Chandler, P. Hagström and Ö. Sölvell, 15-44.
   Oxford, UK: Oxford University Press.
Galli, R., and M. Teubal. 1997. Paradigmatic shifts in national innovation systems. In *Systems of innova-
   tion: Technologies, institutions and organizations*, edited by C. Edquist, 342-70. London: Pinter.
Gassmann, O., and M. von Zedtwitz. 1999. New concepts and trends in international R&D organization.
   Research Policy 28: 231-50.
Gelsing, L. 1992. Innovation and the development of industrial networks. In *National systems of innova-
   tion: Towards a theory of innovation and interactive learning*, edited by B.-Å. Lundvall, 116-28.
   London: Pinter.
Gertler, M. S. 1995. "Being there": Proximity, organization, and culture in the development and adop-
   tion of advanced manufacturing technologies. *Economic Geography* 71: 1-26.

Gourevitch, P., R. Bohn, and D. McKendrick. 2000. Globalization of production: Insights from the hard disk drive industry. *World Development* 28: 301-17.

Granstrand, O. 1998. Towards a theory of the technology-based firm. *Research Policy* 27: 465-89.

Green, K., R. Hull, A. McMeekin, and V. Walsh. 1999. The construction of the techno-economic: Networks vs. paradigms. *Research Policy* 28: 777-92.

Grupp, H. 2000. Learning in a science-driven market. *Industrial and Corporate Change* 9: 143-72.

Guinet, J. 1995. *National systems for financing innovation*. Paris: OECD.

Hägerstrand, T. 1970. What about people in regional science? *Papers of the Regional Science Association* 24: 7-21.

Hampden-Turner, C., and A. Trompenaars. 1993. *The seven cultures of capitalism*. New York: Currency Doubleday.

Harrison, B., M. R. Kelley, and J. Gant. 1996. Innovative firm behavior and local milieu: Exploring the intersection of agglomeration, firm effects, and technological change. *Economic Geography* 72: 233-58.

Hassink, R. 2000. Regional innovation support systems in South Korea and Japan compared. *Zeitschrift für Wirtscharftsgeografie* 44: 228-45.

Henry, N., S. Pinch, and S. Russell. 1996. In pole position? Untraded interdependencies, new industrial spaces and the British motor sport industry. *Area* 28: 25-36.

Hickson, D. J., ed. 1993. *Management in Western Europe: Society, culture and organization in twelve nations*. Berlin: de Gruyter.

Hobday, M. 1994. Technological learning in Singapore: A test case of leapfrogging. *Journal of Development Studies* 30: 831-58.

———. 1995. East Asian latecomer firms: Learning the technology of electronics. *World Development* 23: 1171-93.

Howells, J., and J. Roberts. 2000. From innovation systems to knowledge systems. *Prometheus* 18: 17-31.

Hudson, R. 1999. "The learning economy, the learning firm and the learning region": A sympathetic critique of the limits to learning. *European Urban and Regional Studies* 6: 59-72.

Jacobs, D., and A.-P. de Man. 1996. Clusters, industrial policy and firm strategy: A menu approach. *Technology Analysis and Strategic Management* 8: 425-37.

Jackson, P., and N. Thrift. 1996. Geographies of consumption. In *Acknowledging consumption*, edited by D. Miller, 204-37. London: Routledge.

Julien, P.-A. 1995. Economic theory, entrepreneurship and new economic dynamics. In *The industrial enterprise and its environment: Spatial perspectives*, edited by S. Conti, E. J. Malecki, and P. Oinas, 123-42. Aldershot, UK: Avebury.

Keeble, D., C. Lawson, H. Lawton Smith, B. Moore, and F. Wilkinson. 1998. Internationalisation processes, networking and local embeddedness in technology-intensive small firms. *Small Business Economics* 11: 327-42.

Kelley, M. R., and H. Brooks. 1992. Diffusion of NC and CNC machine tool technologies in large and small firms. In *Computer integrated manufacturing, vol. III. Models, case studies, and forecasts of diffusion*, edited by R. U. Ayres, W. Haywood, and I. Tchijov, 117-35. London: Chapman and Hall.

Kelley, M. R., and S. Helper. 1999. Firm size and capabilities, regional agglomeration, and the adoption of new technology. *Economics of Innovation and New Technology* 8: 79-103.

Kim, L. 1997. *Imitation to innovation: The dynamics of Korea's technological learning*. Boston: Harvard University Press.

Kim, L. 1999. Building technological capability for industrialization: Analytical frameworks and Korea's experience. *Industrial and Corporate Change* 8: 111-36.

Krogh, G. von, K. Ichijo, and I. Nonaka. 2000. *Enabling knowledge creation: How to unlock the mystery of tacit knowledge and release the power of innovation*. Oxford, UK: Oxford University Press.

Kumaresan, N., and K. Miyazaki. 1999. An integrated network approach to systems of innovation—The case of robotics in Japan. *Research Policy* 28: 563-85.

Lave, J., and E. Wenger. 1991. *Situated learning: Legitimate peripheral participation*. Cambridge, UK: Cambridge University Press.

Leonard-Barton, D. 1995. *Wellsprings of knowledge*. Boston: Harvard Business School Press.

Lundvall, B.-Å., ed. 1992. *National systems of innovation: Towards a theory of innovation and interactive learning*. London: Pinter.

MacPherson, A. 1992. Innovation, external technical linkages and small-firm commercial performance: An empirical analysis from western New York. *Entrepreneurship and Regional Development* 4: 165-83.

———. 1995. Product design strategies amongst small- and medium-sized manufacturing firms: Implications for export planning and regional economic development. *Entrepreneurship and Regional Development* 7: 329-48.

———. 1997. A comparison of within-firm and external sources of product innovation. *Growth and Change* 28: 289-308.

Maillat, D. 1995. Territorial dynamic, innovative milieus and regional policy. *Entrepreneurship and Regional Development* 7: 157-65.

Malecki, E. J. 1997. *Technology and economic development: The dynamics of local, regional and national competitiveness*. 2d ed. London: Addison Wesley Longman.

———. 2000. Soft variables in regional science. *Review of Regional Studies* 30: 60-69.

Malecki, E. J., and R. M. Poehling. 1999. Extroverts and introverts: Small manufacturers and their information sources. *Entrepreneurship and Regional Development* 11: 247-68.

Malmberg, A., and P. Maskell. 1997. Towards an explanation of regional specialisation and industry agglomeration. *European Planning Studies* 5: 25-41.

Maskell, P. 1999. Globalisation and industrial competitiveness: The process and consequences of ubiquitification. In *Making connections: Technological learning and regional economic change*, edited by E. J. Malecki and P. Oinas, 35-59. Aldershot, UK: Ashgate.

Maskell, P., and A. Malmberg. 1999a. The competitiveness of firms and regions: "Ubiquitification" and the importance of localized learning. *European Urban and Regional Studies* 6: 9-26.

———. 1999b. Localised learning and industrial competitiveness. *Cambridge Journal of Economics* 23: 167-85.

McKendrick, D. G., R. F. Doner, and S. Haggard. 2000. *From Silicon Valley to Singapore: Location and competitive advantage in the hard disk drive industry*. Stanford, CA: Stanford University Press.

Mendelson, H., and R. R. Pillai. 1999. Information age organizations, dynamics and performance. *Journal of Economic Behavior and Organization* 38: 253-81.

Mody, A., R. Suri, and M. Tatikonda. 1995. Keeping pace with change: International competition in printed circuit board assembly. *Industrial and Corporate Change* 4: 583-613.

Mueller, F., and R. Loveridge. 1995. The "second industrial divide"? The role of the large firm in the Baden-Württemberg model. *Industrial and Corporate Change* 4: 555-82.

Nelson, R. R., ed. 1993. *National innovation systems: A comparative analysis*. New York: Oxford University Press.

———. 1996. The evolution of comparative or competitive advantage: A preliminary report on a study. *Industrial and Corporate Change* 5: 597-617.

Nelson, R. R., and S. G. Winter. 1982. *An evolutionary theory of economic change*. Cambridge, MA: Harvard University Press.

Nonaka, I., and N. Konno. 1998. The concept of "ba": Building a foundation for knowledge creation. *California Management Review* 40: 40-54.

Nooteboom, B. 1999a. Innovation, learning and industrial organisation. *Cambridge Journal of Economics* 23: 127-50.

———. 1999b. *Interfirm alliances*. London: Routledge.

———. 2000. *Learning and innovation in organizations and economies*. Oxford, UK: Oxford University Press.

O'hUallachain, B. 1997. Restructuring the American semiconductor industry: Vertical integration of design houses and wafer fabricators. *Annals of the Association of American Geographers* 87: 217-37.

Oerlemans, L.A.G., M.T.H. Meeus, and F.W.M. Boekema. 1998. Do networks matter for innovation? The usefulness of the economic network approach in analysing innovation. *Tijdschrift voor Economische en Sociale Geografie* 89: 298-309.

Ohmae, K. 1995. *The end of the nation state.* New York: Free Press.

Oinas, P. 1999. Activity-specificity in organizational learning: Implications for analysing the role of proximity. *GeoJournal* 49: 363-72.

———. 2000. Distance and learning: Does proximity matter? In *Knowledge, innovation and economic growth: Theory and practice of the learning region,* edited by R. Rutten, S. Bakkers, K. Morgan, and F. Boekema, 57-69. Cheltenham, UK: Edward Elgar.

———. 2001. Dissecting learning: Cognitive units, activity sets, and geography. Paper presented at the Meetings of the Association of American Geographers, New York, 27 February-3 March.

Oinas, P., and E. J. Malecki. 1999. Spatial innovation systems. In *Making connections: Technological learning and regional economic change,* edited by E. J. Malecki and P. Oinas, 7-33. Aldershot, UK: Ashgate.

Oinas, P., and A. Packalén. 1998. Four types of strategic inter-firm networks—An enrichment of research on regional development [in Finnish]. *Terra* 110: 69-77.

Oinas, P., and H. van Gils. 2001. Identifying contexts of learning in firms and regions. In *Promoting local growth: Process, practice and policy,* edited by D. Felsenstein and M. Taylor. Aldershot, UK: Ashgate. In press.

Oinas, P., and S. Virkkala. 1997. Learning, competitiveness and development. Reflections on the contemporary discourse on "learning regions." In *Regional specialisation and local environment—Learning and competitiveness,* edited by H. Eskelinen, 263-77. Copenhagen, Denmark: NordREFO.

Patchell, J., R. Hayter, and K. Rees. 1999. Innovation and local development: The neglected role of large firms. In *Making connections: Technological learning and regional economic change,* edited by E. J. Malecki and P. Oinas, 109-42. Aldershot, UK: Ashgate.

Patel, P., and K. Pavitt. 1994. The nature and economic importance of national innovation systems. *STI Review* 14: 9-32.

Pinch, S., and N. Henry. 1999. Discursive aspects of technological innovation: The case of the British motor-sport industry. *Environment and Planning A* 31: 665-82.

Porter, A. L., J. D. Roessner, N. Newman, and D. Cauffiel. 1996. Indicators of high technology competitiveness of 28 countries. *International Journal of Technology Management* 12: 1-32.

Porter, M. E. 1998. Clusters and competition: New agendas for companies, governments, and institutions. In *On competition,* edited by M. E. Porter, 197-287. Boston: Harvard Business School Press.

Powell, W. W., K. W. Koput, and L. Smith-Doerr. 1996. Interorganizational collaboration and the locus of innovation: Networks of learning in biotechnology. *Administrative Science Quarterly* 41: 116-45.

Quigley, J. M. 1998. Urban diversity and economic growth. *Journal of Economic Perspectives* 12: 127-38.

Reger, G. 1999. Changes in the R&D strategies of transnational firms: Challenges for national technology and innovation policy. *STI Review* 22: 243-76.

Roberts, E. B. 1990. Evolving toward product and market-orientation: The early years of technology-based firms. *Journal of Product Innovation Management* 7: 274-87.

Roessner, J. D., A. L. Porter, N. Newman, and D. Cauffiel. 1996. Anticipating the future high-tech competitiveness of nations: Indicators for twenty-eight countries. *Technological Forecasting and Social Change* 51: 133-49.

Roobeek, A.J.M. 1990. *Beyond the technology race: An analysis of technology policy in seven industrial countries.* Amsterdam: Elsevier.

Rosenfeld, S. A. 1997. Bringing business clusters into the mainstream of economic development. *European Planning Studies* 5: 3-23.

Rothwell, R. 1992. Successful industrial innovation: Critical factors for the 1990s. *R&D Management* 22: 221-39.

Rycroft, R., and D. E. Kash. 1999. *The complexity challenge: Technological innovation for the 21st century*. London: Pinter.

Sachs, J. 2000. A new map of the world. *The Economist* 24 June, 81-83.

Sanderson, S., and M. Uzumeri. 1995. Managing product families: The case of the Sony Walkman. *Research Policy* 24: 761-82.

Saxenian, A. 1999. *Silicon Valley's new immigrant entrepreneurs*. San Francisco: Public Policy Institute of California.

Scott, A. J. 1998. *New industrial spaces*. London: Pion.

Singh, M. S. 1995. Formation of local skills space and skills networking: The experience of the electronics and electrical sector in Penang. In *Human resources and industrial spaces*, edited by B. van der Knaap and R. Le Heron, 197-226. Chichester, UK: Wiley.

Smith, K. 1995. Interactions in knowledge systems: Foundations, policy implications and empirical methods. *STI Review* 16: 69-102.

Steiner, M. 1998. The discreet charm of clusters: An introduction. In *Clusters and regional specialisation: On geography, technology and networks*, edited by M. Steiner, 1-17. London: Pion.

Storper, M. 1993. Regional "worlds" of production: Learning and innovation in the technology districts of France, Italy and the USA. *Regional Studies* 27: 433-55.

———. 1996. Innovation as collective action: Conventions, products and technologies, *Industrial and Corporate Change* 5: 761-90.

———. 1997. *The regional world*. New York: Guilford.

Storper, M., and R. Salais. 1997. *Worlds of production*. Cambridge, MA: Harvard University Press.

Storper, M., and R. Walker. 1989. *The capitalist imperative: Territory, technology, and industrial growth*. New York: Basil Blackwell.

Sweeney, G. P. 1991. Technical culture and the local dimension of entrepreneurial vitality. *Entrepreneurship and Regional Development* 3: 363-78.

Sweeney, G. P. 1999. *Local and regional innovation: Governance issues in technological, economic and social change*. Report of the Six Countries Programme. Dublin, Ireland: SICA Innovation Associates.

Tidd, J., J. Bessant, and K. Pavitt. 1997. *Managing innovation: Integrating technological, market and organizational change*. Chichester, UK: Wiley.

Tödtling, F. 1995. The innovation process and local environment. In *The industrial enterprise and its environment: Spatial perspectives*, edited by S. Conti, E. J. Malecki, and P. Oinas, 171-93. Aldershot, UK: Avebury.

Tödtling, F., and A. Kaufmann. 1999. Innovation systems in regions of Europe—A comparative perspective. *European Planning Studies* 7: 699-717.

Tsipouri, L. J. 1991. The transfer of technology issue revisited: Some evidence from Greece. *Entrepreneurship and Regional Development* 3: 145-57.

Tushman, M. L., P. C. Anderson, and C. O'Reilly. 1997. Technology cycles, innovation streams, and ambidextrous organizations: Organization renewal through innovation streams and strategic change. In *Managing strategic innovation and change*, edited by M. L. Tushman and P. C. Anderson, 3-23. New York: Oxford University Press.

Utterback, J. M., and A. Afuah. 2000. Sources of innovative environments: A technological evolution perspective. In *Regional innovation, knowledge and global change*, edited by Z. J. Acs, 169-85. London: Pinter.

Vaessen, P., and D. Keeble. 1995. Growth-oriented SMEs in unfavourable regional environments. *Regional Studies* 29: 489-505.

Vaessen, P., and E. Wever. 1993. Spatial responsiveness of small firms. *Tijdschrift voor Economische en Sociale Geografie* 84: 119-31.

Wang, J. 1999. In search of innovativeness: The case of Zhong'guancun. In *Making connections: Technological learning and regional economic change*, edited by E. J. Malecki and P. Oinas, 205-30. Aldershot, UK: Ashgate.

Wenger, E. C., and W. M. Snyder. 2000. Communities of practice: The organizational frontier. *Harvard Business Review* 78: 139-45.

Wong, C. 1998. Determining factors for local economic development: The perception of practitioners in the North West and Eastern regions of the UK. *Regional Studies* 32: 707-20.

Zander, I. 1998. The evolution of technological capabilities in the multinational corporation—Dispersion, duplication, and potential advantages from multinationality. *Research Policy* 27: 17-35.

———. 1999. How do you mean "global"? An empirical examination of innovation networks in the multinational corporation. *Research Policy* 28: 195-213.

[20]

The Economic Geography of the Internet Age

Edward E. Leamer*
ANDERSON SCHOOL OF MANAGEMENT,
UNIVERSITY OF CALIFORNIA, LOS ANGELES

Michael Storper**
SCHOOL OF PUBLIC POLICY AND SOCIAL RESEARCH,
UNIVERSITY OF CALIFORNIA, LOS ANGELES

This paper combines the perspective of an international economist with that of an economic geographer to reflect on how and to what extent the Internet will affect the location of economic activity. Even after the very substantial transportation and communication improvements during the 20th Century, most exchanges of physical goods continue to take place within geographically-limited "neighborhoods." Previous rounds of infrastructure improvement always have had a double effect, permitting dispersion of certain routine activities but also increasing the complexity and time-dependence of productive activity, and thus making agglomeration more important. We argue that the Internet will produce more of the same: certain forces for deagglomeration, but offsetting and possibly stronger tendencies toward agglomeration. Increasingly the economy is dependent on the transmission of complex uncodifiable messages, which require understanding and trust that historically have come from face-to-face contact. This is not likely to be affected by the Internet, which allows long distance "conversations" but not "handshakes."

Will the Internet generate a revolution in the economic geography of the 21st century, creating neighborhoods connected not with streams and roads but with wires and microwave transmissions? History, linguistic theory, and economic theory help to answer this question.

The economic geography of the 18th century was much affected by the costs

---

*Leamer is the Chauncey J. Medberry Professor of Management and Professor of Economics and Statistics at UCLA, where he has been since 1975.

**Storper is Professor of Regional and International Development in the School of Public Policy and Social Research at UCLA. He is also Professor of Social and Human Sciences at the University of Paris/Marne-la-Vallee in France, and Visiting Centennial Professor at the London School of Economics.

Acknowledgement: Supported by a Sage Foundation Grant and edited by Nancy Hsieh

ECONOMIC GEOGRAPHY OF INTERNET AGE

of moving raw materials to production locations where the raw materials could be combined with labor and some capital to make final products. But at the end of the 18th century, home and workshop production were still the rule, and towns and cities were mostly marketplaces and transportation nodes.

A shortcoming of home or workshop production is that most of the capital sits idle most of the time—the hammer and scythe are idle when the spade is used (Leamer, 1999). In the 19th century, the growing importance of mechanization in manufacturing, and hence of physical capital, created pressures to centralize production in factories and in cities where a deeper division of labor allowed capital to operate many more hours during the day. The agglomerations needed to support this division of labor were made possible by great improvements in transportation systems—roads, canals, railroads, clipper ships.

Over the 20th century, improvements in transportation and communication systems allowed increasing geographical fragmentation of production (e.g., Arndt and Kierzkowski, 2001) and increasing global trade in intermediate inputs (e.g., Feenstra and Hansen, 1996). Yet the phenomenon of agglomeration of producers remained quite common. In many manufacturing industries, there still exist clusters of input producers that include both clients and competitors. There are cities specialized in rug making, watch making, and automobile manufacture (Scott, 1988).

Geographical clustering exists for many reasons. Retailing is clustered to save shopping costs for customers. For some types of material production, there are still important transportation advantages to location of the different stages in the division of labor—making the frame,

the doors and the wheels, and doing the assembly—at not too great a distance from each other. Even the 20th-century tendency toward geographical fragmentation of the chain of production was accompanied by the spatial agglomeration of certain parts of the chain, particularly the intellectual/immaterial activities such as accounting, strategy, marketing, finance and legal work. These intellectual activities have increased greatly as a share of value added and are amenable to extremely fine and highly efficient divisions of labor that make it uneconomic for a single firm to employ these specialists on a full time basis. Businesses instead "outsource" many of these functions to specialized firms producing intermediate intellectual inputs.

Since immaterial products can be transported virtually without cost, these intellectual activities are amenable to procurement at a distance: the design in Detroit, advertising in New York and strategy in Chicago. Although the clients of these specialized intellectual firms are sometimes far-flung, their competitors often are not. These specialized firms tend to cluster tightly together in financial "districts" and downtown office buildings, such as Wall Street, the City of London, and Chicago's Loop. Furthermore, it is common for specialized immaterial producers to have branch offices in major cities near the location of deployment of the ideas, suggesting that the "shipping" of an intellectual product may be as costly as shipping a tire or an axle.

From this history we conclude that economic progress over the last three centuries has come with an increasingly fine division of labor, physical labor in the 19th Century and intellectual activities in the 20th. But the finer the division of labor, the greater are the coordination

EDWARD E. LEAMER, MICHAEL STORPER

needs. Routine coordination of standardized intellectual or physical tasks can be done within "markets" that can be extended geographically with communication technologies. But complex and unfamiliar coordination of innovative activities requires long-term relationships, closeness and agglomerations.

The history of economic geography is thus a story of coordination over space and has been shaped by two opposing forces: (1) the constant transformation of complex and unfamiliar coordination tasks into routine activities that can be successfully accomplished at remote but cheaper locations (e.g., commodification), and thus an ongoing tendency toward deagglomeration or dispersion of production; and (2) bursts of innovations that create new activities requiring high levels of complex and unfamiliar coordination, which, in turn, generate bursts of agglomeration (Storper, 1997).

As with previous rounds of innovation in transacting technologies, the Internet offers some of both. It is making routine some coordination tasks ("dumbing down" of computers), but it is also creating a host of new and unfamiliar activities (mass customization). Thus, it creates forces for deagglomeration and forces for agglomeration.

It is widely believed that the Internet will have a more dramatic effect on economic geography than previous rounds of innovation, somehow suspending the force for agglomeration by allowing remote coordination of new and innovative activities.[1] This, we argue, is not likely. Coordination of new and innovative activities depends on the successful transfer of complex uncodifiable messages, requiring a kind of closeness between the sender and receiver that the Internet does not allow. The problem with the Internet is that he cannot look

her in the eye through a screen, and she cannot "feel" or "touch" him. It is a medium that may help to maintain relationships, but does not establish deep and complex contacts.

CLUSTERING OF MATERIAL AND INTELLECTUAL ACTIVITY

An examination of historical data on trade in products between countries reveals that the vast and steady improvements in technologies for transacting across space have not eliminated a strong role for geographical proximity. Pure intellectual activities are even more clustered. This suggests that present or future improvements in communication technologies, such as the Internet, also may not eliminate the role of proximity.

Material Production Remains Highly Clustered

Most of the value-added in the goods and services that we consume originates surprisingly close to home. North America is one trading neighborhood. Over the last fifty years, the share of U.S. trade with Canada has held steady at about 20%, while the share of U.S. trade with Mexico has increased from 5% to 10% (Figure 1). Since 1950, and probably much earlier, Canada has been both the number one destination for U.S. exports and the biggest source of U.S. imports. Mexico, which has always been in the top five U.S. trade partners, slipped in the 1960s and 1970s because of inward-looking policies, which were only partly offset by the rise in petroleum exports. But following the Mexican liberalization of 1986 and the NAFTA agreement, Mexico surpassed Japan to become the number two destination of U.S. exports in 2000, and is edging up to Japan as the second most important source of U.S. imports.

ECONOMIC GEOGRAPHY OF INTERNET AGE

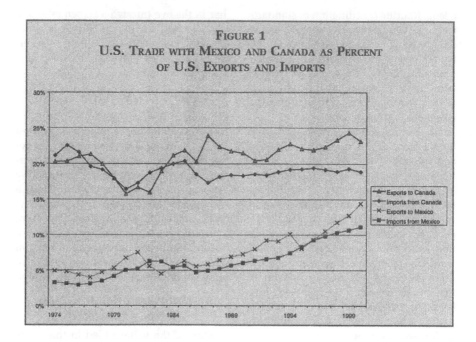

FIGURE 1
U.S. TRADE WITH MEXICO AND CANADA AS PERCENT
OF U.S. EXPORTS AND IMPORTS

North America is not the only trade neighborhood. Table 1 indicates the percent of 1970 and 1985 trade between adjacent countries at the two-digit SIC level of aggregation (the data set excludes trade between pairs of non-OECD countries). The commodities are sorted by the percentage of trade carried out between adjacent countries in 1985. Wood is the most locally-traded commodity, by this measure, with 42.4% of total trade occurring between adjacent countries, followed by printing and publishing. The long-distance products, with less than 20% of trade between adjacent countries, include apparel, footwear, professional equipment and miscellaneous (toys and umbrellas).

Table 1 also shows the percentages of trade involving island nations (especially Japan and the UK) which cannot have adjacent partners. The third column is the residual which, loosely speaking, is long-distance trade. Over 60% of trade in footwear falls into this long-distance category, and almost as much refined petroleum, tobacco and apparel. This contrasts with about 26% long-distance trade in printing and publishing, and in transportation equipment.

The last three columns of Table 1 report the differences between 1970 and 1985. Although trade relative to GDP grew rapidly during this period, long-distance trade and neighborhood trade in many products grew at similar rates. The exceptions include apparel, "other manufacturing industries" and tobacco, which did have rising shares of long-distance trade over this period. We thus pose the rhetorical question: what is there about these products that allows them to be traded over increasingly great distances?

One of the great empirical regularities in economic geography is that the greater

EDWARD E. LEAMER, MICHAEL STORPER

TABLE 1
OECD TRADE (EXPORTS PLUS IMPORTS) BETWEEN ADJACENT COUNTRIES, ISLAND NATIONS AND OTHERS (PERCENT OF TOTAL TRADE PER CATEGORY)

	1970			1985			1985-1970		
	Adjacent (1)	Island (2)	Other (3)	Adjacent (4)	Island (5)	Other (6)	Adj. (4)-(1)	Island (5)-(2)	Other (6)-(3)
TOTAL	30.6%			27.6%			−3%		
Wood	32.7	32.7	34.7	42.4	28.2	29.4	9.7	−4.0%	−5.3
Printing and Publishing	40.4	32.3	27.3	41.0	32.8	26.2	0.6	0.5	−1.1
Paper and Paper Products	35.9	22.5	41.6	37.7	22.7	39.6	1.8	0.2	−2.0
Furniture	50.9	15.0	34.2	37.3	27.0	35.7	−13.6	12.0	1.5
Transport Equipment	41.1	26.3	32.6	36.8	36.5	26.7	−4.3	10.2	−5.0
Misc. Petroleum Products	45.8	16.5	37.7	35.7	26.5	37.8	−10.1	10.0	0.1
Glass and Glass Products	37.1	23.5	39.4	34.4	30.8	34.8	−2.7	7.3	−4.6
Other Non-metallic Minerals	39.5	24.0	36.6	33.9	25.0	41.2	−5.6	1.0	4.6
Metal Scrap	31.8	29.2	39.1	33.2	31.9	34.9	1.4	2.7	−4.2
Other Food	31.7	26.8	41.5	32.5	29.9	37.6	0.8	3.1	−3.9
Fabricated Metal Products	34.6	29.5	35.9	32.3	33.1	34.6	−2.3	3.6	−1.3
Rubber Products	34.1	32.8	33.1	31.9	38.1	30.0	−2.2	5.3	−3.1
Plastic Products	32.4	33.0	34.7	30.1	36.8	33.1	−2.3	3.8	−1.6
Non ferrous Metal Basic Ind.	26.7	31.1	42.3	28.9	32.8	38.3	2.2	1.7	−4.0
Industrial Chemicals	27.9	30.1	42.1	27.8	30.1	42.1	0.1	0.0	0.0
Iron & Steel Basic Industries	33.2	32.3	34.5	26.1	35.0	38.9	−7.1	2.7	4.4
Textiles	30.3	37.0	32.7	25.3	36.2	38.5	−5.0	−0.8	5.8
Food Manufacture	19.6	30.1	50.3	23.5	26.6	49.9	3.9	−3.5	0.4
Beverage	26.9	37.1	36.0	23.2	35.1	41.7	−3.7	−2.0	5.7
Other Chemicals	24.7	30.5	44.8	23.1	36.4	40.5	−1.6	5.9	−4.3
Petroleum Refineries	18.2	29.3	52.6	22.9	22.3	54.8	4.7	−7.0	2.2
Machinery except elec.	27.7	28.6	43.7	21.8	38.2	40.0	−5.9	9.6	−3.7
Tobacco	22.2	27.2	50.6	20.0	21.4	58.6	−2.2	−5.8	8.0
Pottery, China & Earthware	21.9	46.8	31.3	19.0	44.4	36.7	−2.9	−2.4	5.4
Electric Machinery	25.2	33.9	40.9	18.9	46.7	34.3	−6.3	12.6	−6.6
Wearing Apparel	28.6	26.4	45.0	18.8	24.5	56.7	−9.8	−1.9	11.7
Leather	26.5	27.6	45.9	16.9	31.2	51.9	−9.6	3.6	6.0
Footwear	17.7	17.7	64.6	16.4	21.7	61.9	−1.3	4.0	−2.7
Prof., Scientific, & Measuring	23.4	37.6	38.9	16.4	48.2	35.4	−7.0	10.6	−3.5
Other Manufacturing Ind.	14.8	46.1	39.1	12.4	41.7	45.9	−2.4	−4.4	6.8

Source: OECD Compatible Trade and Production database.

Notes: Columns sorted by Adjacent Trade, 1985. Data include only trade flows with other OECD partners. Ireland and UK are included in "island" nations.

ECONOMIC GEOGRAPHY OF INTERNET AGE

the distance between any pair of countries, the less they trade with each other. This is measured by what economists call the "gravity model." According to the familiar gravity model of Newtonian mechanics, the force between any two objects is proportional to the product of their masses divided by the square of the distance between them. In economics, the amount of commerce between two points is equal to the product of the economic masses (GDPs) divided not by the square of the distance between them but by distance itself (or some lower power)[2]:

$$\text{Exports}_{ij} = \alpha GDP_i GDP_j / (DIST_{ij})^\beta$$

Figure 2 illustrates the "force of gravity" on West German trade in 1985. Each point on the graph represents one of Germany's trading partners. On the vertical axis is trade divided by the GDP of the partner [Trade(Germany,j)/GDP(j)]. On the horizontal axis is the distance between Germany and its partner. Both scales are logarithmic. Distance has a clear effect on the intensity of trade – Germany has close trade links with its close neighbors (France, Austria, Switzerland, the Netherlands, etc.) but does not trade much with far-away Asia.

Proximity Is an Important Source of International Competitiveness

Since commerce declines rapidly with distance, closeness to global GDP is an extremely important source of competitive advantage. Countries that are far from global GDP exchange natural resource products or low value-added manufactures for high value-added outputs and these faraway countries have low levels of GDP per capita (Leamer, 1997). The countries that export high value-added manufactures have high per capita GDPs and are overwhelmingly clustered in Europe and North America.

The dramatic effect that market access has on per capita GDP is revealed by Figure 3, which displays a measure of distance to global GDP on the horizontal axis and per capita GDP on the vertical axis. In 1960 only Australia and New Zealand were able to escape the force of gravity—being far away but managing to have a high GDP per capita. By 1990, two

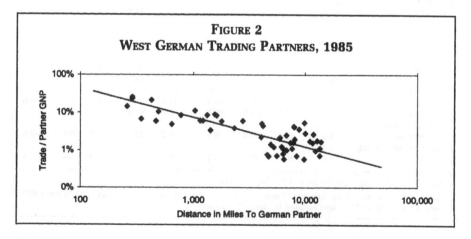

FIGURE 2
WEST GERMAN TRADING PARTNERS, 1985

EDWARD E. LEAMER, MICHAEL STORPER

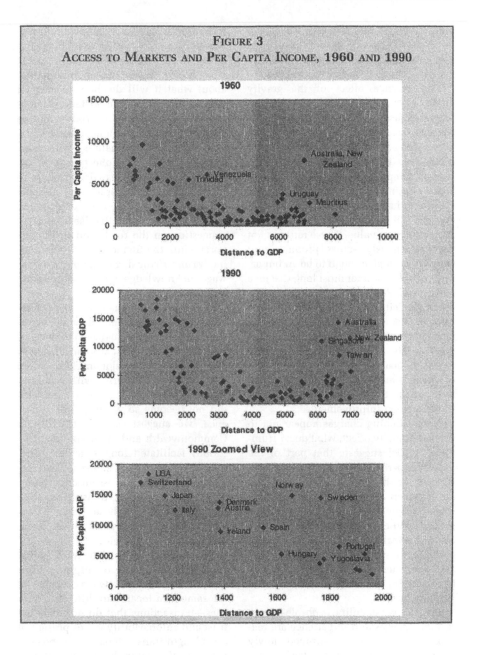

FIGURE 3
ACCESS TO MARKETS AND PER CAPITA INCOME, 1960 AND 1990

ECONOMIC GEOGRAPHY OF INTERNET AGE

other countries had also escaped: Singapore and Taiwan. Except for these, there is a very clear relationship between per capita GDP and distance to markets.

The distance effect of the gravity model captures something more important and more permanent than simple shipping charges. While there have been very substantial reductions in the cost of communicating at long distance and the cost of shipping goods, the role of distance remains very powerful. Hummels (1999a) estimates that the average freight costs of U.S. imports in 1994 were only 3.8% of import value. These rates do not vary enormously across products and they are not high enough to be an important consideration for most long-distance commerce. Rose (1999) estimates distance elasticity of −1.09 with a standard error of 0.05 using 1970 data, and a distance elasticity of −1.12 with a standard error of 0.04 using 1990 data. In other words, in that 20-year period the impact of distance remained unchanged, essentially unaffected by communication and transportation improvements.

Since shipping charges appear not to drive the gravity effect, what does? Hummels (1999c) suggests that part of the answer is perishability, broadly defined. He uses the shift toward more expensive air transport in the last half century, stimulated by a sharp relative decline of air transport costs, to infer the very substantial revealed perishability of traded goods, not just fruits and vegetables, but also computers and high-fashion handbags (it is the "perishable" items that are shipped by air.)

While perishability, obsolescence, fashion changes and impatience are reasons why goods are not shipped slowly over long distances, we emphasize here another reason why commerce is confined to neighborhoods: the technology

for the transmission of complex information has not improved much. This is important because the value of a product is often highly dependent on information about what it will do, how to make it work and with whom to talk when it does not work. While the product may be shipped cheaply over long distances, the accompanying information often needs to be delivered from one person to another. Thus humans remain the containers for shipping complex uncodifiable information. The time costs of shipping these containers is on the rise because of congestion on the roads and in the airports while the financial costs of so doing are also rising due to increases in real wages of knowledge workers who are the human containers.

The four anomalous countries in Figure 3 also help sharpen our understanding. These countries managed to develop high per capita GDPs at great distances from the world's major markets, while other far-away countries with access to the same communication technologies did not become so wealthy. The difference, we suggest, is that the British Commonwealth and other similar institutions facilitated long-term deep relationships over long distances. These relationships create the essential prerequisites of any complex transaction: trust and understanding. These are not the automatic result of transportation and communication technologies.

Furthermore, given that much global trade consists merely of shipping products or components between divisions of the same firm located in different countries—transactions that do not raise the trust and enforceability issues present in arms length transactions—the persistent distance effect on trade strongly suggests that trust alone is insufficient. What must limit trade, even within distant sis-

EDWARD E. LEAMER, MICHAEL STORPER

ter units of the same business entity, is that many information transactions presuppose a high level of mutual understanding. What do you mean when you say you'll do a good job? Do we agree on what is to be done?

Intellectual and Innovative Activities Are also Clustered

The empirical evidence regarding the clustering of intellectual, immaterial production is much more limited than for goods production and trade, in part because there are no widely accepted measures and readily available data which separate intellectual and informational activities from the rest of the economy. But virtually all the descriptive and empirical evidence we have on their locations suggest that—at least up until now—they remain highly, and even increasingly urbanized and located in proximity to GDP. Financial service industries are highly clustered in big cities, and especially in the triad of New York, London and Tokyo, followed by about a dozen other large metropolitan areas (Sassen, 1991). A handful of American metropolitan areas concentrate the intellectual and innovation-based industries (Jacobs, 1960; Pollard and Storper, 1996). Such localized clusters in the informational, intellectual, and innovation-based industries have progressively broadened the reach of their contacts—to other producers in other major localized clusters, and to a widely spread client base—as is suggested by the parallel growth in long-distance telecommunications and travel (Hall, 1998). This is a geography of highly-packed agglomerations which communicate over long distances, not one of dispersal and indifference to distance or proximity.

The Internet economy has produced high densities of dot.com firms in San Francisco, New York, Los Angeles and Seattle, and is following precisely the same geographical pattern as all of its innovative forebears: the establishment of a small number of core agglomerations, characterized by strong inter-firm and firm-labor market network relations, the existence of an "industrial atmosphere," and circular and cumulative advantage from these local external economies. The larger and more globally-linked metropolitan areas are enjoying stronger economic growth than their respective national economies in general, as they reinforce their positions as centers of economic reflexivity: inventiveness, creativity, the management of non-standardized transactions and elements of production and supply chains, i.e., the functions that steer and guide an increasingly elaborate division of labor in modern capitalism as a whole. The economies of these central places are increasingly comprised of core agglomerations of: (a) creative and cultural functions (including industries linked to this, such as fashion, design and the arts); (b) tourism; (c) finance and business services; (d) science, high technology and research; and (e) power and influence (government, corporate headquarters, trade associations and international agencies).

For immaterial intellectual production, there is great value in being at the "center of the action," where the division of labor can be pursued intensively, where specialized talent and "buzz" are important to keeping up with rapidly changing outputs (ideas) (Jaffe et al, 1993; Almeida and Kogut, 1997; Audretsch, 2000), and where complex but understandable contracts can be written with a glance and sealed with a handshake. In addition, many intellectual outputs are not products that can be dropped at the doorstep, but are services

ECONOMIC GEOGRAPHY OF INTERNET AGE

that have to be delivered by one human to another. Value is created jointly by seller and buyer, by coach and student, often involving many hours of direct communication.

Academic office hours, seminars, conferences, and coffees testify to the importance of face-to-face interactions in the production and distribution of new or complex ideas. It is not just union power that has kept the labor-intensive universities operating in more or less the same manner for four centuries. It is the production function itself. While routine learning and training can be done remotely, education is a decidedly up-close and personal activity.

THE JOINT DETERMINATION OF MESSAGES AND RELATIONSHIPS

Many messages can be communicated effectively only if the parties "know" each other. Information and communication technologies (ICTs), such as the Internet, create new possibilities for transmitting messages, and in so doing they may affect the kind of relationship the parties must have in order to send and receive a given kind of message. A first

step toward understanding the role of relationships in the transmission of information is to classify existing technologies according to the form and content of their messages, as we do in Table 2. The rows in this table refer to the form of the message: written words, spoken words, images and "presence," the latter referring to senses of touch and smell, as well as sixth or higher senses. The columns refer to the content of the message, contrasting simple codifiable one-way messages such as a stop sign, with complex context-dependent interactive messages such as the claim that "I love you." It is the complex, context-dependent messages that require the greatest investments in relationships.

Codifiable information has a stable meaning which is associated in a determinate way with the symbol system in which it is expressed, whether it be linguistic, mathematical, or visual. Generally speaking, codifiable information is cheap to transfer because its underlying symbol systems can be widely disseminated through information infrastructure, thus reducing the marginal cost of individual messages. Acquiring the sym-

	TABLE 2	
	MESSAGES	
	Content:	
FORM:	Simple Codifiable One-Way Messages	Complex Context-Dependent Interactive Messages
Written Words	Instructions, Print Media	Exchange of Letters
Spoken Words	Lecture, Command	Telephone Conversation
Images	Blueprint Photograph	Teleconference
Presence: "Feel"/smell		Handshake
EXAMPLES	Stop Sign	"I love you."

EDWARD E. LEAMER, MICHAEL STORPER

bol system may be expensive or slow (language, mathematical skills, etc), as may be building the transmission system, but once acquired using it to communicate information is cheap.

By contrast, much information is only loosely related to the symbol system in which it is expressed. This includes much linguistic, words-based expression (the famous distinction between "speech" and "language"), particularly what might be called "complex discourse" (Searle, 1969). For example, one can master the grammar and the syntax of a language without understanding its metaphors. This is also true for some mathematically expressed information, and for much visual information. If the information is not codifiable, merely acquiring the symbol system or having the physical infrastructure is not enough for the successful transmission of a message. It also takes mutual trust and mutual understanding. The parties therefore need to "know" each other, or have a broad common background which goes well beyond their direct contact, but the existence of which they can verify through their direct contact, using many forms of communication to do so.[3] The transmission of codifiable information has strong network externalities, since once the infrastructure is acquired a new user can plug in and access the whole network. The transmission of uncodifable information may have very limited network externalities, since the successful transmission of the message depends on infrastructure that is largely committed to one specific sender-receiver pair by their mutual trust and understanding.[4]

The distinction between codifiable and uncodifiable messages comes up implicitly in the economics literature on "search" goods and "experience" goods (Nelson, 1974). A "search" good has a transparent value—evident upon initial inspection. An "experience" good has a nontransparent value that depends on the user and that is experienced slowly over time. Markets that match faceless buyers and faceless sellers can mediate the exchange of search goods, but the exchange of experience goods requires trust, understanding and long-term relationships, either directly or through third party certification and enforcement. The persistence of the distance effect on global commerce is due in part to the fact that there are comparatively few search goods and comparatively few market-mediated exchanges. Most transactions require long-term relationships. It is no surprise, therefore, that B2B Internet exchange markets are having a hard time surviving.[5]

Understanding and Trust Come from Relationships

The Wall Street Journal headline "Record Sales Drop" might suggest a small problem for the music industry or a much bigger problem for business overall. Coding and decoding this kind of message involves mutual understanding of the context from which the information emerges: incompleteness is overcome by reasoning through analogy and reading between the lines. The infrastructure needed to accurately establish the context can be very substantial and can be very specific to the sender/recipient pair.

The coding and decoding of a contract (a promise) also depends on the context, which establishes the unstated contract contingencies and also the interests of the parties in honoring the agreement.[6] Understanding the context is not enough if the contract requires future actions that are not in the interests of the parties at the time they promise to take them.

Then the parties either have to align their interests through internalization (Williamson, 1985) or they have to invest in enforcement mechanisms that create mutual trust.

Trust can come from multi-layered relations between the parties to a transaction that can create low-cost enforcement opportunities (Lorenz, 1992). Trust can be created if reputation assets are put at risk. Trust also comes from the bonds that both parties establish to guarantee the truthfulness of the message. One important economic bond is the time and money costs of co-presence (schmoozing), which can far exceed the direct costs of sending the message. These costs, like advertising expenses (Klein and Leffler, 1981), amount to a forfeitable bond that assures the validity of the message.

Also like advertising, there is a special incentive to continue to invest in the relationship in order to maintain the value of the relational asset that was created by earlier encounters (absent the second date, the value of the first date disappears). This will encourage the formation of relatively few long-lasting deep relationships as opposed to many fleeting shallow ones.

To create a relationship bond, the costs must be substantial and transparent. E-mail, paradoxically, can be so efficient that it destroys the value of the message. The e-mail medium greatly reduces the cost of sending a message, somewhat reduces the cost of receiving the message, and it makes the costs mostly nontransparent. The low costs and the non-transparency greatly limit the value of the relationship bond. Mass mailings are indistinguishable from personalized messages. A return receipt only means that the recipient has opened the message, but the sender cannot be sure that enough attention has been devoted to it to absorb the content. Thus, for complex context-dependent information, the medium is the message. An e-mail declaration of "I love you" is more likely to be a computer virus than a credible promise of good things to follow.

THE EFFECT OF THE INTERNET ON THE CLUSTERING OF INTELLECTUAL ACTIVITIES

The Internet might lead to a substantially different economic geography if it allowed communication over long distances those complex, uncodifiable messages that historically have depended on human closeness. Earlier innovations in communication technologies including print media, the telegraph, recordings, the telephone and the television have not eliminated the tendency toward geographical agglomeration. But unlike these earlier technologies, broadband Internet communication will soon allow inexpensive simultaneous real-time interactive visual, oral, numerical and textual messages, creating a much more powerful imitation of closeness than has heretofore been possible.

But the imitation does not have all the properties of the real thing.[7] Face-to-face communication derives its richness and power not just from allowing us to see each other's faces and to detect the intended and unintended messages that can be sent by such visual contact. Co-presence—being close enough literally to touch each other—allows visual "contact" and "emotional closeness," the basis for building human relationships.

To help understand the impact of the Internet, the "face-to-face" metaphor thus needs to be separated into two parts: the "handshake" and the "conversation." The "handshake" refers to information exchanges made while in same

EDWARD E. LEAMER, MICHAEL STORPER

physical space; the "conversation" refers to interactive long-distance exchanges of visual and oral information. One can have a conversation through a computer, but not a handshake.

The Internet is a highly efficient system for the cataloging, accessing and delivery of images such as blueprints and photographs and advertisements. This greatly facilitates the transmission of codifiable visual information. Broadband Internet also promises to support much cheaper teleconferencing, which will allow the exchange of some complex context-dependent messages that heretofore were delivered in person. But the Internet does nothing by itself to put a message in the right context, and does not help in understanding. Moreover, an Internet conversation resembles e-mail in that it involves such low levels of costs to sender and receiver that there is little relationship bond created by the process.

The virtual world of the Internet has no physical neighborhoods, no Starbucks where like-minded people bump into each other for serendipitous handshaking in communities defined by cultural affiliation, language, ideology, desire, mutual identification, and other powerful forms of bonding. If such relationships are important to 21st century capitalism, then the Internet is unlikely to eliminate the need for physical co-presence.

However, unlike e-mail, the costs of this Internet conversation are transparent—one can "see" the time expended by the other party—which is necessary for the creation of the bond. It is possible, but unlikely in our opinion, that there will be Internet teas and cocktail hours, with far-flung parties to the apparently casual conversation making the critical investment in their relationships

(chat rooms do not serve this function, because the participants are anonymous). Therefore, the exchange of uncodifiable ambiguous information that depends on a high level of trust and shared context is likely to continue to require a significant amount of co-presence.

Meanwhile, air and auto transportation technologies have done nothing to decrease the cost of handshakes over long distances for almost 40 years. On the contrary, congestion on the roads and at airports, together with the increasing cost of time, have added considerably to the cost of handshakes beyond our local communities and have been a force for further agglomeration of immaterial commerce.[8]

The Continuing Importance of Relationships

Although the Internet may not do much to facilitate the long-distance communication of complex context-dependent information, it could make these less important. Will the economy of the 21st century be more or less dependent on the kind of complex, context-dependent messages, whose transmission the Internet cannot achieve? Or will most messages become codifiable, allowing immaterial and material commerce at great distances?

Table 3 helps to answer these questions. The two columns separate products according to the character of information needed to use them. At the left are products that come with codified information in the form of simple and understandable manuals/blueprints/specs, or that embody that information with plug and play features. At the right are products that depend for their usefulness on uncodifiable information.

ECONOMIC GEOGRAPHY OF INTERNET AGE

TABLE 3
MESSAGES, TRANSACTIONS AND LOCATION OF STANDARDIZED AND SPECIALIZED PRODUCTS

	Mass-Produced, Standardized Products	Specialized, Customized and Innovative Products
Messages	Codified, transparent	Tacit
Degree of intermediate transacting	Low (high scope economies)	High (low scope economies, high roundaboutness)
Degree of agglomeration of supply chain	Remote/Low agglomeration	Market-centered/agglomerated
Location of production/ distribution in relation to markets	Remote	Indeterminate

New products are constantly entering in the right of this table, and moving to the left. In the 1970s personal computers came with an extensive amount of complex, uncodifiable information about the tasks they could perform and how to get them to do those tasks. Today, these same PCs concentrate on a small range of standardized tasks including word processing, spreadsheets and e-mail. The information needed to perform these tasks is mostly codified in simple instructions expressed in words, drop-down menus and clickable icons.[9]

If the product is standardized, information about its features can be codified and shipped separately from the product in the form of specifications, blueprints, consumer magazines, standards, etc. This allows geographical distance between seller and buyer. If the product is not standardized, its operating features are less likely to be expressed in a codifiable form. The principal way of verifying the product's qualities is then by seeing, touching, feeling or otherwise actually "knowing" the product. This introduces a strong element of hands-on, relational verification of product qualities and thus geographic closeness.

Informational characteristics and the organization of production systems have strong interrelationships. The products in the left-hand column have a high scale of output and a high degree of standardization. This enables simplification of the information required to produce and transact, in the sense that the knowledge and information needed at each step in the process can be codified, set down in blueprints, and hence understood by any sufficiently skilled transacting party. These characteristics—high scale, standardization, routinization and codification of underlying knowledge and information generally permit transactions, whether B2B or B2C, to be carried out over considerable geographical distance.

In the right-hand column of Table 3 are low-volume specialized products and activities. Standardization of the product and routinization of the production process are impeded by fluctuating markets and rapid technological change. In addition, the innovation process itself generally involves the principle of "many heads are better than one," where access to ideas and talent occurring in other specialized, but closely related parts of the economy, heighten the

EDWARD E. LEAMER, MICHAEL STORPER

probability of a successful innovation (Jacobs, 1960; Scott, 1993; Feldman and Audretsch, 1999; Duranton and Puga, 2000). This sort of access is not amenable to codification and routinization. Its unpredictability is precisely its raison d'être. Furthermore, in the case of innovation-based activities, much of the information needed to innovate—even if it is sci-ence- and engineering-based—cannot be entirely codified, at any cost. Much of it is available only through access to the right persons, often few in number, who are working in a given problem area (Zucker, Darby and Brewer, 1998). Most importantly, transforming this information into economically useful knowledge involves recombinations and synergies that are often complex and qualitative. Marshall, in writing about the textile districts of Lancashire in the 19th century, alluded to this information-sharing process with his famous phrase, "the secrets of industry are . . . in the air" (Marshall, 1919). Put another way, these transactional relations are not amenable to complete contracting, and they depend on human relations, involving combinations of social networks, trust, interpretative communities, and reputation effects (Lorenz, 1992).

Reductions in Spatial Transactions Costs Need Not De-agglomerate Production

Technological innovations in transacting, such as the substitution of the Internet for other means of carrying out transactions, can push the outcomes in both directions in Table 3. First, as transportation and communications become better and cheaper, locations formerly foreclosed become economically feasible and activities are freed to relocate according to a field of redefined comparative advantages. They expand the geo-graphical reach of previously existing, usually standardized, activities.

But the second order changes induced by new transportation or IC technologies may reinforce the power of agglomeration. One of the great parables of modern economics is, of course, that "the division of labor is limited by the extent of the market" (Smith, 1776; Young, 1929; Stigler, 1951). In a dynamic framework, there are important feedbacks between technologies of transacting, and what is made (product differentiation) how it is made (production organization), and the geography of the production system. Greater access to suppliers makes possible new combinations of inputs, which, in turn, modifies the costs of producing a given kind of final output. Transportation technologies, in other words, can strongly influence the scale and division of labor in production.

There is substantial, if fragmentary evidence of these feedbacks from previous rounds of infrastructure development (Fishlow, 1965; Pred, 1966 and 1974; DeVries, 1984; Hall, 1998). Relative city sizes remained stable over the 20th century in the United States (Black and Henderson, 1998), and this pattern of stability (parallel growth in cities) is true of other advanced countries such as Japan and France (Eaton and Eckstein, 1997). Moreover, there is mostly persistence of the same activities in the same cities; only a few industries really change their geographical centers or entirely abandon them, once they are initially locked into a location (Storper and Walker, 1989; Brezis and Krugman, 1997; Henderson, 1998). U.S. industrial location patterns at the three-digit level from the mid-nineteenth century onward have been remarkably stable (Kim, 1995; Dumais et al, 1997). Thus, there is strong evidence that innovations in physical transport in-

ECONOMIC GEOGRAPHY OF INTERNET AGE

frastructures such as the canal system, the railroads and the Interstate Highway System, and informational infrastructures such as the postal service, the telegraph and the telephone, have not brought about the end of the urbanizing tendencies of modern capitalism. Quite the contrary, these innovations have tended to reinforce industrial localization and the consequent growth of cities (Teaford, 1986).

They do this by making feasible transactions that open up new, specialized sectors in the economy, where time and product differentiation are essential. This in turn heightens the complexity of B2B and B2C transactions in those sectors, and makes face-to-face contact critical. It is thus reasonable to assume that the Internet will increase product differentiation in the economy and create new forms of complex transactions, even as it simplifies others and permits further spreading out of routinized activities.

Table 4 suggests, by way of examples, this double-edged geography of the Internet age, with its tendencies toward specialization and agglomeration, on the one hand, and spreading out on the other. Three principal changes are being stimulated by the Internet: (1) increases in product *variety*; (2) increases in the fineness of the division of labor (*roundaboutness*); and (3) the automation of intermediation/coordination tasks (*disintermediation*).

Increases in roundaboutness[10] often leads to increased agglomeration because the transactions required to link the separate units or sectors are sometimes sensitive to geographical distance, as we argued above. With the advent of the Internet, greater roundaboutness and agglomeration is probable in such sectors as quality and design-driven products, customer-driven manufacturing,

parts production, innovation-based manufacturing, new consumer services, entertainment, and intellectual, research and managerial activities. Greater variety through recombination of more varied inputs which are sourced from longer distances (with more sophisticated and faster integration and inventory control) is likely in many standardized manufacturing markets, designer retail, consumer-driven manufacturing and parts, engineering and conception, new consumer services (customized take-out food, Internet-ordered home repair), and intellectual, research and managerial inputs to production. The geography of these new "mass variety" sectors will be determined by whether the input-output relations are conversations or handshakes—some will be far-flung, others clustered.

Disintermediation will come with Internet auctions, online consumer banking and finance, online stock transactions, and online medical advice as well as many information services. Here, a general tendency toward spatial disconnection of products and markets will be found, but the organizers of such far-flung systems will very likely be located in clusters.

Thus the ways that a new ICT such as the Internet interacts with production and its geography are many and varied. There appears to be no single new business model that it creates, but complex feedbacks to specialization and divisions of labor within and between sectors. In general, however, the activities which depend on handshaking will continue to be clustered. This was dominated by material production in the age of increasing physical scale and capital-intensity of 19th century manufacturing. It was later dominated by headquarters functions, and in turn by intellectual and innova-

EDWARD E. LEAMER, MICHAEL STORPER

TABLE 4
EXAMPLES OF POTENTIAL CHANGES IN THE INTERNET AGE

	Mass Produced, Standardized Products	Specialized, Customized and Innovative Products
Product differentiation or variety	Less differentiation/variety:	More differentiation/variety: *design-driven retail: *mass customization (design-your-own car;. some consumer services)
Messages: B2C	More transparent: *consumer services (auction/clearinghouse/search model); *consumer banking, finance;	More embedded in product: *mass customization
Messages: B2B	More codified: *Internet auctioning and codified intermediation services; *Some medical advice	More tacit: *quality, design-driven manufacturing; *dot.com firms *consumer mass customized services; *entertainment products
Degree of intermediate transacting:	Unchanged or lower (higher economies of scope):	Higher (lower scope economies): *quality, design-driven manufacturing· *mass customization *entertainment prodn *intellectual, research, managerial, engineering functions
Location of production in relation to intermediate markets:	More remote/dispersed: *Some skilled producer services *codified advice & service linkages, e.g. home repair (delivery still localized)	Market-centered (i.e. agglomerated): *dot.com firms; *customer-driven manufacturing; *innovation-based manufacturing; entertainment production; *intellectual, research, management
Location of production/distribution in relation to final markets:	More remote: *Design-stabilized retailing; *standardized, content-driven retailing (website sales); *electronically-delivered entertainment; consumer banking, finance	Market-centered: *design-driven retail (feel, touch needed); *Customer-driven services (e.g. ordered by internet); *complex consumer services

tion-based producers' complexes. We suggest that the latter, far from having lost their lease on life, are experiencing renewed importance, through the Internet's ability to make possible new specialized divisions of labor in many different activities, and hence to renew the need for handshakes.

THE GEOGRAPHY OF THE NEW ECONOMY

The Internet will probably reinforce the roundaboutness of production and hence of the importance of face-to-face contact, though it will also probably make possible greater linkages between different localized clusters at very long distances. All present signs are that the metropolitan areas that house these activities, which will be increasingly large and internally polycentric, will be the big "global city" winners of the Internet Age, and that these cities will be increasingly interlinked as the sites of these clusters (Scott, 2001).

The consumer service oriented sectors of these metropolitan areas will likely grow in new ways, continuing to adapt to the changing lifestyles of urban residents. One of the great growth sectors has been food and beverages (restaurants), including take-out, which is one more step in the increasingly fine division of urban labor. The Internet may transform the take-out industry into a mass-customized food preparation industry. We will be able to order custom-prepared meals from caterers, who will have a supply structure (possibly by being located within, or close to, supermarkets), an on-command cooking staff and facility, and delivery facilities.

But these cities will certainly lose some other activities. The decades-long tendency for them to shed routine but mobile production activity in the manufactures will now be extended to much routine intellectual labor in other industries, notably the service industries. An illustration of this is architectural services. Architectural firms are currently outsourcing production of shop drawings to developing countries such as China and cheaper developed countries such as Australia. Typically, a large construction project, once it has an accepted architectural design and goes through initial engineering stages, will be defined through shop drawings. These number from something like 30-50 for an average house, to tens of thousands for a concert hall or large office building. Australia's labor costs are considerably lower than those of Europe or the United States. So, many U.S. firms are contracting their shop drawings with Australian firms, and working with them over the net. But China is getting into the act, too. Highly skilled labor in China can be had for $3 per hour. At this price, firms in China working for the world market can afford to have big permanent staffs of shop drawing producers and to work at a large scale, whereas their counterparts in the developed countries bring on and lay off such labor on a project-by-project basis. The real possibility exists of a drop of 50-80% in the labor requirements of many architectural firms. Another possibility is simply outsourcing within the home country, with large-scale shop drawing "factory firms" serving downstream architectural client firms.[11] Such outsourcing is likely to generate important employment changes in the most advanced metropolitan areas. Thus, the products of routine intellectual labor may escape the neighborhood effect. Those developing economies that invest heavily in education and research are likely to become sites for the routine in-

EDWARD E. LEAMER, MICHAEL STORPER

tellectual labor that can now be moved offshore from developed areas, as in the example above.

Other regions of developed countries, which today are home to routine production (manufacturing) and services (e.g., back offices), and whose main appeal is a combination of lower labor and land costs and good access to home markets, will very likely experience mixed effects in the Internet age. On the one hand, they might become the new logistical platforms for the massive transactional web of goods exchange that the Internet will make possible. Insofar as the Internet encourages the further internationalization of manufacturing (facilitating management of operations at a distance, for instance), these routine production regions may also benefit as locations of greater foreign direct investment. The pattern of such investments is likely to be closely associated with the geography of final demand, as it has been in the recent past. There will undoubtedly be complex international sourcing of such industries, generating growth in intra-firm trade in goods and increased information flows. The Internet will make possible more global brands with—at least partially—local production. This is one way in which the neighborhood effect will probably be reproduced in the Internet age.

The Internet may make it feasible for certain types of manufacturing and routine intellectual labor to be more effectively managed at much greater distance than is now the case, both in terms of technical quality, ongoing operation of facilities "on line," and coordination of quantities between far-flung and interconnected units of production systems. The absence of adequate physical infrastructure will be an initial constraint on this (e.g., the state of the road and/or

telecommunication systems in Mexico, India or Eastern Europe). But all in all, non-metropolitan and low-cost regions of developed countries, such as the Intermountain West and the American Southeast, or southern and eastern Europe, are likely to be placed into greater competition with developing countries.

In developing countries, certain very large cities will take their places as platforms for the global transactional economy and as centers of economic reflexivity, alongside their developed world counterparts, as highly skilled technical and managerial labor there is brought into closer operational contact with their homologues in the global city-regions of developed countries. Developing countries will also probably gain in competitiveness as manufacturing sites because they will now be more directly connected to worldwide supply chains, with better technological capacities and quality monitoring than is now possible.

In other words, the Internet age is likely to be highly urban, where global city-regions are the central nodes in world economic geography. The relationship of these cities to their hinterlands will undergo significant change, as we have noted, with the latter no longer serving so much as sites of routine production but as sites of flexible logistics. In turn, far-flung physical supply chains will tie developing country cities and regions into developed country cities, through this logistical surrounding tissue. The aggregate effect, on a world scale, will confirm the existing gravity effect, but with some slow and highly uneven evolutionary tendencies to enlarge its reach, due to the combined effects of more transfer offshore of routine production and routinized intellectual labor.

ECONOMIC GEOGRAPHY OF INTERNET AGE

History Will Matter

Whatever tendencies to relocation are created by the Internet, those forces will matter only if they can overcome the inertia created by the built-in advantages of existing systems of locations. These include physical infrastructure and human network relations which are well-organized, institutionalized, and enjoy the advantages of scale. External economies attached to such patterns make them more efficient than alternatives. Insofar as scale is important to their levels of efficiency, it becomes difficult for alternative locations to break existing patterns, simply because alternatives have to start out at low scale. This can be true even where comparative statics show that an alternative pattern would be more efficient (Arthur, 1989; Krugman and Obstfeld, 1991). A key question to pose of a new transacting technology, then, is whether it can create advantages sufficient to overcome existing external economies and they way they tend to lock-in the winner locations. Overcoming the force of distance involves breaking the existing advantages.

NATIONAL AND REGIONAL COMPETITIVENESS IN THE INTERNET AGE

It has been received wisdom in recent years that infrastructure and education are the keys to competitiveness, in addition to the standard ingredients such as correct tax and property rights policies.[12] This formula may actually bear more fruit than ever before, if the Internet brings about a transformation of the geography of routine intellectual labor, as we have suggested. Then an educated work force and orderly process of doing business could enable less developed regions and countries to "leap over" the problem of distance.

For the higher order activities of invention, innovation and management, however, competitiveness may require more than education. There are cultural and relational dimensions to these activities that cannot be replaced by Internet conversations, as is indicated by the limited number of "faraway" countries who have overcome the force of gravity. These cases, whether the older Anglo-Saxon ones (New Zealand and Australia) or the more recent ones (Taiwan, Singapore and, increasingly, Ireland) strongly suggest that there is a long and difficult, though not impossible, process of creating the relational networks necessary to become part of the world core. The Internet may be a handmaiden of this process, but it will not bring it about in any automatic way; its effects will depend on a wide array of institutionalized human network-building processes. In this light, national and regional competitiveness in the age of the Internet will require "being in the loop" more than ever before, and the loop is only partially wired; it is also in the flesh.

Combining the perspective of an international economist with that of an economic geographer, we have analyzed the historical role that information and communication technologies have had on the location of economic activities, noting that they have always both reinforced agglomeration and urbanization, but also permitted dispersion of economic activity. We have given this history a theoretical interpretation that may help clarify some of the likely impacts of the Internet on the economic geography of the 21st century. In so doing, we have raised a large number of questions whose answers call for additional theory and empirical analysis by international business scholars.

EDWARD E. LEAMER, MICHAEL STORPER

NOTES

1. See Gaspar and Glaeser (1998) for a similar expression of doubt about the deagglomeration effects of the Internet, as well as supporting evidence from previous communications innovations.

2. The gravity model was first applied in the early days of econometric analysis by Beckerman (1956) who used it to study intra-European trade. The gravity model was used also by Poyhonen (1963), Tinbergen (1962) and Linneman (1966). Recent applications can be found in Leamer (1993 and 1997), Frankel and Wei (1993), and Hummels (1999a).

3. This notion has a long history in the social sciences. Michael Polanyi (1966:4) noted that "we know more than we can tell," suggesting that tacit knowledge is deeply rooted in action, commitment and context. Bateson (1973) refers to the "analogue" quality of tacit knowledge: communication between individuals that requires a kind of parallel processing of the complexities of an issue, as different dimensions of a problem are processed simultaneously. Tacit knowledge can often only be successfully communicated as metaphor (Nisbet, 1969), but metaphors are highly context-dependent (Lakoff and Johnson, 1980).

4. Relationships may have externalities (Kogut, Shan and Walker, 1993; Nonaka, 1994; Storper, 1997).

5. Tedeschi (2001) reports the situation as follows: "Earlier this year, Forrester Research, the Internet consulting firm, predicted that the universe of business-to-business e-marketplaces would shrink to a mere 180 in the next two years, from 1,000 or more today. It was a bleak forecast, but one that surprised few who had watched such sites search in vain for customers in 2000." He goes on to quote one industry participant who

states that, "Most of the business-to-business marketplaces . . . were created with the premise that if a corps of powerful buyers in a given market gathered on one site, the suppliers would come running—even if that meant the suppliers had to engage in auctions in which they underbid one another for the right to sell their wares. But attributes that go beyond price, like quality, service, the stability of the brand, warranties—all the things suppliers build around their products—marketplaces haven't allowed them to offer."

6. Knowing what the intentions of another actor are enable us to decode the practical consequences of what they are expressing to us (Husserl, 1968). Speech and action are tightly interrelated, but speech does not automatically reveal to us what another person intends to do (Searle, 1969).

7. Searle (1969) and Austin (1962), in their notion that "language is behavior" provided the basis for showing precisely the limits of strictly conversational interactions: real dialogue, they suggested, is a complex socially-creative activity. Sociologists such as Goffman (1982) and Garfinkel (1987) have shown that complex interaction involves a linguistic and visual "performance," which they liken to being on stage, playing a role, where the visual and corporeal cues are at least as important to knowing what is being "said" as the words themselves.

8. The need for physical co-presence can be met either through locational proximity of the activities, or by using another technology of transacting—transportation—to bring people together. In fact, over the last 20 years, business travel has grown just about as rapidly as electronic transactions (Hall, 1998). The relative merits of each depend on the quantities of physical co-presence re-

ECONOMIC GEOGRAPHY OF INTERNET AGE

quired, the marginal costs in time and transportation of each contact, and the spontaneity, regularity and formality of such contacts. At the moment, we know little theoretically or empirically about the relative merits of, say, occasionally bringing people together and then allowing them to relate via Internet and phone the rest of the time, as opposed to giving them the possibility of immediate, low-cost face-to-face encounters on short notice, via geographical proximity. This brings to mind the software industry which had been moving routine tasks offshore, while keeping the specialized customer-specific tasks in the hands of large focused firms in developed countries (e.g. Scient, Siebel, IBM, etc.) More recently, the Internet has allowed a hybrid model in which designers and architects are spread around the world and do the hand-shaking with customers, while being backed up by a large team of software engineers and programmers in places like Bangalore, India.

9. The notion of a progression from unstandardized, innovative products and services to standardized and mass-produced ones, and its impact on the international location of economic activity, has long been present in such notions as the product life cycle (Vernon, 1966). More recently, it has been reinterpreted in informational terms as the general progression from new, tacit and uncodified information to codified and transparent information, brought about by investments in codification. Nonaka (1994) suggests that firms create knowledge precisely through a dialogue between tacit and explicit knowledge. These notions have to be qualified, however, for the cases of industries that remain permanently artisanal or quasi-artisanal, such as the fashion- or design-based sectors.

10. The classical definition of round-aboutness comes from Allyn Young (1929). Roundaboutness refers to number of intermediate steps required to generate a final output. A modern economy fragments production into an ever-increasing number of different specialized business units and separate sectors, such that a final product emerges in a "roundabout" way through the combination of intermediate products via transactions between these units and sectors. Roundaboutness is measured, in modern terms, through input-output analysis. The more roundabout the organization of production, the more complex the upstream division of labor or input-output system at hand. Greater roundaboutness often leads to the persistence of agglomeration because the transactions required to link the separate units or sectors are sometimes sensitive to geographical distance.

11. Paulo Tombesi, University of Melbourne, Faculty of Architecture, in an interview with Michael Storper, January 2000.

12. See the paper by Oaxley and Yeung in this Symposium.

REFERENCES

Almeida, Paul & Bruce Kogut. 1997. The Exploration of Technological Diversity and the Geographic Localization of Innovation. *Small Business Economics*, 9(1): 21-31.

Arndt, Sven W., & Henryk Kierzkowski, editors. 2001. *Fragmentation: New Production Patterns in the World Economy*. Oxford, England: Oxford University Press.

Arthur, W. Brian. 1989. Competing Technologies, Increasing Returns, and Lock-in by Historical Events. *Economic Journal*, 99: 116-31.

Audretsch, David. 2000. Knowledge, Globalization, and Regions: An Economist's Perspective. In J. Dunning, editors, *Regions, Globalization and the Knowledge-Based Economy*. Oxford: Oxford University Press, 63-82.

Bateson, Gregory. 1973. *Steps Toward an Ecology of Mind*. London: Paladin Press.

Beckerman, W. 1956. Distance and the Pattern of Intra-European Trade. *Review of Economics and Statistics*, 38: 31-40.

Black, D., & Henderson, J.V. 1998. Urban Evolution in the USA. London: London School of Economics, Dept. of Economics, mimeo.

Brezis, E., & Krugman, P. 1997. Technology and the Life Cycles of Cities. *Journal of Economic Growth*. 2: 369-83.

De Vries, Jan. 1984. *European Urbanization 1500-1800*. Cambridge, MA: Harvard University Press.

Dumais, G., G. Ellison & E. Glaeser. 1997. Geographic Concentration as a Dynamic Process. Cambridge, MA: National Bureau of Economic Research, Working Paper No. 6270.

Duranton, G., & D. Puga. 2000. Diversity and Specialization in Cities: Why, Where and When Does It Matter? *Urban Studies*, 37(3): 533-55.

Eaton, J., & Z. Eckstein. 1997. Cities and Growth: Theory and Evidence from France and Japan. *Regional Science and Urban Economics*, 27: 443-74.

Feenstra, Robert, & Gordon Hansen. 1996. Globalization, Outsourcing and Wage Inequality. *American Economic Review*, 86(2): 240-45.

Feldman, M., & D. Audretsch. 1999. Innovation in Cities: Science-Based Diversity, Specialization and Localized Competition. *European Economic Review*, 43: 409-29.

Fishlow, Albert. 1985. *American Railroads and the Transformation of the Antebellum Economy*. Cambridge, MA: Harvard University Press.

Frankel, J., & S. Wei. 1993. Trade Blocks and Currency Blocks. Cambridge, MA: National Bureau of Economic Research, Working Paper No. 4335.

Garfinkel, Harold. 1987. *Studies in Ethnomethodology*. Oxford: Blackwell.

Gaspar, J. & E. Glaeser. 1998. Information Technology and the Future of Cities. *Journal of Urban Economics*, 43: 136-56.

Goffman, Erving. 1982. *Interaction Rituals: Essays on Face-to-Face Behavior*. New York: Pantheon Books.

Hall, Peter. 1998. *Cities in Civilization*. Oxford: Basil Blackwell.

Henderson, J.V. 1997. Medium Sized Cities. *Regional Science and Urban Economics*, 27: 583-612.

Hummels, David. 1999a. Have International Transportation Costs Declined? Purdue University, mimeo.

_____. 1999b. Toward a Geography of Trade Costs, Purdue University, mimeo.

_____. 1999c. Time as a Trade Barrier, Purdue University, mimeo.

Husserl, Edmund. 1968. *The Ideas of Phenomenology*. The Hague, Netherlands: Nijhoff.

Jacobs, J. 1960. *The Economy of Cities*. New York: Random House.

Jaffe, A., M. Trajtenberg & R. Henderson. 1993. Geographic Localization of Knowledge Spillovers as Evidenced by Patent Citations. *Quarterly Journal of Economics*, 63: 577-98.

Kim, S. 1995. Expansion of Markets and the Geographic Distribution of Economic Activities: The Trends in U.S. Manufacturing Structure, 1860-1987. *Quarterly Journal of Economics*, 110: 881-908.

Kogut, B., W. Shan & G. Walker. 1993. Knowledge in the Network and the Network as Knowledge: The Structuring of New Industries. In Gernot Grab-

ECONOMIC GEOGRAPHY OF INTERNET AGE

her, editor, *The Embedded Firm: On the Socioeconomics of Industrial Networks*. London: Routledge.

Klein, Benjamin, & Keith B. Leffler. 1995. The Role of Market Forces in Assuring Contractual Performance. In Oliver Williamson and Scott E. Masten, editors, *Transaction Cost Economics*. Volume 1. Theory and concepts. London: Elgar. 181-207. Originally published in 1981.

Krugman, Paul, & Maurice Obstfeld. 1991. *International Economics: Theory and Policy*. New York: Harper-Collins.

Lakoff, G., & M. Johnson. 1980. *Metaphors We Live By*. Chicago: University of Chicago Press.

Leamer, Edward. 1999. Effort, Wages and the International Division of Labor. *Journal of Political Economy*, 107(6): 1127-63.

_____. 1997. Access to Western Markets and Eastern Effort. In Salvatore Zecchini, editors, *Lessons from the Economic Transition, Central and Eastern Europe in the 1990s*. Dordrecht: Kluwer Academic Publishers, 503-26.

_____. 1993. U.S. Manufacturing and an Emerging Mexico. *The North American Journal of Economics and Finance*, 4(1): 51-89.

Linnemann, H. 1966. *An Econometric Study of International Trade Flows*. Amsterdam: North-Holland

Lorenz, Edward. 1992. Trust and the Theory of Industrial Districts. In M. Storper and A.J. Scott, editors, *Pathways to Industrialization and Regional Development*. London: Routledge.

Marshall, Alfred. 1919. *Industry and Trade*. London: Macmillan.

Nisbet, Robert 1969. *Social Change and History: Aspects of the Western Theory of Development*. Oxford: Oxford University Press.

Nelson, Phillip. 1974. Advertising as Information. *The Journal of Political Economy*, 82(4): 729-54.

Nonaka, Ikujiro. 1994. A Dynamic Theory of Organizational Knowledge Creation. *Organization Science* 5(1): 14-37.

Polanyi, Michael. 1966. *The Tacit Dimension*. London: Routledge.

Pollard, Jane, & Michael Storper. 1996. A Tale of Twelve cities: Metropolitan Employment Change in Dynamic Industries in the 1980s. *Economic Geography*, 72(1): 1-22.

Poyhonen, P. 1963. A Tentative model for the Volume of Trade Between Countries. *Weltwirschaftliches Archiv*, 90: 93-9.

Pred, Allan. 1974. *Urban Growth and the Circulation of Information, 1790-1840*. Cambridge, MA: Harvard University Press.

_____. 1966. *The Spatial Dynamics of Urban Growth in the United States, 1800-1914*. Cambridge, MA: Harvard University Press.

Rose, Andrew K. 1999. One Money, One Market: Estimating the Effect of Common Currencies on Trade. Cambridge, MA: National Bureau of Economic Research, Working Paper 7432 .

Sassen, Saskia. 1991. *Global Cities: New York, London, Tokyo*. Princeton: Princeton University Press.

Scott, A.J., editor. 2001. *Global City-Regions: Trends, Theory, Policy*. Oxford: Oxford University Press.

_____. 1993. *Technopolis: High Technology Industry and Regional Development in Southern California*, Berkeley: University of California Press.

_____. 1988. *Metropolis: From the Division of Labor to Urban Form*. Berkeley: University of California Press.

Searle, John R. 1969. *Speech Acts: An Essay in the Philosophy of Language*. New York: Cambridge University Press.

Smith, Adam. 1776. *The Wealth of Nations*. New York: Modern Library (1937 edition).

Stigler, G.J. 1951. The Division of Labor is Limited by the Extent of the Market. *Journal of Political Economy*, 69: 213-25.

Storper, Michael. 1997. *The Regional World: Territorial Development in a Global Economy*. New York: The Guilford Press.

_____, & Richard Walker. 1989. *The Capitalist Imperative: Territory, Technology and Industrial Growth*. Oxford: Basil Blackwell.

Tadeschi, Robert. 2001. *New York Times*, July 26.

Teaford, John. 1986. *The Twentieth Century American City: Problem, Promise and Reality*. Baltimore: Johns Hopkins University Press.

Tinbergen, J. 1962. *Shaping The World Economy: Suggestions for an International Economic Policy*. New York: Twentieth Century Fund.

Vernon, Raymond. 1966. International Investment and International Trade in the Product Life Cycle. *Quarterly Journal of Economics*, 80: 190-207.

Williamson, Oliver. 1985. *The Economic Institutions of Capitalism*. New York: The Free Press.

Young, Allyn. 1929. Increasing Returns and Economic Progress. *Economic Journal*, 38: 527-42.

Zucker, L.. M. Darby & M. Brewer. 1998. Intellectual Human Capital and the Birth of U.S. Biotechnology Enterprises. *American Economic Review*, 88(1): 290-306.

_____ & J. Armstrong. 1994. Intellectual Capital and the Firm: The Technology of Geographically-Localized Knowledge Spillovers. Cambridge, MA: National Bureau of Economic Research, Working Paper 9496.

Part V
Regulating Economic Spaces

[21]

THE POST-KEYNESIAN STATE AND THE SPACE ECONOMY

RON MARTIN AND PETER SUNLEY

INTRODUCTION: BRINGING THE STATE INTO ECONOMIC GEOGRAPHY

Traditionally, all of the mainstream schools of economics – classical, neo-classical, Marxist and Keynesian – have had difficulty coping with the role of the state in the capitalist economy, and have tended to treat the state in a subsidiary and reductive way (Schott, 1984; Caporaso and Levine, 1992).[1] States may help clear the way for accumulation, and may help alleviate some of the shortcomings and undesirable consequences of the accumulation process, but they do not decisively influence the *forms* of economic organization and co-ordination of economic activities. Recently, a number of new 'state-centred' approaches to political economy have appeared (*see* Caporaso and Levine, 1992), which at last recognize that states may interfere in the economy not simply to readjust it or to compensate for its undesirable externalities or to manage its social tensions, but with the explicit political-ideological intention of shaping its very functioning and organization.[2] However, these still tend to see the state and the market as regulatory institutions that are necessarily antithetical to one another, so that when one increases the other correspondingly declines. As a result, our understanding of the *forms of capitalism* that are likely to be promoted by state intervention remains limited.

Likewise, economic geographers have been remarkably reluctant to integrate the state into their theorizations and analyses of the space economy. To the extent that the state is considered at all (and in some recent key works, it is ignored almost completely), it tends to be viewed simply as a regulatory *deus ex machina* to be lowered on to the economic landscape to resolve this or that specific aspect of uneven development (typically via 'regional policies'). Consequently, we still know little about whether different spatial configurations of production, accumulation and welfare are likely to be associated with different forms of state–economy relations. The precise nature of this association will, of course, depend on various contingent factors, including the state's *capacities*, the particular economic and political *strategies* (both domestic and international) it pursues, its institutional-territorial *organization*, and the prevailing material spatio-economic tendencies and constraints (both domestic and international) that it faces. Nevertheless, it is clear that the state and the space economy are inextricably interrelated: state intervention helps to *constitute* the spatial structure of the economy, and that spatial structure in turn influences the state's economic policy actions and their outcomes.

However, ironically, this argument to bring the state into economic geography comes precisely at a time when there is a rising chorus of opinion that the capitalist state is in fact 'withering away', that its role and influence are being undermined by new socio-economic forces from both within and without. The emerging view is that as we move into a new phase of capitalist accumulation and development that is at once both increasingly more localized and fragmented on the one hand, and more market-driven and globalized on the other, so the state is no longer an appropriate or effective agent of economic regulation and co-ordination, and is itself undergoing fundamental restructuring. The implication of this view would seem to be that the state's influence on the space-economy is evaporating and that the locus of regulation and intervention is shifting both downwards to local and regional institutions and upwards to supranational bodies and organizations.

Our aim in this chapter is to evaluate and move beyond these claims about the 'withering away' and 'retreat' of the state. We begin, in the next section, by outlining the influence of the post-war Keynesian welfare state form on the economic landscape, as a backcloth for then critically examining the nature, extent and spatial implications of the changes in state intervention and regulation that are allegedly being driven by new social-economic-political pressures and forces. We suggest that the argument that the state is being undermined by globalization is overdrawn. Likewise, the extent of the alleged shift from the welfare to the 'workfare' state is argued to be much more problematic and piecemeal than has been claimed. The state may have ceded or lost its powers of intervention in some spheres of socio-economic life, but it has extended or reinforced them in others. A change in the *mode* of state intervention is occurring rather than the withdrawal or weakening of the state. We suggest that future debate surrounding the future form of state–economy relations is likely to revolve around two organizational-regulatory models: a trend towards decentralized local micro-social and economic regulation on the one hand, and the rise of the co-ordinatory or network state on the other. While these models may appear to mark alternative trajectories for the future of state–economy relations, each with different implications for the processes and patterns of regional development, we argue that they should be seen as complementary. Without central orchestration, support and co-ordination, local micro-regulation is likely to accentuate, and ultimately flounder on, problems of regional instability and imbalance. Conversely, central state regulation on its own is unlikely to create the indigenous socio-economic dynamism and flexibility which are needed at local and regional levels. Hence the changing character of state–economy relations is fundamental to the economic geography of contemporary capitalism and vice versa.

THE SPACE ECONOMY UNDER THE KEYNESIAN WELFARE STATE

From its origins in the 1930s (in the UK, the USA and Sweden), the Keynesian welfare state project emerged as the dominant post-war model of economic regulation among the advanced industrialized nations. Its twin goals were the stabilization of the inherent cyclical trend of capitalist growth, and the construction of mass societal support and

harmony through the maintenance of full employment and the provision of a public welfare system. Different capitalist states pursued different variants of this model. According to the particular balance of socio-economic forces and historical political-institutional legacies in each country, the specific policies and forms of intervention employed differed, as did the degree of incorporation of capital and labour interests into the policy-making process.[3] Notwithstanding this diversity, however, there were also sufficient common features to permit some generalization as to the implications of the post-war Keynesian welfare mode of state intervention for the patterns and processes of uneven regional development within nations during this period.

In the Keynesian welfare model, the *national* economic space is the essential geographical unit of economic organization, accumulation and regulation over which the state is sovereign actor. As Radice (1984: 116) points out, 'The national economy is privileged in Keynesian theory for the purely practical reason that the nation-state system defines geopolitical space with the necessary features convenient for the theory: a common currency, common laws, and shared institutions.' Indeed, the Keynesian welfare project required a high degree of *closure* of the national economic space in order for the state's domestic policy measures to have their desired effects. Of course, extensive flows of money, capital and goods took place across the borders of the national space-economy, but under the post-war Keynesian regime these flows were controlled by and negotiated between nation-states, through such international bodies as the IMF, World Bank and the General Agreement on Tariffs and Trade, precisely in order to guarantee the stability of each *national* economy.

At the same time, as an accumulation strategy the Keynesian welfarist mode of intervention necessarily involved a high degree of *spatial centralization* of political regulation of the domestic economy. The management of aggregate national demand – the key innovation of Keynesianism – itself required new, centralized powers of co-ordination and manipulation of basic fiscal and monetary measures, and of concertation with the representatives of large national organizational interests, especially of capital and labour (Regini, 1995). Likewise, the use of a wide range of regulatory and legislative controls, on markets, prices, corporate organization, wages, labour, unions and the like, all part of the panoply of *extensive* intervention that characterized the Keynesian state, brought regions and localities within the economy under much greater central state control and dependence. To be sure, the degree of spatial centralization of political power over the economy varied from country to country, reflecting the spatio-organizational structure of different state systems (especially whether federal or unitary), although in most nations the administration of certain aspects of taxation and welfare spending was decentralized to subnational regional or local levels, as was responsibility for other aspects of allocation. But in every instance, the regions were subordinated to the macro-economic and macro-redistributive imperatives of the centre.

The counterpart of this centralized political regulation was *spatial socio-economic integration*. One of the direct consequences of the development of the post-war welfare state has been the dramatic growth in state spending, on production, allocation and redistribution (Cochrane. and Clarke, 1993; Corny, 1990).[4] Whereas in 1937 government spending averaged only 21 per cent of GDP among the industrialized countries, by 1980 this had grown to 43 per cent (Tanzi and Schuknecht, 1996). In all of the advanced economies, public spending increased on utilities, social and physical infrastructures, and various collective goods, especially housing and nation-wide education, health care and social benefit systems. The effect of these programmes was to foster consistent standards of social welfare and social infrastructure provision across regions and localities, thereby incorporating them into an increasingly *collective* or *public* space-economy, which in some countries extended to large-scale state ownership and management of key industries, in addition to utilities and other collective goods.

Finally, the Keynesian welfare state model of economic intervention and regulation was a *spatially redistributive and stabilizing* one. A defining feature of the post-war Keynesian welfare state – though again one that varied in intensity between countries – has been its redistributive role. Progressive tax regimes combined with benefit programmes aimed at low-income groups had the effect of redistributing income from rich sections of society to the poorer sections.[5] Similarly, the scale and nature of public finance under the Keynesian state has involved substantial fiscal transfers between regions and localities. Most of these transfers are 'automatic', in the sense that they derive from the way in which national (and federal) tax and benefit systems operate, and from the public expenditure stabilizers that are automatically activated by the economic cycle (especially unemployment and related social security payments). In addition, various other forms of expenditure, for example on infrastructure, industrial support and military and other procurement, also involve inter-regional fiscal transfers. The same is true of the explicit regional and urban policies adopted by

most Western advanced countries over the post-war period, aimed at reducing spatial income and employment inequalities. The limited evidence that exists on inter-regional fiscal transfers suggests that they have played a critical role in stabilizing regional economies. For example, one of the most comprehensive studies, the McDougall Report (Commission of the European Communities, 1977), found that inter-regional fiscal transfers reduced inter-regional per capita income differentials by an average of 46 per cent in unitary states like the UK, France and Italy, and by 35 per cent in federal systems such as the USA, West Germany and Canada (*see* MacKay, 1995).[6] These regional redistributive-stabilizing effects were probably one of the most significant spatial consequences of the growth of the post-war Keynesian welfare state, the more so because they have largely resulted from the workings of public finance without any explicit government intent.

While it is difficult to isolate the precise regional impacts of post-war Keynesian policies in different individual countries, especially in the absence of any counterfactual evidence as to what would have happened in the absence of this form of state intervention and the state-led post-war economic boom, there can be little doubt that the direct and indirect impact has been substantial. Keynesianism helped to *underwrite* regional economies in a variety of ways. By stimulating and maintaining mass consumer demand, it helped to support those manufacturing regions specializing in the mass production industries of the period.[7] By intervening in or acquiring direct ownership of key industries, such as coal, steel, power, aerospace, shipbuilding, even motor vehicle manufacture, it influenced the fortunes of those local communities built around such sectors. And by promoting the substantial expansion of public-sector employment, it introduced a whole new layer to the spatial division of labour. Further, in those countries where Keynesianism became entwined with large-scale state spending on military production and research, as in the USA and UK, this spending itself produced specific geographies of investment and jobs, in some cases boosting economically backward areas but in others reinforcing pre-existing spatial inequalities (*see*, for example, Markusen *et al.*, 1991).[8] In many countries, Keynesian welfare policies became allied with more explicit attempts at reshaping the space-economy through various 'regional policies' aimed at directing the location of industry and jobs (as in the UK), or through more strategic spatial economic planning (as in Japan). More generally, post-war state policies and expenditures on housing, transportation and public utilities encouraged massive investments in particular spatial patterns of work, consumption and residence. In the USA and UK, for example, one expression of this process was a wave of suburbanization, which in turn helped to maintain aggregate mass demand for consumer durables. Once in place, these spatial configurations of industry, employment, infrastructure and population shaped the alternatives available to policy-makers on a range of fiscal, welfare and regulatory issues. On balance, as a result of these various policies and mechanisms, in most countries post-war Keynesian interventionism was a key factor behind the steady process of *regional convergence* in per capita incomes that characterized most advanced capitalist nations until the late 1970s.[9]

Over the past two decades, this model of the regulated and managed space economy has been undergoing radical change. The Keynesian welfare state has been undermined from without and from within, and is now widely viewed as obsolete, a project no longer relevant to the changing imperatives and contours of socio-economic development, and perhaps inherently flawed even during its apparent heyday (Janicke, 1990). On the domestic front, the inability of the Keynesian state to resolve supply-side economic rigidities as evidenced by inflation, unemployment and 'lame-duck' industries, its endemic 'fiscal crisis' of rising costs of social welfare and public resistance to higher taxation, and its inability to control the demands of organized labour had culminated by the end of the 1970s in a crisis both of economic management and social legitimation.[10] In addition, the very bases of national Keynesian interventionism have, it is widely argued, been completely undermined by changes in the international economic context, in particular the collapse of the international regulatory regime (Bretton Woods) that underpinned national financial stability and, more recently, the 'globalization' of economic activities and economic spaces. These internal and external challenges have thus raised the fundamental issue of whether and in what sense there is a future economic role for the state. The political response to this question, as embodied in the widespread shift to New Right neo-liberal state models throughout much of the advanced capitalist world during the 1980s and early 1990s, epitomized by Thatcherism in the UK and Reaganism in the USA (*see* King, 1987; Thompson, 1990), has been to roll back national systems of regulation, intervention and welfare support in an attempt to give national economies the 'flexibility' needed to compete in today's global markets (Regini, 1995). This reorientation of state–economy relations has in turn had equally profound implications for the space-economy.

GLOBALIZATION VERSUS THE STATE: THE END OF NATIONAL ECONOMIC SPACE?

The idea that *globalization* is undermining the economic sovereignty of the nation-state is now widespread. The term 'globalization' itself is a 'chaotic concept'. For some it conveys the notion of increasing time–space *compression* of economic activity, symbolized by the almost instantaneous hypermobility of money, capital and information around the globe without reference to national boundaries (what Ohmae, 1990, labels the 'borderless world'). For others, it denotes an increasing process of socio-economic *integration* and *convergence*, in the sense of the replacement of nationally distinct products, firms, services and markets by truly global products, markets, trade and corporations. In yet other accounts, the term 'globalization' is a process of increasing time–space *distanciation*; that is, 'stretching' and interdependency of socio-economic events and processes across the globe irrespective of geographical separation. Driven by the conjoint forces of competitive deregulation by nation-states themselves, the market-driven global strategies of firms, internationally convergent consumer tastes, and by the thrust of new revolutionary, supranational information technologies, globalization, it is claimed, renders conventional notions of 'national' economic space, 'national' money, 'national' firms and 'national' technologies increasingly irrelevant (*see* Reich, 1995a, 1995b; Ohmae, 1995a, 1995b).

According to this view, then, globalization is 'decentring' national economic space, thereby undermining one of the basic tenets of Keynesian interventionism, indeed of all the mainstream schools of political economy. National economic boundaries are now so 'porous' that attempts to manipulate domestic demand by fiscal or monetary means are almost certain to be futile, even counterproductive. In particular, states have lost their economic sovereignty, their control over their exchange rates, money supplies and currencies to 'stateless' financial institutions and global markets, and these have no overriding obligations of national interest (Allen, 1994; Banuri and Schor, 1992; Camilleri and Falk, 1992; O'Brien, 1992; Martin, 1994b).[11] Likewise, states are no longer able to exercise control over the investment, employment and location decisions of firms, an increasing proportion of which are no longer committed to their home nations, but see themselves as global, and willing to switch production, jobs and investment between countries in response not only to market opportunities and cost advantages but also in search of less regulated environments. Nor, in today's world of global corporations and global information flows, can states easily confine the external economies associated with domestic R&D activity within their national boundaries (Comor, 1994). Under these conditions, it is claimed, states have no option but to withdraw from extensive regulation and intervention in their domestic economic spaces, to remove all barriers to the free flow of money, goods and capital, and to try to reduce their levels of corporate and individual taxes in order to create the low-tax 'enterprise' spaces required by the competitive forces of the new global economy (*see* Tanzi, 1995). The New Right response to globalization has thus been a sort of transnational economic liberalism or neo-classicism, in which the need for states to cede economic power to global markets and corporations is seen not only as inescapable but optimally efficient.[12] For extreme exponents of this view, nation-states are dinosaurs waiting to die, a 'cartographic illusion'; once efficient engines of wealth creation, nation-states today have lost that role and have become reduced to inefficient engines of wealth (re)distribution (Ohmae, 1995a).

The flip-side of this loss of economic sovereignty and power by nation-states is that the individual regions within nations have been exposed to the intense competition and uncertainties associated with globalization and the global economy. In a sense the national economic space is becoming a 'glocalized' composite, and individual regions within it increasingly linked into and integrated with diffuse webs of overseas markets, suppliers, technology and competitors, and less and less with 'domestic' ones. This makes individual regions and localities more prone to idiosyncratic demand and technology shocks, as well as to externally originating decisions with respect to investment and disinvestment, employment and production. Some states have actively encouraged this increased exposure of their regions as part of a 'shock therapy' to force their economies to restructure away from rigid, old, inefficient industries and activities towards new and more flexible ones capable of competing in world markets. The running down of regional aid, the deregulation of industry and labour markets, legislative attacks on unions, and the privatization and marketization of public-sector activities – these and other strategies pursued by free-market conservative governments since the early 1980s, especially in the UK and USA, but

also elsewhere – have all been directed at increasing the flexibility of, and reducing central support for, the space-economy. While this state-sponsored retreat from the regions has certainly heightened the 'shock' felt by localities, whether and in what sense it has been 'therapeutic' remains debatable. If the Keynesian state was concerned to integrate its constituent regional and local economies and to cushion them from economic instability, the approach of its successor, the neo-liberal conservative state, has been to *dismantle* and *fragment* those systems of central support in deference to the restructuring forces of global competition, destabilizing its regions in the process.

Both because of this withdrawal of central state intervention in the space-economy, and because of the local impact of globalization itself, individual regions and localities have increasingly sought to assume responsibility over their own economic destinies. *Indigenous growth* and *local economic governance* are the new buzz-words in regional economics and politics. As economic geographers and other regional analysts endeavour to unravel the structural, technological and socio-economic foundations of 'indigenously generated' localized growth and adjustment within a globalized economy, regional states and authorities themselves are busy trying to pursue their own local enterprise strategies, whether these be based on local deregulation, local technology parks, or local small-business initiatives. In some accounts, the contemporary counterpart of the demise of the national economy is the rebirth of 'regional economies'; the emergence of a global network of specialized local industrial districts, varying in the type of activity involved but sharing the key trait of being based on localized networks of small, flexible firms (Amin and Thrift, 1994). Some commentators, again led by Ohmae (1995a), take this vision of current trends one step further, and see new *region-states* as the emerging successors to nation-states.[14] According to Ohmae, these 'region-states', defined by economic activity rather than by political borders, are emerging as the growth regions of the global economy precisely because they embrace the very characteristics demanded by the logic of that economy: deregulated, flexible and multinationalized accumulation, and little of the centralized and collectivized interventionist baggage of the nation-state.

However, compelling though these arguments and accounts may seem, their validity and generality are questionable. The 'death of the nation-state' by globalization has been much exaggerated.[15] While globalization has indeed weakened the economic sovereignty and power of nation-states in certain spheres – most notably in the realm of monetary policy – individual states still exercise substantial independence and authority in the regulation and management of their domestic political economies (Porter, 1990; Pooley, 1991; Hirst and Thompson, 1992, 1996). States still possess a large measure of autonomy over their fiscal policy, still control large sections of industry and services, still set much of the regulatory framework governing economic markets, and still exert considerable influence through their public spending programmes. Indeed, some policy areas, such as public spending, have been characterized by intensified central state regulation. Likewise, although regions and localities within nations are now subject to considerable competitive pressures and internal disarticulation from global economic forces, their economic fortunes and prosperity still depend in fundamental ways on the economic policies and expenditures of their central states. In fact, despite the political rhetoric of monetarist supply-side governments of the 1980s that 'Keynesianism was dead', political reality has been somewhat different: Reagan's expansion of the US budget deficit, especially through increased military spending, and Thatcher's tax cuts in the UK both had the classic Keynesian effect of boosting national consumer demand. And in these and other industrialized economies, programmes of privatization and restructuring of public sector activities and post-Cold War demilitarization are having highly significant impacts on regions and localities.

There is little doubt that the globalization process is changing the nature and meaning of 'national economies', in the sense that this concept has traditionally been understood, but it would be wrong to assume that state intervention is rendered obselete in a world of post-Keynesian globalized capitalism (*see also* Drache and Gertler, 1991). Rather, the challenge is one of rethinking conventional notions of 'economic nationalism', 'national' economic policy and the division between the 'private' and the 'public' in ways that are more appropriate to new economic realities. The New Right political response, of seeking to cut the size of the state, to squeeze and privatize the public sector, and to reduce taxes and workers' real wages, all so as to compete with low-wage, low-welfare newly industrializing countries, is one that threatens to lock the advanced nations into a vicious circle of progressively declining living standards (*see* Wood, 1994). Globalization does not justify less state intervention, but a *redirection* of that intervention. There is a growing connection between the kind of investment and spending that the state sector undertakes and the capacity of a country to attract worldwide capital and technology. Although states may now exert less influence

over economic development through conventional fiscal and monetary policies, their role in moulding the socio-institutional embeddedness of that development is, if anything, becoming more important (Amin and Tomaney, 1995a). For it is the specific nature of the socio-institutional framework that in large part accounts for the differences between 'national capitalisms' and for the differences in national growth performance within them (*see* Hutton, 1995a; Berger and Dore, 1996). In each case the economic, political and social interlock in different, distinctive ways.

Thus even if money and technology have 'gone global', different nation-states continue to differ in their relationship between finance and industry (Cox, 1986), and the generation of new technologies (Archibugi and Michie, 1995). Perhaps above all, as the advanced economies become global, each state's most important competitive asset becomes the skills and cumulative learning of its workforce and the quality and efficiency of its public infrastructure. Unlike capital, technology, raw materials and information, all of which have become much more mobile, even 'borderless', the workforce and public infrastructure are unique to a state, and are vital to the prosperity of its constituent regions.[16] The extent to which different states invest in and upgrade their workforces and their social and infrastructural capital is likely to be a decisive factor in determining the ability of their regions and localities to compete in the global market-place. The state is in a pivotal position to promote these investments in education, training, R&D and in all the infrastructure that moves people and goods and facilitates communication. These are the investments that distinguish one state's economy from another: they are the relatively non-mobile factors in global competition, and key determinants of the local outcomes of that competition.

FROM THE WELFARE STATE TO GEOGRAPHIES OF WORKFARE?

The New Right attempt to refashion the state has been designed not only as a response to perceived globalization but also as an attack on the whole ideology of welfarism and state support. This has involved a displacement in the focal point of politics from the maximization of general welfare to the promotion of enterprise, innovation and profitability in both public and private sectors (Cerny,

1993). Jessop (1994a, 1994b) argues that the transition from Fordism to post-Fordism, the rise of new technologies and the trend to continental regionalism have all acted to hasten the succession of the Keynesian welfare state (KWS) by the Schumpeterian workfare state (SWS). In prioritizing innovation and competitiveness, the SWS furthers the 'hollowing out' of the nation-state as powers and responsibilities are transferred to smaller regional and local governments whose intervention is closer to the sites of competitiveness (*see also* Goodwin *et al.*, 1993). In this view, social policy is becoming subordinate to the needs of labour market flexibility (Geddes, 1994) so that welfare systems are being replaced by policies designed to increase the skills and flexibility of those already in work (hence the label 'workfare').[17] Others see this shift towards a competition focus not as an inevitable institutional logic caused by the crisis of Fordism but as a reflection of deliberate policy choices by neo-liberal governments (Peck and Tickell, 1994b; Amin and Tomaney, 1995b).

There is already some consensus on the geographical implications of this reorientation from welfare to competition and enterprise. The disengagement of the state from its former spatially distributive role has stark implications for less favoured regions as it removes an important mainstay of their socio-economic welfare and weakens their protective framework. Thus according to Amin and Tomaney,

> Most obviously, the logic implies a reduction in the level of direct state support for industry in the less favoured regions, as the commitment to regional incentives is reduced, as the state disengages from and restructures industry under its ownership, as it deregulates public utilities and services, and as it ceases to direct public procurement contracts to firms located in the less favoured regions. (1995a: 39)

Furthermore, any retrenchment of welfare expenditure would particularly threaten the less favoured regions and poorer localities in which low-income and marginalized groups are disproportionately concentrated. The geographical danger is that the state will increasingly, perhaps inadvertently, support the most competitive regions and cities and that the less prosperous will be left to the market.[18] It could be argued that this general move to a type of neo-liberal workfare or competition state will undo the integrative and stabilizing effects associated with the Keynesian welfare state, with the result that geographical inequalities will widen and social cohesion will fracture.

This picture of radical discontinuity in state–economy linkages, however, should be viewed with circumspection. It relies on a selective emphasis and tends to impart too much coherence and logical neatness to contemporary experiments with new policies.[19] Studies which have examined the restructuring of welfare states have found that they have proved surprisingly resilient. Both in the UK and the USA, the state has experienced difficulty in retrenching the welfare system (Pierson, 1994). In the UK, the expenditure effects of cuts in programmes such as housing, the freezing of real benefit levels and wider means-testing have been cancelled out by the increased costs of persistent mass unemployment and population ageing, so that total welfare expenditure has been broadly static. While the Keynesian goal of full employment has undoubtedly been abandoned by most states, the Beveridgian aim of maintaining a minimum standard of living has proved more robust (Mishra, 1990). Welfare state retrenchment is difficult as welfare institutions have become fixed over a *longue durée* and exert considerable inertial power.[20] Moreover, welfare state reform has been restrained by its political sensitivity and unpopularity: governments are reluctant to alienate their social and regional support, and maintaining some socio-spatial coherence is still vital.[21] States' responses to international economic conditions will therefore be conditioned by political struggles and their need to sustain the social bases of their autonomy (Gourevitch, 1986), and in this respect universal benefits, in particular, are harder to cut back.

The argument that welfare is being subordinated to labour market flexibility is also problematic in that it portrays welfare states and labour market flexibility as incompatible alternatives. This seems to concede too much to the view that there is an inevitable trade-off between equity and efficiency. In fact, there are a large number of micro-economic flexibilities and efficiencies which are secured by the operation of welfare systems (Barr, 1993; Esping-Andersen, 1990).[22] Indeed, this may be one of the reasons why there is no easy cross-national correlation between the size of states' expenditures on welfare and their relative economic performances (Morley and Schmid, 1993; Pfaller *et al.*, 1991). It also needs to be emphasized that welfare reform has operated from very different starting-points and, despite its widespread incidence, these differences have not by any means been erased.[23] A great deal of confusion and uncertainty over the future direction of welfare remains, and it would be premature to assume that all states are converging on a post-welfare enterprise model (Cochrane and Clarke, 1993).[24] In general, despite the rhetoric of the 1980s, public spending has continued to increase, to an average of 47 per cent of GDP in the industrial countries by 1994 (Tanzi and Schuknecht, 1996). Furthermore, if there is a new convergence in welfare provision, then it seems more likely to centre on the use of '*quasi-markets*' through which social services are centrally financed but provided by contractors, quangos and private agencies (Taylor Gooby and Lawson, 1993). The creation of more decentralized and pluralistic systems may be in tune with demands for greater consumer responsiveness but its implications for resources are unclear. The geography of service provision will play an increasingly important role in determining the outcomes for public spending, for where suppliers are able to exploit a local monopoly the costs of provision may increase.[25] It is also likely to exacerbate differences between localities in the resourcing and provision of services (*see*, for example, Mohan, 1995).

It is important, then, not to exaggerate the logical and practical coherence of the 'post-welfare' agenda as it is not easy to explain precisely what the new focus of state policy is to be. According to Jessop (1994b), for example, 'The growing importance of structural competitiveness is the mechanism which leads me to believe that we will witness the continuing consolidation of the "hollowed-out" Schumpeterian workfare state in successful capitalist economies' (p. 36). The key problem here is that *competitiveness* is a vague and in some ways obfuscatory idea. As Krugman (1994a, 1994b) argues, the notion is often misleading as it implies that international trade is a zero-sum game in which one country's gain is always at the expense of another's loss.[26] Competitiveness cannot simply be reduced to balances of trade and, to be meaningful, it must refer to comparative levels of productivity and living standards (Porter, 1990). But it is then difficult to argue that contemporary state policies are coalescing on raising productivity and living standards. Raising productivity undoubtedly depends on significant investment in public infrastructure and on a well-motivated and trained workforce, both of which are contradicted by neo-liberalism's determination to reduce public expenditure and lower real wages. It is clear that there are different ways in which states can try to construct competitiveness, including both social democratic and neoliberal policy models (Garrett and Lange, 1991). Moreover, conflictual processes of social learning within institutions are as important as economic pressures in determining which model is adopted.

This means that the geographical implications of the current changes in state–economy relations are graduated and complex. On the one hand, the departure from Keynesianism has encouraged the widening of socio-spatial inequalities. The reluc-

tance to cushion the impact of economic adversity has exposed regions to international trade and a greater risk of instability and decline. Simultaneously, the neo-liberal shift from a fiscal reliance on direct taxation to one on indirect taxation has acted in a spatially regressive way by falling hardest on areas containing relatively poor tax-payers. Conversely, policies of tax cuts for higher earners have no doubt acted to boost leading regions. On the other hand, many of the geographical consequences of recent shifts in state policy are indeterminate and difficult to predict in an abstract and generalized fashion. For instance, allowing regions to specialize to a greater extent on the basis of international trade also increases the risks faced by hitherto successful regions.[27] More generally, if some policies do come to focus more effectively on supply-side improvements then there is no inherent reason why these policies should be of less benefit to less favoured regions and cities. Finally, in the context of persistent mass unemployment, automatic fiscal transfers to poorer regions have remained substantial. MacKay (1995), for example, shows that in the UK since the early 1980s these transfers have come to account for a greater proportion of the total income of the peripheral regions. While some foresee the decline of fiscal transfers to poorer regions under post-Fordism (Dunford and Perrons, 1994), there is little evidence of this as yet.

We should also treat the claim that economic powers are being devolved to regional-level governments with caution. Experiences in Germany, Italy and Spain demonstrate that while regional governments may well foster dynamic economic efficiency, their autonomy is constrained by the operation of nation-wide redistributional mechanisms (Newlands, 1995; Zimmerman, 1990). The contradiction typically faced by regional institutions is that while their economies are increasingly vulnerable to depressions and shocks, their ability to respond is typically strongly constrained by limited resources and national regulatory and legislative structures (Anderson, 1992). The German *Länder*, for example, are often seen as models of the new economic regionalism, but have been subject to greater central controls and fiscal redistribution over the post-war period (Newlands, 1995; Amin and Tomaney, 1995a). The technological and industrial policies developed by the *Länder* have not been an *alternative* to national policies, for the latter have also been expanded (Esser, 1989). Indeed, it appears that regional-level initiatives work best when there is an effective integration of regional and national policies (Gertler, 1992; Weiss, 1989). Moreover, there is a great deal of difference between genuine local economic capability and the local implementation of central state directives, as the case of the British 'quango boom' (there are now over 5000 of these bodies) testifies (Holtham and Kay, 1994).[28]

While there are aspects of state regulation which have undoubtedly changed radically, others have been more resistant to reform, producing a complex assemblage of new and old institutions, pressures and experiments. Rather than all industrialized states shifting to a single post-Keynesian model of intervention in the space economy, different types of policy innovation and experimentation are being pursued in different states. The austerity and marketization associated with neo-liberal policies over the past two decades, and the trend to greater spatial inequities brought in their wake, are undoubtedly common experiences, but it is misleading to represent this primarily as a transition to a locally based, supply-side and micro-interventionist state. For example, privatization programmes have been witnessed in a wide range of countries,[29] but, in many cases, this has not meant a corresponding decline in state intervention but rather a shift to *intensive* regulation in response to the need to protect public interests in the privatized provision of public goods (Thompson, 1990; Helm, 1994; Parker, 1993). Within different national frameworks and models of intervention, there appears to be a growing use of a *contract style* of administration and service supply, separating finance from service provision, and this may well fracture the national uniformity associated with Keynesian welfarism and exacerbate spatial differences in regulatory environments and service provision.[30]

CONCLUSION: MICRO-REGULATION AND THE NETWORK STATE

It is clear, then, that nation-states continue to exert a profound influence on their space-economies and it is also clear that the question of the scale of state intervention and regulation will be fundamental to its future economic role. There appear to be two emerging models of the future of state–economy relations. The first refers to local social regulation modelled on the Italian experience. Both Regini (1995) and Locke (1995), for example, argue that the economic success of Italy during the 1980s was due not to a central state strategy but rather to the 'vibrancy' of local associational networks which

encourage entrepreneurialism. The most successful regions are distinguished by their polycentric networks of egalitarian associationalism, interest group organization, and local institutions which facilitate the pooling of information and scarce resources, mediate conflict, and generate trust among local economic actors. Hence micro social regulation provides collective goods such as co-operative industrial relations, the co-ordination of wage dynamics, training and the development of human resources. In this view, what has happened in Italy is indicative of more general trends sweeping across all advanced industrial nations and, in an era when governments are losing macro-economic control and divesting themselves of micro-economic control through deregulation, established national and *étatist* models are of little help (*see also* Warren, 1994; Schmidt, 1996). This Italian model is proving to have a very wide impact, not only through the industrial districts literature but also by stimulating debate on the wider regional possibilities of associational networks in civil society (Amin and Thrift, 1995). What these accounts imply, however, is a much more pronounced and dramatic experience of regional disparities as those regions and cities with effective local regulatory orders outperform those which lack 'vibrant' micro-regulatory structures.

The second emerging model of state–economy relations is predicated on the view that both the wealthiest states and the fastest-growing economies (particularly the East Asian newly industrialized countries) owe their success, in part, to what might be called an *institutionalized and co-ordinatory* type of state intervention. States such as Germany, Sweden, Austria, Switzerland and Japan have employed co-ordinatory policies at both macro and micro scales (Soskice, 1990; Matzner and Streeck, 1991). At the macro scale the development of co-operative institutions between industry, finance and labour has restrained real wage growth and provided relatively cheap capital, while at the micro scale, competition has been balanced by long-term co-ordination centred on high-skilled labour forces and co-operative, long-term relationships with suppliers. As Holtham and Kay (1994) have argued, 'In these successful capitalist economies it is not simply that the rough edges of capitalism are disguised by a veneer of collective activity. It is that such *collective action is essential to making capitalism work*' (p. 2; emphasis added). At the same time, explanations of the phenomenal growth of the East Asian 'late industrializers' have stressed the *co-ordinated model* of state involvement in the economy. In South Korea and Taiwan, for example, the dense networks which exist between industrial groups and the state have facilitated the rapid development of new sectors and the promotion of export-led industrialization (Amsden, 1989; Wade, 1990). Similarly, Japan has also been described as a 'network state' (Sheridan, 1993; Wilks and Wright, 1991). In the view of many, this type of co-ordination is not so much bureaucratic as based on formal and informal policy networks which connect public, intermediate and private actors. In this model, state capacity is reconceived as the ability to sustain a dense structure of *policy networks*,[31] which allows the state to work through and in co-operation with other organizations (Hall and Ikenberry, 1989; Weiss and Hobson, 1995). Moreover, such institutionalized networks are seen as increasingly important for economic growth (*see*, for example, Lazonick, 1991).

However, the view which we have tried to develop in this chapter is that, contrary to the arguments of some, these two models of public regulation and co-ordination are not necessarily alternatives. Rather, the need for *state-organized* networks to provide the facilitatory infrastructure, co-ordination and redistributed resources in which more local and regional types of social actor co-operation and micro-regulation can flourish is likely to continue.[32] Without central orchestration, micro-regulation is likely to flounder on the problems caused by intensified regional instability and imbalance. Conversely, central state regulation, on its own, is likely to prove incapable of creating and making use of the wide range of external economies and increasing returns which operate at a regional level. One result of the Anglo-American dominance of economic geography, however, is that we know very little about the ways in which macro- and micro-co-ordinatory networks can be made to work symbiotically and to reinforce one another. It appears that this fusion can not only benefit leading industrial regions but can also be strikingly effective in managing the decline of older industries (for example, *see* Young, 1991) and may well increase the adaptability of declining industrial regions.[33] Yet we also know little about the geography of national and local effective policy networks: the extent to which they can be cultivated in states currently dominated by neo-liberalism and whether and how these co-ordinatory networks can be created in differing legal and administrative cultures and within different regional social and political traditions. In short, the whole question of the changing character of state–economy relations remains central to understanding the economic geography of the post-Keynesian era.

NOTES

1. Even regulationist political economy, which stresses the role of the state as a key component and mechanism of the capitalist 'mode of regulation' (Boyer, 1990) fails to escape this limitation (Goodwin *et al.*, 1993; Jessop, 1990b). Most regulationists simply adopt an already available account of the state to fill out their model of political economy. And despite their efforts to the contrary, their treatment of state intervention is close to being a functionalist one, in which state regulation serves the needs of the accumulation regime; hence the regulationists' idea of the 'Fordist' state, and now its successor the 'post-Fordist' state.

2. For example, 'transformational state' theories of economic governance (*see*, for example, Lindberg and Campbell, 1991), Jessop's (1990b) strategic-relational theories of the state and 'state projects', and the new, 'historical institutionalist', comparative political economy (Steinmo *et al.*, 1992). Though these various new approaches differ in their specifics, all share what has been termed a 'relative autonomy' or autonomous view of the state.

3. There was, in fact, a (largely unresearched) international geography to post-war Keynesian interventionism, involving such versions as 'military' Keynesianism in the USA, consensual Keynesianism or 'Butskellism' in the UK, the 'social market' Keynesianism of Germany, the 'regulatory' Keynesianism of France, 'pragmatic' Keynesianism in Italy, and 'social democratic' Keynesianism in Sweden (*see* Hall, 1989, for detailed accounts of these different national forms).

4. While there is nothing inherent in the theory or practice of Keynesian macro-economic management that requires an accompanying welfare state, the two are certainly complementary. By providing a social wage, the welfare system helps to support aggregate demand and mass consumption, and through its education, housing and health components it helps in the reproduction of the labour force needed for full-employment production. By the same token, the maintenance of full employment through demand management ensures the tax take required to fund the welfare state. An extensive welfare state, then, is predicated on the economy being run at or near full employment.

5. To the extent that some social benefits have been universal, and indirect taxes have been regressive, critics of the post-war welfare state argue that contrary to what has been intended, the poor have not always been the main beneficiaries.

6. As Krugman (1993) argues, such 'non-market' fiscal stabilizers are as important as market forces in limiting the impact of regional economic downturns and crises (*see also* Sala-i-Martin and Sachs, 1991). Sala-i-Martin and Sachs (1991) have shown for the USA that declining federal tax payments and rising

federal transfers provide roughly a one-third offset to a regional-specific decline in economic activity.

7. In this sense the Keynesian state played a key role in the growth of Fordist industry in the 1950s to 1970s.

8. Keynes himself only made one reference to the applicability of his policies to the regional development question. Interestingly, this was when he advocated directing the British Government's rearmament programme in the late 1930s to the depressed regions of the north of the country, where surplus labour and capital could be utilized without encouraging national inflationary pressures.

9. For empirical evidence on the convergence of regional incomes in the USA, European countries and Japan, *see* Barro and Sala-i-Martin (1995). Although those authors tend to explain this convergence in terms of market forces and technological diffusion, they admit that government policy is likely to have been important. Our argument here is that Keynesianism was indeed of fundamental importance.

10. The Keynesian welfare state was based on two assumptions concerning both the *level* and the *composition* of employment: that the economy was kept at or near full employment, and that the typical worker was a male who normally earned sufficient to support himself, his wife and a family of at least one child. In the labour market of the 1980s and 1990s, however, not only is high unemployment endemic but employment has become much less secure, less full-time and increasingly feminized. The implications of these changes for the social security system are profound.

11. It is significant in this context that Keynes himself feared the 'globalization' of finance and the rise of 'stateless monies'. Of all aspects of economic and social life, he argued, 'let finance be primarily national' (Keynes, 1933: 758). In his view, 'economic internationalisation embracing the free movement of capital and loanable funds as well as of traded goods may condemn [a] country to a much lower level of material prosperity than could be attained under a different system' (ibid.: 762–3). The contemporary loss of financial autonomy by the leading industrialized states is, in large part, of their own doing, the result of the wave of competitive financial deregulation that they embarked upon during the 1980s.

12. Indeed, some see the dichotomy between nation-states and global capitalism as ultimately unsustainable and as necessitating a shift from a system of national regulation to one of transnational regulation (McMichael and Myhre, 1991).

13. Ohmae has fallen victim to a common complaint among management 'gurus' and 'airport economists', namely gross over-simplification and exaggeration. Ohmae's work is certainly long on sensationalism, but it is much shorter on detailed analytical exegesis and empirical evidence.

14. He cites numerous instances, for example: northern Italy, Baden-Württemburg, Hong Kong/southern China, the Silicon Valley–Bay Area in California, Fukuoka–Kitakyushu in the north of the Japanese island of Kyushu, the growth triangle of Singapore–Johore–Riau, Triangle Park in North Carolina, and the Rhône-Alps region of France.

15. It can equally be argued that the challenge to the nation-state and state intervention has more to do with the expanding trend towards *regional integration* than with globalization; that is, the formation of continental regional free-trade blocs and customs unions (*see* Anderson and Blackhurst, 1993; De Melo and Panagariya, 1993; Cable and Henderson, 1994; Hirst and Thompson, 1996). Although the European Union represents the most advanced example of this new regionalism, there are more than 30 of these regional initiatives worldwide, one of the most recent being NAFTA, the North American Free Trade Agreement. As the case of the EU illustrates only too clearly, regional integration agreements raise critical questions concerning the scope for and nature of economic intervention and regulation by the nation-states that are signatories to such agreements, and concerning the need to ensure that individual regions and localities adjust to the economic shocks and changes that integration agreements bring. Although some believe that integration leads to a reduction of spatial economic disparities within the regional bloc, most observers expect such disparities to be exacerbated as a result of locational shifts in investment and jobs towards the core growth regions. Thus it is often argued that a necessary corollary of the advanced stages of economic and monetary union is the establishment of a corresponding integrated system of spatial fiscal transfers and stabilizers. It is arguably within these blocs that Ohmae's 'region-state' economies assume their real significance.

16. Critics of the globalization thesis have repeatedly emphasized that labour was more internationally mobile during the period of mass migration in the nineteenth century.

17. Jessop argues that there are neo-liberal, neo-statist and neo-corporatist forms of the SWS but insists that we will witness its continuing consolidation in successful capitalist economies. The reason is that this new state form could help to resolve the crises of Fordism, so that 'Thus it seems that the hollowed out Schumpeterian workfare state could prove structurally congruent and functionally adequate to post-Fordist accumulation regimes' (1994b: 27).

18. This conforms with the vision of writers on globalization who also argue that the weakened governance capability of the nation-state leads to new inequalities and dualisms between those who can compete in the global economy and those who cannot (Sassen, 1994).

19. This exaggerated coherence may reflect the influence of regulation theory on these ideas; the regulationist notion of a coherent mode of regulation may be misleading (Painter and Goodwin, 1995).

20. For example, contrasting labour market policies have been shaped and constrained by patterns of institutions fixed early in the twentieth century (King, 1995).
21. While maintaining social consensus may have been given lower priority in recent years, it remains one of the key imperatives of modern states. In this context, the dramatic ageing of the population structures of many OECD poses a twin problem for welfare reform, for while it will escalate costs, it will also endow moves to privatize pension and health systems with enormous political sensitivity.
22. As Barr (1993) argues, 'The welfare state is much more than a saftey net; it is justified not simply by a redistributive aim one may (or may not) have, but because it does things which private markets for technical reasons either would not do at all, or would do inefficiently. We need a welfare state of some sort for efficiency reasons, and would do so even if all distributional problems had been solved' (p. 433).
23. For example, the liberal, social assistance model characteristic of the UK and the USA, the conservative, corporatist regimes of Germany, France and Austria and the social-democratic regimes of the Scandinavian countries (Esping-Andersen, 1990).
24. For instance, the dramatic reduction of state spending under New Zealand's 'Rogernomics' from 46 per cent of GDP in 1988 to 36 per cent in 1994 is clearly different from Australia's combination of stringency with increases in benefit levels for low-income families. Thus those like Bennett (1990) who talk of a 'post-welfare' agenda are thus guilty of *pars pro toto.*
25. This argument is supported by the example of the welfare system in The Netherlands, which is highly decentralized and pluralistic but relatively costly (Glennerster and Le Grand, 1994).
26. Krugman (1994a) also argues that, because of the domestic importance of service industries, raising productivity in these 'non-traded' sectors would have an enormous effect on standards of living, but this has little to do with international competitiveness.

27. For example, Buck *et al's* (1992) comparison of London and New York argues, 'A significant decline in activity in the international financial system, or a competitive loss to other cities . . . would now pose as severe a threat for the two cities as the decline of manufacturing and goods circulation did in the past' (p. 103).
28. While some regulation theorists have distinguished between local government and local governance, regulation theory has failed to explain why these different local political trends prevail in different states (Hay, 1995).
29. These include not only the UK and USA, but also France, Japan, Italy, New Zealand and Sweden.
30. For example, Clark (1992) describes a move in the USA towards an administrative state based on regulatory agencies and argues that this type of regulation is inherently geographical as it is enforced by, and enmeshed in, a legal environment shaped by local precedents and cultural practice. His argument clearly reflects the legalism of the US regulatory environment, and we need to know much more about the different national forms, the local consequences and the incomplete geographies of regulation of this administrative state model.
31. Policy networks in this view represent the linking processes within policy communities and are constituted by complexes of organizations and actors connected to each other by resource dependencies.
32. Regini (1995) himself notes that 'To what extent micro-regulation processes will become an effective alternative to macro-regulation in the production of other public goods is, as I have said, difficult to predict. The boundaries between micro and macro, between their respective ranges of action, are still uncertain. But if macro-political regulation continues to be scaled down in favour of micro-social regulation, this will not necessarily mean the increasing irrelevance of institutions and the greater weight of the deregulated market in determining economic outcomes' (p. 145).
33. A key issue for research is how, and under what conditions, dense policy networks become obstacles to, rather than facilitators of, structural economic change (Katzenstein, 1987; *see also* Grabher, 1993c).

BIBLIOGRAPHY

Allen, R. 1994: *Financial crises and recessions in the global economy.* London: Edward Elgar.

Amin, A. and Thrift, N. (eds) 1994: *Globalization, institutions and regional development in Europe.* Oxford: Oxford University Press.

Amin, A. and Thrift, N. 1995: Institutional issues for the European regions: from markets and plans to socioeconomics and powers of association. *Economy and Society* 24, 41–66.

Amin, A. and Tomaney, J. 1995a: The challenge of cohesion. In Amin, A. and Tomaney, J. (eds), *Behind the myth of European union: prospects for cohesion.* London: Routledge, 10–50.

Amin, A. and Tomaney, J. 1995b: The regional dilemma in a neo-liberal Europe. *European Urban and Regional Studies* 2, 171–88.

Amsden, A. 1989: *Asia's next Giant: South Korea and late industrialization.* Oxford: Oxford University Press.

Anderson, J. 1992: *The territorial imperative: pluralism, corporatism and economic crisis.* Cambridge: Cambridge University Press.

Anderson, K. and Blackhurst, R. (eds) 1993: *Regional integration and the global trading system.* London: Harvester Wheatsheaf.

Archibugi, D. and Michie, J. 1995: The globalisation of technology: a new taxonomy. *Cambridge Journal of Economics* 19, 121–40.

Banuri, T. and Schor, J. 1992: *Financial openness and national autonomy.* Oxford: Clarendon Press.

Barr, N. 1993: *The economics of the welfare state,* 2nd edition. Oxford: Oxford University Press.

Barro, R. and Sala-i-Martin, X. 1995: *Economic growth.* New York: McGraw-Hill.

Bennett, R. 1990: *Decentralization, local governments and markets: towards a post-welfare agenda.* Oxford: Clarendon.

Berger, S. and Dore, R. (eds) 1996: *National diversity and global capitalism.* Ithaca, NY: Cornell University

Boyer, R. 1990: *The regulation school: a critical introduction.* New York: Columbia University Press.

Buck, P., Drennan, M. and Newton, K. 1992: Dynamics of the metropolitan economy. In Fainstein, S., Gordon, I. and Harloe, M. (eds), *Divided cities: New York and London in the contemporary world.* Oxford: Blackwell, 68–103.
Paris: Syros.

Cable, V. and Henderson, D. (eds) 1994: *Trade blocs: the future of regional integration.* London: Royal Institute of International Affairs.

Camilleri, J. and Falk, J. 1992: *The end of sovereignty? The politics of a shrinking and fragmenting world.* Aldershot: Edward Elgar.

Caporaso, J. and Levine, D. 1992. *Theories of political economy.* Cambridge: Cambridge University Press.

Cerny, P.G. 1990: *The changing architecture of politics: structure, agency and the future of the state.* London: Sage.

Clark, G.L. 1992: 'Real' regulation: the administrative state. *Environment and Planning A* 24, 615–27.

Cochrane, A. and Clarke, J. 1993: *Comparing welfare states: Britain in international context.* London: Sage.

Commission of the European Communities 1977: *McDougall Report.* Report of the Study Group on the Role of Public Finance in European Integration, vols 1 and 2. Brussels: CEC.

Comor, E. (ed.) 1994: *The global political economy of communication.* New York: St Martin's Press.

Cox, A. 1986: *The state, finance and industry: a comparative analysis of post-war trends in six advanced industrial economies.* Brighton: Harvester.

de Melo, J. and Panagariya, A. (eds) 1993: *New dimensions in regional integration.* Cambridge: Cambridge University Press.

Drache, D. and Gertler, M. (eds) 1991: *The new era of global competition: state policy and market power.* Montreal: McGill-Queen's University Press.

Dunford, M. and Perrons, D. 1994: Regional inequality, regimes of accumulation and economic development in contemporary Europe. *Transactions of the Institute of British Geographers* NS 20, 163–82.

Esping-Andersen, G. 1990: *The three worlds of welfare capitalism.* Cambridge: Polity.

Esser, J. 1989: Does industrial policy matter? Land governments in research and technology policy in federal Germany. In Crouch, C. and Marquand, D. (eds), *The new centralism: Britain out of step in europe.* Oxford: Blackwell, 94–108.

Garrett, G. and Lange, P. 1991: Political responses to interdependence: what's 'left' for the left? *International Organization* 45, 539–64.

Geddes, M. 1994: Public services and local economic regeneration in a post-Fordist economy. In Burrows, R. and Loader, B. (eds), *Towards a post-Fordist wel-*

fare state? London: Routledge.

Gertler, M. 1992: Flexibility revisited: districts, nation-states, and the forces of production. *Transactions of the Institute of British Geographers* NS 17, 259–78.

Glennerster, H. and Le Grand, J. 1994: *The development of quasi-markets in welfare provision.* London: Suntory Toyota Centre Discussion Paper, London School of Economics.

Goodwin, M., Duncan, S. and Halford, S. 1993: Regulation theory, the local state, and the transition of urban politics. *Environment and Planning D: Society and Space* 11, 67–88.

Gourevitch, P. 1986: *Politics in hard times: comparative responses to international economic crises.* Ithaca, NY: Cornell University Press.

Grabher, G. 1993c: The weakness of strong ties: the lock-in of regional development in the Ruhr area. In Grabher, G. (ed.), *The embedded firm: on the socio-economics of industrial networks.* London: Routledge, 255–77.

Hall, J. and Ikenberry, G. 1989: *The state.* Milton Keynes: Open University Press.

Hall, P. (ed.) 1989: *The political power of economic ideas: Keynesianism across countries.* Princeton, NJ: Princeton University Press.

Hay, C. 1995: Re-stating the problem of regulation and re-regulating the local state. *Economy and Society* 24, 387–407.

Helm, D. 1994: British utility regulation: theory, practice and reform. *Oxford Review of Economic Policy* 10, 17–39.

Hirst, P. and Thompson, G. 1992: The problem of 'globalisation': international economic relations, national economic management and the formation of trading blocs. *Economy and Society* 21, 357–96.

Hirst, P. and Thompson, G. 1996: *Globalization in question: the international economy and the possibilities of governance.* Cambridge: Polity.

Holtham, G., and Kay, J. 1994: The assessment: institutions of economic policy. *Oxford Review of Economic Policy* 10, 1–16.

Hutton, W. 1995a: *The state we're in.* London: Jonathan Cape.

Janicke, M. 1990: *State failure: the impotence of politics in industrial society.* Cambridge: Cambridge University Press.

Jessop, B. 1990b: *State theory: putting capitalist states in their place.* Cambridge: Cambridge University Press.

Jessop, B. 1994a: Post-Fordism and the state. In Amin, A. (ed.), *Post-Fordism: a reader.* Oxford: Blackwell, 251–79.

Jessop, B. 1994b: The transition to post-Fordism and the Schumpeterian workfare state. In Burrows, R. and Loader, B. (eds), *Towards a post-Fordist welfare state?* London: Routledge, 13–37.

Katzenstein, P. 1987: *Policy and politics in West Germany: the growth of a semi-sovereign state.* Philadelphia: Temple University Press.

Keynes, J. 1933: National self-sufficiency. *Yale Review* 22(4).

King, D. 1987: *The New Right: politics, markets and citizenship.* London: Macmillan.

King, D. 1995: *Actively seeking work? The politics of unemployment and wefare policy in the United States and Great Britain.* London: University of Chicago Press.

Krugman, P. 1993: Lessons of Massachusetts for EMU. In Torres, F. and Giavazzi, F. (eds), *Adjustment and growth in the European Union.* Cambridge: Cambridge University Press, 241–60.

Krugman, P. 1994a: *Peddling prosperity: economic sense and nonsense in the age of diminished expectations.* New York: Norton.

Krugman, P. 1994b: Competitiveness: a dangerous obsession. *Foreign Affairs* 73, 28–44.

Lazonick, W. 1991: *Business organization and the myth of the market economy.* Cambridge: Cambridge University Press.

Locke, R. 1995: *Remaking the Italian economy.* Ithaca, NY: Cornell University Press.

MacKay, R. 1995: Non-market forces, the nation state and the European Union. *Papers in Regional Science* 74, 209–31.

McMichael, P. and Myhre, D. 1991: Global regulation versus the nation state, *Capital and Class* 43, 83–106.

Markusen, A., Hall, P., Campbell, I. and Dietrick, S. 1991: *The rise of the gunbelt: the military remapping of industrial America.* Oxford University Press.

Martin, R. 1994b: Stateless monies, global financial integration and national economic autonomy: the end of geography? In Corbridge, S., Martin, R. and Thrift, N. (eds), *Money, power and space.* Oxford: Blackwell. 253–78.

Matzner, E. and Streeck, W. 1991: *Beyond Keynesianism: the socio-economics of production and full employment.* Aldershot: Edward Elgar.

Mishra, R. 1990: *The welfare state in capitalist society.* Brighton: Harvester Wheatsheaf.

Mohan, J. 1995: *A National Health Service? The restructuring of health care in Britain since 1979.* London: St Martin's Press.

Morley, H., and Schmid, G. 1993: Public services and competitiveness. In Hughes, K. (ed.), *European competitiveness.* Cambridge: Cambridge University Press.

Newlands, D. 1995: The economic role of regional governments in the European Community. In Hardy, S., Hart, M., Albrechts, L. and Katos, A. (eds) *An enlarged Europe: regions in competition?* London: Jessica Kingsley, 70–80.

O'Brien, R. 1992: *Global financial integration: the end of geography.* London: Pinter.

Ohmae, K. 1990: *The borderless world: power and strategy in the interlinked economy.* London: Fontana.

Ohmae, K. 1995a: *The end of the nation state: the rise of regional economies.* London: HarperCollins.

Ohmae, K. 1995b: Putting global logic first. In Ohmae, K. (ed.), *The evolving global economy.* New York: Harvard Business Review, 129–40.

Painter, J. and Goodwin, M. 1995: Local governance and concrete research: investigating the uneven development of regulation. *Economy and Society* 24, 334–56.

Parker, D. 1993: Privatisation: Ten years on. In Healey, N. (ed.), *Britain's Economic miracle: myth or reality?* London: Routledge, 174–94.

Peck, J.A. and Tickell, A. 1994b: Searching for a new institutional fix: the after Fordist crisis and global–local disorder. In Amin, A. (ed.), *Post-Fordism: a reader.* Oxford: Blackwell, 280–316. School of Economic:

Pfaller, A., Gough, I., and Therborn, G. 1991: *Can the welfare state compete? A comparative study of five advanced capitalist countries.* London: Macmillan.

Pierson, C. 1994: *Dismantling the welfare state? Reagan, Thatcher and the politics of retrenchment.* Cambridge: Cambridge University Press.

Pooley, S. 1991: The State rules, OK? The continuing political economy of nation-states. *Capital and Class* 43, 65–82.

Porter, M. 1990: *The competitive advantage of Nations.* London: Macmillan.

Radice, H. 1984: The national economy: a Keynesian myth. *Capital and Class* 22, 111–40.

Regini, M. 1995: *Uncertain boundaries: the social and political construction of European economies.* Cambridge: Cambridge University Press. Princeton University

Reich, R. 1995a: Who is us? In Ohmae, K. (ed.), *The evolving global economy.* New York: Harvard Business Review, 141–60.

Reich, R. 1995b: Who is them? In Ohmae, K. (ed.), *The evolving global economy.* New York: Harvard Business Review, 161–82.

Sala-i-Martin, X. and Sachs, J. 1991: *Fiscal federalism and optimum currency areas: evidence for Europe from the United States.* Working Paper 3855. Cambridge, MA: National Bureau of Economic Research.

Sassen, S. 1994: *Cities in a world economy.* London: Pine Forge Press.

Schmidt, V. 1996: Industrial policy and policies of industry in advanced industrialized nations. *Comparative Politics* 28, 225–48.

Schott, K. 1984: *Policy, power and order: the persistence of economic problems in capitalist states.* New Haven, CT: Yale University Press.

Sheridan, K. 1993: *Governing the Japanese economy.* Cambridge: Polity.

Soskice, D. 1990: Reinterpreting corporatism and explaining unemployment: coordinated and uncoordinated market economies. In Brunetta, R. and Dell Aringa, C. (eds) *Labour relations and economic performance.* London: Macmillan.

Steinmo, S., Thelen, K. and Longstreth, F. 1992: *Structuring politics: historical institutionalism in comparative analysis.* Cambridge: Cambridge University Press.

Tanzi, V. 1995: *Taxation in an integrating world.* Washington DC: Brookings Institution.

Tanzi, V. and Schuknecht, L. 1996: The growth of government and the reform of the state in industrial countries. IMF Working Paper (December).

Taylor Gooby, P. and Lawson, R. 1993: *Markets and managers: new issues in the delivery of welfare.* Buckingham: Open University Press.

Thompson, G. 1990: *The political economy of the New Right.* London: Pinter.

Wade, R. 1990: *Governing the market: economic theory and the role of government in East Asian industrialization.* Princeton, NJ: Princeton University Press.

Warren, M. 1994: Exploitation or co-operation: the political basis of regional variation in the Italian informal economy. *Politics and Society* 22, 89–122.

Weiss, L. 1989: Regional economic policy in Italy. In Crouch, C. and Marquand, D. (eds), *The new centralism.* Oxford: Blackwell, 109–24.

Weiss, L. and Hobson, J. 1995: *States and economic development: a comparative historical analysis.* Cambridge: Polity.

Wilks, S. and Wright, M. 1991: *The promotion and regulation of industry in Japan.* Basingstoke: Macmillan.

Wood, A. 1994: *North–South trade, employment and inequality: changing fortunes in a skill-driven world.* Oxford: Clarendon Press.

Young, M. 1991: Structural adjustment of mature industries in Japan: legal institutions, industry associations. In Wilks, S. and Wright, P. (eds), *The promotion and regulation of industry in Japan.* Basingstoke: Macmillan.

Zimmerman, H. 1990: Fiscal federalism. In Bennett, R. (ed.), *Decentralization, local governments and markets: towards a post-welfare agenda.* Oxford: Clarendon, 245–64.

[22]

Neoliberalizing Space

Jamie Peck

Department of Geography, University of Wisconsin-Madison, WI, US;
jpeck@geography.wisc.edu

and

Adam Tickell

School of Geographical Sciences, University of Bristol, Bristol, UK;
a.tickell@bristol.ac.uk

This paper revisits the question of the political and theoretical status of neoliberalism, making the case for a process-based analysis of "neoliberalization." Drawing on the experience of the heartlands of neoliberal discursive production, North America and Western Europe, it is argued that the transformative and adaptive capacity of this far-reaching political-economic project has been repeatedly underestimated. Amongst other things, this calls for a close reading of the historical and geographical (re)constitution of the process of neoliberalization and of the variable ways in which different "local neoliberalisms" are embedded within wider networks and structures of neoliberalism. The paper's contribution to this project is to establish a stylized distinction between the destructive and creative moments of the process of neoliberalism—which are characterized in terms of "roll-back" and "roll-out" neoliberalism, respectively—and then to explore some of the ways in which neoliberalism, in its changing forms, is playing a part in the reconstruction of extralocal relations, pressures, and disciplines.

Neoliberalism seems to be everywhere. This mode of free-market economic theory, manufactured in Chicago and vigorously marketed through principal sales offices in Washington DC, New York, and London, has become the dominant ideological rationalization for globalization and contemporary state "reform." What began as a starkly utopian intellectual movement was aggressively politicized by Reagan and Thatcher in the 1980s before acquiring a more technocratic form in the self-styled "Washington consensus" of the 1990s. Neoliberalism has provided a kind of operating framework or "ideological software" for competitive globalization, inspiring and imposing far-reaching programs of state restructuring and rescaling across a wide range of national and local contexts. Crucially, its premises also established the ground rules for global lending agencies operating in the crisis-torn economies of Asia, Africa, Latin America, and the former Soviet Union, where new forms of "free-market" *dirigisme* have been constructed. Indeed, proselytizing the virtues of free trade, flexible labor, and active individualism has become so commonplace in contemporary politics—from Washington to Moscow—that they hardly even warrant a comment in many quarters.

The new religion of neoliberalism combines a commitment to the extension of markets and logics of competitiveness with a profound antipathy to all kinds of Keynesian and/or collectivist strategies. The constitution and extension of competitive forces is married with aggressive forms of state downsizing, austerity financing, and public-service "reform." And while rhetorically antistatist, neoliberals have proved adept at the (mis)use of state power in the pursuit of these goals. For its longstanding advocates in the Anglo-American world, neoliberalism represents a kind of self-imposed disciplinary code, calling for no less than monastic restraint. For its converts in the global south, neoliberalism assumes the status of the Latinate church in medieval Europe, externally imposing unbending rule regimes enforced by global institutions and policed by local functionaries. Meanwhile, if not subject to violent repression, nonbelievers are typically dismissed as apostate defenders of outmoded institutions and suspiciously collectivist social rights.

Although Margaret Thatcher was never right to claim that "there is no alternative" to the neoliberal vision of a free economy and a minimalist state, two decades later the global hegemony of this mode of political rationality means that the burden of proof has shifted: neoliberalism is no longer a dream of Chicago economists or a night-mare in the imaginations of leftist conspiracy theorists; it has become a commonsense of the times. Hence Bourdieu and Wacquant's (2001:2) portrayal of neoliberalism as a "new planetary vulgate" and Beck's (2000:122) characterization of the same nebulous phenomena as an ideological "thought virus." It is revealing, perhaps, that such resorts to metaphor are not unusual in attempts to develop proximate conceptualizations of neoliberalism, the power of which would seem to have become as compelling as it is intangible. Confronted with an apparently extant neoliberal hegemony, the new challenges are simultaneously theoretical and political. They concern the ways that neoliberalism is conceived and characterized, how it is imposed and reproduced, and the identification of its command centers and its flanks of vulnerability.

Attempts to conceive neoliberalism in specifically geographical terms also call for a careful mapping of the neoliberal offensive—both in its heartlands and in its zones of extension—together with a dis-cussion of how "local" institutional forms of neoliberalism relate to its more general (ideological) character. This means walking a line of sorts between producing, on the one hand, overgeneralized accounts of a monolithic and omnipresent neoliberalism, which tend to be insufficiently sensitive to its local variability and complex internal constitution, and on the other hand, excessively concrete and contingent analyses of (local) neoliberal strategies, which are inadequately attentive to the substantial connections and necessary characteristics of neoliberalism as an extralocal project (see Larner 2000). It is our contention here that critical discussions of neoliberalism have tended to lean in the latter direction of closely specified, institutionally contingent accounts, typically focused on more concrete forms of neoliberalism, such as particular Thatcherite restructuring strategies. While conscious of the risks of overreaching, we seek here a preliminary way to push out from this useful starting point to explore some of the more generic and abstract features of the neoliberalization process. Our explicit focus is on the neoliberal heartlands of North America and Western Europe, which have been at the same time its principal centers of discursive production and sites of intensive institutional reconstruction. Paradoxically, perhaps, critical analyses of the extralocal characteristics of neoliberalism have been somewhat underdeveloped in these its "home" spaces, in contrast to the compelling work that has been carried out on the extension of neoliberalism into the second and third world (see Bond 2000; DeMartino 2000; Peet with Hartwick 1999; Veltmeyer, Petras and Vieux 1997; Weiss 1998).

Like the globalization rhetorics with which they are elided, discourses of neoliberalism have proved to be so compelling because, in representing the world of market rules as a state of nature, their prescriptions have a self-actualizing quality. Even as they misdescribe the social world, discourses of globalization and neoliberalism seek to remake it in their own image (Bourdieu 1998; Piven 1995). Discourses of neoliberalism are "strong discourses" in part by virtue of this self-actualizing nature and in part because of their self-evident alignment with the primary contours of contemporary political-economic power.

> [T]his initially desocialized and dehistoricized "theory" has, now more than ever, the means of *making itself true* ... For neoliberal discourse is not like others. Like psychiatric discourse in the asylum ... it is a "strong discourse" which is so strong and so hard to fight because it has behind it all the powers of a world of power relations which it helps to make as it is, in particular by orienting the economic choices of those who dominate economic relations and so adding its own—specifically symbolic—force to those power relations. (Bourdieu 1998:95; emphasis in original)

There are, in fact, many parallels between analytical treatments of globalism and those of neoliberalism. Both have been associated with a mode of exogenized thinking in which globalism/neoliberalism is presented as a naturalized, external "force." Both ascribe quasiclimatic, extraterrestrial qualities to apparently disembodied, "out there" forces, which are themselves typically linked to alleged tendencies towards homogenization, leveling out, and convergence. And both have attributed to them immense and unambiguous causal efficacy: while conservative commentators emphasize the (ostensibly ubiquitous) benign effects of globalization, critics focus instead on the (just as pervasive) malign effects of neoliberalism. Yet their common flaw is that they have tended to naturalize and exogenize their object of study—be this in the form of an all-powerful globalization process or the all-encompassing politics of neoliberalism. Certainly, critical analyses do have the virtue of underscoring the inescapably *political* character of the globalization project and the hegemonic position of neoliberalism in global agencies and discourses. However, there is more to be done, both theoretically and empirically, on the specification and exploration of different processes of neoliberalization. This would need to take account of the ways in which ideologies of neoliberalism are themselves produced and reproduced through institutional forms and political action, since "actually existing" neoliberalisms are always (in some way or another) hybrid or composite structures (see Larner 2000). On the other hand, though, analyses of neoliberalism in general cannot afford to reduce this multifaceted process to its concrete manifestations.

With the benefits of hindsight, and particularly in light of neoliberalism's "long boom" of the 1990s, our objective here is to explore the (subtly transformed) process of neoliberal*ization* in the context of recent history, and in doing so to refine, develop, and modestly revise the received conception of neoliberal*ism*. Taking certain cues from the globalization debate, we propose a processual conception of neoliberalization as both an "out there" and an "in here" phenomenon whose effects are necessarily variegated and uneven, but the incidence and diffusion of which may present clues to a pervasive "metalogic" (Amin 1997; Dicken, Peck and Tickell 1997). Like globalization, neoliberalization should be understood as a process, not an end-state. By the same token, it is also contradictory, it tends to provoke countertendencies, and it exists in historically and geographically contingent forms. Analyses of this process should therefore focus especially sharply on *change*—on shifts in systems and logics, dominant patterns of restructuring, and so forth—rather than on binary and/or static comparisons between a past state and its erstwhile successor. It also follows that analyses of neoliberalization must be sensitive to its contingent nature—hence the nontrivial differences, both theoretically and politically, between the actually existing neoliberalisms of, say, Blair's Britain, Fox's Mexico, or Bush's America. While processes of neoliberalization are clearly at work in all these diverse situations, we should not expect this to lead to a simple convergence of outcomes, a neoliberalized end of history and geography.

The process of neoliberalization, then, is neither monolithic in form nor universal in effect. However, in the course of the last quarter-century there have been significant *internal* shifts in its institutional form, its political rationality, and its economic and social consequences. Focusing on the changing situation in the neoliberal "heartlands," we wish to outline—and simultaneously problematize—the complex evolution that has taken place, from the experimental proto-neoliberalisms of the 1970s through the constitution of neoliberalism as an explicit political-economic project during the 1980s to the "deep neoliberalisms" of the past decade. Perhaps the most controversial aspect of this process of complex evolution concerns the status of neo-liberalism as a *regulatory* "project" or "regime," given the (contested and uneven) shift that has taken place from the Thatcher/Reagan era of assault and retrenchment to the more recent experience of normalized neoliberalism which has been associated with, inter alia, the technocratic embedding of routines of neoliberal governance, the aggressive extension of neoliberal institutions and their seeming robust-ness even in the face of repeated crises, and the continuing erosion of pockets of political and institutional resistance to neoliberal hegemony, including the "soft neoliberalisms" most clearly epitomized by the Third Way.

In short, in this North Atlantic zone at least, there seems to have been a shift from the pattern of deregulation and dismantlement so dominant during the 1980s, which might be characterized as "roll-back neoliberalism," to an emergent phase of active state-building and regulatory reform—an ascendant moment of "roll-out neoliberalism." In the course of this shift, the agenda has gradually moved from one preoccupied with the active *destruction and discreditation* of Keynesian-welfarist and social-collectivist institutions (broadly defined) to one focused on the purposeful *construction and consolidation* of neoliberalized state forms, modes of governance, and regulatory relations. It is this more recent pattern of institutional and regulatory restructuring, which we characterize here as a radical, emergent com-bination of neoliberalized economic management and authoritarian state forms, that demands both analytical and political attention. This may represent a critical conjuncture, since it reflects *both* the contradictions/limitations of earlier forms of neoliberalization *and* the attainment of a more aggressive/proactive form of contemporary neo-liberalization. In the process, forms of "shallow" neoliberalization— during the Thatcher/Reagan years—that were understood by many at the time as a moment of destructive and reactionary "antiregulation" (see Peck and Tickell 1994a) might now need to be confronted as a more formidable and robust pattern of proactive statecraft and per-vasive "metaregulation." Granted, it may still be (politically) inappro-priate to regard the contemporary manifestation of neoliberalization

as a form of regulatory "settlement," but its multifaceted normalization during the 1990s and beyond surely calls for more sharply focused
analyses both of actually existing neoliberalisms and of the disciplinary
force-fields in which they are embedded. In its diffuse, dispersed,
technocratic, and institutionalized form, neoliberalism has spawned a
free-market in social regression, but simultaneously it is becoming
vulnerable—from the inside as well as the outside—in wholly new ways.

From Jungle Laws to Market Rules

In an earlier attempt to understand neoliberalism, we presented it as
the political essence of the (after-Fordist) crisis, a form of "jungle law"
that broke out in the context of the exhaustion of the postwar social
contract (Peck and Tickell 1994a; Tickell and Peck 1996). Shaped
by the experiences of the "Thatcher decade" (Jessop et al 1988;
Overbeek 1990), many of the ramifications of which were genuinely
international in scope, this conception of neoliberalism was
antagonistic to those analyses that presented the deregulation and
retrenchment of the 1980s as the precursor to some kind of "post-
Fordist" mode of social regulation (see Jessop 1989), contending that
the politics of neoliberalism were—by definition—disorderly and
destructive. Under such conditions, social relations were being
reconstituted in the image of a brutal reading of competitive-market
imperatives, while their geographical corollary—interlocal relations—
were also being remade in competitive, commodified, and monetized
terms, with far-reaching consequences for local political conditions
and regulatory settlements (whether neoliberal or not).

   Thus, in contrast to the Fordist-Keynesian golden age, when the
national-state became the principal anchoring point for institutions of
(gendered and racialized) social integration and (limited) macro-
economic management, neoliberalization was inducing localities to
compete by cutting social and environmental regulatory standards and
eroding the political and institutional collectivities upon which more
progressive settlements had been constructed in the past (and might
again be in future). In this context, our earlier analysis portrayed
the 1980s and early 1990s as a period of institutional searching and
experimentation within restrictive (and ultimately destructive)
neoliberal parameters. Extralocal rule regimes at this time seemed
critically to undermine the potential of non-neoliberal projects at
the local scale, while engendering a lemming-like rush towards urban
entrepreneurialism, which itself would only serve to facilitate, encourage, and even publicly subsidize the accelerated mobility of circulating
capital and resources (see Harvey 1989). As if to add insult to injury,
regimes of public investment and finance, too, would increasingly
come to mimic these marketized conditions (see Cochrane, Peck and
Tickell 1996; Jessop, Peck and Tickell 1999).

For all the implied pessimism of this analysis, however, a decade ago it appeared to us that neoliberalism was unsustainable. In both its American and British heartlands, the early 1990s witnessed significant economic retrenchment, as the irrationalities and externalities of fundamentalist neoliberalism began to take their toll. Well documented at the time were the tendencies, for example, to exacerbate and deepen the macroeconomic cycle, to license short-termist, plundering strategies on the part of competing capitals, to widen social, economic, and spatial inequalities, to undermine the production of public goods and collective services, to degrade social and environmental resources, to constrain and weaken socially progressive alternatives to (or even conservative ameliorations of) market liberalism, and so forth. The transatlantic recessions of the early 1990s, while certainly also having "local" causes, seemed at the time to be vivid precursors of these inherent and substantially "internal" crisis tendencies (Peck and Tickell 1995) The politics of the crisis, it seemed, were working their natural course.

The flexible recessions of the early 1990s and the mounting popular unease with the process of state withdrawal—which were to fatally undermine the Thatcher/Major and Reagan/Bush I projects—appeared, then, to be highlighting at least two key essences of neoliberalism. First, the reactive politics of the after-Fordist downswing seemed to lack the capacity for sustainable reproduction in the medium term. Second, they were evidently more concerned with the *de(con)struction* of "anticompetitive" institutions like labor unions, social-welfare programs, and interventionist arms of the governmental apparatus than with the purposeful *construction* of alternative regulatory structures. In the process of this phase of active deconstruction, however, the neoliberal offensive also helped to usher in, and to legitimize and enforce, a new regime of highly competitive interlocal relations, such that just about all local social settlements were becoming tendentially subject in one way or another to the disciplinary force of neoliberalized spatial relations. The conclusion that we drew at the time was that—in the absence of a more stable, socially/spatially redistributive, and supportive extralocal framework —the neoliberal constitution of competitive relations between localities and regions placed real limits on the practical potential of localized or "bottom-up" political action (Amin 1999). In the asymmetrical scale politics of neoliberalism, local institutions and actors were being given responsibility without power, while international institutions and actors were gaining power without responsibility: a form of regulatory dumping was occurring at the local scale, while macrorule regimes were being remade in regressive and marketized ways. This represented a deeply hostile macro- or extralocal environment, one that even the most robust of progressive localisms seemed ill placed to overturn through unilateral action. So, while there was (and is) certainly a crucial role beyond mere amelioration for progressive projects at the local scale, it seemed that any adequate response to neoliberalism had to be framed in substantially *extra*local terms. Only this could stall and circumvent the *neoliberalization of interlocal relations*—a more nebulous and a more daunting adversary.

A fundamental problem for progressive localisms—such as, for example, the Scandinavian welfare state or "high-road" regional economic initiatives—was that they did not confront neoliberalism on a level playing field of "regime competition," because neoliberalism always represented more than "just" an alternative regulatory project or program. It was not that neoliberalism was simply a creature of, say, California or the Southeast of England (although these were clearly amongst its strongholds), but that the very social and spatial relations in which such regions were embedded had *themselves* become deeply neoliberalized. Neoliberalism was therefore qualitatively different because it inhabited not only institutions and places but also *the spaces in between*. In other words, neoliberalism was playing a decisive role in constructing the "rules" of interlocal competition by shaping the very metrics by which regional competitiveness, public policy, corporate performance, or social productivity are measured—value for money, the bottom line, flexibility, shareholder value, performance rating, social capital, and so on. Neoliberalism therefore represented a form of regulation of sorts, but not a form commensurate with, say, the Keynesian-welfarism that preceded it in many (though not all) cases. In Gough's (1996:392) terms, neoliberalism was a form of "regulation by value," one that he sought explicitly to contrast with our more explicit focus on its manifest *illogics*. While there may have been more than a hint of wishful thinking on our part in the denigration of the Thatcherite project (Peck and Tickell 1995), our reading did emphasize the macrolevel disciplinary effects of neoliberalism, which were clearly part of the "logic" of the system, even whilst we may have been somewhat reluctant to acknowledge it as such (Peck and Tickell 1994b). In the final analysis, though, neoliberalism proved too robust to be brought down by the Anglo-American recessions of the early 1990s, or indeed by the Asian meltdown later in the decade. Instead, these economic crises proved to be important moments in its ongoing transformation.

Mutations of Neoliberalism
Although neoliberalism privileges the unitary logic of the market while advocating supposedly universal cures and one-best-way policy strategies, it is in reality much more variegated than such self-representations suggest. Neoliberalism remains variegated in character, certainly when assessed according to criteria like the scale and scope

of state intervention, forms of capital and labor market regulation, the constitution of institutions of social regulation, patterns of political resistance and political incorporation, and so forth. However, as there are evidently powerful family resemblances, adequate conceptualizations must be attentive to *both* the local peculiarities *and* the generic features of neoliberalism. While not wishing to downplay local and national differences in the constitution of the neoliberal project, we focus here on some of the broader patterns and connections evident in the process of neoliberalization as it has taken shape in the North Atlantic zone. More specifically, we want to draw attention to what seem to us to be important *historical* shifts in the constitution of the neoliberal project in this transnational space (see also Tickell and Peck forthcoming).

The first of these shifts occurred in the late 1970s, as neoliberalism underwent a transformation from the abstract intellectualism of Hayek and Friedman to the state-authored restructuring projects of Thatcher and Reagan. This can be characterized as a movement from "proto-" to "roll-back" neoliberalism: a shift from the philosophical project of the early 1970s (when the primary focus was on the restoration of a form of free-market thinking within the economics profession and its subsequent [re]constitution as the theoretical high ground) to the era of neoliberal conviction politics during the 1980s (when state power was mobilized behind marketization and deregulation projects, aimed particularly at the central institutions of the Keynesian-welfarist settlement). The backdrop to this shift was provided by the macroeconomic crisis conditions of the 1970s, the blame for which was unambiguously laid at the door of Keynesian financial regulation, unions, corporatist planning, state ownership, and "overregulated" labor markets. In this context, the neoliberal text—freeing up markets, restoring the "right to manage," asserting individualized "opportunity rights" over social entitlements—allowed politicians the right to be both conservative and radical. It was they, not their social-democratic adversaries, who were insisting on the need for root-and-branch change; and while labor unions, social advocates, and left-of-center parties were often forced to defend (apparently failing) institutions, neoliberals made the argument for a clean break.

The second neoliberal transformation occurred in the early 1990s, when the shallow neoliberalisms of Thatcher and Reagan encountered their institutional and political limits as evidence of the perverse economic consequences and pronounced social externalities of narrowly marketcentric forms of neoliberalism became increasingly difficult to contest. However, the outcome was not implosion but reconstitution, as the neoliberal project itself gradually metamorphosed into more socially interventionist and ameliorative forms, epitomized by the Third-Way contortions of the Clinton and Blair administrations. This most recent phase might be portrayed as one of "roll-out" neoliberalism, underlining the sense in which new forms of institution-building and governmental intervention have been licensed within the (broadly defined) neoliberal project. No longer concerned narrowly with the mobilization and extension of markets (and market logics), neoliberalism is increasingly associated with the political foregrounding of new modes of "social" and penal policy-making, concerned specifically with the aggressive reregulation, disciplining, and containment of those marginalized or dispossessed by the neoliberalization of the 1980s. Commenting on this shift in the means and methods of neoliberalization, Wacquant (1999:323) pointedly observes that

> [t]he same parties, politicians, pundits, and professors who yesterday mobilized, with readily observable success, in support of "less government" as concerns the prerogatives of capital and the utilization of labor, are now demanding, with every bit as much fervor, "*more* government" to mask and contain the deleterious social consequences, in the lower regions of social space, of the deregulation of wage labor and the deterioration of social protection.

This does not mean, of course, that economic policy concerns have somehow slipped off the agenda, but rather that—as modes of neoliberal economic management have been effectively normalized—the frontier of active policymaking has shifted and the process of state-building has been reanimated. More now than merely a deregulatory political mindset or a kind of ideological software, neoliberalism is increasingly concerned with the roll-out of new forms of institutional "hardware." In the neoliberal heartlands, this is associated with a striking coexistence of technocratic economic management and invasive social policies. Neoliberal processes of economic management—rooted in the manipulation of interest rates, the maintenance of noninflationary growth, and the extension of the "rule" of free trade abroad and flexible labor markets at home—are increasingly technocratic in form and therefore superficially "depoliticized," acquiring the privileged status of a taken-for-granted or foundational policy orientation. Meanwhile, a deeply interventionist agenda is emerging around "social" issues like crime, immigration, policing, welfare reform, urban order and surveillance, and community regeneration. In these latter spheres, in particular, new technologies of government are being designed and rolled out, new discourses of "reform" are being constructed (often around new policy objectives such as "welfare dependency"), new institutions and modes of delivery are being fashioned, and new social subjectivities are being fostered. In complex simultaneity, these social and penal policy incursions represent both the advancement of the neoliberal project—of extending and bolstering market logics, socializing individualized subjects, and disciplining the noncompliant—and a recognition of sorts that earlier manifestations of this project, rooted in dogmatic deregulation and marketization, clearly had serious limitations and contradictions. Consequently, what we characterize here as "roll-out" neoliberalism reflects a series of politically and institutionally mediated responses to the manifest failings of the Thatcher/Reagan project, formulated in the context of ongoing neoliberal hegemony in the sphere of economic regulation. In a sense, therefore, it represents *both* the frailty of the neoliberal project *and* its deepening.

This is not the place to explore the historiography of neoliberalism in detail. Instead, we focus on the principal analytical and political implications of this broad movement in relation to neoliberalism's status as a regulatory project, its scalar constitution, and its substantive policy foci. First, with respect to the changing character of neoliberalism as regulatory project, it should be noted that whereas its initial ascendancy during the 1970s was associated with crisis conditions "external" to the project itself, the shifts of the 1990s were substantially triggered by "internal" contradictions and tensions in the project. The macroeconomic turbulence of the 1970s, and the stresses this placed on the central institutions of Keynesian-welfarism, established both the context for neoliberalism's transition from intellectual to state project and the political conditions conducive to a decisive shift to the right. In contrast, the mutations of the 1990s essentially represented responses to previous market, state, and governance failures partly (or even largely) initiated by neoliberalism itself. While it had been ideologically presumed during the 1980s that the spontaneous operation of market forces and disciplines would alone be sufficient to the task of economic regulation as long as government got out of the way, by the 1990s it had become clear that recurrent failures of a quasisystemic nature in areas like transport, food systems, and pollution, and even in financial and labor markets, called for responses outside the narrow repertoire of deregulation and marketization. Hence the deliberate stretching of the neoliberal policy repertoire (and its associated rhetorics) to embrace a range of extramarket forms of governance and regulation. These included, inter alia, the selective appropriation of "community" and nonmarket metrics, the establishment of social-capital discourses and techniques, the incorporation (and underwriting) of local-governance and partnership-based modes of policy development and program delivery in areas like urban regeneration and social welfare, the mobilization of the "little platoons" in the shape of (local) voluntary and faith-based associations in the service of neoliberal goals, and the evolution of invasive, neopaternalist modes of intervention (along with justifications for increased public expenditure) in areas like penal and workfare policy.

Second, these shifts have been accompanied by—and partially achieved through—changes in neoliberalism's scalar constitution. These have involved complex (and often indirect) extensions of national state power, most notably in the steering and management of programs of devolution, localization, and interjurisdictional policy transfer. In welfare reform, for example, the downloading of resources, responsibilities, and risks to local administrations and extrastate agencies has occurred in the context of a close orchestration of the processes of institutional reform and policy steering by national states (Peck 2001). And while some social policies have been (superficially) decentralized and localized, just as compelling have been the movements in economic policy, where issues once deeply politicized (such as interest-rate setting, currency valuation, trade policy, corporate regulation and taxation) have since been variously parlayed into technocratic structures and routinized conventions, absorbed by transnational agencies and metaregulatory frameworks, or exposed to the "markets." At the level of the national state, neoliberalized forms of macro-economic management—based on low inflation, free trade, flexible job markets, regressive taxation, downsized government, and central-bank (relative) autonomy—now constitute the taken-for-granted context for political debate and policy development (Bluestone and Harrison 2000). Meanwhile, international institutions such as the International Monetary Fund (IMF) and the World Trade Organization (WTO) establish and police neoliberalized "rules of the road" that promulgate free(r) trade, free(r) markets, and increasingly unrestricted access to a wide range of markets (including public services under the General Agreement on Trade in Services) to trans-national corporations, while potentially more progressive institutions and agreements (the International Labour Organisation, the United Nations Conference on Trade and Development, the Kyoto Protocols) are allowed to wither.

Third, in relation to the substantive foci of neoliberal policies, we draw attention to the twin processes of financialization in the realm of economic policy and activation in the field of social policy. In economic policy, the excesses of roll-back neoliberalism found vivid expressions in the financial crises of the 1990s, which took both geographical (Mexico, Russia, East Asia, Turkey) and institutional (Barings, Long Term Capital, Orange County, Equitable Life) forms. In response, new financial architectures are taking shape, under-girding a type of "global stability" by normalizing Anglo-American rationalities and, paradoxically, creating deeper systemic integration in ways that are seriously procyclical (Tickell 1999; Tickell and Clark 2001). In social policy, the (re)criminalization of poverty, the normal-ization of contingent work, and its enforcement through welfare retrenchment, workfare programming, and active employment policies represent a comprehensive reconstitution of the boundary institutions of the labor market. Following the blue-collar shakeouts of the 1980s and the white-collar downsizings of the 1990s, the attention of policymakers has focused with increasing insistency on the challenges of reproducing regimes of precarious work and mobilizing the poor for low-wage employment. Market discipline, it seems, calls for new modes of state intervention in the form of large-scale incarceration, social surveillance, and a range of microregulatory interventions to ensure persistent "job readiness" (Piven and Cloward 1998).

Of course, these shifts in the macroconstitution of neoliberalism have not been reproduced homogenously across space. In fact, they have been associated with a marked intensification of spatially uneven development, which itself has produced new opportunities for—and challenges to—the neoliberal project. Neoliberalism's persistent vulnerability to regulatory crises and market failures is associated with an ongoing dynamic of discursive adjustment, policy learning, and institutional reflexivity. As long as collateral damage from such breakdowns can be minimized, localized, or otherwise displaced across space or scale, it can provide a positive spur to regulatory reinvention. One of neoliberalism's real strengths has been its capacity to capitalize on such conditions. And because little store is set by local loyalties or place commitments in neoliberalized regulatory regimes, which favor mobility over stability and short- over long-term strategies, dynamics of persistent reform and extralocal policy learning assume critical roles in the reconstruction of institutions and the maintenance of legitimacy. Such regimes are characterized, then, by the perpetual reanimation of restless terrains of regulatory restructuring. Thus, the deep neoliberalization of spatial relations represents a cornerstone of the project itself.

Spaces of Neoliberalism
Neoliberalism does indeed seem to be everywhere. And its apparent omnipresence is at the same time a manifestation of and a source of political-economic power. It is no coincidence, in this respect, that viral metaphors are so often deployed in critical analyses of the spread of neoliberalism, because notions of contagion, carrier populations, and susceptibility seem so apposite. Viruses are dangerous, of course, because they spread, and bodies politic—while they may exhibit differing degrees of resistance—are rarely immune to all the strains of neoliberalism. Crucially, any analysis of the diffusion of neoliberalism must pay attention to the nature of these movements *between* sites of incorporation and imposition, because neoliberalism cannot be reduced to an "internal" characteristic of certain institutions or polities; it also exists as an extralocal regime of rules and routines, pressures and penalties. This is not to deny the valuable role of concrete analyses

of actually existing neoliberal projects—for example, in regional manifestations such as the Southeast of England (see Allen et al 1998; Peck and Tickell 1995), or in the more nebulous form of putative "reform models" like learning/flexible/networked regions or modes of zero tolerance/workfarist/activist social and penal policy (see Peck and Theodore 2001; Wacquant 1999). Rather, it is to make the point that such analyses can usefully be complemented by explorations of the webs of interlocal and interorganizational relations in which they are embedded. Moreover, these accounts of the structural contexts—which are both a product of and produced by "local" neoliberalisms—may help to draw attention to the causally substantial connections and telling family resemblances between different forms of neoliberalization.

As David Harvey so persuasively argued of urban entrepreneurialism, the serial reproduction of cultural spectacles, enterprise zones, waterfront developments, and privatized forms of local governance is not simply an aggregate outcome of spontaneous local pressures, but reflects the powerful disciplinary effects of interurban competition: "[I]t is by no means clear that even the most progressive urban government can resist [social polarization] when embedded in the logic of capitalist spatial development in which competition seems to operate not as a beneficial hidden hand, but as an external coercive law forcing the lowest common denominator of social responsibility and welfare provision within a competitively organised urban system" (Harvey 1989:12). Because signature cultural events, prestige corporate investments, public resources, and good jobs are in such short supply, cities (perhaps the most visibly denuded victims of roll-back neoliberalism) are induced to jump on the bandwagon of urban entrepreneurialism, which they do with varying degrees of enthusiasm and effectiveness. And ultimately, their persistent efforts and sporadic successes only serve to further accelerate the (actual and potential) mobility of capital, employment, and public investment. In selling themselves, cities are therefore actively facilitating and subsidizing the very geographic mobility that first rendered them vulnerable, while also validating and reproducing the extralocal rule systems to which they are (increasingly) subjected (Cochrane, Peck and Tickell 1996). The logic of interurban competition, then, turns cities into accomplices in their own subordination, a process driven—and legitimated—by tales of municipal turnaround and urban renaissance, by little victories and fleeting accomplishments, and ultimately also by the apparent paucity of "realistic" local alternatives. Thus, elite partnerships, mega-events, and corporate seduction become, in effect, both the only games in town *and* the basis of urban subjugation. The public subsidy of zero-sum competition at the interurban scale rests on the economic fallacy that every city can win, shored up by the political reality that no city can afford principled noninvolvement in the game.

Clearly, this regime of accelerated interurban competition was not simply a product of neoliberalism, nor can it be reduced entirely to its logic, but the parallel ascendancy of neoliberalism has been crucial in reinforcing, extending, and normalizing these transurban tendencies towards reflexive and entrepreneurial city governance in (at least) seven ways:

- Neoliberalism promotes and normalizes a "growth-first" approach to urban development, reconstituting social-welfarist arrangements as anticompetitive costs and rendering issues of redistribution and social investment as antagonistic to the overriding objectives of *economic* development. Social-welfarist concerns can only be addressed *after* growth, jobs, and investment have been secured, and even then in no more than a truncated and productivist fashion.
- Neoliberalism rests on a pervasive naturalization of market logics, justifying on the grounds of efficiency and even "fairness" their installation as the dominant metrics of policy evaluation. In this analysis, urban policy measures should anticipate, complement, and in some cases mimic the operation of competitive markets.
- As the ideology of choice for both major funding agencies and "the markets," neoliberalism not only privileges lean government, privatization, and deregulation, but through a combination of competitive regimes of resource allocation, skewed municipal-lending policies, and outright political pressure undermines or forecloses alternative paths of urban development based, for example, on social redistribution, economic rights, or public investment. This produces a neoliberal "lock-in" of public-sector austerity and growth-chasing economic development.
- Neoliberalism licenses an extrospective, reflexive, and aggressive posture on the part of local elites and states, in contrast to the inward-oriented concerns with social welfare and infrastructure provision under the Keynesian era. Today, cities must actively—and responsively—scan the horizon for investment and promotion opportunities, monitoring "competitors" and emulating "best practice," lest they be left behind in this intensifying competitive struggle for the kinds of resources (public and private) that neoliberalism has helped make (more) mobile.
- Despite its language of innovation, learning, and openness, neoliberalism is associated with an extremely narrow urban-policy repertoire based on capital subsidies, place promotion, supply-side intervention, central-city makeovers, and local

boosterism. The very familiarity of this cocktail is a reflection both of the coercive pressures on cities to keep up with—or get a step ahead of—the competition and of the limited scope for genuinely novel local development under a neoliberalized environment.

- Neoliberal regimes are unforgiving in the face of incompetence or noncompliance, punishing cities that fail in the unyielding terms of competitive urbanism through equally unambiguous measures such as malign neglect, exclusion from funding streams, and the replacement of local cadres. Reinforcing these pressures, national and transnational government funds increasingly flow to cities on the basis of economic potential and governance capacity rather than manifest social need, and do so through allocation regimes that are competitively con-stituted (in contrast to the Keynesian pattern, where resources were secured on basis of social entitlements, bureaucratic channels, or redistributive mechanisms or through the oper-ation of automatic stabilizers). Yet while zones of deeply impacted poverty and social exclusion may have been no-go areas for neoliberals during the 1980s, in its roll-out guise neoliberalism is increasingly penetrating these very places, animated by a set of concerns related to crime, worklessness, welfare dependency, and social breakdown.

- As key sites of economic contradiction, governance failure, and social fall-out, cities find themselves in the front line of *both* hypertrophied after-welfarist statecraft *and* organized resistance to neoliberalization. Regressive welfare reforms and labor-market polarization, for example, are leading to the (re)urbanization of (working and nonworking) poverty, positioning cities at the bleeding edge of processes of punitive-institution building, social surveillance, and authoritarian governance.

Neoliberalization cannot, therefore, be reduced to an outcome or a side effect of this after-Keynesian environment. Instead, it can be seen to have exercised a cumulatively significant and somewhat autonomous influence on the structure and dynamics of interurban competition and intraurban development. This has not been a period in which inherently competitive forces have been spontaneously "liberated" by the state's withdrawal. Rather, neoliberalism's ascendancy has been associated with the political *construction* of markets, coupled with the deliberate extension of competitive logics and privatized management into hitherto relatively socialized spheres. "Entrepreneurial" regimes of urban governance are, therefore, not simply local manifestations of neoliberalism; their simultaneous rise across a wide range of national, political, and institutional contexts suggests a systemic connection with neoliberalization as a *macro* process. In other words, the remaking of the rules of interlocal competition and extralocal resource allocation—or the deep neoliberalization of spatial and scalar relations—fundamentally reflects the far-reaching macropolitical realignment that has taken place since the 1970s. Harvey (1989:15) hints at this in remarking that the managerialist cities of the Keynesian era were rather less exposed to macropressures: "[U]nder conditions of weak interurban competition … urban governance [was rendered] less consistent with the rules of capital accumulation," while in contrast, during the subsequent phase of neoliberal entrepreneurialism, "urban governance has moved more rather than less into line with the naked requirements of capital accumulation." Yet, even stripped of its Keynesian clothing, this underlying capitalist logic does not automatically secure some kind of Hayekian order of competitive urbanism. Instead, the subsequent neoliberal "settlement" had to be *engineered* through explicit forms of political management and intervention and new modes of institution-building designed to extend the neoliberal project, to manage its contradictions, and to secure its ongoing legitimacy.

There is, then, a space for politics here, even if these are tendentially neoliberalized politics, or limited forms of resistance to this ascendant (dis)order. And where there is politics there is always the scope for geographical unevenness and path-transforming change. Somewhat paradoxically, given the strength of the coercive extralocal forces mobilized and channeled by neoliberalism, the outcome is not homogeneity, but a constantly shifting landscape of experimentation, restructuring, (anti)social learning, technocratic policy transfer, and partial emulation. Even if the overriding dynamics of neoliberalized spatial development involve regulatory undercutting, state downsizing, and races to the bottom, the outcomes of this process are more variegated than is typically assumed to be the case. Rather than some rapidly accomplished "bottoming out" of minimalist regulatory settlements, it tends to result in ongoing institutional restructuring and externally leveraged "reform" around new sets of axes.

While earlier manifestations of spatial restructuring may indeed have reflected the privileged place of destructive, "antiregulatory" moments of the neoliberalization process (emphasized in Peck and Tickell 1994a), the rather different reading outlined here has intentionally drawn attention to the complex ways in which these logics of deregulation became systematically entwined, during the 1990s, with the *roll out* of neoliberalized state forms (themselves partly the outcome of previous tensions and contradictions in the early neoliberalist project). Such forms of "deep" neoliberalization and post-Keynesian statecraft are associated with an especially restless landscape of urban competition, narrowly channeled innovation, and policy emulation, as the processes of institutional searching that have become such compelling features of the after-Fordist era have been (increasingly?) been played out at the urban and subnational scales (see Brenner 1999; Tickell and Peck 1995).

Even if the resultant repertoire of restructuring strategies is hardly more inspiring now than it was in the early 1990s—following the partial accommodations that have been engineered with various social capital, associative economy, Third-Way, and social economy approaches—it must be acknowledged that neoliberalism has demonstrated a capacity variously to spawn, absorb, appropriate, or morph with a range of local institutional (re)forms in ways that speak to its *creatively* destructive character (see Brenner and Theodore [paper] this volume). Certainly this potential has been recognized for some time, whether in the regulationist emphasis on "chance discoveries" (Jessop 2001; Lipietz 1987; Peck and Tickell 1994a), or indeed in Harvey's (1989:15) original formulation of the urban entrepreneurialism thesis:

> Competition for investments and jobs ... will presumably generate all kinds of firmaments concerning how best to capture and stimulate development under particular local conditions ... From the standpoint of long-run capital accumulation, it is essential that different paths and different packages of political, social and entrepreneurial endeavours get explored. Only in this way is it possible for a dynamic and revolutionary social system, such as capitalism, to discover new forms and modes of social and political regulation suited to new forms and paths of capital accumulation.

In contrast to the generalized skepticism that was evident a decade or so ago concerning the potentially transformative potential of neoliberalism, which we were not alone in associating irretrievably with processes of institutional destruction and a climate of crisis politics, there has been a growing recognition of neoliberalism's potential to evolve, mutate, and even "learn." If the Greenspan boom of the 1990s tells us anything, it is surely that, under a particular configuration of geoeconomic power, neoliberal modes of growth management can be brutally effective, at least over the medium term and in the context of largely uncontested US hegemonic dominance (Peck 2002; Setterfield 2000). And the exuberance of ("new") economic growth in the US during the 1990s, in turn, served to validate American policies across a wide range of fields, not least those relating to welfare reform and labor-market flexibility (Peck 2001).

Indeed, if there is a policy-development analog for the coercive regime of interurban competition described by Harvey, it is perhaps to be found in the multifaceted process of "fast policy transfer" that has become an increasingly common characteristic of (inter)local institutional change (Jessop and Peck 2001). In concrete terms, this reflects two interconnected features of the contemporary policy development process, at least within (and between) neoliberalized national and local states. First, this process is increasingly dominated by "ideas from America," both in the direct sense of the modeling of transnational reform projects on the basis of the (technocratically stylized) experience of US cities and states (see Deacon 2000; Peck and Theodore 2001) and, more generally, in the form of the growing influence of US-inflected neoliberal restructuring strategies in fields like policing and incarceration policy, financial management, and corporate governance. Second, and in more functional terms, there is suggestive evidence that policy cycles—the elapsed time between the (re)specification of policy "problems," the mobilization of reform movements, and the selection and implementation of strategies—are being deliberately shortened in the context of a growing propensity to adopt "off-the-shelf," imported solutions in the place of the (usually slower) process of in situ policy development.

Together, these twin processes are resulting in an acceleration in policy "turnover time," as reform dynamics become effectively endemic and as extralocal learning and emulation become more or less continuous processes. This, in turn, is leading to a deepening and intensification in the process of neoliberalization, US-style. Such mechanisms of international and interlocal policy transfer— which take place along channels that have been created, structured, and lubricated by technocratic elites, think tanks, opinion-formers, consultants, and policy networks—have been rapidly established as one of the principal modes of policy development in strategically critical fields such as systemic financial stability, the management of urban "underclasses," the regulation of contingent labor markets, and the displacement of welfare entitlements with socially authoritarian packages of rights and responsibilities (Peck 2001; Tickell and Clark 2001; Wacquant 1999). For Wacquant (1999:319–320), this represents an aggressive and deliberate internationalization of a "new penal commonsense," rooted in the recriminalization of poverty and the resultant normalization of contingent labor:

> [Neoliberal nostrums] did not spring spontaneously, ready-made, out of reality. They partake of a vast constellation of terms and theses that come from America on crime, violence, justice, inequality, and responsibility—of the individual, of the "community," of the national collectivity—which have gradually insinuated themselves into European public debate to the point of serving as its framework and focus, and which owe the brunt of their power of persuasion to their sheer omnipresence and to the prestige recovered by their originators.

A conspicuously important outcome of this increased "transcontinental traffic in ideas and public policies" (Wacquant 1999:322) is the marked deepening of the process of neoliberalization on an international, if not global, scale. Crucially, this has massively enlarged the space for extensive forms of neoliberalized accumulation and policy formation, in ways that were distinctly favorable to the United States through most of the 1990s (see Wade 2001). At the same time, however, it has also shifted the scales and stakes of the attendant crisis tendencies in potentially very serious ways.

In fact, having been aggressively upscaled to the transcontinental level, the neoliberal "settlement" is now surely more, not less, vulnerable to systemic crisis. Even though the successive remaking of economic and social institutions in US-inflected forms is not leading to any kind of simple "convergence," it may nevertheless be serving to degrade the institutional "gene pool" essential for subsequent forms of neoliberal mutation and crisis containment. The (incomplete and uneven) global extension of neoliberalism may make large parts of the world more receptive (if not "safer") places for Americanized forms of accumulation and regulation, but by the same token it also amplifies their structural vulnerabilities, while reducing the scope for creative (and self-renewing) institutional learning. The specter is therefore raised that the very same channels through which the neoliberal project has been generalized may subsequently become the transmission belts for rapidly diffusing international crises of overaccumulation, deflation, and serial policy failure. Indeed, as we hover on the brink of a global recession—the first since the 1930s, when an earlier form of liberalism was the commonsense of the time—neoliberalism may be about to face its sternest test of credibility and legitimacy.

Finally, as the deleterious social consequences and perverse externalities of neoliberal economic policies become increasingly evident and widespread, the foundations may be inadvertently created for new forms of translocal political solidarity and consciousness amongst those who find themselves marginalized and excluded on a *global* basis (Hardt and Negri 2000). The globalization of the neoliberal project has therefore been tendentially (though not necessarily) associated with the partial globalization of networks of resistance. And while one of the conspicuous strengths of roll-back neoliberalism was its capacity to disorganize sources of (actual and potential) political opposition, roll-out neoliberalism is becoming just as conspicuously associated with disruption and resistance, as the process of deep neoliberalization has created new basing points, strategic targets, and weak spots. New forms of strategically targeted resistance may therefore represent both obstacles to neoliberalization and spurs to its continuing transformation.

Conclusion: Neoliberal(ism) Rules?

In many respects, it would be tempting to conclude with a teleological reading of neoliberalism, as if it were somehow locked on a course of increasing vulnerability to crisis. Yet this would be both politically complacent and theoretically erroneous. One of the most striking features of the recent history of neoliberalism is its quite remarkable transformative capacity. To a greater extent than many would have predicted, including ourselves, neoliberalism has demonstrated an ability to absorb or displace crisis tendencies, to ride—and capitalize upon—the very economic cycles and localized policy failures that it was complicit in creating, and to erode the foundations upon which generalized or extralocal resistance might be constructed. The transformative potential—and consequent political durability—of neoliberalism has been repeatedly underestimated, and reports of its death correspondingly exaggerated. Although antiglobalization protests have clearly disrupted the functioning of "business as usual" for some sections of the neoliberal elite, the underlying power structures of neoliberalism remain substantially intact. What remains to be seen is how far these acts of resistance, asymmetrical though the power relations clearly are, serve to expose the true character of neoliberalism as a *political* project. In its own explicit politicization, then, the resistance movement may have the capacity to hold a mirror to the process of (ostensibly apolitical) neoliberalization, revealing its real character, scope, and consequences.

Just as neoliberalism is, in effect, a form of "high politics" that expressly denies its political character (Beck 2000), so it also exists in a self-contradictory way as a form of "metaregulation," a rule system that paradoxically defines itself as a form of *anti*regulation. In their targeting of global rule centers like the WTO, the IMF, and the G8, resistance movements seek not only to disrupt the transaction of neoliberal business but also to draw attention to the inequities and perversities of these rule regimes themselves. At the same time, however, these rule systems cannot unproblematically be reduced to institutional condensates like the WTO, because one of the funda-mental features of neoliberalism is its pervasiveness as a system of *diffused power* (Hardt and Negri 2000). Contemporary politics revolve around axes the very essences of which have been neoliberalized. As such, neoliberalism is qualitatively different from "competing" regulatory projects and experiments: it shapes the environments, contexts, and frameworks within which political-economic and socio-institutional restructuring takes place. Thus, neoliberal rule systems are perplexingly elusive; they operate between as well as within specific sites of incorporation and reproduction, such as national and local states. Consequently, they have the capacity to constrain, condition, and constitute political change and institutional reform in far-reaching and multifaceted ways. Even if it may be wrong-headed to characterize neoliberalization as some actorless force-field of extra-local pressures and disciplines—given what we know about the decisive purposive interventions of think-tanks, policy elites, and experts, not to mention the fundamental role of state power itself in the (re)production of neoliberalism—as an *ongoing ideological project* neoliberalism is clearly more than the sum of its (local institutional) parts.

Thus, it is important to specify closely—and challenge—the extralocal rule systems that provide a major source of neoliberalism's reproductive and adaptive capacity. Local resistance—especially strategically targeted local resistance—is a necessary but perhaps insufficient part of this task. Crucially, neoliberalism has been able to make a virtue of uneven spatial development and continuous regulatory restructuring, rendering the macro power structure as a whole partially insulated from local challenges. In addition, progressive local alternatives are persistently vulnerable, in this turbulent and marketized environment, to social undercutting, institutional over-loading, and regulatory dumping. This is not to say that the hegemony of neoliberalism must necessarily remain completely impervious to targeted campaigns of disruption and "regime competition" from progressive alternatives, but rather to argue that the effectiveness of such counterstrategies will continue to be muted, absent a phase-shift in the constitution of extralocal relations. This means that the strategic objectives for opponents of neoliberalism must include the reform of macroinstitutional priorities and the remaking of extralocal rule systems in fields like trade, finance, environmental, antipoverty, education, and labor policy. These may lack the radical edge of more direct forms of resistance, but as intermediate and facilitative objectives they would certainly help to tip the macroenvironment in favor of progressive possibilities. In this context, the defeat (or failure) of local neoliberalisms—even strategically important ones—will not be enough to topple what we are still perhaps justified in calling "the system." It will continue to be premature to anticipate an era of "push-back" neoliberalism, let alone the installation of a more progressive regulatory settlement, until extralocal rule regimes are remade in ways that contain and challenge the forces of marketization and commodification—until there is a far-reaching *deliberalization* of spatial relations.

Acknowledgments

An earlier version of this paper was given at the seminar "Neoliberalism and the City" at the Center for Urban Economic Development, University of Illinois at Chicago, September 2001. We would like to thank all of the participants for the incisive discussion, which helped to clarify our thinking and also to thank Neil Brenner, Nik Theodore, Mark Goodwin, and Nigel Thrift for their perceptive comments and suggestions. The usual disclaimers apply.

References

Allen J, Massey D and Cochrane A, with Charlesworth J, Court G, Henry N and Sarre P (1998) *Rethinking the Region.* London: Routledge

Amin A (1997) Placing globalisation. *Theory, Culture, and Society* 14:123–137

Amin A (1999) An institutionalist perspective on regional economic development. *International Journal of Urban and Regional Research* 23:365–378

Beck U (2000) *What is Globalization?* Cambridge, UK: Polity Press

Bluestone B and Harrison B (2000) *Growing Prosperity.* Boston: Houghton Mifflin

Bond P (2000) *Elite Transition: From Apartheid to Neoliberalism in South Africa.* London: Pluto

Bourdieu P (1998) *Acts of Resistance.* Cambridge, UK: Polity Press

Bourdieu P and Wacquant L (2001) NeoLiberalSpeak: Notes on the new planetary vulgate. *Radical Philosophy* 105:2–5

Brenner N (1999) Globalisation as reterritorialisation: The rescaling of urban governance in the European union. *Urban Studies* 36:431–451

Cochrane A, Peck J and Tickell A (1996) Manchester plays games: Exploring the local politics of globalization. *Urban Studies* 33:1319–1336

Deacon A (2000) Learning from the US? The influence of American ideas upon "New Labour" thinking on welfare reform. *Policy and Politics* 28:5–18

DeMartino, G F (2000) *Global Economy, Global Justice: Theoretical Objections and Policy Alternatives to Neoliberalism.* London: Routledge

Dicken P, Peck J and Tickell A (1997) Unpacking the global. In R Lee and J Wills (eds) *Geographies of Economies* (pp 158–166). London: Arnold

Gough J (1996) Neoliberalism and localism: Comments on Peck and Tickell. *Area* 28:392–398

Hardt M and Negri A (2000) *Empire.* Cambridge, MA: Harvard University Press

Harvey D (1989) From managerialism to entrepreneurialism: The transformation of urban governance in late capitalism. *Geografiska Annaler* 71B:3–17

Jessop B (1989) *Thatcherism: The British road to post-Fordism.* Essex Papers in Politics and Government 68. Essex: Department of Government, University of Essex

Jessop B (ed) (2001) *Regulation Theory and the Crisis of Capitalism 1—The Parisian Regulation School.* Cheltenham: Elgar

Jessop B and Peck J (2001) Fast policy/local discipline. Mimeograph. Lancaster: Department of Sociology, Lancaster University

Jessop B, Bonnett K, Bromley S and Ling T (1988) *Thatcherism.* Cambridge, UK: Polity

Jessop B, Peck J and Tickell A (1999) Retooling the machine: Economic crisis, state restructuring, and urban politics. In A Jonas and D Wilson (eds) *The Urban Growth Machine* (pp 141–159). New York: SUNY Press

Larner W (2000) Theorising neoliberalism: Policy, ideology, governmentality. *Studies in Political Economy* 63:5–26

Lipietz A (1987) *Mirages and Miracles.* London: Verso

Overbeek H (1990) *Global Capitalism and National Decline.* London: Unwin Hyman

Peck J (2001) *Workfare States.* New York: Guilford

Peck J (2002) Labor, zapped/growth, restored? Three moments of neoliberal restructuring in the American labor market. *Journal of Economic Geography* 2(2): 179–220

Peck J and Theodore N (2001) Exporting workfare/importing welfare-to-work: Exploring the politics of Third Way policy transfer. *Political Geography* 20:427–460

Peck J and Tickell A (1994a) Jungle law breaks out: Neoliberalism and global-local disorder. *Area* 26:317–326

Peck J and Tickell A (1994b) Searching for a new institutional fix: The *after*-Fordist crisis and global-local disorder. In A Amin (ed) *Post-Fordism* (pp 280–316). Oxford: Blackwell

Peck J and Tickell A (1995) The social regulation of uneven development: "Regulatory deficit," England's South East and the collapse of Thatcherism. *Environment and Planning A* 27:15–40

Peet R, with Hartwick E (1999) *Theories of Development*. New York: Guilford

Piven F F (1995) Is it global economics or neo-laissez-faire? *New Left Review* 213: 107–114

Piven F F and Cloward R (1998) *The Breaking of the American Social Compact*. New York: New Press

Polanyi K (1944) *The Great Transformation*. Boston: Beacon Press

Setterfield M (2000) Is the new US social structure of accumulation replicable? Mimeograph. Hartford, CT: Department of Economics, Trinity College

Tickell A (1999) Unstable futures: Controlling and creating risks in international money. In L Panitch and C Leys (eds) *Socialist Register* (pp 248–277). Rendleshem: Merlin Press

Tickell A and Clark G L (2001) New architectures or liberal logics? Interpreting global financial reform. Mimeograph. Bristol: School of Geographical Sciences, University of Bristol

Tickell A and Peck J (1995) Social regulation *after* Fordism: Regulation theory, neoliberalism and the global-local nexus. *Economy and Society* 24: 357–386

Tickell A and Peck J (1996) Neoliberalism and localism: A reply to Gough. *Area* 28:398–404

Tickell A and Peck J (forthcoming) Making global rules: Globalisation or neoliberalisation? In J Peck and H W-C Yeung (eds) *Making Global Connections*. London: Sage

Veltmeyer H, Petras J and Vieux S (eds) (1997) *Neoliberalism and Class Conflict in Latin America: A Comparative Perspective on the Political Economy of Structural Adjustment*. New York: St. Martin's Press

Wacquant L (1999) How penal common sense comes to Europeans: Notes on the transatlantic diffusion of the neoliberal *doxa*. *European Societies* 1:319–352

Wade R (2001) Showdown at the World Bank. *New Left Review* 7:124–137

Weiss L (1998) *The Myth of the Powerless State: Governing the Economy in a Global Era*. Ithaca, NY: Cornell University Press

Jamie Peck is a Professor of Geography and Sociology at the University of Wisconsin, Madison. He is author of *Work-Place: The Social Regulation of Labor Markets* (New York: Guilford, 1996), *Workfare States* (New York: Guilford, 2001), and a number of articles on issues related to labor-market restructuring and policy, welfare reform, urban political economy, and theories of regulation. He is currently coeditor of *Antipode* and joint editor of *Environment and Planning A*. His research at the present time is concerned with two issues: the reconstitution of temporary and low-wage labor markets and the political economy of policy transfer.

Adam Tickell is a Professor of Human Geography at the University of Bristol and has previously lectured at the universities of Leeds, Manchester and Southampton. He is editor of *Transactions, Institute of British Geographers* and review editor of the *Journal of Economic Geography*. His work explores the geographies and politics of international financial reform, governance structures in the UK, and the reconfiguration of the political commonsense.

[23]

Globalization and the politics of local and regional development: the question of convergence

Kevin R Cox

Globalization, and the increased exposure to international competition that it has supposedly induced, has led to expectations of institutional convergence in, among other things, local and regional development policy and the politics surrounding it. There have been changes in the United Kingdom, but not of the decentralizing, neo-liberalizing form anticipated. A comparison of the British with the very different, highly decentralized, American case seeks to shed light on this. Emphasis is placed on both the strongly embedded nature of institutions and on misunderstandings about the strength of the forces of globalization.

key words politics of local and regional development globalization class struggles state form United Kingdom United States

Department of Geography, The Ohio State University, Columbus, OH 43210, USA
email: kcox@geography.ohio-state.edu

revised manuscript received 8 April 2004

Introduction

The point of departure for this paper is the question of convergence – institutional, technical, macro-economic policy – that emerged in the context of the debate about globalization. The claim was that the pressures exerted on corporations and state agencies, through international competition for markets and for inward investment, would result in the adoption of similar practices, institutional forms and the like. An extreme version of this was 'the race to the bottom' argument. This was later tempered by a recognition that the increased openness of national economies could only affect policy in combination with national specificities; specificities of both a limiting and enabling variety (Kenworthy 1997; Henderson 1999). This did not mean that there might not be some homogeneity of response. As Storper (1987) argued, the local does indeed make the global in many respects, and we should be alert to convergences as much as to continuing difference. The idea of convergence has also found a home in discussions of the theory and practice of local and regional development policy. This has in turn generated expectations of some

convergence in the sorts of politics that emerge around these policies. The paper provides a critical examination of ideas of convergence with respect to the policy and politics of local and regional development, and draws on the evidence provided by two cases: those of the United Kingdom and the United States.

What convergence might look like is unclear. Reading the tea leaves suggests that an important element of a local and regional development policy appropriate to the times and circumstances would be one that decentralizes powers and responsibilities to very local levels and induces competition between them, either through local governments or growth coalitions. There is evidence for this in some of the initiatives adopted by British governments, of both political stripes, over the last 20 years. The idea of glocalization (Jessop 1994 1999 2000; Swyngedouw 1997a 1997b; N Brenner 1998) explicitly links changes in the state's scale division of labour with respect to local and regional development to both the pressures unleashed by globalization and to the possibilities that it opens up. This is complemented by the evident popularity of promoting industrial clusters – an idea developed in the US – as the vehicle of choice for bringing about a revival of regional economies.[1]

One virtue of this conception is that it is consistent with the neo-liberal agenda and with the ultimate purpose of these policies, the reimposition of the law of value. Arguably, this is what underlies the globalization that has occurred since the early 1970s, and it in turn, through increased international competition for trade and investment, has served the original goal of disciplining capital. It does not seem unreasonable, therefore, to expect powerful pressures in the direction of a neo-liberalized scalar division of labour of the state as it confronts issues of local and regional development.

In this respect, the post-war US might be regarded as an appealing template. It exhibits a high level of decentralization to local and regional levels of powers and responsibilities appropriate to the economic development function. These include powers over land use planning, financial incentives and also that fiscal responsibility which enervates the search for taxables. Not uncoincidentally, localities and states engage in an intense competition for inward investment and enhanced positions in wider geographic divisions of labour. Historically, the British case, as it has been defined for most of the post-war period, might be regarded as the antithesis of the American one: a scale division of labour characterized by a very high level of top-down control, working in particular through a centrally regulated planning system and regional selective assistance. The assumption in this paper, therefore, is that a comparison of the two instances could shed light on the possibilities and limits of institutional convergence. To what extent is convergence apparent, either in ways unanticipated, or in ways which might bring the British policy and politics of local and regional development closer to the American version? And how might one account for these empirical trends, or alternatively, their absence?

An important context for this discussion is provided by Harvey's (1985a 1985b) theorizing of what he called 'the geopolitics of capitalism'. He outlines a necessary tension between fixity and mobility in the politics of capitalist development, or what one might call in the current parlance, globalizing and localizing tendencies. Struggles among the classes that, in virtue of their relative immobility, cannot escape one another, are the condition for 'settlements' which then provide the basis for what he calls territorial structured coherences. These would embrace production technologies, patterns of consumption, physical and social infrastructures, the division of the product between capital and labour, along with policies regulating the labour process and labour organizing.

Given the rise of competitors elsewhere, shifts in the circulation of value, however – what Storper and Walker (1989, chap 1) called capital's 'inconstant geography' – are inevitably unstable. This initiates a politics of restructuring as classes and class fractions struggle to shift its costs onto others. Existing conditions are reworked, new models of state, labour or corporate practice imported from elsewhere, the redistribution of value through the state challenged, and so on.

We should remark here, however, on just how problematic this is likely to be. If one is to take the notion of 'coherence' seriously, then *any* change has wide social ramifications. It is not just the redistribution of values that is likely to be at stake. In addition, there are ingrained practices, deeply held convictions, structures of cooperation to be overturned. None of this can be easy. There are also scale aspects to this. What underpins structured coherences at more local levels is likely some more general condition having to do with policy or state structure, and over which power can be exerted only through a politics of scale that would require much wider alliances. Local and regional development policy is one aspect of structured coherences. Like other aspects, it is conditioned by class struggles, as well as the state form within which struggle occurs. It is unlikely that it would prove any more malleable when the question of a 'necessary' restructuring arises.

The paper now divides into three major sections. In the first of these I provide an outline of the two politics of local and regional development in their classical forms and their necessary conditions of existence. A second section situates them with respect to arguments about globalization and identifies some of the – very modest – changes that have occurred in local and regional development policy and what they might have to do with globalization. A final section then addresses the British case in the context of the American with a view to understanding why the changes that have occurred in the former have not been more dramatic.

Contrasting politics of local and regional development

At a very high level of abstraction, the fundamental process around which local economic development occurs is not that different between the US and Britain. In both instances it is a relation between local government and private investors within a set of limits and possibilities laid down by central branches of the state. Businesses negotiate with local government about planning permissions, re-zonings, some financial aid, infrastructural changes perhaps. It is, however, those centrally defined conditions that have varied quite starkly between the two instances. In addition, and historically, it was not always the case in Britain that it was a relation between local government and private investors, since nationalized industries also made their location decisions, and central government could be involved in them. This involvement, more-over, could be on specifically local and regional development grounds.

In the United Kingdom, the central government presence has been much stronger, and that remains the case to the present day, albeit with some recent, quite mild, weakening. Elements of this contrast include, in the first place, policies aimed at steering new investment into areas of persistently high unemployment. These have worked in the British case through grants of various sorts to firms mak-ing new investments in areas that turned out to be of quite limited geographic extent, although during the 1960s there was an increase in their magnitude (Rees and Lambert 1985, 104). These policies were supplemented by attempts to limit investment in both industry and office employment in more buoyant local economies.

Local economic development is conditional in its geography on land use plans and on the permissions made within the context of those plans. In both coun-tries local governments have their land use planning departments, their plans and their policy positions. But in the American case, these are almost wholly unconstrained by state interventions. For sure, there is a body of law in each state governing land use zoning, subdivision regulations, building standards and the like. In contrast to the British instance, however, there is nothing corresponding to the structure plans, and regional planning guid-ance, with which local government plans must be consistent. Nor is there any possibility of recourse to a higher authority, other than a judicial one, when the necessary land use permits are refused. This is a conception of planning that, it is widely believed, should start at the bottom rather than at the top.

In addition to providing an overall regulatory context for the local economic development process,

successive British governments have also played a more active role. An early instance of this was the new town programme. In addition to relieving congestion in the major conurbations, in some instances it also provided new poles for develop-ment by concentrating labour resources. This was particularly the case with new towns in the old coalfield areas, like Peterlee and Washington in the Northeast,[2] Glenrothes in Fifeshire and Cwmbran in South Wales. Without the leverage provided over housing choices by a large public housing sector and the institution of local housing authority waiting lists, it seems unlikely that these projects could have come to easy fruition. Public ownership of industry also provided the government with tools for altering the geography of the space econ-omy and for stimulating local economic develop-ment.[3] And although the new town programme ended 30 years ago, and through denationalization of industry the government has lost some of its leverage over local economic development, analogous initiatives in the form of the Urban Development Corporations continued until quite recently.

There is virtually nothing comparable to this in the US. Certainly, there are 'new towns' – typically originating in small settlements on the fringes of major metropolitan areas that, through the private development process can quickly acquire the trap-pings of the urban and which are as 'new' in the sense of mushroom growth as Britain's new towns. But they have never been part of a concerted government policy, subject to goals of relieving congestion by, for example, choosing sites which would minimize commuting back to major cities. Likewise, even if state and federal governments have never taken industries into public ownership, they have controlled huge amounts of technically 'footloose' investment which can alter the topogra-phy of development: military bases, federal office buildings, branches of the state university, federal and state prisons, state and federal research facili-ties. But again, this has never been used to impose some top-down vision of what the map of develop-ment should look like; redistributing in order to create some measure of interregional equality or possibly to underpin a settlement policy. Rather what state and federal branches have had under their control has been up for grabs for local govern-ments, the local growth coalitions in which they typically play a part, and their friends in state and federal legislative branches looking for new forms of patronage. In consequence, it should not come as a surprise to learn that federal expenditures across the states have virtually no equalizing effect in terms of per capita incomes.

Historically, therefore, the scalar organization of the state, as it pertains to local economic development in the United States and Britain, has been quite different; in fact, starkly so. In the British case there have been clear central government attempts to control and regulate the local economic development process in order to achieve some objectives for the country as a whole; objectives which can be referred to some conception of 'the national interest'. In the American case, however, central control has been very limited indeed. Local economic development policy has been formulated locally and with minimal central oversight. It has been accompanied, for reasons which will become clearer later on, by an intense territorialization of its concomitant politics. Local governments are part of local growth coalitions, made up largely of private businesses with a strong interest in the expansion of 'their' local economies. Under these conditions, state and federal government become objects of lobbying for various expenditures and regulatory relief that will redound to local economic growth in some way. Where in the post-war period British policy and politics was led from the centre, American policy and politics started in the localities.

It is this which, in conjunction with the form of the American state, accounts for the strongly centrifugal effects to which *any* central state initiatives are subjected there. Accordingly, seemingly *any* attempts to channel government money into areas of significant economic distress along the lines of British development area policy have been severely compromised. The same tendencies can be observed in respective enterprise zone legislation. Whereas in the British instance local authorities were asked to bid for a very limited number of enterprise zone designations, in the American case the legislation of the individual states has been extraordinarily permissive. While in Britain there have only ever been 18, there are currently 39 in California, 91 in Illinois, 52 in New York State and, in the State of Ohio, with just under a sixth of the population of the United Kingdom, there are well in excess of 300.

So how might one understand these very considerable differences in local and regional development policies and the politics surrounding them? My fundamental assumption here is that country-specific policies of local and regional development, along with their scalar organization and forms of politics, are constructed over long periods of time, and in the context of geohistorically very specific conditions. The accumulation process and the class struggles that constitute it, and which it deepens, occur within a (geographically) constituted field of limits and possibilities that varies greatly: one of state forms, national imaginaries, geographically uneven development and historically embedded interests and identities, among other things. It is within this field of limits and possibilities that contending, class-pertinent forces, with interests in places at different geographic scales, try to organize and forge coalitions and develop supporting imaginaries around particular state projects – again at a variety of scales – aimed at furthering those interests.[4] At the level of form, the variety of possibilities cannot be overemphasized. Class has been a significantly more overt organizing principle in Britain than in the US, while territory has been more prominent in the latter. But projects can also be organized around ethnicity, some putative national affiliation or, to infer from the success of the BNP in particular labour markets in Britain and Le Pen's National Front in France, racial attributions. Moreover, particular understandings of local and regional development, of interests in the process, of its conditions, both limiting and facilitating, and to the extent that they are realized through practices that at least in some instances are effective, tend to endure and acquire the status of a received wisdom. Accordingly they become very difficult to change when subjected to new forces of, for example, a neo-liberalizing kind.

State forms have provided an essential context for these struggles and an object of their transformative power. The American state, as befits the checks and balances for which it is typically celebrated, is defined by a high degree of fragmentation of state power: parties are weak relative to individual legislators, the executive branch is checked not just by Congress but by the committee system, which in turn provides a vehicle for delivering to those individual legislators or factions of them, district- or state-specific fiscal flows or regulatory relief, and in ways often contrary to party programmes. The internal organization of the state, including its federal form and the powers that the states delegate to local governments, reinforces that fragmentation as does the penchant for more commodified forms of intervention in which state agents rely heavily on markets in order to achieve their purposes. The latter is reinforced by the divided nature of state power, including its territorial division, since this has tended to further the commodification of the state itself.

The British state is quite different. State power is highly concentrated in that overlapping of the national legislative and executive branches known as 'the government'. Strong party discipline prevails. Committees are relatively weak. The state for most of the post-war period has been highly unitary, with local government subjected to strong central oversight and limited in its ability to create fiscal space of its own. Modes of intervention, at least in form, have leaned towards the decommodified, seemingly challenging the wisdom of the market, as we noted in the discussion of attempts to regulate the location of industrial and office investments.

Working within, with, against, these very different state forms we find, again, quite sharply differentiated labour movements. The greater strength of the labour movement in Britain is widely acknowledged. It has been apparent both institutionally and, one might argue, in terms of results. Union membership has always been significantly higher in Britain and, since the Second World War, at least twice as high. In the United States, however, there is no social democratic party, even of the relatively conservative caste that has been true of the Labour Party, to contend for state power. Policy and changes in state form in the British case bear out the greater strength of the popular forces. The welfare state is altogether stronger, a relatively large fraction of the population still lives in public housing, industries were taken into public ownership, access to health services is universal, and so on. Accordingly, the discursive environment, the relative balance of beliefs between the competing claims of individualism and collectivism, and of growth and distribution as policy priorities, remains different. This is so even after the markedly privatizing, pro-growth impetus of the Thatcher years.

One should recognize at the outset, however, the clear possibility of a link between state form and the strength of respective labour movements. This is not to attribute causality to any one direction. Rather one could argue that state forms were a condition for the development of respective labour movements as much as they were a result of their respective transformative powers. For sure, there is a tradition in British historiography linking the

growth of the central branches of the state to the intense challenges often faced by an emergent industrial capitalism at the local level (for example, Foster 1974; Corrigan and Sayer 1985). But to what extent was it the centralization of power that encouraged class organization at a national scale to begin with? In the US, the obstacle that the constitution has provided to working class organization frequently has been recognized. As Madison famously argued in the *Federalist Papers* (no 10; http://lcweb2.loc.gov/const/fed/fed_html, accessed 18 April 2004), the 'majority . . . must be rendered, by their numbers and local situation, unable to concert and carry into effect schemes of oppression' (see also Lazare 1998). On the other hand, American labour has had its moments, moments at which national politics might have taken a decisive turn in its favour, if conjunctural forces had not conspired against it, as Katznelson (1989) has argued. So there remains an unresolved tension between state form and mass politics.

In the event, the combination of a weak labour movement and a fragmented state has been a divided capital, its diverse forces inimical to state initiatives attempting to unite them around some grand vision. American capital has never felt the need to unite around a state project because it has never felt threatened in the way in which British capital has.[5] Rather, and as Lowi (1969) has argued, national politics in the US has been an interest group politics, with labour as one interest among many, rather than a class one. The federal state has been parcelled out and, to the extent that there is a national interest, it is simply defined as the outcome of the struggle between these various 'special interests'. This is a process that has clearly had its territorial aspects, though they cannot be read off from either state form or the relative weakness of the labour movement.

It is the dispersion of state power that has tended to select in, in more direct, unmediated ways, the representation of territorial coalitions of various sorts. Given the organization of the state, these commonly include state officials anxious about revenue streams or simply votes from an electorate primed by an incessant drumbeat of business-orchestrated concern about the local economy and its significance for 'jobs'. The, often quite specific, demands of these coalitions can then get funnelled upwards, again in a fairly direct way, through local legislators. The primary system of nominating candidates has provided incentives to put together specifically local coalitions of forces for electoral purposes. Weak party discipline,[6] the ability of congress-persons and senators to then get on the committees whose deliberations have the most serious import for their respective districts or states, have subsequently facilitated delivery to those same local coalitions. In short, territorial coalitions achieve their goals at higher levels of the state – both federal and in the constituent states – through the networking of 'their' legislative representatives in committees and with the state agencies anxious for affirmative votes from those committees,[7] and through – often – bipartisan voting blocs on house and senate floors.

But in no way can the presence or absence of territorializing imaginaries and practices be *reduced* to state forms. Local governments with strong interests in expanding respective tax bases are an important condition for them, but their appeals will have limited effect if they are restricted to concerns about tax burdens. In this regard, one of the absolutely crucial differences between the American and British cases has been the presence or otherwise of strong, local capitalist classes. American cities, metro areas, suburbs, almost invariably have sizeable capitalist classes – groups of firms, individual operators, franchise holders, perhaps, who have major stakes in particular local economies and in their expansion. These have been the heart of growth coalition politics and have consisted, with some variation depending on scale, of developers, local newspaper and media empires, landlords, the public utilities, sometimes a bank or two that have yet to take advantage of the changes in the law permitting inter-county and inter-state branching, the owners of auto dealership and beverage franchises. All of these experience some sort of local dependence by virtue of some combination of local knowledge, property investments, dependence on a specifically local market, fixed capital assets, which limit movement into other, geographically defined, markets. This is a local dependence, moreover, that can deepen as a result of the way the agents involved constitute a local investment community that commits funds to development projects simply by virtue of their local knowledge and the super-profits available to them as a result of their monopoly of that information. As we will see, there has been nothing quite like this in Britain and that remains so to the present day.

Adding an extra wrinkle to the role of local initiatives and their ability to make a difference has been the relative co-optability of the masses. Local growth coalitions have enjoyed a good deal of success in achieving popular support for, or at least acquiescence in, the policies they pursue, even though they often have adverse effects on the popular classes. The ability to co-opt has been particularly evident in plant closure issues where local growth coalitions have been instrumental in putting together rescue packages but where union locals have often gone along with plans for so-called 'worker givebacks' as part of the package. Certainly one can say that, in trying to explain this, labour has indeed its own forms of local dependence, as was discussed earlier in the paper and this could make them susceptible to growth coalition overtures. But this is true of Britain too.

In significant part it has to do with the strengths of respective labour movements and the discursive environments that have been built up in their wake. Arguments about property taxes, property values, competition with other localities or regions, appeal to the individualizing element in popular ideology, just as claims about moving up the ladder to 'major league city' status appeal to the search for some sort of community not available in other forms (Cox 1999, 29–34). But in addition, and particularly fuelling the co-optability of labour in the workplace, there has been a difference in degrees of uneven development of the labour movement. In the American case, union density has always been geographically highly variable, with organization typically much stronger in the industrialized states of the Northeast, Midwest and West Coast. As firms sought out locations in less unionized states, and ones usually with lower wages, such as the Sunbelt states of the South, the Great Plains and the Mountain West, this was a challenge to a clearly localized faction of labour and made it vulnerable to appeals from capital to join a cross-class alliance which would stem the losses.

The contrast provided by the British case is considerable. It is not just that a state form in which power is highly concentrated at the centre makes it an unsuitable vehicle for initiatives aimed at securing advantages for particular localities or regions. It is also the weakness of local interests, either in the state or in civil society. Unlike the American case, banks and utilities have lacked the sort of local dependence that has put them in the forefront of local economic development initiatives there. Historically, banks have been extraordinarily concentrated in their ownership, and electricity and gas were brought under public ownership shortly after the Second World War. Local growth coalitions have therefore lacked the strong business base that they have historically enjoyed in the US. Furthermore, there has been only very, very limited incentive for local governments themselves to get involved. This stems in large part from the virtual absence of local government dependence on its own tax base. Local governments have limited interest in bolstering property values – which, through the rates, is the main source of locally raised government revenue – because, and in contrast to the American case, it won't do them much good. Rather, a central feature of local government finance has been the centrally provided 'rate support grant'. This doles out by far the larger proportion of local government revenue – 75 per cent on average – through a formula that takes into account both local resources, including the rate base, and local needs; so holding needs constant, increasing the resource base will simply result in a lowering of the rate support grant that local government receives. In addition, any local initiative along American lines aimed at securing concessions from local workers would quickly encounter limits imposed by the relative strength of the labour movement and the fairly even geographical spread of that strength. There might be variations between urban and rural areas, but not the sort of interregional variation one finds in the US.[8]

In sum, in explaining the peculiarities of the American politics of local and regional development, one has to look to the way in which strong local and regional interests have been inserted in a state whose highly fragmented form facilitates their expression; and to a national balance of class forces, at both material and discursive levels, which has tended to work to the disadvantage of the vast majority. In consequence of the weakness of the labour movement and the enabling features of the state, American business has remained divided, not just unable but uninterested in uniting around a national programme which would subordinate the interests of some to the advantage of American capital as a whole.

In the United Kingdom, on the other hand, a different disposition of class forces, along with a highly centralized state, has resulted in a greater susceptibility to central state definitions of what the national economy, its sectoral disposition, its balance between financial and industrial fractions and its geography should look like. This does not mean that those national visions have always been translated into practice effectively. But in contrast to the US, they have existed and there has been enduring appeal to them.

Urban and regional policy provides an illuminating case in point. For most of the post-war period, the United Kingdom had a relatively coherent urban and regional policy designed to take the labour- and housing-market pressure off major urban centres by redistributing employment to new towns and to areas of relatively high unemployment. This was always interpreted in terms of national goals that capital could find acceptable. For sure, the initial justification of depressed area policy was in terms of the welfare of the unemployed, but in the immediate post-war period of shortages of building materials, making use of existing infrastructure made sense from the standpoint of a national accumulation strategy. Later, in the 1960s, the rationale shifted. The rapidity with which economic expansion in the 1950s led to overheating and subsequent deflationary measures suggested that regional policy could work in a counter-inflationary manner, and that was the way it was defined.

In the US no such rationalization was required because there was no strong capitalist interest in, or willingness to go along with, a federal urban and regional policy. Nowhere is the disunity of American capital more apparent than in the cannibalization of those measures that the federal governments *have* proposed. Indicative is a programme in the 1960s, orchestrated by the Department of Commerce's Economic Development Administration, which aimed to provide various soft loans and grants for infrastructural investments in areas of high unemployment. But at the end of the legislative process some 80 per cent of the population found themselves living in eligible areas (Barnekov *et al.* 1989, 111).[9]

Recent changes and the question of 'globalization'

These are the classical forms assumed by local and regional policy and politics in the post-war period. From the early 1970s onwards, however, there were signs of change; ones that have picked up momentum over the period since then. Some of this was common across different countries, including the United Kingdom and the United States, and can be linked to changes in the overall policy climate, particularly its neo-liberal character and the increasing internationalization of the economy that was subsequent to this. There is, however, only very limited sign of convergence on a decentralized model for local and regional policy. In this section the changes that have occurred are described and their links to globalization, and more particularly the neo-liberal policies underlying it, exposed. In the penultimate section of the paper I return to consider why British policy continues to be so centralized, despite the continuing challenges of intensified international competition in product markets and for inward investment.

Change in respective policies and politics of local and regional development

One of the changes that has occurred in both countries is a heightened level of territorial conflict around development issues. This has been apparent since the late 1970s. In Britain, during the 1980s, the support for the two major political parties took on a distinctly regionalized character. The Labour Party was dominant in the relatively depressed labour and housing markets of Wales, the North and Scotland, with the Conservative Party dominant in the (relatively) booming London and the Southeast (Savage 1987). This changed somewhat in the 1990s as it became evident to the Labour Party, in the guise of 'New Labour', that in order to regain power their programme had to be modified in a more private-sector friendly direction in order to carry some of the constituencies in the Southeast.

These tensions, ongoing to the present day, have been manifest in specific territorial issues.[10] The so-called Barnett formula, according to which local authorities in Scotland and Wales get considerably higher per capita grants than local authorities in England – something introduced in 1976 in order to dampen down separatist sentiment – has come under attack.[11] Along similar lines, cities are now making calculations of the difference between the share of national taxes that their residents provide and the share of national domestic spending that they receive. This was an issue raised by all candidates for the recent mayoral election in London: an historic first in British local politics. This new interest in 'territorial justice' has been picked up by the regional studies literature (for example, Mackay 2001; Morgan 2001). Also symptomatic has been the establishment of the Core Cities group by the cities of Birmingham, Bristol, Leeds, Liverpool, Manchester, Newcastle, Nottingham and Sheffield. According to their mission statement their goal is

to work in partnership with Government and other key stakeholders to promote and strengthen Core Cities as drivers of regional and national competitiveness and prosperity with the aim of creating internationally competitive regions. (http://www.corecities.com/coreDEV/about.html, accessed 18 April 2004)

The territorialization of the politics of local and regional development could not be clearer.[12]

Similar tensions are evident in the American case. In the mid to late 1970s a regional question got defined around an emerging differential in terms of income and employment growth. Central to this was the distinction between a so-called Sunbelt and Coldbelt (Markusen 1987). In an attempt to shift the channels through which state economies are irrigated with federal money, a group of senators and congress-persons, with the backing of local governments and growth coalitions, organized the Northeast Midwest Congressional Coalition. Much of their focus was a demand for greater equality of treatment based on the notion of some sort of balance between what a state sends to the federal government in the form of taxes and what it gets back in the form of expenditures. There have also been calls for a 'levelling of the playing field', meaning abolition of the right-to-work clause of the Taft-Hartley Act which gives some Sunbelt states an advantage in attracting the most seriously union-averse of corporations; and a federalization of the welfare state with standards defined as those presently prevailing in the more generous Coldbelt states. The objective has been to take away some of the so-called 'business climate' advantages enjoyed by Sunbelt states and their localities.

The interest in 'business climate' is also expressive of some change in the institutional character of local and regional development policy in the United States. It is, in fact, a concern for honing it, comparing it with that of other states, which seems to be one element of what Jessop (2002) has referred to as a refunctionalization of the central government's role in local and regional development. Other elements include enhanced financial assistance in the case of investments seen as having strategic significance: American states have been very active in easing the way financially for foreign automobile manufacturers, for example, or even in attracting the headquarters of major American firms, as was apparent in the recent bidding war for Boeing. And finally there is the turn to the so-called 'knowledge economy': part of the effort to create new niches in an international division of labour that is increasingly inhospitable for them in what have come to be known as lotech industries. This has been particularly evident in the states of the old Manufacturing Belt, where a number of them, including Michigan, Ohio and Pennsylvania, have all sponsored research and development agencies funded by public–private partnerships with the aim of providing a technical leg-up to existing industries and nurturing new ones.[13] These are typically specialized in particular lines and located so as to complement local specializations.[14]

Similar changes are evident in the British case. Although still presiding over a, now diminished, programme of aid to the distressed regions, the establishment of enterprise zones, and other relics of the old regional development regime, the British state now plays a central role in orchestrating the knowledge or learning economy. It also provides the necessary assistance for investments, like those of the German and Japanese automobile and electronics companies, regarded as of strategic importance, and ensures, through its labour policies, that Britain remains attractive to foreign investors.

The particular niche that British governments have tried to develop over the past 20 years or so has been as a relatively low cost platform for inward investors producing for the West European market and keen to get behind EU trade barriers. To the extent that the government has had a 'regional policy', one of regenerating the zones of de-industrialization outside of London and the Southeast, this has been it. But in order to carry it through, the issue of 'low costs' has had to be addressed. The struggle around whether or not Britain should opt into the EU's Social Chapter, requiring the introduction of management–labour councils to confer on issues like redundancies and plant closures, was central to this. Of all EU members, Britain was the only one that initially opted out.[15] The failure of successive British governments to agree to the Social Chapter remained a thorn in the side of the union movement until the decision to join in 2001. Nevertheless, continuing differences in labour law remain an issue. Firms often have plants both in Britain and elsewhere in the EU. The closure of a British plant and the retention of one in, say, Belgium or France is invariably interpreted in terms of the greater ease of closing plants in Britain.

This is not to ignore what is supposed to have been the big change in local and regional development policy in England: the creation of the (nine)

Regional Development Agencies (RDAs) in 1998 and the earlier emergence of sub-regional partnerships among local government authorities and with some business participation.[16] Each of the RDAs has broad responsibility for economic development within respective regions, orchestrating inward investment from outside the country, developing new industrial estates in areas of the country where private investment has been reluctant, and exercising power over regional planning through its own regional economic strategy: something which Regional Planning Guidance is statutorily obliged to take into account. However, ambitions for the funding and responsibilities of the RDAs were much greater before some central government departments stepped in and opposed the transfer of powers. Secretaries of State for Trade and Industry and Education and Employment resisted transfer of apparently relevant functions such as Regional Selective Assistance and funding for Training and Enterprise Councils (Mawson 1999). Moreover, the centralized planning apparatus remains in place, slowing down change in the national space economy and often frustrating local economic development plans and projects. Regional Development Agencies take the lead in the construction of regional economic plans, but these are subject to central monitoring and intervention. In other words, the significance of the RDAs should not be exaggerated.[17]

The role of globalization

Turning to how we might understand these changes, modest as they might be, one can certainly see why globalization as it is conventionally understood might be implicated. The dramatic changes in the international division of labour are a clear condition for the refunctionalization of central branches of the state towards a learning economy. The emergence of the idea of 'business climate' and correlative benchmarking likewise make little sense outside of the increase in FDI and foreign trade and the increased exposure of local economies to competition from outside.

Yet things are more complex than this. One can get a sense of what might be at stake through a critical examination of those theories of globalization which abstract from broader social changes; which see it, for example, as an emanation of technical changes or contingent factors such as the oil shock and the need to recycle petrodollars. One competing explanation would understand globalization as conditioned by the long downturn (R Brenner 1998) and the measures taken by Western governments and businesses to re-establish the conditions for profitability. The general tendency of these measures, of course, has been what has come to be known as neo-liberal. Some of the measures undertaken under that banner have had direct consequences for 'globalization' as it is conventionally understood. These would include the move to eliminate capital controls, and an easing of trade restrictions, associated, *inter alia*, with the rise of common markets and free trade areas. In addition, there have been moves whose relation to globalization are more mediated and complex. Among these are the privatization of previously nationalized industries exposing them to foreign ownership, once their more serious loss-creating assets had been eliminated in order to make them attractive purchases; the attempt to rein in the welfare state and the power of labour unions; and the increased commodification of the state through competitive tendering.

As numerous others have noted, there have been clear consequences for the politics of local and regional development. It would be difficult to contest that there has been an intensification of territorial competition as the economic bases of local and regional economies have been subject to increased challenges from elsewhere and local/regional/national quasi-monopolies have been undermined. Harvey (1989) wrote about the rise of urban entrepreneurialism, and while I think this is an overgeneralization, since American cities have been entrepreneurial going back at least to the 1960s,[18] it does have applicability to Western Europe, including Britain.

Equally important, however, and less emphasized, is the fact that the stakes for some localities/regions/countries have been intensified as a result of increased geographically uneven development.[19] This has had diverse origins. Old sectors have dramatically declined as new ones have – equally dramatically – risen. The removal of barriers to trade and investment has facilitated offshore relocation of lower skill, less knowledge-intensive manufacturing, though this has also been enabled by deskilling of labour processes and changes in transportation. Privatization and the selling off of nationalized industries has also had striking effects on local and regional employment as Ray Hudson (1986) has documented for the British case; to paraphrase him, the veritable 'wrecking of regions'. At the same time, there have been new and revivified localities and regions. The movement of money into speculative activity since 1970, the enhanced financialization of the global economy, along with the growth in trade of currencies, has stimulated the emergence of the so-called 'world cities'. Alongside rustbelts there have been silicon valleys, silicon fens and silicon prairies. The macroeconomic austerity that has been part of the neo-liberal medicine has accelerated these changes, driving low profit businesses, and the regions in which they are disproportionately located, to the wall.

It is in the context of this intensification of geographically uneven development and the lowering of barriers to movement and to spatial substitutability that I think we can make sense of recent changes in both the politics and policy of local and regional development. Central branches of the state have engaged in a twofold strategy in their refunctionalizations of the regional development role. On the one hand, they have tried to save what they can through a newfound emphasis on 'business climate'. On the other, they have moved to stimulate the creation of new, knowledge-intensive firms and sectors or to revivify existing ones, as in the case of London's financial services: sectors, in other words, that will be more immune to competition from firms in developing countries.

The heightening of territorial tensions and widespread notions of territorial exploitation have been common to both the United States and to Britain, as we have seen. The central branches of the state have become the foci of struggles around territorial redistribution and relief for lagging regions. In some cases there has been pressure for change in the state's scalar division of labour. As John Mawson (1999) has pointed out, these tensions have been one of the conditions for redrawing the institutional map of local and regional development policy in Britain. There is evidence of particularly strong support for the creation of the RDAs from a group of about 40 Northern MPs, along with rumours of a threat to disrupt the legislation governing devolution to Scotland and Wales unless the RDAs were speedily introduced after Labour's accession to power in 1997. Likewise, antagonism to the Barnett formula is especially strong in the North of England, and also in London where, among other things, the high cost of living, and a context of relatively uniform public sector wages across the country, has made hiring public service workers noticeably more difficult.[20]

Convergence revisited

So, while there has been some convergence between US and British policies of local and regional development, and while this can be linked to the tensions set up by globalization, or more specifically neo-liberal policy, the sort of decentralization of responsibility anticipated by glocalization theorists, and looked for by others in the academic industry, that is the study of local and regional development, is still a rather weedy growth. The question is: why?

To be sure, inserting such a scale division of labour within the British state appears to have been one objective of British governments, of both the major parties, since the early 1980s. This has taken a number of different forms, including the top-down creation of competition for discretionary government funds, as in the Single Regeneration Budget, and the attempts to assemble local, largely business-led, local coalitions for that purpose (Peck 1995); and most recently the introduction of the RDAs. But, despite the hype that has accompanied these, if indeed something appropriate to the realization of a neo-liberal vision is to be accomplished, then much remains to be done.

It is not just a matter of the extremely modest character of the RDA initiative: the almost minute share of regional government spending that they command – less than 1 per cent (Morgan 2001, 345) – and the, related, resistance of central government departments to transferring powers and resources to them. There are also the rigours of a still quite centralized planning system. The relation between local economic development and the local planning function historically has been a fraught one. Only within the last 25 years has the relationship begun to change, as planning departments acquired officers in charge of local economic development. Nevertheless, and compared with the American case, local planning regulation in Britain remains highly stringent, often making it hard to find a place for new developments, either industrial or commercial, or the housing to support them.

The underlying problem, however, is one of altering the material conditions so as to release bottom-up forces that would then push, of their own initiative, for changes in the scale division of labour through which local economic development occurs. This cannot be overemphasized. Business interests in local economies remain weak. Equally,

so do those of local governments. *The Economist* for 14–20 April 2001 (US edition) included a leader with the provocative title 'Does Britain want to be rich?' (pp. 18, 20). Like most of their leaders, it linked up with a news article in the body of the magazine, in this case a discussion of planning for growth in the Cambridge area and the discouragement that the entrepreneurial element there was experiencing as a result of adverse planning decisions. As it went on to argue:

> Many of the companies (in the Cambridge area) say it is a struggle to grow . . . They say that the main reason they cannot grow is the planning system. Applications to erect a new building grind slowly through the system; appeals, if an application is turned down, take years. High-tech businesses do not have years to waste. (*The Economist* 14–20 April 2001, 53–4)

The underlying problem, according to the leader, was that 'while councils have so much power to determine the level of growth in their area they have little interest in fostering it'. The reason for this is that their worries about revenues, in contrast to the US case, are minor ('The state of the local economy . . . makes little difference to local government coffers').

This is slightly exaggerated. Without the support of a local business interest in expanding local economies – and strong business interests in the expansion of particular local economies clearly do not exist as they do in the US – local government would be lacking allies in confronting local opposition. But it does point to one important precondition for a locally generated, bottom-up politics of economic development of the sort characteristic of the United States. Having said that, the obstacles to bringing about this decentralization of fiscal responsibility are huge. In practice, it would certainly challenge widely held ideals regarding equality in provision of public services. It would also, and inevitably, threaten the system of national wage bargaining for schoolteachers and municipal workers, and invite, in consequence, not just their opposition but that of the whole labour movement.[21] Government concern that the RDAs limit their competition with one another reflects a broader discursive hostility towards the idea of territorial competition that local fiscal responsibility would almost certainly unleash: a hostility that is far less evident in the US. Nevertheless, changing the basis for the funding of local government in the United Kingdom is now in the air, and a review of local government funding is promised for next year. The watchword, apparently, is a 'new localism', which to judge from some of the proposals being circulated, could include local income and sales taxes. However, as *The Economist* again concludes:

> The new localism is a big idea all right, but the government has yet to discover whether it is one whose time has come. Still, as they say, full marks for bravery. (*The Economist* 2004, 51)[22]

This is to come at the question of moving towards a neo-liberal model for structuring local and regional development policy purely in terms of the forces of resistance to it. I want to conclude, however, by also shedding some doubt on the strength of the forces that, according to some, should be encouraging it. Just what 'complete' globalization would look like is, of course, unclear. Is it the elimination of factor price differentials between countries, a situation in which all of a country's product is exported and all that it consumes imported, or what? We can, however, see what the obstacles are to increasing the value of international trade as a percentage of gross domestic products, and they are very, very substantial indeed. It is not just the continued protectionism of national economies, including those protections afforded to member countries by the EU.

There is also the fact that the fields of competition for many firms are *not* international. Rather they exist at all manner of geographical scales below that of the national and their competitors are national rather than international. This also sheds light on the issue of inward investment. The idea of local economies in different countries competing for investment by the same MNC attracts the media headlines, but it is a small minority of all the cases of site selection that come through the door. Rather, it is a matter of competing for the headquarters, the branch plant or back office of a national firm.

Quite what scale actually signifies when talking about the internationalization of exchange relations is also an issue. In terms of their geographical extent, some national economies are much more expansive than others and this affects the balance between intra- and inter-national exchange. The US economy, as is well known, is a much less open one than say that of Belgium or The Netherlands.[23] As a result, in the US, and historically, it is reasonable to assume that the markets to which local growth coalitions have been oriented in terms of

attracting inward investment have tended to be more national in scope than in less closed economies. As, in the post-war period, corporations separated off different functions in the form of branch plants, corporate headquarters, distribution centres, etc., so it was possible for localities to compete for them and to start thinking of a future as a corporate headquarters city, a branch plant town or a distribution centre. This was occurring long before talk of globalization and, as we have seen, continues to the present day.

The United Kingdom has a much more open economy, but an important question is, 'open to where'? Its field of competition is defined increasingly by the EU. With respect to the institutional forms through which it struggles to ensure the development of localities and regions, it is other European countries rather than the United States with which it is competing. The EU provides a major market, not just for European firms, but for others, the Americans included. For various reasons they need to be within the EU. The institutional mechanisms that the member countries of the EU have adopted in order to stimulate local and regional development are all quite similar: certainly more similar to each other in this respect than they are to the US. There is no reason why the sort of resistance to change that has been experienced in Britain should not be replicated elsewhere within the EU.

Conclusions

Anticipation of some convergence of the institutional forms underpinning local and regional development policy is common. This expectation is based on assumptions about globalization and how it is affecting the limits on, and possibilities open to, agents at different geographical scales, including those defined by the various, territorially specific, branches of the state. If in fact they are converging on a decentralized form, then some countries have further to go than others. As Michael Mann has stated, '. . . it is difficult to see much of a weakening of US (federal and state) government powers, since these were never exercised very actively' (1997, 484). In Britain, on the other hand, local and regional development policies have been subject to quite stringent central control. Furthermore, the British case shows very little movement to the decentralized American model, despite all the hype. This does not mean to say that there has been no change, but for the most part it is not of a decentralizing nature, either in the US or in Britain. Rather there has been what Jessop has called a refunctionalization of the British central state and the American states around the knowledge economy, facilitating strategic investments and ensuring a competitive business climate (Jessop 2002, chaps 5, 7).

How, therefore, can one make sense of what has been happening? Harvey's concept of territorial structured coherences is suggestive (Harvey 1985a, 146 1985b, 139–44). Located in the contradiction between capital's necessary fixity and mobility, they express the form and effects of local class 'settlements' arrived at within the context of an existing set of local conditions and extra-local linkages of diverse sorts. These 'settlements' are inevitably temporary and subject to challenge as the limits and possibilities confronting the different classes and class fractions shift. Local and regional development policy, the underlying disposition of state powers structuring such policies, are an intrinsic aspect of territorial structured coherences. As such, they too reflect the outcomes of class struggles, their discursive effects and, as I have suggested, the enabling and limiting effects of state structures.

The British and American versions of the politics of local and regional development are strikingly different. The British one has unfolded in the context of tight regulation from the centre, while in the US there is a heightened element of local discretion. A politics constituted by very different balances of class forces at both national local levels, is at the root of these, albeit conditioned by very different state structures.

The turn to neo-liberalism that emerged in the wake of what Robert Brenner has called 'the long downturn' and the increased distanciation of commodity exchanges that it unleashed has provided a challenge to territorial structured coherences and their associated politics of local and regional development throughout the advanced capitalist societies. This is particularly the case in those, like the United Kingdom, where departure from a neo-liberal ideal of decentralized, competitive institutional forms was greatest. This in turn has generated anticipation of a convergence on a more locally driven and regulated politics, much like that of the US.

Change, however, has been very modest and the central question is why this might be. Attention focuses on two complementary explanations. The first is the strength of the forces of resistance to

change within a territorially structured coherence. The second is the strength of those external forces of internationalization which are supposedly providing a challenge to existing institutional forms. Structured coherences define a totality of internally related elements, discursive, spatial, practical, institutional, cooperative, having to do with power, among others. Disturbing any one aspect of a structured coherence, and of the state structures underpinning it, threatens values in place, the institutional forms and beliefs underpinning those values and is therefore likely to incite strong resistance. The RDAs and the refunctionalization of the central state role have provided minimal threat, though the struggle of organized labour against central government attempts to turn much of the country into a low wage platform for firms exporting to the EU is an indicator of what is at stake. Any move towards shifting fiscal responsibilities towards local government as a prelude to a more territorialized form of local economic development policy would bring in its wake other consequences of a highly inflammatory sort. For a start, inequalities in public provision would sharply increase and it seems unlikely that national wage bargaining for local government employees could survive.

On the other hand, it is very possible that the challenge of globalization has been overestimated and, according to some, deliberately so for reasons of an entirely political nature.[24] A number of different things could be mentioned here. It is not just countertendencies in a protectionist direction or the fact that so much of the competition for local economic development is oriented to national rather than international investors. There is also the relatively closed nature of both the US economy and the increasing orientation of the British towards the EU, which is itself more closed than the American one.[25] British competition for inward investment, for example, is increasingly with other members of the EU rather than with the US.

Capital has both homogenizing and differentiating, universalizing and particularizing tendencies. The idea of convergence in institutional forms picks up only on the universalizing and then exaggerates their geography and their significance. The challenge in understanding institutional change is a critical scrutiny of the tension between the universalizing on the one hand and the particularizing on the other. It has been the objective of this paper to do just that with respect to debates about the changing form of the politics of local and regional development.

Notes

1 For a strong statement on the significance of these and their political expression in what he calls 'regional directorates', see Scott (1998).

2 See Hudson (1982) on the Washington case.

3 An example of this was the – ultimately abortive – attempt to build up the iron and steel and chemical industries on Teesside in the 1970s (Beynon *et al.* 1986).

4 Compare Harvey: 'The tension between free geographical mobility and organized reproduction processes within a confined territory exists for both capitalists and labourers alike. And how that tension is resolved for either depends crucially on the state of class struggle between them' (1985a, 149).

5 For a good discussion of the British case, see Leys (1985).

6 A function of the separation of powers, and therefore of state form.

7 For an excellent case, see Chapter 5 of Berkman and Viscusi (1971) on the politics of the Central Arizona (irrigation) Project.

8 A suggestive example of how the very different conditions of the British case, in terms of levels of geographically uneven development, affect the politics of local and regional development comes from Foster and Woolfson (1989). They have discussed an instance that erupted in 1987 around plans on the part of Ford to locate a plant in Dundee, Scotland, an area of persistent and high unemployment. This was a plant projected to eventually employ over 400. Part of the agreement was that the new plant would operate outside the collective bargaining structures of the rest of the company in Britain. Existing unions in the Ford worker representation structure were sidestepped by the completion of an agreement with the Amalgamated Electrical Workers' Union which would be the sole organizer at the plant. The plan was to exempt the plant from the Ford National Joint Negotiating Committee (FNJNC) that bargained on wages and conditions for Ford's 22 other plants in the UK, and that it would be exempt from all existing agreements incorporated into the FNJNC's 'blue book'. This proved highly controversial. There was strong opposition from those unions party to the FNJNC and a later agreement by the FNJNC to black all parts coming from the new plant. Amidst widespread recriminations against the unions on the part of the Scottish Office, which had midwifed the agreement, this led to a decision on the part of Ford not to proceed with the plant. The claims of a national wage bargaining agreement took precedence in this instance over the claims of locality.

9 Ellwood and Patashnik (1993) comment in a similar vein on the outcome of the Model Cities programme of the mid-1960s. Originally intended to funnel billions in demonstration grants to the nation's ten most severely distressed cities to see if comprehensive aid could alleviate urban poverty, by the time the bill became law the number of eligible cities had increased by a factor of fifteen.

10 One such was the decision to award a major, 500 million pound addition to a research facility in the Oxford area rather than to one at Daresbury in the Manchester area.

11 These are not insignificant differences. Scotland receives 25 per cent more per capita in public expenditure per year than the English regions, and Wales 15 per cent.

12 Yet another concern has been interest rate policy which it is believed in many areas outside the Southeast, including Scotland, is driven by inflationary pressures arising in the Southeast and to the detriment of areas elsewhere that have unused resources. Note in addition, however, that the territorialization of the politics of local and regional development was never entirely absent prior to the Thatcher revolution. Some useful case studies are provided by Cooke (1983) and Bassett and Hoare (1984).

13 Compare Friedmann and Bloch: 'On the terrain of state economic development planning, the narrow focus on attracting industry has shifted to a broader concern with fostering economic competitiveness, primarily through the creation of an entrepreneurial regional milieu in which innovation can flourish... Through the 1980s, the economic development agencies of many states, often in partnership with business and academia, and sometimes with labour, have crafted a large number of programmes and tools to serve these ends. Increased funding for universities, university research parks, attempts at improving educational systems, business assistance centres, small business assurance centres, small-business incubators, improved sources of venture and seed capital, subcontracting assistance programmes, consultancy services, mediation programmes and upgraded vocational training are all mechanisms that have been used here' (1990, 593).

14 In the case of the Edison Center programme in Ohio, the Center in Cincinnati is devoted to research in machine tools, while that in Cleveland specializes in medical technologies and the one in Akron – onetime centre of the rubber tyre industry – on polymers.

15 The reason given at the time: 'There is a clear difference between the British and continental models of industrial and employment policy. The British approach is one of low burdens on business and voluntarism. The continental approach is based on regulation and statutory obligation. The Social Chapter embodies the continental approach ... it would not help either businesses or employees in the United Kingdom. It would lead to higher costs, more restrictions and higher unemployment' (Ian Lang 'The Social Chapter – Blank Cheque and Competitive Disaster.' Press release from the Department of Trade and Industry P96/725, 30 September 1996).

16 From the 1970s on, both Scotland and Wales acquired their own regional development capacities through the Welsh and Scottish Offices – agencies of the central state – respectively and so they were not affected by the 1998 legislation.

17 As Kevin Morgan has written: 'The RDAs may be a step in the right direction so far as the under-endowed English regions are concerned, but their ambitions seem to have raced ahead of their resources, indeed they control less than 1% of total public expenditure within their areas' (2001, 345).

18 For example: Salisbury (1964) and Mollenkopf (1976).

19 On the British case, see Dunford (1997). In the American case there has been convergence, but this has been due to a simultaneous increase in incomes and employment in the South and West and a relative decline, along with quite severe local pockets of unemployment, in the old Manufacturing Belt.

20 There is a degree of universality in these changes across the advanced capitalist countries. Jessop has made a strong case for the refunctionalization of central states around learning economy agendas (2002, chap 5). The competition between the states for inward investment on the basis of 'business climate' now finds a parallel in similar anxieties among the member states of the EU. The role of uneven development in the emergence of interregional tensions and drives to rescale the state is likewise echoed among other member countries of the EU. In Belgium the adoption of a federal constitution in 1993 is hard to imagine outside of the widening differences between the Walloon-speaking rustbelt in the South, where the Socialists were anxious to take control of recapitalization in their own way; and a booming Flanders keen to divert tax flows to internal uses rather than support the profligate South. Italy provides a case that is similar in some respects. Uneven development has always been an issue in the country, but the movement towards a single market, concern about the impending euro and the emergence of the Third Italy, all weighed heavily in the balance in the growth of the Northern League and its aspirations for separation or at least a measure of federalism. The South was seen as a fiscal drag on the North, and an impediment, through the demands it made on national spending, on the ability of the Italian government to meet the requirements for entry into the euro-zone and the advantages of access to capital that it was believed that would bring. And in Germany, according to Jeffery, 'Increasingly, competition, rather than cooperation, has become the organizing principle of territorial politics' (1999, 153).

21 Significantly in the US these are determined at the local level with individual municipalities and school districts.

22 For the position of the Labour left on this, see Walker (2002).

23 The average of imports and exports as a percentage of GDP for select countries in 1998: Belgium: 62 per cent; Canada: 38.5 per cent; France: 21.5 per cent; Germany: 23.5 per cent; Italy: 25.5 per cent; Japan: 7.25 per cent; The Netherlands: 45.5 per cent; United Kingdom: 31 per cent; and the US: 12.5 per cent.

24 As Piven has put it: 'The key fact of our historical moment is said to be the globalization of national economies which, together with "post-Fordist" domestic restructuring, has had shattering consequences for the economic well-being of the working class, and especially for the power of the working class. I don't think this explanation is entirely wrong but it is deployed so sweepingly as to be misleading. And right or wrong, the explanation itself has become a political force, helping to create the institutional realities it purportedly merely describes' (1995, 108).

25 According to Kleinknecht and ter Wengel (1998), in 1995 external trade amounted to less than 10 per cent of the total GDP of the EU.

References

Barnekov T, Boyle R and Rich D 1989 *Privatism and urban policy in Britain and the United States* Oxford University Press, Oxford

Bassett K and Hoare A 1984 Bristol and the sage of Royal Portbury: a case study in local politics and municipal enterprise *Political Geography Quarterly* 3 223–50

Berkman R L and Viscusi W K 1971 *Damming the West* Grossman Publishers, New York

Beynon H, Hudson R and Sadler D 1986 Nationalised industries and the destruction of communities: some evidence from northeastern England *Capital and Class* 29 27–57

Brenner N 1998 Global cities, glocal states: global city formation and state territorial restructuring in contemporary Europe *Review of International Political Economy* 5 1–37

Brenner R 1998 The economics of global turbulence *New Left Review* 229 1–265

Cooke P 1983 Regional restructuring: class politics and popular protest in South Wales *Environment and Planning D: Society and Space* 1 265–80

Corrigan P and Sayer D 1985 *The Great Arch* Blackwell, Oxford

Cox K R 1999 Ideology and the growth coalition in Jonas A E G and Wilson D eds *The urban growth machine* State University of New York Press, Albany chap 2

Dunford M 1997 Divergence, instability and exclusion: regional dynamics in Great Britain in Lee R and Wills J eds *Geographies of economies* Edward Arnold, London chap 20

Ellwood J W and Patashnik E M 1993 In praise of pork *The Public Interest* 110 19–33

Foster J 1974 *Class struggle and the industrial revolution* Weidenfeld and Nicholson, London

Foster J and Woolfson C 1989 Corporate reconstruction and business unionism: the lessons of Caterpillar and Ford *New Left Review* 174 51–66

Friedmann J and Bloch R 1990 American exceptionalism in regional planning, 1933–2000 *International Journal of Urban and Regional Research* 14 576–601

Harvey D 1985a The geopolitics of capitalism in **Gregory D and Urry J** eds *Social relations and spatial structures* Macmillan, London chap 7

Harvey D 1985b *The urbanization of capital* Blackwell, Oxford

Harvey D 1989 From managerialism to entrepreneurialism: the transformation of urban governance in late capitalism *Geografiska Annaler* 71B 3–17

Henderson J W 1999 Uneven crises: institutional foundations of East Asian economic turmoil *Economy and Society* 28 327–68

Hudson R 1982 Accumulation, spatial policies, and the production of regional labor reserves: a study of Washington New Town *Environment and Planning A* 14 665–80

Hudson R 1986 Nationalized industry policies and regional policies: the role of the state in capitalist societies in the deindustrialization and reindustrialization of regions *Environment and Planning D: Society and Space* 4 7–28

Jeffery C 1999 Party politics and territorial representation in the Federal Republic of Germany *West European Politics* 22 130–66

Jessop B 1994 Post-Fordism and the state in **Amin A** ed *Post-Fordism: a reader* Blackwell, Oxford chap 8

Jessop B 1999 Narrating the future of the national economy and the national state? Remarks on remapping regulation and reinventing governance in **Steinmetz G** ed *State/culture* Cornell University Press, Ithaca NY

Jessop B 2000 The crisis of the national spatio-temporal fix and the tendential ecological dominance of globalizing capitalism *International Journal of Urban and Regional Research* 24 323–60

Jessop B 2002 *The future of the capitalist state* Blackwell, Oxford

Katznelson I 1989 Was the great society a lost opportunity? in **Fraser S and Gerstle G** eds *The rise and fall of the new deal order, 1930–1980* Princeton University Press, Princeton NJ 185–211

Kenworthy L 1997 Globalization and economic convergence *Competition and Change* 2 1–64

Kleinknecht A and ter Wengel J 1998 The myth of economic globalisation *Cambridge Journal of Economics* 22 637–47

Lazare D 1998 America the undemocratic *New Left Review* 232 3–40

Leys C 1985 Thatcherism and British manufacturing *New Left Review* 151 1–25

Lowi T J 1969 *The end of liberalism* W W Norton, New York

Mackay R R 2001 Regional taxing and spending: the search for balance *Regional Studies* 35 563–75

Mann M 1997 Has globalization ended the rise and rise of the nation-state? *Review of International Political Economy* 4 472–96

Markusen A R 1987 *Regions: the economics and politics of territory* Rowman and Littlefield, Totowa NJ

Mawson J 1999 Devolution and the English regions *The Source Public Management Journal* (http://www sourceuk net/columnists/johnmawson-articles html) Accessed 18 April 2004

Mollenkopf J H 1975 The post-war politics of urban development *Politics and Society* 5 257–95

Morgan K 2001 The new territorial politics: rivalry and justice in post-devolution Britain *Regional Studies* 35 343–8

Peck J 1995 Moving and shaking: business elites, state localism and urban privatism *Progress in Human Geography* 19 16–46

Piven F F 1995 Is it global economics or neo-laissez-faire? *New Left Review* 213 107–15

Rees G and Lambert J 1985 *Cities in crisis: the political economy of urban development in post-war Britain* Edward Arnold, London

Salisbury R H 1964 Urban politics: the new convergence of power *Journal of Politics* 26 775–97

Savage M 1987 Understanding political alignments in contemporary Britain: do localities matter? *Political Geography Quarterly* 6 53–76

Scott A J 1998 *Regions and the world economy* Oxford University Press, Oxford

Storper M 1987 The post-enlightenment challenge to Marxist urban studies *Environment and Planning D: Society and Space* 6 418–26

Storper M and Walker R 1989 *The capitalist imperative* Blackwell, Oxford

Swyngedouw E 1997a Neither global nor local: 'glocalization' and the politics of scale in **Cox K R** ed *Spaces of globalization* Guilford, New York chap 6

Swyngedouw E 1997b Excluding the other: the production of scale and scaled politics in **Lee R and Wills J** eds *Geographies of economies* Arnold, London chap 13

The Economist 2001 Silicon Fen strains to grow *The Economist* 14–20 April 53–4

The Economist 2001 Does Britain want to be rich? *The Economist* 14–20 April 18, 20

The Economist 2004 A little local difficulty *The Economist* 24 January 51 (US edition)

Walker D 2002 *In praise of centralism* A Catalyst working paper (http://www.catalystforum.org.uk/ pdf/walker.pdf) Accessed 18 April 2004

[24]

The global trend towards devolution and its implications

Andrés Rodríguez-Pose, Nicholas Gill
Department of Geography and Environment, London School of Economics, Houghton Street,
London WC2A 2AE, England; e-mail: A.Rodriguez-Pose@lse.ac.uk; N.M.Gill@lse.ac.uk
Received 14 June 2002; in revised form 7 October 2002

Abstract. Globalisation has been accompanied by an equally global tendency towards devolution of
authority and resources from nation-states to regions and localities that takes on various forms,
depending upon which actors are driving the decentralisation efforts. The existence of a general trend
towards devolution also has significant implications for efficiency, equity, and administration. The
authors outline first the general drive towards devolution and then proceed to examine which
countries are experiencing which forms of decentralisation. A theoretical argument emphasising the
role of governmental legitimacy across various tiers of government is used to explain the diversity of
devolution initiatives, drawing on examples that include Brazil, Mexico, India, China, the USA,
and some European countries. Having supported their model of decentralisation, the authors then
examine the implications of the widespread downward transfer of power towards regions. Some of the
less widely discussed pitfalls of decentralisation are presented; caution in promoting devolutionary
efforts is the prescription of this paper.

1 Introduction

Research on globalisation has tended to stress the role global processes are playing in
undermining the importance of nationally based policymaking, politics, culture, and
society. Such trends are underpinned by the proliferation of communication and transportation media, the emergence of dominant forms of international, brand-based
capitalism, and the standardisation of various modes of interaction, from the convergence of languages to digitisation (Dicken, 1998). Accordingly, globalisation tends to
promote what Agnew (2000, page 101) has called a new international homogeneity
across the global order. Such homogeneity implies a certain erosion of the importance
of spatiality at the national, and by extension the regional, level as global processes
succeed in diluting and internationalising the traditional 'nexus of interactions' associated with local and regional spaces (Gray, 1998; see also Castells, 1996; Massey, 1999;
Storper, 1997).

Yet, in spite of this, globalisation is failing to obliterate the importance of the
local dimension across the world. In many ways, recent developments point in an
opposite direction: towards a greater relevance of place, space, and regions. The
growing visibility of the local and regional dimension has many manifestations.
Although some argue that the demise of the nation-state is continuing, the rise
in regional political activism (Rodríguez-Pose, 1998, pages 215f), the increasing
importance of regionalism in government (Keating, 1998), and the regionally based
competition that mobile capital is inducing (Cheshire and Gordon, 1998) have given
rise to a renewed interest in the role of regions. In this paper we will concentrate on
one of the most significant recent developments at the global level: the widespread
transfer of power downwards towards regions. This process, which in some cases
involves the creation of new political entities and bodies at a subnational level

and in others an increase in their content and power, is known as devolution (Prud'homme, 1994).[1]

There is now enough evidence to claim that since the outbreak of the process of globalisation—and perhaps as a result of it—subnational units have increased their demands for power. This process has numerous positive aspects but also raises important issues regarding national equity and welfare, public finance, and territorial competition.

The focus of this paper is thus fourfold. First, in section 2, in response to the often simplistic conceptions of devolution, a theoretical model based upon Donahue's (1997, pages 7–15) threefold classification of the mechanisms of devolution—legitimacy and the decentralisation of authority and of resources—is introduced. Subsequently, in section 3, we expose the devolutionary process in a selection of countries, including Brazil, China, India, Mexico, the USA, and countries of the European Union. These represent a cross-section of some of the largest areas of the globe and cover the developed and the developing worlds. In section 4, different forms of devolution are compared and examined. In section 5 we critically assess the global implications of the devolutionary trend in light of its evident diversity. In the final section we conclude that a greater awareness of the benefits and drawbacks of devolution is required in order to prevent the escalation of some of the downsides associated with it (section 6).

2 The theoretical framework

Devolution is a complex and heterogeneous process. From the high level of decentralisation of certain federal states, such as Germany, and of some Spanish regions, to the more limited power of regions in France or, until recently, Mexico, decentralisation processes across the world have taken on a variety of forms. Consequently, conceptualisation of devolution is far from simple. Looking for a minimum common denominator, Donahue (1997, pages 7–15) characterises the process as being made up of three separate factors: legitimacy, the decentralisation of resources, and the decentralisation of authority. Any form of devolution implies some degree of subnational legitimacy and some form of decentralisation of authority and resources; consequently, any analysis of devolution should take these three factors into consideration.

There is, however, a need for caution in examining evidence, because a simple list-based approach may overlook the interaction between the elements. The complexity of the devolution process derives from the interest conflicts of the actors involved and the differences in legitimacy that they share. Most importantly, the interests of subnational and national governments tend to be at odds across the component factors of devolution. Although national governments would prefer, ceteris paribus, to devolve responsibilities (authority) to their regional or state governments with as few accompanying resources as possible, the subnational governments would prefer the opposite case. The balance between these extremes will depend upon the relative strength, or, in political terms, legitimacy, of the two tiers of government.[2] In figure 1 we illustrate this approach.

[1] Prud'homme's (1994) classification of decentralisation into spatial, market, and administrative decentralisation provides a useful conceptualisation. For the work in this paper—concerned with administrative decentralisation—the redistribution of decisionmaking to lower government tiers (deconcentration) and the closer involvement of semiautonomous organisations (delegation) hold less importance than the third type of administrative decentralisation that Prud'homme outlines—that of power transfer to lower government tiers (devolution). As a consequence, throughout this text decentralisation will be taken to refer to devolution.

[2] Donahue conceptualises legitimacy as incorporating 'popular support' and 'citizen cooperation'. He states that, "ultimately the most important asset that government can command ... is not legal authority, or fiscal resources, or even talented personnel, but legitimacy" (Donahue, 1997, page 12).

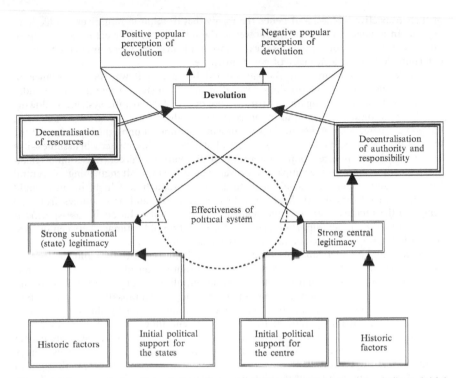

Figure 1. The complexity of devolution. Note: double-lined boxes and arrows indicate initial factors and processes, respectively; single-lined boxes and arrows indicate subsequent factors and processes, respectively.

If we begin at the bottom of the diagram, we can see that the legitimacy of subnational and national governments is determined for the most part by processes of history and political support. Regarding historical processes, culture, language, religion, and ethnicity have traditionally been the factors behind a strong regional identity and determine the legitimacy of subnational claims (see Litvack et al, 1998, page 1). Economic development has also recently been added to the list (Allmendinger and Tewdwr-Jones, 2000; Keating, 1998). Regarding political support, the inclusion of legitimacy into the analysis brings into play a wealth of political factors that shape the power and room for manoeuvre of governmental tiers. An important caveat here, however, is that a reasonably effective political and democratic system must be in place to facilitate the influence of the electorate. This being the case, and all other things being equal, poor political support for the regional cause would translate into a relatively weak regional legitimacy and tend to promote a devolutionary process in which the central government holds the upper hand, favouring progressive decentralisation of responsibilities and often forcing regional and local governments "to undertake increasing expenditure responsibilities on a static, and often narrowing, financial base" (Bennett, 1997, page 330). In contrast, a strong regional legitimacy, underpinned by high political support for the regional lobby, would favour a more rapid decentralisation of resources, as there would be strong demand for such transfers to subnational tiers of government. In general, the combination of historical and political factors in democratic countries shapes the legitimacy of governmental tiers, the relative strength of their respective lobbies, and hence the forms that devolution initiatives are likely to assume. In nondemocratic systems, however, the influence of the electorate is compromised and other factors, which are less transparent, take on more importance in the determination of the legitimacy of government tiers.

As depicted in figure 1 by the single-lined boxes and flows, the importance of legitimacy and popular support does not end at the initial stage. There is also an endogenous role (if we assume again an operative political and democratic system). Following initial moves towards devolution, in terms of decentralisation of resources or authority, the popular perception of devolution will subsequently impact upon the legitimacy of the two government tiers. It is difficult to predict what direction this impact will assume—it is just as feasible to envisage a popular approval or disapproval of decentralisation originating from the centre, for example. Approval would mean a strengthening of central legitimacy and disapproval would lead to a relative reduction, which, in turn, could be translated into greater or lesser transfers of authority and/or resources from the centre to the regions. Although the political reaction to devolution is case-specific it nevertheless remains an important driver in the evolution of legitimacy between the two tiers, wherever a reasonable democracy has taken root.

In general, an understanding of the interaction between legitimacy and the transfer of resources and authority is imperative to the examination of devolutionary trends (Donahue, 1997). A case that depicts strong decentralisation of resources displays no more evidence for devolution than one showing strong decentralisation of responsibilities. It merely indicates a different type of devolution, driven by different levels of government, and deriving, ultimately, from a different allocation of legitimacy across governmental tiers. Moreover, following from this, we should not necessarily expect cases to depict high levels of both resource and responsibility devolution, because forces often operate to promote their mutual exclusivity. It is with this conceptualisation in mind that we approach recent devolutionary efforts.

3 The global trend towards devolution

Before the onset of globalisation the world was dominated by strong national governments, and regional governments tended to be either weak or nonexistent. Hence in Europe, with the exceptions of Austria, Germany, Switzerland, and Yugoslavia, as well as in Africa and Asia, central governments dominated throughout the postwar era. Latin American countries also had centralised states. Although some countries officially maintained federalist or regionalist constitutions, such as Brazil, Mexico, and Venezuela, they tended to be federalist on paper, with regions and states representing little more than administrative units. In the former Soviet Union (USSR) the situation was similar to that of Latin America. The USSR consisted of a union of sovereign states on paper, but, in reality, was heavily centralised and controlled from Moscow. Outside this framework, only the above-mentioned European countries, Australia, Canada, India, and the USA had systems in which the regional tier of government played any significant role, and even in some of these cases the role of regional governments had been waning. This was, for example, the case of the USA, where the power of the states had been declining with respect to those of the federal government since at least the reforms introduced by Franklin Roosevelt during the Great Depression (Donahue, 1997); it was also the case in India, where the centrally based mechanism of planned economic development undermined the power of the regions, as the central government dictated financial and economic goals to the states (Sury, 1998).

At the beginning of the 21st century this panorama has radically changed. A devolutionary trend has swept the world. In some cases, subnational turbulence has led to the demise of former countries and to the emergence of new states. The fifteen constituent republics of the former Soviet Union have become independent states; Czechoslovakia peacefully split into the Czech Republic and Slovakia; four new states have emerged out of war-torn Yugoslavia, and Kosovo and Montenegro may follow suit. This phenomenon is not, however, exclusive to former Central and Eastern European socialist countries. Eritrea achieved independence after a long guerrilla war, and East Timor has recently become an independent state after twenty-five years of Indonesian occupation.

The emergence of new independent states is an extreme form of a more general, global trend in the transference of power, authority, and resources to subnational levels of government. Few spaces around the world have remained untouched by this trend. Eight out of the fifteen members of the current European Union—accounting for 87% of its population (Rodríguez-Pose, 2002, page 174)—have seen some level of decentralisation. In addition to the already federalised Austria and Germany, Belgium became a federal state in the early 1990s, and Italy is in the process of federalisation. Spain, despite not being a federal state, is arguably the most decentralised state in Western Europe. France has taken limited steps towards regionalisation—with ongoing debates over the granting of regional autonomy to Corsica. The United Kingdom and Portugal have also transferred a considerable amount of power to some of their regions. Similarly, Poland has recently followed the path towards regional devolution.

Outside Europe, devolution has also been widespread, especially in large and heterogeneous states. In some cases, regional autonomy has been granted ex nihilo. This is the case for Indonesia, which passed autonomy laws aimed at undoing decades of extremely centralised government and at appeasing separatist tendencies in 1999 (Aspinall and Berger, 2001). In China, although political devolution has not formally occurred and the Chinese Communist party still keeps a tight grip on political developments, there has been widespread fiscal decentralisation that has provided regional and local governments with considerable powers and that has encouraged policy innovation at the regional level (Ma, 1996, page 5).

In other cases, preexisting levels of regional autonomy have been enhanced. The most striking cases have been those of Latin American countries. In Mexico the collapse of the economic system in 1982 and the political uncertainty that followed led to extensive changes in territorial politics. Although Mexico's constitution has officially been one of federalism since at least the revolution of 1910 (Rodríguez, 1998, pages 236–238), extreme presidentialism and the dominance of the executive branch of government ensured seventy subsequent years of centralism and an enduring centralist culture (Rodríguez, 1998, pages 235–236). But, as Ward and Rodríguez (1999, page 28) assert, the last two decades have seen a dramatic improvement in the political systems of representation, accountability, flexibility, and democracy and have led to a profound reform of territorial politics in support of greater federalism.

In Brazil, the power of the states has been reinforced since the passing of the 1988 Constitution. The 'regional interest' lobby was extremely influential during the drafting process and was well placed to capitalise on the anti-central-government sentiment that had been developing during military rule (Coutinho, 1996, page 7). The regions were able to secure residual powers of legislation and maintain a lack of any clear constitutional demarcation of responsibilities between the state, the centre, and the local governments (Dillinger and Webb, 1999, pages 9–11). This has subsequently afforded them a hitherto unprecedented level of discretion over their own financing, administration, and responsibilities.

Among states that had considerable regional autonomy before the onset of globalisation, the trend has also been towards even greater decentralisation. In India, the overconcentration of power in the hands of a few national elites until the early 1980s brought about a reaction that started to redress the balance from the centre to the regions (Sharma, 1999). In the USA, the trend towards centralisation, which some trace back to the American Civil War or to the Great Depression, started to be reverted during Nixon's presidency in the early 1970s (Donahue, 1997), but it was really under Reagan's New Federalism in the 1980s when states began to recover greater freedom of action.

The global trend towards devolution is based on subnational legitimacy and implies greater transfers of authority and resources from the centre to the states or regions. In most cases, and as in previous waves of decentralisation, regional legitimacy has historic, linguistic, religious, and/or cultural roots. Regions and states with their own ethnic, historical, cultural, or linguistic identity have paved the way for decentralisation. That has been the case for Catalonia and the Basque Country in Spain, for Scotland in the United Kingdom, for Brittany and Corsica in France, and even for Chiapas in Mexico and for Tibet and Xinjiang in China, which have 'brandished' ethnic, cultural, or historical arguments as the source of their demands for greater autonomy. Economic arguments are also increasingly becoming a source of subnational legitimacy (Keating, 1998). Uneven regional economic development, alongside the achievement of greater economic efficiency through decentralisation, are coming to the fore and gradually starting to occupy the bulk of the regionalist discourse in favour of decentralisation. The Northern Italian Leagues were the first to base their devolutionary claims heavily on economic demands after their failure to gain visibility by highlighting traditional ethnic or linguistic issues (Diamanti, 1993). Nationalist and regionalist parties in Spain have increasingly resorted to similar arguments, as indeed have the Zapatistas in Chiapas.

Additional factors also contribute to boost the legitimacy of calls for decentralisation. In some circumstances, decentralisation goes hand in hand with democracy. This is the case for Brazil and most of Latin America, where the advent of democracy and decentralisation are intrinsically related (Shah, 1991; Souza, 1997). Spain represents a similar case: forty years of dictatorship generated greater legitimacy for the devolutionary cause and contributed to the profound territorial transformation of the Spanish state after the return to democracy (Pérez Díaz, 1990). In other circumstances, decentralisation tends to accompany changes in the economic regime—especially moves towards the marketisation of national economies. This trend has been followed in India, China (Ping, 2000, page 180; Da-dao and Sit, 2001, page 29), and, to a lesser extent, in Mexico and Brazil, since the opening of these countries to trade.

The process of devolution operates through transfers of authority and resources. Subnational governments across the globe currently enjoy greater authority and powers than they did a few decades ago. The trend is widespread. The powers of Italian regions have progressively increased since the late 1970s, and today they exercise considerable control in the fields of agriculture, tourism, regional planning, environment, and economic development (Rodríguez-Pose, 2002). All Spanish regions now have competence for health and education, and some have also secured powers over policy areas such as policing, taxation, and fiscal affairs, which have traditionally been the prerogative of the nation-state (Castells, 2001). The Scottish Parliament enjoys tax-raising and law-making powers, and the Northern Ireland Assembly, the National Assembly for Wales, and the Greater London Authority have taken over varying levels of central government activities (Tomaney, 2000; Tomaney and Ward, 2000).

Outside Europe, a similar transfer has occurred. The USA has, for example, witnessed devolutionary efforts centred around two key areas: welfare and medical insurance (Schram and Soss, 1998). In Mexico, the transfer of power from the centre to the states includes: the increased ability of states to raise revenue; greater control for the states over development funds; a strengthened administrative capacity for the municipalities; and a clarification of the divisions of responsibility between different tiers of government (Rodríguez, 1998, pages 251–252). In India, state responsibilities embrace public order, police, prisons, irrigation, agriculture and related activities, land, public health, industries other than those centrally assigned, and trade and commerce. In addition, states share with the central government authority over economic and social planning, education, labour, and forestry (Bagchi et al, 1992; RBI, 2000). Chinese reforms since 1980 have been aimed at transferring to provinces greater responsibility for budgets (Ma, 1996, page 5). The Brazilian Constitution of 1988 grants, as mentioned above, state governors all those powers not otherwise prohibited by the Constitution (Dillinger and Webb, 1999).

Last, decentralisation has also implied a substantial transfer of resources from the centre to the regions. In figure 2 we illustrate the growth in subnational government expenditure as a proportion of total government expenditure in our case-study countries between 1982 and 1999. Readers should be aware of the limitations of these data from the International Monetary Fund as they do not convey the degree of local spending autonomy of subnational governments, do not distinguish between sources of tax and nontax revenues, intergovernmental grants, and other grants, and do not disclose what proportion of intergovernmental transfers are conditional or discretionary (Ebel and Yilmaz, 2002, pages 6–7). However, given the lack of alternative, more detailed, and credible data sources, as well as the strength of the general trends identified here, these data must suffice for our purposes. Two points are apparent. First, in the group of countries as a whole, there is an average increase of around 15% in the proportion of subnational government expenditures. Second, as the framework in figure 1 implies, not all countries are party to this trend. Some countries (that is, Brazil, China, and Spain) have witnessed a considerable decentralisation of resources, which, in the case of Brazil and China, has not been accompanied by similar

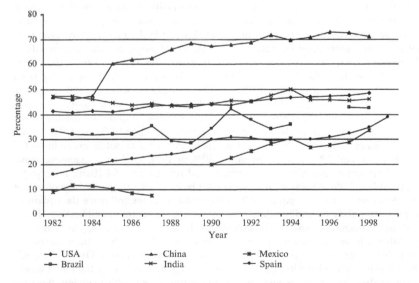

Figure 2. Subnational expenditure as a percentage of total Government expenditure. Note: in the case of Brazil, Mexico, Spain, and the USA, figures include local government expenditure as well as state or provincial expenditure (source: IMF, various years).

levels of decentralisation of authority. In other cases, such as India, the decentralisation of authority has not been matched by a similar decentralisation of resources, as the share of central government expenditure increased slightly at the expense of that of regional government during the period of analysis.

4 Differing forms of devolution

Having established the existence of a widespread trend towards devolution in this section we address its diversity, with reference to the theoretical discussion in section 2. Recall we examined the interrelationship between the legitimacy of governmental tiers and the form of devolution we can expect to find in a given country. In general, all other things being equal, where subnational governments have enjoyed relatively strong legitimacy [as in the case of Spain (Rodríguez-Pose, 1996)], devolution biased towards the decentralisation of resources can be expected; in contrast, where the central government has dominated [as in the case of the English regions (Jones, 2001; Morgan, 2002, page 802)], greater decentralisation of responsibility may well ensue. In this section we present support for this contention by examining the cases of China, India, Brazil, Mexico, and the USA.

China's efforts towards decentralisation began in conjunction with its marketisation initiatives in the late 1970s. The recognition that marketisation warranted a shift in government outlook, from a commanding role towards facilitating functions, brought about a series of fiscal reforms that saw the centre's influence over public resources reduced markedly across the 1980s (Ping, 2000, pages 180 – 181). By 1993 the provinces held control over revenue collection, with the centre's share in total revenue falling from 35.8% in 1983, to 22% in 1993 (Yi and Chusheng, 2001, page 86). In terms of figure 1, therefore, provincial legitimacy had soared to unprecedented levels, largely because of the enthusiastic nature of devolutionary initiatives. Furthermore, the importance of the provinces was compounded by the introduction of negotiation-based fiscal contracting between the provinces and the centre at the end of the 1980s (Lee, 2000, pages 1009 – 1015), which allowed certain states, especially the richer ones, to exploit their bargaining positions to a far greater extent than under the previous formula-based revenue-sharing contracts. The impacts of this rise in the power of the provinces in relation to the centre soon became evident. Alongside the stated increase in subnational revenue claims, the provinces were able to extract far more aid from the centre than the centre itself could afford, contributing to the national deficit spiralling from parity to 600 billion yuan between 1985 and 1995 (Yi and Chusheng, 2001, page 90). Inevitably, this effect was skewed towards the richer provinces, whose bargaining positions were stronger, and Ma (1996, pages 22 – 23) documents a dramatic decline in the progressivity of transfers from the centre to the regions between 1983 and 1991. Alongside this, as figure 2 illustrates, the central government's share in public expenditure fell steadily, from around 53% in 1982, to under 30% in 1998, hindering macroeconomic control and prompting emergency widespread fiscal reforms in 1994 (Bahl, 1999, pages 105 – 129). Although these fiscal reforms succeeded in increasing the centre's involvement in revenue raising, Lee (2000, pages 1009 – 1015) documents the influence the provinces had in preserving their expenditure levels, undermining the centre's equilibrating efforts. But, although resources have been decentralised, many subnational governmental responsibilities, in accordance with the marketisation of China, have been discontinued, including large areas of responsibility over state-run enterprises. Overall, China's provinces can be seen to have increased their legitimacy markedly over the past twenty-five years, leading, as figure 1 would predict, to a surge in regional financing in excess of any accompanying devolution of responsibility.

The 1988 Brazilian constitution was written in an environment of pronounced hostility towards central government control (Hagopian, 1996, pages 259 – 262 and 268 – 269). Constitutional provisions to curb national party dominance resulted in a lack of consistent central politics, a persistent feature of the Brazilian system since the return to democracy. In 1990 the twenty-seven states were represented at the central level by eleven different parties (Kraemer, 1997, page 35), and in such a climate it comes as no surprise that state legitimacy and power have outweighed that of the centre. As Dillinger and Webb point out, "even when the President seems to have strong political support on a roll call vote, that support is the result of extensive prior negotiations and concessions to regional interests" (1999, page 11). The power of the states has been translated into unruly fiscal behaviour. States overspent massively during the late 1980s and 1990s, to the extent of printing their own money, borrowing from their own banks, and running up huge state-level debts (Dillinger and Webb, 1999, pages 23 – 25). Subsequently, the central government has serviced these debts and has faced little choice but to offer financial assistance to struggling states (Montero, 2000, page 67).

More recently, from the mid-1990s onwards, large debtor states have repeatedly threatened the central government with default (Dillinger and Webb, 1999, pages 25 – 26) and in January 1999 seven state governments declared themselves bankrupt, contributing with their action to an economic crisis. Hence, state legitimacy can be viewed as so strong in Brazil that not only have devolution initiatives been biased towards resource devolution but also resource devolution has been paid for by the centre itself. Moreover, when individual fiscal packages with the states were eventually negotiated, beginning with São Paulo in 1997, the lack of extra debt-servicing responsibilities for large debtor states confirmed that states had secured such a degree of legitimacy as to be able to promote resource-biased decentralisation yet avoid responsibility (Dillinger and Webb, 1999, pages 25 – 26). Brazil therefore epitomises both the rapid decentralisation of resources and the gradual decentralisation of responsibility that we would expect, from figure 1, to proceed from high subnational governmental legitimacy.

India presents a contrasting case. Here, the continued legislative and administrative dominance of the central government has engendered a very different decentralisation trajectory. Whereas the Chinese and Brazilian central governments yielded readily to the provinces and states, the Indian central government maintained its influence and managed devolutionary initiatives largely to its own advantage. Historically, although India is a highly decentralised country, the influence of the central government has been consistently high. Following British rule, which emphasised the centre as a means of coordination of export and political control, the regimented system of five-year economic plans has guaranteed the preservation of central legitimacy. So when, during the 1980s, fiscal imbalance began seriously to affect the national budget, with the internal debt of the central government rising from 15.9% of GDP in 1980/81, to 33.6% in 1993/94 (Buiter and Patel, 1997, page 36), the temptation to exploit this legitimacy advantage and to decentralise responsibilities began to manifest itself.

The Indian federal system is based upon three lists of responsibilities—accruing to the centre, the centre and the states, and the states alone. The centre's progressive redefining of the contents of the state and joint lists has allowed the central government subtly to raise the responsibilities of Indian states over the past twenty years (see Bagchi et al, 1992; RBI, 2000). Simultaneously, aid to the states as a proportion of state expenditure has actually fallen, from 54.3% in 1990/91 to 50.6% in 1998/99 (ICSO, 1999). To find that central expenditure as a proportion of total public expenditure has risen slightly over the period (as figure 2 illustrates) therefore comes as no surprise. As discussed in section 2, this fact should not be taken as evidence of centralisation but, as the preceding discussion confirms, a symptom of the form of devolution India has experienced, underpinned by central legitimacy and consequently dominated by the central government's agenda.

Mexico and the USA represent intermediate cases. Political currents in Mexico have also had a decentralising effect. Through the presidencies of de la Madrid (1982 – 88), Salinas (1988 – 94), and Zedillo (1994 – 2000) the historical dominance of the centre has been progressively undermined. This is owed in large part to political developments that served to make the centre more accountable and at the same time providing regional politics and politicians with ever-increasing legitimacy. These developments have taken various forms. The first set of changes took place in national politics. The incorporation in 1977 of 100 (out of 400) seats based on proportional representation, as opposed to the original, and remaining, first-past-the-post allocated seats, gave opposition parties some institutional voice at the central level (Ward and Rodríguez, 1999, pages 23 – 27). In 1986 this representation was increased again as seats allocated on the basis of proportional representation swelled to account for 200 of the 500 congressional seats.

Regional politics provided opposition politicians with a second forum. The central government, suffering from a legitimacy deficit in the mid-1980s (Rodríguez, 1998, page 241), took the first steps towards greater decentralisation, but, with time, the process led to a rise in state legitimacy. As one would predict from figure 1, devolution initiatives have graduated from emphasising the decentralisation of responsibilities to the states initially, towards a much more well-funded programme as states have risen in political terms and the fallibility of the centre has developed. Hence, under de la Madrid, although devolution initiatives emphasised the 'autonomy' of regions and included many initiatives designed to give the states and municipalities of Mexico 'greater freedom' over their own governance, these policies could be dismissed as either lip service or thinly veiled excuses to dump expenditure responsibilities on the states (Ward and Rodríguez, 1999, pages 51 – 53). During Salinas's presidency further responsibilities were devolved to the states and localities, including comprehensive welfare duties. During this period the election of opposition governors began to shift the legitimacy balance in favour of the states. Finally, Zedillo's ascendancy to the presidency brought with it the renunciation of presidential metaconstitutional powers and the initiation of a cross-party forum on intergovernmental relations that has defined drives towards a new federalism in the country since 1996. The level of transparency and cogovernance these initiatives have achieved have guaranteed the states at least a reasonable accompaniment of resources with each devolutionary effort (Ward and Rodríguez, 1999). In general, then, the 1980s and 1990s have witnessed an increase in devolution of resources alongside devolution of responsibility as a result of political changes that have acted to equalise the relative legitimacy of governmental tiers, in line with the mechanisms illustrated in figure 1.

The case of the USA, in spite of significant political and economic differences, bears some similarities to the case of Mexico. As in Mexico, it was the centre, under the presidencies of Nixon and Reagan, that initiated drives towards devolution (Donahue, 1997). Indeed, Nixon coined the phrase 'New Federalism' in his bid to involve the states more closely in national governance. Popular disillusionment with the centre during the 1970s and 1980s, following the Vietnam War, the oil crises, and a shift in ideological principles away from macroeconomic management, served as a convenient political platform from which to launch various responsibility-shifting policies. The 1990s, during Clinton's presidency, saw a continuation of this trend, with the introduction of block grants for welfare shifting much of the risk and decisionmaking capacity for social security to the states, but including only nominal, and potentially short-lived, increases in funding (Powers, 1999). Central legitimacy, therefore, had so far served to emphasise responsibility-biased devolution in the USA.

As with Mexico, however, there have been forces operating to dilute this dominance, which have gained greater momentum over the past ten years. Whereas Mexico experienced an exogenous strengthening state-level legitimacy, driven by the political revolution, America's states have been able to bolster their own bargaining and negotiating positions as a result of endogenous factors—precisely because their responsibilities and authority have increased throughout the reforms. The states have benefited from a positive popular perception of devolution, which has increasingly held the central government to account. As Donahue (1997, page 13) asserts,

"Contemporary opinion surveys ... show dwindling faith in the federal government and (at least in relative terms) rising state legitimacy ... [T]he fraction of respondents identifying the federal government as 'the level from which you feel you get least for your money' rose by 10 points (to 46%) between 1989 and 1994 alone."

This rise in legitimacy of the states has acted to curb the emphasis on responsibility so prominent in previous devolutionary efforts. As an illustration, President Clinton's reluctance to devolve responsibilities for medical health, although trumpeted as a moralistic decision in order to safeguard equity, was prompted by widespread resistance from the states, whose fears over defective funding procedures stemmed from their experience of welfare reform (Offner, 1999). As regional and state-level politics continues to become more important in the USA, such checks and balances will act to equalise the devolution of responsibilities and resources. In the future, greater devolution of resources to the states may become a political necessity for the centre.

5 The impacts of the devolutionary trend

At this point we are able to draw two conclusions. First, as evidenced in section 3, the devolutionary trend is a common and general one. Second, however, this generality should not be mistaken for homogeneity. As we revealed in section 4, there are different forms of devolution that arise from different legitimacy distributions. In this section we reassess the implications of the trend in light of its virtual universality and complexity.

From a strategic political perspective we should not be surprised to learn that awareness of the advantages of devolution tend to be more developed than awareness of its drawbacks. Whether devolution is driven mostly by national or subnational governments, each potential driver has an inherent interest in defending its policies to the electorate and bolstering its legitimacy and popularity. The most common supporting arguments for devolutionary policies draw upon the efficiency advantages that lower level governance can engender (Donahue, 1997; Keating, 1999; Oates, 1972; Tanzi, 1996, pages 297–300). Given that the population in any country has a diverse preference structure, which varies across geographical space, this efficiency has three major sources, the themes of which recur time and again (Litvack et al, 1998, page 5; Oates, 1972). First, a smaller democratic and financial base should result in a managerial reform that will lead to a heightened degree of accountability, bolstered by the reduced administrative distance between the electorate and the politicians (Bennett, 1990; Hatry, 1994). Second, the lack of diluting influences—experienced by central government having responsibility for alternative, diverse regions—tends to allow local governments the flexibility to respond to the preferences of their 'customers' or electorates (Bennett, 1990). Third, the local nature of governance implies a greater chance of the election of local politicians with the specialist knowledge necessary to detect and react to the wishes of the electorate and defend its interests at higher levels (Putnam, 1993) as well as increasing the ability to implement policy innovations that would have been more difficult to pursue at the central or federal level (Bennett, 1990; Donahue, 1997). Devolutionists therefore acknowledge these three factors act both to ensure that local governments are more representative of and responsive to the interests of a given locality or region and to allow them to fulful this role. Thus, public policy in general is brought closer into line with the diverse preference sets of a nation, and welfare efficiency increases. Moreover, for those whose interests are not represented in their original location, there is in theory an increasing opportunity to choose and move between regional and local governmental systems as devolution progresses and as the diversity of public systems available to a nation as a whole increases. Arguments in favour of devolution are therefore underpinned by free-market assumptions, such as easy mobility, the politicoeconomic machinery of choice and democracy, and rational models of public sector behaviour under a democratic framework (see Tanzi, 1996, page 300).

There are areas, however, where developments associated with the global decentralising trend we have identified above cast doubt on the picture of greater efficiency, greater democracy, and greater welfare painted by devolutionists. As Prud'homme (1994, pages 1.2–1.3) outlines, the enthusiasm for devolution is driven not only by a commonly ill-considered faith in economic arguments but also by considerations of strategy on the part of ruling elites, the often 'fashionable' aura surrounding the concept of devolution, and the encouragement of international organisations, such as the World Bank, whose support for devolution is at best based upon a 'weak analytical basis' (Prud'homme, 1995). Indeed, on such a basis, over 12% of all World Bank project activity in 1998 included a decentralisation component (Litvack et al, 1998, page 1). The need to expound the merits of caution with regard to devolution is therefore manifest. Consequently, in the following subsections, we explore some of the less encouraging developments associated with devolution by focusing on three aspects—efficiency issues, equity issues, and administrative issues.

5.1 Efficiency issues

5.1.1 *Devolution and debt*

As discussed in earlier sections, devolutionary initiatives in our case-study countries have often included some separation of responsibilities from resources. It is rare to find simultaneous decentralisation of responsibility and resources, as the driving actor behind devolutionary efforts will invariably have incentives to separate these two factors one way or another (Rodden, 2002a, page 684; also, see figure 1). Under these circumstances it is common to discover a mismatch between responsibilities and financing, which, as Bennett (1997, page 331) underlines, have inevitably led to the development of debts either at the central or at the local level, to a steady erosion of the linkage of local decisions to their financial consequences, and to growing tension between central governments and subnational administrations. The cases of India, Brazil, and Spain illustrate various ways in which devolution can encourage debt formation.

In India, central government debt has been one of the major drivers behind devolutionary efforts. Between 1980/81 and 1993/94 the internal debt of the central government doubled, and total internal state debts rose from 4.6% to 7% of GDP (Buiter and Patel, 1997, page 36). It is easy, therefore, to envisage a situation where the pressure on central government finances at least contributed to a decentralisation of expenditure responsibilities. With a strong centre and calls for greater regional autonomy, the temptation to engineer favourable forms of devolution that will assist in limiting the central deficit may well become overwhelming.

In cases where the states have played a key role in the decentralisation effort, the decentralisation of resources can contribute towards large central deficits and developing regional debts—the former because of the de facto decentralisation of resources and the latter through the moral-hazard problem of central government effectively underwriting the expenditure of regions [for a general exposition of this commitment problem, see Rodden (2002a); for an examination of the Brazilian case, see Rodden (2002b)]. Perhaps the most apparent case of this type of development is found in Brazil. In Brazil, where decentralisation initiatives have been driven more by the states than by the centre, states were allowed to accumulate huge debts, often resulting from the development of short-term populist policies by Brazilian governors, aimed at securing their grip on power (Dillinger and Webb, 1999; Rodríguez-Pose and Arbix, 2001). The generation of regional debts was traditionally perceived as a low-risk strategy by local politicians, because, first, the burden of the debt would fall on future generations and not on current voters and, second, states expected to be bailed out in the medium term by the federal government. This, in combination with a weak centre, the political survival of which during much of the second half of the 1990s depended on the support of regional politicians, created conditions where several Brazilian states defaulted on their debts (Rodden, 2002a, page 670).

A devolutionary process driven by the so-called 'historical' regions in Spain has led to the establishment of a system where, although in most regions regional finances are still largely based on central government grants, autonomous regions hold the upper hand (Rodríguez-Pose, 1996). The consequence of this situation has been an erosion of national control over regional spending and a weak fiscal accountability, leading to a spiralling of regional debt across Spain (Castells, 2001; Montero, 2001).

5.1.2 Inefficient competition for industry

Another factor that threatens to proceed from efforts towards administrative devolution concerns the phenomenon of territorial competition. The proliferation and greater powers of subnational governments are shifting the focus of development policies away from achieving greater equality or national cohesion, towards securing greater economic efficiency at the local level (Cheshire and Gordon, 1998). Consequently, there has been an increasing tendency for subnational governments to engage in competition for the attraction of foreign direct investment (FDI) (Scott, 1998; Vernon, 1998). When subnational governments offer incentives for mobile industry to locate within their region, and the private sector firm chooses between the most attractive packages offered by the states, the impact upon national efficiency can be damaging. Cheshire and Gordon (1996) conceptualise this situation in terms of a zero-sum game—one in which the aggregate payoff of the game for all the actors involved is independent of the final outcome of the game across the actors. Specifically, the total gain of the arrival of a firm in a given country will be the same, or very similar, no matter where within the country the firm chooses to locate. But there are costs involved in trying to affect the locational decision of the firm that stem from the sort of measures offered to the mobile firms—such as tax grants, incentives and subsidies, or specific policy arrangements for foreign investors—that contribute to the spiralling of subnational debt (Rodríguez-Pose and Arbix, 2001). So, although the nation as a whole has nothing to gain as regions compete for investment it has much to lose as competition for mobile investment increases and regional advertising and marketing, government grants and loans, and interest and debt concessions increase. There are numerous examples of this sort of zero-sum competition for industry across Europe and the USA (Donahue, 1997; Mytelka, 2000; Phelps and Tewdwr-Jones, 2001) but it is perhaps in Brazil and China where the level of inefficiency associated with this sort of territorial competition is more apparent.

Of all the industries in Brazil, the automobile industry represents the clearest illustration of a general tendency. The Brazilian automobile industry expanded rapidly during the relatively stable period between 1995 and 1999 at the start of President Cardoso's term in office. During that time, car manufacturers invested over US$12 billion in Brazil (Rodríguez-Pose and Arbix, 2001, page 134). Although this influx of FDI appears to be beneficial to the whole country, industrial mobility has created the conditions for fierce rivalry between Brazilian states hoping to attract investment and consequently it has paved the way for wasteful expenditure designed to influence companies' locational decisions. As has been pointed out elsewhere, "Tax and bidding wars have become the norm in the motor industry" (Rodríguez-Pose and Arbix, 2001, page 145), with concessions routinely including the donation of land, the provision of infrastructure, state and local tax breaks, loans, and a series of financial cautions and guarantees.

In China, a similar pattern of territorial competition has become commonplace. Under the marketisation drive of the 1980s provinces, along with thousands of counties and townships, launched their 'opening up' programmes with the establishment of 'zones'. These were of differing varieties but each shared the common theme of offering significant concessions to attract industry, including, as a typical example, a 15% flat rate on corporation tax, a two-year tax exemption on profits, and a further three-year 50% tax reduction. Ma (1996, page 15) documents the nature of these zones, which, by mid-1993, had swelled in number to around 1800 across China. Crucially, there were no official standards in the level of concessions that could be offered by the zones. The lack of an official standard of concessions has led to the development of intense competition, with concessions often extended well beyond the example above, towards extremes such as five tax-free years and a further five years of 50% tax payment.

The impact on China as a whole of the development of these zones may well have been detrimental, for the reasons Cheshire and Gordon (1996; 1998) provide. Ma (1996, page 15) points out that not only are concessions available to internationally footloose companies but also to indigenous companies in an attempt to attract them from one region to another. No discernible national economic gain from the resources expended on inducing these movements, which boil down to a simple spatial reallocation of industry within the nation, can be expected. Furthermore, McKenney (1993, pages 20 – 21) points out that in Beijing, one of the richer and more tax-effective areas in China, the ability of the subnational government to extract the taxes rightfully owed to them in their economic zones has been extremely poor. All in all, although much industry has been attracted to China through the use of economic zones, especially from Hong Kong, Taiwan, and Singapore, the efficiency cost through competitive concession making across provinces detracts markedly from the advantages.

5.2 Equity issues

One of the traditional roles of national government is the redistribution of resources in order to safeguard minimum levels of welfare throughout the country. Decentralisation of authority and resources undermines a central government's ability to achieve this, in two fundamental ways. First, devolution of decisionmaking authority progressively transfers the responsibility for devising ways in which redistribution will occur to subnational government. Second, this tier of government is multifarious, and it is often the case that larger or more prosperous regions are overrepresented at this level. Hence, following devolution, a smaller role for national transfers and a larger voice for the regions in deciding how transfers are allocated is likely to result in a less progressive system of fiscal redistribution than would be the case under a centralist system (Thompson, 1989) unless an explicit and transparent interterritorial fiscal transfer system is established. The political and economic muscle of stronger regions is likely to skew public expenditure in their favour, regardless of whether the greatest legitimacy is based in the centre or in the regions and of whether the financing system of regions is based on local tax revenue or on grants from the centre. Where the finance system is locally based, the devolution of fiscal powers will inevitably favour wealthier areas, and where the finance system is centrally based the greater political muscle of larger and richer regions may be reflected in a greater capacity to secure transfers from the centre and to impede the evolution of a more centralised regional system of transfers or regional policy, as Markusen (1994) demonstrated in the US case.

In figure 3 we illustrate the regressiveness of regional budgets in three of our case-study countries: Mexico, the USA, and Spain. The three cases represent three different forms of devolution and three different forms of regional financing, yet, in all cases, regions with the highest GDP per capita enjoy, as a general rule, a greater capacity for expenditure than do poorer regions. The positive regression lines in the graphs confirm this trend (figure 3).

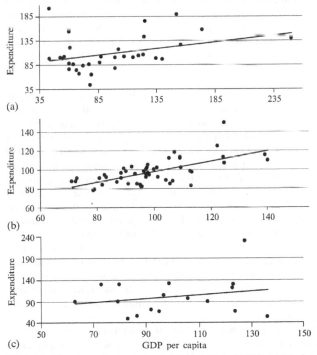

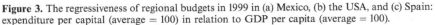

(c)   GDP per capita

Figure 3. The regressiveness of regional budgets in 1999 in (a) Mexico, (b) the USA, and (c) Spain: expenditure per capital (average = 100) in relation to GDP per capita (average = 100).

5.3 Institutional and political issues

Three further issues concerning the understated downside of the devolutionary trend outlined in section 2 merit attention. First, and most obviously, devolution involves an increase in the number and a decrease in the size of administrative units, with accompanying costs. Second, more subtly, lobbying costs under a decentralised system are likely to exceed those under a central system. In much the same way that regions compete for mobile capital, subnational governmental units also compete for central financial assistance (see Milgrom and Roberts, 1992), and, as for competitive bidding between regions, the expense incurred during the process of competition for government financing is a deadweight loss to the nation as a whole. Moreover, we can expect these losses to become greater as subnational governments become more powerful and complex. The proliferation of intragovernmental conventions and reviews of fiscal relations—for example, in Brazil (1989), China (1994), and Mexico (1995)—underscore the importance, complexity, and expense associated with public redistribution. Third, devolution of authority carries with it the threat of increased corruption (Rodden, 2002a, page 684). As Prud'homme (1994, pages 2.17–2.21) points out, although corruption is hard to measure, several reasons can be given for suspecting corruption rates to be higher at local levels. On the one hand, more opportunities for corruption probably exist, along with greater pressure from interest groups resulting from their proximity to local decisionmakers. On the other hand, fewer obstacles exist to prevent corruption. Local officials typically enjoy more discretion over funds than do those at the national level as well as more long-standing and personal relationships with stakeholders, creating the conditions for unethical relations that are promoted further by a relative lack of media scrutiny at lower governmental levels (Prud'homme, 1995). Furthermore, in Tanzi's (1996, page 301) view, contiguity—the fact that officials and citizens live and work close to one another and have often known each other for some time and may even be related—serves to undermine the ethics of local governmental institutions.

6 Conclusions

The purpose of this paper has been to bring to attention the general trend in decentralisation and to outline the complexity of its forms and implications. To this end, some of the key issues surrounding the mechanisms of devolution were addressed and incorporated into an informal model at the start of the paper. In this model we identified three factors of decentralisation as composing the devolutionary trend—legitimacy, and the decentralisation of resources and authority. Some of the interest conflicts arising from the coexistence of these attributes of devolution were addressed in order to deepen our understanding of devolutionary mechanics. The relevance of this understanding became clear in section 3, in which we outlined the depth and breadth of the global trend towards devolution. In those countries with centralised systems twenty or thirty years ago, decentralisation has been widespread, and in those countries with initially more vertically dispersed government systems further decentralisation from their respective starting points has become the norm. This global tendency towards the devolution of authority and resources elevates the importance of the need to understand this phenomenon. To this end, in section 4 the theoretical arguments given in section 2 were then applied to our group of case-study countries, exposing the diversity of devolutionary efforts across the globe.

In section 5, given the understated heterogeneity of devolutionary processes, the parallel heterogeneity of devolutionary implications was addressed. The expectation that devolution leads to greater efficiency, as the devolutionists and many policymakers appear to hold, can be called into question on the grounds that the process tends to engender both debt and territorial competition which are harmful to national efficiency. The gains from devolution through the matching of public services to a heterogeneous population preference structure is a static argument that may overlook dynamic alterations in the behaviour of the actors involved. From either perspective, the incentive structure facing the national and subnational governments alter and present the potential for opportunistic interaction that is damaging for the economy as a whole. Although the matching argument remains strong it should be weighted against the expected losses resulting from these factors before any devolutionary processes are undertaken.

In terms of equity, evidence was present to support the case that decentralisation of resources is often regressive from a territorial point of view. The combination of dwindling central government outlays in relative terms with the greater bargaining power of the richer and/or larger subnational authorities frequently leaves weaker and poorer regions in a worse financial state than under a centralised system. In section 5 we provided a brief discussion of the administrative, lobbying, and corruption costs that devolution is also likely to entail.

In summary, it is imperative for policymakers to recognise varying forms of decentralisation and to be aware of the vested interests of national and subnational governments across these alternative devolutionary forms. It is also imperative that commentators, policymakers, and analysts remain aware of the context of debates and opinions surrounding devolution. It is no coincidence that devolution tends to be supported by national electorates, because powerful and influential actors seek to muster the support of the electorate to facilitate their own initiatives. At the same time, however, this situation might well be damaging if enthusiasm for devolution is not tempered with, first, an awareness of the context of any debates that occur and, second, an awareness of not just the benefits but also the understated drawbacks that devolution can engender. As Tanzi (1996, page 314) has emphasised, successful devolution is highly dependent upon a long list of preexistent circumstances, the generation and, equally, consideration of which are by no means assured. The prescription of this paper is therefore cautionary with respect to both the interpretation and the promotion of devolutionary efforts.

Acknowledgements. The authors would like to thank Robert Bennett, Gilles Duranton, Murray Low, and two anonymous referees for their comments to earlier drafts of this paper. Andrés Rodríguez-Pose gratefully acknowledges the financial support of the Royal Society – Wolfson Research Merit Award and Philip Leverhulme Prize during this research.

References
Agnew J, 2000, "From the political economy of regions to regional political economy" *Progress in Human Geography* **24** 101 – 110
Allmendinger P, Tewdwr-Jones M, 2000, "Spatial dimensions and institutional uncertainties of planning and the 'new regionalism'" *Environment and Planning C: Government and Policy* **18** 711 – 726
Aspinall E, Berger M T, 2001, "The break-up of Indonesia? Nationalisms after decolonisation and the limits of the nation-state in post-cold war Southeast Asia" *Third World Quarterly* **22** 1003 – 1024
Bagchi A, Bajaj J L, Byrd W (Eds), 1992 *State Finances in India* (Vikas, New Delhi)
Bahl R, 1999 *Fiscal Policy in China* (1990 Institute, San Francisco, CA)
Bennett, R J, 1990 *Decentralization, Local Governments and Markets: Towards a Post-welfare Agenda* (Clarendon Press, Oxford)
Bennett R J, 1997, "Administrative systems and economic spaces" *Regional Studies* **31** 323 – 336
Buiter W H, Patel U R, 1997, "Solvency and fiscal correction in India", in *Public Finance: Policy Issues for India* Ed. S Mundle (Oxford University Press, Oxford) pp 30 – 75
Castells A, 2001, "The role of intergovernmental finance in achieving diversity and cohesion: the case of Spain" *Environment and Planning C: Government and Policy* **19** 189 – 206
Castells M, 1996 *The Rise of the Network Society* (Blackwell, Cambridge, MA)
Cheshire P C, Gordon I, 1996, "Territorial competition and the predictability of collective (in)action" *International Journal of Urban and Regional Research* **20** 383 – 400
Cheshire P C, Gordon I, 1998, "Territorial competition: some lessons for policy" *Annals of Regional Science* **32** 321 — 346
Coutinho M, 1996, "The Brazilian fiscal system in the 1990s: equity and efficiency under inflationary conditions", RP 41, Institute of Latin American Studies, University of London, London

Da-dao L, Sit V, 2001, "China's regional development policies: a review", in *China's Regional Disparities: Issues and Policies* Eds V Sit, L Da-dao (Nova Science Publishers, Hauppauge, NY) pp 19–37

Diamanti I, 1993 *La Lega: Geografia, Storia e Sociologia di un Nuovo Soggetto Politico* (Donzelli, Rome)

Dicken P, 1998 *Global Shift: Transforming the World Economy* 3rd edition (Guilford Press, New York)

Dillinger W, Webb S B, 1999, "Fiscal management in federal democracies: Argentina and Brazil", Policy Research Working Paper 2121, World Bank, Washington, DC

Donahue J D, 1997 *Disunited States* (HarperCollins, New York)

Ebel R, Yilmaz S, 2002, "On the measurement and impact of fiscal decentralization", http://econ.worldbank.org/files/13170.wps2809.pdf

Gray J, 1998 *False Dawn: The Delusions of Global Capitalism* (Granta Books, London)

Guigale M M, Webb S B, 2000 *Achievements and Challenges of Fiscal Decentralisation: Lessons from Mexico* (World Bank, Washington, DC)

Hagopian F, 1996 *Traditional Politics and Regime Change in Brazil* (Cambridge University Press, Cambridge)

Hatry H, 1994, "Accountability for service quality, performance measurement and closeness to customs", in *Local Government and Market Decentralisation* Ed. R J Bennett (United Nations University Press, Tokyo) pp 163–181

IMF, various years *Government Finance Statistics* (International Monetary Fund, Washington, DC)

ICSO, 1999 *Statistical Abstract of India* (Indian Central Statistical Organisation, Delhi)

Jones M, 2001, "The rise of the regional state in economic governance: 'partnerships for prosperity' or new scales of state power?" *Environment and Planning A* **33** 1185–1211

Kraemer M, 1997, "Intergovernmental transfers and political representation: empirical evidence from Argentina, Brazil and Mexico", WP 345, Inter-American Development Bank, 1300 New York Avenue NW, Washington, DC 20577

Keating M, 1998 *The New Regionalism in Western Europe* (Edward Elgar, Northampton, MA)

Keating M, 1999, "Asymmetrical government: multinational states in an integrating Europe" *Publius Journal of Federalism* **19** 71–87

Lee P K, 2000, "Into the trap of strengthening state capacity: China's tax assignment reform" *The China Quarterly* **164** 1008–1024

Litvack J, Ahmad J, Bird R, 1998 *Sector Studies Series 21491: Rethinking Decentralisation in Developing Countries* (World Bank, Washington, DC)

Ma J, 1996 *Intergovernmental Relations and Economic Management in China* (St Martin's Press, New York)

McKenney K I, 1993, "An assessment of China's special economic zones", The Industrial College of the Armed Forces, National Defence University, Washington, DC

Markusen A, 1994, "American federalism and regional policy" *International Regional Science Review* **16** 3–15

Massey D, 1999, "Power-geometrics and the politics of space–time", Department of Geography, University of Heidelberg, Heidelberg

Milgrom P, Roberts J, 1992 *Economics, Organisation and Management* (Prentice-Hall, Englewood Cliffs, NJ)

Montero A P, 2000, "Devolving democracy? Political decentralisation and the New Brazilian Federalism", in *Democratic Brazil: Actors, Institutions and Processes* Eds P R Kingstone, T J Power (University of Pittsburgh Press, Pittsburgh, PA)

Montero A P, 2001, "Decentralizing democracy: Spain and Brazil in comparative perspective" *Comparative Politics* **33** 149–172

Morgan K, 2002, "The English question: regional perspectives on a fractured nation" *Regional Studies* **36** 797–810

Mytelka L K, 2000, "Location tournaments for FDI: inward investment into Europe in a global world", in *The Globalization of Multinational Enterprise Activity and Economic Development* Eds N Hood, S Young (Macmillan, London) pp 278–302

Oates W E, 1972 *Fiscal Federalism* (Harcourt Brace Jovanovich, New York)

Offner P, 1999 *Medicaid and the States: A Century Foundation Report* (Century Foundation Press, New York)

Pérez Díaz V, 1990, "Governability and the scale of governance: mesogovernments in Spain", WP 1990/6, Instituto Juan March de Estudios e Investigaciones, Madrid

Phelps N A, Tewdwr-Jones M, 2001, "Globalisation, regions and the state: Exploring the limitations of economic modernisation through inward investment" *Urban Studies* **38** 1253–1272

Ping X, 2000, "The evolution of Chinese fiscal decentralisation and the impacts of tax reform in 1994" *Hitotsubashi Journal of Economics* **41** 179 – 191

Powers E T, 1999, "Block granting welfare: fiscal impact on the states", OP23, Urban Institute, Washington, DC

Prud'homme R, 1994, "On the dangers of decentralization", WP-1252, World Bank, Washington, DC

Prud'homme R, 1995, "The dangers of decentralization" *World Bank Research Observer* **10** 201 – 220

Putnam, R J, 1993 *Making Democracy Work: Civic Traditions in Modern Italy* (Princeton University Press, Princeton, NJ)

RBI, 2000, "State finances: a study of budgets of 1999 – 2000", Reserve Bank of India, Mumbai

Rodden J, 2002a, "The dilemma of fiscal federalism: grants and fiscal performance around the world" *American Journal of Political Science* **46** 670 – 687

Rodden J, 2002b, "Bailouts and perverse incentives in the Brazilian states" http:// www1.worldbank.org/publicsector/decentralization/cd/Brazil.pdf

Rodríguez V E, 1998, "Recasting federalism in Mexico" *Publius Journal of Federalism* **28** 235 – 254

Rodríguez-Pose A, 1996, "Growth and institutional change: the influence of the Spanish regionalisation process on economic performance" *Environment and Planning C: Government and Policy* **14** 71 – 87

Rodríguez-Pose A, 1998 *Dynamics of Regional Growth in Europe: Social and Political Factors* (Clarendon Press, Oxford)

Rodríguez-Pose A, 2002 *The European Union: Economy, Society, and Polity* (Oxford University Press, Oxford)

Rodríguez-Pose A, Arbix G, 2001, "Strategies of waste: bidding wars in the Brazilian automobile sector" *International Journal of Urban and Regional Research* **25** 134 – 154

Schram S F, Soss J, 1998, "Making something out of nothing: welfare reform and a new race to the bottom" *Publius Journal of Federalism* **28** 67 – 88

Scott A J, 1998 *Regions and the World Economy: The Coming Shape of Global Production, Competition, and Political Order* (Oxford University Press, Oxford)

Shah A, 1991, "The new fiscal federalism in Brazil", DP 124, World Bank, Washington, DC

Sharma S D, 1999 *Development and Democracy in India* (Lynne Rienner, Boulder, CO)

Souza C, 1997 *Constitutional Engineering in Brazil: The Politics of Federalism and Decentralization* (Macmillan, London)

Storper M, 1997 *The Regional World: Territorial Development in a Global Economy* (Guilford Press, New York)

Sury M M, 1998 *Fiscal Federalism in India* (Indian Tax Institute, Delhi)

Tanzi V, 1996, "Fiscal federalism and decentralization: a review of some efficiency and macroeconomic aspects", in *Annual World Bank Conference on Development Economics* (World Bank, Washington, DC) pp 295 – 316

Thompson C, 1989, "Federal expenditure-to-revenue ratios in the United States of America, 1971 – 85: an exploration of spatial equity under the 'new federalism'" *Environment and Planning C: Government and Policy* **7** 445 – 470

Tomaney J, 2000, "End of the empire state? New labour and devolution in the United Kingdom" *International Journal of Urban and Regional Research* **24** 675 – 688

Tomaney J, Ward N, 2000, "England and the 'new regionalism'" *Regional Studies* **34** 471 – 478

Vernon R, 1998 *In the Hurricane's Eye: The Troubled Prospects of Multinational Enterprises* (Harvard University Press, Cambridge, MA)

Ward P M, Rodríguez V E, 1999, "New federalism and state government in Mexico: bringing the states back in", AS – Mexican Policy Report 9, University of Texas at Austin, Austin, TX

Yi L, Chusheng L, 2001, "The tax division scheme: central – local financial relations in the 1990s", in *China's Regional Disparities: Issues and Policies* Eds V Sit, L Da-dao (Nova Science Publishers, Hauppauge, NY) pp 85 – 104

Name Index